GEOLOGICAL SOCIETY SPECIAL PUBLICATION NO. 177

The Evolutionary Biology of the Bivalvia

EDITED BY

E. M. HARPER
Cambridge University, UK

J. D. TAYLOR
The Natural History Museum, UK

&

J. A. CRAME
British Antarctic Survey, UK

2000
Published by
The Geological Society
London

THE GEOLOGICAL SOCIETY

The Geological Society of London was founded in 1807 and is the oldest geological society in the world. It received its Royal Charter in 1825 for the purpose of 'investigating the mineral structure of the Earth' and is now Britain's national society for geology.

Both a learned society and a professional body, the Geological Society is recognized by the Department of Trade and Industry (DTI) as the chartering authority for geoscience, able to award Chartered Geologist status upon appropriately qualified Fellows. The Society has a membership of 9099, of whom about 1500 live outside the UK.

Fellowship of the Society is open to persons holding a recognized honours degree in geology or a cognate subject, or not less than six years' relevant experience in geology or a cognate subject. A Fellow with a minimum of five years' relevant postgraduate experience in the practice of geology may apply for chartered status. Successful applicants are entitled to use the designatory postnominal CGeol (Chartered Geologist). Fellows of the Society may use the letters FGS. Other grades of membership are available to members not yet qualifying for Fellowship.

The Society has its own Publishing House based in Bath, UK. It produces the Society's international journals, books and maps, and is the European distributor for publications of the American Association of Petroleum Geologists (AAPG), the Society for Sedimentary Geology (SEPM) and the Geological Society of America (GSA). Members of the Society can buy books at considerable discounts. The Publishing House has an online bookshop (http://bookshop.geolsoc.org.uk).

Further information on Society membership may be obtained from the Membership Services Manager, The Geological Society, Burlington House, Piccadilly, London W1V 0JU (E-mail: enquiries@geolsoc.org.uk; Tel: +44 (0) 207 434 9944).

The Society's Web Site can be found at http://www.geolsoc.org.uk/.The Society is a Registered Charity, number 210161.

Published by The Geological Society from:
The Geological Society Publishing House
Unit 7, Brassmill Enterprise Centre
Brassmill Lane
Bath BA1 3JN, UK

(*Orders*: Tel. +44 (0)1225 445046
Fax +44 (0)1225 442836)
Online bookshop: http://bookshop.geolsoc.org.uk

British Library Cataloguing in Publication Data
A catalogue record for this book is available from the British Library.

ISBN 1-86239-076-2
ISSN 0305-8719

Typeset by E & M Graphics, Bath, UK
Printed by Alden Press, Oxford, UK.

Distributors

USA
AAPG Bookstore
PO Box 979
Tulsa
OK 74101-0979
USA
Orders: Tel. + 1 918 584-2555
Fax +1 918 560-2652
E-mail *bookstore@aapg.org*

Australia
Australian Mineral Foundation Bookshop
63 Conyngham Street
Glenside
South Australia 5065
Australia
Orders: Tel. +61 88 379-0444
Fax +61 88 379-4634
E-mail *bookshop@amf.com.au*

India
Affiliated East-West Press PVT Ltd
G-1/16 Ansari Road, Daryaganj,
New Delhi 110 002
India
Orders: Tel. +91 11 327-9113
Fax +91 11 326-0538
E-mail *affiliat@nda.vsnl.net.in*

Japan
Kanda Book Trading Co.
Cityhouse Tama 204
Tsurumaki 1-3-10
Tama-shi
Tokyo 206-0034
Japan
Orders: Tel. +81 (0)423 57-7650
Fax +81 (0)423 57-7651

Contents

It is recommended that reference to all or part of this book should be made in one of the following ways:

HARPER, E. M., TAYLOR, J. D. & CRAME, J. A. (eds) 2000. *The Evolutionary Biology of the Bivalvia.* Geological Society, London, Special Publications, **177**.

MORTON, B. 2000. The function of pallial eyes within the Pectinidae, with a new description of those present in Patinopecten yessoensis *In*: HARPER, E. M., TAYLOR, J. D. & CRAME, J. A. (eds) *The Evolutionary Biology of the Bivalvia.* Geological Society, London, Special Publications, **177**, 247–257.

Preface

Bivalves are key components of Recent marine and freshwater ecosystems, and have been so for most of the Phanerozoic. Bivalves have undergone an almost relentless increase in diversity and have exploited a number of varied life habits. Many have abandoned the primitive burrowing habit to attach to surfaces by flexible byssal threads, or rigid attachment by cementation, or to bore into rock, shells and wood, or simply to rest on the sediment surface. Others swim, at least if danger threatens. In the deep oceans the septibranchs are predatory. The shells of bivalves – through their shape, arrangement of muscle scars, hinges, ligaments, growth lines, microstructure and chemistry – reveal more information about the life habits of the animal than for any other molluscan class. The shell provides a continuous and detailed record of shell growth from the larval phase to maturity, a factor allowing the study of the both chemical and structural signatures, which may, in turn, be used to interpret both ontogenetic and palaeoenvironmental histories. It is for these reasons that palaeontologists have long been to the forefront of evolutionary and functional studies of bivalves.

The long history of both zoological and palaeontological research on the bivalves has revealed a startling degree of convergence and parallel evolution which has hampered the interpretation of their evolutionary history. However, new discoveries and developments make it timely for a new concerted effort to integrate zoological and palaeontological techniques to improve comprehension of the evolutionary history of the class. Molecular phylogenies, coupled with multi-character morphological studies, promise to improve the understanding of relationships, along with new discoveries of early Palaeozoic faunas which illuminate the scant knowledge of the early radiation of the class. The discovery of chemo-symbiotic relationships, and the location of diverse and specialized bivalves in the deep sea, hydrothermal vents, cold seeps and submarine caves, are continuously expanding conceptions of the morphological range of the class.

The Biology and Evolution of the Bivalvia meeting, organized on behalf of the Malacological Society of London, was held in Cambridge (UK) in September 1999. Attended by nearly 120 scientists, it was the first international bivalve meeting of zoologists and palaeontologists to be held for over 20 years. This book contains a selection of the papers presented at the meeting.

We are extremely grateful to the following people who reviewed papers for the present volume: L. Adamkewicz, M. Amler, E. B. Andrews, M. Angel, A. E. Bogan, J. Buckland-Nicks, C. A. Campbell, J. G. Carter, A. Chartrosse, A. Checa, J. C. W. Cope, J. Davenport, M. Foote, G. Giribet, E. Glover, S. Gofas, E. Gosling, G. Haszprunar, A. Hodgson. J. D. Hudson, J. Hutchinson, D. Jablonski, A. L. A. Johnson, P. A. Johnston, D. S. Jones, H. Jones, M. B. Jones, D. Lees, D. T. J. Littlewood, C. Lydeard, B. Morton, L. S. Peck, H. O. Pörtner, P. Rainbow, C. A. Richardson, L. R. Saul, J. Schneider, P. W. Skelton, A. B. Smith, G. Steiner, T. Steuber, A. R. H. Swan, J. A. Todd and T. R. Waller.

Liz Harper, John Taylor & Alistair Crame

Unravelling the evolutionary biology of the Bivalvia: a multidisciplinary approach

E. M. HARPER[1], J. D. TAYLOR[2] & J.A. CRAME[3]

[1] *Department of Earth Sciences, Downing Street, Cambridge CB2 3EQ, UK (e-mail: emh21@cus.cam.ac.uk)*

[2] *Department of Zoology, The Natural History Museum, London SW7 5BD, UK*

[3] *British Antarctic Survey, Madingley Road, Cambridge CB3 0ET, UK*

Bivalves have been important members of marine communities since the early Palaeozoic, in terms of both their numerical abundance and diversity. They are particularly prevalent in shallow shelf sediments, but they have also conquered the intertidal zone as well as the deep sea, where they are successful predators and key components of some vent communities. They have also invaded freshwater systems a number of times, where today they are important (and costly) foulers. In terms of general community structure, bivalves are important as prey items for a range of different predatory groups, and as major space occupiers, particularly on hard substrata where space may be limited.

The abundance and diversity of both Recent and fossil bivalves have made them attractive subjects for both zoologists and palaeontologists, and both disciplines have contributed to the present system of classification and the understanding of their phylogenetic relationships. However, somewhat inevitably, the focus of the two groups has been rather different, with zoologists concentrating on anatomical characters such as those associated with the gills and stomach, whilst palaeontologists have necessarily dwelt on hard-part characters such as dentition and shell microstructure. It is becoming increasingly clear, however, that convergence and parallelism is rife within the class and more integrated approaches are necessary to unravel these.

The last major attempt to integrate the palaeontological and zoological approaches to bivalve evolution was more than 20 years ago, at the Royal Society of London meeting in 1977. The resulting volume of the *Philosophical Transactions of the Royal Society of London*, Volume **284**, is still a benchmark volume. In particular, an early attempt at a cladistic analysis of the Pteriomorphia by Waller (1978) and papers on the origin and taxonomic diversification of the bivalves (Pojeta 1978) and the rostroconchs (Runnegar 1978) are still widely cited. However, in 1977 the *Treatise* volumes (Cox *et al.* 1969; Stenzel 1971) were still very much in vogue as a reliable data source, although even then there was a feeling that it was in need of a comprehensive revision (Yonge 1978). This sentiment has been echoed ever since, most strongly by Johnston & Haggart (1998) in their introduction to *Bivalves: An Eon of Evolution. Paleobiological Studies Honoring Norman D. Newell.* The Royal Society volume was also written at a time when cladistic studies were virtually unknown and there was not the wealth of molecular techniques at hand to tackle questions of phylogeny. The enormous changes which have taken place in the way in which evolutionary processes are thought about has changed during the last 20 years, as have the number of taxa, both Recent and fossil, that have been discovered during this period, both of which have inspired this volume of multidisciplinary papers from both palaeontologists and zoologists.

Bivalve classification and phylogeny

Despite the long scientific interest in bivalves, the relationships between the higher taxa are perhaps surprisingly unresolved. Certainly the *Treatise* classification is now dated, but even more recent schemes are often deeply conflicting, probably because of their reliance on single or small groups of characters (Starabogatov 1992; Morton 1996; Salvini-Plawen & Steiner 1996; Waller 1998; Amler 1999). The relatively recent boom in molecular techniques is of great potential here and two papers in this volume have used them to try to elucidate the relationships between higher taxa. **Steiner** & **Hammer** have used an analysis of 18S rDNA sequences in order to tackle the phylogenetic

From: HARPER, E. M., TAYLOR, J. D. & CRAME, J. A. (eds) *The Evolutionary Biology of the Bivalvia*. Geological Society, London, Special Publications, **177**, 1–9. 1-86239-076-2/00/$15.00

relationships of the pteriomorphs, both in terms of the major pteriomorph taxa (such as the ostreoids, pinnoids and pterioids, etc.), and between the Pteriomorphia as a whole and the other major bivalve clades (e.g. protobranchs, heteroconchs, anomalodesmatans). Their researches supports the monophyly of the Protobranchia, Heteroconchia and Pteriomorphia, and float the idea that the Heteroconchia and Pteriomorphia may be sister taxa. Within the pteriomorphs, there is strong support for two major clades, ([Pinnoidea (Ostreoidea + Pterioidea)] and [(Anomioidea + Plicatuloidea) + (Limioidea + Pectinoidea)]. Further molecular studies are presented by **Campbell**, also using the 18S gene, who investigated the relationships between a wide range of bivalve superfamilies, orders and subclasses. Gratifyingly, he found all superfamilies, which have been established on morphological grounds, appeared monophyletic, as were most orders. Only the Myoida appear polyphyletic, thus confirming the suspicions of some previous workers (e.g. Morris *et al.* 1991).

Palaeontologists face the fascinating but difficult challenge of trying to test various models of bivalve phylogeny with the known fossil record. At face value the task should be relatively simple as the bivalves have, by virtue of their robust calcareous valves, an excellent fossil record, but this has not proved to be the case. It is particularly difficult because following the good fossil record of the earliest widely accepted bivalves (*Fordilla* and *Pojetaia*), which are relatively abundant from the Tommotian of North America, Europe and Australasia, there is an unfortunate gap in the fossil record of the class that spans from the middle Middle Cambrian–Early Ordovician, some 4% of the entire evolutionary history of the class (Harper 1998). Unhappily, this gap in the fossil record appears to cover a period of major evolutionary changes within the class, for on their reappearance the bivalves are larger and more diverse both taxonomically and in life habits. None the less, new discoveries in the last decade have improved current knowledge of early Palaeozoic bivalve faunas. These findings have inspired two papers in this volume which tackle the problem in two rather different ways.

Carter *et al.* have undertaken an ambitious cladistic analysis of 62 bivalve taxa of Palaeozoic age, including a number of equivocal forms such as *Tuarangia*, using 117 characters derived almost exclusively from the hard-parts. The large number of characters (and states) and taxa has limited this study to a heuristic analysis, and the analysis has revealed 411 most parsimonious trees. Doubtless an increase in computer power in the next few years will allow more exhaustive searches and further character refinement may reduce the numbers of possible models. The paper is accompanied by full character descriptions and the data matrix which will allow such future re-analysis. Despite the large number of trees produced, the consensus trees do introduce a number of interesting conclusions, some of which support the conclusion of previous studies, such as Waller (1998), on the basal position of the Mytiloidea amongst the pteriomorphs, and the basal dichotomy within the palaeotaxodonts between the Nuculanoidea and the Nuculoidea + Solemyoida. They also support the view of shared ancestry of the Nuculoidea and the palaeoheterodonts. In other areas, however, the authors come to rather different conclusions to those reached by others. The most striking of these is the rejection of the popular notion that the Bivalvia were derived from the rostroconchs, preferring instead a stenothecoid monoplacophoran origin. Elsewhere they are at odds with Waller (1998), rejecting a link between the Anomalodesmata and the Heteroconchs, preferring to place the former as a sister group to the pteriomorphs. Again, there are striking anomalies between the results of this analysis and those of molecular studies. **Cope** has used a more traditional approach, based on the culmination of detailed research of a range of early Ordovician bivalve faunas. He bases his scheme on dentition, shell microstructure and inferred gill grade. Cope identifies the evolution of the filibranch gill, which he infers from dental changes, as providing the initial impetus for the diversification of the bivalves in either the latest Cambrian or earliest Ordovician. He recognizes two subclasses; (1) the Protobranchia – including the Nuculoidea and chemosymbiotic forms such as the solemyoids; (2) the Autolamellibranchiata – from which he derives three principal stocks, the Trigonioida, the Anomalodesmata and the Heteroconchia, which he believes then gave rise to the Pteriomorphia.

Two other papers in the volume deal with the phylogenetic relationships of enigmatic higher bivalve taxa. In a major new synthesis, **Skelton & Smith** present a first attempt at a comprehensive phylogeny for the rudists, an extinct group of sessile, epifaunal bivalves that flourished in warm temperate and tropical seas from the Late Jurassic to the Late Cretaceous. Their cladistic study was based on 33 skeletal characters and 31 species, representing each of the higher taxa recognized in previous classification schemes. One of the ten most parsimonious trees obtained from their study was selected as a working hypothesis. This showed many of the basic phylogenetic relationships established by early workers, such as Douvillé and MacGillavry, were perfectly sound. An outer shell layer of fibrillar prismatic calcite is a diagnostic synapomorphy for all rudists and attachment to the substrate by one or the other valve provides a fundamental division into two stocks. Other features shown by the cladogram include the monophyletic status of the uncoiled rudists, as well as established families such as the Radiolitidae and Hippuritidae. Important new light is shed on the origin of the latter family by its sister-group relationship with the polyconitid genus *Tepeyacia*, from the Albian of Mexico. There is every prospect that future iterations of this study will refine rudist phylogeny even further.

Deep-sea carnivory is just one of some rather unusual lifestyles found amongst members of the bivalve subclass Anomalodesmata. This group comprises 14 extant families of morphologically disparate bivalves, many from the deep sea, whose classification and phylogenetic relationships have been highly confused. **Harper *et al.*** use a cladistic analysis of a set of anatomical and shell characters in an attempt to determine relationships amongst this diverse group. One conclusion from this analysis is that carnivory has arisen independently within two separate clades. Trees generated from shell characters alone were incongruent with those derived from the total evidence of anatomy and shell, suggesting that although anomalodesmatans have a long fossil record it may be difficult to integrate fossils into such an analysis.

Bivalves play a major role in freshwater communities, where they are significant in water filtering, and their larval phases are important parasites of fish, whilst their propensity to spread has made them important biofoulers (Aldridge 1999). Because of the nature of freshwater bodies, many taxa are endemic to specific regions and hence it is probable that they, above all other bivalves, may show extreme degrees of morphological convergence. Similarly, many geographically restricted taxa are at great danger from extinction. **Lydeard *et al.*** have used the mitochondrial cytochrome C oxidase subunit I and 16S rRNA gene sequences to examine the phylogenetic relationships of five species of Gulf Coastal unionid generally accepted as belonging to three different genera. These are particularly important taxa as they are all considered endangered. Their analyses found that the five taxa were all distinct and the 'genera' in which they had previously been dispersed in were polyphyletic. The threat of morphological convergence hampering classification is an issue with the three unionid genera of cementing 'freshwater oysters' [*Etheria* (Africa), *Acostaea* (South America) and *Pseudomulleria* (India)]. Despite the fact that the cemented habit within the Bivalvia as a whole is well known to be polyphyletic (Yonge 1979) and that these freshwater cementers have highly disjunct geographic distributions, they have traditionally been placed within a monophyletic family, the Etheriidae. **Bogan** & **Hoeh** have integrated new cytochrome C oxidase subunit I DNA sequences from these three taxa with existing data from other unionid taxa from around the world. Their findings do not support a close relationship between the three genera, suggesting rather that the cemented habit has arisen at least twice. They also pose the rather interesting question as to why the cemented habit might have evolved in the first place, noting that there are no recorded freshwater cementers in either North America or China, the major hubs of unioinoid diversity.

Comparative morphology continues to be an important tool in our efforts to determine phylogenetic relationships amongst bivalves. A major benefit from the introduction of cladistic methods in phylogenetic reconstruction has been in the objective definition of characters and their states. Concomitantly, there has been a search for new types of characters and two papers show how detailed morphological studies of bivalve sperm, using transmission electron microscopy, can generate sets of phylogenetically important characters. In the first paper, **Healy *et al.*** use sperm characters to explore relationships within the large bivalve subclass Pteriomorphia, e.g. showing how the Ostreoidea and Limoidea are probably closely linked, as are the Pterioidea, Pinnoidea and Pectinoidea. By contrast, sperm characters suggest that the Arcoidea and Limopsoidea are less closely related than generally supposed. At a more detailed level, **Keys** & **Healy** use characters of sperm ultrastructure to explore relationships amongst the living species of giant clams – Tridacninae. Sperm data supports a close relationship between giant clams and cardiid bivalves, but perhaps indicates a separate origin of *Hippopus* and the other *Tridacna* species. Amongst the species of *Tridacna*, *T. squamosa* shares more characters with *T. gigas* than with the other species in the subgenus, *Chametrachea*, where it is usually placed; a conclusion supported by some molecular data.

Bivalve form and function

The great diversity of bivalve life habits is reflected in a wide range of morphological variation in terms of both shell and anatomy. The seminal work of Stanley (1970) provided a firm basis for the use of shell characters to infer the life habits of extinct taxa and few have approached the elegance of the work of C. M. Yonge in demonstrating the variation in anatomical features across the class. Despite the established interest in bivalve form and function, there remains a surprising amount to do and, in particular, to synthesize the information. It is important to document the morphological diversity displayed by the class and to investigate the constraints and preadaptations which have favoured some body plans and not others.

Taylor & **Glover** report that all the species of Lucinidae studied to date harbour sulphide-oxidizing, chemosymbiotic bacteria within their gills. Although chemosymbiosis in living lucinids has been recognized for some time, it seems that little thought has gone into its mechanics – if sulphide-laden water is drawn into the mantle cavity alongside the inhalant respiratory current, how is oxidization prevented before reaching the chemosymbionts? Taylor & Glover's careful work revealed a number of adaptations by which the two sites of respiration and chemosynthesis are partitioned, e.g. by the detached anterior muscle which separates the oxidized water that come in anteriorly from the sulphide-rich waters which are drawn in posteriorly. The presence of mantle gills on the internal surface of the anterior of the mantle of many taxa underlines this 'switch' in the location of normal respiratory activity. Mantle gills had been described previously, but only superficially and their complexity and true significance had never been realized. This underlines the continued importance of anatomical studies of seemingly familiar bivalves. Such is the importance of the chemosymbiotic relationship, that evolutionary syntheses of the lucinoids should take account of it and it is therefore extremely useful to now have morphological characters which may be used as markers for the habit. Many of these features are demonstrable in the Silurian *Ilionia* suggesting that the chemosymbiosis within the lucinoids is extremely ancient.

In recent years it has become apparent that the very distinctive chemosymbiotic bivalve communities associated at the present day with both hot vents and cold seeps can be traced back through considerable periods of time. Such faunas seem to have been particularly widespread in the Cretaceous and **Kelly *et al.*** record a further occurrence in the Early Cretaceous (Late Barremian) of Wollaston Forland, Northeast Greenland. Sedimentological evidence indicates that this is a methane cold-seep complex which accumulated in a mid- to outer-shelf setting. It is characterized by a large number of limestone mounds, 1–3 m in diameter and up to 1.8 m high, that occur within an otherwise monotonous, mudstone-dominated sequence. The mounds reveal an internal structure of calcite-cemented tube systems, laminated calcite crusts and void fills. A large new species of the lucinid genus *Cryptolucina* is a dominant element within the fauna, and there are occasional specimens of the cryptodont *Solemya* too. However, the most distinctive bivalve within the assemblage is a large, new taxon which is referred to the anomalodesmatan family Modiomorphidae. This new bivalve reaches *c.* 300 mm in length by 120 mm in height, and has a shell which is, in places, up to 28 mm thick. It seems to be yet another new giant bivalve taxon associated exclusively with cold seeps. Members of the Pectinidae possess the anatomically most advanced eyes of any bivalve family and it would seem only logical to conclude that they have developed these to provide an advanced early warning system of vicinal predators. However, in a wide-ranging review of the functional significance of pectinid pallial eyes, **Morton** suggests that their true purpose is still largely unknown. He first adds to our anatomical knowledge of pectinid eyes by describing those of the boreal species *Patinopecten yessoensis* and then goes on to show that in almost all taxa studied an escape response can only be elicited by either tactile or chemical stimuli. Visual clues rarely, if ever, provoke active swimming. Is the pectinid eye an example of evolutionary overdesign? If pectinid and spondylid eyes are indeed homologous, it would suggest that they evolved before the two families diverged in the Early Jurassic. Sophisticated pallial eyes may have been retained since then for a number of poorly understood functions.

Few organisms have evolved the capability of digesting wood, but the teredinid bivalves, the shipworms, have been successful in this mode of feeding, dating back to the earliest Cretaceous, a time when angiospermous wood would have become abundant in coastal habitats. Indeed, in Recent ecosystems teredinids have been the scourge of human-made coastal structures and boats made of wood. However, relatively little is known about the evolution of this very specialized habit. **Lopes *et al.*** have described and compared the functional anatomy of the digestive system of two wood boring taxa, *Neoteredo reynei* (Bartsch, 1920) and *Psiloteredo healdi* (Bartsch, 1931), which are common in Brazilian mangroves. Important differences between the two suggest that *N. reyenei*, with its enlarged wood-packed appendix and anal canal, is more dependant on wood for its nutrition than is *P.healdi*, in which these structures are small and where active palps have been retained, suggesting the importance of suspension feeding. There is evidence that during ontogeny wood may increase in importance in the diet of the latter.

Gill structure in bivalves has long been the focus of functional and evolutionary studies, and **Beninger** & **Dufour** describe and discuss the fine structure of the abfrontal surface of the gill filaments. From the supposed primitive function as a mucociliary cleaning surface in protobranch bivalves, two evolutionary trajectories involving cilia and mucocytes are proposed. One involves the progressive reduction in density of cilia and mucocytes, and the other involves reduction of cilia but retention or increase in acid mucopolysaccharide-secreting mucocytes.

Many arcoid bivalves are characterized by a duplivincular ligament, wherein the lamellar layer is embedded in the fibrous one in a distinctive series of chevrons. The one prominent exception to this type of organization occurs in the family Noetiidae, which is characterized by vertical, chevron-shaped strips of lamellar ligament. At first sight it would appear that these two types of ligament must be the product of two quite distinct groups, but chance observations on a series of *Limopsis* ligaments suggested to **Thomas *et al.*** that striking differences in the form of the arcoid ligament can be produced by only a very modest change in the developmental process. To develop this idea further they have hypothesized that, along the mantle isthmus, alternate zones of activation and inhibition control the production of lamellar ligament that forms either oblique sheets or vertical layers; only a single, simple change of instructions to a narrow field of cells is required to change from one pattern to the other. Tests provided by two contrasting sets of mathematical models indicate that this may indeed be the case and it may be concluded that disparate growth patterns can in fact be produced by only subtle character shifts.

With the current revolution in the understanding of which taxonomic groups may or may not be assigned to the subclass Cryptodonta (Johnston & Collom 1998), considerable importance is now attached to a proper understanding of the Palaeozoic order Praecardioida. **Yancey** & **Heaney** present some timely information on the Lunulacardiidae, the youngest praecardioid family in North America. Their material comes from the Pennsylvanian (Upper Carboniferous) Buckhorn Asphalt biota of south-central Oklahoma, and comprises a unique assemblage of small bivalves. Some of these are definitely juveniles but others are adults that can be assigned to an unusual new taxon. Characterized by a distinct change in shape from a veneriform juvenile to a subtrigonal adult, this new form possesses both stout teeth and a prominent ligament. Indeed, were it not for these latter characters, it could easily be confused with a conocardioid rostroconch!

The morphology of larval shells of both living and fossil bivalves has the potential to supply useful phylogenetic information, particularly where, as in the case of the oysters, the post-larval shell is prone to a large degree of ecophenotypic variation. **Malchus** has recognized and studied a number of prodissoconchs from Middle Jurassic oysters. Careful examination of their morphology revealed that the ancestral larval hinge most likely evolved from a mytilid-like provinculum, to produce a secondarily symmetrical hinge line. He also suggests that the small size of the prodissoconch I, along with the small P I/P II ratios, confirm that the planktotrophic development style seen in modern oysters is in fact the plesiomorphic state for the entire superfamily.

The tube-dwelling clavagellid bivalves are some of the most enigmatic of all bivalves and as indicated by Harper *et al.*, their position within the subclass Anomalodesmata is extremely equivocal, their extreme morphological adaptations apparently masking useful phylogenetic signals. **Savazzi** has investigated the little-known clavagellid *Bryopa*. Unlike most clavagellids, which are thought to be facultative semi-endolithic tube dwellers, *Bryopa* appears to be fully and obligatorily endolithic. Despite the fact that the left valve is permanently attached to the tube, the bivalve appears to migrate forwards within the substratum throughout growth. Despite the fact that no living animals, or indeed soft parts, were available for study, Savazzi was able to use morphodynamic principles to suggest how this could occur. He envisages a mechanism by which the shell elongates anteriorly, sliding the soft parts forward, whilst the posterior portion of the right valve is continually being resorbed. The strong inequivalvy which results is seen in none of the other clavagellids and, unfortunately, it seems that *Bryopa* is too derived to provide the much needed information as to the affinities of the superfamily.

Although the mineralization of the external shells of bivalves has been extensively studied and used as a phylogenetically useful character (Taylor *et al.* 1969, 1973; Carter 1990), some internal tissues are also mineralized and freshwater mussels (Unionoida) often have large accumulations of calcium granules in their tissues. **Byrne** describes the fine structure and chemistry of abundant orange, iron-rich calcium granules in the Australian freshwater mussel *Hyridella*. Comparison of granule distribution and composition between different unionoids suggests the existence of phylogenetically significant differences.

Teranota is a relatively newly described taxon, known only from the Middle Devonian of Germany. These scimitar-shaped bivalves are interpreted by **Rogalla** & **Amler** as belonging to the enigmatic family Orthonotidae, whose affinities are debated as belonging to either the modiomorphids, anomalodesmatans or even the heteroconchs. Rogalla & Amler are persuaded of an anomalodesmatan affinity but support a close link with the modiomorphids. The discovery of specimens in (apparently) life position, embedded anterior-end downwards at angles of 60–80° to the sediment surface, has allowed them to reconstruct these as endobyssate bivalves with no true siphons.

Biodiversity and biogeography

The documentation and analysis of patterns of biodiversity on regional and global scales, and, in particular, the latitudinal diversity gradient, have generated much interest, a plethora of hypotheses and considerable controversy. Marine bivalves have been often used in such analyses by both biologists and palaeontologists. A new analysis of regional bivalve faunas by **Crame** shows that both latitudinal and longitudinal gradients are not as regular in form as supposed. There is a distinct step between 20 and 30°N in the latitudinal diversity gradient for the northern hemisphere and in the southern hemisphere the bivalve fauna of Australia forms a distinct hotspot of diversity. The causes of these large-scale patterns are likely multiple and complex, and Crame critically reviews various hypotheses emphasizing that historical processes in the Neogene, such as the closure of Tethys and the northward movement of Australia into the tropics, have been important in shaping the biodiversity patterns of a region.

Jablonski *et al.* have also been investigating the latitudinal gradient in marine bivalve diversity, paying particular attention to that present on the eastern Pacific continental shelf. Here, their very comprehensive data set also picks out a sharp (50%) fall in the number of species at the edge of the tropics, followed by a much gentler decline into Arctic regions. They believe that this gradient may be a positive function of available energy and, using a residuals analysis to counter the effects of spatial autocorrelation within the data set, demonstrate that mean sea-surface temperature is a highly significant predictor of bivalve diversity. It is also apparent that both infaunal and epifaunal taxa increase in diversity from the poles to the tropics, although at somewhat different rates. Along the Pacific coast of North America their ratio changes significantly with latitude and there is preliminary evidence from the Mesozoic fossil record to suggest that it may also change with time. Clearly, this is a dynamic gradient driven by the differential diversification of the two functional groups. Do factors such as the mass extinction event at the *K–T* boundary and global climatic change have a role to play in driving these clade dynamics?

Analysis of global patterns of marine biodiversity as exemplified in the Crame and Jablonski *et al.* papers is highly dependent on the quality and completeness of the source data from regional levels. An excellent example of how these data are acquired is the analysis by **Mikkelson** & **Bieler** of the bivalve fauna of the Florida Keys. Prior to their survey only 163 species were recorded from the area. Using a combination of original collecting, search of museum collections and scouring of literature records this list has now been increased to 325 species. Interestingly, but rather worrying, primary literature sources recovered only 44% of the species, although this increased to 73% when 'grey' literature was included. Museum collections were the most important source of records, picking up 77% of the species including 62 not found by other methods. The message from this paper is that a diversity database for a particular area based solely on literature records is likely to be a severe underestimate of the true diversity.

Molecular techniques are now being used in biogeographical analysis to test hypotheses concerning the area of origin and movements of species. The smooth blue mussel, *Mytilus galloprovincialis*, has an unusually widespread distribution in temperate and subpolar waters of both the northern and southern hemispheres. Introduction by humans may account for some of the distribution pattern but mussels are present in shell middens predating European arrival in southern Australia and natural dispersal also seems likely. **Daguin** & **Borsa** used nuclear DNA markers to determine the genetic relationships and tease apart the history of *Mytilus galloprovincialis* populations around the world. They demonstrate the genetic distinctiveness of the Australian populations and reject the hypothesis that these mussels were recently introduced by humans. By contrast, the close similarity of samples from the Mediterranean, the North Pacific and Chile supports the hypothesis that mussels in the latter two areas have been introduced recently.

Ecological and evolutionary trends

Bivalves have a crucially important role to play in the study of evolutionary lineages. **Crampton** & **Maxwell** argue that outline shape should be a key morphological character and proceed to develop a new method for studying it based on Fourier shape analysis. An outline trace of either a growth stage or the complete adult valve is digitized, yielding a suite of Fourier coefficients that describe a spectrum of harmonically related trigonometric curves, or harmonics. These coefficients can then be used to compare synthetic average morphologies at a variety of ontogenetic and evolutionary stages. In a detailed study of the shallow-burrowing crassatellid genus, *Spissatella*, it was shown that growth is strongly allometric. Ontogenetic changes in shape of individuals are invariably far greater than total evolutionary changes within the genus over *c.* 20 Ma. An overall strong correspondence between individual species ontogenies and evolutionary changes in adult form strongly suggest that heterochrony has been the dominant evolutionary mechanism throughout the lifespan of *Spissatella*. There is good evidence to suggest that paedomorphosis and peramorphosis occurred in approximately equal frequencies from the latest Oligocene onwards.

Johnson *et al*. have highlighted the potential of the Queen Scallop, *Aequipecten opercularis*, as a tool to investigate Late Cenozoic palaeoenvironmental change. Occurring at the present day from Norway to the Adriatic, it has rapid, year-round growth rates that ensure recovery of well-resolved records of seasonal variation, and a fossil record that stretches back into the Miocene. Somewhat surprisingly, an oxygen isotope palaeothermometry study on mid-Pliocene shells indicated seasonal temperatures similar to those of the present day. This stands in marked contrast to a variety of other isotope and general assemblage data which suggests that the mid-Pliocene was globally warmer than now. Perhaps the southern North Sea Basin was in some way exempt from this general trend, or maybe even these well-preserved fossils suffered a degree of cryptic diagenetic alteration. More promising in this particular respect may be the study of microgrowth increments, for these do show markedly higher maxima and minima (i.e. related to extreme summer and winter temperatures) in the mid-Pliocene. Carbon isotope and various trace element data also offer considerable potential for studying variation in seasonality patterns.

The evolution of physiological characters has been investigated experimentally by **Peck & Conway**. They were particularly concerned with the concept of metabolic cold adaption (MCA), which is the hotly debated subject of whether the metabolic rates of polar ectotherms are elevated to compensate for the physiological constraints imposed by living at low temperatures. The recent demonstration that the muscles of certain polar fish contain twice the number of mitochondria as temperate species is taken by some to be evidence for the operation of MCA. The approach adopted in this study was to look at rates of oxygen consumption in two Antarctic bivalve species and then to compare the results with those obtained previously from a range of lower latitude taxa. In this way it was possible to show, quite conclusively, that rates of oxygen consumption decline with increasing latitude. Together with a similar study of Q_{10} coefficients, the data indicate that there is no significant elevation of whole-organism metabolism at low temperatures. The restricted temperature envelope and low metabolic rates of polar ectotherms are adaptations that have evolved over at least the last 15 Ma, if not considerably longer.

Edelaar has been investigating whether the depth of burrowing in the deposit-feeding tellinid *Macoma balthica* might be regarded as a trade-off between safety from predators and access to its food source. Obviously, the deeper *Macoma* burrows the safer it is from surface-dwelling predators, but the more remote it is from its food. The key question of interest here is whether organisms such as *Macoma* can adjust to optimum values by modifying their phenotype (such as depth of burrowing). To try and answer this, a series of laboratory experiments was devised to investigate depth of burrowing with, firstly, food (finely chopped spinach) either present or absent, and then a common predator (the shorecrab, *Carcinus maenas*) either present or absent. The results showed a significant decrease in burrowing depth when food was added but then a significant increase when a predator was introduced. On average, individual *Macoma* burrowed 57% deeper in the presence of the shorecrab. Such behavioural plasticity may have a number of important ecological and evolutionary implications. For example, pronounced changes in the shape or density of growth rings might reflect predation intensity as much as a low food supply.

Ensis directus, a recent ballast-water invader along European North Sea coasts, shows periodic mass mortalities that are still poorly understood. **Cadée** has noticed how, at certain times during the winter, large numbers of shells become exposed by at least half their length above the sediment surface and then cannot re-burrow. At such times they become easy prey for herring gulls, who have perfected a technique of alternate shaking and dropping to break open the shell. This invariably causes damage to just the middle part of the valve and the production of a unique type of shell fragment. Many types of shell fragment are now known to be specific to individual predators and thus have considerable potential for palaeoecological studies. Presumably, avian predators exert a significant predation pressure on intertidal and shallow subtidal bivalves and, therefore, the recognition of diagnostic shell damage caused by these predators, which is demonstrably different from taphonomic damage, is extremely useful in any future attempts to document this activity from the fossil record.

Marine mytilid mussels are probably the most familiar and well-studied of all bivalves. They have a huge biomass and are major space-occupying organisms upon rocky intertidal shores throughout the world, where they are often important components of intertidal food webs. **Seed *et al.*** review the attributes that contribute towards the ecological success of mytilid mussels. They then go on to show how mussel beds themselves form major habitats for many other organisms, with the composition and structure of these communities remarkably similar in widely separated parts of the world. Because of their abundance and accessibility, as well as their importance as food, mussels have been widely used as sentinel indicators of environmental change and pollution. Seed *et al.* demonstrate, by detailed studies of shell growth and trace element analysis, how mussel shells can be used as chronometers of environmental change. Their work provides a model for the interpretation of growth patterns in fossil shells.

Future directions

It would be easy to presume that the long tradition of bivalve study by both zoologists and palaeontologists, and their basic familiarity with the class, has led to a comprehensive understanding of their evolutionary history. However, this is far from the case. The start of the twenty-first century offers many exciting challenges to the understanding of the evolutionary biology of the Bivalvia.

Exciting discoveries continue to be made amongst living bivalves, and two in particular have changed the concepts of trophic adaptations and evolution of the class. Bivalves had long been regarded as having two main feeding strategies – suspension and deposit feeding – with a few species nutritionally dependent on having photosymbiosis with unicellular algae. It is only 20 years since the discovery of predatory behaviour amongst bivalves and this is now recognized as a major feeding strategy amongst at least three groups of deeper water bivalves (Morton 1981). Similarly, exploration of hydrothermal vents and cold-seep sites over the last 20 years has revealed communities of bivalves nutritionally dependent on symbiotic sulphur- and methane-oxidizing bacteria. It is remarkable that this chemosymbiosis has now been recognized in at least six different clades and many species of bivalves, living not only in the deep sea but also in shallow water and even intertidal habitats (Distel 1998). Comparative morphology suggests that this chemosymbiosis is an ancient nutritional strategy which may date back to the Ordovician.

The biology of these humble animals continues to surprise, as witnessed by the extraordinary behavioural and morphological adaptations employed by freshwater unionid bivalves to ensure transfer of their larvae to host fish species. These remarkable adaptations involve extensions of the mantle edges to mimic fish, complete with fin-like structures and 'eyes', and also extruded structures containing the glochidian larvae which mimic fish and arthropod larvae (Haag *et al.* 1995).

A remarkable recent discovery is that the worm-like animal *Xenoturbella*, a ciliated, virtually organless bag from northern Europe, is in fact a highly modified shell-less protobranch bivalve (Israelsson 1999). This umasking of *Xenoturbella* extends present conceptions of the morphological range of bivalves.

Palaeontologists also continue to unearth wonderful and extraordinary animals which truly extend our concepts of morphological disparity amongst bivalves, e.g. the discovery of a new family, the Wallowaconchidae, of large-vaned and chambered Triassic bivalves (Yancey & Stanley 1999), which may have harboured microbial symbionts. Whilst debate still continues over the life habits of enigmatic extinct taxa such as the rudists and inoceramids, new discoveries are gradually extending our knowledge of early Palaeozoic bivalve evolution and continually extending, or otherwise modifying, the geological ranges of known groups. The most recent published compilation of familial ranges (Skelton & Benton 1993) is strikingly different from that published in the *Treatise* and even then its authors admit that its quality is rather 'patchy'.

One of the most obvious short-term challenges is to produce a comprehensive compilation of all the available phylogenetic information and geological ranges and the present authors echo the calls by Yonge (1978) and Johnston & Haggart (1998) for an updated version of the *Treatise*.

References

ALDRIDGE, D. C. 1999. The morphology, growth and reproduction of Unionidae (Bivalvia) in a Fenland waterway. *Journal of Molluscan Studies*, **65**, 47–60.

AMLER, M. R. 1999. Synoptical classification of fossil and Recent Bivalvia. *Geologica et Palaeontologica*, **33**, 237–248.

CARTER, J. G. 1990. *Skeletal Biomineralization: Patterns, Processes and Evolutionary Trends*. Van Nostrand Reinhold, New York.

COX, L. R. & 24 others. 1969. *In:* MOORE, R. C. (ed.) *Treatise on Invertebrate Paleontology. Part N. Mollusca (1–2), Bivalvia.* Geological Society of America, Boulder, CO, and University of Kansas Press, Lawrence, KS.

DISTEL, D. L. 1998. Evolution of chemautotrophic endosymbioses in bivalves. *Bioscience*, **48**, 277–286.

HAAG, W. R., BUTLER, R. S. & HARTFIELD, P. D. 1995. An extraordinary reproductive strategy in freshwater bivalves: prey mimicry to facilitate larval dispersal. *Freshwater Biology*, **34**, 471–476

HARPER, E. M. 1998. The fossil record of bivalve

molluscs. *In:* DONOVAN, S. K. & PAUL, C. R. C. (eds) *The Adequacy of the Fossil Record.* John Wiley and Sons, Chichester, 243–267.

ISRAELSSON, O. 1999. New light on the enigmatic *Xenoturbella* (phylum uncertain): ontogeny and phylogeny. *Proceedings of the Royal Society of London,* **266B**, 835–841.

JOHNSTON, P. A. & COLLOM, C. J. 1998. The bivalve heresies – inoceramidae are Cryptodonta, not Pteriomorphia. *In:* JOHNSTON, P. A. & HAGGART, J. W. (eds) *Bivalves: An Eon of Evolution. Paleobiological Studies Honoring Norman D. Newell.* Calgary University Press, Calgary, 347–360.

—— & HAGGART, J. W. 1998. *In:* JOHNSTON, P. A. & HAGGART, J. W. (eds) *Bivalves: An Eon of Evolution. Paleobiological Studies Honoring Norman D. Newell.* Calgary University Press, Calgary, xi–xii.

MORRIS, N. J., DICKINS, J. M. & ASTAFIEVA-URBAITIS, K. 1991. Upper Palaeozoic anomalodesmatan bivalves. *Bulletin of the British Museum of Natural History (Geology Series)*, **47**, 51–100.

MORTON, B. 1981. Prey capture in the carnivorous septibranch *Poromya granulata* (Bivalvia: Anomalodesmata: Poromyacea). *Sarsia,* **66**, 241–256.

—— 1996. The evolutionary history of the Bivalvia. *In:* TAYLOR, J. D. (ed.) *Origin and Evolutionary Radiation of the Mollusca.* Oxford University Press, Oxford, 337–356.

POJETA, J. 1978. The origin and taxonomic diversification of pelecypods. *Philosophical Transactions of the Royal Society of London, Series B*, **284**, 225–243.

RUNNEGAR, B. 1978. Origin and evolution of the Class Rostroconchia. *Philosophical Transactions of the Royal Society of London, Series B,* **284**, 319–333.

SALVINI-PLAWEN, L. v. & STEINER, G. 1996. Synapomorphies and plesiomorphies in higher classification of the Mollusca. *In:* TAYLOR, J. D. (ed.) *Origin and Evolutionary Radiation of the Mollusca.* Oxford University Press, Oxford, 29–51.

SKELTON, P. W. & BENTON, M. J. 1993. Mollusca: Rostroconchia, Scaphopoda and Bivalvia. *In:* BENTON, M. J. (ed.) *The Fossil Record 2*. Chapman & Hall, London, 237–263,

STAROBOGATOV, Y. I. 1992. Morphological basis for phylogeny and classification of Bivalvia. *Ruthenica,* **2**, 1–25.

STANLEY, S. M. 1970. Relation of shell form to life habits of the Bivalvia. *Geological Society of America Memoir*, **125**, 1–296.

STENZEL, H. B. 1971. Oysters. *In:* MOORE, R. C. (ed.) *Treatise on Invertebrate Paleontology. Part N. Mollusca (3), Bivalvia.* Geological Society of America Boulder, CO, and University of Kansas Press, Lawrence, KS.

TAYLOR, J. D., KENNEDY, W. J. & HALL, A. 1969. The shell structure and mineralogy of the Bivalvia. Introduction, Nuculacea–Trigonacea. *Bulletin of the British Museum (Natural History), Zoology Series, Supplement*, **3**, 1–125.

——, —— & —— 1973. The shell structure and mineralogy of the Bivalvia. II. Lucinacea–Clavagellacea, Conclusions. *Bulletin of the British Museum (Natural History), Zoology Series, Supplement*, **22**, 253–284.

WALLER, T. R. 1978. Morphology, morphoclines and a new classification of the Pteriomorphia (Mollusca: Bivalvia). *Philosophical Transactions of the Royal Society of London, Series B,* **284**, 345–365.

—— 1998. Origin of the molluscan Class Bivalvia and a phylogeny of the major groups. *In:* JOHNSTON, P. A. & HAGGART, J. W. (eds) *Bivalves: An Eon of Evolution. Paleobiological Studies Honoring Norman D. Newell.* Calgary University Press, Calgary, 1–45.

YANCEY, T. E. & STANLEY, G. D. 1999. Giant alatoform bivalves in the Upper Triassic of western North America. *Palaeontology*, **42**, 1–23.

YONGE, C. M. 1978. Introductory remarks. *Philosophical Transactions of the Royal Society of London, Series B*, **284**, 201.

—— 1979. Cementation in bivalves. *In:* VAN DER SPOEL, S., VAN BRUGGEN, A. C. & LEVER, J. (eds) *Pathways in Malacology.* Bohn, Scheltema, Holkema and Junk, Utrecht, 83–106.

Molecular phylogeny of the Bivalvia inferred from 18S rDNA sequences with particular reference to the Pteriomorphia

GERHARD STEINER & SABINE HAMMER

Institute of Zoology, University of Vienna, Althanstr. 14, A-1090 Vienna, Austria
(e-mail: gerhard.steiner@univie.ac.at)

Abstract: The classification of the Pteriomorphia, a major division of the Bivalvia with Ordovician origins, is controversial both among palaeontologists and neontologists. To elucidate phylogenetic relationships new near-complete 18S rDNA sequences of 26 Pteriomorphia, three Protobranchia, three Heterodonta, one Anomalodesmata and three Scaphopoda were obtained, aligned with 71 other published molluscan sequences, and analysed with parsimony, maximum likelihood and spectral analysis. Although Bivalvia appear diphyletic due to heterogeneity of substitution rates among lineages, monophyly of Protobranchia, Heteroconchia and Pteriomorphia is supported. The heteroconch Lucinida, Myoida and Venerida are not monophyletic, and Anomalodesmata arise from within Heteroconchia. The basal nodes of Pteriomorphia have little support but two major clades, [Pinnoidea (Ostreoidea + Pterioidea)] and [(Anomioidea + Plicatuloidea) + (Limoidea + Pectinoidea)], are resolved with more confidence. The strongly supported clade of Anomioidea + Plicatuloidea, the separation of Pinnoidea from Pterioidea and most of the intrafamiliar relationships are not in accordance with morphological classifications. Combining these results with selected morphological characters, a phylogenetic hypothesis is proposed showing Mytiloidea and Arcoidea as the basal pteriomorph groups, the latter giving rise to the clade uniting the pinnoid–ostreoid–pterioid and the anomioid–limoid–pectinoid lines.

Molecular characters are potentially useful for phylogenetic studies, their usefulness depending on the time past since the cladogeneses or splits in question. Nucleotide sequences characteristically provide high numbers, usually hundreds or even thousands, of characters of little but equal complexity. They form data sets independent from morphological data and are subject to different selective constraints: DNA sequences, especially of ribosomal genes, are less selected by lifestyle and habitat than is morphology. This is especially important and needed when many morphological characters and their underlying homology decisions are equivocal due to convergent and/or parallel evolution, as is the case in the Bivalvia. Thus, it may be expected that molecular phylogenetic studies on bivalves, a group notorious for parallel evolution, will provide valuable insights into the branching pattern of its higher taxa.

Sequences of the nuclear-encoded small ribosomal subunit, the 18S rDNA, are particularly popular for studying relationships among and within phyla (e.g. Riutort *et al.* 1993; Philippe *et al.* 1994; Winnepenninckx *et al.* 1994, 1996; Mackey *et al.* 1996; Aguinaldo *et al.* 1997). The main reason for this popularity of this gene, as summarized by Winnepenninckx *et al.* (1994), is the alternation of variable and highly conserved sequences, facilitating amplification and providing phylogenetic information for various systematic levels. The first bivalve 18S rDNA sequences were used for investigations of relationships of Mollusca to other phyla and among molluscan classes (Field *et al.* 1988; Winnepenninckx *et al.* 1994, 1996). Studies on bivalve phylogeny inferred from 18S rDNA were initiated by Littlewood *et al.* (1991), Rice *et al.* (1993) and Kenchington *et al.* (1994, 1995), focusing on the families Ostreidae, Mactridae, Pectinidae and Mytilidae. Further complete sequences were included in the analyses by Steiner & Müeller (1996), Frischer *et al.* (1998), Maruyama *et al.* (1998) and Canapa *et al.* (1999), whereas Adamkewicz *et al.* (1997) and Campbell *et al.* (1998) relied on partial 18S rDNA sequences. Monophyly of the Bivalvia was unambiguously supported only by the analysis of Frischer *et al.* (1998) and, in a re-analysis of Steiner & Müeller (1996), by Giribet & Carranza (1999) and Steiner (1999). All other analyses showed Gastropoda or Polyplacophora separating Pteriomorphia from Heteroconchia. One reason for this unconventional topology was attributed to higher substitution rates

From: Harper, E. M., Taylor, J. D. & Crame, J. A. (eds) *The Evolutionary Biology of the Bivalvia*. Geological Society, London, Special Publications, **177**, 11–29. 1-86239-076-2/00/$15.00

in the heterodont and gastropod branches by Steiner & Müeller (1996). However, their study also found the phylogenetic signal for a bivalve clade almost equally strong as that for competing clades.

There are considerable discrepancies in the molecular studies also in regard to the relationships among the higher bivalve taxa, although protobranchs are generally the first bivalve branch. Frischer *et al.* (1998), with a limited set of species, show Pteriomorphia as sister group to the Heterodonta. Adamkewicz *et al.* (1997) and Campbell *et al.* (1998) present paraphyletic Pteriomorphia with respect to the Heterodonta or Heteroconchia, albeit with different topologies. A clade of Ostreoidea, Pinnoidea and Pterioidea takes a basal position in the tree of Adamkewicz *et al.* (1997), the Pectinoidea are the sister group to the Heteroconchia, and the anomalodesmatan species roots in the basal polytomy. In contrast, Campbell *et al.* (1998), using a different species set, have the Pectinoidea as the first autobranch clade and a polytomy of Pterioidea, Ostreoidea and Heterodonta, excluding the Pinnoidea. Furthermore, Adamkewicz *et al.* (1997) found no support for the heteroconch groups Anomalodesmata, Veneroida and Myoida nor for the Protobranchia.

The present study focuses on the phylogenetic connection of the Pteriomorphia to the other bivalve taxa and on the relationships among the major pteriomorph groups. Pteriomorphia appear in the fossil record no later than the early Ordovician period, where mytiloid, arcoid and pterioid forms are also recognized. These groups experienced periods of radiating diversification in the Late Devonian and both in the beginning and at the end of the Mesozoic (Carter 1990). Similar to the molecular studies, phylogenetic hypotheses based on morphological evidence differ considerably in the status of the Pteriomorphia and in the sister-group relationships proposed. Figure 1 summarizes five recently published hypotheses. Pteriomorphia appear monophyletic in Cope (1996), Waller (1998) and Morton (1996), although the latter includes the Unionida. Polyphyletic Pteriomorphia are represented in Carter (1990), with separate origin of the Mytiloidea, and in Starobogatov (1992), with Unionida and Heterodonta separating two pteriomorph groups. The aim of this investigation is to provide a database of near-complete 18S rDNA sequences with at least two representatives of all major pteriomorph groups and at least one of the Protobranchia, Heterodonta and Anomalodesmata as potential sister-group taxa. The tree(s) resulting from phylogenetic reconstruction are expected to elucidate sister group relationships of the Bivalvia (by the inclusion of scaphopod sequences), the origin of the Pteriomorphia as well as relationships within Pteriomorphia. These results can subsequently be of aid in evaluating homology decisions of morphological characters.

Materials and Methods

Most specimens sequenced in this study were collected in Safaga Bay (Red Sea, Egypt), Rovinj (Northern Adriatic Sea, Croatia) and Trondheim Fjord (Norway), either by scuba diving or dredging. The species used for the phylogenetic analyses are listed in Table 1 with their EMBL/GenBank accession numbers (sequences obtained herein are in bold).

DNA Extraction

Total DNA was extracted from adductor muscle, gills or, in small specimens, the total soft body of 30 bivalves (23 pteriomorph, three heterodont, one anomalodesmatan and three protobranch species) and three scaphopods. Minced tissue samples were incubated for 2 h at 65°C in 0.2 mL of sterile water, 5% Chelex® 100 resin (Sigma), 25 μL of proteinase K (20 mg mL^{-1}), 10 μL RNAse A (0.5 mg mL^{-1}) and 10 μL dithiothreitol (1 M) on a shaker and vortexed several times for 5 s during incubation. Following centrifugation for 2 min at 14 000 rpm, the remaining lysate was transferred to a fresh tube and stored at 4°C. Depending on the quality and amount of genomic DNA, 1–5 μL or appropriate dilutions of the sample lysate were used in the amplification procedure.

Polymerase Chain Reaction (PCR)

Near-complete sequences of the 18S rDNA gene were obtained by a direct PCR product-sequencing approach. In the first step, the complete 18S rDNA was amplified using the primers F (5′-CCG AAT TCG TCG ACA ACC TGG TTG ATC CTG-3′) and R (5′-CCC GGG ATC CAA GCT TGA TCC TTC TGC AGG TTC ACC TAC-3′) (Steiner & Mueller 1996). The PCR was performed on a Robocycler 96 Gradient (Stratagene) in a 25 μL reaction mix containing 1 mM $MgCl_2$, each dNTP at 200 μM, each primer at 50 pM, 2.5 units of *TAQ* polymerase (BioTherm, Genecraft), and the supplied reaction buffer at × 1 concentration. After an initial denaturation step of 2 min at 94°C, five PCR cycles of 40 s at a 50°C annealing temperature, 40 s at a 72°C primer extension, and 40 s at a 94°C denaturation were performed, followed by 35 cycles at a 61°C annealing temperature and a final primer extension step of 10 min at 72°C. In the second step, a nested PCR was performed to amplify two overlapping 18S rDNA fragments of 1050 and 900 base pair (bp)

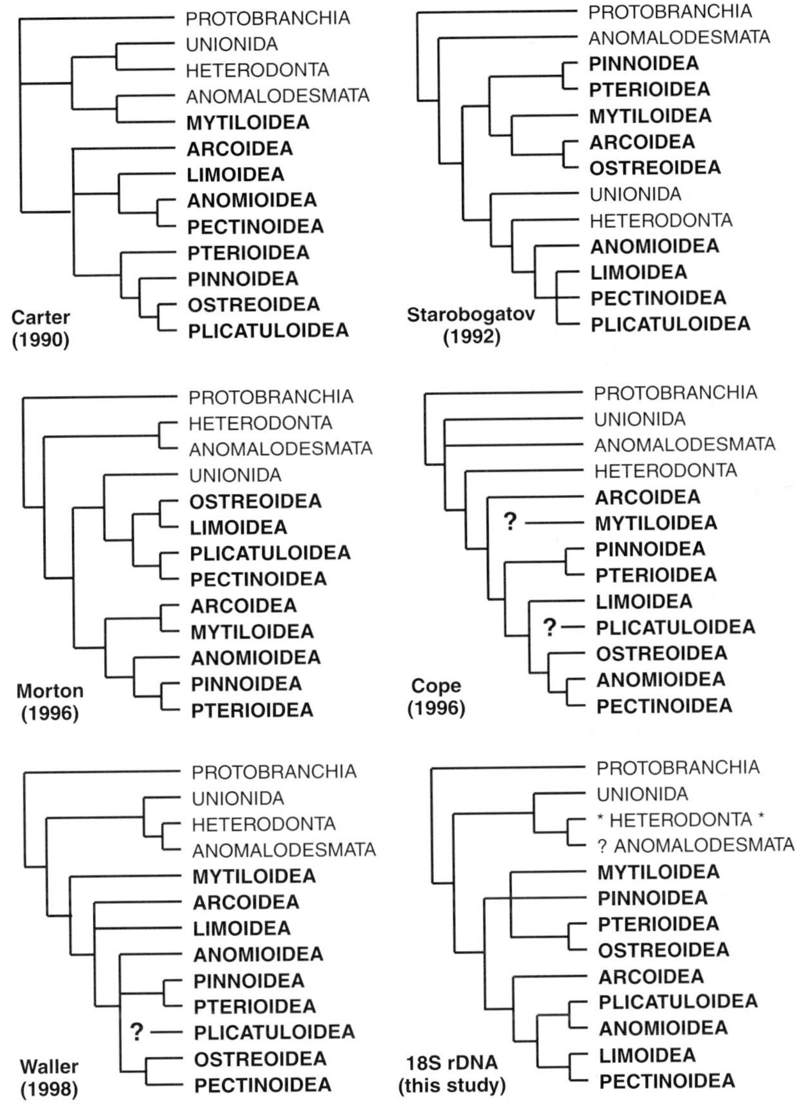

Fig. 1. Comparison of five recently published morphology based phylogenetic hypotheses of pteriomorph relationships with the results obtained in the present study.

length using the tagged primer pairs: pair 1: F*-T3 (5′-att aac cct cac taa agC AAC CTG GTT GAT CCT G-3′) and R2-STI2 (5′-cga tga aga acg cag cgA GAA CTR CGA CGG TAT C-3′); pair 2: F2-STI2 (5′-cga tga aga acg cag cgT CAG AGG YTC GAA GAC G-3′) and R*-T3 (5′-att aac cct cac taa agC CTT CTG CAG GTT CAC CTA C-3′). Lower case letters indicate the tag sequences (T3 and STI2). Nested PCR was carried out in a 75 μL reaction mix containing 1 mM $MgCl_2$, each dNTP at 100 μM, primers R2-STI2 and F2-STI2 at 75 pM, primers F*-T3 and R*-T3 at 37.5 pM, 7.5 units of *TAQ* polymerase (BioTherm, Genecraft), and the supplied reaction buffer at × 1 concentration. Each of the 35 PCR cycles consisted of 40 s at 94°C, 40 s at 52°C and 60 s at 72°C, followed by a final primer extension step of 10 min at 72°C. PCR products were purified with either Cleanmix purification kit (Talent) or QIAquick-spin PCR purification kit (Qiagen), and sequenced automatically with T3 and STI2 sequencing primers on a LI-COR 4000 (VBC-GENOMICS Bioscience Research GmbH).

Table 1. *Species used in the phylogenetic analysis*

Systematic position	Species	EMBL/GENBANK accession number
BIVALVIA		
Pteriomorphia		
Arcoidea		
Arcidae	*Arca noae* Linné, 1758	X90960
	Acar plicata (Dillwyn, 1817) §	AJ389630
	Barbatia virescens *(Reeve, 1844)*	X9197
	Barbatia cancellaria (Lamarck, 1819) [p]	AF022470, AF022471
	Anadara ovalis (Bruguiére, 1789) [p]	L78852
Noetiidae	*Striarca lactea* (Linné, 1758)	AF120531
Glycymerididae	***Glycymeris pedunculus*** (Linné, 1758) §	AJ389631
	Glycymeris sp.	X91978
Mytiloidea		
Modiolinae	***Modiolus auriculatus*** Krauss, 1848 §	AJ389644
Mytilinae	***Brachidontes variabilis*** (Krauss, 1848) §	AJ389643
	Septifer cf. ***bilocularis*** (Linné, 1758) §	AJ389645
	Mytilus galloprovincialis Lamarck, 1819	L33451
	Mytilus californianus Conrad, 1837	L33449
	Geukensia demissa (Dillwyn, 1817)	L33450
Lithophaginae	*Lithophaga lithophaga (Linné, 1758)*	AF120530
Pinnoidea		
Pinnidae	***Pinna muricata*** Linné, 1758 §	AJ389636
	Atrina pectinata (Linné, 1767)	X90961
	Atrina rigida (Lightfoot, 1786) [p]	L78850
Pterioidea		
Pteriidae	***Pteria macroptera*** (Lamarck, 1819) §	AJ389637
	Pteria hirundo (Linné, 1758)	AF120532
	Pteria brevialata (Dunker, 1872) [p]	L78849
	Pinctada margaritifera (Linné, 1758) §	AJ389638
	Pinctada imbricata Röding, 1798 [p]	AF022472, AF022473
	Electroma alacorvi (Dillwyn, 1817) §	AJ389641
Isognomonidae	***Isognomon legumen*** (Gmelin, 1791) §	AJ389639
	Isognomon alatus (Gmelin, 1791) [p]	AF022468, AF022469
Malleidae	***Malvifundus regulatus*** (Forsskal, 1775) §	AJ389640
	***Vulsella* sp.**§	AJ389642
Ostreoidea		
Gryphaeidae	***Hyotissa*** cf. ***hyotis*** (Linné, 1758) §	AJ389632
	Hyotissa cf. ***numisma*** (Lamarck, 1819) §	AJ389633
Ostreidae, Ostreinae	*Ostrea edulis* Linné, 1758	L49052
	Ostrea equestris Say, 1834 [p]	AF022466, AF022467
Ostreidae, Lophinae	***Lopha cristagalli*** (Linné, 1758) §	AJ389635
Ostreidae, Crassostreinae	*Crassostrea virginica* (Gmelin, 1791)	Z29549
	Saccostrea cucullata (Born, 1778) §	AJ389634
Anomioidea		
Anomiidae	***Anomia ephippium*** Linné, 1758 ‡	AJ389661
	Pododesmus caelata (Reeve, 1859) §	AJ389650
Plicatuloidea		
Plicatulidae	***Plicatula plicata*** (Linné 1767) §	AJ389651
Limoidea		
Limidae	***Lima lima*** (Linné, 1758) §	AJ389652
	Limaria hians (Gmelin, 1791)	AF120534
	Ctenoides annulatus (Lamarck, 1819) §	AJ389653
Pectinoidea		
Spondylidae	***Spondylus crassisquamatus*** Lamarck, 1819 §	AJ389646
	Spondylus hystrix Röding, 1798 §	AJ389647
Pectinidae	*Pecten maximus* (Linné, 1758)	L49053
	Placopecten magellanicus (Gmelin, 1791)	X53899
	Flexopecten glaber (Linné, 1758) ‡	AJ389662
	Argopecten irradians (Lamarck, 1819)	L11265
	Chlamys islandica (Müller O.F., 1776)	L11232
	Chlamys hastata (Sowerby, 1843)	L49049
	Mimachlamys varia (Linné, 1758)	L49051

Table 1. *Continued*

Systematic position	Species	EMBL/GENBANK accession number
	Crassadoma gigantea (Gray, 1825)	L49050
	Exellichlamys spectabilis (Reeve, 1853) §	AJ389648
	Pedum spondyloideum (Gmelin, 1791) §	AJ389649
Protobranchia		
Solemyida		
Solemyidae	***Solemya togata*** (Poli, 1795) * [p]	AJ389658
Nuculida		
Nuculidae	*Nucula proxima* Say, 1822 [p]	AF022476, AF022477
Nuculanidae	*Yoldia limatula* (Say, 1831) [p]	L78848
	Yoldiella nana (Sars M., 1865) †	AJ389659
	Nuculana pella (Linné, 1767) ‡	AJ389665
Heteroconchia		
Unionida		
Unionidae	*Anodonta imbecilis* Say, 1829 [p]	L78858
	Elliptio complanata (Lightfoot, 1786) [p]	L78857
Lucinida		
Lucinidae	***Cardiolucina semperiana*** (Issel, 1869) §	AJ389655
	Ctena divergens (Philippi, 1850) §	AJ389656
Ungulinidae	***Diplodonta*** cf. ***subrotundata*** Issel, 1869 §	AJ389654
Veneroida		
Veneridae	*Venus verrucosa* Linné, 1758	AJ007614
	Callista chione (Linné, 1758)	AJ007613
	Dosinia discus (Reeve,1850) [p]	L78863
	Mercenaria mercenaria (Linné, 1758) [p]	L78864
Tellinidae	*Donax variabilis* Say, 1822 [p]	L78867
	Tellina versicolor De Kay, 1843 [p]	U93556
Arcticidae	*Arctica islandica* (Linné, 1767)	U93555
Corbiculidae	*Corbicula leana* Prime, 1864 [p]	L78861
Galeommatidae	*Divariscintilla yoyo* Mikkelsen & Bieler, 1989 [p]	L78869
	Galeomma takii Kuroda, 1945	X91969
Cardiidae	*Tridacna gigas* (Linné, 1758)	D84189
	Tridacna squamosa Lamarck, 1819	D84190
	Hippopus hippopus (Linné, 1758)	D84660
	Hippopus porcellanus Rosewater, 1982	D84661
	Fragum unedo (Linné, 1758)	D84664
	Corculum cardissa (Linné, 1758)	D88909
	Vasticardium flavum (Linné, 1758)	D88910
	Fulvia mutica (Reeve, 1844)	D88911
Mactridae	*Mactromeris polynyma* (Stimpson, 1860)	L11230
	Mulinia lateralis (Say, 1822)	L11268
	Spisula solida (Linné, 1758)	L11266
	Tresus nutalli (Conrad, 1837)	L11269
Solenidae	*Ensis directus* Conrad, 1843 [p]	L78871
Myoida		
Myidae	*Mya arenaria* Linné, 1758 [p]	AF022478, M20094, M21541, M21175, J03774
Corbulidae	*Corbula contracta* Say, 1822 [p]	L78860
Pholadidae	*Cyrtopleura costata* (Linné, 1758) [p]	L78865
Hiatellidae	*Panopea japonica* A. Adams, 1850 [p]	L78870
Anomalodesmata		
Cuspidariidae	***Tropidomya abbreviata*** (Forbes, 1843) †	AJ389657
CAUDOFOVEATA		
Limifossoridae	*Scutopus ventrolineatus* Salvini-Plawen, 1968	X91977
POLYPLACOPHORA		
Ischnochitonina		
Chitonidae	*Liolophura japonica* (Lischke, 1873)	X70210
	Acanthopleura granulata (Gmelin, 1791) [p]	L78872
Ischnochitonidae	*Lepidochitona corrugata* (Reeve, 1848) [p]	X91975
Acanthochitonina		
Acanthochitonidae	*Crytptochiton stelleri* (Middendorf, 1847) [p]	L78873

Table 1 *continued overleaf*

Table 1. *Continued*

Systematic position	Species	EMBL/GENBANK accession number
GASTROPODA		
Neritopsina		
Neritidae	*Nerita albicilla* Linné, 1758	X91971
Vetigastropoda		
Trochidae	*Monodonta labio* (Linné, 1758)	X94271
Caenogastropoda		
Nassariidae	*Nassarius siquinjorensis* (A. Adams) = crematus (Hinds, 1844)	X94273
Tonnidae	*Bursa rana* (Linné, 1758)	X94269
Calyptraeidae	*Crepidula adunca* Sowerby, 1825	X94277
Gymnomorpha		
Onchidiidae	*Onchidella celtica* (Cuvier, 1817)	X70211
CEPHALOPODA		
Coleoidea		
Loliginidae	*Loligo pealei* Lesueur, 1821 [p]	no acc. number
SCAPHOPODA		
Dentaliida		
Dentaliidae	*Antalis vulgaris* (Da Costa, 1778)	X91980
	Antalis inaequicostata (Dautzenberg, 1891) ‡	AJ389660
	Dentalium bisexangulatum Sowerby, 1860 §	AJ389663
	Dentalium laqueatum Verrill, 1885 ‖	AJ389664

Sequences obtained in this study are in bold. Partial sequences are indicated with [p].
* Greece, Mediterranean; † Trondheim Fjord, Norway; ‡ Northern Adriatic; § Safaga Bay, Red Sea; ‖ Charleston, South Carolina

Sequence alignment, choice of taxa and phylogenetic analysis

The sequences obtained were aligned to published ones using CLUSTAL W (Thompson *et al.* 1994) with subsequent manual adjustment in DCSE (De Rijk & De Wachter 1993), taking into account secondary structural information when possible. Nearly all bivalve 18S rDNA sequences were included in the alignment to achieve a dense taxonomic representation of species, as advised by Lecointre *et al.* (1993). Some heteroconch partial sequences of doubtful quality and origin were excluded [e.g. the anomalodesmatan sequences of Adamkewicz *et al.* (1997) that are clustering with the Gastropoda]. Only two species of the genera *Mytilus, Tridacna* and one representative of *Fragum,* and each mactrid genus were used. Because sequence divergence, branch lengths and resolution among these taxa are minimal, this helped reduce the number of taxa in the analyses and computing time without losing valuable information. All available pteriomorph sequences were included covering most major groups. Unfortunately, no representative of the Dimyidae was available for sequencing. Gastropod sequences were selected mainly from the streptoneuran grade. Thus, the data set includes 91 Bivalvia (five Protobranchia, 33 Heteroconchia, 53 Pteriomorphia), six Gastropoda, four Scaphopoda, one Cephalopoda, four Polyplacophora and one Caudofoveata, totalling 107 species. Of these, 27 are represented by partial sequences. The alignment and input file is available by anonymous FTP from *ftp.ebi.ac.uk*/pub/databases/embl/align as alignment file DS40811.

A relative rate test for the complete 18S rDNA sequences in the data set was made using the output of the branch length test of LINTRE (Takezaki *et al.* 1995). The mean of the root-to-tip distances in a neighbour-joining tree (using *Scutopus ventrolineatus* or *Liolophura japonica* as outgroup) was calculated and the distribution of the individual differences from the mean recorded for each taxon in question. These values were tested by non-parametric tests (Kruskal-Wallis and Mann-Whitney-U-Test in SPSS for Windows 6.0.1; SPSS Inc.) and a permutation test [software described in Nemeschkal (1999)].

Parsimony and maximum likelihood were used for phylogenetic reconstruction because they both allow for the inclusion of partial sequences. Results of neighbour-joining analyses with the near-complete pteriomorph sequences were similar to the other results and are not shown. Unweighted parsimony analyses were performed with *PAUP**

4.0b2 (Swofford 1998) under Windows 95 and with *PAUP* 4d65* under UNIX. The high number of species made exhaustive and branch-and-bound searches unfeasible. For the heuristic searches uninformative characters were excluded, gaps were treated as missing and *Scutopus ventrolineatus* (Caudofoveata) was used as outgroup. The search strategy consisted of an initial search with 50 random taxon addition sequences, TBR branch swapping, MULPARS option in effect, no steepest descent, keeping 100 most parsimonious trees (MPT) of each replicate. This search was repeated three times with different random seed numbers. In the second step, all MPT and trees up to 1% longer than the MPT were submitted to TBR branch swapping to find all MPT. Because the overall transition/transversion ratio was found to be only slightly > 1, no weighted parsimony searches were made.

Bremer support indices (Bremer 1988, 1994) were calculated for the strict consensus tree to assess robustness of each branch in *PAUP** using a batch file produced by *Tree-Rot* (Sorensen 1996) with ten random addition sequences and keeping 100 trees per replicate for each search. The Bremer support index (or decay index) gives the number of additional steps needed to collapse a clade. Bootstrap values were calculated for the pteriomorph subtree. They are derived from 1000 bootstrap replicates with five random taxon addition sequences, TBR branch swapping, keeping 50 most parsimonious trees. It should be pointed out that the high proportion (*c.* 25%) of partial sequences in the alignment may affect both measures of support for the concerned nodes.

For the maximum-likelihood analyses (ML) both *PAUP** and *fastDNAml 1.0.6* (Olsen *et al.* 1992) were used. The large data set and a limitation in computer resources forced elimination of tree building by stepwise addition. Instead, the strict consensus tree was used as the starting tree for the calculation of ML parameters and subsequent branch swapping. In *PAUP**, empirical nucleotide frequencies were used, and the values for the transition/transversion ratio, proportion of invariable sites and gamma shape parameter were estimated under the Hasegawa–Kishino–Yano (HKY85) model with rate heterogeneity and four categories of substitution rates following a gamma distribution. The resulting values were then set for the nearest-neighbour interchange (NNI) branch swapping. This procedure was iterated as long as better trees were found. The options set for *fastDNAml* included the use of empirical nucleotide frequencies, a transition/transversion ratio of 1.2 (as estimated in *PAUP**), four categories of substitution rates (0.5, 1, 2 and 4) and tree rearrangements crossing five branches. Each alignment position was assigned to a rate category manually by the following criteria: category 1, sites with no more than two species having a substitution; category 2, no more than three different nucleotides and substitutions, mainly among family groups; category 3, all nucleotides present, substitutions also within families; category 4, hypervariable regions, single or multiple insertions/deletions. As arbitrary as this procedure is, it reflects the best estimate of substitution rates and yields trees with better ML scores than by using uniform rates. The ML trees obtained by both programmes were mutually tested and submitted to branch swapping in an iterative way until no further improvement was achieved. In addition to the complete data set, subsets containing only Arcoidea, Pectinoidea, Pinnoidea, Pterioidea and Ostreoidea (each with the nearest outgroup species), and Pteriomorphia + Protobranchia were used in both parsimony and ML analyses (including stepwise addition) to check the influence of the number of outgroup organisms on the topology. Resulting trees were handled with *TREEVIEW* (Page 1996).

The programs *PREPARE* and *HADTREE* (Hendy & Penny 1993) were used for spectral analysis of the Pteriomorphia using the same settings as in Steiner (1999), i.e. recoding the data to two-state characters and the 'sum of 7' option for calculating the frequencies of splits. Because the number of species is limited to 20 in these programs, only two species with complete sequences of each family group were included by chosing the least and the most derived ones in the ML tree. The protobranch *Yoldiella nana* was used as outgroup.

Results

The 18S rDNA sequences obtained in this study range in length from 1687 (*Flexopecten glaber*) to 1929 (*Ctenoides annulatus*) base pairs, except for *Solemya togata* where only a 778 base pairs long fragment of the 3′ end could be sequenced. This species is nevertheless included in the analyses because it is the only solemyid sequence available clustering with bivalves. The alignment with the other molluscan sequences submitted to the phylogenetic analyses has 2026 positions of which 875 are parsimony informative. The distribution of tree lengths of one million random trees is highly skewed with a g_1 index of –0.74 for the complete data set and –0.60 for the Protobranchia + Pteriomorphia subset. This indicates a highly hierarchical data structure both for the complete data and the subset.

Relative rate test

The relative rates are expressed in Fig. 2 by the difference from the average branch length in a NJ tree. Pteriomorphia have the lowest rates in this data set and differ significantly ($p < 0.05$) both from the Heteroconchia and all other outgroups. Among the Bivalvia only the two protobranch sequences show similar rates to the fastest pteriomorphs, the Limidae. Although Pteriomorphia are represented by 46 sequences, their range of rates is very small, especially when compared to the Heteroconchia.

Relationships of higher taxa

The parsimony analysis returned 30 571 MPT of 4617 steps (CI = 0.376, RCI = 0.259). The strict consensus tree (Figs 3 and 4) is one of the MPT and fairly well resolved, although the deep nodes especially have low support indices. Monophyly of the Bivalvia is not supported as the Heteroconchia are separated from the Protobranchia + Pteriomorphia clade by the Gastropoda and the Polyplacophora. The partial sequence of *Loligo* clusters with the gastropod *Monodonta*. The Scaphopoda appear paraphyletic at the base of the tree with *Dentalium laqueatum* joining the caudofoveat outgroup. Except for *Yoldiella* and *Nuculana*, the Protobranchia form an unresolved polytomy with the pteriomorph clade. The Unionida are the sister group to the remainder of the Heteroconchia clade. The family group taxa Galeommatidae, Lucinidae, Tridacnidae and Mactridae are monophyletic with high branch support between 8 and 60. Monophyly of Veneridae, Donacidae and Cardioidea is supported with Bremer indices from 2 to 4, whereas the order-level taxa Lucinida, Venerida and Myoida are polyphyletic. The branches connecting the family group taxa generally have little support (1 or 2) and are unstable with respect to the ML analysis (see below). Only the Cardioidea + Tridacnidae branch and the sister-group relation of Myidae and Corbulidae are robust.

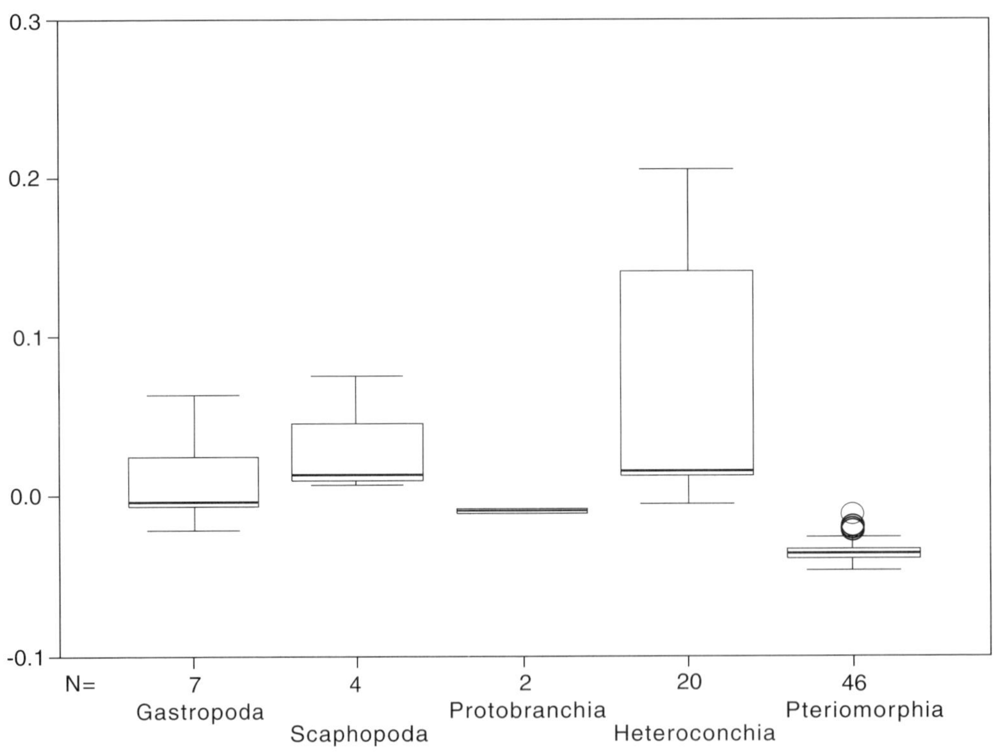

Fig. 2. Relative rate test comparing the differences from the mean root-to-tip distance (ordinate) in an NJ tree with *Liolophura japonica* as outgroup by taxonomic group (abscissa) using LINTRE. Pteriomorphia have significantly lower values ($p < 0.05$) than both Heteroconchia alone and all outgroup organisms pooled. The pteriomorph outliers (circles) correspond to the three species of the Limidae and *Lopha cristagalli* (Ostreidae) with the highest substitution rates.

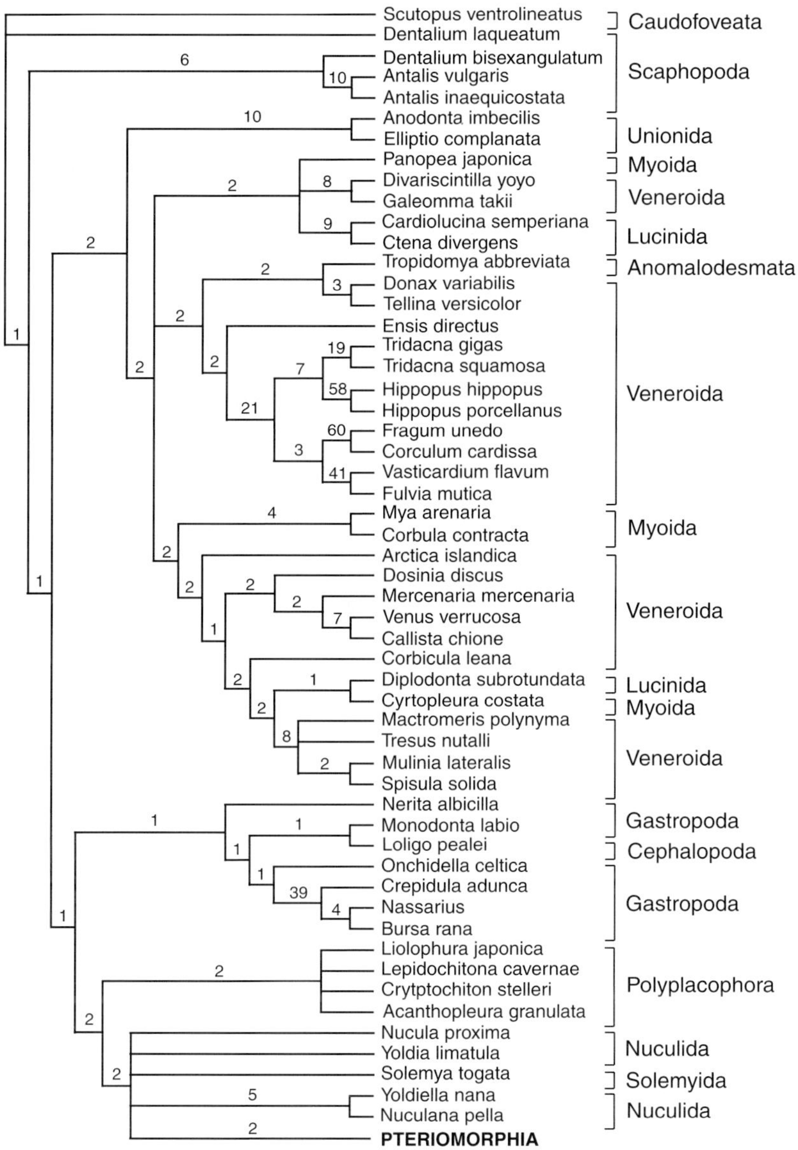

Fig. 3. Strict consensus tree and one of the 30 571 most parsimonious trees (length = 4617, CI = 0.376, RCI = 0.259) found by unweighted parsimony using a heuristic search with 50 replicates of random sequence addition and TBR branch swapping. The pteriomorph subtree is shown in detail in Fig. 4. Bremer support indices are indicted above the branches.

The branch of the Pteriomorphia has the same support (Bremer index 2) as that of the Heteroconchia. It splits up into two lines: one consists of the Arcoidea as the basal branch followed by a clade of Anomioidea and *Plicatula*, being the sister group to Limoidea and Pectinoidea. The second line shows a trichotomy of the Pinnoidea and Mytiloidea, and a clade uniting Pterioidea and Ostreoidea. As in the heteroconch clade, the highest branch support values are detected for some of the family rank taxa. Mytiloidea, Anomioidea, Limoidea, Spondylidae and Pectinidae have support indices of between 8 and 12, and bootstrap values of ≥ 97; Arcoidea,

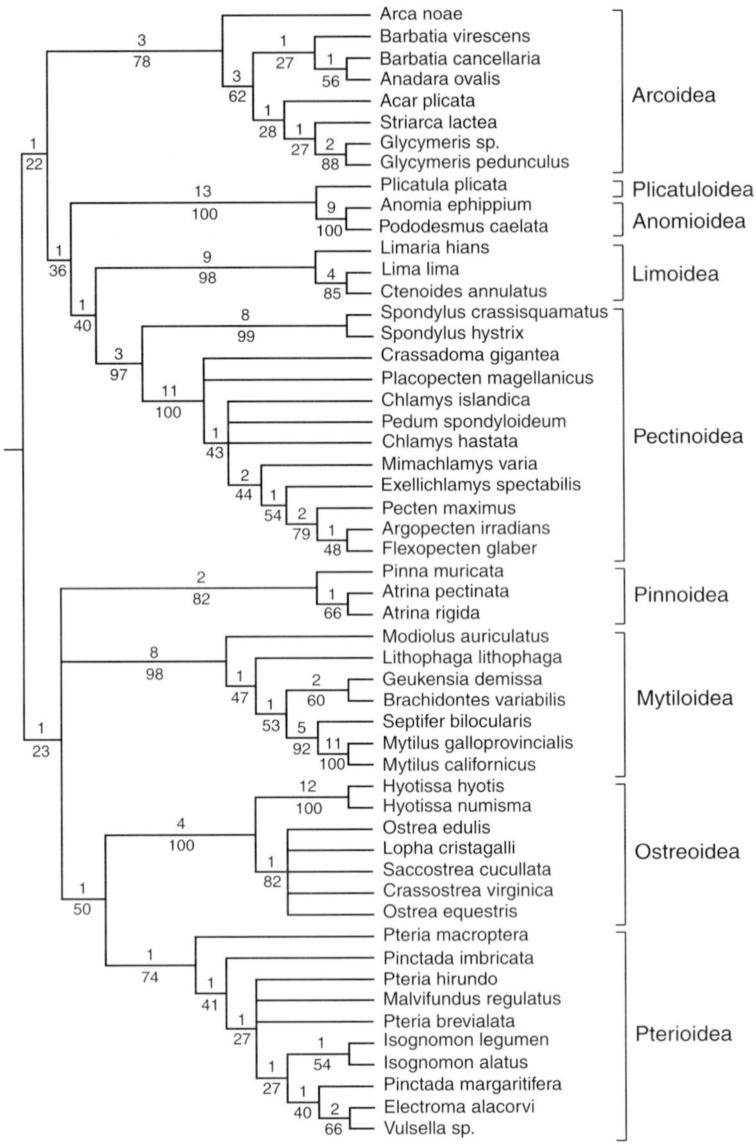

Fig. 4. Pteriomorph subtree of the strict consensus tree, continuing from the subtree of the outgroup species in Fig. 3. Bremer support indices are indicted above, and bootstrap values below, the branches.

Ostreioidea and Pectinoidea are intermediate, with support indices of 3–4 and bootstrap values between 78 and 100. Pinnoidea and Pterioidea have branch support indices of 2 and 1, and bootstrap values of 82 and 74, respectively. Again, internal branches connecting these family group taxa have low support and bootstrap values, the exception being the clade of *Plicatula* and the Anomioidea with an index of 13 and full bootstrap support, the highest score of the entire pteriomorph part of the tree. In contrast to the Heteroconchia, however, all internal pteriomorph branches are also present in the ML tree.

The ML tree (Figs 5 and 6) has log likelihood values of –26 224.810 55 in *PAUP** and –26 510.053 72 in *fastDNAml*. Although its overall topology is similar to the strict consensus tree, there are differences both in the relationships of the

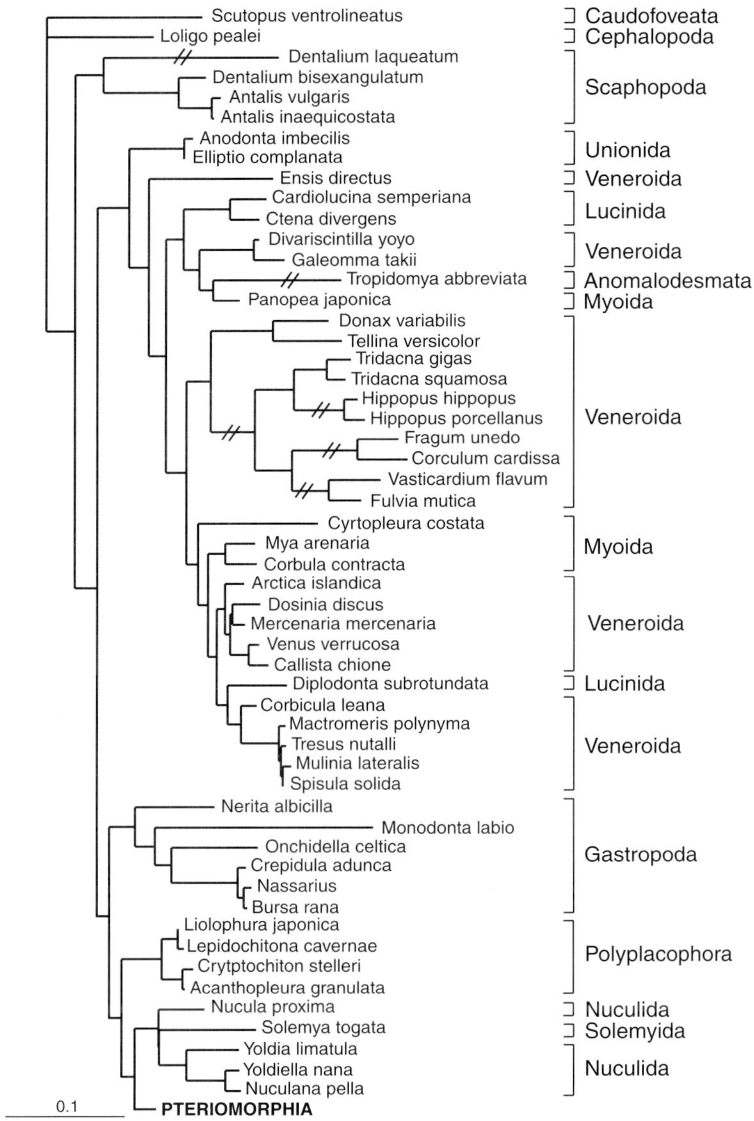

Fig. 5. Maximum likelihood (ML) tree found by *PAUP** (log L = –26 224.810 55) and fastDNAml (log L = –26 510.053 72). Details on the search strategy are given in the text. For better readability the lengths of the marked branches are reduced by a factor 1.5 in *Dentalium laqueatum* and *Tropidomya abbreviata*, and by factors 2 or 4 in the Tridacnidae and Cardiidae, respectively. This is the subtree of the outgroup species to the Pteriomorphia. The pteriomorph subtree is shown in detail in Fig. 6.

major molluscan groups and within the Heteroconchia and Pteriomorphia. The analysis yields monophyletic Scaphopoda and Gastropoda with *Loligo* here being closest to the outgroup. Protobranchia form a clade containing monophyletic Nuculanidae. The differences in the Heteroconchia clade concern the position of *Ensis*, now at the base of the Heterodonta, as well as the Anomalodesmata and Myoida. *Tropidomya* and *Panopea* are the sister group of the Galeommatidae. *Cyrtopleura* is close to the Myidae + Corbulidae clade forming a paraphyletic group to a part of the Veneroida and Lucinida.

The ML topology for the pteriomorph family groups is the same as in the parsimony strict consensus tree, except for resolving the trichotomy

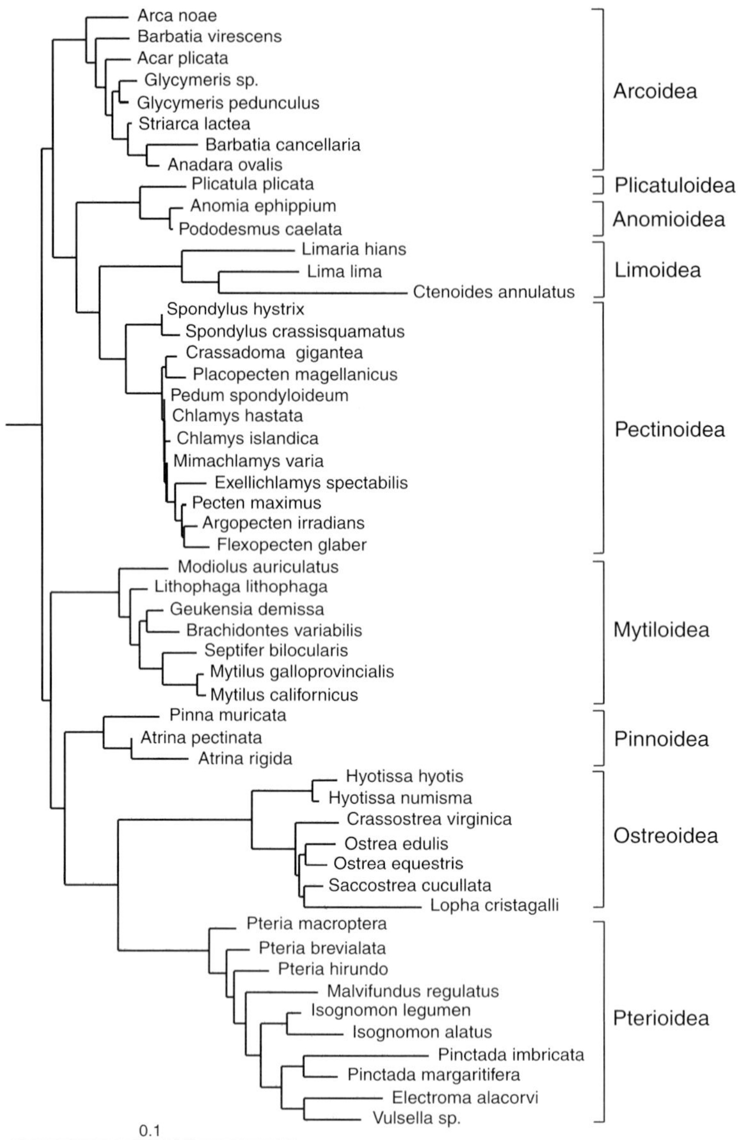

Fig. 6. Pteriomorph subtree of the maximum likelihood tree, continuing from the subtree of the outgroup species in Fig. 5.

by connecting the Pinnoidea as the sister group to the Pteriodea + Ostroidea clade. Analyses with the subsets Pteriomorphia + Protobranchia and with each of the family groups yield identical results, although the estimations of parameters for the transition/transversion ratio, proportion of invariant sites and gamma shape differ from the analysis of the complete data set.

Neighbour-joining analyses (HKY85 model, gamma distributed rate heterogeneity) with the complete pteriomorph sequences only (results not shown) agree with these results in regard to the Anomioidea + Plicatuloidea being the sister group to Limoidea + Pectinoidea, in clustering Ostreoidea to Pterioidea, and to very short branch lengths of the basal clades. Differences concern the Mytiloidea and the Ostreoidea + Pterioidea clade forming the basal branches of the Pteriomorphia varying in order according to the parameter used in the analysis. The position of the Pinnoidea also varies but never clusters with the Ostreoidea + Pterioidea clade.

Spectral analysis for selected pteriomorph species (Fig. 7) ranks the signals (frequency of occurrence of a split minus normalized frequency of conflicting splits) for the family group taxa highest, as could be expected. There are two exceptions: the node of Anomioidea + *Plicatula* as the most robust node uniting different family groups takes rank 4, thus being better supported than the Anomioidea; the Pectinoidea (rank 66), represented by *Spondylus hystrix*, *Placopecten magellanicus* and *Argopecten irradians*, shows the weakest signal of the family group taxa, and even of all nodes found in the strict consensus and ML trees. However, it also has very low conflict. The other branches connecting family groups take ranks 11 (Ostreoidea + Pterioidea), 20 (Ostreoidea + Pterioidea + Pinnoidea), 39 (Pectinoidea + Limoidea) and 55 (Pectinoidea + Limoidea + Anomioidea + *Plicatula*). The numerous nodes in between are mostly those uniting single species of either Limoidea or Ostreoidea to other families which can be explained by the high conflict or noise for these two groups. Some of the nodes in this area that are not in the ML tree (not indicated in Fig. 6) represent the strongest signal for connecting the Mytiloidea and Arcoidea to other family groups. Node 27 links Mytiloidea with Ostreoidea and *Yoldiella*, node 38 unites Arcoidea and Pterioidea, node 42 Arcoidea with Anomioidea and *Plicatula*, and node 50 Arcoidea with Mytiloidea and Pinnoidea. The latter is of interest because it is also present in some of the neighbour-joining trees and underlines the separated origin of the Pinnoidea and Pterioidea. The clade connecting these two groups ranks only 87.

Relationships within pteriomorph family groups (Figs 4 and 6)

The results for those pteriomorph groups with more than three species in the data set are reported in detail.

Arcoidea. In both analyses, *Arca noae* is the basal offshoot in the arcoid tree, and the Arcidae are not monophyletic. The strict consensus tree shows Arcidae paraphyletic to the sister groups *Striarca lactea* (Noetiidae) and Glycimerididae, whereas the ML tree roots these families within the Arcidae. The basal position of *Arca noae* has the highest Bremer support of all arcoid internal branches.

Mytiloidea. The representatives of the Modiolinae and Lithophaginae appear ancestral to the monophyletic Mytilinae. Although branch support is low, both parsimony and ML analyses return the same topology. *Geukensia* and *Brachidontes* are the

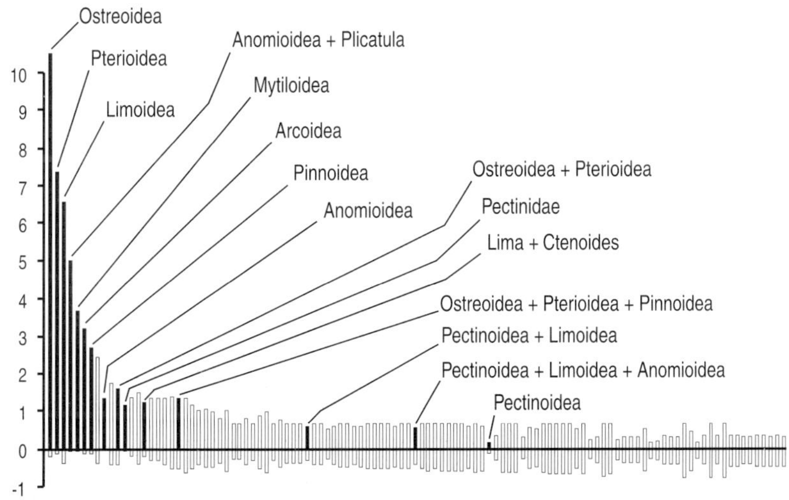

Fig. 7. Spectral analysis of selected pteriomorph species. Histogram of support (positive values on ordinate) and corrected conflict (negative values on ordinate) for splits (abscissa) in the alignment with net signal > 0, ranked by net signal (signal minus conflict). Shaded bars indicate nodes present in the ML tree. Species selected for each taxon: Ostreoidea = *Hyotissa hyotis* + *Lopha cristagalli*; Pterioidea = *Pteria hirundo* + *Malvifundus regulatus*; Mytiloidea = *Modioilus auriculatus* + *Geukensia demissa*; Arcoidea = *Arca noae* + *Glycymeris pedunculus*; Pinnoidea = *Atrina pectiniata* + *Pinna muricata*; Pectinidae = *Placopecten magellanicus* + *Argopecten irradians*, Pectinoidea = Pectinidae + *Spondylus hystrix*; all species of Anomioidea and Limoidea in the data set.

sister group to the clade of *Septifer* and *Mytilus* that is characterized by exceptionally high support values.

Ostreoidea. The sister-group relationship of Gryphaeidae and Ostreidae is supported by high bootstrap values, although the Bremer index for the ostreid branch is low. The resolution within the Ostreidae is poor, resulting in a polytomy of all five species in the strict consensus tree and very short internal branch lengths in the ML tree. However, the ML tree recovers the genus *Ostrea* as a clade. The subfamily Crassostreinae appears polyphyletic with *Crassostrea* at the base of the clade and *Saccostrea* clustering with *Lopha*.

Pterioidea. This taxon name is used here in a sense excluding the Pinnoidea. Of the three pterioid families treated in this data set only Isognomonidae, represented by two *Isognomon* species, appear monophyletic, whereas both Pteriidae and Malleidae are polyphyletic. The genus *Pteria* forms a paraphyletic stem group of the Pterioidea in the ML tree. The two species of *Pinctada* are recovered as a clade by ML only and not set in close relationship to *Pteria*. The pteriid *Electroma alacorvi* clusters with the malleid *Vulsella* as the sister group of *Pinctada*. With *Malvifundus* rooting just above the *Pteria* grade, this renders Malleidae polyphyletic. Branch support is low within the Pterioidea, the most robust clade being that of *Electroma* and *Vulsella*. However, the topologies of the parsimony and ML analyses are very similar.

Pectinoidea. The two *Spondylus* species comprise a well supported sister group to the Pectinidae. Branch support values within the Pectinidae are low but the topology of parsimony and ML trees are compatible. *Crassadoma* and *Placopecten* take the basal position in the pectinid subtree. The branching pattern of the *Chlamys* species and *Pedum* at the base of the remaining pectinids is unresolved in the strict consensus and marked by very short branch lengths in the ML tree. While *Mimachlamys varia* is part of this assemblage in the ML tree the support value of 2 connects it more to the top part of the tree by parsimony criteria. The top part consists of a graded series of *Excellichlamys*, *Pecten*, *Argopecten* and *Flexopecten*.

Discussion

Potential limits of methods

The data set and phylogenetic analysis presented here is one of the largest of bivalve 18S sequences and the largest of pteriomorph species published to date, and considerable significance of the results can be expected. However, a number of methodological limitations and known potential sources of error must be heeded to evaluate the results adequately.

The alignment is a difficult step in the analysis of the 18S rDNA gene due to the variations in sequence length in certain loop regions. This results in many alternative alignments of which only a few were tested. Information about secondary structure can be of assistance but is in itself ambiguous. If, for example, one end of a stem has a conserved sequence on one strand and substitutions on the other strand, the loop is elongated relative to the stem in the species concerned. Aligning strictly to secondary structure yields a less parsimonious alignment in this case. Dense taxonomic sampling should improve alignment quality as intermediate forms are expected to connect the extremes. This is one of the reasons why the partial sequences are included in the data set.

The low overall transition/transversion (s/v) ratio of 1.2 indicates a certain saturation of observed distances, as was already shown with a smaller data set (Steiner & Müeller 1996). The erosion of the phylogenetic signal by multiple substitutions concerns mostly ancient speciation events. The deep nodes are consequently not well supported and need to be viewed with greater caution than nodes higher up in the tree. On the other hand, the low overall s/v parameter estimated by and for ML analyses could lead to erroneous topologies in younger clades where the observed parameter value is much higher. Several analyses with subsets of pteriomorph families yield the same topology as the entire set. However, the tree topology of these speciation events depends less on the s/v parameter and more on the outgroup.

Another potential source of error is the strategy of the tree search in the computationally demanding ML analysis of this large data set. Taking the strict consensus tree of the parsimony analysis as the starting tree and allowing rearrangements crossing not more than five branches, significantly narrows the tree space covered by the search. That this range of branch swapping can alter and optimize a tree considerably is demonstrated by the removal of *Loligo pealei* from within the gastropods in the strict consensus to a position between the caudofoveat outgroup and the scaphopods. In any case, the swapping range is sufficiently wide to check on all topological variations of the major molluscan classes and subclasses represented in the study.

A frequently encountered problem is the 'long-branch attraction' effect causing parsimony and distance algorithms to cluster distantly related species or clades, either because they share a higher substitution rate than the remaining lines, increasing their random sequences similarities, or because the long undivided branches bear a high number of unobservable multiple hits (Swofford *et al.* 1996). The high degree of similarity of the parsimony and ML results indicates that the 'long-branch attraction' is not an important problem in the analysis. The taxa with the longest branches (Cardiidae and Tridacnidae, *Tropidomya abbreviata*, *Dentalium laqueatum*, *Monodonta labio*) are found in the outgroups, while Pteriomorphia have shorter than average branch lengths.

The error expected to be introduced by the partial sequences in the data set (*c.* 25% of the species) seems to be of little significance judged by their systematically orthodox position in the trees in most cases. Congeners of *Glycymeris*, *Atrina*, *Ostrea*, *Isognomon* and *Pinctada*, represented by a complete and partial sequence, are monophyletic, at least in the ML tree. Only the genera *Barbatia* and *Pteria* do not form clades, although *Pteria* is represented by two complete sequences. More problematic are the systematically questionable positions of *Loligo pealei* and *Ensis directus*, two of the shortest sequences used, varying greatly among the analyses. These findings need confirmation by more and more complete sequences of the respective taxa.

Relationships of higher taxa

The sister group of Bivalvia remains obscure, partly due to their apparently diphyletic origin. None of the conchiferan groups are close to the Protobranchia + Pteriomorphia clade, and Heteroconchia stand isolated between Scaphopoda and Gastropoda. Thus, a sister group of Bivalvia and Scaphopoda (e.g. Salvini-Plawen & Steiner 1996) is not supported here. Monophyletic Bivalvia are difficult to resolve with the 18S rDNA (Kenchington *et al.* 1994; Steiner & Mueller 1996; Winnepenninckx *et al.* 1996; Adamkewicz *et al.* 1997; Campbell *et al.* 1998), although Steiner & Müeller (1996) found the phylogenetic signal for Bivalvia to be only slightly weaker than conflicting ones. It was shown, however, by Frischer *et al.* (1998) and Giribet & Carranza (1999) that alignments yielding bivalve monophyly do exist. However, Steiner (1999) demonstrated that the alignment of Giribet & Carranza (1999) supports the bivalve clade only by a single non-homoplastic position. A common feature of the trees with monophyletic Bivalvia is the para- or polyphyly of Pteriomorphia with Ostreidae joining the heterodont branch. However, in Frischer *et al.* (1998) both Pteriomorphia and Heterodonta form clades but Ostreidae are the first pteriomorph offshoot. Apparent bivalve polyphyly is caused by separate origin of the Pteriomorphia and Heterodonta and/or *Solemya velum* clustering with the gastropods (Adamkewicz *et al.* 1997; Campbell *et al.* 1998). In the latter cases, this seems due to an aberrant sequence of this *Solemya* species, as *S. togata* in the present study clusters with the other Protobranchia. A reason for the autobranch groups not to cluster may be their differences in substitution rates as suspected by Steiner & Mueller (1996). The relative rate test showing pteriomorph substitution rates to be significantly lower than those of the other molluscan groups, and that of the Heteroconchia in particular, corroborates this explanation. The homogeneity of rates among the Pteriomorphia is remarkable and lends additional weight to the significance of these results.

All analyses presented here indicate monophyletic Pteriomorphia and Heteroconchia (Heterodonta + Anomalodesmata). The Protobranchia clade is supported in the ML tree only. The results for the Heteroconchia confirm the findings of Adamkewicz *et al.* (1997) showing Unionida as the basal branch, polyphyly of Myoida and Venerida, and generally little support for internal nodes. Not in agreement is the position of the Anomalodesmata: instead of joining the gastropod branch, *Tropidomya abbreviata* emerges from within the Heterodonta together with the hiatelloid *Panopea* as sister group to the Galeommatidae. Thus, if the probably artificial position of *Ensis* is not taken into account, the basal heterodont lineage consists of the Lucinidae, Galeommatidae, Hiatellidae and Cuspidariidae. This is an extraordinary taxonomic mix but corroborates McAlester (1966) and Morton (1996) in regard to the ancient origin of Lucinidae and Galeommatidae. It also agrees with Morton (1996), Salvini-Plawen & Steiner (1996) and Waller (1998) on the heterodont roots of the Anomalodesmata, although only with a part of the Myoida. On the other hand, the wide separation of the ungulinid *Diplodonta* from the Lucinidae challenges all systematic concepts. From these results, it appears highly unlikely that any of the Heteroconchia lines is a candidate for a sister group of the Pteriomorphia.

Pteriomorph relationships

The Pteriomorphia are supported as a clade in all analyses in a similar way to the Heteroconchia, both in terms of Bremer indices and of branch length in the ML tree. However, the basal nodes of the pteriomorph tree are not well supported and have to be taken *cum grano salis*. Two major clades showing stability in the analyses, in that all algorithms converge on the same topology, though with low branch support, are the lines of (Pinnoidea [Pterioidea + Ostreoidea)] and [(Anomioidea + Plicatuloidea) + (Limoidea + Pectinoidea)]. Among these, the position of the Pinnoidea varies most, suggesting an origin as ancient as that of the Arcoidea and Mytiloidea, lending support to Starobogatov´s (1992) concept of an early separation of this line from the other Pteriomorphia. Among the hypotheses shown, only Carter (1990) considers a non-monophyly of Pinnoidea and Pterioidea. None of the hypotheses depicted in Fig. 1 are fully supported, instead this study corroborates certain groupings in each of them.

The species set used for this study contains only single representatives of many pteriomorph families and some are not represented at all. The results, therefore, permit only restricted inference of intra family relationships. However, it is striking that the Arcidae and Pteriidae as the basal families of their respective family groups appear para- or even polyphyletic. The pterioid family Malleidae is also not supported, with *Vulsella* and *Malvifindus* being clearly separated on the tree. The common origin of Gryphaeidae and Ostreidae is strongly supported, contradicting the views of Stenzel (1971) and Carter (1990). Resolution within the Ostreidae is not satisfactory with the 18S rDNA, paralleling the findings with 28S rDNA sequences (Littlewood 1994). The branching order of subfamilies in the Mytiloidea has Modiolinae as the basal line followed by Lithophaginae and Mytilinae. Within the Mytilinae, *Septifer* and the *Mytilus* species are set off from *Geukensia* and *Brachidontes* species by a strongly supported and comparatively long branch. This is not congruent with the recent results of comparative sperm morphology (Kafanov & Drozdov 1998), suggesting a closer relationship of *Brachidontes* and *Septifer*. The pectinid topology recovered here differs in one aspect from the one in Frischer *et al.* (1998). The clade of *Crassadoma* and *Placopecten* is confirmed but instead of being derived from the *Chlamys* + *Mimachlamys* groups it takes the basal position in the pectinid tree. Instead, the *Chlamys* + *Mimachlamys* groups give rise to the *Decatopecten* – *Pecten* – *Aequipecten* groups. The difference in branching sequence may be attributed to the rather distant outgroup (*Crassostrea* or *Geukensia*) Frischer *et al.* (1998) used. This demonstrates the importance of using not too distantly related outgroups in molecular studies as the resulting long branches connecting it to the ingroup may bear 'so many changes that the sequences have become effectively randomized' (Swofford *et al.* 1996, p. 478). The present topology lends support to the systematic scheme of Waller (1991), deriving the pectinoid shell morphology with more or less equal size of shell auricles from the chlamydoid form with pronounced unequal auricles (Frischer *et al.* 1998), although Waller (1991) proposed that this occurred independently in the *Aequipecten* group.

Morphological aspects and proposed phylogeny

Morphological characters important for relationships among the major pteriomorph groups were selected from Carter (1990) and Waller (1998) and projected on the ML tree topology for these groups (Fig. 8a). The ML tree is preferred to the strict consensus parsimony tree because it resolves the mytiloid–pinnoid branching sequence. Recoding the characters given by these authors to a data matrix for a combined molecular and morphological analysis has been attempted but is now postponed. At the time of preparation of this paper, too many uncertainties in homology and gaps in character states for all the species in the molecular data set were evident, so that a more comprehensive analysis will be presented elsewhere.

Non-homoplastic synapomorphies (in bold) for the Pteriomorphia in Fig. 8a are the discontinuity of larval and post-larval ligaments, stomach morphology of type III of Purchon, rectum with a flattened cross-section, connected cardic auricles and a branchial ocellus in the juvenile. The two latter characters listed by Waller (1998) are somewhat doubtful; branchial eyes are not known from all pteriomorph groups (Cragg 1996). A fusion of auricles was reported by Atkins (1938) for all pteriomorphs except for *Anomia*, expanding on the results of Pelseneer (1911) but never confirmed on a broad scale. At least in *Arca noae*, this fusion must occur late in ontogeny as juveniles have clearly separate auricles (Steiner, pers. comm.). Another potential synapomorphy, though a homoplastic one, is the loss of the opisthodetic ligament. The reconstruction of this character involves either a reversal in the Mytiloidea or an independent loss in the Ostreioidea + Pterioidea + Pinnoidea line and on the branch leading to the clade including the Arcoidea.

Special attention is drawn to the five other characters that evolve in parallel along these two branches: the calcitic prisms in the outer shell layer, the inequivalve shell, the presence of a byssal notch, the palp–gill connection of type III and the dominant right abdominal sense organ (ASO). This collection of parallelisms, caused by the weakly supported and short-branched basal nodes of the 18S trees, prompt us to propose a preliminary phylogenetic hypothesis combining the more reliable branches of the molecular tree and the

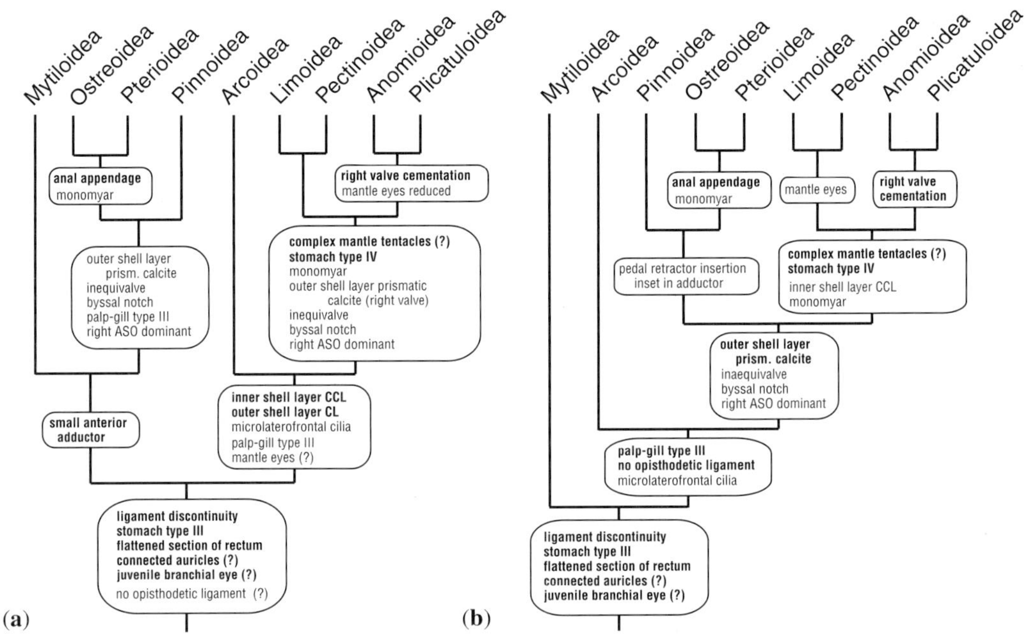

Fig. 8. (**a**) Morphological characteristics of major groups of the Pteriomorphia plotted on the idealized ML tree shown in Fig. 6. (**b**) Proposed phylogenetic tree of the Pteriomorphia combining molecular and morphological data. Non-homoplastic synapomorphies (bold) and homoplastic synapomorphies are shown. Reversals and convergences on terminal branches not shown; (?), questionable homologies.

morphological data (Fig. 8b). The proposed phylogeny recognizes the Mytiloidea as the first pteriomorph branch. Its sister group is defined by the loss of the opisthodetic ligament, microlaterofrontal cilia and a palp–gill connection of type III. The sister group to the Arcoidea now includes the ostreoid–pterioid–pinnoid clade and shows the calcitic prisms in the outer shell layer as non-homoplastic synapomorphy.

The other synapomorphies, inequivalve shells, a byssal notch and the dominant right ASO suffer reversals in the limoid branch. A possible synapomorphy for the ostreoid–pteroid–pinnoid clade is the inset of the posterior pedal retractor insertion, with the posterior adductor muscle experiencing reversal in the Ostreoidea. Ostreoidea and Pteriodea can be defined by their monomyarian condition and the presence of an anal appendage, although Harry (1985) and Waller (1998) express certain doubts about its homology. Monomyary is also a synapomorphy of the clade uniting Limoidea, Pectinoidea, Anomioidea and Plicatuloidea, together with the complex cross-lamellar inner shell layer and the type IV stomach. The complex mantle tentacles as classified by Waller (1978, 1998) unfortunately lack a sound morphological definition. Depending on the definition, their presence may be synapomorphic for the entire clade or for the Limoidea + Pectinoidea only. The presence of mantle eyes in the mantle margins diagnoses the latter (though paralleled in the Arcoidea); the cementation of the right valve is synapomorphic for the Anomioidea + Plicatuloidea. It is again emphasized that the proposed tree is still preliminary and has to be subjected to rigorous cladistic testing both by morphological and additional molecular characters as well as by the incorporation of palaeontological data.

Briefly summarizing the results for the relationships on a family level, the following can be pointed out without going into the detail of morphological characters: the monophyly of Mytilinae and Pectinidae is supported; Arcidae are paraphyletic; Pteriidae and Malleidae are polyphyletic.

The authors are indebted to Werner Piller and Martin Zuschin (Institute of Paleontology, University of Vienna) for the possibility to collect in Safaga, Egypt. Additional tissue samples were provided by David Campbell (University of South Carolina) and Argyro Zenetos (National Center of Marine Research, Greece). Robert Felber and Manfred Mueller (VBC Genomics Bioscience Research GmbH, Vienna) gave valuable technical support. Gonzalo Giribet permitted the use of four sequences prior to publication. This study was funded by the Fonds zur Föderung der wissenschaftlichen Forschung (FWF project no. P11846-GEN).

References

ADAMKEWICZ, S. L., HARASEWYCH, M. G., BLAKE, J., SAUDEK, D. & BUL, C. J. 1997. A molecular phylogeny of the bivalve mollusks. *Molecular Biology and Evolution*, **14**, 619–629.

AGUINALDO, A. M. A., TURBEVILLE, J. M., LINFORD, L. S., RIVERA, M. C., GAREY, J. R., RAFF, R. A. & LAKE, J. A. 1997. Evidence for a clade of nematodes, arthropods and other moulting animals. *Nature*, **387**, 489–493.

ATKINS, D. 1938. On the ciliary mechanisms and interrelationships of Lamellibranchia. Part VII. Latero-frontal cilia of the gill filaments and their phylogenetic value. *Quarterly Journal of Microscopic Science*, **80**, 345–436.

BREMER, K. 1988. The limits of amino acid sequence data in angiosperm phylogenetic reconstruction. *Evolution*, **42**, 795–803.

—— 1994. Branch support and tree stability. *Cladistics –The International Journal of the Willi Hennig Society*, **10**, 295–304.

CAMPBELL, D. C., HOEKSTRA, K. J. & CARTER, J. G. 1998. 18S ribosomal DNA and evolutionary relationships within the Bivalvia. *In:* JOHNSTON, P. A. & HAGGART, J. W. (eds) *Bivalves: An eon of Evolution – Paleobiological Studies honoring Norman D. Newell*. University of Calgary Press, Calgary, 75–85.

CANAPA, A., MAROTA, I., ROLLO, F. & OLMO, E. 1999. The small-subunit rRNA gene sequences of venerids and the phylogeny of bivalvia. *Journal of Molecular Evolution*, **48**, 463–468.

CARTER, J. G. 1990. Evolutionary significance of shell microstructure in the Paleotaxodonta, Pteriomorhpia and Isofilibranchia (Bivalvia: Mollusca). *In:* CARTER, J. G. (ed.) *Skeletal Biomineralization: Patterns, Processes and Evolutionary Trends. Vol.ume1.* Van Nostrand Reinhold, New York, 135–296.

COPE, J. C. W. 1996. The early evolution of the Bivalvia. *In:* TAYLOR, J. D. (ed.) *Origin and Evolutionary Radiation of the Mollusca.* Oxford University Press, Oxford, 361–370.

CRAGG, S. M. 1996. The phylogenetic significance of some anatomical features of bivalve veliger larvae. *In:* TAYLOR, J. D. (ed.) *Origin and Evolutionary Radiation of the Mollusca.* Oxford University Press, Oxford, 371–380.

DE RIJK, P. & DE WACHTER, R. 1993. DCSE, an interactive tool for sequence alignment and secondary structure research. *CABIOS*, **9**, 735–740.

FIELD, K. G., OLSEN, G., LANE, D., GIOVANNONI, ST., GHISELIN, M., RAFF, E., PACE, N. & RAFF, R. 1988. Molecular phylogeny of the animal kingdom. *Science*, **239**, 748–753.

FRISCHER, M. E., WILLIAMS, J. & KENCHINGTON, E. 1998. A molecular phylogeny of some major groups of Pectinidae inferred from 18S rDNA gene sequences. *In:* JOHNSTON, P. A. & HAGGART, J. W. (eds) *Bivalves: An eon of Evolution – Paleobiological Studies honoring Norman D. Newell*. University of Calgary Press, Calgary, 213–221.

GIRIBET, G. & CARRANZA, S. 1999. Point counter point: What can 18S rDNA do for bivalve phylogeny? *Journal of Molecular Evolution*, **48**, 256–258.

HARRY, H. W. 1985. Synopsis of the supraspecific classification of living oysters (Bivalvia: Gryphaeidae and Ostreidae). *Veliger*, **28**, 121–158.

HENDY, M. D. & PENNY, D. 1993. Spectral analysis of phylogenetic data. *Journal of Classification*, **10**, 5–24.

KAFANOV, A. I. & DROZDOV, A. L. 1998. Comparative sperm morphology and phylogenetic classification of recent Mytiloidea (Bivalvia). *Malacologia*, **39**, 129–139.

KENCHINGTON, E., LANDRY, D. & BIRD, C. J. 1995. Comparison of taxa of the mussel *Mytilus* (Bivalvia) by analysis of the nuclear small-subunit rRNA gene sequence. *Canadian Journal of Fisheries & Aquatic Sciences*, **52**, 2613–2620.

——, RODDICK, D. L., SINGH, K. R. & BIRD, C. J. 1994. Analysis of small-subunit rRNA gene sequences from six families of molluscs. *Journal of Marine Biotechnology*, **1**, 215–217.

LECOINTRE, G., PHILIPPE, H., LÊ, L. H. V. & LE GUYADER, H. 1993. Species sampling has a major impact on phylogenetic inference. *Molecular Phylogeny and Evolution*, **2**, 205–224.

LITTLEWOOD, D. T. J. 1994. Molecular phylogenetics of cupped oysters based on partial 28S rRNA gene sequences. *Molecular Phylogenetics and Evolution*, **3**, 221–229.

——, FORD, S. E. & FONG, D. 1991. Small subunit rRNA gene sequence of *Crassostrea virginica* (Gmelin) and a comparison with similar sequences from other bivalve molluscs. *Nucleic Acids Research*, **19**, 6048.

MACKEY, L. Y., WINNEPENNINCKX, B., DE WACHTER, R., BACKELJAU, T., EMSCHERMANN, P. & GAREY, J. R. 1996. 18S rDNA suggests that Entoprocta are protostomes, unrelated to Entoprocta. *Journal of Molecular Evolution*, **42**, 552–559.

MARUYAMA, T., ISHIKURA, M., YAMAZAKI, S. & KANAI, S. 1998. Molecular phylogeny of zooxanthellate bivalves. *Biological Bulletin*, **195**, 70–77.

MCALESTER, A. L. 1966. Evolutionary and systematic implications of a transitional Ordovician lucinoid bivalve. *Malacologia*, **3**, 433–439.

MORTON, B. 1996. The evolutionary history of the Bivalvia. *In:* TAYLOR, J. D. (ed.) *Origin and Evolutionary Radiation of the Mollusca.* Oxford University Press, Oxford, 337–360.

NEMESCHKAL, H. L. 1999. Morphometric correlation patterns of adult birds (Fringillidae: Passeriformaes and Columbiformes) mirror the expression of developmental control genes. *Evolution*, **53**, 899–918.

OLSEN, G. J., MATSUDA, H., HAGSTROM, R. & OVERBEEK, R. 1992. *Documentation for fastDNAml 1.0.* Electronic file distributed with the program.

PAGE, R. D. M. 1996. TREEVIEW: An application to display phylogenetic trees on personal computers.

Computer Applications in the Biosciences, **12**, 357–358.

PELSENEER, P. 1911. Les lamellibranches d'l'expédition du Siboga. Partie anatomique. *Siboga Expéditie Siboga Expédition Monograph*, **71**, 1–125.

PHILIPPE, H., CHENUIL, A. & ADOUTTE, A. 1994. Can the Cambrian explosion be inferred through molecular phylogeny? *Development*, **Supplement 1994**, 15–25.

RICE, E. L. 1990. Nucleotide sequence of the 18S ribosomal RNA gene from the Atlantic sea scallop *Placopecten magellanicus*. *Nucleic Acids Research*, **18**, 5551.

——, RODDICK, D. & SINGH, R. K. 1993. A comparison of molluscan (Bivalvia) phylogenies based on palaeontological and molecular data. *Molecular Marine Biology and Biotechnology*, **2**, 137–146.

RIUTORT, M., FIELD, K. G., RAFF, R. A. & BAGUÑA, J. 1993. 18S rRNA sequences and phylogeny of the Platyhelminthes. *Biochemical Systematics and Ecology*, **21**, 71–77.

SALVINI-PLAWEN, L. & STEINER, G. 1996. Synapomorphies and plesiomorphies in higher classification of Mollusca. *In:* TAYLOR, J. D. (ed.) *Origin and Evolutionary Radiation of the Mollusca*. Oxford University Press, Oxford, 29–52.

SORENSON, M. D. 1996. *Tree-Rot*. University of Michigan, Ann Arbor.

STAROBOGATOV, Y. I. 1992. Morphological basis for the phylogeny and classification of Bivalvia. *Ruthenica*, **2**, 1–26.

STEINER, G. 1999. Point counter point: What can 18S rDNA do for bivalve phylogeny? Response. *Journal of Molecular Evolution*, **48**, 258–261.

—— & MÜLLER, M. 1996. What can 18S rDNA do for bivalve phylogeny? *Journal of Molecular Evolution*, **43**, 58–70.

STENZEL, H. B. 1971. Oysters. *In:* MOORE, R. C. (ed) *Treatise on Invertebrate Paleontology, Part N. Volume 3, Mollusca 6, Bivalvia*. Geological Society of America, Boulder, CO, and Kansas University Press, Lawrence, KS, N953-N1224.

SWOFFORD, D. L. 1998. *PAUP*. Phylogenetic Analysis Using Parsimony (*and Other Methods). Version 4.* Sinauer Associates Inc., Sunderland, MA.

——, OLSEN, G. J., WADDELL, P. J. & HILLIS, D. M. 1996. Phylogenetic inference. *In:* HILLIS, D. M., MORITZ, C. & MABLE, B. K. (eds) *Molecular Systematics*, 2nd Edition. Sinauer Associates Inc., Sunderland, MA, 407–514.

TAKEZAKI, N., RZHETSKY, A. & NEI, M. 1995. Phylogenetic test of the molecular clock and linearized trees. *Molecular Biology and Evolution*, **12**, 823–833.

THOMPSON, J. D., HIGGINS, D. G. & GIBSON, T. J. 1994. CLUSTAL W: improving the sensitivity of progressive multiple alignment through sequence weighting, position specific gap penalties and weight matrix choice. *Nucleic Acid Research*, **22**, 4673–4680.

WALLER, T. R. 1978. Morphology, morphoclines and a new classification of the Pteriomorphia. *Philosophical Transactions of the Royal Society, London*, **B284**, 345–365.

—— 1991. Evolutionary relationships among commercial scallops (Mollusca: Bivalvia: Pectinidae). *In:* SHUMWAY, S. E. (ed.) *Scallops: Biology, Ecology and Aquaculture. Volume 21.* Elsevier, New York, 1–73.

—— 1998. Origin of the molluscan class Bivalvia and a phylogeny of major groups. *In:* JOHNSTON, P. A. & HAGGART, J. W. (eds) *Bivalves: An Eon of Evolution – Paleobiological Studies honoring Norman D. Newell*. Calgary University Press, Calgary, 1–47.

WINNEPENNINCKX, B., BACKELJAU, T. & DE WACHTER, R. 1994. Small ribosomal subunit RNA and the phylogeny of Mollusca. *Nautilus*, Suppl. **2**, 98–110.

——, —— & —— 1996. Investigation of molluscan phylogeny on the basis of 18s rRNA sequences. *Molecular Biology and Evolution*, **13**, 1306–1317.

Molecular evidence on the evolution of the Bivalvia

D. C. CAMPBELL

Department of Geological Sciences, University of North Carolina, Chapel Hill, NC 27599-3315, USA

Present address: Biology Department, St Mary's College of Maryland, 18952 E. Fisher Rd, St Mary's City, MD 20686–3001, USA (e-mail: dcampbell@osprey.smcm.edu)

Abstract: Despite widespread agreement on the monophyly of several major taxa of bivalves, others remain uncertain and the relationships among them are debated. The present study compares new and published morphological phylogenies with new analyses based on 18S gene sequences.

All but one family and all superfamilies in the Bivalvia were monophyletic in all the analyses. Several higher taxa, including most subclasses and orders, were also resolved as monophyletic. Only Myoida shows strong evidence for polyphyly, with at least two origins from Veneroida. Autobranchia was supported as monophyletic in the parsimony analyses. Within Pteriomorphia, Ostreoida is the sister taxon of Pterioida, if not derived from within it, rather than closest to Pectinoida. The numerous points of agreement with morphology based analyses suggests that both types of evidence are converging on a common phylogeny; however, differences remain to be resolved by further study.

The Class Bivalvia has been recognized in essentially its modern form since Linnaeus (Cox *et al.* 1969), but the relationships of its constituent superfamilies, orders and subclasses remain contested. Most families and superfamilies seem valid, but 'real difficulties arise...when attempts are made to group well-defined superfamilies into orders and subclasses' (Newell *in* Cox *et al.* 1969, p. N208). The present research found that molecular data from the 18S gene can address many of these problems.

Most phylogenetic analyses of the Bivalvia so far published are based on what the author considered to be significant morphological synapomorphies rather than computerized analyses. Several morphological characters are known to have arisen convergently in different bivalve lineages (eulamellibranch gill grade, porcelaneous microstructure, taxodont dentition, etc.), resulting in scepticism as to the ability of parsimony analyses to satisfactorily resolve bivalve classification. However, Salvini-Plawen & Steiner (1996) attempted a computerized parsimony analysis while noting difficulties with homoplasy. A new phylogenetic analysis (Carter *et al.* 2000) was also compared to the DNA results. Other recent morphology based bivalve phylogenies include Starobogatov (1992), Morton (1996), Cope (1997) and Waller (1998); these summarize much previous work on the class. Additional phylogenetic models for specific taxa were also considered. The proposed phylogenies in these papers were compared to the results of the present DNA analyses and to published DNA-based analyses.

The 18S gene was chosen for the present study because it has been useful in prior studies of higher level phylogeny of the Bivalvia and because the gene was already at least partially sequenced for several taxa (Field *et al.* 1988; Rice 1990; Holland *et al.* 1991; Littlewood *et al.* 1991; Kenchington *et al.* 1993; Rice *et al.* 1993; Kenchington *et al.* 1995; Steiner & Müller 1996; Winnepenninckx *et al.* 1996; Adamkewicz *et al.* 1997; Bell & Grassle 1998; Campbell *et al.* 1998; Frischer *et al.* 1998; Maruyama *et al.* 1998; Canapa *et al.* 1999). Although many taxonomically significant results were obtained, anomalies and uncertainties remain. Thus, further study of 18S gene sequences promises important data for resolving the relationships of higher taxa within the Bivalvia.

The only other genes for which sequences have been obtained for many bivalve superfamilies are the 16S gene, COI (both mitochondrial) and portions of the 28S gene. Mitochondrial genes risk reflecting an organelle phylogeny rather than the organismal phylogeny because many bivalves have biparental inheritance of mitochondria, with occasional loss and re-evolution of one lineage

From: HARPER, E. M., TAYLOR, J. D. & CRAME, J. A. (eds) *The Evolutionary Biology of the Bivalvia*. Geological Society, London, Special Publications, **177**, 31–46. 1-86239-076-2/00/$15.00

Table 1. *Taxa and sources*

Taxon	Reference	Accession number
Brachiopoda		
Lingula anatina Lamarck, 1801 [*'lingua'*]	Mackey *et al.* 1996	X81631
Glottidia pyramidata (Stimpson, 1860)	Halanych *et al.* 1995	U12647
Discinisca tenuis (G. B. Sowerby II, 1846)	Cohen *et al.* unpublished	U08327
'Neocrania' anomala (Müller, 1776)	Cohen *et al.* unpublished	U08328
Terebratalia transversa (G. B. Sowerby, 1846)	Halanych *et al.* 1995	U12650
Thecidellina blochmanni Dall, 1920 [*'blochmanii'*]	Cohen *et al.* 1998	AF025935
Neorhynchia sp. D1157	Cohen *et al.* 1998	AF025937
Eohemithiris grayii (Woodward, 1855) [*'Eohemithyris'*]	Cohen *et al.* 1998	AF025936
Phoronida		
Phoronis architecta Andrews, 1890	Mackey *et al.* 1996	U36271
Polyplacophora		
Cryptochiton stelleri (Middendorff, 1847)	Field *et al.* 1988	M20056-58
Liolophura japonica (Lischke, 1873)	Winnepenninckx *et al.* 1993	X70210
Lepidochitona corrugata (Reeve, 1848)	Winnepenninckx *et al.* 1996	X91975
Gastropoda		
Nerita albicilla Linnaeus, 1758	Winnepenninckx *et al.* 1996	X91971
Monodonta labio (Linnaeus, 1758)	Winnepenninckx *et al.* 1998*b*	X94271
Littorina littorea (Linnaeus, 1758)	Winnepenninckx *et al.* 1998*a*	Y91970
Crepidula adunca G. B. Sowerby I, 1825	Winnepenninckx *et al.* 1998*b*	X94277
Thais clavigera (Küster, 1860)	Winnepenninckx *et al.* 1996	X91979
Anisodoris nobilis (MacFarland, 1905)	Field *et al.* 1988	M20097-99
Onchidella celtica (Cuvier, 1817)	Winnepenninckx *et al.* 1994	X70211
Limicolaria kambeul (Bruguière, 1792)	Winnepenninckx *et al.* 1992	X66374
Bivalvia: Nuculoidea		
Nucula proxima Say, 1822	new	AF120526
Acila castrensis (Hinds, 1843)	new	AF120527
Nuculanoidea		
Nuculana minuta (Müller, 1776)	new	AF120529
Neilonella subovata (Verrill & Bush, 1897)	new	AF207645
Mytiloidea		
Brachidontes exustus (Linnaeus, 1758)	new	AF229623
Geukensia demissa (Dillwyn, 1817)	Kenchington *et al.* 1995	L33451
Modiolus americanus (Leach, 1815)	new	AF229624
Musculus lateralis (Say, 1822)	new	AF229625
Mytilus californianus Conrad, 1837	Kenchington *et al.* 1995	L33449
Mytilus edulis Linnaeus, 1758	Kenchington *et al.* 1995	L24489
Septifer bilocularis (Linnaeus, 1758)	new	AF229622
Arcoidea		
Arca noae (Linnaeus, 1758)	Steiner & Müller 1996	X90960
Barbatia virescens (Reeve, 1844)	Winnepenninckx *et al.* 1996	X91974
Glycymeris glycymeris (Linnaeus, 1758) ['sp.']	Winnepenninckx *et al.* 1996	X91978
Anomioidea		
Pododesmus macrochisma (Deshayes, 1839)	new	AF229627, AF229628
Plicatuloidea		
Plicatula australis (Lamarck, 1819)	new	AF229626
Pectinoidea		
Chlamys islandica (Müller, 1776)	Kenchington *et al.* 1993	L11232
Argopecten irradians (Lamarck, 1819)	Rice *et al.* 1993	L11265
Pecten maximus (Linnaeus, 1758)	Frischer *et al.* 1998	L49053
Placopecten magellanicus (Gmelin, 1791)	Rice 1990	X53899
Spondylus sinensis Schreibers, 1793	new	AF229629
Pinnoidea		
Atrina pectinata (Linnaeus, 1767)	Steiner & Müller 1996	X90961
Pterioidea		
Pinctada margaritifera (Linnaeus, 1758)	new	AF229620
Isognomon isognomum (Linnaeus, 1758)	new	AF229621
Ostreoidea		
Ostrea edulis Linnaeus, 1758	Frischer *et al.* 1998	L49052
Crassostrea virginica (Gmelin, 1791)	Littlewood *et al.* 1991	Z29549
Hyotissa hyotis (Linnaeus, 1758)	new	AF229618, AF229619

Table 1. *Continued*

Taxon	Reference	Accession number
Unionoidea		
Elliptio complanata (Lightfoot, 1786)	Adamkewicz *et al.* 1997	L78857
Utterbackia imbecilis (Say, 1829) ['*Anodonta*']	Adamkewicz *et al.* 1997	L78858
Margaritifera margaritifera (Linnaeus, 1758)	new	AF229612
Carditoidea		
Carditamera floridana Conrad, 1838	new	AF229617
Cyamioidea		
Basterotia elliptica (Récluz, 1850)	new	AF229616
Galeommatoidea		
Galeomma takii (Kuroda, 1945)	Winnepenninckx *et al.* 1996	X91969
Divariscintilla yoyo Mikkelsen & Bieler, 1989	Adamkewicz *et al.* 1997	L78869
Gastrochaenoidea		
Gastrochaena stimpsonii Tryon, 1861	new	AF229615
Solenoidea		
Ensis directus Conrad, 1843	Adamkewicz *et al.* 1997; Bell & Grassle 1998	L78871; U93557
Ensiculus cultellus (Linnaeus, 1758)	new	AF229614
Tellinoidea		
Tellina versicolor DeKay, 1843	Bell & Grassle 1998	U93556
Asaphis deflorata (Linnaeus, 1758)	Adamkewicz *et al.* 1997	L78868
Donax variabilis Say, 1822	Adamkewicz *et al.* 1997	L78867
Cardioidea		
Fulvia mutica (Reeve, 1844)	Maruyama *et al.* 1998	D88911
Vasticardium flavum (Linnaeus, 1758)	Maruyama *et al.* 1998	D88910
Tridacna gigas (Linnaeus, 1758)	Maruyama *et al.* 1998	D84189
Tridacna crocea Lamarck, 1819	Maruyama *et al.* 1998	D88908
Hippopus hippopus (Linnaeus, 1758)	Maruyama *et al.* 1998	D84660
Fragum fragum (Linnaeus, 1758)	Maruyama *et al.* 1998	D84662
Corculum cardissa (Linnaeus, 1758)	Maruyama *et al.* 1998	D88909
Veneroidea		
Venus verrucosa Linnaeus, 1758	Canapa *et al.* 1999	AJ007614
Mercenaria mercenaria (Linnaeus, 1758)	Frischer *et al.*, unpublished	AF106073
Callista chione (Linnaeus, 1758)	Canapa *et al.* 1999	AJ007613
Arcticoidea		
Arctica islandica (Linnaeus, 1767)	Bell & Grassle 1998, unpublished	U93555
Corbiculoidea		
Corbicula leana Prime, 1864	Adamkewicz *et al.* 1997	L78861
Mactroidea		
Mactromeris polynyma (Stimpson, 1860)	Rice *et al.* 1993	L11230
Spisula solida (Linnaeus, 1758)	Rice *et al.* 1993	L11266
Mulinia lateralis (Say, 1822)	Rice *et al.* 1993	L11268
Tresus nuttalli (Conrad, 1837)	Rice *et al.* 1993	L11269
Myoidea		
Mya arenaria Linnaeus, 1758	Field *et al.* 1988; Campbell *et al.* 1998	M20094, M21541, M21175; AF022478
Pholadoidea		
teredinid (pallets not preserved)	new	AF229613
Pandoroidea		
Pandora arenosa Conrad, 1834	new	AF120539
Lyonsia floridana Conrad, 1849	new	AF120540
Poromyoidea		
Myonera species	new	AF120544

For Bivalvia, the superfamily is given; for other Mollusca, the class; and for non-molluscs, the phylum. Taxonomic changes or corrections are noted in square brackets. Most of these have also been noted by the original author or other reviewers. However, *Neocrania* is an apparently overlooked junior homonym. Bell & Grassle (1998) in their text give corrected, more complete sequences for *Tellina versicolor* and *Ensis directus* than the GenBank files, but only use part of the *Arctica islandica* sequence; the most complete sequences were used in the present analyses.

(Hoeh *et al.* 1997). Also, they generally evolve more rapidly than nuclear genes and so are less useful for earlier evolutionary events (Hillis & Dixon 1991). Finally, the 18S gene has the most available data.

Despite its potential, molecular phylogenetics has often yielded problematical results, conflicting with previous morphological studies. In some cases, review of morphological data has provided support for the new classification, whereas other instances seem to reflect problems with the molecular analyses. Although exceptions are known (Felsenstein 1978), usually increasing either the number of taxa or the amount of data will improve accuracy (Poe & Swofford 1999). In particular, it is important to avoid wide phylogenetic gaps among taxa to avoid spurious associations ('long-branch attraction') (Felsenstein 1978). By including additional taxa, the present study sought to address some of these problems.

Materials and Methods

Taxa

The present study sought to include as many higher level taxa as possible. However, various higher level taxa are used differently by different authors. A consensus of recent systematic works suggests recognizing five extant subclasses, although other classifications have been advocated (e.g. Carter 1990; Cope 1997; Carter *et al.* 2000). The subclass Protobranchia includes the orders Nuculoida and Solemyoida; the subclass Pteriomorphia includes the orders Arcoida, Limoida, Ostreoida, Pectinoida, Pterioida and Mytiloida; the subclass Heterodonta includes the orders Veneroida and Myoida; the subclass Palaeoheterodonta includes the orders Unionoida and Trigonioida; and the subclass Anomalodesmata includes the order Pholadomyoida. The subclasses other than Protobranchia are often grouped as the Autobranchia. Within the Autobranchia, two major groups have been recognized: the Pteriomorphia and the Heteroconchia (including Palaeoheterodonta, Heterodonta and Anomalodesmata). Other classifications split the Mytiloida from Pteriomorphia as the subclass Isofilibranchia. Suborders of Veneroida are herein referred to as Lucinoidei and Veneroidei, following the endings used by Starobogatov (1992) in the light of variation in ending and assigned rank.

The present analyses included all published complete bivalve 18S sequences, except in the case of multiple sequences for the same genus. New sequences focused on additional higher taxa. Published partial sequences were included if they represented superfamilies for which no more than one complete sequence was available and if their position in analyses was relatively well-constrained. Table 1 lists the taxa and sources. Accession numbers allow location of the data in GenBank. In preliminary analyses, Brachiopoda, Phoronida and other molluscs were the closest relatives of Bivalvia, so these were retained as outgroups for the final analyses. Only a single sequence was available for Scaphopoda and Chaetodermomorpha, so they were excluded due to the long branch (at least Cambrian–Recent). This left multiple sequences for Polyplacophora, Gastropoda and Brachiopoda, along with the phoronid, for outgroups.

Molecular Techniques

Molecular procedures followed standard protocols for cetyltrimethylammonium bromide (CTAB) extraction, polymerase chain reaction (PCR) and dye-labelled cycle sequencing. PCR and sequencing primers were designed using published primers (Rice 1990; Winnepenninckx *et al.* 1994; Giribet *et al.* 1996; Whiting *et al.* 1997) and results of preliminary research by Campbell *et al.* (1998). PCR primers were CAACCTGGTTGATCCTG (forward) and CTGATCCTTCTGCAGGTTC (reverse). Additional sequencing primers were GTTCGATTCCGGAGAGGGA and ATGGTTGCAAAGCTGAAAC (forward), and AGGCTCCCTCTCCGGAATCGAAC and TTGGCAAATGCTTTCGC (reverse). The PCR started with 4 min at 94°C, followed by 26 cycles with 1 min 94°C, 1 min at the annealing temperature and 2 min at 72°C. Annealing temperatures were 56, 54, and 52°C for the first three cycles, 50°C for the next three cycles and 48°C for the remainder. This was followed by 10 min at 72°C. The sequencing reaction was held at 96°C for 4 min, followed by 25 cycles at 96°C for 30 s and then at 49°C for 4 min 30 s. The protobranch and anomalodesmatan sequences were obtained by Gonzalo Giribet of the American Museum of Natural History from specimens obtained by the present author. New sequences for Pteriomorphia were largely obtained while working at the American Museum of Natural History. The new complete sequences have both strands sequenced. A preliminary partial sequence was included for Teredinidae, as no complete sequences are available. It agrees well with unpublished preliminary data for Pholadidae in the same superfamily. Attempts at further sequencing from the teredinid were unsuccessful. The shells from analysed specimens were deposited in the University of North Carolina at Chapel Hill collections as vouchers.

Analytical Methods

The DNA sequences were aligned by hand, with reference to the secondary structure model of Winnepenninckx *et al.* (1994). Alignment procedures remain problematical (Giribet & Wheeler 1999) and variations in alignment may generate different cladograms (Winnepenninckx & Backeljau 1996). Although computerized alignments are more replicable than manual alignments, there is no guarantee that they are better. Repeated alignment of the same sequences was used to identify areas of uncertain alignment, which were revised to ensure consistency. The present alignment was submitted to EMBL at http:/www.ebi.ac.uk/ and is alignment DS41611. Maximum parsimony, neighbour-joining and bootstrap analyses were performed using *PAUP 3.1.1* (Swofford 1993) and *PAUP*4.0* (Swofford 1999). Decay indices (also called Bremer indices) were calculated using *TreeRot* (Sorenson 1999), *PAUP 3.1.1* and *PAUP*4.0*. Multiple outgroups were used as there is disagreement on the closest relatives of the Bivalvia (e.g. Runnegar 1996; Waller 1998). After preliminary analyses, including several outgroups, Gastropoda, Polyplacophora and Brachiopoda were selected because they consistently placed close to Bivalvia, if not within it, and because multiple complete sequences were available, decreasing the risk of long-branch attraction. The first 20 and last 23 bases were excluded because they represent primer sequence rather than actual data for most sequences. Multistate characters were treated as uncertain.

Maximum-parsimony analyses were used to generate phylogenetic models. The heuristic analyses used random addition of taxa, with 100 replicates. Tree bisection–rejoining was used for branch swapping (Swofford & Begle 1993). The best treatment of gaps is debated (Giribet & Wheeler 1999). Although ignoring them can omit informative data, the presence of long indels made other methods more problematical. The first analysis treated gaps as missing data. As a test of the effect of gaps, a second analysis was performed with gaps treated as a fifth character state. This represents an extreme weighting of gaps, so clades supported by both analyses are probably relatively stable with respect to gap weighting.

The trees from the heuristic searches were combined to generate strict consensus trees. Bootstrap analysis used 100 replicates, each of which was a heuristic search with closest addition, holding ten trees at each step and using tree bisection–rejoining. Decay indices were calculated using heuristic searches of 20 random replicates for trees incompatible with the constraint (Sorenson 1999). Although no absolute criteria are known for evaluating decay indices (Swofford *et al.* 1996), higher indices suggest stronger support.

A neighbour-joining tree was also calculated (Saitou & Nei 1987). Neighbour joining is better regarded as an algorithm for generating trees to use in further searching rather than as a source for final trees, as it has no optimality criterion (Swofford *et al.* 1996). However, it has been widely used to generate final trees. It is included for comparison and as a demonstration of nodes that are robust to multiple methods of analysis. It used a one-parameter model with gaps coded as missing data.

Hillis & Bull (1993) found that, for unbiased data sets, bootstrap values above 70% correlate with a 95% probability of phylogenetic accuracy. Asymmetry in the present cladograms and variation in evolutionary rates would tend to raise the bootstrap value necessary for 95% confidence, but unless the data are extremely divergent, bootstrap percentages tend to underestimate the phylogenetic probability.

Results

Complete sequence lengths range from 1794 to 2192 bases, with the alignment length at 2330 bases. The longest sequences were the anomalodesmatans, ranging from 1919 to 2192. The next longest were the fragines, from the data of Maruyama *et al.* (1998). Other new sequences ranged from 1794 to 1820 bases, although some of this variation reflects variation in the number of bases resolved near the primers.

With gaps coded as missing data, 764 included positions are parsimony informative. The nine shortest trees found by parsimony with gaps coded as missing were 3744 steps and were found by 97 of the replicates. The g1 value was –0.730 689, based on 100 000 random trees. Other tree statistics include a consistency index of 0.4581 (0.3945 excluding uninformative characters), a retention index of 0.6999 and a rescaled consistency index of 0.3206. Figure 1 shows the strict consensus tree with decay index values indicated and Fig. 2 shows the bootstrap results. Decay values range from 1 to 57, suggesting that long-branch attraction may cause problems. The highest decay indices (over 20) were confined to Cardiidae, Anomalodesmata, Nuculanoidea and Caenogastropoda. The highest values are within Cardiidae.

The neighbour-joining tree was 3780 steps long, evaluated using maximum parsimony. The consistency index was 0.4537 (0.3903 excluding uninformative characters), the retention index was 0.6946 and the rescaled consistency index was 0.3151. Figure 3 shows the neighbour-joining tree. Differences with the first analysis are noted.

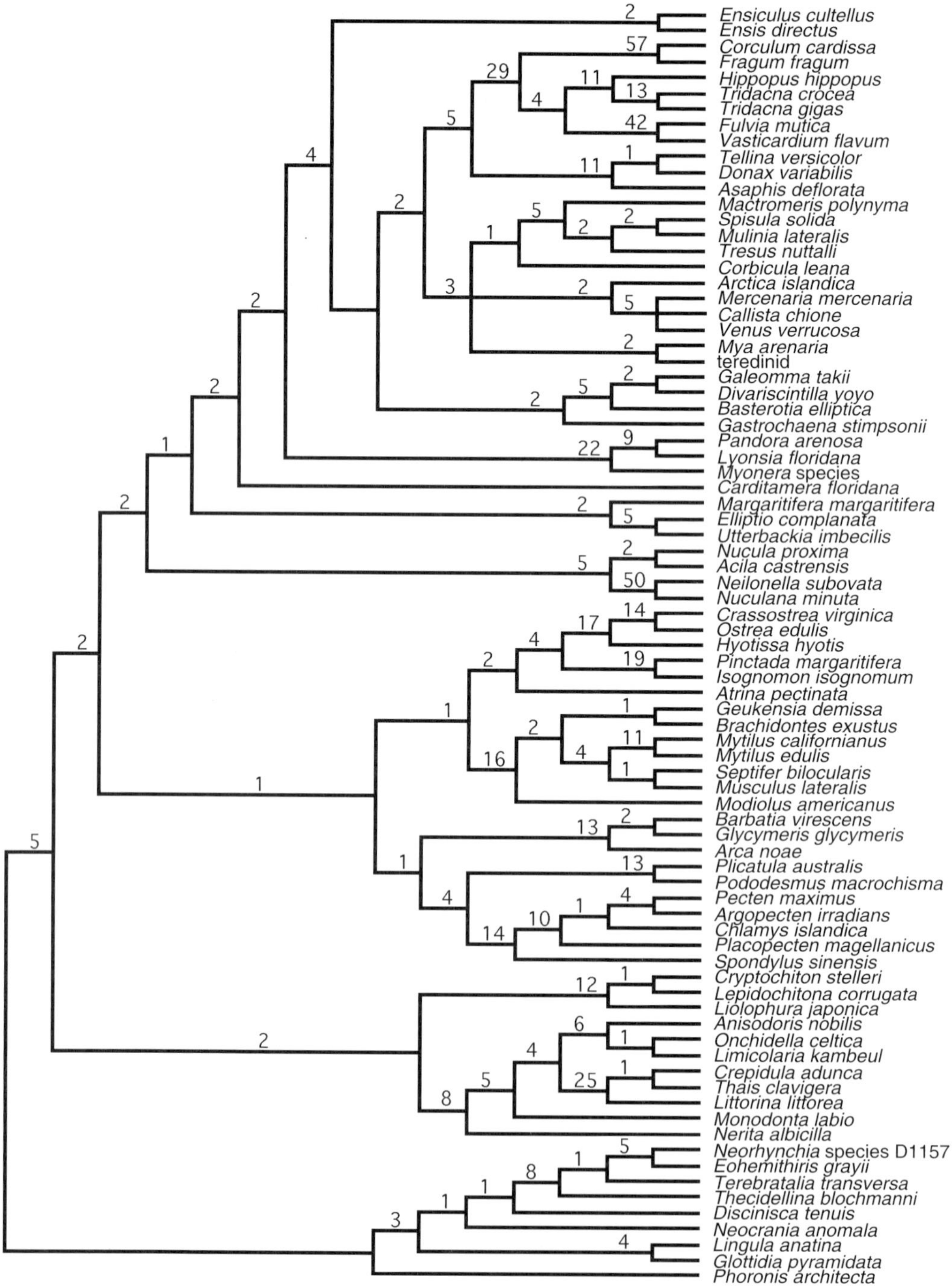

Fig. 1. Strict consensus with gaps coded as missing data. Numbers above branches are the decay index or Bremer support values.

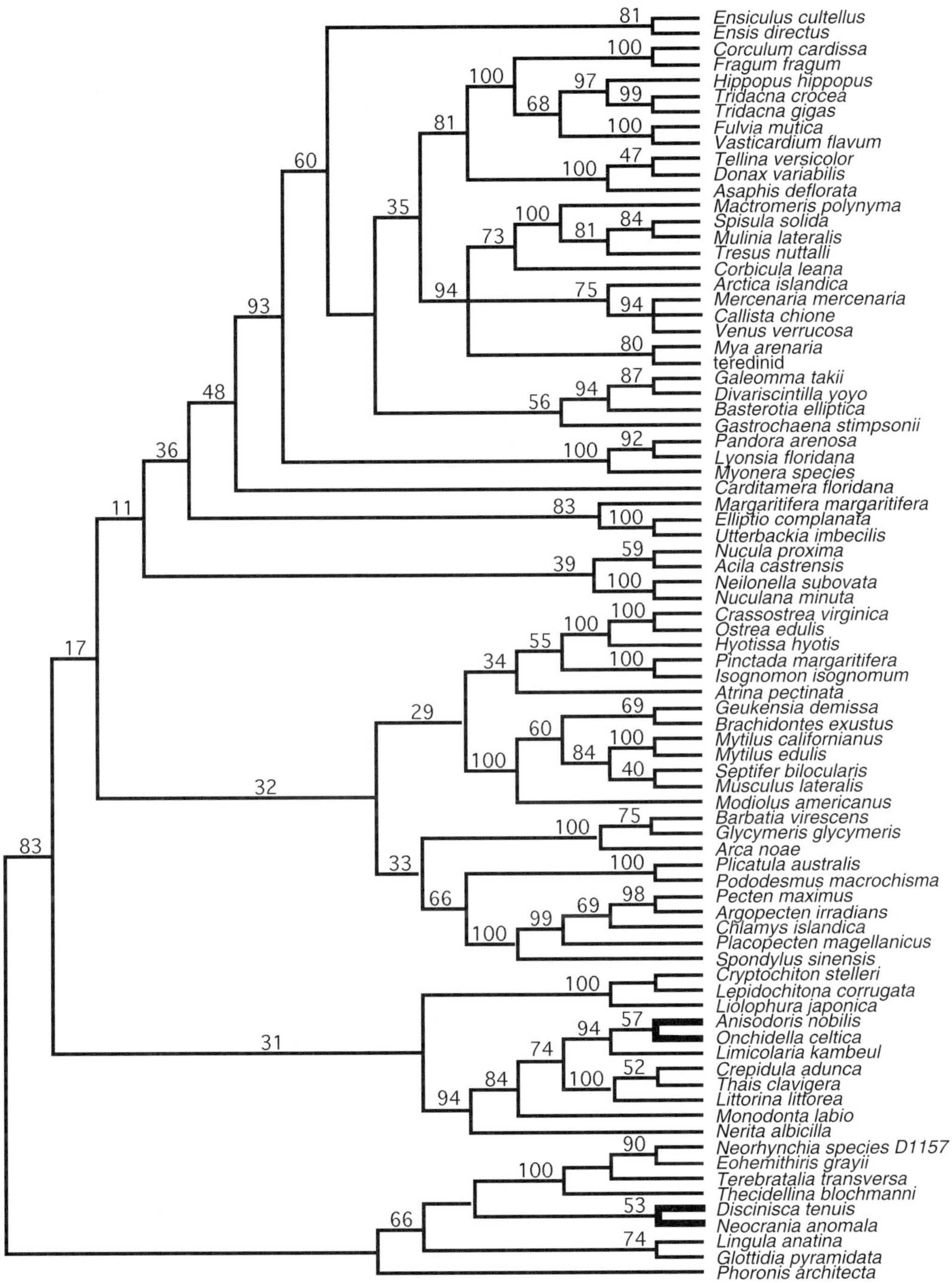

Fig. 2. Strict consensus tree with gaps coded as missing data. Numbers above branches are the bootstrap percentages for all clades with least 50% support. Also, the bootstrap support for other ingroup nodes from Fig. 1 is given. Note that conflicting clades may have higher bootstrap support if support for the clade is under 50%. The two bold clades had over 50% bootstrap support but conflict with the strict consensus tree.

With gaps coded as a fifth character state, there were 1024 parsimony informative characters. The 122 shortest trees found by parsimony with gaps coded as missing were 5672 steps. This represents two groups of trees, found by 54 and 21 searches. The remaining 25 searches found three longer local minima. The g1 value was –0.801 573, based on 100 000 random trees. Other tree statistics include

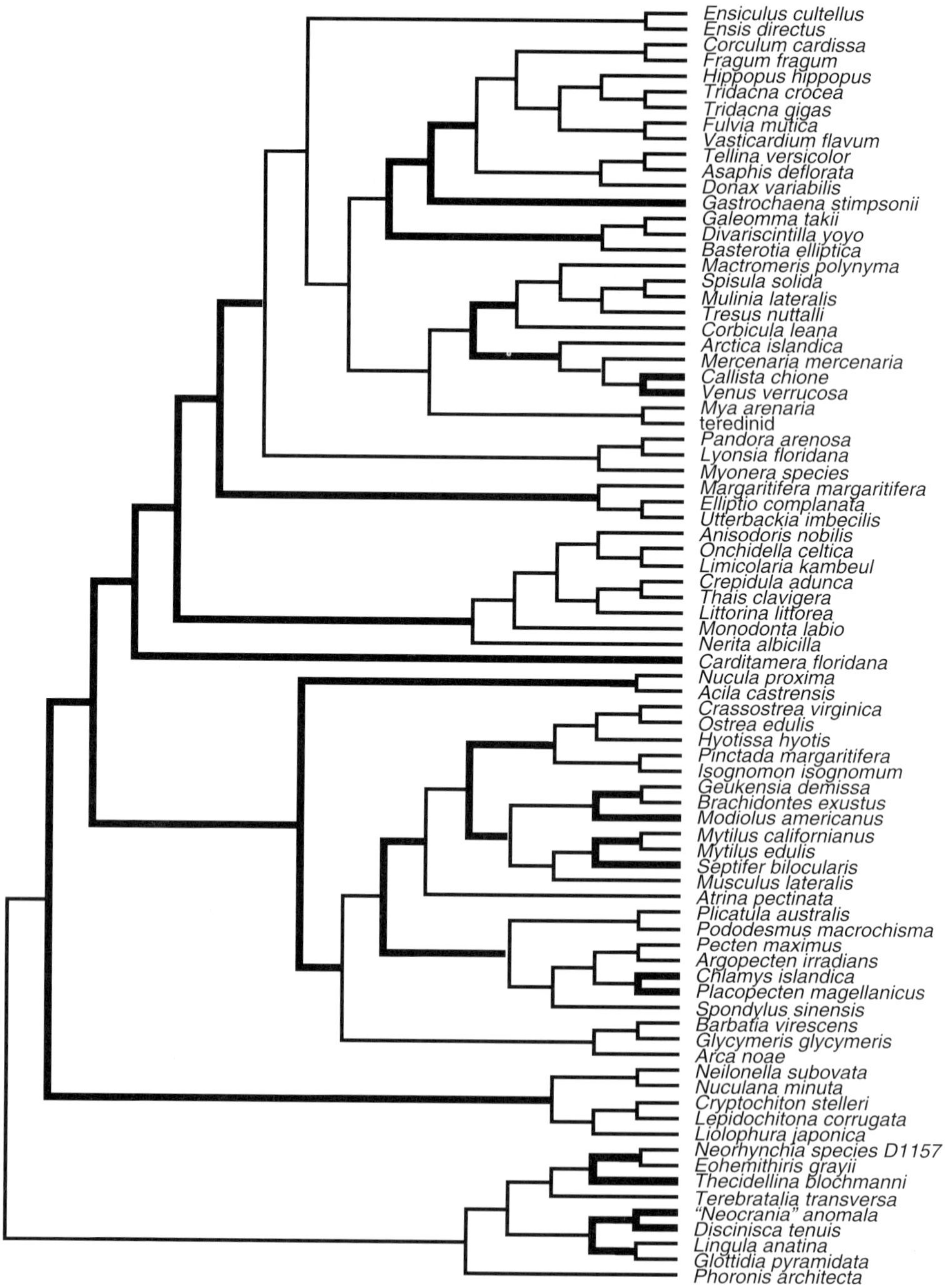

Fig. 3. Neighbour-joining tree, gaps coded as missing data. Clades not supported in Fig. 1 are shown with thicker branches, including polytomies in Fig. 1 resolved by neighbour joining.

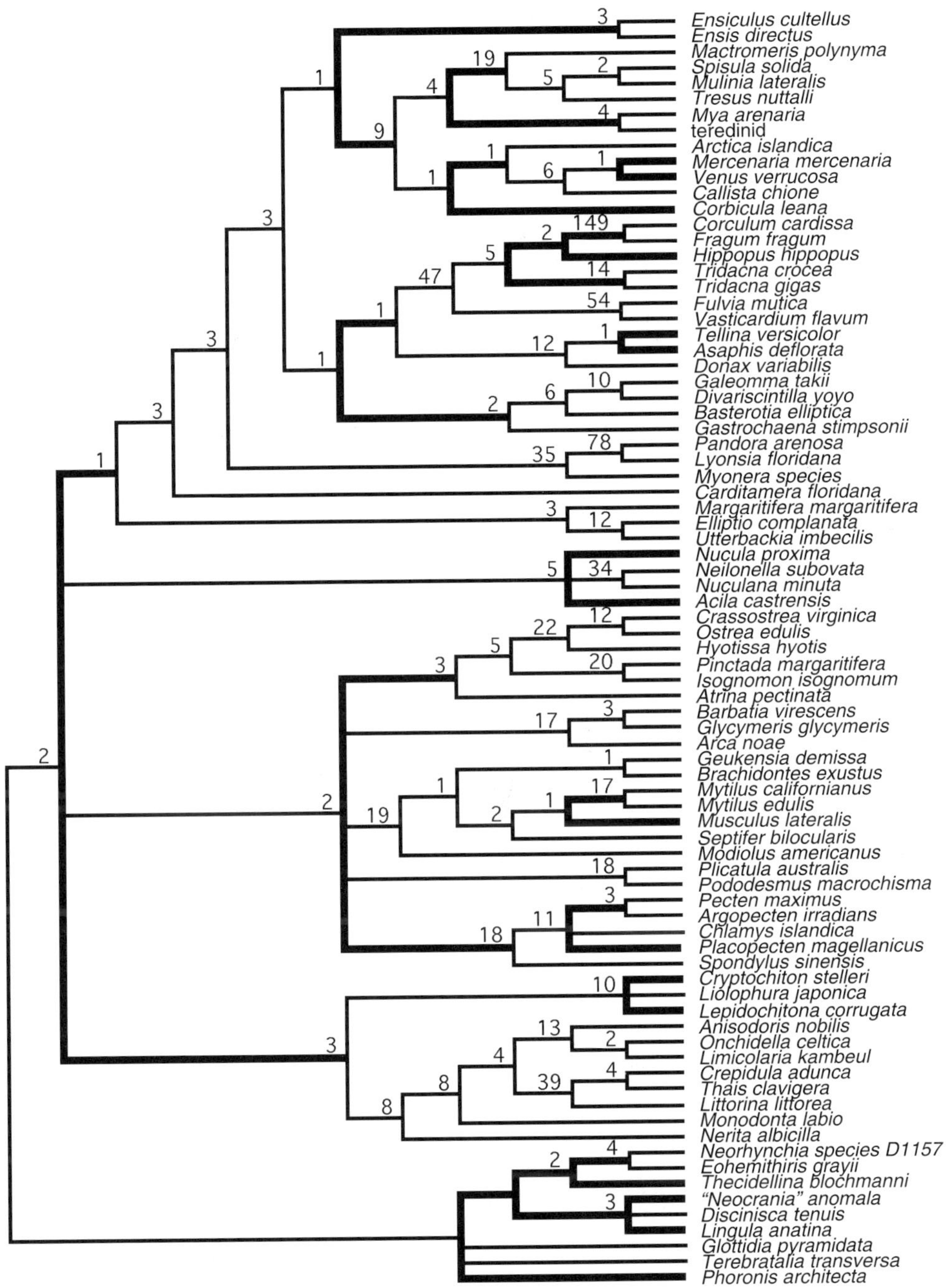

Fig. 4. Strict consensus with gaps coded as a fifth character state. Numbers above branches are the decay index or Bremer support values. Clades not supported in Fig. 1 are shown with thicker branches, including polytomies in Fig. 1 resolved in this second analysis. Polytomies in this consensus that were resolved in Fig. 1 are also indicated with thicker branches.

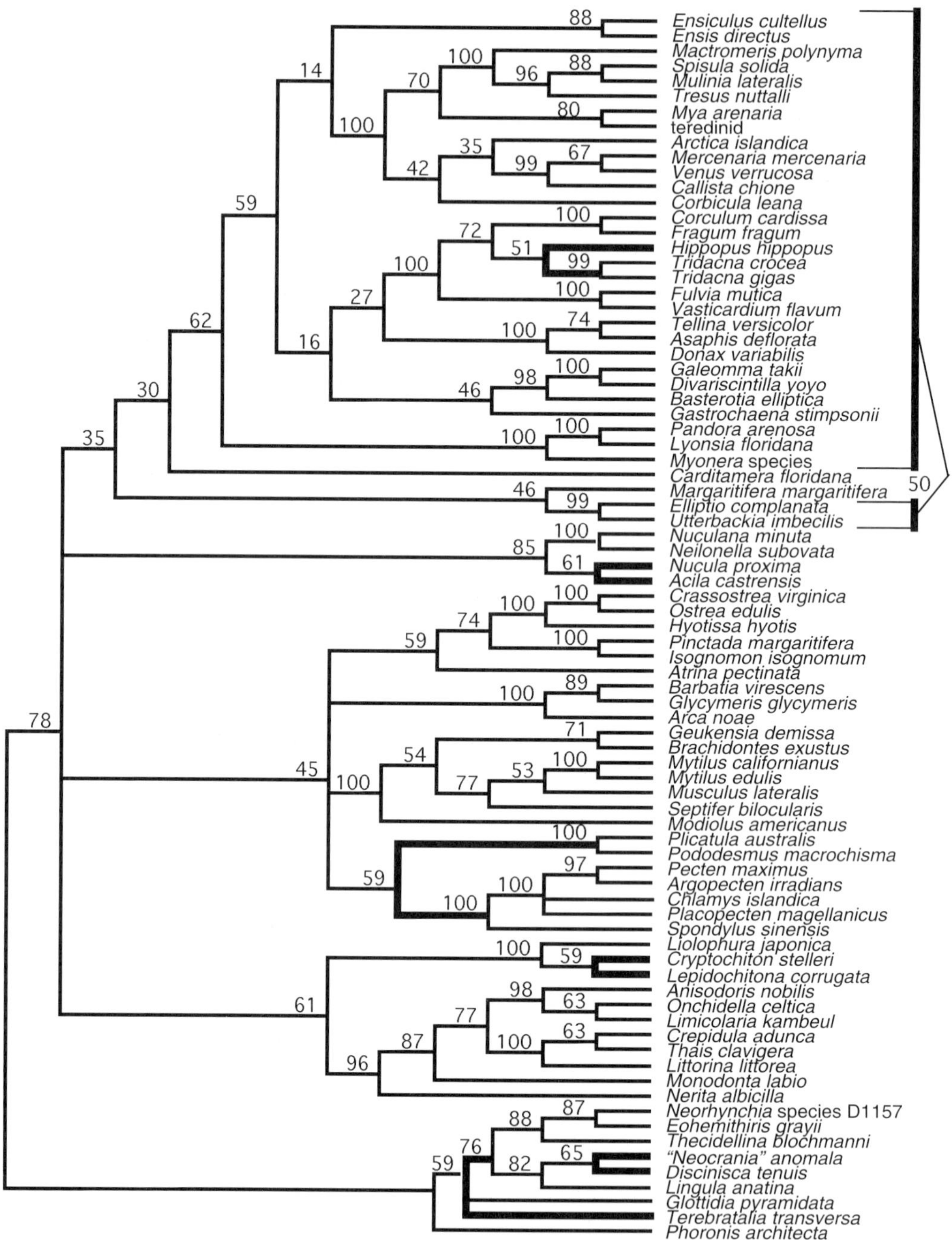

Fig. 5. Strict consensus tree with gaps coded as a fifth character state. Numbers above branches are the bootstrap percentages for all clades with least 50% support. Also, the bootstrap support for other ingroup nodes from Fig. 4 is given. Note that conflicting clades may have higher bootstrap support if support for the clade is under 50%. The six bold clades were either unresolved in the strict consensus or had over 50% bootstrap support but conflict with the strict consensus. The clade of *Hippopus*, *Fragum* and *Corculum*, supported in the strict consensus, had 49% support. A clade with 50% bootstrap support also conflicted with the strict consensus; it includes Veneroida, except *Carditamera*, Anomalodesmata and Unionidae, as indicated by the lines to the right.

a consistency index of 0.4318 (0.3837 excluding uninformative characters), a retention index of 0.6642 and a rescaled consistency index of 0.2868. Figure 4 shows the strict consensus tree with decay index values indicated and Fig. 5 shows the bootstrap results; differences from the first analysis are highlighted in Fig. 4. Decay values range from 1 to 149, suggesting that long-branch attraction may cause problems. Decay indices of over 20 were confined to Cardiidae, Anomalodesmata, Nuculanoidea, Ostreoidea, Pterioidea and Caenogastropoda. Again, the highest values are within Cardiidae.

Discussion

In general, agreement between the molecular and morphological results is good, ranging from the class to the genus level. However, there are some points of disagreement between them, as well as disagreements among models based on either data source. Higher taxa are used in these discussions as defined above, although some papers used the same names differently [e.g. Cope (1997) excluded Arcoida from Pteriomorphia and Campbell *et al.* (1998) placed Myoida in Anomalodesmata].

Despite the wide range of decay indices and sequence lengths, taxa with high values generally grouped into clades compatible with morphological evidence. Only the neighbour-joining tree and the internal topology of Cardiidae conflict with morphological evidence in the placement of these probable long branches.

Most morphology based analyses agree on the monophyly of the Bivalvia, apart from uncertainties about poorly known Cambrian forms. However, molecular support for this has been weak. In fact, Hoeh *et al.* (1998) suggested, based on molecular data, that the class is diphyletic and other molecular analyses have had similar results (e.g. Kenchington *et al.* 1994; Steiner & Müller 1996; Winnepenninckx *et al.* 1996; Campbell *et al.* 1998), usually interpreted as error. This may reflect a limited range of taxa, especially with limited representation of protobranchs. Such wide phylogenetic spacing of taxa makes long-branch attraction a likely problem. If, as suggested by the phylogeny of Waller (1998), the Bivalvia are relatively distantly related to other conchiferans, this problem will be particularly acute, with the divergence between them and other extant molluscs no later than the earliest Cambrian (Runnegar 1996). Modern protobranchs diverged from other bivalves by the early Ordovician (Cope 1996; Carter *et al.* 2000). Thus, limited sampling of outgroups and protobranchs produces very long branches. This remains a problem in the present study, with only four protobranchs, no solemyoids and only two of the other molluscan classes represented. Preliminary analyses in the present study showed that excluding outgroups with only a single available sequence and adding additional representatives of remaining outgroup taxa led to stronger support for bivalve monophyly. The results of Hoeh *et al.* (1998) may also have been affected by the higher mutation rate of the 16S gene (Hillis & Dixon 1991), which could lead to greater long-branch attraction. In the present study, the first analysis (gaps coded as missing data) supported bivalve monophyly, though with weak decay and bootstrap support. The second analysis (gaps coded as a fifth character state) did not resolve bivalve monophyly relative to other molluscs, yet bootstrap support for bivalve monophyly was higher (44.3% with no more than 18.5% against, as opposed to 17.0% in the first analysis with 21.8% and 15.8% supporting the highest supported clades with bivalve paraphyly or polyphyly). None of these bootstrap values are high enough to view as strong support. The neighbour-joining tree places Gastropoda near the base of Heteroconchia and Polyplacophora with Nuculanoidea as basal molluscs. As Nuculanoidea was an exceptionally long branch (decay index of 50 with gaps coded as missing) and Gastropoda moderately long (decay index of 8), these can be dismissed as probable errors, in addition to the inherent problems of evaluating neighbour-joining trees. Monophyly for Gastropoda, Polyplacophora and Mollusca had at least 75% bootstrap support in both analyses, and other outgroup clades also received strong support.

The parsimony analyses supported all five subclasses of the Bivalvia as monophyletic, in agreement with most morphological studies, except for the position of *Carditamera*. The neighbour-joining analysis had Protobranchia as diphyletic. However, the bootstrap and decay support for the subclasses was often weaker. Previous molecular analyses have frequently failed to find monophyly of the subclasses, again probably because of poor representation of some subclasses (both in number of taxa and completeness of sequences). Another source of anomalous results has been the tendency for *Crassostrea virginica* (one of the first species sequenced and thus included in most analyses) to place outside of Pteriomorphia or make Pteriomorphia paraphyletic, with *Crassostrea* as the sister taxon to heterodonts (Steiner & Müller 1996; Winnepenninckx *et al.* 1996; Campbell *et al.* 1998). This species seems to have a relatively high rate of molecular evolution relative to other pteriomorphians (Steiner & Müller 1996; Canapa *et al.* 1999), making it a likely candidate for long-branch attraction. The inclusion of other ostreoids [*Ostrea* in Campbell *et al.* (1998), Frischer *et al.* (1998) and Canapa *et al.* (1999), and *Hyotissa hyotis*, herein] helps solve this problem. Most previous molecular analyses only include Pteriomorphia and Heterodonta. Some of these have supported monophyly for both included subclasses (Rice *et al.* 1993; Frischer *et al.* 1998; Giribet & Carranza 1998). Also, some analyses in Hoeh *et al.* (1998) supported monophyly of all four subclasses included in their analyses, although not supporting monophyly for the Bivalvia. Adamkewicz *et al.* (1997) published the only prior molecular analysis to include all five subclasses, though some of their sequences were very short. Pseudogenes, or errors by the sequencing laboratory, may also have caused problems with some of their sequences (Adamkewicz pers. comm.). Only Heterodonta and Palaeoheterodonta appeared monophyletic in their analyses. The relatively short sequences and limited taxonomical coverage (though much wider than any previous study) probably affected their resolution of higher taxa.

Although many morphological analyses support the monophyly of all included subclasses (e.g. Cope 1997; Waller 1998), others disagree. Salvini-Plawen & Steiner (1996) separated Trigonioida from Unionoida and Myoida from Veneroida (which they divided into Veneroida *sensu stricto* and Lucinoidea), although they noted the risk of convergent evolution in some supporting characters (compare the apomorphies for Trigonioida with the synapomorphies for Eulamellibranchia, nodes 14 and 16 in fig. 2.10). Morton (1996) had Pteriomorphia paraphyletic to Palaeoheterodonta and Heterodonta paraphyletic to Anomalodesmata. Although *Neotrigonia* is not represented in the present analyses, Hoeh *et al.* (1998) found close similarity between *Neotrigonia* and unionids for the 16S gene, and morphological similarities are also strong, with plausible fossil transitions (Newell & Boyd 1975). Likewise, the similarities between Myoida and Anomalodesmata are prone to convergent evolution (Salvini-Plawen & Steiner 1996), and the similarities between Trigonioida and Pteriomorphia are probably plesiomorphies (Waller 1998). As discussed below, molecular evidence strongly supports placing Myoida in Heterodonta rather than in Anomalodesmata.

The relationships among these subclasses are much more contentious. Almost all morphological studies agree that the Protobranchia are basal to the remaining Bivalvia. Most see Pteriomorphia as the next group to diverge (Waller 1998), possibly along with the Palaeoheterodonta (Morton 1996). However, Cope (1997) grouped Heterodonta with Pteriomorphia (divided into Pteriomorphia *sensu stricto* and Neotaxodonta) and Anomalodesmata with Palaeoheterodonta. Previous molecular studies provide some resolution of subclass-level relationships, although often not well-supported. Adamkewicz *et al.* (1997) had protobranchs and anomalodesmatans basal, and tending to place outside Bivalvia, and Pteriomorphia paraphyletic to a sister grouping of Palaeoheterodonta and Heterodonta. Similarly, Campbell *et al.* (1998) had Protobranchia basal, with tendencies to place outside Bivalvia, and Pteriomorphia paraphyletic to Heterodonta. Hoeh *et al.* (1998) again had protobranchs basal, tending to place outside of Bivalvia, and Palaeoheterodonta next most basal, with Pteriomorphia and Heterodonta as sister taxa. In the present study, Protobranchia, Anomalodesmata and Palaeoheterodonta are each represented by, at most, four complete sequences, making their affinities less certain. However, the association between Anomalodesmata and Heterodonta seems well-supported by the present molecular data as well as morphological evidence (Waller 1998). Monophyly for Heteroconchia is supported by all the analyses, although the neighbour-joining tree includes Gastropoda. Bootstrap and decay support are lower than for the Heterodonta and Anomalodesmata (without *Carditamera*) clade. The relative positions of Protobranchia and Pteriomorphia, as well as their position relative to Heterodonta and Anomalodesmata, vary from analysis to analysis. No relationships among subclasses had a decay index higher than two in the first analysis or three in the second.

Relationships similar to those proposed by Waller (1998) seem most likely from all available evidence. The morphological similarities between Palaeoheterodonta, Heterodonta and Anomalodesmata are numerous (Salvini-Plawen & Steiner 1996; Waller 1998), and molecular evidence weakly supports grouping Palaeoheterodonta with the other two (Adamkewicz *et al.* 1997; present study). Cope (1997) grouped Heterodonta and Pteriomorphia largely on the basis of crossed-lamellar microstructure; however, this is convergent and the basal microstructure in Pteriomorphia is nacreous (Carter 1990; Carter *et al.* 2000).

Relationships between orders within the subclasses are also debated. Within the Pteriomorphia, Mytiloida and Arcoida are generally agreed to be basal [not universally, though, e.g. Morton (1996)]. Mytiloida is usually considered basal (Waller 1998; Carter *et al.* 2000), but Arcoida is viewed as basal in some studies (Cope 1997). Among the remaining orders, several phylogenetic arrangements have been proposed. Limoida was traditionally associated with Pectinoida (Cox *et al.* 1969), but Waller (1978, 1998) moved it to a more basal position and Morton (1996) grouped it with Ostreoida. The relationships between Ostreoida, Pterioida, and Pectinoida are particularly debated, with about equal support for closer relationships between Ostreoida and Pectinoida (Morton 1996; Waller 1998), and between Ostreoida and Pterioida (Carter 1990). Assignments of superfamilies to these orders has also varied. Cox *et al.* (1969) associated Plicatulidae with Pectinoidea and Pinnoidea with Mytiloida, but Waller (1978) and most subsequent work group Plicatulidae with Ostreoidea and Pinnoidea with Pterioida. Anomioidea is usually associated with Pectinoidea, but Waller (1998) put them in a polytomy relative to Ostreoidea and Morton (1996) grouped them with Pterioida.

Previous molecular studies have supported some of these ideas. Mytiloida and Arcoida were generally relatively basal, although the position of Ostreoida varied (Kenchington *et al.* 1994; Steiner & Müller 1996; Winnepenninckx *et al.* 1996; Adamkewicz *et al.* 1997; Campbell *et al.* 1998; Frischer *et al.* 1998; Giribet & Carranza 1998; Steiner 1999; Canapa *et al.* 1999). Ostreoida was consistently distant from Pectinoida; when Pterioidea was represented, it grouped with Ostreoida. Pinnoidea also tended to appear basal, but with some tendency to group with Pterioidea and Ostreoidea. The slow evolutionary rate of *Atrina*, as shown by low percentage differences with the outgroup, probably affected its apparent affinities (Steiner & Müller 1996). Only Adamkewicz *et al.* (1997) and Campbell *et al.* (1998) had multiple families for any pteriomorphian order (Pinnidae and Pteriidae, and those plus Isognomonidae, respectively). These produced weak to strong support for monophyly of Pterioida and Pterioidea.

With several additional superfamilies and families represented (Gryphaeidae, Plicatuloidea, Anomioidea, Spondylidae and Glycymerididae, as well as complete sequences for Pteriidae and Isognomonidae), and better representation of other superfamilies, the present analyses were able to address additional morphology based hypotheses. Ostreoida and Pterioida show a close relationship, with Pterioida paraphyletic to Ostreoida. The association of Pterioidea and Ostreoida has the highest decay support (4 in the first analysis, 5 in the second) of any supraordinal group within the Bivalvia. The present analyses also strongly support monophyly for Ostreoidea, including both Gryphaeidae and Ostreidae, in agreement with Waller (1998) but contrasting with the uncertainty expressed in Carter (1990). Plicatuloidea is strongly supported as the sister taxon of Anomioidea and they are both generally associated with Pectinoidea (including Spondylidae and Pectinidae). This association of Plicatuloidea and Anomioidea is unexpected from morphological studies. Also, Arcidae appears paraphyletic relative to Glycymerididae; this is not entirely surprising because both *Arca* and *Barbatia* had arisen by the Jurassic, whereas Glycymerididae extends back only to the Cretaceous (Cox *et al.* 1969). However, relationships between Pectinoida, Arcoida, Mytiloida and the Ostreoida–Pterioida clade are poorly resolved in the present study. Ancestors of all four were present in the Ordovician (Carter *et al.* 2000), so the radiation may have been relatively rapid, and the basal branches are moderately long, with very few branches within the clades represented before Late Palaeozoic.

In contrast to the Pteriomorphia, there are many fewer phylogenies proposed for relationships among superfamilies in Heterodonta. Many examples of convergence are known, and many authors have proposed that Veneroida, Myoida and Heterodonta may be polyphyletic (e.g. Newell 1965). Although Veneroida is sometimes split into lucinid- and venerid-centred groups, the assignment of taxa to these two divisions (orders or suborders, depending on the reference) is not well-established. For example, Newell (1965) placed Tellinoidea with the Lucinoidei, but Morton (1996) put them closer to venerids and split off some of Newell's lucinoideines basally, relative to the rest of Heterodonta and Anomalodesmata.

Despite the uncertainties from morphology, the DNA evidence provides resolution of various groups within the Heterodonta above the superfamily level. All parsimony analyses supported a clade including Mactroidea, Veneroidea, Corbiculoidea, Arcticoidea, Myoidea, and Pholadoidea; this had 94.2% and 100% bootstrap support and 3 and 9 for decay index values. Previous molecular studies have found support for this grouping (Adamkewicz *et al.* 1997; Canapa *et al.* 1999) but most studies have not included enough representatives of both this group and other heterodonts, preventing assessment of its monophyly. This group is similar to the traditional suborder Veneroidei of morphological analyses, but also includes Mactroidea, Myoidea and Pholadoidea. Saul (1973) proposed evolution of Mactroidea from Arcticoidea, which would put it into this group. Other molecular studies suggest that Dreissenoidea, Chamoidea and Pisidiidae may belong here (Adamkewicz *et al.* 1997; Rosenberg *et al.* 1997; Hoeh *et al.* 1998). Another veneroid group, found by all parsimony analyses but with weaker bootstrap and decay values (55.7% and 45.9%, 2 and 2), included Gastrochaenoidea, Galeommatoidea and Sportellidae. Strong support (93.6% and 98.3%, 5 and 6 decay index) for a clade of Sportellidae and Galeommatoidea agrees with recent morphological studies (Salas & Gofas 1998). This clade resembles Erycinoinei of Starobogatov

(1992), but morphological studies suggest that Cyrenoidea is polyphyletic with the other families probably in the suborder Veneroidei. Splitting Myoida among Vereroida was also supported by Adamkewicz *et al.* (1997) and some morphological studies. Although Gastrochaenoidea has been recognized as very different from Myoidea (Carter 1978), its association with Galeommatoidea and Sportellidae is unexpected and only weakly supported by bootstrap and decay (not supported by neighbour-joining). The Gastrochaenoidea does consistently place within Heterodonta but outside Veneroidei, which contains Myoidea and Pholadoidea. Another grouping supported by the present parsimony analyses is Tellinoidea plus Cardioidea, similar to the suborder Cardioidei of Starobogatov (1992) but including Donacidae and excluding Ungulinoidea. However, this may be a case of long-branch attraction, as Tellinoidea and Cardioidea both have high decay indices. Although this clade received 80.6% bootstrap support in the first analysis, it only received 27.3% support in the second, less than the 43.0% support for grouping Cardioidea, Gastrochaenidae and Erycinoinei. Solenoidea and the Galeommatoidea–Sportellidae–Gastrochaenoidea clade seem to be relatively basal, but with poor bootstrap support. Finally, Carditoidea is basal not only to all other sampled heterodonts but also to Anomalodesmata. This is unexpected; although they appear to be relatively primitive heterodonts, they are morphologically similar to other heterodonts exclusive of anomalodesmatans. More molecular data for other primitive heterodonts are needed to test this result, especially Crassatelloidea and Lucinoidea.

All bivalve superfamilies and families appear monophyletic except Arcidae. Support for monophyly of Nuculidae and Nuculoidea was weak. Although the strict consensus in the second analysis did not resolve Nuculidae relative to Nuculanoidea, bootstrap support for Nuculidae was slightly higher than in the first analysis (61.0% v. 59.4%).

Conclusions

Analysis of 18S rDNA helps resolve several issues in the phylogeny of the Bivalvia. In particular, conflicts between different morphology based interpretations can often be resolved by use of this new data source. Further study promises to continue to improve our understanding of bivalve evolution.

The numerous areas of agreement between the present study and previous morphological and molecular studies suggest that we are converging on a common phylogeny of the Bivalvia. In particular, the traditional subclasses, orders, superfamilies and families generally seem to be monophyletic. However, many important taxa remain poorly known and areas of disagreement remain to be resolved, especially in the relationships between taxa of equal rank. Data from additional genes and further morphological analyses should resolve many of these issues.

Relationships among the subclasses are not yet clear, apart from the grouping of Anomalodesmata and Heterodonta. Heteroconchia (Palaeoheterodonta, Heterodonta and Anomalodesmata) received weaker support. Monophyly of the subclasses is supported, with the possible exception of Carditoidea placing basally in the Heterodonta–Anomalodesmata clade, as is monophyly of most orders, superfamilies and families.

In addition, three clades not corresponding to named higher taxa are strongly supported. The clade of advanced veneroids (Mactroidea, Arcticoidea, Corbiculoidea, Veneroidea, Myoidea, and Pholadoidea), the Sportellidae–Galeommatoidea clade and the Plicatuloidea-Anomioidea clade had over 93% bootstrap support in both analyses. Also supported by both parsimony analyses and receiving over 75% bootstrap support in at least one analysis were a clade of Tellinoidea and Cardioidea, a clade of Arcticoidea and Veneroidea, a clade of Anomalodesmata and Heterodonta excluding Carditoidea, and a clade of Myoidea and Pholadoidea. Plicatuloidea and Anomioidea belong in Pectinoida rather than Ostreoida or a separate order.

Myoida is polyphyletic, with at least two separate origins from within Veneroida, and should not be considered as an order. However, two 'myoid' superfamilies, Myoidea and Pholadoidea, form a clade within Veneroida.

K. Hoekstra, in her master's thesis, established the usefulness of 18S gene sequence data for suprafamilial systematics in the Bivalvia and experimented with various molecular techniques. G. Giribet and K. Hoekstra obtained some sequence data used in the present study and helped with molecular techniques. Work on several pteriomorph sequences was performed in W. Wheeler's laboratory, working with G. Giribet. The Paleontological Society, Sigma Xi, the Geological Society of America, the Conchologists of America, a Royster Dissertation Fellowship and a Martin grant helped fund this research. Several people helped obtain specimens. G. Giribet, E. Harper and an anonymous reviewer provided helpful comments and corrections.

References

ADAMKEWICZ, S. L., HARASEWYCH, M. G., BLAKE, J., SAUDEK, D. & BULT, C. J. 1997. A molecular phylogeny of the bivalve mollusks. *Molecular Biology and Evolution*, **14**, 619–629.

BELL, J. L. & GRASSLE, J. P. 1998. A DNA probe for

identification of larvae of the commercial surfclam (*Spisula solidissima*). *Molecular Marine Biology and Biotechnology*, **7**, 127–137.

CAMPBELL, D. C., HOEKSTRA, K. T. & CARTER, J. G. 1998. 18S ribosomal DNA and evolutionary relationships within the Bivalvia. *In:* JOHNSTON, P. A. & HAGGART, J. W. (eds) *Bivalves: an Eon of Evolution – Paleobiological Studies Honoring Norman D. Newell*. University of Calgary Press, Calgary, 75–85.

CANAPA, A., MAROTA, I., ROLLO, F. & OLMO, E. 1999. The small-subunit rRNA gene sequences of venerids and the phylogeny of Bivalvia. *Journal of Molecular Evolution*, **48**, 463–468.

CARTER, J. G. 1978. Ecology and evolution of the Gastrochaenacea, with notes on the evolution of the endolithic habitat. *Peabody Museum of Natural History, Yale University, Bulletin*, **41**, 1–92.

—— 1990. Evolutionary significance of shell microstructure in the Palaeotaxodonta, Pteriomorphia, and Isofilibranchia (Bivalvia: Mollusca). *In*: CARTER, J. G. (ed.) *Skeletal Biomineralization: Patterns, Processes, and Evolutionary Trends, Vol. 1*. Van Nostrand Reinhold, New York, 135–296.

——, CAMPBELL, D. C. & CAMPBELL, M. R. 2000. Cladistic perspectives on early bivalve evolution. *This volume.*

COHEN, B. L., GAWTHROP, A. B. & CAVALIER-SMITH, T. 1998. Molecular phylogeny of brachiopods and phoronids based on nuclear-encoded small subunit ribosomal RNA gene sequences. *Philosophical Transactions of the Royal Society of London, Series B*, **353**, 2039–2061.

COPE, J. C. W. 1996. Early Ordovician (Arenig) bivalves from the Llangynog Inlier, South Wales. *Palaeontology*, **39**, 979–1025.

—— 1997. The early phylogeny of the class Bivalvia. *Palaeontology*, **40**, 713–746.

COX, L. R., NEWELL, N. D., BOYD, D. W. *ET AL.* 1969. *In*: MOORE, R. C. (ed.) *Treatise on Invertebrate Paleontology. Part N, Mollusca 6, Bivalvia*. Geological Society of America, Boulder, CO, and University of Kansas Press, Lawrence, KS, N1–N952.

FELSENSTEIN, J. 1978. Cases in which parsimony or compatibility measures will be positively misleading. *Systematic Zoology*, **27**, 401–410.

FIELD, K. G., OLSEN, G. J., LANE, D. J. *ET AL.* 1988. Molecular phylogeny of the animal kingdom. *Science*, **239**, 748–753.

FRISCHER, M. E., WILLIAMS, J. & KENCHINGTON, E. L. R. 1998. A molecular phylogeny of some major groups of Pectinidae inferred from 18S rRNA gene sequences. *In:* JOHNSTON, P. A. & HAGGART, J. W. (eds) *Bivalves: an Eon of Evolution – Paleobiological Studies honoring Norman D. Newell*. University of Calgary Press, Calgary, 213–221.

GIRIBET, G. & CARRANZA, S. 1998. What can 18S rDNA do for bivalve phylogeny? *Journal of Molecular Evolution*, **48**, 256–258.

—— & WHEELER, W. 1999. On gaps. *Molecular Phylogenetics and Evolution*, **13**, 132–143.

——, CARRANZA, S., BAGUÑA, J., RIUTORT, M. & RIBERA, C. 1996. First molecular evidence for the existence of a Tardigrada + Arthropoda clade. *Molecular Biology and Evolution*, **13**, 76–84.

HALANYCH, K. M., BACHELLER, J. D., AGUINALDO, A. M., LIVA, S. M., HILLIS, D. M. & LAKE, J. A. 1995. Evidence from 18S ribosomal DNA that the lophophorates are protostome animals. *Science*, **267**, 1641–1643.

HILLIS, D. M. & BULL, J. J. 1993. An empirical test of bootstrapping as a method for assessing confidence in phylogenetic analysis. *Systematic Biology*, **42**, 182–192.

—— & DIXON, M. T. 1991. Ribosomal DNA: molecular evolution and phylogenetic inference. *The Quarterly Review of Biology*, **66**, 411–453.

HOEH, W. R., STEWART, D. T., SAAVEDRA, C., SUTHERLAND, B. W. & ZOUROS, E. 1997. Phylogenetic evidence for role-reversal of gender-associated mitochondrial DNA in *Mytilus* (Bivalvia: Mytilidae). *Molecular Biology and Evolution*, **14**, 959–967.

——, BLACK, M. B., GUSTAFSON, R., BOGAN, A. E., LUTZ, R. A. & VRIJENHOEK, R. C. 1998. Testing alternative hypotheses of *Neotrigonia* (Bivalvia: Trigonioida) phylogenetic relationships using cytochrome *c* oxidase subunit 1 DNA sequences. *Malacologia*, **40**, 267–278.

HOLLAND, P. W. H., HACKER, A. M. & WILLIAMS, N. A. 1991. A molecular analysis of the phylogenetic affinities of *Saccoglossus cambrensis* Brambell & Cole (Hemichordata). *Philosophical Transactions of the Royal Society of London, Series B*, **332**, 185–189.

KENCHINGTON, E. L. R., LANDRY, D. & BIRD, C. J. 1995. Comparison of taxa of the mussel *Mytilus* (Bivalvia) by analysis of the nuclear small-subunit rRNA gene sequence. *Canadian Journal of Fisheries and Aquatic Sciences*, **52**, 2613–2620.

——, RODDICK, D. L., SINGH, R. K. & BIRD, C. J. 1994. Analysis of small- subunit rRNA gene sequences from six families of molluscs. *Journal of Marine Biotechnology*, **1**, 215–217.

——, NAIDU, K. S., RODDICK, D. L., COOK, D. I. & ZOUROS, E. 1993. Use of biochemical genetic markers to discriminate between adductor muscles of the sea scallop (*Placopecten magellanicus*) and the Iceland scallop (*Chlamys islandica*). *Canadian Journal of Fisheries and Aquatic Sciences*, **50**, 1222–1228.

LITTLEWOOD, D. T. J., FORD, S. E. & FONG, D. 1991. Complete nucleotide sequence of the small subunit ribosomal RNA gene of the American oyster *Crassostrea virginica* (Gmelin) and a comparison with similar sequences from other bivalve molluscs. *Nucleic Acids Research*, **19**, 6048.

MACKEY, L. Y., WINNEPENNINCKX, B., BACKELJAU, T., DE WACHTER, R., EMSCHERMANN, P. & GAREY, J. R. 1996. 18S rRNA suggests that Entoprocta are protostomes, unrelated to Ectoprocta. *Journal of Molecular Evolution*, **42**, 552–559.

MARUYAMA, T., ISHIKURA, M., YAMAZAKI, S. & KANAI, S. 1998. Molecular phylogeny of zooxanthellate bivalves. *Biological Bulletin*, **195**, 70–77.

MORTON, B. 1996. The evolutionary history of the Bivalvia. *In*: TAYLOR, J. (ed.) *Origin and Evolutionary Radiation of the Mollusca*. Oxford University Press, Oxford, 337–359.

NEWELL, N. D. 1965. Classification of the Bivalvia. *American Museum Novitates*, **2206**, 1–25.

—— & BOYD, D. W. 1975. Parallel evolution in early trigoniacean bivalves. *Bulletin of the American Museum of Natural History*, **154**, 53–162.

POE, S. & SWOFFORD, D. L. 1999. Taxon sampling revisited. *Nature*, **398**, 299–300.

RICE, E. L. 1990. Nucleotide sequence of the 18S ribosomal RNA gene from the Atlantic sea scallop *Placopecten magellanicus* (Gmelin, 1791). *Nucleic Acids Research*, **18**, 5551.

——, RODDICK, D. & SINGH, R. K. 1993. A comparison of molluscan (Bivalvia) phylogenies based on palaeontological and molecular data. *Molecular Marine Biology and Biotechnology*, **2**, 137–146.

ROSENBERG, G., TILLIER, S., TILLIER, A., KUNCIO, G. S., HANLON, R. T., MASSELOT, M. & WILLIAMS, C. J. 1997. Ribosomal RNA phylogeny of selected major clades in the Mollusca. *Journal of Molluscan Studies*, **63**, 301–309.

RUNNEGAR, B. 1996. Early evolution of the Mollusca: the fossil record. *In*: TAYLOR, J. (ed.) *Origin and Evolutionary Radiation of the Mollusca*. Oxford University Press, Oxford, 77–87.

SAITOU, N. & NEI, M. 1987. The neighbor joining method: a new method for reconstructing phylogenetic trees. *Molecular Biology and Evolution*, **4**, 406–425.

SALAS, C. & GOFAS, S. 1998. Description of four new species of *Neolepton* Monterosato, 1875 (Mollusca: Bivalvia: Neoleptonidae), with comments on the genus and on its affinity with the Veneracea. *Ophelia*, **48**, 35–70.

SALVINI-PLAWEN, L. V. & STEINER, G. 1996. Synapomorphies and plesiomorphies in higher classification of Mollusca. *In*: TAYLOR, J. (ed.) *Origin and Evolutionary Radiation of the Mollusca*. Oxford University Press, Oxford, 29–51.

SAUL, L. R. 1973. Evidence for the origin of the Mactridae (Bivalvia) in the Cretaceous. *University of California Publications in Geological Sciences*, **97**, 1–59.

SORENSON, M. D. 1999. *TreeRot*, version 2a. Boston University, Boston, MA.

STAROBOGATOV, YA. I. 1992. Morphological basis for phylogeny and classification of Bivalvia. *Ruthenica*, **2**, 1–25.

STEINER, G. 1999. What can 18S rDNA do for bivalve phylogeny? *Journal of Molecular Evolution*, **48**, 258–261.

—— & MÜLLER, M. 1996. What can 18S rDNA do for bivalve phylogeny? *Journal of Molecular Evolution*, **43**, 58–70.

SWOFFORD, D. L. 1993. *PAUP 3.1.1*. Smithsonian Institution, Washington, DC.

—— 1999. *PAUP**. Sinauer Associates, Sunderland, MA.

—— & BEGLE, D. P. 1993. *User's Manual for PAUP: Phylogenetic Analysis Using Parsimony*. Laboratory of Molecular Systematics, Smithsonian Institution, Washington, DC.

——, OLSEN, G. J., WADDELL, P. J. & HILLIS, D. M. 1996. Phylogenetic inference. *In:* HILLIS, D. M. & MABLE, B. K. (eds) *Molecular Systematics*, 2nd Edition. Sinauer Associates, Sunderland, MA, 407–514.

WALLER, T. R. 1978. Morphology, morphoclines and a new classification of the Pteriomorphia (Mollusca: Bivalvia). *Philosophical Transactions of the Royal Society of London, Series B*, **284**, 345–365.

—— 1998. Origin of the molluscan Class Bivalvia and a phylogeny of major groups. *In:* JOHNSTON, P. A. & HAGGART, J. W. (eds) *Bivalves: an Eon of Evolution - Paleobiological Studies Honoring Norman D. Newell*. University of Calgary Press, Calgary, 1–45.

WHITING, M. F., CARPENTER, J. C., WHEELER, Q. D. & WHEELER, W. C. 1997. The Strepsidura problem: phylogeny of the holometabolous insect orders inferred from 18S and 28S ribosomal DNA sequences and morphology. *Systematic Biology* **46**, 1–68.

WINNEPENNINCKX, B. & BACKELJAU, T. 1996. 18S rRNA alignments derived from different secondary structure models can produce alternative phylogenies. *Journal of Zoological Systematics and Evolutionary Research*, **34**, 135–143.

——, —— & DE WACHTER, R. 1993. Complete small ribosomal subunit RNA sequence of the chiton *Acanthopleura japonica* (Lischke, 1873) (Mollusca, Polyplacophora). *Nucleic Acids Research*, **21**, 1670.

——, —— & —— 1994. Small ribosomal subunit RNA and the phylogeny of the Mollusca. *Nautilus*, Suppl 2, **108**, 98–110.

——, —— & —— 1996. Investigation of molluscan phylogeny on the basis of 18S rRNA sequences. *Molecular Biology and Evolution*, **13**, 1306–1317.

——, REID, D. G. & BACKELJAU, T. 1998*a*. Performance of 18S rRNA in littorinid phylogeny (Gastropoda: Caenogastropoda). *Journal of Molecular Evolution*, **47**, 586–596.

——, BACKELJAU, T., VAN DE PEER, Y. & DE WACHTER, R. 1992. Structure of the small ribosomal subunit RNA sequence of the pulmonate snail *Limicolaria kambeul*, and phylogenetic analysis of the Mollusca. *FEBS Letters*, **309**, 123–126.

——, STEINER, G., BACKELJAU, T. & DE WACHTER, R. 1998*b*. Details of gastropod phylogeny inferred from 18S rRNA sequences. *Molecular Phylogenetics and Evolution*, **9**, 55–63.

Cladistic perspectives on early bivalve evolution

J. G. CARTER[1], D. C. CAMPBELL[1] & M. R. CAMPBELL[2]

[1] *Department of Geological Sciences, University of North Carolina at Chapel Hill, Chapel Hill NC 27599-3315, USA (e-mail: clams@email.unc.edu)*

[2] *Department of Geological Sciences, Indiana University, 1005 East Tenth Street, Bloomington IN 47401, USA*

Abstract: Parsimony analysis suggests derivation of the Bivalvia from monoplacophorans rather than from rostroconchs, and additionally indicates that a phylogenetic classification of the Bivalvia can be achieved by erecting the superorder Nuculaniformii *nov.* and the order Nuculanoida *nov.* for the superfamily Nuculanoidea; relegating all other palaeotaxodonts to the superorder Nuculiformii; restricting the order Nuculoida to the families Nuculidae and Pristiglomidae; expanding the order Solemyoida to include ctenodontid genera as basal plesions; restricting the superorder Heteroconchia to palaeoheterodonts and heterodonts, exclusive of the Modiomorphidae; relegating the new family Evyanidae, the Colpomyidae, Matheriidae and Modiolodontidae to near-basal plesion status within the superorder Pteriomorphia; restricting the Mytiloida to the superfamily Mytiloidea, inclusive of modiolopsid genera as basal plesions; placing *Ortonella* as a basal plesion within the Cyrtodontoida; expanding the order Pectinoida to include the Myodakryotidae and the suborders Limina and Pectinina; and expanding the superfamily Arcoidea to include the frejid genera and *Catamarcaia* as basal plesions, and the family Glyptarcidae. Modiomorphid anomalodesmatans appear to be more closely related to the Pteriomorphia than to the Heteroconchia, and *Evyana* lies close to the common ancestry of modiomorphids and colpomyid pteriomorphians. Arcoids may have evolved from left–right symmetrical but otherwise rhombopteriid-like ancestors, rather than from actinodontoids or directly from cyrtodontids. The new family Eodonidae is proposed to distinguish the nacreous genus *Eodon* from the non-nacreous Astartidae within the superfamily Crassatelloidea.

Bivalves appear to have evolved from laterally compressed, stenothecid monoplacophorans similar to early Cambrian *Anabarella* Vostokova and *Watsonella* Grabau (Runnegar & Pojeta 1974). These two genera are morphologically intermediate between monoplacophorans and both rostroconchs and bivalves. They differ from rostroconchs in lacking a true pegma (Runnegar 1996) and they differ from bivalves in lacking a microstructurally differentiated ligament, although they do have a ligament precursor (Kouchinsky 1999). *Watsonella* also resembles bivalves in having a divided larval shell (Dzik 1994). *Anabarella*, *Watsonella* and the Cambrian bivalve *Pojetaia* Jell have a columnar prismatic outer shell layer with the apices of the prisms mutually isolated by thick, interprismatic organic matrices, together with an inner shell layer consisting of imbricated laminae with fibrous second order structural units. These fibres constitute matted and lamello-fibrillar structures ['type-2 nacre' of Mutvei (1970)] in stenothecids, but a wide-tablet nacreous structure in *Pojetaia* and *Fordilla* Barrande (Runnegar & Bentley 1983; Carter 1990; Runnegar & Pojeta 1992; Kouchinsky 1999). Modern bivalve nacre consists of smaller, more horizontal tablets, which only rarely show traces of second order fibres (Mutvei 1983). The evolutionary transition from stenothecid monoplacophoran to true bivalve required microstructural differentiation of the ligament, the evolution of at least one adductor muscle, loss of permanent shell gapes and rearrangement of imbricated, matted/lamello-fibrillar laminae into imbricated nacreous laminae as in fordilloids, or imbricated calcitic foliated laminae as in tuarangioids.

The early differentiation of the bivalve subclasses and orders is less well understood and remains controversial. When *Fordilla* was the only known Cambrian bivalve, Pojeta & Runnegar (1974) suggested that it gave rise to cycloconchid palaeoheterodonts, and palaeoheterodonts then gave rise to the Pteriomorphia, Isofilibranchia and Palaeotaxodonta. Pojeta (1975) then offered an alternative scenario in which *Fordilla* gave rise directly to the subclass Isofilibranchia. Babin

From: HARPER, E. M., TAYLOR, J. D. & CRAME, J. A. (eds) *The Evolutionary Biology of the Bivalvia*. Geological Society, London, Special Publications, **177**, 47–79. 1-86239-076-2/00/$15.00

(1977) suggested that *Fordilla* gave rise to *Babinka* Barrande and *Coxiconcha* Babin, and *Coxiconcha* then gave rise to the Isofilibranchia. Pojeta (1978) agreed with Babin (1977) that *Babinka* evolved directly from a *Fordilla*-like ancestor, but he favoured an independent ancestor for the Isofilibranchia. Pojeta (1985) presented a phylogeny in which *Pojetaia* and *Fordilla* are ancestral to palaeotaxodonts and isofilibranchs, respectively. Runnegar & Bentley (1983) allied *Fordilla* with *Modiolodon* Ulrich, and suggested that *Fordilla* evolved into mytiloids, orthonotids and pholadomyoids, whereas *Pojetaia* evolved into praenuculid palaeotaxodonts and, through them, actinodonts and non-mytiloid pteriomorphians. Runnegar & Pojeta (1992) and Runnegar (1996) offered still another possibility, that *Fordilla* and *Pojetaia* were stem-group bivalves predating the last common ancestor of the extant Bivalvia. Runnegar & Pojeta (1992) noted that it is unlikely that the distinctive shell microstructure of *Pojetaia* and *Fordilla* was lost independently in lines leading to palaeotaxodonts and autolamellibranchs.

MacKinnon (1982, 1985) and Berg-Madsen (1987) suggested that the Pteriomorphia evolved directly from Cambrian *Tuarangia* MacKinnon. Runnegar & Bentley (1983), Runnegar (1983, 1985) and Runnegar & Pojeta (1992) believed that *Tuarangia* is a laterally compressed monoplacophoran closely related to *Pseudomyona*, noting that these two genera have similarly foliated shells. According to their interpretation, *Pseudomyona* has a univalved protoconch and an adductor muscle positioned differently than in early bivalves [see also Runnegar & Jell (1976)]. *Tuarangia* and *Pseudomyona* differ from undoubted Cambrian and Ordovician bivalves in their D-shaped, largely calcitic shells. Specimens of Middle Cambrian *Tuarangia gravgaerdensis tenuiumbonata*, described by Hinz-Schallreuter (1995), show short, anterior and posterior ligament insertion areas; differentiated left and right beaks; and possible anterior and posterior adductor muscle scars, suggesting closer affinity with the Bivalvia than with *Pseudomyona*.

Starobogatov (1992) utilized evolutionary trends in shell and soft anatomical features, plus larval structure and development, to divide the Bivalvia (including the Rostroconchia) into three superorders, 15 orders, 30 suborders, 15 infraorders and 119 superfamilies. Starobogatov (1992, fig. 4) indicated that the superfamily Phaseoloidea gave rise to the superorders Nuculiformii Dall, Mytiliformii Férussac and Conocardiformii Neumayr. *Phaseolus* Monterosato is a Recent, resiliated nuculoid, but Starobogatov nevertheless suggested that its pattern of few, radially disposed, lamelliform hinge teeth is a good ancestral model for nuculoid, nucinelloid and lyrodesmoid hinges. The lyrodesmoid hinge would then have been transformed into schizodont, actinodont, and other palaeoheterodont and heterodont dentitions. Starobogatov included in the Conocardiformii not only fordilloids and various anatomically derived anomalodesmatans, but also rostroconchs, some of which he claimed had a ligament and adductors. Within the Mytiliformii, the suborder Lyrodesmatoidei was identified as the ancestral stock for the order Unioniformes (lyrodesmatids, modiomorphids, gastrochaenids, trigonioids, unionoids and various actinodonts), the order Pholadomyiformes (Pholadomyoidea, Clavagelloidea, Pandoroidea, Thracioidea and ancient pholadomyoid groups) and the order Mytiliformes (mytiloids, pterioids, cyrtodontoids, arcoids and early pectinoids). Starobogatov separated the Ctenodontidae from the Solemyidae at the ordinal level. He also separated advanced pectinoids (including limoids) from other pteriomorphians, implying that advanced pectinoids are more closely related to unionoids and lucinoids. Starobogatov divided the remaining pteriomorphians into an ancestral pterioid–early pectinoid clade and a more derived cyrtodontoid–arcoid–ostreoid–mytiloid clade, with arcoids evolving from cyrtodontoids through the addition of medial hinge teeth, and mytiloids, ostreoids and gryphaeoids evolving from arcoids through replacement of normal hinge teeth with shell marginal denticles.

Waller (1998) maintained that bivalves are more closely related to rostroconchs than to stenothecid monoplacophorans, by virtue of sharing a strong pallial line and a similar lateral profile, i.e. with the ventral, apertured part of the shell exceeding the dorsal part in height. However, the Lower Cambrian stenothecid *Watsonella crossbyi* Grabau, 1900, has a profile similar to bivalves and one of the oldest rostroconchs, Middle Cambrian *Ribeiria junior* Runnegar, 1996, has a typical monoplacophoran profile. Waller (1998) regarded fordilloids as stem-group bivalves evolving from an unnamed proto-bivalve prior to mineralization of the ligament. He summarized features supporting a basal dichotomy between the Protobranchia and Autobranchia, all of which are non-palaeontological except for the autobranch pattern of 'umburrid-like dental ontogeny with elongate, shallowly inclined primary teeth and sockets characterizing early growth stages' [citing Johnston (1991, p. 315)]. Waller (1998) also summarized the features, non-palaeontological except for predominance of nacreous shell microstructure, indicating a basal split in the Protobranchia between the Nuculanoidea and the Nuculoidea–Solemyoidea, the latter superfamily expanded to include the Ctenodontidae and Nucinellidae. Within the Autobranchia, a basal dichotomy allegedly separates the Pteriomorphia from the Heteroconchia (the latter inclusive of the Palaeoheterodonta, Heterodonta and Anomalodesmata). The Pteriomorphia were said to be characterized by four soft anatomical features plus discontinuous fibrous ligament ontogeny and anisomyarian adductors. The Heteroconchia were characterized as having ligament nymphs and a core homology of 3a/2/3b hinge dentition.

Salvini-Plawen & Steiner (1996) performed a phylogenetic analysis of the Bivalvia based largely on soft anatomy. They utilized 42 characters generalized for three families, three superfamilies, three orders and three subclasses. Unfortunately, they incorrectly assumed that amphidetic ligaments and combined external/internal ligament are less derived than opisthodetic and external ligaments in their polarizations, and they incorrectly indicated the absence of an abdominal sense organ in the Unionoida (see Herbers 1914). They concluded that the Palaeotaxodonta is monophyletic, with the Solemyidae closer to the Nucinellidae than to the Nuculoidea or Nuculanoidea; that the Trigoniidae are closely related to the Pteriomorphia; and that the Unionoida, Veneroida, Lucinoidea and Myoida–Anomalodesmata–Septibranchia are united in a tetratomy. Our analysis of their data, with corrected character states and polarizations, produced 18 equally parsimonious trees. All 18 trees show the Palaeotaxodonta basal to the Autolamellibranchiata, but the Nucinellidae, Solemyidae, Nuculoidea and Nuculanoidea comprise a paraphyletic assemblage. The Septibranchia are shown as basal autolamellibranchs; above the Septibranchia half of the trees unite the Unionoida with a Pteriomorphia–Trigoniidae clade and the other half unite the Unionoida with the remaining autolamellibranchs. In the latter subclade, the Veneroida, Lucinoidea and Myoida–Anomalodesmata differentiate in that sequence.

Cope (1996*a*, *b*, 1997) indicated that *Fordilla* and *Pojetaia* are palaeotaxodonts, and he and Ratter & Cope (1998) favoured ultimate palaeotaxodont ancestry for all bivalve subclasses. However, Cope (1996*b*) acknowledged that the extremely small shell sizes of *Pojetaia* and *Fordilla* (2–4 mm) make it difficult to distinguish between palaeotaxodont and actinodont hinges. Cope (1996*b*) followed Newell (1969, p. N393) and Bailey (1983) in uniting modiomorphoideans and actinodontians in the Palaeoheterodonta, although he distinguished between the orders Modiomorphoida (Modiolopsidae and Colpomyidae) and Actinodontoida (Actinodontoidea and Glyptarcoidea). Cope (1996*b*) assumed that modiomorphoids evolved from actinodontoids through loss of hinge teeth but he did not address the distinction between the families Modiomorphidae and Modiolopsidae. Cope (1996*a*, 1997) suggested that the Anomalodesmata evolved from weakly dentate 'modiomorphoids' such as colpomyids by reducing the hinge dentition. Ratter & Cope (1998) indicated that frejids and parallelodontids share common ancestry in a descendant of *Catamarcaia*, and that *Catamarcaia* and the Cyrtodontoidea (plus the ambonychioidean and pterioidean descendants of cyrtodontids) share common ancestry in a hypothetical descendant of *Glyptarca* Hicks. *Glyptarca* presumably evolved from the nuculoid *Cardiolaria* Munier-Chalmas (Cope 1996*a*; Ratter & Cope 1998). Cope (1996*a*, *b*, 1997) and Ratter & Cope (1998) subdivided the Pteriomorphia into the subclass Pteriomorphia *sensu stricto* [orders Cyrtodontoida, Limoida, Pterioida (Ambonychioidea and Pterioidea), Mytiloida, Ostreoida and Praecardioida] and Neotaxodonta (*Catamarcaia*, Arcoidea and Limopsoidea). Cope (1996*a*, p. 366) believed that the Limoida, Pterioidea and Ambonychioidea evolved independently from the Cyrtodontidae. In 1997, he indicated that the Neotaxodonta is characterized by a unique combination of duplivincular ligaments, an ancestral crossed-lamellar or complex crossed lamellar structure, and a tendency to develop taxodont dentition. According to Cope (1997), neotaxodonts gave rise to the Heterodonta, whereas palaeotaxodonts gave rise to the Palaeoheterodonta. Cope's (1997) suggested porcelaneous neotaxodont ancestry for the Heterodonta is incompatible with the observation by Carter & Lutz (1990, pl. 104) that the Devonian crassatelloidean *Eodon tenuistriata* (Hall, 1870) is nacreous.

Geyer & Streng (1998) regarded their new Middle Cambrian taxon *Arhouriella opheodontoides* as a bivalve of undetermined subclass and order. However, the holotype has overhanging hinge teeth undercut by the hinge plate; a single muscle scar far from the shell margins; a hinge bar underlying a putative anterior, submarginal ligament insertion area; and a lancet-shaped hinge flange (putative posterodorsal ligament attachment area), all of which are reminiscent of ostracodes (see Scott 1961). The paratype has a more typical bivalve hinge (possibly edentulous, reportedly abraded), but much more elevated beaks and a possible ligament nymph. It is not obviously the same taxon as their holotype.

Other hypotheses for early bivalve phylogeny have focused on whether nuculoids gave rise to actinodontoids or vice versa (Dechaseaux 1952; Babin 1966; Newell 1969; Desparmet *et al.* 1971; Pojeta 1971, 1978; Morris & Fortey 1976; Scarlato

& Starobogatov 1978; Morris 1979, 1980; Babin & Le Pennec 1982; Liljedahl 1984*a*, *b*; Waller 1990, 1998; Johnston 1991, 1995; Sánchez 1995; Cope 1996*a*, *b*, 1997; Johnston & Zhang 1998; Ratter & Cope 1998). Nuculoid ancestry is almost universally favoured because nuculoids are common early Ordovician fossils, they retain a plesiomorphic protobranch ctenidium and they show certain early ontogenetic features similar to other molluscan classes. However, actinodontoid ancestry for the Nuculoida was suggested by Morris & Fortey (1976) and Morris (1980) on the basis of the existence of early Palaeozoic bivalves with transitional taxodont/actinodont hinges. Sánchez & Babin (1998) offered a third alternative, that palaeotaxodonts and palaeoheterodonts originated independently.

The present study is a phylogenetic analysis of a wide diversity of Palaeozoic bivalves, based on characters of shell morphology, hinge dentition, shell musculature, shell mineralogy, and shell and ligament microstructure, together with four soft anatomical features inferred from Recent taxa. The purpose of this study is to evaluate whether bivalves are more closely related to monoplacophorans or rostroconchs and to establish a preliminary phylogenetic classification for this class.

Analysed taxa

The phylogenetic analysis includes 62 Palaeozoic bivalve species with one species *Lyrodesma majus* (Ulrich 1894), represented by separate entries for adult and juvenile stages (Appendix 1). We also analysed: the Lower Cambrian, laterally compressed, stenothecid monoplacophorans *Anabarella plana* Vostokova 1962 and *Watsonella crossbyi* Grabau 1900; the problematic, Middle Cambrian monoplacophoran *Pseudomyona queenslandica* (Runnegar & Jell 1976); and the Middle Cambrian, ribeirioid rostroconch *Ribeiria junior* Runnegar, 1996. Shell muscle scars for the latter species were inferred from the Lower Ordovician ribeirioid rostroconch *Ribeiria lucan* (Walcott 1924). The two stenothecids served as the outgroup for the analysis. Both species have been implicated in the origin of both the Bivalvia and the Rostroconchia (see Pojeta & Runnegar 1985 and refs cited therein). In selecting bivalves for study, first preference was given to the best known species of each genus; preference was otherwise given to the type species. All of the studied genera have Palaeozoic origins, and all but *Acharax* and *Modiolus* are extinct. The represented families are also extinct, except for the Nuculidae, Acharacidae, Malletiidae, Limidae and Parallelodontidae.

Character definitions

The characters defined for this study (Appendix 2) are all palaeontological in nature except for inferences of protobranch or non-protobranch ctenidia, and the presence or absence of abdominal, adoral ('cephalic') and anterior mantle sense organs. It is assumed that protobranch ctenidia characterized all Cambrian monoplacophorans, rostroconchs, fordilloids and undoubted palaeotaxodonts, whereas non-protobranch ctenidia were present in all undoubted pteriomorphians, palaeoheterodonts, heterodonts and pholadomyoids. The ctenidial grade of Cambrian *Pseudomyona*, Cambrian *Tuarangia* and all bivalves with intermediate palaeotaxodont/actinodont hinges (e.g. tironuculids and cardiolariids) were designated as uncertain. It was assumed that an abdominal sense organ was present in all undoubted Trigonioida, Unionoida and Pteriomorphia, and absent in all Monoplacophora, Rostroconchia, Tuarangioida, Fordilloida, Tironuculidae, Nuculoida, Solemyoida and Heterodonta. Lucinoids, actinodontoids (Cyloconchidae, Redoniidae, Nyassidae, and Lyrodesmatidae), cardiolariids, modiomorphids, *Thoralia* Morris, *Babinka* and *Coxiconcha*, were scored as uncertain for this feature. The reader is referred to Herbers (1914), Yonge (1977), Haszprunar (1983, 1985) and Waller (1998) for discussion of this feature. An adoral sense organ is present in modern nuculids and solemyids, but is apparently absent in the Nuculanoidea, Nucinellidae, Pristiglomidae and in autolamellibranch bivalves (Waller 1998, p. 16, and refs cited therein). It was assumed that an adoral sense organ is present in nuculids and praenuculids as well as in the solemyid *Acharax* Dall, but this feature was indicated as uncertain for the Ctenodontidae, Tironuculidae and Cardiolariidae. The adoral sense organ was scored as absent in the Monoplacophora, Rostroconchia, Tuarangioida, Fordilloida and Autolamellibranchiata. An anterior mantle sense organ is apparently restricted to the Nuculanoidea (Yonge 1939; Allen 1985; Allen & Hannah 1986; Waller 1998). This feature is indicated as present in the Nuculanoidea but uncertain in the Tironuculidae, Cardiolariidae and Ctenodontidae. It was scored as absent in the Monoplacophora, Rostroconchia and in all other analysed bivalves.

A few of the characters in this study are continuous in nature, e.g. hinge arch, shell elongation, beak position and beak direction. Most of these continuous characters, however, are known to show quasi-discontinuous distributions. All unknown character states were coded as uncertain, and all intermediate character states and intraspecific-variable characters as polymorphisms. Polymorphisms comprised < 2% of the observations (see separate listings in Appendix 3). Polymorphic characters, while not ideal for phylogenetic analyses, can nevertheless increase phylogenetic accuracy (Wiens & Servedio 1997). Although many of the characters have missing observations, it is important to note that missing data *per se* do not distort phylogenetic analyses. Rather, they merely mimic the effects of limited taxon sampling (Wiens 1998). Also, highly significant characters with considerable missing data may still increase phylogenetic accuracy (Smith 1994; Wiens 1998).

Current terminology of shell morphology and shell musculature follows the definitions in Cox *et al.* (1969–1971). Use of the term 'hinge' implies the presence of a divided larval shell, whereas 'pseudohinge' implies an undivided larval shell (e.g. in *Pseudomyona*). No distinction is made between a 'primary' pallial line, which maintains contact with the posterior adductor, and a 'secondary' pallial line (Yonge 1953), which does not. These two modes of pallial attachment are intergradational, and the relationship between the pallial line and the posterior adductor muscle scar is dealt with in other characters. The shell microstructure terminology follows Carter (1990).

Ligament terminology is presently modified from Waller (1990) and Carter (1990) to define a number of new varieties. We distinguish between true ligaments and pseudoligaments. Ligaments and pseudoligaments are both morphologically specialized to enhance dorsal flexing of the shell, but only true ligaments differ microstructurally from the adjacent shell. For example, the foliated structure comprises the pseudoligament and the adjacent shell in *Pseudomyona* (see Runnegar 1983, fig. 6B), whereas lamellar and fibrous structures comprise the ligament, and nacreous, porcelaneous and/or foliated structures comprise the adjacent shell in pteriomorphians. The terminology of ancestrally dorsal ligaments is summarized in Table 1. Nuculoid and solemyoid internal resilia are excluded from Table 1 because they are not strictly homologous with dorsal ligaments. Despite the physical continuity of the dorsal and resilial ligaments in modern nuculids (Waller 1990), these two ligaments have ancestrally separate fibrous layers. This is shown by the Triassic malletiid *Palaeoneilo elliptica* (Goldfuss 1838), which has a well-mineralized dorsal, simple ligament separate from its laterally mineralized, internal resilium.

It is tentatively assumed that discontinuous fibrous ligament ontogeny characterizes all members of the Modiolopsoidea as defined by Fang & Morris (1997) (i.e. families Modiolopsidae, Colpomyidae and Modiolodontidae), as well as all mytilids and other pteriomorphians, including frejids and glyptarcids [see discussions by Waller (1990, 1998), and Carter & Seed (1998)].

The term 'lateral tooth', from the definition by Cox *et al.* (1969, p. N106), is restricted to exclude pseudolateral teeth. A lateral tooth, as presently defined, is partly or entirely located some distance from the beak and its proximal end does not extend to the beak. A pseudolateral tooth is partly located some distance from the beak but extends all the way to the beak. Rather than utilizing inferences of dental homologies for specific subumbonal/cardinal teeth, the pivotal (major or central) subumbonal/cardinal tooth has been identified, and the shape and orientation of this tooth and the subumbonal/cardinal teeth immediately anterior and posterior to it have been described. In counting the number of anterior, subumbonal/cardinal and posterior teeth, only socketed teeth (i.e. as opposed to unsocketed crura) are included.

Some cyrtodontids present a problem for classifying teeth as subumbonal or anterior because what is probably an homologous cluster of relatively anterior teeth may be positioned subumbonally or anterior to the umbones. In such cases the more posterior and posteriorly dipping of any transitional subumbonal/anterior teeth in cyrtodontids have been arbitrarily counted as subumbonal, and the more horizontal and anteriorly dipping of these teeth as anterior. Unfortunately, it is possible that some of these 'subumbonal' teeth are homologous with anterior frejid, glyptarcid and/or parallelodontid teeth.

Edentulous hinges also present a problem for phylogenetic studies because edentuly has clearly evolved several times in this class. Many bivalves have apparently lost ancestral dentition in connection with life habits and/or biomechanical adaptations reducing the need to precisely guide the valves. Therefore, five categories of subumbonal edentuly are recognized on the basis of associations with particular life habits or biomechanical adaptations: (1) wide, thick, pleated, periostracal margins (e.g. solemyids); (2) wide, flexible, calcitic simple prismatic margins (e.g. pterineids); (3) thick, gasket-like periostracal margins (e.g. some mytilids); (4) deep burrowing life habits with very thin, commonly spiculose periostracal extensions of the shell margins (many pholadomyoids); (5) subumbonal edentuly apparently resulting from a posterior shift in the umbones rather than from loss of teeth (e.g. possibly falcatodontids).

'Taxodonty' (Cox 1969, p. N108) is here subdivided into palaeotaxodonty, pretaxodonty, pseudotaxodonty and heterotaxodonty. Palaeotaxodonty refers to five or more laterally adjacent, regularly shaped taxodont teeth that are neither strongly ventrally divergent nor ventrally bifid.

Table 1. *Dorsal ligaments and their evolutionary derivatives*

1. Simple narrow (SN): simple ligament with narrow insertion area.
(a) SNweak: SN ligament with plesiomorphically weakly or non-mineralized medial portion of inner sublayer, e.g. *Pojetaia* Jell.
(b) SNstrong: SN ligament with strongly mineralized lateral and medial portions of inner sublayer, e.g. *Cycloconcha milleri* (Meek, 1871); also the dorsal ligament but not the resilium in *Nuculoidea pinguis* (Lindström, 1880).

2. Simple wide (SW): simple ligament with wide insertion area, e.g. *Paracyclas* Hall.

3. Parivincular narrow (PN): parivincular ligament with short, narrow nymph parallelling dorsal shell margin.
(a) PNweak: PN ligament with nymphs not strongly projecting above dorsal shell margin, e.g. *Astarte* Sowerby.
(b) PNstrong: PN ligament with nymphs strongly projecting above dorsal shell margin, e.g. *Lyrodesma* Conrad.
(c) PNincip: PN ligament with incipient nymph flanked by very weakly differentiated lamellar ligament groove, e.g. *Silurozodus* Liljedahl.

4. Parivincular wide (PW): short to intermediate length, barrel-shaped parivincular ligament with wide, strongly dorsally projecting nymph in which anterior nymph margin rises abruptly above the hinge line.
(a) PWext: external PW ligament, e.g. *Acharax* Dall.
(b) PWint: internal PW ligament, partially covered by the prismatic outer shell layer, e.g. *Solemya* Lamarck.

5. Parivincular ramped (PR): medium to long, barrel-shaped parivincular ligament with wide, strongly dorsally projecting nymph in which nymph margin increases gradually in height toward the posterior.
(a) PRthick: PR ligament attached to thick, flat to concave hinge plate, with dorsal margin of nymph at or below dorsal edge of dorsal shell margin.
(i) PRthick-shallow: with nymph at or shallowly submerged below dorsal edge of dorsal shell margin, e.g. *Modiomorpha* Hall & Whitfield.
(ii) PRthick-deep: with nymph deeply submerged below dorsal edge of dorsal shell margin, e.g. *Lahillia larseni* Sharman & Newton, 1897.
(b) PRthin: PR ligament attached to thin hinge plate, with dorsoposterior nymph margin projecting above the dorsal edge of dorsal shell margin ('thin' refers to the hinge plate thickness, not necessarily to the ligament thickness).
(i) PRthin-weak: relatively weak PRthin ligament, e.g. *Edmondia gibbosa* M'Coy, 1844.
(ii) PRthin-strong: relatively strong PRthin ligament, e.g. *Goniophora onyx* Liljedahl, 1984.

6. Parivincular submerged (PS): relatively short, parivincular ligament with nymph dipping distinctly posteroventrally relative to dorsal edge of dorsal shell margin.
(a) PSshallow: PS ligament with fibrous layer shallowly submerged below dorsal edge of dorsal shell margin, e.g. *Eodon tenuistriata* (Hall, 1870).
(b) PSdeep: PS ligament with fibrous layer deeply submerged below the dorsal edge of dorsal shell margin, e.g. *Montanaria* Spriestersbach.

7. Duplivincular horizontal (DH): duplivincular ligament with adult insertion grooves nearly parallel to growth lines on hinge; early growth stage may or may not have inclined insertion grooves.
(a) DHopis: opisthodetic DH ligament with ventralmost insertion groove extending further posteriorly than the dorsalmost insertion groove.
(i) DHopis1: DHopis ligament with only one adult fibrous layer but with another ontogenetically attenuated, early post-larval, fibrous layer, e.g. *Mytilus* Linné.
(ii) DHopis2: DHopis ligament with two or more adult fibrous layers at least in some individuals of the species. In some DHopis2 ligaments the insertion area is very narrow and there are few, widely separated ligament grooves (e.g. *Aleodonta burei* Liljedahl, 1994). In others, there are more numerous ligament grooves and the early-formed fibrous layers may be largely restricted to the anterior ligament insertion area (e.g. *Cyrtodonta huronensis* Billings, 1858). A third variety shows only growth lines on the ligament insertion area, overlain by relatively thick, fibrous ligament ligament layers, each covering several growth lines [e.g. *Gosseletia triquetra* Conrad, 1838; see Carter (1990, p. 198)].
(iii) DHopis1–2: DHopis ligament with normally only one adult fibrous layer but with additional adult fibrous layers formed in some individuals as a consequence of traumatic withdrawal of the mantle margins [e.g. *Pinna* Linné; see Yonge (1953)].
(b) DHopis/amph: slightly prosodetic but still predominantly opisthodetic DH ligament, e.g. *Actinopteria* Hall.
(c) DHamph: strongly amphidetic DH ligament, e.g. *Leiopecten* Khalfin.

8. Duplivincular inclined (DI): duplivincular ligament with adult insertion grooves distinctly dipping relative to the hinge margin.
(a) DIopis: entirely opisthodetic DI ligament, e.g. *Ischyrodonta* Ulrich.
(b) DIopis/amph: slightly prosodetic, but predominantly opisthodetic DI ligament, e.g. *Ptychodesma* Hall & Whitfield.
(c) DIamph: strongly amphidetic DI ligament.
(i) DIamph-incl: DIamph ligament with all ligament grooves distinctly inclined relative to growth lines, e.g. *Pseudaviculopecten* Newell.
(ii) DIamph-aliv: DIamph ligament with posterior ligament grooves all distinctly inclined relative to growth lines, but anterior part of ligament changing to a broad, flat, insertion area covered by weakly or strongly mineralized fibrous ligament, e.g. *Cosmetodon obsoletus* (Meek, 1871).

Table 1. *Continued*

9. Alivincular (ALIV): single, triangular to slightly trapezoidal area of fibrous ligament (resilifer) flanked anteriorly and posteriorly by lamellar ligament.
(a) ALIVnar: ALIV ligament with angle of divergence < 20°, e.g. *Spondylus* Linné.
(b) ALIVmod: ALIV ligament with moderate angle of divergence (20–110°).
(i) ALIVmod-trian: ALIVmod ligament with triangular resilifer, e.g. *Lentipecten hochstetteri* (Zittel, 1864).
(ii) ALIVmod-trap: ALIVmod ligament with nearly trapezoidal resilifer, e.g. *Eumorphotis multiformis* (Bittner, 1899).
(c) ALIVwide: ALIV ligament with angle of divergence > 110°, e.g. *Acanthopecten coloradoensis* (Newberry, 1861).

10. Multivincular (MULT): serially repeated alivincular ligaments.

Palaeotaxodont teeth may be orthomorphodont, concavodont (distally concave) or convexodont (distally convex), or a combination thereof [terminology after Babin (1966)]. In palaeotaxodont hinges, the tooth number increases with shell growth above *c.* 1 mm length. Pretaxodonty is presently applied to minimally taxodont hinges with only two to four stout, orthomorphodont teeth. Unlike palaeotaxodont teeth, pretaxodont teeth do not increase in number with continued shell growth beyond *c.* 1 mm length. For example, *Pojetaia sarhroensis* Geyer & Streng, 1998, has only three or four pretaxodont teeth in shells 2.25 mm in length, whereas the Recent nuculoid *Microgloma yongei* Sanders & Allen, 1973 has six to eight palaeotaxodont teeth in a shell 1.0 mm in length (Sanders & Allen 1973, fig. 10) and *Ennucula aegeensis* (Forbes 1843) has ten palaeotaxodont teeth in a shell 0.98 mm in length (Ockelmann & Warén 1998, fig. 4C). Pseudotaxodonty (cf. Cox 1969, p. N107) presently refers to five or more closely adjacent, more or less irregularly shaped, taxodont teeth in one or two rows. Pseudotaxodont teeth may be strongly ventrally divergent and/or ventrally bifid, but they are not regularly concave or convex. Heterotaxodonty refers to hinges in which the anterior teeth, which may be palaeotaxodont or pseudotaxodont, are abruptly much larger than posterior palaeotaxodont teeth (e.g. in *Praeleda subtilis*, Cope 1999). Cope's (1996*a*) term 'gradidentate' has been adopted for taxodont teeth more or less gradually increasing in size anteriorly and posteriorly from the subumbonal region.

Methods of analysis

PAUP Version 4b2a* (Swofford 1998) was used with random addition, tree bisection–reconnection (TBR), delayed transformation, and one 425 replicate and two 1000 replicate runs. The decay index (D.I.) for each node in the strict consensus tree was calculated using a 20-replicate heuristic search (random addition, TBR, delayed transformation) for trees incompatible with each node. The D.I. is the difference between the strict consensus tree and the shortest tree incompatible with the node. With reference to Fig. 4 strict consensus tree nodes with D.I. other than zero include: D.I. = 41 (node 2); D.I. = 6 (node 4); D.I. = 5 (node 21); D.I. = 4 (nodes 3, 16–20, 31, 55, 64); D.I. = 3 (nodes 7, 26, 27, 30, 34); D.I. = 2 (nodes 5, 6, 29, 33, 41); and D.I. = 1 (nodes 9 22–25, 28, 32, 38–40, 43, 48, 52, 53, 56, 61–63, 65). Appendix 3 shows the database for these analyses.

Results

The 425 replicate and two 1000 replicate heuristic searches found the same 411 most parsimonious trees (length 1103), suggesting that all of the most parsimonious trees were found. These trees had a consistency index of 0.5313, a homoplasy index of 0.5830, a retention index of 0.6372 and a rescaled consistency index of 0.3385. The strict, Adams and majority rule consensus trees are shown in Figs 1–3.

The strict consensus tree (Fig. 1) shows that the Bivalvia (taxa 5–67) resolve closer to the laterally compressed monoplacophorans *Watsonella* and *Pseudomyona* than to the laterally compressed monoplacophoran *Anabarella* or the rostroconch *Ribeiria*. The strict consensus tree also indicates monophyly for the subclass Palaeotaxodonta (taxa 8–19); a clade consisting of the Anomalodesmata (taxon 41) together with the Pteriomorphia (taxa 42–67); a clade comprising the Tironuculidae (taxon 8), Praenuculidae (taxon 9), Nuculidae (taxon 10) and Solemyoida (taxa 17–19); a clade of the Thoraliidae (taxon 21) and Trigonioida (taxa 22–25); a clade of three cycloconchid genera (taxa 26–28); a monophyletic order Mytiloida as presently defined (taxa 44–46); a monophyletic Cyrtodontoida (taxa 50–52); monophyletic

Pectinoida as presently defined (taxa 55–57); and a monophyletic order Arcoida (taxa 62–67). The Adams consensus tree (Fig. 2) additionally resolves a clade comprising the Palaeoheterodonta and Heterodonta (i.e. several actinodonts plus the orders Lucinoida, Trigonioida and Veneroida; taxa 20–40); a clade within the Pteriomorphia comprising the Colpomyidae, Modiolopsidae,

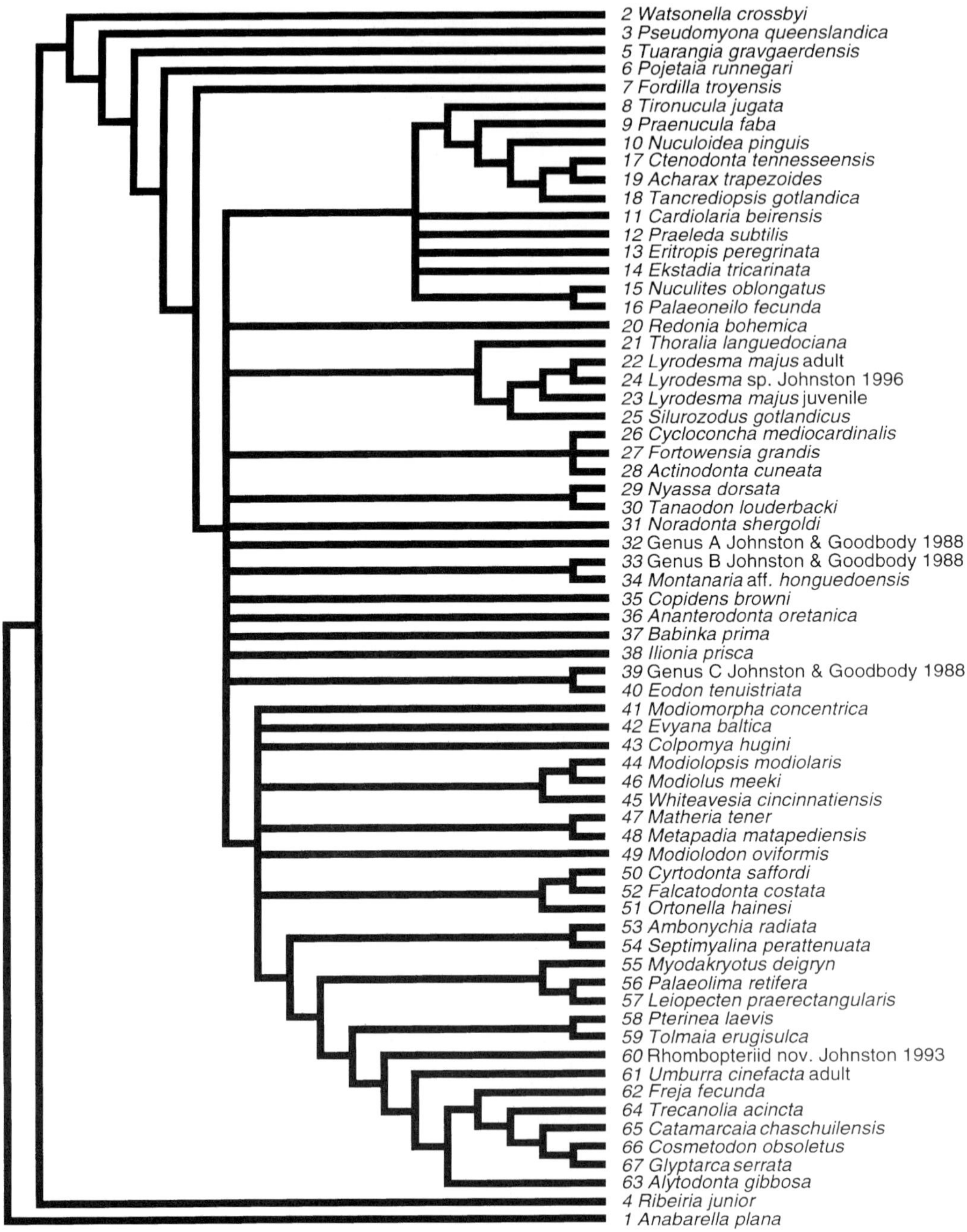

Fig. 1. Strict consensus tree for 411 most parsimonious trees from a 425 replicate, heuristic search.

Mytilidae, Matheriidae, Modiolodontidae, Cyrtodontidae and Falcatodontidae, taxa 43–52; and a second pteriomorphian clade represented by the Ambonychioidea, Pectinoida, Pterioida, Ostreoida and Arcoida. The majority rule consensus tree (Fig. 3) adds a weakly supported (53%), poorly resolved association of the Palaeotaxodonta together with the Palaeoheterodonta and Heterodonta.

From the 411 most parsimonious trees, selected for further analysis were those trees in which the

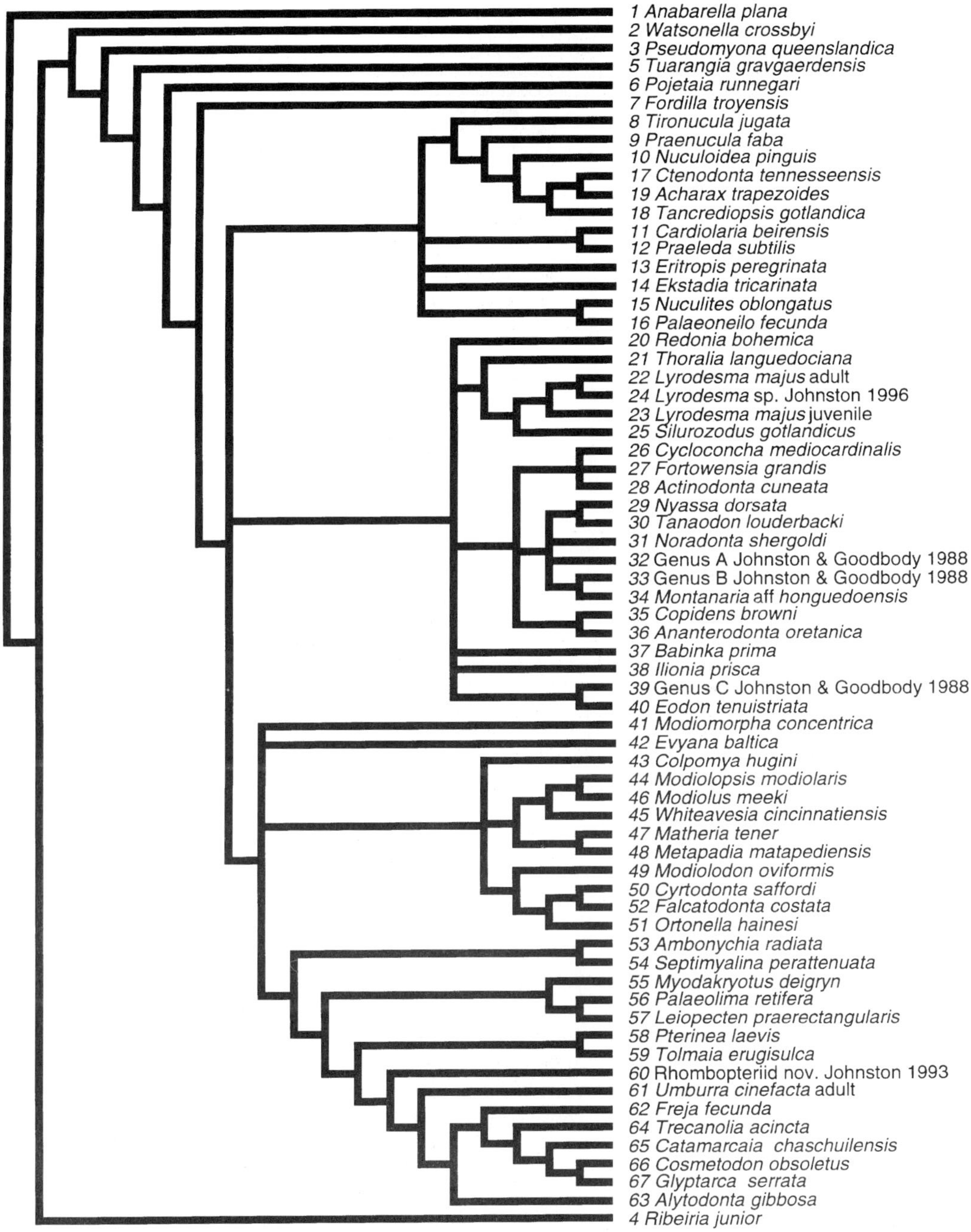

Fig. 2. Adams consensus tree for 411 most parsimonious trees from a 425 replicate, heuristic search.

clade of the Palaeoheterodonta plus Heterodonta was resolved as monophyletic and exclusive of the Palaeotaxodonta and Pteriomorphia. This filtering did not interfere with evaluation of the hypothesis of actinodont ancestry to arcoids because all of the most parsimonious trees excluded this hypothesis. The filtering excluded all but 16 trees. In these 16 trees, a palaeoheterodont–heterodont clade is

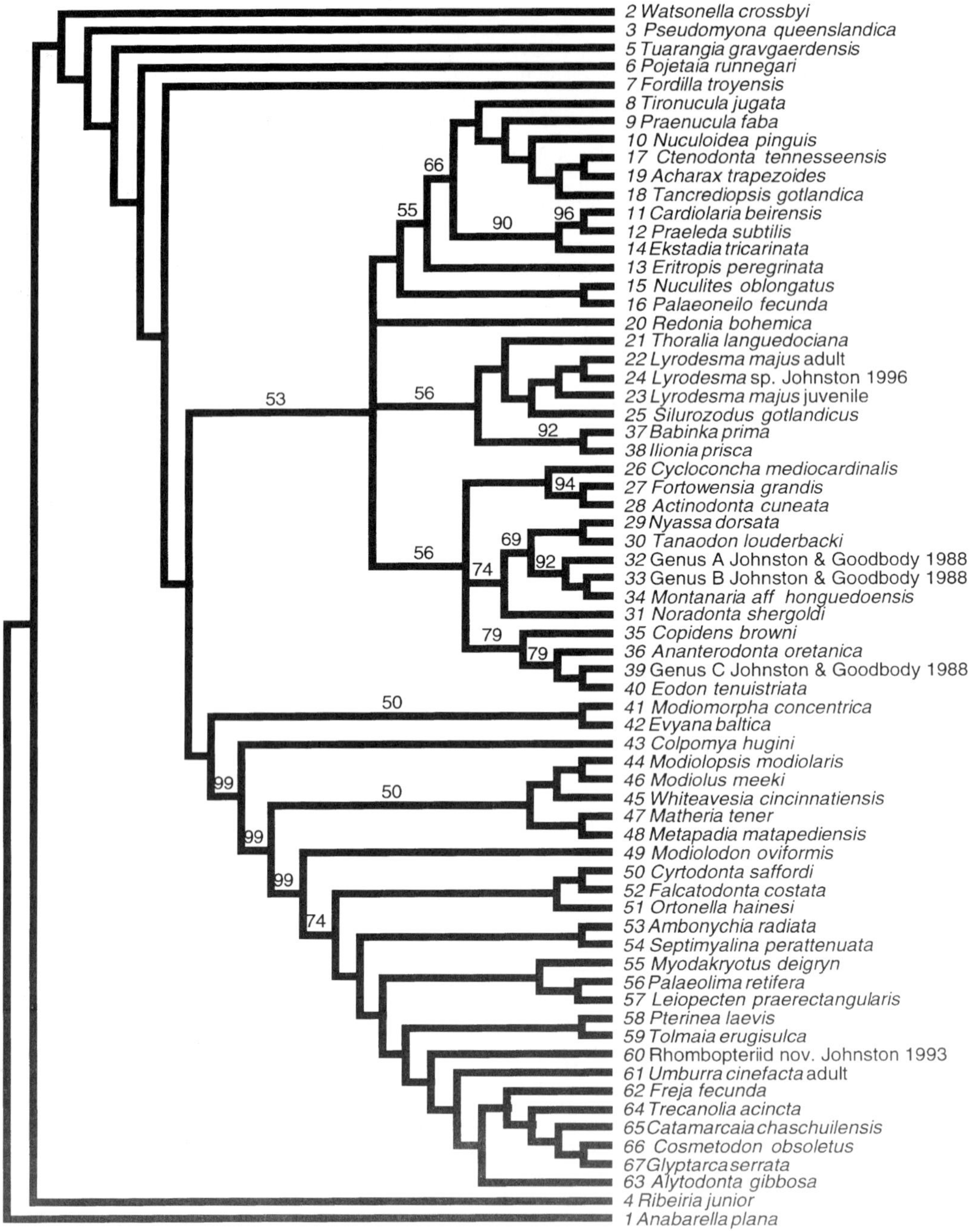

Fig. 3. Majority rule consensus tree for 411 most parsimonious trees from a 425 replicate, heuristic search. All nodes are 100% except where indicated otherwise.

consistently basal to an Anomalodesmata–Pteriomorphia clade. The 16 trees differ only in the position of *Modiomorpha* basal to *Evyana*, or as a sister group to *Evyana*, and in the position of the Matheriidae (*Matheria* and *Metapadia*) as a sister group to *Modiolopsis–Whiteavesia–Modiolus*, basal to these three genera or basal to *Modiolodon*. From these 16 trees the reconstruction (Fig. 4) in which *Modiomorpha* is basal to the Pteriomorphia and in which the Matheriidae are closer to the Cyrtodontidae was selected. This selection was justified on the basis that *Modiomorpha* lacks the multiple ligament insertions characteristic of *Evyana* and many other basal or near-basal pteriomorphians (including some colpomyids and modiolopsids) and because matheriids have been regarded as closely related to cyrtodontids (Desbiens 1994). All character state changes for Fig. 4 are given in Appendix 4 and the more significant of these are reviewed briefly below. This review is largely limited to unambiguous apomorphies and non-terminal nodes.

The clade of *Pojetaia–Fordilla*–higher Bivalvia (node 5) is characterized by the appearance of a subumbonal socket in the right valve and a corresponding tooth in the left valve, and the loss of radial costae. The clade of *Fordilla*–higher Bivalvia (node 6) begins with a more anteriorly situated but non-terminal beak and loss of an ancestral posterior auricle. The common ancestor of the subclasses Palaeotaxodonta and Autolamellibranchiata (node 7) is characterized by separation of the left valve (LV) pedal retractor scar from the adductor muscle scar, a shift in the position of the posterior adductor muscle scar closer to the hinge, a more solidly inserted pallial line, small tablet nacreous and/or porcelaneous [crossed lamellar (CL), complex crossed lamellar (CCL), and fine CCL and/or homogeneous (FCCL/HOM)] structures, and more extensive mineralization of the simple, narrow ligament. The Palaeotaxodonta (node 55) then differentiates by evolving an anterior mantle sense organ and palaeotaxodont hinge teeth, the latter anteriorly concavodont and posteriorly orthomorphodont, with the anterior tooth row less than one third as long as the posterior tooth row and with the posterior teeth increasing in size toward the posterior. The Autolamellibranchiata (node 8) differentiates by changing from protobranchiate to autolamellibranchiate. Within the Autolamellibranchiata, the Heteroconchia as presently defined (node 35) then separates from the clade of the Anomalodesmata and Pteriomorphia (node 9). The Heteroconchia is characterized by a more discoidal, less dorsally indented anterior shell margin and, possibly (ambiguous), by a change from anisomyarian to nearly isomyarian. The Anomalodesmata–Pteriomorphia clade has stronger posteroventral obliquity and possibly (ambiguous) a stronger inflection in the posteroventral shell margin. *Modiomorpha* has numerous apomorphies, including a medioventral or posteroventral sinus, proximity of the anterior pedal retractor and anterior adductor muscle, placement of the anterior adductor close to the hinge, partial overgrowth of the hinge dentition by the advancing ligament insertion area and a shallowly submerged, parivincular ramped ligament. The suborder Pteriomorphia (node 10) differs from the Anomalodesmata in having discontinuous fibrous ligament ontogeny and multiple ligament grooves.

Focusing on the Palaeotaxodonta, there is a basal dichotomy between the superfamily Nuculanoidea (*Nuculites* and *Palaeoneilo*) and other taxa traditionally placed in the orders Nuculoida or Solemyoida. The Nuculanoidea (node 56) is characterized by a shallow pallial sinus and possibly also (ambiguous) by a change from nacreous to finely textured porcelaneous microstructure. The remaining Palaeotaxodonta (node 57) are characterized by an increase in umbonal elevation and, except for *Eritropis*, also a change from anterior to medial beaks and from posterior orthomorphodont to posterior concavodont dentition (node 58). Except for the basal genus *Eritropis*, this clade is divided into the family Cardiolariidae Cope 1997, and a sister group (node 61) which includes *Tironucula* (family Tironuculidae), *Praenucula* (family Praenuculidae), *Nuculoidea* (family Nuculidae) and the order Solemyoida. The Cardiolariidae (node 59) is characterized by a heterotaxodont hinge. The sister group to the Cardiolariidae (node 61) corresponds with a change from posteroventral to anteroventral shell obliquity, nuculiform shell shape, more uniform size in the posterior palaeotaxodont hinge teeth and an increase in the relative length of the anterior tooth row. Autapomorphies for *Tironucula jugata* include an increase in the hinge arch and the addition of a single posterior lamellar tooth above the posterior palaeotaxodont teeth. Autapomorphies for *Praenucula faba* include reduction in the number of dorsal accessory muscle scars to one or two and a change in the posterior hinge teeth from concavodont to orthomorphodont. The latter, however, is not widespread for *Praenucula*, as concavodont posterior dentition is present in *Praenucula infirma* Tunnicliff, 1982, among other species. The family Nuculidae is presently represented by *Nuculoidea pinguis*. Autapomorphies for this species include an internal resilium and increase in size of the anterior adductor muscle scars relative to the posterior ones.

The order Solemyoida (node 64) is characterized by: a triangular posteroventral shell margin; medial or posterior beaks; a greatly elongated shell anterior with nearly parallel dorsal and ventral margins; at most, only a slight indentation anterior to opisthogyrate beaks; an external parivincular wide ligament; and anterior palaeotaxodont teeth with variable shapes. The hinge dentition is secondarily reduced in *Acharax* and other solemyoideans.

The subclass Autolamellibranchiata (node 8) is characterized by non-protobranch ctenidia and perhaps also (ambiguous) by non-pretaxodont, non-

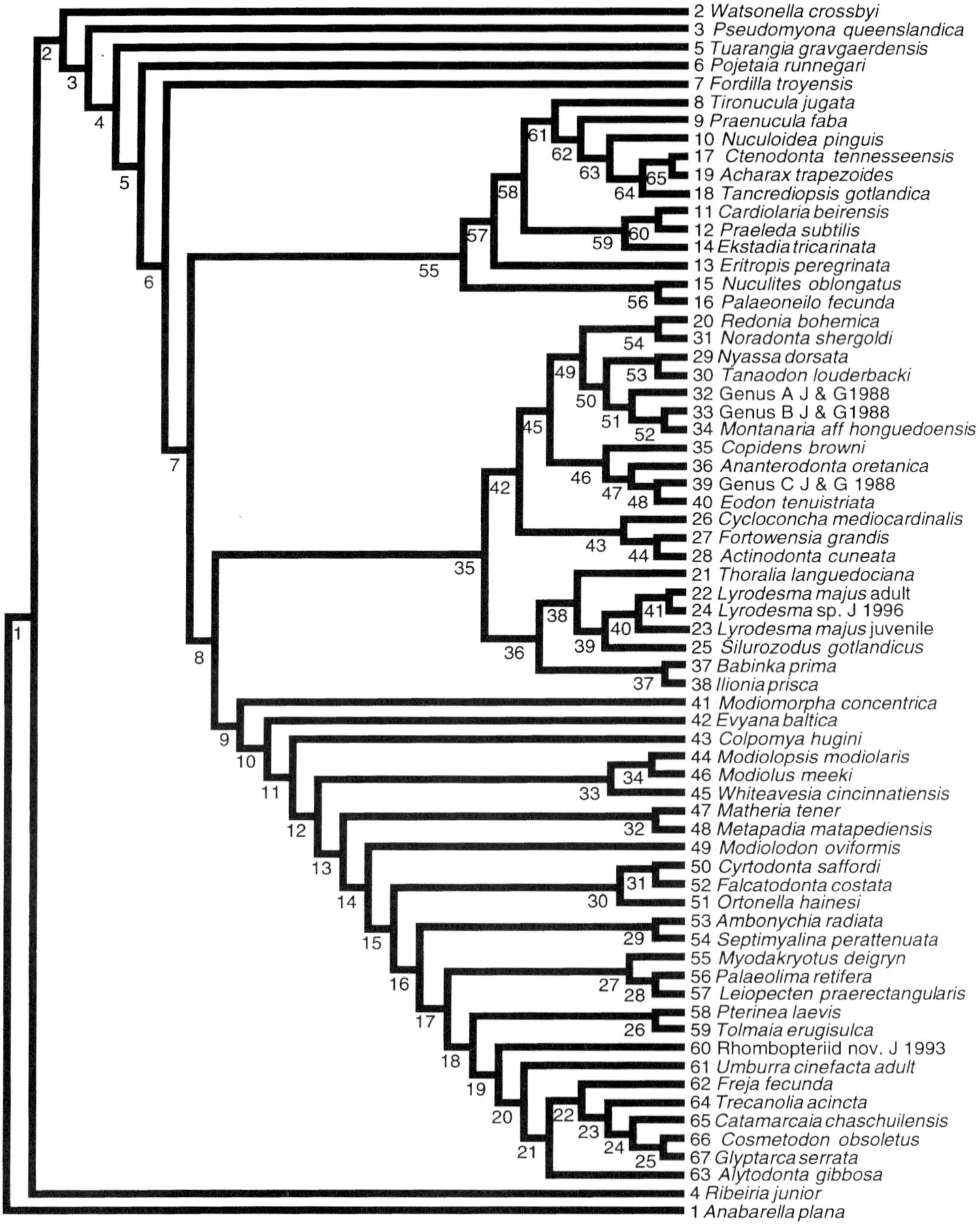

Fig. 4. Cladogram selected from 411 most parsimonious trees to base the present phylogenetic classification.

palaeotaxodont hinge teeth. The superorder Heteroconchia (node 35) is characterized by a change in the shape of the anterior shell margin from indented anterior to the beaks to discoidal, with at most a very slight indentation. This node may also correspond with a change from unequal to nearly equal adductor muscle scars (ambiguous). The Heteroconchia shows a basal dichotomy between an unnamed clade (node 36) comprising the Lucinoida (node 37), Thoraliidae and Trigonioida (node 39), and a second unnamed clade comprising the Cycloconchidae *sensu stricto* (node 43), Redoniidae (node 54), Nyassidae (node 50) and basal or near-basal heterodonts (node 46). The lucinoid–thoraliid–trigonioid clade (node 36) is characterized by a shift in the beak position from anterior to medial and, possibly, also by a slender, non-bifid or slightly bifid, right pivotal cardinal tooth. Within this clade, the order Lucinoida (node 37) shows a reduction in the shell length/height ratio to < 1.25, a more nearly equilateral shell shape, and a wider, simple ligament. The Thoraliidae shows an increase in umbonal elevation, changes from posteroventrally to anteroventrally oblique, evolves a more robust right pivotal cardinal tooth and adds a slender, posteroventrally directed cardinal tooth behind the pivotal cardinal tooth in the left valve. The order Trigonioida (node 39) is characterized by the change from a simple to an incipient parivincular narrow ligament. It may be noted that *Lyrodesma* did not place close to *Silurozodus* until its juvenile stage was included as a separate taxon.

An unnamed cycloconchid–redoniid–nyassid–heterodont clade (node 42) is characterized by posteriorly dipping pseudotaxodont hinge teeth located posteriorly to the pivotal cardinal tooth in each valve. Within this clade, the Cycloconchidae (as presently restricted; node 43) shows two anterior lateral or pseudolateral teeth, the stronger of these long, slender, lamellar and continuous with the anterior cardinal tooth. The remaining taxa in this clade (node 45; redoniids, nyassids and near basal or basal heterodonts) are united by subumbonal pseudotaxodonty and by posteroventrally dipping pseudotaxodont teeth anterior to the pivotal cardinal tooth.

The near basal or basal heterodont clade (node 46) is characterized by a shell length/height ratio of < 1.25; three or more posterior lateral or pseudolateral teeth, the strongest long, slender, and lamellar; and a wider posterior hinge plate. Within this clade, Genus C of Johnston & Goodbody (1988) shares with *Eodon tenuistriata* (Hall, 1870) (node 48) slight posterior rostrateness, a broadly concave area immediately anterior to the beaks and reduction of ancestral subumbonal pseudotaxodonty to two cardinal teeth.

The genera *Redonia* and *Noradonta* (node 54) are united by having a single posterior lateral or pseudolateral tooth, a wider anterior hinge plate and, possibly, also separation of the posterior pedal retractor scar from the posterior adductor scar (the latter remains uncertain for *Noradonta*).

The family Nyassidae (node 50), represented by *Nyassa*, *Tanaodon*, *Montanaria*, and genera A and B of Johnston & Goodbody (1988), is characterized by the addition of a posteroventral shell marginal sinus. Possible synapomorphies include a slightly triangular posteroventral margin, a wider posterior hinge plate and very wide CL and/or CCL structure in the inner shell layer.

Within the superorder Pteriomorphia, *Evyana baltica* (Liljedahl, 1989) is transitional between the Modiomorphidae and early 'mytiloid' pteriomorphians. *Evyana* combines the deeply incised lunule and rather crudely shaped cardinal dentition of *Modiomorpha* (although more distinct than in *Modiomorpha*) with the opisthodetic, duplivincular ligament of other near-basal pteriomorphians such as colpomyids. Nodes 11–15 join the various 'mytiloid' families with the order Cyrtodontoida. Node 11 corresponds with the appearance of a *Fordilla*-like shell shape and a horizontal, opisthodetic, duplivincular ligament with few ligament grooves. Node 12 involves reduction in the convexity of the left valve, reduction of the left pivotal subumbonal hinge tooth and closer spacing of the ligament grooves. Node 13 is supported by only one ambiguous synapomorphy – a change in the right pivotal cardinal/subumbonal tooth from bulbous or tuberculiform to more distinctly shaped and slender. Node 14 coincides with shortening of the shell to a length/height ratio of < 1.25 and reduction of posteroventral shell obliquity. Node 15 is characterized by a decrease in left valve convexity and partial overgrowth of the hinge plate by the ligament insertion area.

Nodes 16–20 correspond with the differentiation of the Ambonychioidea, Pectinoida, Pterioida, Ostreoida and the Umburridae–Arcoida clade. Node 16 is supported by the evolution of a posterior auricle without significant posterior extension, a byssal sinus, posteroventral shift of the posterior adductor muscle scar and calcitic, simple prismatic structure. At node 17 the following are observed: a commissural shelf; an angular anterior auricle; an anteriorly dipping anterior shell margin; and a straight, proximally restricted gill wheal, the latter extending from the umbonal cavity posteroventrally below the posterior adductor. At node 18 the left valve becomes more convex than the right one; the duplivincular ligament changes from symmetrical to asymmetrical; the subumbonal area has posteriorly dipping pseudotaxodont teeth; the posterior hinge acquires three or more lateral and/or pseudolateral teeth, with the lower ones aligned en echelon closer to the beak, and at least one of these long and slender; and the tooth closest to the subumbonal area is short and nearly horizontal.

The Ostreoida–Arcoida clade (node 19) corresponds with a reduced byssal sinus; slightly more elevated umbones; a change from *Pterinea*form to *Umburra*form shell shape; a gill wheal changing from straight and proximally restricted, to distally posteriorly curving; two anterior lateral teeth, the stronger of these short and tabular; and pseudotaxodont teeth filling the gap between these teeth and the beak. At node 20, the duplivincular ligament changes from opisthodetic horizontal to amphidetic inclined, and the ligament symmetry changes from asymmetrical to symmetrical with a wide interfacial angle between the ligament insertion areas.

The clade of the order Mytiloida as presently restricted (node 33) is characterized by a change from *Fordilla*form to modioliform or mytiliform shell shape, and loss of subumbonal teeth in association with a thick, gasket-like periostracum. Within this clade, the family Mytilidae, represented by Cretaceous *Modiolus*, differs from *Modiolopsis* and *Whiteavesia* in reducing the convexity of its valves; shifting posteriorly the contact of the pallial line with the posterior adductor muscle scar; shifting the posterior adductor scar ventrally and posteriorly; and adding abundant irregular simple prismatic structure in the inner shell layer.

The superfamily Cyrtodontoidea (node 30) is characterized by two long, lamellar, posterior lateral teeth, not uniformly aligned en echelon, and separated by a wide, edentulous gap from the subumbonal hinge teeth. Within this superfamily, *Ortonella* is basal to the superfamilies Cyrtodontoidea and Falcatodontoidea. The Cyrtodontidae, as represented by *Cyrtodonta saffordi* (Hall 1859), has a more discoidal ventral shell margin; more medially situated beaks; more equal adductor muscle scars; a gill wheal proximally dorsoventral and distally posteriorly curving; a shortened posterior lateral tooth closest to the beak; a thicker anterior hinge plate; and a deeply bifid, ventrally or anteroventrally open subumbonal tooth anterior to the pivotal subumbonal tooth. In contrast, the Falcatodontidae has a posterior auricle; strong, radial costae; a less elliptical, more triangular posterior adductor muscle scar; two posterior lateral teeth in one valve and three in the other; subumbonal edentuly resulting from a posterior shift in the umbones; and a reduction in overgrowth of the hinge dentition.

The clade of the Pectinoida (node 27) is defined by reduction in the convexity of the left valve, loss of hinge overgrowth and a change in the ancestral duplivincular, horizontal ligament from opisthodetic to amphidetic. Ambiguous synapomorphies include the adoption of valve-indifferent pleurothetic habits and a shift from anterior to medial beaks. Within this clade, suborder Limina (represented by *Palaeolima*) is characterized by a change from slight posteroventral to slight anteroventral shell obliquity; loss of the commissural shelf; a change from prosogyrate to orthogyrate; closely to medium-spaced radial costae; and the change from duplivincular, horizontal, amphidetic ligament to an alivincular ligament. The suborder Pectinina (represented by *Leiopecten*) shows a slight increase in extension of the posterior auricle beyond the subauricular sinus; a slight increase in the shell length/height ratio; a more discoidal, anteriorly inclined anterior shell margin; a more convex left valve; and a hinge with one or two anterior and posterior, elongate teeth that both approximate the subumbonal area without strong, intervening teeth.

Synapomorphies for the order Arcoida (node 21) include a stronger inflection in the posteroventral shell margin; posteroventral inclination of the anterior shell margin; reduction of the commissural shelf; secondarily equivalve shells; movement of the posterior adductor closer to the hinge; a more elliptical anterior adductor; increase in the number of anterior lateral or pseudolateral hinge teeth from two to three or more; and a change in the orientation of the pseudotaxodont subumbonal teeth from posteroventrally inclined to nearly vertical or dorsally divergent.

Discussion

The present analysis suggests that the Bivalvia are more closely related to *Watsonella* and *Pseudomyona* than to *Anabarella* and the Rostroconchia. This analysis also suggests that the common ancestor of *Tuarangia* and *Pojetaia* (Fig. 4, node 4) had a non-rostrate posterior; a moderately to slightly elongate ventral shell margin; no shell marginal sinuses; slight posteroventral obliquity; low, medial beaks with, at most, a slight indentation anterior to them; an angular posterior auricle; no lunule or escutcheon; no pegma or pegma-like shelf; three or more subumbonal, dorsal accessory muscle scars; a non-sinuate pallial line; a straight, edentulous hinge; an aragonitic shell; and a non-mineralized or weakly mineralized, narrow, dorsal, simple ligament. Some of the features indicated for this ancestor are mutually inconsistent. For example, presence of a protoconch (character 35, beak direction) conflicts with medial division of the larval shell (character 98). Here, PAUP reverted to a more distant ancestral state for character 35 because *Tuarangia* and *Pojetaia* differ in this character. For a similar reason, a pseudoligament and true ligament are indicated as present under different characters. Contrary to these indications, the immediate common ancestor of *Tuarangia* and *Pojetaia* probably had a divided larval shell and a true ligament, although whether it had differentiated anterior and posterior adductor muscles remains uncertain.

The present data reinforce the observation by Waller (1998, fig. 3) that the subclass Palaeotaxodonta shows a basal dichotomy between the superfamily Nuculanoidea and the superfamilies Nuculoidea and Solemyoidea. This is significant because Waller's evidence was largely soft anatomical and the present is principally palaeontological. In recognition of this dichotomy, the subclass Palaeotaxodonta is divided into the superorders Nuculaniformii *nov.* (containing the order Nuculanoida *nov.*) and Nuculiformii (*nom. transl.* Starobogatov, 1992, *ex* Nuculidae Gray, 1824). The Nuculiformii comprises all non-nuculanoidean palaeotaxodonts. It is also suggested that *Eritropis* is excluded from the Cardiolariidae and that the 'malletiid' *Ekstadia* Soot-Ryen, 1964, is included in this family. The Cardiolariidae thus defined is uniquely characterized by a heterotaxodont hinge. In order to preserve the order Nuculoida as non-paraphyletic, it should be limited to the superfamily Nuculoidea, families Nuculidae and Pristiglomidae (Sanders & Allen, 1973). The Pristiglomidae share numerous apomorphies with the Nuculidae, including a sharply separated heel, multiple loops in the hindgut and a laterally mineralized resilium (Waller 1998).

Contrary to Cope (1996*a*, 1997), the Solemyoidea appear to be closely related to the Palaeotaxodonta through the family Ctenodontidae Wöhrmann, 1893. As constituted by McAlester (1969), the family Ctenodontidae is presently shown to be paraphyletic with respect to the Solemyoidea. Pojeta (1988) indicated that the Ctenodontidae is poorly conceptualized, so he tentatively assigned *Ctenodonta* and *Tancrediopsis*, among other genera, to the nuculanoidean family Malletiidae. The present analysis suggests that *Ctenodonta* and *Tancrediopsis* should be considered near-basal plesions within the order Solemyoida.

The present analysis supports the hypothesis of Sánchez & Babin (1998) that nuculoids and palaeoheterodonts merely share common ancestry. In this regard, it is important to note that *Fordilla* was diagnosed as non-pretaxodont. One might argue that *Fordilla* has instead a reduced pretaxodont dentition, considering its close affinity with *Pojetaia*. In either case, however, fordilloid dentition, and hence the dentition immediately preceding the differentiation of the subclasses Palaeotaxodonta and Autolamelibranchiata, was sparsely dentate and neither palaeotaxodont nor actinodont. Actinodonty has apparently evolved convergently among palaeotaxodonts, palaeoheterodonts and pteriomorphians, just as taxodonty *sensu lato* has evolved convergently among palaeotaxodonts, actinodonts, unionoids and pteriomorphians. The origin of the subclass Autolamellibranchiata may correlate with a change from pretaxodont to non-pretaxodont and non-palaeotaxodont dentition, as well as with the change from protobranch to non-protobranch ctenidia.

The superorder Heteroconchia, as presently restricted to palaeoheterodonts and heterodonts, ancestrally differs from the Anomalodesmata and Pteriomorphia in having a more discoidal anterior and more equal adductor muscle scars. These changes suggest a shift from relatively inactive to active burrowing life habits.

Within the Heteroconchia, the thoraliid–trigonioid–lucinoid clade (node 36) is not supported by the strict or Adams consensus trees and it is only supported at 56% by the majority consensus tree. Its significance is therefore questioned. The other heteroconch clade, consisting of actinodonts and heterodonts, is supported by the Adams consensus tree and therefore more likely reflects a natural association.

Within the actinodont–heterodont clade, *Noradonta* does not place close to *Lyrodesma* in the Adams or majority consensus trees, but it is resolved close to *Redonia* in Fig. 4. Pojeta & Gilbert-Tomlinson (1977) and Babin (1982*a*) placed *Noradonta shergoldi* Pojeta & Gilbert-Tomlinson, 1977, in the family Lyrodesmatidae. However, some species of *Noradonta*, such as *Noradonta redoniaeformis* (Thoral, 1935), are strikingly similar to *Redonia bohemica* Barrande, 1881, differing in little more than the presence of dental microcrenulations. *Redonia* and *Noradonta* lack ligament nymphs (Pojeta 1978, pl. 4, figs 1 and 2; Babin 1982*a*, pl. 9, fig. 2), but nymphs are well developed in *Lyrodesma*. The present study is in agreement with Johnston & Goodbody (1988) and Babin & Gutiérrez-Marco (1991) in that the taxonomic significance of dental microcrenulations has been overemphasized in the definition of palaeoheterodont families.

Eodon is one of the earliest known representatives of the superfamily Crassatelloidea, and hence of the order Veneroida and group Heterodonta. Chavan (1969, p. N565) placed this genus in the family Astartidae d'Orbigny, but it differs from astartids in retaining a partially nacreous interior [Carter & Lutz 1990, pl. 104; see Taylor (1973) for Astartidae] and in having a ligament transitional between simple wide, as in

Table 2. *Classification*

Class Bivalvia Linné, 1758

Order Tuarangioida Mackinnon, 1982
    Genus *Tuarangia* Mackinnon, 1982
Genus *Pojetaia* Jell, 1980
Genus *Fordilla* Barrande, 1881
Subclass Palaeotaxodonta Korobkov, 1954
    Superorder Nuculaniformii *nov.*
        Order Nuculanoida *nov.*
            Superfamily Nuculanoidea Adams & Adams, 1858
                Family Malletiidae Adams & Adams, 1858
                    Genus *Palaeoneilo* Hall & Whitfield, 1869
                    Genus *Nuculites* Conrad, 1841
    Superorder Nuculiformii Gray, 1824
        Genus *Eritropis* Pojeta & Gilbert-Tomlinson, 1977
        Family Cardiolariidae Cope, 1997
            Genus *Ekstadia* Soot-Ryen, 1964
            Genus *Cardiolaria* Munier-Chalmas, 1876
            Genus *Praeleda* Pfab, 1934
        Unnamed clade
            Family Tironuculidae Babin, 1982
                Genus *Tironucula* Morris & Fortey, 1976
            Family Praenuculidae McAlester, 1969
                Genus *Praenucula* Pfab, 1934
            Order Nuculoida Gray, 1824
                Superfamily Nuculoidea Gray, 1824
                    Family Nuculidae Gray, 1824
                        Genus *Nuculoidea* Williams & Breger, 1916
            Order Solemyoida Gray, 1840
                Genus *Tancrediopsis* Beushausen, 1895
                Genus *Ctenodonta* Salter, 1852
                Superfamily Solemyoidea Gray, 1840
                    Family Acharacidae Scarlato & Starobogatov, 1979
                        Genus *Acharax* Dall, 1908

Subclass Autolamellibranchiata Grobben, 1894
    Superorder Heteroconchia Hertwig, 1895
        Unnamed clade
            Order Lucinoida Fleming, 1828
                Superfamily Babinkoidea Horný, 1960
                    Family Babinkidae Horný, 1960
                        Genus *Babinka* Barrande, 1881
                Superfamily Lucinoidea Fleming, 1828
                    Family uncertain
                        Genus *Ilionia* Billings, 1875
            Unnamed clade
                Family Thoraliidae Morris, 1980
                    Genus *Thoralia* Morris, 1980
                Order Trigonioida Lamarck, 1819
                    Family Lyrodesmatidae Ulrich, 1894
                        Genus *Lyrodesma* Conrad, 1841
                    Superfamily Trigonioidea Lamarck, 1819
                        Family Schizodidae Newell & Boyd, 1975
                            Genus *Silurozodus* Liljedahl, 1992

        Unnamed clade
            Superfamily Cycloconchoidea Ulrich, 1893
                Family Cycloconchidae Ulrich, 1893
                    Genus *Cycloconcha* Miller, 1874
                    Genus *Fortowensia* Cope, 1996
                    Genus *Actinodonta* Phillips, 1848
            Unnamed clade
                Unnamed clade
                    Genus *Copidens* Pojeta & Gilbert-Tomlinson, 1977
                    Genus *Ananterodonta* Babin & Gutiérrez-Marco, 1985
                    Unnamed clade
                        Genus C Johnston & Goodbody, 1988
                        Order Veneroida Rafinesque, 1815
                            Superfamily Crassatelloidea Férussac, 1822
                                Family Eodonidae *nov.*
                                    Genus *Eodon* Hall in Miller, 1877
                Unamed clade
                    Family Redoniidae Babin, 1966
                        Genus *Redonia* Rouault, 1851
                        Genus *Noradonta* Pojeta & Gilbert-Tomlinson, 1977

Table 2. *Continued*

- Family Nyassidae Hall, 1885
  - Unnamed clade
    - Genus *Nyassa* Hall & Whitfield, 1869
    - Genus *Tanaodon* Kirk, 1927
  - Unnamed clade
    - Genus A Johnston & Goodbody, 1988
    - Genus B Johnston & Goodbody, 1988
    - Genus *Montanaria* Spriestersbach, 1909

- Superorder Anomalodesmata Dall, 1889
  - Order Modiomorphoida Miller, 1877
    - Superfamily Modiomorphoidea Miller, 1877
      - Family Modiomorphidae Miller, 1877
        - Genus *Modiomorpha* Hall & Whitfield, 1869
- Superorder Pteriomorphia Beurlen, 1944
  - Family Evyanidae *nov.*
    - Genus *Evyana* Liljedahl, 1990
  - Family Colpomyidae Pojeta & Gilbert-Tomlinson, 1977
    - Genus *Colpomya* Ulrich, 1894
  - Order Mytiloida Rafinesque, 1815
    - Superfamily Mytiloidea Rafinesque, 1815
      - Genus *Modiolopsis* Hall, 1847
      - Genus *Whiteavesia* Ulrich, 1893
      - Family Mytilidae Rafinesque, 1815
        - Subfamily Modiolinae Keen, 1958
          - Genus *Modiolus* Lamarck, 1799
  - Family Matheriidae Scarlato & Starobogatov, 1979
    - Genus *Matheria* Billings, 1858
    - Genus *Metapadia* Desbiens, 1994
  - Family Modiolodontidae Fang & Morris, 1997
    - Genus *Modiolodon* Ulrich, 1894
  - Order Cyrtodontoida Ulrich, 1894
    - Genus *Ortonella* Ulrich, 1894
    - Superfamily Cyrtodontoidea Ulrich, 1894
      - Family Cyrtodontidae Ulrich, 1894
        - Genus *Cyrtodonta* Billings, 1858
    - Superfamily Falcatodontoidea Cope, 1996b
      - Family Falcatodontidae Cope, 1996b
        - Genus *Falcatodonta* Cope, 1996b
  - Superfamily Ambonychioidea Miller, 1877
    - Family Ambonychiidae Miller, 1877
      - Genus *Ambonychia* Hall, 1847
    - Family Myalinidae Frech, 1891
      - Genus *Septimyalina* Newell, 1942
  - Order Pectinoida Rafinesque, 1815
    - Family Myodakryotidae Tunnicliff, 1987
      - Genus *Myodakryotus* Tunnicliff, 1987
    - Suborder Limina Rafinesque, 1815
      - Superfamily Limoidea Rafinesque, 1815
        - Family Limidae Rafinesque, 1815
          - Genus *Palaeolima* Hind, 1903
    - Suborder Pectinina Rafinesque, 1815
      - Superfamily Leiopectinoidea Krasilova, 1959
        - Family Leiopectinidae Krasilova, 1959
          - Genus *Leiopecten* Khalfin, 1940
  - Order Pterioida Gray, 1847
    - Superfamily Pterioidea Gray, 1847
      - Family Pterineidae Miller, 1877
        - Genus *Pterinea* Goldfuss, 1826
        - Genus *Tolmaia* Williams, 1908
  - Order Ostreoida Férussac, 1822
    - Family Rhombopteriidae Korobkov, 1960
      - New genus, new species Johnston, 1993
  - Family Umburridae Johnston, 1991
    - Genus *Umburra* Johnston, 1991
  - Order Arcoida Stoliczka, 1871
    - Superfamily Arcoidea Stoliczka, 1871
      - Genus *Alytodonta* Cope, 1997
      - Genus *Freja* Liljedahl, 1984
      - Genus *Trecanolia* Ratter & Cope, 1998
      - Genus *Catamarcaia* Sánchez & Babin, 1993
      - Family Glyptarcidae Cope, 1996b
        - Genus *Glyptarca* Hicks, 1873
      - Family Parallelodontidae Dall, 1898
        - Genus *Cosmetodon* Branson, 1942

Genus C of Johnston & Goodbody (1988), and parivincular submerged, as in astartids. In recognition of these distinctions, *Eodon* is placed in the new crassatelloidean family Eodonidae.

The present phylogenetic reconstruction conflicts with Cope's (1997) hypothesis that palaeoheterodonts gave rise to pteriomorphians and heterodonts through arcoid intermediates. The Heterodonta, as represented by *Eodon*, is a sister group to the Redoniidae and Nyassidae, with all three taxa most likely evolving from cycloconchid-like ancestors independently of the Arcoida. In order to exclude paraphyly with the Crassatelloidea, the superfamily Cycloconchoidea and the family Cycloconchidae are restricted to *Cycloconcha*, *Actinodonta* and *Fortowensia* among the genera presently studied.

This analysis also conflicts with Waller's (1990, 1998) hypothesis that the Heteroconchia should include the Anomalodesmata. Waller (1998) cited a 3a/2/3b core dental homology and ligament nymphs as apomorphies, separating the Heteroconchia (inclusive of the Anomalodesmata) from the Pteriomorphia. However, most early anomalodesmatans are edentulous, including some modiomorphids (e.g. *Phorinoplax striata* Johnston, 1993). *Modiomorpha*, regarded as an anomalodesmatan by Waller (1990), Morris *et al.* (1991), Johnston (1993) and Fang (1998), has only crudely formed subumbonal hinge teeth that are not obviously consistent with a 3a/2/3b dental homology. Evidence is observed for a close relationship between the Anomalodesmata and the Pteriomorphia. For this reason, the superorder Heteroconchia is restricted to the Palaeoheterodonta and Heterodonta, exclusive of the Modiomorphidae and probably most other anomalodesmatans. However, the possibility that some non-modiomorphids currently classified as anomalodesmatans are actually heteroconchs cannot presently be excluded. It is also noteworthy that ligament nymphs are absent in many early heteroconchs. Waller (1998, p. 34) acknowledged their absence in *Copidens* and no evidence is observed for them in *Actinodonta cuneata* Philips, 1848 and *Cycloconcha milleri* (Meek, 1871). The latter species has a dorsally open, simple, narrow ligament fossette with, at most, only a very weakly developed lateral groove for the lamellar ligament. The redoniids *Redonia* and *Noradonta*, the nyassids *Nyassa* and *Tanaodon*, and the lucinoids *Babinka* and *Ilionia* also lack distinct nymphs. Thus, ligament nymphs are clearly not a synapomorphy for the superorder Heteroconchia, and these structures have probably evolved convergently in trigonioids, veneroids and among later lucinoids.

Waller (1998) appears to be correct in regarding mytiloids (*sensu lato*) as near-basal pteriomorphians. In fact, the superfamily Modiolopsoidea Fischer, 1887, as defined by Fang & Morris (1997) to include the Modiolopsidae Fischer, Colpomyidae Pojeta & Gilbert-Tomlinson and Modiolodontidae Fang & Morris, appears to be paraphyletic with respect to the remaining Pteriomorphia. Also, the family Modiolopsidae is paraphyletic with respect to the superfamily Mytiloidea. The family Matheriidae Scarlato & Starobogatov, 1979, which Desbiens (1994) included in the superfamily Cyrtodontoidea, is apparently more closely related to the Modiolopsidae and Modiolodontidae than to the Cyrtodontidae.

Within the order Cyrtodontoida, the superfamily Cyrtodontoidea is paraphyletic with respect to the superfamily Falcatodontoidea Cope, and within the superfamily Arcoidea, the family Frejidae Ratter & Cope (represented by *Freja*, *Alytodonta* and *Trecanolia*) is paraphyletic with respect to *Catamarcaia*, the family Glyptarcidae Cope, and the family Parallelodontidae Dall.

In order to resolve paraphyletic relationships within the Pteriomorphia, it is recommended that the superfamily Modiolopsoidea is abandoned relegating the modiolopsid genera to basal plesion status within the superfamily Mytiloidea; relegating *Ortonella* to basal plesion status within the Cyrtodontoida; relegating the frejid genera to basal plesion status within the superfamily Arcoidea; and including the family Glyptarcidae within the superfamily Arcoidea. Cope (1997) suggested that *Catamarcaia* belongs in the Parallelodontidae. However, this placement would make the Parallelodontidae paraphyletic with respect to the Glyptarcidae. The present analysis supports the suggestion by Sánchez (1995) that *Catamarcaia* is basal to the Parallelodontidae, but also that *Catamarcaia* is basal to the Glyptarcidae. Therefore, *Catamarcaia* is here placed in the superfamily Arcoidea as an isolated plesion basal to both the Parallelodontidae and Glyptarcidae.

The genus *Myodakryotus* Tunnicliff was regarded by Tunnicliff (1987) and Pojeta & Runnegar (1985; as *Prolobella*) as a possible link between the families Cyrtodontidae and Limidae. Pojeta & Runnegar (1985) placed this genus in the Limidae, but Tunnicliff (1987) assigned it to the family Myodakryotidae. The present study suggests a basal relationship with both the superfamily Limoidea and superfamily Leiopectinoidea. Therefore, the Myodakryotidae, the superfamily Limoidea, the superfamily Leiopectinoidea and the various other pectinoid superfamilies are placed in the order Pectinoida. Within this order, the suborders Limina Rafinesque and Pectinina Rafinesque are recognized.

The present study confirms that the Ostreoida are

more closely related to the Pterioida than to the Pectinoida. This relationship is also compatible with data of shell microstructure (Carter 1990; McRoberts & Carter 1994) and DNA (Campbell *et al.* 1998). The major anatomical and shell microstructural features reportedly linking ostreoids with pectinoids (Waller 1998), such as tentacle grade and calcitic foliated structure, are either plesiomorphic or convergent.

Umburra apparently bridges the morphological gap between the early ostreoid family Rhombopteriidae and the order Arcoida. This placement is compatible with the suggestion by Johnston (1991) that *Umburra* is closely related to the Rhombopteriidae, but it contradicts Johnston's hypothesis that *Umburra* is basal to the Rhombopteriidae, Pterineidae and Pectinoida. The present analysis suggests placement of the family Umburridae between the orders Ostreoida and Arcoida.

Johnston (1991) and Waller (1998) defined the Eupteriomorphia of Boss (1982) to include the family Umburridae, the order Pterioida and the order Ostreoida, the latter containing both oysters and scallops. They also regarded arcoids, ambonychiids and limids as basal to the Eupteriomorphia. According to the present analysis, arcoids are derived, rather than basal, relative to the Pectinoida, Pterioida and Ostreoida as presently defined. Consequently, any phylogenetic definition of the Eupteriomorphia that includes the latter three orders must also include the Arcoida. It is therefore recommended that the name Eupteriomorphia be abandoned. The position of the Arcoida in Fig. 4 contradicts most current hypotheses for the origin of this group. It suggests that arcoids evolved from left–right symmetrical relatives of rhombopteriids rather than from cyrtodontids or actinodonts. Verification of mineralogical data for rhombopteriids, umburrids, actinodonts and early arcoids, much of which is presently tentative, should help resolve this issue. It may be noted, in this regard, that irreversibly polarizing outer layer calcite (character 90), as suggested by Carter (1980), reduced the proportion of clades supporting many of the nodes in the majority rule consensus tree, but resulted in no change in the topology of the Pteriomorphia. This polarization added two steps, i.e. replacing the appearance of calcite at node 16 and its secondary loss at node 21 (Arcoida), with the appearance of calcite at nodes 26, 28, 29 and taxon 60. The taxonomic revisions suggested by the present phylogenetic analysis are summarized in Table 2.

This research was supported by the Nation Science Foundation (NSF GB36048). The Microstructure data for the Nyassidae are based on a paper in preparation by the senior author and Dr P. A. Johnson.

Appendix 1: Species analysed

1. Class Monoplacophora, family Stenothecidae: *Watsonella crossbyi* Grabau, 1900, Lower Cambrian, Siberian Platform. 2. Class Monoplacophora, family Stenothecidae: *Anabarella plana* Vostokova, 1962, Lower Cambrian, Siberian Platform. 3. Class Monoplacophora, family uncertain: *Pseudomyona queenslandica* (Runnegar & Jell, 1976), Middle Cambrian, Queensland, Australia. 4. Class Rostroconchia, order Ribeirioida, family Ribeiriidae: *Ribeiria junior* Runnegar, 1996, Middle Cambrian, Queensland, Australia [muscle scars based on *Ribeiria lucan* (Walcott, 1924), Lower Ordovician, Canada]. **CLASS BIVALVIA:** 5. *Tuarangia gravgaerdensis* Berg-Madsen, 1987, Middle Cambrian, Denmark. 6. *Pojetaia runnegari* Jell, 1980, Lower Cambrian, South Australia and New South Wales, Australia. 7. *Fordilla troyensis* Barrande, 1881, Lower Cambrian, Greenland. 8. *Tironucula jugata* Morris & Fortey, 1976, Lower Ordovician, Spitsbergen. 9. *Praenucula faba* Liljedahl, 1994, Upper Silurian, Gotland. 10. *Nuculoidea pinguis* (Lindström, 1880), Upper Silurian, Wenlock, Gotland. 11. *Cardiolaria beirensis* (Sharpe, 1853), Middle Ordovician, Spain. 12. *Praeleda subtilis* Cope, 1999, Middle Ordovician, Wales. 13. *Eritropis peregrinata* Cope, 1999, Middle Ordovician, Wales. 14. *Ekstadia tricarinata* Soot-Ryen, 1964, Upper Silurian, Gotland. 15. *Nuculites oblongatus* Conrad 1841, Middle Devonian, New York. 16. *Palaeoneilo fecunda* (Hall, 1862), Upper Ordovician, Dubuque, Iowa. 17. *Ctenodonta tennesseensis* Pojeta, 1988, Middle Ordovician, Tennessee. 18. *Tancrediopsis gotlandica* (Soot-Ryen, 1964), Upper Silurian, Gotland. 19. *Acharax (Nacrosolemya) trapezoides* (Meek, 1874), Upper Carboniferous, eastern Kentucky. 20. *Redonia bohemica* Barrande, 1881, Middle Ordovician, Bohemia, Czech Republic. 21. *Thoralia languedociana* (Thoral, 1935), Lower Ordovician, southern France. 22. Adult *Lyrodesma majus* (Ulrich, 1894), Ordovician, Ohio. 23. Juvenile *Lyrodesma* (Ulrich, 1894), as in adult *Lyrodesma majus* except for juvenile hinge and ligament based on *Lyrodesma* sp. from Lower Silurian of Ohio (Harrison & Harrison, 1975, pl. 2, figs 14–17). 24. *Lyrodesma* sp., Middle Ordovician, British Columbia, from Johnston (1996, pl. 3, figs 1–5). 25. *Silurozodus gotlandicus* Liljedahl, 1992, Upper Silurian, Gotland. 26. *Cycloconcha mediocardinalis* Miller, 1874, Upper Ordovician, Ohio. 27. *Fortowensia grandis* Cope, 1996*b*, Lower Ordovician, Arenig, South Wales. 28. *Actinodonta cuneata* Philips, 1848, Lower Silurian, Llandovery, Wales. 29. *Nyassa dorsata* (Goldfuss, 1840), Middle Devonian, Hamilton Group, New York. 30. *Tanaodon louderbacki* Kirk, 1927, Middle Devonian, Sichuan Province, China. 31. *Noradonta shergoldi* Pojeta & Gilbert-Tomlinson, 1977, Ordovician, Georgina Basin, Australia. 32. New genus A of Johnston & Goodbody (1988), Middle Devonian, Melville Island, Arctic Canada. 33. New genus B of Johnston & Goodbody (1988), Middle Devonian, Melville Island, Arctic Canada. 34. *Montanaria* aff. *M. honguedoensis* Desbiens, 1994, Lower Devonian, Emsian, Gaspé, Québec, Canada. 35. *Copidens browni* Pojeta & Gilbert-Tomlinson, 1977, Ordovician, Australia. 36. *Ananterodonta oretanica* Babin & Gutiérrez-Marco, 1985, Middle Ordovician, Llanvirn,

Spain. 37. *Babinka prima* Barrande, 1881, Lower Ordovician, Europe. 38. *Ilionia prisca* (Hisinger, 1837), Upper Silurian, Ludlow, Gotland. 39. New genus C of Johnston & Goodbody (1988), Middle Devonian, Melville Island, Arctic Canada. 40. *Eodon tenuistriata* (Hall, 1870), Middle Devonian, Moscow Shale, Morrisville, New York. 41. *Modiomorpha concentrica* (Conrad, 1838), Middle Devonian, Chanango Valley, New York. 42. *Evyana baltica* (Liljedahl, 1989), Upper Silurian, Wenlock, Gotland. 43. *Colpomya hugini* Liljedahl, 1994, Upper Silurian, Wenlock, Gotland. 44. *Modiolopsis modiolaris* (Conrad, 1838), Upper Ordovician, New York State. 45. *Whiteavesia cincinnatiensis* (Hall & Whitfield, 1875), Upper Ordovician, Kentucky. 46. *Modiolus meeki* (Evans & Shumard, 1857), Upper Cretaceous, Montana. 47. *Matheria tener* Billings, 1858, Middle or Upper Ordovician, Quebec, Canada; ligament based on *Matheria rugosa* Ulrich, 1892, Middle or Upper Ordovician, Minnesota. 48. *Metapadia matapediensis* Desbiens, 1994, Lower Devonian, Canada. 49. *Modiolodon oviformis* (Ulrich, 1890), Middle or Upper Ordovician, Trenton, Kentucky. 50. *Cyrtodonta saffordi* (Hall, 1859), Middle or Upper Ordovician, Nashville, Tennessee. 51. *Ortonella hainesi* (Miller, 1874), Upper Ordovician, eastern North America. 52. *Falcatodonta costata* Cope, 1996, Lower Ordovician, Arenig, South Wales. 53. *Ambonychia radiata* Hall, 1847, Upper Ordovician, New York and Kentucky. 54. *Septimyalina perattenuata* (Meek & Hayden, 1858), Upper Carboniferous, Kentucky. 55. *Myodakryotus deigryn* Tunnicliff, 1987, Upper Ordovician, Wales. 56. *Palaeolima retifera* (Shumard, 1858), Upper Carboniferous, Kentucky. 57. *Leiopecten praerectangularis* Krasilova, 1959, Lower Devonian, Kazakhstan. 58. *Pterinea laevis* Goldfuss, 1826, Lower Devonian, Germany. 59. *Tolmaia erugisulca* Johnston, 1993, Lower Devonian, southeast Australia. 60. Rhombopteriid new genus and species of Johnston (1993, fig. 16), Lower Devonian, southeast Australia. 61. *Umburra cinefacta* Johnston, 1991, Upper Silurian, Wenlock, New South Wales, Australia. 62. *Freja fecunda* Liljedahl, 1984*a*, Upper Silurian, Gotland. 63. *Alytodonta gibbosa* Cope, 1997, Lower Silurian, Scotland. 64. *Trecanolia acincta* Ratter & Cope, 1998, Upper Silurian, Wenlock, South Wales. 65. *Catamarcaia chaschuilensis* (Acenolaza & Toselli, 1977), Lower Ordovician, Argentina. 66. *Cosmetodon obsoletus* (Meek, 1871), Upper Carboniferous, Kentucky. 67. *Glyptarca serrata* Cope, 1996*b*, Lower Ordovician (Arenig), South Wales.

Appendix 2: Character definitions

1. Arch of LV ventral hinge margin: 1, straight; 2, weak to medium; 4, strong; 5, no hinge or pseudohinge. **2. LV convexity:** 1, strong; 2, moderate; 3, weak. **3. Extension of LV posterior auricle beyond sinus:** 1, no LV posterior auricle; 2, slight with angular termination; 3, moderate with angular termination; 5, angular termination but no posterior extension. **4. LV posterior auricle:** 1, absent; 2, present. **5. LV posterior auricular sinus:** 1, absent; 2, present. **6. Extension of LV anterior auricle beyond sinus:** 1, no LV anterior auricle; 3, moderate with angular termination; 5, with angular termination but no extension. **7. LV anterior auricle:** 1, absent; 2, present. **8. Shape of RV anterior auricle:** A, no RV anterior auricle; not secondarily absent; B, dorsal margin dips toward the anterior; C, anterior margin nearly straight and vertical; D, anterior margin nearly straight but dipping toward the posterior; E, anterior margin is nearly straight antero-ventrally and < 45° to hinge axis; dipping toward the posterior; with an angular to slightly rounded shoulder. **9. Shape of LV posterior auricle:** 1, primitively absent or none of the following; 2, straight, horizontal; 3, slopes upward posteriorly; 4, straight; slopes downward posteriorly; 5. curves downward posteriorly. **10. Relative length of LV anterior and posterior auricles:** 1, both primitively absent; 2, equal; 3, anterior longer, posterior shorter or absent; 4, posterior longer, anterior shorter or absent. **11. RV byssal sinus:** 1, absent; 2, present. **12. LV umbonal elevation:** 1, absent or slight; 3, moderate; 4, strong. **13. LV posterior rostrateness:** 1, absent; 2, slight to moderate; 3, strong. **14. LV elongation (L/H):** 1, intermediate (1.25–2.00); 3, long (> 2.00); 5, short (< 1.25). **15. LV posteroventral shell margin:** 1, rounded, oval or slightly truncate without strong inflection; 2, like 1 but with strong inflection but neither triangular nor rectangular; 3, triangular posteroventrally; 4, triangular medioposteriorly; 5. rectangular. **16. LV obliquity (predominant growth gradient relative to beak and hinge axis):** 1, equilateral; 2, slightly posteroventral; 3, slightly anteroventral; 4, strongly anterior; 5, strongly posterior; 6, strongly posteroventral. **17. Shape of anterior shell margin:** *Beak terminal anterior, no anterior auricle:* G, beak anterior to anteroventral aperture, the latter oriented at high angle to length axis of aperture. 2, anterior obsolescent and anterior margin below the byssal sinus, if the latter is present, is strongly inclined toward the posterior; E, anterior obsolescent, anterior margin below byssal sinus, if the latter is present, is broadly ovate to nearly straight and nearly perpendicular to hinge axis; B, anterior margin slopes posteriorly and is broadly ovate, without anterior auricle. *Beak anterior but not terminal, no anterior auricle*: 1, entire anterior is discoidal with, at most, a very slight indentation anterior to beaks; 4, shallow but distinct indentation or no indentation anterior to the beaks, anterior shell margin is convex with a tight curvature in comparison with the maximum posterior height of the shell and the anteroventral margin is broadly ovate to straight; 5, dorsal anterior margin broadly concave immediately anterior to beaks and convex anteroventrally; 6, dorsoanterior margin is distinctly indented anterior to beaks, not broadly concave; anterior margin is discoidal to moderately ovate to slightly triangular, with a convexity that is nearly as wide as the maximum posterior shell height; anterior margin is not greatly extended anteriorly nor is it triangular; anterior passes into moderately to broadly ovate margin anteroventrally; 7, presence or absence of indentation immediately anterior to beak, with shell anterior elongated and with nearly horizontal dorsal and ventral areas, and with discoidal to moderately oval anteriormost termination; A, no indentation anterior to beaks, anterior margin very broadly ovate and nearly vertical; C, anterior margin slopes dorsally, not forming a straight-edged auricle, and then swings sharply downwards to make a broadly ovate to discoidal anterior; H, shallow, distinct indentation or no indentation anterior to beaks, anterior margin smoothly and broadly convex anterodorsally but slightly straight anteroventrally, and shell is not elongate enough to be modioliform. *Beak anterior but not terminal, with anterior auricle*: 3, anterior margin ventral to the byssal sinus (if the latter is present) is broadly ovate to nearly straight and posteroventrally inclined; D, anterior margin below auricle and auricular sinus (if the latter is present) is broadly ovate or straight and nearly perpendicular or anteriorly slightly inclined to hinge axis; F, anterior margin below auricle and auricular sinus (if the latter is present) is moderately ovate to discoidal and anteriorly inclined. *Beak medial or posterior:* 8, at most a slight indentation anterior to beaks and anterior margin is narrowly ovate to smoothly triangular but not extremely elongated, with no anterior auricle; 0, as in option 8, but with distinct indentation anterior to beaks; K, anterior slightly truncate; L, slight but distinct indentation immediately anterior to beaks, anterior margin nearly discoidal; P, at most a slight indentation anterior to beaks; anterior greatly elongated with nearly parallel dorsal and ventral margins. **18. Slope of LV sharply angular dorsoanterior shell margin relative to hinge axis (using line drawn from the anteriormost hinge tangent to anteroventral shell margin):** 0, no sharply angular dorsoanterior shell margin; 1, nearly right angle; 2, slightly < 90° between ligament axis and tangent, angled posteriorly; 5, > 90° between ligament axis and tangent, angled anteriorly; 6, hinge, pseudohinge, ligament and pseudoligament all absent. **19. LV adult ventral shell shape:** 1, nearly discoidal without posteroventral sinus; 2, moderately to broadly ovate or straight, moderately to slightly elongate, without a medial ventral or posteroventral sinus; 4, not elongate, anteroventrally broadly ovate, nearly straight, or straight, with posteroventral sinus; 6, anteroventrally concave and posteroventrally spatulate; 7, anteriorly nearly straight or convex, moderately elongate to elongate, with medial ventral or posteroventral sinus; 8, very elongate, virtually straight or only slightly convex, without medial or posteroventral sinus. **20. Permanent anterior shell gape:** 1, absent; 2, anteroventral gape oblique to hinge axis in bivalved shell; 4, far anterior gape in bivalved shell; 6, anterior to anteroventral gape in univalved shell; 7, anterior gape closable by flexing the shell margins and periostracum. **21. Permanent posterior shell gape:** 1, absent; 6, present. **22. Posteroventral sinus:** 1, absent; 2, present. **23. LV lunule:** 1, absent but hinge or pseudohinge present; 2, weak and shallow; 5, sharply defined, very deep; 6, no hinge or pseudohinge (*Ribeiria lucan*). **24. LV escutcheon:** 1, absent but hinge or pseudohinge present; 2, weakly developed; 4, no hinge or pseudohinge. **25. Topography of LV lunule:** 1, lunule absent but hinge or pseudohinge present. 2, prominent growth lines anterior to ligament area on nearly commarginal, deeply inset lunular surface or on the hinge margin immediately ventral to beaks; 4, no hinge or pseudohinge; 5, smooth. **26. LV commissural shelf:** 1, absent; 2, present. **27. Shell morphologic evidence for branchitellum:** 1, no; 2, yes. **28: Valve down in pleurothetic taxa:** 1, non-pleurothetic; 2, LV down; 4, either valve down indifferently. **29. LV beak position:** 2, posterior; 3, medial; 4, anterior but not terminal; 5, terminal anterior. **30. Beaks are posteriorly situated:** 1, no; 2, yes. **31. Shell valve equality:** 1, equivalve; 2, RV slightly larger or more convex; 4, LV slightly larger or more convex; 5, LV much larger or more convex; 6, no hinge or pseudohinge. **32. General trend of shell valve equality:** 1, equivalve; 2, LV larger or more convex; 3, RV larger or more convex; 4, no hinge and pseudohinge. **33. Medial subumbonal septum or buttress:** 1, absent but a hinge or pseudohinge present; 3, prominent septum deep in umbonal cavity; 4, no hinge or pseudohinge; but septum or pegma possibly present. **34. Umbonal-anteroventral buttress or umbonal-anteroventral deck:** 1, absent but hinge or pseudohinge present; 2, non-auricular, ridge-like buttress, passing just posterior to anterior adductor scar, possibly continuous with anterior hinge; 3, buttress passing from hinge line to anteroventral margin on proximal margin of auricle; 5, anterior pegma-like structure but no true pegma; 6, true pegma. **35. Predominant adult beak direction:** 1, prosogyrate with prodissoconch; 2, orthogyrate with prodissoconch; 3, opisthogyrate with prodissoconch; 4, protoconch present. **36. Distinctive shell shapes:** 0, none of the following; A, mytiliform or modioliform; B, *Ambonychia*form; C, *Pterinea*form; D, *Silurozodus-gotlandicus*form; E, *Arca*form; F, *Falcatodonta*form; G, *Cyrtodonta*form; H, *Fordilla*form; I, *Umburra*form; J, *Nyassa*form; K, *Eodon*iform; L, *Solemya*form or nearly *Solemya*form; M, *Palaeoneilo*form; N, *Pinna*form; P, *Crassatella*form; Q, *Redonia*form; R, *Leiopecten*iform, *Leiopectinella*form, or *Pecten*iform; S, *Pholadomya*form; T, *Orthonota*form; U, *Nuculoidea*form; 1, *Pojetaia*form; 2, *Anabarella*form; 3, *Nucula*form; 4, Permophoriform; 5, *Matheria*form; 7, *Cycloconcha*form; 8, *Lima*form; 9, *Gryphaea*form–*Lopha*form. **37. Distinctive prosopon:** 1, neither of the following; 2, edmondiid type sharply reflected commarginal flanges; 3, late ontogenetic change from regular commarginal ribs to irregular growth lines. **38. Adult LV antimarginal prosopon:** 1, absent; 2, fine, regular, closely spaced costae or costellae; 3, strong, regular, closely to medium-spaced costae or plicae; 5, very fine, regular, closely spaced lineations not associated with marginal macrocrenulations; 9, largely smooth but with some radial costellae anteriorly and/or posteriorly; B, very fine, irregular radial striae. **39. Adult LV ranks of antimarginal prosopon:** 1, absent; 2, one size rank; 3, more than one size rank. **40. Adult RV antimarginal prosopon:** 1, absent; 2, fine, regular, closely spaced costae or costellae; 3, strong, regular, close to medium spaced costae or plicae; 5, very fine, regular, closely spaced lineations not associated with marginal macrocrenulations; 6, fine, regular, widely spaced costae or costellae; 9, radial costellae only anteriorly and/or posteriorly; B, very fine, irregular radial striae. **41. Anterior LV pedal retractor/levator muscle scar position:** 1, separate from adductor; 2, just touching adductor dorsally or dorsoposteriorly; 3, overlapping adductor; 6, just touching adductor posteroventrally; 7, primitively no anterior adductor but anterior left and right pedal retractors present. **42. Special support for anterior pedal retractor:** 1, absent; 4, pegma-like shelf; 5, true pegma. **43. Posterior LV pedal pedal retractor/levator position:** 1, separate from adductor and dorsal or dorsoanterior to it; 2, just touching adductor and dorsal or dorsoanterior to it; 3, overlapping adductor and dorsomedial or dorsoanterior to it; 4, confluent with adductor scar and dorsomedial or dorsoanterior to it, or secondarily absent; 8, no adductor but posterior pedal retractors present; 9, no adductor but median posterior pedal retractor present. **44. Junction of LV pallial line and posterior adductor muscle scar:** 1, midventral to anteroventral edge of adductor; 2, posteroventral edge of adductor; 3, posterior edge of adductor; 4, adductor is entirely interior to trend of the pallial line but close to it; 7, primitively no posterior adductor; 8, primitively only one adductor. **45. Position of LV posterior adductor:** 1, just below hinge; 2, moderately ventral and posterior to hinge; 6, primitively no posterior adductor; 7, primitively only one adductor. **46. Buttress or deck for LV posterior adductor:** 1, absent or adductor secondarily reduced; 2, distinct ridge or non-auricular deck; 3, primitively no adductors; 4, primitively only one adductor; 5, radial, platform-like thickening of shell along muscle migration pathway. **47. Shape of LV posterior adductor scar:** 1, round to subcircular; 2, elliptical to subtrapezoidal with length axis nearly parallel or only slightly oblique to nearest shell margin; 3, subtriangular; 5, primitively no adductors; 6, primitively only one adductor. **48. Position of LV anterior adductor (inferred ancestral position if secondarily reduced):** 1, on hinge plate or cutting into hinge plate or adductor deck; 2, off hinge plate; 4, primitively no adductors; 5, primitively only one adductor. **49. LV anterior adductor buttress:** 1, adductor present but not buttressed, or slight buttress lacks a posterior slope break; 3, buttress with distinct posterior slope break; 4, adductor supported by hinge plate; 5, adductor secondarily reduced and there is no trace of a former supporting buttress or deck; 7, primitively no anterior adductor; 8, primitively only one adductor. **50. Shape of LV anterior adductor:** 2, point of attachment; 3, round to subcircular; 4, elliptical to subtrapezoidal; 5, subtriangular; 6, elongate; 7, primitively no anterior adductor; 8, primitively only one adductor. **51. Relative size of adductor scars:** 1, anterior one secondarily reduced; 2, posterior one larger by at least 100%; 3, nearly isomyarian; 4, anterior much larger; 5, primitively no adductors; 6, primitively only one adductor. **52: Gill retractor, suspensor and mantle blood vessel scars:** 1, none apparent; 2, line of scars proximally restricted, straight, extending from umbonal cavity posteroventrally, below posterior adductor; Quenstedt scar may or may not be differentiated; may be expressed as a ridge or as a distinct break in the slope of the shell interior; 3, proximally dorsoventral and distally posteriorly curving scar, usually represented by a groove on the shell interior; Quenstedt scar may or may not be differentiated; rarely expressed as the extension of an umbonal septum extending dorsoventrally from the medial or anterior umbonal cavity, then curving slightly toward the posterior; not an anterior adductor buttress; 4, anterior–posterior, nearly horizontal line of scars on flanks of umbonal shell interior. **53. Umbonal to posteroventral gill wheal:** 1, absent; 2, present (2 or 3 in preceding category). **54. Anterior pallial, protractor or other muscle scar not dorsal to anterior adductor:** 1, none evident; 3, single or multiple scars medioposterior or posteroventral to anterior adductor; adjacent to anterior adductor and not extending posteriorly from the pallial line; 4, single, discrete scar at junction of anterior adductor with anteroventral pallial line; 5, primitively no anterior adductor scars and no anterior pallial line; 6, primitively no anterior adductors but anterior pallial or cephalic feeding organ retractors present. **55. Dorsal accessory muscle scars (excluding one anterior and one posterior pedal retractor or pedal levator scar closest to each adductor and excluding scars on ventral margin of a posterior internal ridge, scored elsewhere):** 1, absent; 2, only one or two closely spaced, medial subumbonal, inserting on surface other than posteroventral margin of hinge plate and other than apex of umbonal cavity; 3, three or more, subumbonal, inserting on concave interior surface, not on ventral edge of a posterior internal ridge and not only near anterior adductor muscle scar, or between umbonal cavity and posterior adductor; or one or more scars between umbonal apex and anterior pedal retractor, not present posteriorly except possibly on ventral margin of posterior internal ridge or on ventral margin of posterior hinge plate; 5, one or two closely spaced, small, isolated muscle scars just ventral to the hinge, between beak and posterior adductor, distinct from the posterior pedal retractor; 6, several small, closely spaced muscle scars just ventral to the hinge, between umbonal cavity and posterior adductor, distinct from the posterior pedal retractor; 7, only one or two closely spaced scars at the far apex of umbonal cavity, excluding anterior pedal retractor; 8, only one or two closely spaced scars on ventral edge or umbonal cavity side of hinge plate, not at far apex of umbonal cavity. **56. Depth of posterior pallial sinus:** 1, absent; 2, shallow; 3, deep. **57. Pallial line structure:** 1, solid or with small, closely spaced scars except possibly discontinuous at anterodorsal and posterodorsal extremities; 3, entirely broken; 4, entirely of long, radial scars, not a solid pallial line with proximal radial extensions; 5, solid anteriorly, broken posteriorly. **58. Pallial line continuity:** 1, solid or consisting of closely spaced, small scars (1 above); 2, at least partially discontinuous (3–5 above). **59. Non-shell marginal posterior lateral or pseudolateral teeth in LV or RV (excluding palaeotaxodont teeth):** 1, absent and none of the following; 2, only one such tooth in either or both valves, not more than one; 3, two such teeth in one or both valves, not more than two; 4, three or more such teeth in either or both valves; 5, no hinge or pseudohinge. **60. Shape of non-palaeotaxodont posterior lateral or pseudolateral tooth closest to subumbonal area:** 1, absent; none of the following; 3, long, lamellar lateral tooth; 4, lamellar tooth immediately posterior to a palaeotaxodont subumbonal tooth, derived ontogenetically from palaeotaxodont teeth; 6, no hinge or pseudohinge; 7, short, nearly horizontal, lamellar or tabular tooth either separated from the cardinal area by an edentulous gap, separated from the anterior area by an edentulous gap, or adjacent to cardinal pseudotaxodont teeth with a distinctly different orientation than the lamellar/tabular tooth; 8, short, distinctly posteriorly dipping, lamellar or tabular tooth adjacent to similarly posteriorly dipping cardinal pseudotaxodont teeth. **61. At least one posterior tooth not reaching beak in either or both valves, excluding palaeotaxodont and pseudolateral teeth:** 1, no hinge or pseudohinge present; 2, one and only one solid lamellar or tuberculiform posterior lateral in either or both valves; not more than one in either valve; 4, more than one in either or both valves, uppers en echelon closer to beak; 5, more than one in either or both valves, not uniformly aligned en echelon, with wide edentulous gap from beak; 7, more than one in either or both valves, lowers en echelon closer to beak, and no cardinolateral nor transitional palaeotaxodont/lamellar tooth; 8, more than one in either or both valves, lowers en echelon closer to beak; cardinolateral present; A, no hinge or pseudohinge. **62. Shape of strongest posterior lateral, non-palaeotaxodont tooth:** 1, does not apply; 3, strong, long, slender, lamellar; 4, strong but short and tabular or tuberculiform; 7, no hinge or pseudohinge. **63. More than one posterior lateral with lower teeth en echelon closer to beak:** 1, no, but hinge or pseudohinge present; 2, yes; 3, no hinge or pseudohinge. **64. Posterior palaeotaxodont teeth, not transitional to lamellar:** 1, no, but hinge or pseudohinge present; 2, yes, with little increase in tooth size posteriorly; 3, yes, with great increase in tooth size posteriorly; 4, no hinge or pseudohinge. **65. Distinctive posterior hinge dentitions:** 1, hinge or pseudohinge present but none of the following apply; 2, lateral, pseudolateral or pseudotaxodont teeth above separate posterior palaeotaxodont or pseudotaxodont teeth; 3, numerous posterior teeth dipping towards posterior, increasing in number ontogenetically by branching and/or intercalation; 5, posterior lamellar tooth with pseudotaxodont branches, without accompanying separate pseudotaxodont or palaeotaxodont teeth; 6, no hinge or pseudohinge; 8, two long posterior pseudolateral teeth, one above the other, immediately posterior to subumbonal palaeotaxodont teeth. **66. Glyptarcid-type hinge (subumbonal pseudotaxodont teeth represent subdivisions of posterior pseudolateral tooth, dorsal and posterior to other subumbonal pseudotaxodont teeth):** 1, neither of the following; 5, primitively no hinge or pseudohinge; 7, present. **67. Hinge or pseudohinge subumbonally edentulous:** ?, yes, but associated life habit and biomechanical factors unknown; 1, no; 2, yes and associated with a gasket-like, periostracal margin; 3, hinge or pseudohinge present and ancestral edentulous condition is retained (*Watsonella*, *Pseudomyona*); 5, no hinge or pseudohinge; 6, yes, associated with wide, flexible, periostracal margins; 7, yes, associated with rigid shell margins and shallow burrowing or semi-infaunal life habits, with subumbonal edentuly apparently an artifact of posterior shift in the umbones rather than actual tooth reduction; 8, yes, associated with deep burrowing habits and extension of rigid shell margins by a thin periostracum studded with aragonitic spikes or granules. **68. Subumbonal pseudotaxodonty, palaeotaxodonty or pretaxodonty:** 1, hinge or pseudohinge present but none of the following apply; 2, pseudotaxodont in one row, teeth parallel or ventrally divergent, possibly ventrally bifid; 3, pseudotaxodont in two rows, one over the other; 4, palaeotaxodont; 5, no hinge or pseudohinge; 6, pseudotaxodont in one row, teeth dorsally divergent; 7, pretaxodont. **69. Subumbonal or anterior tooth row relationship with posterior palaeotaxodont or pseudotaxodont tooth row:** 1, hinge or pseudohinge present but not exclusively palaeotaxodont nor exclusively pseudotaxodont; 2, tooth rows in line; 3, posterior row overlaps anterior row; 4, no hinge or pseudohinge. **70. Subumbonal dentition bulbous or tuberculiform, excluding indistinct teeth as in *Modiomorpha* and *Ambonychia*:** 1, no, but hinge or pseudohinge present; 2, yes; 3, primitively no hinge or pseudohinge. **71. Any non-palaeotaxodont, non-pretaxodont subumbonal teeth, including indistinct teeth as in *Fordilla*, *Modiomorpha*, and *Ambonychia*, but excluding crural teeth:** 1, absent but hinge or pseudohinge present; 2, present; 3, no hinge or pseudohinge. **72. Distinctive cardinal-subumbonal dentitions:** 0, no hinge or pseudohinge; 1, hinge or pseudohinge present but none of the following apply; 2, no anterior hinge teeth, but at least three non-bulbous, short subumbonal teeth, and anteriormost dips towards the posterior or is orthocline; 3, strongly posteriorly concave, distinct, non-taxodont, anterior subumbonal tooth in either valve; 5, no anterior hinge teeth but at least three non-bulbous, short subumbonal teeth, deeply grooved, and tightly fan-like, with anterior tooth dipping towards the anterior; A, dorsally divergent teeth, subumbonal or anterior and posterior to subumbonal area, possibly with an edentulous gap below the beak; C, overlap of anterior palaeotaxodont or pseudotaxodont dentition by medial and posterior palaeotaxodont dentition; D, pretaxodont; E, overlap of medial pseudotaxodont or transitional pseudotaxodont/lamellar dentition by posterior pseudotaxodont or transitional pseudotaxodont/lamellar dentition. **73. Anterior, non-palaeotaxodont, pseudolateral but not anterior lateral teeth in either or both valves (excluding crura).** 1, hinge or pseudohinge present but does not apply; 2, one and only one solid, long, pseudolateral anterior tooth in either or both valves; not more than one; 4, more than one solid, long, anterior pseudolateral tooth in either or both valves; 6, no hinge or pseudohinge. **74. At least one adult anterior non-palaeotaxodont, non-pseudotaxodont, lateral tooth (i.e. not reaching beak) in either or both valves, excluding crura, but including small, tuberculiform teeth:** 1, no, but hinge or pseudohinge is present; 5, more than one such tooth in at least one valve, not uniformly aligned en echelon, with wide edentulous gap to beak; 6, more than one such tooth in at least one valve, not necessarily aligned en echelon, with pseudotaxodont teeth filling gap to beak; 7, more than one such tooth in at least one valve, lowers en echelon closer to beak; 8, one or more transitional palaeotaxodont–pseudotaxodont teeth in either or both valves, with palaeotaxodont or pseudotaxodont teeth filling gap to beak; 0, no hinge or pseudohinge. **75. Anterior lateral or pseudolateral, non-pseudotaxodont, non-palaeotaxodont tooth in LV or RV (minute, tuberculiform teeth are scored as 1 and 2) (this excludes mytilid anterior shell marginal teeth):** 1, absent or does not apply but hinge or pseudohinge is present; 2, only one such anterior tooth in either or both valves; not more than one; 3, two such teeth in either or both valves; not more than two; 4, three or more such teeth in either or both valves; 5, no hinge or pseudohinge. **76. Relationship of strongest non-shell marginal, non-palaeotaxodont, non-pseudotaxodont, anterior lateral or pseudolateral tooth with anterior cardinal-subumbonal tooth:** 1, no such anterior tooth, but hinge or pseudohinge present; 2, continuous with cardinal-subumbonal tooth; 3, separate from cardinal-subumbonal tooth; 4, separated from cardinal-subumbonal tooth by palaeotaxodont teeth; 5, no hinge or pseudohinge. **77. Strongest non-shell marginal, anterior lateral or pseudolateral, non-palaeotaxodont tooth:** 1, none of the following; 2, very small, short, possibly tuberculiform and impersistent; 3, strong, long, slender lamellar; 4, strong but short and tabular; 5, no hinge or pseudohinge; 6, short, derived ontogenetically from arm of anterior, concavodont tooth. **78. Distinctive hinges:** 0, none of the following; 1, evyanid-type; 2, cyrtodont-type; 3, rhombopteriid–umburrid-type; 4, modiomorphid-type; 5, thoraliid-type; 6, gryphaeid–ostreid-type; 7, right pivotal cardinal-subumbonal tooth nearly parallel with dorsoanterior shell margin; 8, colpomyid-type; 9, subumbonal teeth pseudotaxodont and nearly vertical or dorsally divergent in at least one valve; A, only one or two anterior and posterior elongate hinge teeth approximating the subumbonal area without strong, intervening teeth in adult or early juvenile shell; B, montanariid-type; C, babinkid-type; D, redoniid-type; F, nyassid-type; G, cycloconchid-type; H, secondarily edentulous from immediate palaeotaxodont ancestors; I, subumbonal teeth pseudotaxodont and only posteroventrally inclined, not ventrally divergent; J, tironuculid-type; K, astartid–crassatellid-type; L, at least partially palaeotaxodont with anterior tooth row at least one-third as long as posterior tooth row; M, lyrodesmatid-type; O, matheriid-type; P, at least partially palaeotaxodont with anterior tooth row less than one-third as long as posterior tooth row and not tironuculid-type hinge; R, schizodian type; T, pretaxodont-type. **79. RV pivotal cardinal-subumbonal tooth:** 0, differentiated cardinal-subumbonal teeth absent in RV but hinge or pseudohinge present, or otherwise none of the following; 1, amorphous, crudely wedge-shaped, typically with growth lines on articulating surface; 2, bulbous or tuberculiform, rounded to elliptical, horizontally elongate or oblique, generally non-bifid; 3, distinctly shaped, heavy, subtrigonal, with only a faint indication of bifidness, although the tooth may be smooth or irregularly grooved; 4, distinctly shaped, heavy, subtrigonal, only superficially bifid; 5, distinctly and regularly shaped, fully bifid, with bisectrix oriented toward posteroventral, ventral or anteroventral direction; 6, distinctly shaped, slender, non-bifid or slightly bifid only at the far distal end, possibly slightly lobate distally; 7, palaeotaxodont; A, no hinge or pseudohinge; B, pseudotaxodont; G, pretaxodont. **80. LV cardinal-subumbonal tooth immediately anterior to LV pivotal cardinal-subumbonal tooth:** 0, hinge or pseudohinge present but cardinal-subumbonal teeth absent; 1, no tooth in this position; 2, slender, not pseudotaxodont; directed anteroventrally; not parallel with dorsoanterior shell margin; 3, slender, not pseudotaxodont; parallel or nearly parallel with dorsoanterior shell margin; 4, slender, not pseudotaxodont, not palaeotaxodont, directed posteroventrally; 6, palaeotaxodont; 8, pseudotaxodont and nearly vertical or dipping towards the anterior; A, pseudotaxodont and dipping toward the posteroventral; B, no hinge or pseudohinge; D, pretaxodont (*Pojetaia*); E, strongly and deeply bifid, opening ventrally or anteroventrally. **81. LV cardinal-subumbonal tooth immediately posterior to LV pivotal cardinal-subumbonal tooth:** 0, hinge or pseudohinge present but cardinal-subumbonal teeth absent in LV; 1, no cardinal-subumbonal tooth in this position; 2, slender, not pseudotaxodont; directed posteroventrally; 3, slender, not pseudotaxodont; nearly parallel with dorsoposterior shell margin and base of ligament nymph; 4, slender, nearly vertical to slightly anteroventrally directed, not palaeotaxodont or pseudotaxodont; 5, palaeotaxodont; 6, pseudotaxodont and nearly vertical; A, pseudotaxodont and dipping towards the posterior; D, pseudotaxodont and dipping towards the anterior; C, no hinge or pseudohinge; E, pretaxodont. **82. Pivotal RV cardinal-subumbonal socket:** 1, hinge or pseudohinge present but no socket; 2, proximally floored; 3, unfloored; 4, completely floored; 5, no hinge or pseudohinge. **83. Presence or absence of dental striations (not longitudinal grooves)**. 1, absent or hinge is edentulous, but hinge or pseudohinge present; 2, present; 3, no hinge or pseudohinge. **84. Heterotaxodonty:** 1, absent but hinge or pseudohinge present; 2, present; 3, no hinge or pseudohinge. **85. Predominant anterior palaeotaxodont tooth shape.** 1, non-palaeotaxodont but hinge or pseudohinge present; 2, orthomorphodont; 3, concavodont; 6, concavoconvex, convexoconcave or diconvex; 7, no hinge or pseudohinge. **86. Predominant posterior palaeotaxodont tooth shape:** 1, non-palaeotaxodont but hinge or pseudohinge present; 2, orthomorphodont; 3, concavodont; 6, concavoconvex, convexoconcave or diconvex; 7, no hinge or pseudohinge. **87. Pseudohinge or hinge overgrowth:** 1, none of the following; 2, partial overgrowth of hinge dentition by advancing ligament insertion area or deeply incised lunule; 4, no hinge or pseudohinge; 5, pseudohinge present but no true hinge. **88. Width of LV posterior hinge or pseudohinge excluding ligament insertion area and auricle:** 1, narrow to medium, obsolescent or absent; 3, wide; 4, no hinge or pseudohinge. **89. Width of LV anterior hinge or pseudohinge excluding ligament insertion area, auricle and deeply impressed lunule:** 1, narrow to medium, obsolescent or absent; 3, wide; 4, no hinge or pseudohinge. **90. Mineralogy LV outer shell layer:** 1, aragonite or predominantly aragonite with isolated calcite; 2, all calcite or all calcite in a continuous outer sublayer. **91. Mineralogy LV adult inner shell layer:** 1, aragonite; 3, calcite. **92. General microstructural grade of middle/inner shell layers:** 1, exclusively matted to lamello-fibrillar, or large tablet nacre, or Cambrian and entirely regularly foliated, the latter convergent on some post-Ordovician pteriomorphians; 2, small tablet nacreous; CL, CCL and/or FCCL/HOM. **93. Predominant left outer layer microstructure:** 1, regular simple prismatic; 2, mixed regular simple prismatic, irregular simple prismatic, HOM, or just ISP and HOM; 6, porcelaneous; 7, non-denticular composite prismatic and/or aragonitic regular simple prismatic; the NDCP may grade laterally into irregular spherulitic prismatic, regular spherulitic prismatic and/or homogeneous; 8, RSP to ISP to HOM (aragonite) overlying a fibrous prismatic sublayer; 9, aragonitic FP or denticular composite prismatic, possibly with minor ICCL, FCCL or irregular spherulitic and irregular simple prisms. **94. Calcitic regular simple prisms present near adult shell margins:** 1, no; 2, yes. **95. Calcitic foliated structure in LV:** 1, absent; 2, present. **96: Predominant porcelaneous structure in inner shell layer:** 1, absent; 2, fine complex crossed-lamellar and/or HOM (undifferentiated); 5, irregular complex crossed-lamellar possibly also with cone CCL, medium and/or coarse textured; 6, matted to complex crossed lamellar; 7, very wide CL and CCL. **97.**

Aragonitic prisms in middle and/or inner shell layer, excluding adductor myostraca: 1, minor ISP sublayers; 2, abundant ISP sublayers; 3. vertical ISP pillars; 6, fibrous prisms and matted structure comprise most of the shell interior. **98. Medial division of larval shell:** 1, absent; 2, present. **99. Hinge or pseudohinge:** 1, absent; 2, present. **100. Discontinuous fibrous ligament ontogeny, or derived from such ancestors:** 1, no; 2, yes. **101. Nuculoid/solemyoid internal ligament:** 1, absent; 2, present as a resilium (nuculoids), as a RV ligamental chondrophore (Palaeozoic solemyoids) or as a possible internal lamellar ligament homolog (modern solemyoids). **102. Spacing of elongated ligamental grooves:** 1, only one present; 2, multiple grooves closely spaced; 3, multiple grooves moderately spaced; 4, multiple grooves very widely spaced, including pteriomorphian alivincular and multivincular ligaments; 5, no ligament or pseudoligament; 6, pseudoligament present. **103. Mineralization of ancestrally elongated dorsal ligament (including lamellar remnant of dorsal ligament in resiliated nuculoids):** 1, non-mineralized or weakly mineralized without accompanying resilium; 2, mineralization extends from valve to valve, whether the ligament is still external or now submerged; 6, no ligament or pseudoligament; 7, only pseudoligament present. **104. General ligament grade:** 1, no ligament, or only pseudoligament, or primitively weakly mineralized or unmineralized dorsal ligament; 2, well mineralized true ligament or derived from such ancestors. **105. General categories of ancestrally dorsal ligaments (excluding nuculoid and solemyoid internal ligaments):** 0, neither ligament nor pseudoligament present; 1, pseudoligament; A, DH (duplivincular horizontal); B, DI (duplivincular inclined); D, SN (simple narrow); E, SW (simple wide); F, PN (parivincular narrow); G, PW (parivincular wide); H, PR (parivincular ramped); J, PS (parivincular submerged); K, ALIV (alivincular). **106. Major varieties of ancestrally dorsal ligaments (excluding nuculoid and solemyoids internal ligaments):** 1, ligament present but none of the following; 2, neither ligament nor pseudoligament is present; 3, pseudoligament; A, DHopis; B, DHopis/amph; C, DHamph; D, DIopis; E, DIopis/amph; F, DIamph; G, DIpros; K, SNweak; L, SNstrong; M, PNincip; N, PNweak; O, PNstrong, P, PWext; Q, PWint; R, PRthick; S, PRthin; T, PSshallow; 4, PSreduced; 5, PSdeep; 7, ALIVnar; 8, ALIVmod; 9, ALIVwide; 0, ALIVsplit. **107. Detailed ligament varieties (excluding nuculoid and solemyoid internal ligaments):** 0, neither ligament nor pseudoligament; 1, pseudoligament; 2, true ligament but none of the following; A, DHopis2; B, DHopis1; C, DHopis1–2; D, DIamph-incl; 3, DIamph-aliv; I, PRthick-shallow; J, PRthick-deep; K, PRthin-weak; L, PRthin-strong; M, PSdeep-sub; N, PSdeep-sub/resil; O, PSdeep-resil; P, ALIVmod-trian; Q, ALIVmod-trap; R, SNweak-opis; S, SNweak-amph. **108. Alternative groupings of similar dorsal ligament, excluding palaeotaxodont internal ligaments:** 0, pseudoligament or no ligament; 1, ligament present but none of the following apply; 2, DHopis or DIopis (opisthodetic duplivincular, regardless of inclinations); 3, DHopis/amph or DIopis/amph (slightly amphidetic duplivincular regardless of inclinations); 4, DHamph or DIamph (strongly amphidetic duplivincular regardless of inclinations); 5, SN or SW (all simple ligaments); 6, PN, PW, PR or PS (all parivincular ligaments except PF and PW, which clearly evolved separately). **109. Additional dorsal ligament varieties:** 0, ligament present but none of the following; 5, pseudoligament or no ligament; 7, DHopis2 ligament with several ligament grooves and with early formed fibrous layers largely restricted to the anterior of the ligament insertion area; 8, DHopis2 ligament with very few, widely spaced ligament grooves. **110. Dorsal shell plate kinetics:** 1, flexible by virtue of microstructural differentiation; 3, cracking of medial dorsal shell plate during growth to keep valves apart ventrally; 4, pseudoligament; 5, none of the above (most Monoplacophora). **111. Combined interfacial angle of L–R ligament insertion areas for duplivincular ligaments and their alivincular, multivincular and planivincular derivatives, measured near the adult hinge margin (this excludes nuculoid and solemyoid internal ligaments):** 1, symmetrical and < 70°, or ligament is not duplivincular nor derived therefrom; 2, L–R asymmetrical; 3, alivincular and submerged below shell margins; 4, symmetrical and > 70°; 5, pseudoligament or no ligament; 6, adult portion of elongate, simple ligament inserts on ventromedial surface of dorsal hinge margin with no true nymph but possibly with a pseudonymph. **112. Support for palaeotaxodont internal ligament:** 0, no ligament or only pseudoligament present; 1, ligament present but no palaeotaxodont internal ligament; 2, resilifer in one valve opposes a tooth or tooth-like structure in other valve; 3, opposing resilifers, not strongly projecting beneath hinge rotational axis; may be symmetrical or asymmetrical; 5, ligamental chondrophore attaching below beak in RV, with ligament arching upwards without a chondrophore in LV. **113. L–R asymmetrical palaeotaxodont resilifer (excluding solemyoid ligamental chondrophore):** 0, neither ligament nor pseudoligament present; 1, absent, but a ligament is present; 2, present. **114. Ctenidial grade:** ?, uncertain; 1, protobranchiate; 2, autolamellibranchiate. **115. Abdominal sense organ:** 1, absent; 2, present. **116. Adoral and anterior mantle sense organs:** 1, both absent; 2, adoral sense organ present; 3, anterior mantle sense organ present. **117. Shell height above dorsal limits of aperture exceeds shell height below dorsal limits of aperture:** 1, yes; 2, no.

Appendix 3: Database for phylogenetic analysis

Vertical, species number; horizontal, character number. *Polymorphisms, indicated separately as follows. Note that 6-15[1,2] signifies species 6, character 15, character states 1 and 2, etc. Polymorphisms, 5-5[1,2], 6-15[1,2], 6-18[0–5], 6-41[2,3], 6-50[4,6], 6-55[2–7], 7-79[1,2], 8-64[1,2], 8-78[J,L], 8-86[1–3], 9-14[1,5], 10-38[1,5], 10-40[1–5], 11-15[1-2], 11-29[3,4], 13-2[1-2], 13-85[2,3], 14-29[3,4], 16-69[2,3], 16-72[1,C], 16-86[2,3], 18-22[1,2], 21-17[7,8], 22-35[2,3], 23-35[2,3], 24-35[2,3], 25-15[3,5], 26-29[3,4], 26-50[4,5], 26-59[2,3], 26-60[1–3], 26-61[1,2], 26-62[1,3], 26-68[1,2], 27-15[2,4], 27-29[3,4], 27-60[1,3], 27-61[1,2], 27-62[1,3], 27-79[0,B], 28-15[1,4], 28-59[3,4], 28-60[1,3], 28-61[1,8], 28-62[1,3], 28-68[1,2], 28-88[1,3], 29-17[1,6], 30-60[1,8], 32-78[B,F], 34-15[2,3], 34-19[2,4], 36-15[1,2], 36-17[1,4], 36-34[1,2], 36-61[4,5], 37-16[1,3], 37-17[1,6], 37-29[2,3], 37-30[1,2], 37-50[4–6], 37-80[1,2], 37-81[1,4], 39-17[5,6], 40-15[2,5], 40-17[5,6], 40-55[4,5], 40-78[B,K], 41-11[1,2], 41-13[1,2], 41-17[6,9], 41-36[A,4], 41-37[1,3], 41-49[1,4], 43-15[2,5], 43-43[1,3], 43-93[2,9], 44-12[1,3], 44-13[1,2], 44-43[1,2], 44-56[1,2], 45-12[1,3], 45-13[1,2], 45-16[5,6], 45-36[4,A], 46-12[1,3], 46-13[1,2], 48-36[A,H], 48-48[1,2], 49-50[3,4], 50-17[4,H], 50-49[1,4], 51-23[1,2], 51-25[1,5], 51-49[3,4], 51-61[5,7], 51-63[1,2], 51-109[7,8], 52-50[3,4], 52-63[1,2], 53-3[1,5], 53-4[1,2], 53-18[0,1], 54-18[0,3], 54-36[A,C], 55-1[1,2], 55-5[1,2], 55-9[2,4], 55-10[2,4], 55-16[1,2], 55-18[1,5], 56-10[2,3], 56-93[1,2], 57-16[2,5], 57-36[R,C], 58-17[3,D], 58-48[1,2], 58-59[3,4], 58-106[A,B], 58-108[2,3], 59-17[2,E], 59-34[1,3], 59-48[1,2], 59-61[5,7], 60-78[3,I], 61-59[3,4], 61-65[1,3], 61-78[3,I], 62-16[1,2], 62-19[1,2], 62-88[1,3], 63-74[6,7], 63-88[1,3], 65-29[3,4], 65-60[1,7], 65-75[3,4], 66-14[1,3], 66-60[1,7], 67-60[1,7], 67-78[9,G].

```
 1234567890123456789012345678901234567890123456789012345678901234567890123456789012345678901234567890123456789012345678901234567
15311111A11111136G6266164411151644542151574876354775115312?56A7346555433060555OABC533774441117111611115610200555001111
22211111A111111426026611111114111151O1?1?74876354775115312?11111111311111111110000111115111117111622116711310545001112
31352152C24142113D121111111113111114O1?1??1?8746588611??1??2323111131111115312000011111511231??211121167113105450O?112
45211111A11111136G6266164411151644642l???759763547751163l??56A7346555433060555OABC5337744411??11111111561020053500l111
51252*11A341121528521111111113111112O1515?1??21?21?311??1??1111131111211111111000011122111231??211221111DKS501111?112
62252111A441111*18*21111111113111111111*11121221*2111*1321111111117111D11111TGDE411111111111?1111221111DKR501111112
72211111A111111126021111111114111111H111131312122142221315211111111111211111O*114111111111?1111221111DKR501111112
84211111A1113151360211111111131111113111111?1122143111?1??24111*2114211118146*7654113*113112?11??22111?2DL250111?1?2
92211111A11131*130021111111122111113111111?111221431112l??11111211142111111111L7654113211111291111221111 22DL25011111122
02211111A11131114002111111111122111111U1*1*1111122144111311111111211142111111111L765411331111129111122121 22DL25011311122
12211111A111315*26021111111111*11111101111111?11221431113111111111311143 11C11111P76541213111112?11??22111?2DL2501111?1?2
22252111A54132522802111111113111110111111?112214311131111111113111421111111P7654121211111 2?11??22111?2DL2501111?1?2
32*11111A111331564041111111111411111J1111111?1122143111?1?111111311142111111111P765411*2111112?11??22111?2DL2501111?1?2
42211111A11131512H0211111111*111121011112122112234341133111111113111421111111P76541233111112911212221122DL25011111132
52211111A11111315102111121111411121M1111111112234311362111111311142111111111P76541132111112611212211112DL25011111132
62211111A1111111260211111111141111171111111?1122143111?211111112111 4*11*11111P7654112*11111221121222111 22DL25011111132
72211111A111321338081111111113111113L1111?1??11?2143111?1??11111211142111111111L0001116611111 2?11??2211122GP210111111?2
82211111A11112134P0811*2251112211113L111111?122214311131111111131114211111111L765411631111129111122111 22GP210111111?2
92211111A11111334P0871111111112211123L1B1B111?1122144114314211111111611111111111H000111111111112211112212122GP21011511122
02111111A1113111260211111111114111121Q1111?112111143311l?1112123111111122311111D2114111111113112?11??22111?2DL25011112?12
12211111A11132143*02111111111131111130111111?11221431113l??1111111111112111111533242111111112?11??22111?2DL2501111??12
22211111A11112155102111111111411112*P19291111121433111821111111111121125111111MB8A4211113311 2?11??2211122FO26011112?12
32211111A11112155102111111111411111*P1929111112234311182111111111111112111111R63332111111112?11??2211122FM26011112?12
42211111A11112155102111111111411112*P192911111??14331118???281111111211251111 1MB8A4211113311 2?11??2211122FO26011112?12
52211111A111125*2102111111111311111 3D1111312111221432217111111111111112111111R6313111111112?11??2211122FM26011112212
62211111A1111111210211121511 1*111111711111??111?21*311?3111****11111*112141323GB8A41111111112?11??22111?2DL25011112?12
72211111A111311*560211111111*1111110????1?1113213211 1?1113***1111111121413 23G*33411111111112?11??22111?2DL25011112?12
```

```
82211111A111111*5102111??11114111110111121311122133111?111****11111*112117424GB8A411111*1112?11??22111?2DL25011112?12
92211111A11111136*041121111114111121J1111???1112234311??1114873211112112211111FBAA41111231112611212211122DL25011112?12
02211111A11111136B041121111115111121J1111?1??11?144?111?1114*73211112112211111FBA3421111311126117122111?2DL25011112?12
12211111A111111366021111111114111121J1111?11?1121443111????2111111112112211111MBAA42111113112?11??22111?2DL25011112?12
22211111A111111325041121111114111111J111111??11?146?11??1??1111111112112511111*B8A41111131112611712211122E125011112?12
32211111A111121326041121111114111111J111111??11?143??????11111111111111221111 1B54342111111112?11712211122E125011112?12
42311111A111121*250*1112151114111111J11111133112143311181111111111111112211111B54341111131112911712211122J5M6011112?12
52211111A11111522102111111111141111110111 1???1112214311 1?1114873211112112111111FBAA41111133112?11??22111?2DL25011112?12
62311111A111115*2*01111215111411111*1K?11111?11122143111?11143*3111112112111111FBAA41111131112?11??22111?2DL25011112?12
72211111A1113151**02111111111**11111011112141111 21*3411311111111111111122111111C4**32111111112?11??22111?2E125011112?12
82211111A11111521C041121251113111111041113134112216422131421111111111111211111106004111111112?11??22111?2E125011112?12
92211111A11112522*021112151114111111K31111141112214311 1?11123231111111121111 31B6???211113111 2?11112211122E125011112112
02311111A111121*2*021112151114111121K311111311121343221*111232311111111212122 3*22141111131112911512211122JT26011112112
12211111A11*1*126*071115121114111111**1112131112 1*42111311111111111111211111141114111121111291111221 1122HRI6011112?12
22211111A111111266021115121114111111011111?4111221423233111111111111122111111211111111111129111122213 22BD22011112212
32211111A111321*6H021111111114123111H111111*?112214?221311111111111112211111182112111111111 2*111122213 22AAA2811112212
42311111A111**1264071111111114111111A111111*111221321113*1111111111211111111111000011111111112?111122211 22AAB2116112212
52311111A111**32*402111111111141111 11*3111???1112214311 1??11111111112111111111100001111111111 2?111122212 22AAA2811112212
62211111A111**12640711111111114111111A1111?1332122142???1111111111211111111110000111111111129111222211 22AAB2116112212
72311111A11111126402111215111411111151111?1??11?2???????1??111111111111211111O613411111111112??1??22212?2AAA2?11112212
82311111A111111164021112151114111111*1111???11?*142111???11111111111121111 11O600411111111112??1??22212?2AAA2?11112212
92311111A111115124021111111114111121H1111?12211213*21?13111111111111111121111 11O65141111111111 2?111122212 22AAA2811112212
02211111A11131512*01111111111311111 1G1111?1321121*4332??11137541111111112114334 26E141111213112?11??22212?2AAA2?11112212
12211111A11112552402111*1*111411112 1H11121??11?1*4????11133*3*111111121111112512411112311 12?11??22212?2AAA2*11112212
22255111A1113155240211111111114111111F1323?1??11321*2111?1114373*111711111143342000111111111112?11??22212?2AAA2?11112212
312**111A24211516E*141151211151111 11B132311332122511??15132475411111111 22111112642411112112121 2?1122212?2AAA2?11112212
41152111A242115162*6111512112514214 1*1111113221116111115111111111111112111111010041111211212221112221222AAA2?11112212
5*322*52C**21151*D*11111112143111111816 2631?32121152221?1111111111111112111111051241111111??2?????2221??2AC2401?112212
61352152B2*211513D521111111114311111281323?1???1?????????1??11111111011111111110000111111111212*213122214?2K8P1011112212
71332232E2221111*F52111111121?3142111*1111?1???1?2????????1??1111111101111111111A00011111111212?2???22212?2AC24011112212
82232232B24211512*222111112124152111C1111?1?1211*132??1?111*873211112112211111IBAA411112112 1222?1122212?2A*A*?1?112212
92332252C24211526*121111112124 1421*1C1336111 2211*122221311147*321311211221111 1IBAA4111123321 2?????22212?2AB23012112212
02252152B24131512D511111112223152111I1331?1?32122132321?1113764211112112116334*B8A4111121321 2?2???22212?2AAA2?12112212
12252152C24131512D111111112123142121I1111112121221323233111*76321*112112116334*B8A41111233?12??1??22212?2BFD4014112212
22152152B2413152*35*11111111131111 12G11113131112214211 3?11147631111121121164349B86411112*1112?11??22213?2BFD4014112212
32252152B2413252230111111?111311111 1F1222?13?1122142??1????477321111211211*4329B86411112*3112?11??22212?2BFD4014112212
41252152B44131512351111111111311111 2G????1??11?2142111?11137931151161 12A163349BAD42111211112?11??22212?2BFD4014112212
52252152B441?2??635?111111111*111121E1111?1??1122342??13???4*93215116112A16*339B8641111231?12?11??22212?2BD22014112212
61122252D24233*263271111111214111112E0222?1??11?2143????1??4*43211116112A164349BAD421112111129115122212 22BF34014112212
71252152D24132356352111111111411111 1E1111?1??1112?33111??112*11111713112E16434*B8642111111112?11??22212?2?????11112212
```

Appendix 4: Character state changes for nodes in Fig. 4

Key: 34(5–6) indicates character 34 changing from state 5 to state 6; 34(5–6,7) indicates character 34 changing from state 5 to state 6 or state 7 (indeterminate). *, State change of uncertain hierarchical position due to uncertainty of character state in one or more adjacent sister groups.

Node 1-*Anabarella*: 2(2–3), *97(1–6), *110(3–5). **Node 1-2**: 16(6–2), 18(6–0), 23(6–1), 24(4–1), 25(4–1), 31(6–1), 32(4–1), 33(4–1), 36(2–0), 59(5–1), 60(6–1), 61(A–1), 62(7–1), 63(3–1), 64(4–1), 65(6–1), 66(5–1), 67(5–3), 68(5–1), 69(4–1), 70(3–1), 71(3–1), 72(0–1), 73(6–1), 74(0–1), 75(5–1), 76(5–1), 77(5–1), 79(A–0), 80(B–0), 81(C–0), 82(5–1), 83(3–1), 84(3–1), 85(7–1), 86(7–1), 87(4–5), 88(4–1), 89(4–1), 99(1–2), 102(5–6), 103(6–7), 105(0–1), 106(2–3), 107(0–1), *110(3–4), 117(1–2). **Node 2-*Watsonella***: *1(5–2), *15(3–4), *17(G–6), *29(5–4), *35(4–1), *97(1–6), *98(1–2). **Node 2–3**: *1(5–1), 3(1–5), 4(1–2), 10(1–4), *15(3–1), 20(6–1), 21(6–1), *29(5–3), 34(5–1), 42(4–1). **Node 3-*Pseudomyona***: 2(2–3), 6(1–5), 7(1–2), 8(A–C), 9(1–2), 12(1–4), *13(1–2), 16(2–3), *17(G–D), 18(0–1), *44(7–8), *45(6–7), *46(3–4), *47(5–6), *48(4–5), *49(7–8), *50(7–8), *51(5–6), 59(1–2), 60(1–3), 61(1–2), 62(1–3), 74(1–5), 75(1–3), 77(1–2), *90(1–2), *91(1–3), *95(1–2). **Node 3–4**: *17(G–8), *45(6–2), *46(3–1), *48(4–2), *49(7–1), 67(3–1), 87(5–1), *98(1–2), 102(6–1), 103(7–1), 105(1–D), 106(3–K), 108(0–5), 109(5–0), 110(4–1), 111(5–1), 112(0–1), 113(0–1). **Node 4-*Tuarangia***: 9(1–3), *13(1–2), 15(1–5), 18(0–5), *35(4–2), *51(5–3), 64(1–3), 69(1–2), 85(1–2), 86(1–2), *90(1–2), *91(1–3), *95(1–2), *107(1–S). **Node 4–5**: *1(1–2), *35(4–1), 38(5–1), 40(5–1), *41(7–3), *44(7–1), *47(5–2), *50(7–4), *51(5–2), *54(5–1), 82(1–4), *107(1–R). **Node 5-*Pojetaia***: 9(1–4), 16(2–1), 36(0–1), *43(8–1), 55(3–2,7), *57(1–3), 68(1–7), 72(1–D), 78(0–T), *79(0–G), *80(0–D), *81(0–E). **Node 5–6**: 3(5–1), 4(2–1), 10(4–1), *17(8–6), 29(3–4), *43(8–3), *79(0–1), *80(0–1), *81(0–1). **Node 6-*Fordilla***: 36(0–H), 52(1–2), 53(1–2), *57(1–5), *71(1–2). **Node 6–7**: 41(3–1), 45(2–1), 58(2–1), 92(1–2), *93(7–9), 103(1–2), 104(1–2), 106(K–L), 107(R–2). **Node 7–55**: *43(3–1), *51(2–3), 64(1–3), 68(1–4), 69(1–2), 78(0–P), 79(1–7), 80(1–6), 81(1–5), 85(1–3), 86(1–2), 116(1–3). **Node 55–57**: 12(1–3). **Node 57–58:** 29(4–3), 86(2–3). **Node 58–61:** 16(2–3), 36(0–3), 64(3–2), 78(P–L). **Node 61-*Tironucula***: 1(2–4), *14(1–5), 59(1–2), 60(1–4), 65(1–2), 74(1–8), 76(1–4), 77(1–6), 89(1–3). **Node 61–62**: 17(6–0), 29(3–2), 30(1–2), *116(3–2). **Node 62-*Praenucula***: 55(3–2), 86(3–2). **Node 62–63**: 16(3–4). **Node 63-*Nuculoidea***: *36(3–U), 51(3–4), 101(1–2), 112(1–3). **Node 63–64**: 15(1–3), 17(0–P), 19(2–8), 35(1–3), *36(3–L), 85(3–6), 105(D–G), 106(L–P), 108(5–1). **Node 64–65**: 79(7–0), 80(6–0), 81(5–0), 82(4–1). **Node 65-*Ctenodonta***: *13(1–2), 16(4–3), 17(P–8), 29(2–3), 30(2–1), *86(3–6). **Node 65-*Acharax***: *12(3–1), 14(1–3), 20(1–7), 34(1–2), 38(1–B), 40(1–B), 51(3–4), 54(1–4), *57(1–4), *58(1–2), 64(2–1), 67(1–6), 68(4–1), 69(2–1), 78(L–H), 85(6–1), *86(3–1), *93(9–2), 101(1–2), 112(1–5). **Node 64-*Tancrediopsis***: *12(3–1), *13(1–2), 23(1–2), 24(1–2), 25(1–5), 46(1–2), 64(2–3). **Node 58–59**: *14(1–5), 84(1–2). **Node 59–60**: 85(3–1). **Node 60-*Cardiolaria***: 69(2–3), 72(1–C). **Node 60-*Praeleda***: 3(1–5), 4(1–2), 9(1–5), 10(1–4), 13(1–2), *15(1–2), 17(6–8), 86(3–2). **Node 59-*Ekstadia***: 17(6–H), 34(1–2), 41(1–2), 43(1–2), *44(1–2), 49(1–3), 52(1–4), 56(1–3), *96(1–2). **Node 57-*Eritropis***: 13(1–3), 15(1–5), 16(2–6), 17(6–4), 19(2–4), 36(0–J). **Node 55–56**: 56(1–2), *96(1–2). **Node 56-*Nuculites***: 14(1–3), 16(2–5), 17(6–1), 24(1–2), 34(1–2), *36(0–M), 49(1–3), 54(1–3), *55(3–6), *93(9–6), 103(2–1). **Node 56-*Palaeoneilo***: *36(0–7), 64(3–2), 85(3–2), *93(9–2). **Node 7–8**: *71(1–2), 114(1–2). **Node 8–35**: 17(6–1), *51(2–3). **Node 35–42**: *79(1–B), 81(1–A). **Node 42–45**: 68(1–2), *78(0–F), *80(1–A). **Node 45–49**: 36(0–J), 48(2–1), 49(1–4), 72(1–2). **Node 49–54**: *17(1–6), *34(1–2), 43(3–1), 59(1–2), 89(1–3). **Node 54-*Redonia***: 2(2–1), 12(1–3), 36(J–Q), *44(1–2), 47(2–1), 50(4–3), 61(1–2), *62(1–3), 68(2–1), 70(1–2), 72(2–3), *78(F–D), 79(B–2), 80(A–1), 81(A–1). **Node 54-*Noradonta***: *15(1–3), 16(2–6), *78(F–M), 83(1–2). **Node 49–50**: *15(1–3), 19(2–4), 22(1–2), *88(1–3), *96(1–7). **Node 50–53**: 16(2–6), *34(1–2), *59(1–4), 61(1–7), *62(1–3), 63(1–2), *93(9–6). **Node 53-*Nyassa***: 48(1–2), 49(4–3), *60(1–8), 87(1–2), 96(7–2). **Node 53-*Tanaodon***: 17(1–B), 29(4–5), 81(A–3), 83(1–2). **Node 50–51**: *17(1–5), 105(D–E), 106(L–1). **Node 51-Genus A Johnston & Goodbody 1988**: *50(4–6), 72(2–5), *80(A–8), *93(9–6). **Node 51–52**: 13(1–2), *50(4–3), 68(2–1), *78(F–B), 79(B–F), *80(A–4), 81(A–3). **Node 52-Genus B Johnston & Goodbody 1988**: *17(5–6), 83(1–2), 88(3–1). **Node 52-*Montanaria***: 2(2–3), 22(2–1), 23(1–2), 25(1–5), *44(1–3), *55(3–8), 105(E–J), 106(1–5), 107(2–M), 108(5–6). **Node 45–46**: 14(1–5), *15(1–2), *59(1–4), *62(1–3), *88(1–3). **Node 46-*Copidens***: *60(1–8), *61(1–7), 63(1–2), 89(1–3). **Node 46–47**: 23(1–2), 25(1–5), 36(0–K), *60(1–3). **Node 47-*Ananterodonta***: *2(2–3), 19(2–1), *61(1–4,5). **Node 47–48**: 13(1–2), 17(1–5), *37(1–3), 59(4–2), *61(1–2), 68(2–1), 78(F–B). **Node 48-Genus C Johnston & Goodbody 1988**: 43(3–4), *76(1–3), *79(B–6), 83(1–2), *105(D–E), *106(L–1). **Node 48-*Eodon***: *2(2–3), 14(5–1), 34(1–2), 48(2–1), 49(1–3), 52(1–2), 53(1–2), *55(3–4,5), 73(1–2), 75(1–2), *76(1–2), 77(1–3), *79(B–2), *80(A–2), *81(A–1), 96(1–5), *105(D–J), *106(L–T), 108(5–6). **Node 42–43**: *59(1–3), 75(1–3), 76(1–2), 77(1–3), *78(0–G), *80(1–8). **Node 43-*Cycloconcha***: 23(1–2), 25(1–5), 36(0–7), *73(1–4). **Node 43–44**: 16(2–5), 50(4–3). **Node 44-*Fortowensia***: 12(1–3), *15(1–2,4), 17(1–6), 47(2–3), 51(3–2), *73(1–4), 80(8–3), 81(A–3). **Node 44-*Actinodonta***: *41(1–2), 74(1–7), 75(3–4), 77(3–4). **Node 35–36**: 29(4–3), *79(1–6). **Node 36–38**: 13(1–2), 35(1–3), 43(3–1), 80(1–3). **Node 38-*Thoralia***: 12(1–3), *15(1–4), 16(2–3), 17(1–7,8), *78(0–5), 79(6–3), 81(1–2), *83(1–2). **Node 38–39**: *15(1–5), *78(0–R), 105(D–F), 106(L–M), 108(5–6). **Node 39–40**: 16(2–5), 29(3–4), *36(0–P), 38(1–9), 39(1–2), 40(1–9), *55(3–8), 56(1–2), *83(1–2). **Node 40–41**: 34(1–2), 48(2–1), *49(1–4), 50(4–3), 68(1–2), 72(1–5), 78(R–M), 79(6–B), 80(3–8), *81(1–A), 88(1–3), 89(1–3), 106(M–O). **Node 41-*Lyrodesma majus* adult:** no autapomorphies. **Node 41-*Lyrodesma* sp. Johnston 1996**: 59(1–2), 60(1–8). **Node 40-*Lyrodesma majus* juvenile**: *49(1–3), *81(1–3), *82(4–3). **Node 39-*Silurozodus***: 14(1–5), *36(0–D), 41(1–3), 43(1–2), 52(1–2), 53(1–2), *55(3–7), *82(4–3), *115(1–2). **Node**

36–37: 14(1–5), 16(2–1), 105(D–E), 106(L–1). **Node 37-*Babinka***: 12(1–3), *41(1–2), 43(3–4), 47(2–1), *52(1–4), 70(1–2), 78(0–C), 79(6–4), 82(4–3), *83(1–2). **Node 37-*Ilionia***: *15(1–2), 17(1–C), 19(2–4), 22(1–2), 24(1–2), 25(1–5), 37(1–4), *41(1–3), 44(1–4), *50(4–6), 51(3–4), *52(1–2), 53(1–2), 57(1–4), 58(1–2), 80(1–0), 81(1–0). **Node 8–9**: *15(1–2), 16(2–6). **Node 9-*Modiomorpha***: 19(2–7), *23(1–5), *25(1–2), 36(0–4,A), 41(1–2), 48(2–1), 78(0–4), 87(1–2), *105(D–H), *106(L–R), 107(2–I), *108(5–6), *79(1–2). **Node 9–10**: *79(1–2), 100(1–2), 102(1–3), *108(5–2), *115(1–2). **Node 10-*Evyana***: *23(1–5), *25(1–2), 43(3–4), 52(1–3), *53(1–2), 54(1–3), *70(1–2), 78(0–1), *82(4–1), *105(D–B), *106(L–D). **Node 10–11**: 36(0–H), *105(D–A), *106(L–A), 107(2–A), 109(0–8). **Node 11-*Colpomya***: 12(1–3), 13(1–2), *17(6–H), 31(1–2), 32(1–3), 52(1–2), *53(1–2), *70(1–2), 78(0–8), 82(4–2). **Node 11–12**: 2(2–3), *17(6–4), 80(1–0), 81(1–0), 102(3–2). **Node 12–33**: 36(H–A), 67(1–2), 71(2–1), *79(2–0), *82(4–1). **Node 33–34**: 19(2–7), 102(2–1), 107(A–B), 109(8–1), 111(1–6). **Node 34-*Modiolopsis***: 43(3–1,2), 50(4–3). **Node 34-*Modiolus***: 2(3–2), 44(1–3), 45(1–2), 97(1–2). **Node 33-*Whiteavesia***: 14(1–3), 37(1–3), 51(2–3). **Node 12–13**: *79(2–6). **Node 13–32**: 23(1–2), 25(1–5), 78(0–O). **Node 32-*Matheria***: 36(H–5), 80(0–1), 81(0–3). **Node 32-*Metapadia***: *15(2–1). **Node 13–14**: 14(1–5), *15(2–1), 16(6–2), *44(1–2). **Node 14-*Modiolodon***: 34(1–2), 43(3–2), *48(2–1), 49(1–3), 80(0–5), 81(0–1). **Node 14–15**: 2(3–2), 87(1–2). **Node 15–30**: 59(1–3), 60(1–3), 61(1–5), 62(1–3), 78(0–2). **Node 30–31**: 12(1–3), 74(1–4), 75(1–3), 76(1–3), 77(1–4). **Node 31-*Cyrtodonta***: 19(2–1), 29(4–3), *36(H–G), *48(2–1), 51(2–3), 52(1–3), 53(1–2), 60(3–7), 62(3–4), 80(0–E), 81(0–1), 89(1–3). **Node 31-*Falcatodonta***: 3(1–5), 4(1–5), *15(1–5), *36(H–F), 38(1–3), 39(1–2), 40(1–3), 47(2–3), 59(3–4), *61(5–7), 67(1–7), 71(2–1), 79(6–0), 82(4–1), 87(2–1). **Node 30-*Ortonella***: 13(1–2), *15(1–5), 34(1–2), *41(1–2), *48(2–1), 49(1–3,4), 79(6–5), 80(0–1), 81(0–2), 88(1–3). **Node 15–16**: 3(1–5), 4(1–2), 9(1–2), 10(1–4), 11(1–2), 36(H–C), 45(1–2), 90(1–2), *93(9–2), 94(1–2). **Node 16–29**: *1(2–1), 16(2–6), 23(1–5), 25(1–2), *29(4–5), *50(4–1), 51(2–1), 55(3–5). **Node 29-*Ambonychia***: *17(4–E), *19(2–1), 20(1–4), 36(C–B), 38(1–3), 39(1–2), 40(1–3), *44(2–3), *49(1–5), 57(1–3), 58(1–2), 59(1–4), 60(1–7), 61(1–5), 62(1–4), 72(1–2), 78(0–2), 80(0–4), 81(0–2), 93(2–1). **Node 29-*Septimyalina***: 2(2–1), *17(4–2), *19(2–6), *28(1–2), 31(1–4), 32(1–2), 34(1–4), 47(2–1), *48(2–1), *49(1–6), 79(6–1). **Node 16–17**: 6(1–5), 7(1–2), 8(A–B), *17(4–D), 18(0–5), 26(1–2), 52(1–2), 53(1–2), *109(8–0). **Node 17–27**: 2(2–3), *28(1–4), *29(4–3), 87(2–1), 106(A–C), 107(A–2), *108(2–4). **Node 27-*Myodakryotus***: 3(5–2), 8(B–C), *19(2–1), *36(C–8), 38(1–6), *39(1–2), 40(1–6), *41(1–3), *44(2–3), *48(2–1), *50(4–5), *79(6–5), 80(0–1), 81(0–2). **Node 27–28**: *1(2–1), *10(4–2), 67(1–0), 71(2–1), *79(6–0), 82(4–1). **Node 28-*Palaeolima***: 16(2–3), 26(2–1), 35(1–2), *36(C–8), 38(1–3), *39(1–2), 40(1–3), *96(1–3), 102(2–4), 105(A–K), 106(C–8), 107(2–P), 108(4–1). **Node 28-*Leiopecten***: 3(5–3), 5(1–2), 6(5–3), 8(B–E), 14(5–1), 17(D–F), 31(1–4), 32(1–2), 78(0–A). **Node 17–18**: *28(1–2), 32(1–2), *50(4–3), 59(1–4), 60(1–7), 62(1–3), 63(1–2), 68(1–2), 78(0–I), *79(6–B), 81(0–A), 111(1–2). **Node 18–26**: 3(5–3), 5(1–2), 47(2–1), *61(1–7), 72(1–2), *80(0–A). **Node 26-*Pterinea***: 6(5–3), *18(5–2), 20(1–2), *31(1–5), *44(2–1), 60(7–8). **Node 26-*Tolmaia***: 2(2–3), 8(B–C), 15(1–2), 16(2–6), 17(D–2,E), *18(5–1), *31(1–4), 38(1–3), 39(1–3), 40(1–6), *43(3–1), 50(3–2), 65(1–3), 88(1–3), *89(1–3), *106(A–B), 107(A–2), *108(2–3). **Node 18–19**: 11(2–1), 12(1–3), *19(2–1), *29(4–3), 36(C–I), 52(2–3), *61(1–6), 74(1–6), 75(1–3), 76(1–3), 77(1–4), *80(0–8), *89(1–3). **Node 19-Rhombopteriid new Genus Johnston 1993**: 27(1–2), *31(1–5), 38(1–3), 39(1–3), *44(2–3), 59(4–3), 62(3–4). **Node 19–20**: *44(2–1), 105(A–B), 106(A–F), 107(A–D), *108(2–4), 111(2–4). **Node 20-*Umburra***: 8(B–C), 18(5–1), *31(1–4), 34(1–2), 43(3–2), 54(1–3), 88(1–3). **Node 20–21**: 15(1–2), 17(D–3), 26(2–1), 28(2–1), 32(2–1), 45(2–1), 50(3–4), 75(3–4), 78(I–9), 81(A–6), *90(2–1), *94(2–1). **Node 21–22**: *36(I–G), *52(3–1), *53(2–1), 89(3–1). **Node 22-*Freja***: 2(2–1), *35(1–2), *41(1–3), 54(1–3), *63(2–1), 102(2–3). **Node 22–23**: 61(6–9), 68(2–6), 72(1–A). **Node 23-*Trecanolia***: *1(2–1), *9(2–4), 15(2–1), *35(1–2), 59(4–3), *63(2–1), *65(1–5), 75(4–3), 80(8–A), 81(6–D), *83(1–2). **Node 23–24**: 13(1–2), 16(2–6), 36(G–E). **Node 24-*Catamarcaia***: *9(2–4), 34(1–2), 49(1–3), *65(1–5), 77(4–3), 88(1–3), 106(F–D), *107(D–2), 108(4–2). **Node 24–25**: *1(2–1), 8(B–D), *14(5–3), *29(3–4), 51(2–3), *83(1–2). **Node 25-*Cosmetodon***: 2(2–1), 3(5–2), 5(1–2), 11(1–2), 13(2–3), 18(5–2), *19(1–7), 27(1–2), *35(1–2), 37(1–0), 38(1–2), 39(1–2), 40(1–2), *61(9–4), 80(8–A), 81(6–D), *93(2–9), *96(1–5), *107(D–3). **Node 25-*Glyptarca***: 15(2–5), *19(1–2), *47(2–1), 50(4–3), 59(4–2), *61(9–1), 62(3–1), *63(2–1), 66(1–7), 68(6–3), 72(A–E), 87(2–1), 111(4–1). **Node 21-*Alytodonta***: 13(1–2), 18(5–0), *36(I–F), 38(1–2), 39(1–2), 40(1–2), 61(6–7), 77(4–2). **Node 1-*Ribeiria***: 34(5–6), 42(4–5), 43(8–9), 54(5–6).

References

Adams, H. & Adams, A. 1854–1858. *The Genera of Recent Mollusca Arranged According to their Organization*. J. van Voorst, London, **2**, 1–661.

Allen, J. A. 1985. The Recent Bivalvia, their form and evolution. *In*: Trueman, E. R. & Clarke, M. R. (eds) *The Mollusca, Volume 10, Evolution*. Academic Press, New York, 337–403.

—— & Hannah, F. J. 1986. A reclassification of the Recent genera of the subclass Protobranchia (Mollusca, Bivalvia). *Journal of Conchology*, **32**, 225–249.

Babin, C. 1966. *Mollusques bivalves et céphalopodes du Paléozoique Armoricain*. Imprimerie Commerciale et Administrative, Brest.

—— 1977. Étude comparée des genres *Babinka* Barrande et *Coxiconcha* Babin (Mollusques bivalves de l'Ordovicien). Intérêt phylogénétique. *Geobios*, **10**, 51–79.

—— 1982*a*. Mollusques bivalves et rostroconches. *In*: Babin, C., Courtessole, R., Melou M., Pillet, J., Vizcaino, D. & Yochelson, E. (eds) Brachiopodes (articulés) et Mollusques (bivalves, rostroconches, monoplacophores, gastropodes) de l'Ordovicien inférieur (Trémadocien–Arenigien) de la Montagne Noire (France méridionale). *Mémoire de la Société des Études Scientifiques de l'Aude*, 37–49.

—— 1982*b*. *Tromelinodonta nov. gen.*, bivalve lyrodesmatide (Mollusca) de l'Arenigien (Ordovicien Inférieur). *Geobios*, **15**, 423–427.

—— & GUTIÉRREZ-MARCO, J. 1991. Middle Ordovician bivalves from Spain and their phyletic and palaeogeographic significance. *Palaeontology*, **34**, 109-147.

—— & LE PENNEC, M. 1982. Ontogénèse et phylogénèse, à propos de quelques caractères dentaires des mollusques bivalves. *Malacologia*, **22**, 709-720.

BAILEY, J. B. 1983. Middle Devonian Bivalvia from the Solsville Member (Marcellus Formation), central New York State. *Bulletin of the American Museum of Natural History*, **174**(3), 193–326.

BERG-MADSEN, V. 1987. *Tuarangia* from Bornholm (Denmark) and similarities in Baltoscandian and Australasian late Middle Cambrian faunas. *Alcheringa*, **11**, 245–259.

BEURLEN, K. 1944 [1947]. Beiträge zur Stammesgeschichte der Muscheln. *Sitzungsberichte der Mathematisch-Naturwissenschaftlichen Abteilung der Bayerischen Akademie der Wissenschaften zu München*, **1944**, 133–145.

BOSS, K. J. 1982. Mollusca. *In*: PARKER, S. P. (ed.) *Synopsis and Classification of Living Organisms. Volume 1.* McGraw-Hill, New York, 945–1166.

CAMPBELL, D. C., HOEKSTRA, K. J. & CARTER, J. G. 1998. 18S ribosomal DNA and evolutionary relationships within the Bivalvia. *In*: JOHNSTON, P. A. & HAGGART, J. W. (eds) *Bivalves, an Eon of Evolution — Paleobiological Studies Honoring Norman D. Newell.* University of Calgary Press, Calgary, 75–85.

CARTER, J. G. 1980. Environmental and biological controls of bivalve shell mineralogy and microstructure. *In*: LUTZ, R. A. & RHOADS, D. C. (eds) *Skeletal Growth of Aquatic Organisms.* Plenum, New York, 69–113.

—— 1990. Evolutionary significance of shell microstructure in the Palaeotaxodonta, Pteriomorphia and Isofilibranchia. *In*, CARTER, J. G. (ed.) *Skeletal Biomineralization, Patterns, Processes, and Evolutionary Trends, Volume I.* Van Nostrand Reinhold, New York, 135–296.

—— & LUTZ, R. A. 1990. Bivalvia (Mollusca), Plates. *In*: CARTER, J. G. (ed.) *Skeletal Biomineralization, Patterns, Processes, and Evolutionary Trends, Volume II.* Van Nostrand Reinhold, New York, 5–28, pls 1–121.

—— & SEED, R. 1998. Thermal potentiation and mineralogical evolution in *Mytilus* (Mollusca; Bivalvia). *In*: JOHNSTON, P. A. & HAGGART, J. W. (eds) *Bivalves, an Eon of Evolution — Paleobiological Studies Honoring Norman D. Newell.* University of Calgary Press, Calgary, 87–117.

CHAVAN, A. 1969. Superfamily Crassatellacea. *In*: COX, L. R. & 24 others (eds) 1969–1971 *Treatise on Invertebrate Paleontology. Part N. Mollusca 6. Bivalvia.* Geological Society of America, Boulder, CO, and University of Kansas Press, Lawrence, KS, N562–N580.

COPE, J. C. W. 1996*a*. The early evolution of the Bivalvia. *In*: TAYLOR, J. (ed.) *Origin and Evolutionary Radiation of the Mollusca.* Oxford University Press, Oxford, 361–370.

—— 1996*b*. Early Ordovician (Arenig) bivalves from the Llangynog inlier, South Wales. *Palaeontology*, **39**, 979–1025.

—— 1997. The early phylogeny of the class Bivalvia. *Palaeontology*, **40**, 713–746.

—— 1999. Middle Ordovician bivalves from mid-Wales and the Welsh Borderland. *Palaeontology*, **42**, 467–499.

COX, L. R. & 24 OTHERS. 1969–1971. *In*: MOORE, R. C. (ed.) *Treatise on Invertebrate Paleontology. Part N. Mollusca 6. Bivalvia.* Geological Society of America, Boulder, CO, and University of Kansas Press, Lawrence, KS, **1**, 1969, xxxvii, N1–N489; **2**, 1969, ii, N491–N952; **3**, 1971, iv, N953–N1124.

DALL, W. H. 1889. On the hinge of pelecypods and its development, with an attempt toward a better subdivision of the group. *American Journal of Science*, **38**, 445–462.

—— 1898. Contributions to the Tertiary fauna of Florida. *Transactions of the Wagner Free Institute of Science of Philadelphia*, **3**(4), 571–948.

DECHASEAUX, C. 1952. Classe des Lamellibranches. *In*: PIVETEAUX, J. (ed.) *Traité de Paléontologie*, Tome II. Masson et Cie, Paris, 220–364.

DESBIENS, S. 1994. Les bivalves Cyrtodontacea Ulrich 1894 de la Formation de York River, Emsien de Gaspésie, Québec. *Geobios*, **27**, 449–458.

DESPARMET, R., TERMIER, G. & TERMIER, H. 1971. Sur un bivalve protobranche ante-Arenigien trouvé au nord de Wardak (Afghanistan). *Geobios*, **4**, 143–150.

DZIK, J. 1994. Evolution of 'small shelly fossils' assemblages. *Acta Palaeontologica Polonica*, **39**, 247–313.

FANG, Z. 1998. Revision and taxonomic position of the aberrant Devonian bivalve *Beichuania*. *In*: MORTON, B. (ed.) *The Bivalvia – Proceedings of a Memorial Symposium in Honour of Sir Charles Maurice Yonge, Edinburgh, 1986.* Hong Kong University Press, Hong Kong, 185–191.

—— & MORRIS, N. J. 1997. The genus *Pseudosanguinolites* and some modioliform bivalves (mainly Palaeozoic). *Palaeoworld*, Press of University of Science and Technology of China, **7**, 49–69.

FÉRUSSAC, A. E. DE 1822. *Tableaux Systématiques des Animaux Mollusques classés en familles naturelles, dans lesquels on a établis la concordance de tous les systèmes.* A. Bertrand, London.

FISCHER, P. H. 1880–1887. *Manuel de Conchyliologie et de Paléontologie Conchyliologique.* F. Savy, Paris.

FLEMING, J. 1828. *A History of British Animals, Exhibiting the Descriptive Characters and Systematical Arrangement of the Genera and Species of Quadrupeds, Birds, Reptiles, Fishes, Mollusca, and Radiata of the United Kingdom.* Bell & Bradfute, Edinburgh.

FRECH, F. 1891. Die devonischen Aviculiden Deutschlands, ein Beitrag zur Systematik und Stammesgeschichte der Zweischaler. *Abhandlungen z. Geolologischen Specialkarte von Preussen und den Thuringischen Staaten,* **9**(3).

GEYER, G. & STRENG, M. 1998. Middle Cambrian

pelecypods from the Anti-Atlas, Morocco. *Revista Espanola de Paleontología, num. extr. Homenaje al Prof. Gonzalo Vidal*, 83–96.

GRAY, J. E. 1824. A natural arrangement of Mollusca according to internal structure. *Repository*, **15**, 229–239.

—— 1840. *Synopsis of the contents of the British Museum*, 42nd Edition. British Museum, London.

—— 1847. A list of the genera of Recent Mollusca, their synonyms and types. *Proceedings of the Zoological Society of London*, **15**, 129–219.

GROBBEN, K. 1894. Zur Kenntniss der Morphologie, der Verwandschaftsverhältnisse und des Systems der Mollusken. *Sitzungsberichte der Mathematisch-Naturwissenschaftlichen Classe der Kaiserlichen Akademie der Wissenschaften, Wien*, **103**, 61–86.

HALL, J. 1885. *Lamellibranchiata II, Text and Plates. Containing Descriptions and Figures of the Dimyaria of the Upper Helderberg, Hamilton, Portage, and Chemung Groups. Geological Survey of the State of New York, Paleontology, Vol. 5, Pt. 1.* Van Benthuysen & Sons, Albany, New York, 269–562, pls 34–80, 93–96.

HARRISON, W. B. III & HARRISON, L. K. 1975. A Maquoketa-like molluscan community in the Brassfield Formation (early Silurian) of Adams County, Ohio. *Bulletins of American Paleontology*, **67**, 193–234.

HASZPRUNAR, G. 1983. Comparative analysis of the abdominal sense organs of Pteriomorphia (Bivalvia). *Journal of Molluscan Studies, Supplement*, **12A**, 47–50.

—— 1985. The fine structure of the abdominal sense organs of Pteriomorphia (Mollusca, Bivalvia). *Journal of Molluscan Studies*, **51**, 315–319.

HERBERS, K. 1914. Entwicklungesgeschichte von *Anodonta cellensis* Schröter. *Zeitschrift für Wissenschaftliche Zoologie*, **108**, 1–174.

HERTWIG, R. 1895. *Lehrbuch der Zoologie*, 3rd Edition. Jena.

HINZ-SCHALLREUTER, I. 1995. Muscheln (Pelecypoda) aus dem Mittelkambrium von Bornholm. *Geschiebekunde Aktuel*, **11**, 71–84.

HORNÝ, R. 1960. On the phylogeny of the earliest pelecypods (Mollusca). *Vestník Ústředního Ústavu Geologického*, **35**, 479–482, pl. 1.

JOHNSTON, P. A. 1991. Systematics and ontogeny of a new bivalve, *Umburra cinefacta*, from the Silurian of Australia, implications for pteriomorphian evolution. *Alcheringa*, **15**, 293–319.

—— 1993. Lower Devonian pelecypods from southeastern Australia. *Association of Australasian Palaeontologists, Memoir*, **14**, 1–134.

—— 1995. Evolutionary pathways in the bivalve subclass Pteriomorphia. *In*: JOHNSTON, P. A. (ed.) *Canadian Paleontology Conference Program and Abstracts*, **5**, 16–17.

—— 1996. Middle Ordovician (Caradoc) pelecypods from the Advance Formation, British Columbia. *Geological Survey of Canada Bulletin*, **491**, 113–129.

—— & GOODBODY, Q. H. 1988. Middle Devonian bivalves from Melville Island, Arctic Canada. *In*: MCMILLAN, N. J., EMBRY, A. F. & GLASS, D. J. (eds) *Devonian of the World, Volume III: Paleontology, Paleoecology and Biostatigraphy*. Canadian Society of Petroleum Geologists, Calgary, Alberta, Canada, 337–346.

—— & ZHANG, R. 1998. Re-interpretation of *Sinodora* Pojeta and Zhang, 1984, an unusual bivalve (Trigonoida) from the Devonian of China. *In*: JOHNSTON, P. A. & HAGGART, J. W. (eds) *Bivalves, An Eon of Evolution — Paleobiological Studies Honoring Norman D. Newell*. University of Calgary Press, Calgary, 361–375.

KEEN, A. M. 1958. *Marine Shells of Tropical West America*. Stanford University Press, Stanford, CA.

KOROBKOV, I. A. 1954. *Handbook and Methodical Guide to the Tertiary Mollusca, Lamellibranchiata*. Gosud. Nauchno-tecnoï Isledov. Nefti. Gorno-toplivnoi lit-ry, Leningradskoi Otdelenie [in Russian].

—— 1960. Rhombopteriidae *fam. nov. In*: EBERZIN, A. G. (ed.) *Osnovui Paleontologhii. Spravochnik dlya paleontologov i geologov SSSR. 3, Mollyuski, Pantsirnye. Dvustvorchatye, Lopatonogie.* Moskva, Izdatel'stvo Akademii Nauk SSSR, 81.

KOUCHINSKY, A. V. 1999. Shell microstructures of the Early Cambrian *Anabarella* and *Watsonella* as new evidence on the origin of the Rostroconchia. *Lethaia*, **32**, 173–180.

KRASILOVA, I. N. 1959. A new family Leiopectinidae from the Upper Silurian and Lower Devonian of the northeastern part of the Balkhash region. *Paleontologicheskiai Zhurnal*, **1959**(2), 41–46 [in Russian].

LAMARCK, J. B. P. A. DE M. DE 1815–1822. *Histoire Naturelle des Animaux sans Vertèbres, Volumes 1–7.* Verdière & Deterville, Paris [see Volume 6, Part 1, 1819]

LILJEDAHL, L. 1984*a*. Silurian silicified bivalves from Gotland. *Sveriges Geologiska Undersökning, Avhandlingar och Uppsater*, Serie C, Nr. **804**, **78**, 1–82.

—— 1984*b*. Silicified Silurian bivalves from Gotland. *Lund Publications in Geology*, **24**, 1–28.

—— 1994. Silurian nuculoid and modiomorphid bivalves from Sweden. *Fossils and Strata*, **33**, 1–89.

LINNÉ, K. 1758. *Systema Naturae per Tria Regna Naturae. Edit. decima, 1.* Laurentii Salvii, Stockholm.

MACKINNON, D. I. 1982. *Tuarangia paparua* n. gen. and n. sp., a late Middle Cambrian pelecypod from New Zealand. *Journal of Paleontology*, **56**, 589–598.

—— 1985. New Zealand late Middle Cambrian molluscs and the origin of Rostroconchia and Bivalvia. *Alcheringa*, **9**, 65–81.

MCALESTER, A. L. 1969. Family Praenuculidae. *In*: COX, L. R. & 24 others (eds) 1969–1971. *Treatise on Invertebrate Paleontology. Part N. Mollusca 6, Bivalvia.* Geological Society of America, Boulder, CO, and University of Kansas Press, Lawrence, KS, N229–N230.

MCROBERTS, C. A. & CARTER, J. G. 1994. Nacre in an early gryphaeid bivalve (Mollusca). *Journal of Paleontology*, **68**, 1405–1408.

MILLER, S. A. 1877. *The American Palaeozoic Fossils, a Catalogue of the Genera and Species, with Names of Authors, Dates, Places of Publication, Groups of*

Rocks in Which Found, and the Etymology and Significance of the Words, and an Introduction Devoted to the Stratigraphical Geology of the Palaeozoic Rocks. Published by the author, Cincinnati Times Co. Book and Job Rooms, Cincinnati, Ohio.

MORRIS, N. J. 1979. On the origin of the Bivalvia. *In*: HOUSE, M. R. (ed.) *The Origin of Major Invertebrate Groups*. Systematics Association Special Volume No. **12**, Academic Press, London, 381–413.

—— 1980. A new Lower Ordovician bivalve family, the Thoraliidae (?Nuculoida), interpreted as actinodont deposit feeders. *Bulletin of the British Museum of Natural History (Geology)*, **34**, 265–272.

—— & FORTEY, R. A. 1976. The significance of *Tironucula gen. nov.* to the study of bivalve evolution. *Journal of Paleontology*, **50**, 701–709.

——, DICKINS, J. M. & ASTAFIEVA-URBAITIS, K. 1991. Upper Palaeozoic anomalodesmatan Bivalvia. *Bulletin of the British Museum (Natural History), Geology*, **47**, 51–100.

MUTVEI, H. 1970. Ultrastructure of the mineral and organic components of molluscan nacreous layer. *Biomineralization Research Reports*, **2**, 48–61.

—— 1983. Ultrastructural evolution of molluscan nacre. *In*: WESTBROEK, P. & DE JONG, E. W. (eds) *Biomineralization and Biological Metal Accumulation*. D. Reidel Co., Boston, 267–271.

NEWELL, N. D. 1942. Late Paleozoic pelecypods, Mytilacea. *State Geological Survey of Kansas*, **10**, 1–115.

—— 1969. Classification of Bivalvia, and Superfamily Modiomorphacea. *In*: COX, L. R. & 24 others (eds) 1969–1971. *Treatise on Invertebrate Paleontology. Part N. Mollusca 6, Bivalvia*. Geological Society of America, Boulder, CO, and University of Kansas Press, Lawrence, KS, N205–N224, and N393–N399.

—— & BOYD, D. W. 1975. Parallel evolution in early Trigoniacean bivalves. *Bulletin of the American Museum of Natural History*, **154(2)**, 53–162.

OCKELMANN, K. & WARÉN, A. 1998. Taxonomy and biological notes on the bivalve genus *Microgloma*, with comments on protobranch nomenclature. *Ophelia*, **48**, 1–24.

ORBIGNY, A. D', 1844 [1843]. *Paléontologie Français terrains crétacés, Volume 3, Mollusques*. G. Masson, Paris.

POJETA, J. JR 1971. Review of Ordovician pelecypods. *US Geological Survey Professional Paper*, **695**, 1–46, pls 1–20.

—— 1975. *Fordilla troyensis* Barrande and early pelecypod phylogeny. *Bulletins of American Paleontology*, **67**, 363–384.

—— 1978. The origin and early taxonomic diversification of pelecypods. *Philosophical Transactions of the Royal Society of London, Series B*, **284**, 225–246.

—— 1980. Molluscan phylogeny. *Tulane Studies in Geology and Paleontology*, **16**, 55–80.

—— 1985. Early evolutionary history of diasome mollusks. *In*: BROADHEAD, T. W. (ed.) *Mollusks, Notes for a Short Course*. University of Tennessee Department of Geological Sciences, Studies in Geology, **13**, 102–130.

—— 1988. The origin and Paleozoic diversification of solemyoid pelecypods. *New Mexico Bureau of Mines and Mineral Resources Memoir*, **44**, 201–271.

—— & GILBERT-TOMLINSON, J. 1977. Australian Ordovician pelecypod molluscs. *Bulletin of the Bureau of Mineral Resources, Geology and Geophysics*, **174**, 1–64, pls 1–29.

—— & RUNNEGAR, B. 1974. *Fordilla troyensis* and the early history of pelecypod mollusks. *American Scientist*, **62**, 706–711.

—— & —— 1985. The early evolution of diasome molluscs. *In*: WILBUR, K. M. (ed.) *The Mollusca, Volume 10, Evolution*. Academic Press, London, 295–336.

RAFINESQUE, C. S. 1815. *Analyse de la Nature ou Tableau de L'Univers et des Corps Organisés*. Published by the author, Palerme, 1–214.

RATTER, V. A. & COPE, J. C. W. 1998. New neotaxodont bivalves from the Silurian of South Wales and their phylogenetic significance. *Palaeontology*, **41**, 975–991.

RUNNEGAR, B. N. 1983. Molluscan phylogeny revisited. *Memoirs of the Association of Australasian Palaeontologists*, **1**, 121–144.

—— 1985. Shell microstructures of Cambrian molluscs replicated by phosphate. *Alcheringa*, **9**, 245–257.

—— 1996. Early evolution of the Mollusca, the fossil record. *In*: TAYLOR, J. (ed.) *Origin and Evolutionary Radiation of the Mollusca*. Oxford University Press, Oxford, 77–87.

—— & BENTLEY, C. 1983. Anatomy, ecology and affinities of the Australian early Cambrian bivalve *Pojetaia runnegari* Jell. *Journal of Paleontology*, **57**, 73–92.

—— & JELL, P. A. 1976. Australian Middle Cambrian molluscs and their bearing on early molluscan evolution. *Alcheringa*, **1**, 109–138.

—— & POJETA, J. JR 1974. *Fordilla troyensis* and the early history of pelecypod mollusks. *American Scientist*, **62**, 706–711.

—— & —— 1992. The earliest bivalves and their Ordovician descendants. *American Malacological Bulletin*, **9**, 117–122.

SALVINI-PLAWEN, L. V. & STEINER, G. 1996. Synapomorphies and plesiomorphies in higher classification of Mollusca. *In*: TAYLOR, J. (ed.) *Origin and Evolutionary Radiation of the Mollusca*. Oxford University Press, Oxford, 29–51.

SÁNCHEZ, T. M. 1995. Comments on the genus *Catamarcaia* Sánchez and Babin and the origin of the Arcoida. *Geobios*, **28**, 343–346.

—— & BABIN, C. 1998. The origin of actinodont from taxodont dentition or vice versa, an unnecessary controversy? *In*: JOHNSTON, P. A. & HAGGART, J. W. (eds) *Bivalves, An Eon of Evolution — Paleobiological Studies Honoring Norman D. Newell*. University of Calgary Press, Calgary, 409–412.

SANDERS, H. L. & ALLEN, J. A. 1973. Studies on deep-sea Protobranchia (Bivalvia); prologue and the

Pristiglomidae. *Bulletin of the Museum of Comparative Zoology*, **145**, 237–362.

SCARLATO, O. A. & STARABOGATOV, Y. I. 1978. Phylogenetic relations and the early evolution of the class Bivalvia. *Philosophical Transactions of the Royal Society of London, Series B*, **284**, 217–224.

—— & —— 1979. General evolutionary patterns and the system of the class Bivalvia. *Trudy Zoologicheskogo Instituta, Akademiya Nauk, SSSR*, **80**, 5–38 [in Russian].

SCOTT, H. W. 1961. Shell morphology of Ostracoda. *In*: BENSON, R. H. & 16 others. *In:* MOORE, R. C. (ed.) *Treatise on Invertebrate Paleontology, Part Q, Arthropoda 3, Crustacea, Ostracoda*. Geological Society of America, Boulder, CO, and University of Kansas Press, Lawrence, KS, Q21–Q37.

SMITH, A. B. 1994. *Systematics and the Fossil Record, Documenting Evolutionary Patterns*. Blackwell Scientific Publications, Oxford.

STAROBOGATOV, Y. I. 1992. Morphological basis for phylogeny and classification of Bivalvia. *Ruthenica*, **2**, 1–25.

STOLICZKA, F. 1870–1871. Cretaceous fauna of southern India. 3. The Pelecypoda, with a review of all known genera of this class, fossil and Recent. *Geological Survey of India, Palaeontologica Indica, Series 6*, **3**, 1–537.

SWOFFORD, D. L. 1998. *PAUP*. Phylogenetic Analysis Using Parsimony (*and Other Methods). Version 4b2a*. Sinauer Associates, Sunderland, MA.

TAYLOR, J. D. 1973. The structural evolution of the bivalve shell. *Palaeontology*, **16**, 519–534.

TUNNICLIFF, S. P. 1987. Caradocian bivalve molluscs from Wales. *Palaeontology*, **30**, 677–690, pls 76–77.

ULRICH, E. O. 1893. New and little known Lamellibranchiata from the Lower Silurian rocks of Ohio and adjacent states. *Report of the Geological Survey of Ohio*, **7**, 627–693, pls 45–56.

—— 1894. The Lower Silurian Lamellibranchiata of Minnesota. *In*: *Final Report of the Geological and Natural History Survey of Minnesota, Volume 3*, 475–628 [published separately in 1894 prior to the entire Volume 3 in 1897].

WALLER, T. 1990. The evolution of ligament systems in the Bivalvia. *In*, MORTON, B. (ed.) *The Bivalvia – Proceedings of a Memorial Symposium in Honour of Sir Charles Maurice Yonge, Edinburgh, 1986*. Hong Kong University Press, Hong Kong, 47–91.

—— 1998. Origin of the molluscan class Bivalvia and a phylogeny of major groups. *In*: JOHNSTON, P. A. & HAGGART, J. W. (eds) *Bivalves, An Eon of Evolution — Paleobiological Studies Honoring Norman D. Newell*. University of Calgary Press, Calgary, 1–45.

WIENS, J. J. 1998. Does adding characters with missing data increase or decrease phylogenetic accuracy? *Systematic Biology*, **47**, 625–640.

—— & SERVEDIO, M. R. 1997. Accuracy of phylogenetic analysis including and excluding polymorphic characters. *Systematic Biology*, **46**, 332–345.

WÖHRMANN, S. F. VON 1893. Ueber die systematische Stellung der Trigoniden und die Abstammung der Nayaden. *Jahrbuch der Kaiserlich-Koniglichen Geologischen Reichsanstalt, Wien,* **43**, 1–28.

YONGE, C. M. 1939. The protobranchiate Mollusca; a functional interpretation of their structure and evolution. *Philosophical Transactions of the Royal Society of London, Series B*, **230**, 79–147.

—— 1953. Form and habit in *Pinna carnea* Gmelin. *Philosophical Transactions of the Royal Society of London, Series B*, **237**, 335–374.

—— 1977. Form and evolution in the Anomiacea (Mollusca, Bivalvia) – *Pododesmus*, *Anomia*, *Patro*, *Enigmonia* (Anomiidae), *Placunanomia*, *Placuna* (Placunidae *fam. nov.*). *Philosophical Transactions of the Royal Society of London, Series B*, **276**, 453–523.

A new look at early bivalve phylogeny

JOHN C. W. COPE

Department of Earth Sciences, Cardiff University, PO Box 914, Cardiff CF10 3YE, UK

(e-mail: copejcw@cardiff.ac.uk)

Abstract: In the 30 years since publication of the bivalve *Treatise*, (Moore, R. C. (ed.) 1969. *Treatise on Invertebrate Paleontology. Part N. Mollusca 6, Bivalvia*, Geological Society of America and University of Kansas) important new faunas have been described from the early and mid Cambrian and from the early and mid Ordovician. These contain significant new forms, including some long-ranging intermediate groups, that indicate the relationships between the principal bivalve clades, but lack of fossils from the late Cambrian and earliest Ordovician is a major hindrance. The principal phase of bivalve diversification followed on from the evolution of the filibranch gill in the latest Cambrian or earliest Ordovician. The fundamental division of the class is into two subclasses, Protobranchia and Autolamelli branchiata; links between the two can be demonstrated in the early Ordovician. Major divisions of each subclass are recognized as superorders.

Within the Protobranchia, the Nuculoida developed specialist food-gathering palps and an enlarged foot. Diverging early from the protobranch stock were other bivalves that lived symbiotically with sulphur-oxidizing chemoautotrophic bacteria; this allowed colonization of anaerobic substrates and produced two distinct stocks: the deeply infaunal anteriorly elongate Solemyoida and the shallower infaunal Nucinelloida.

The Autolamellibranchiata, initially identified by strongly asymmetrical hinges, diversified in three directions, each characterized by distinctive hinges. The Trigonioida were characterized by ligamental nymphs and frequently denticulate teeth, and rapidly regained greater symmetry; the Anomalodesmata also developed a strong ligamental insertion within nymphs and largely lost their dentition, whilst the Heteroconchia, principally with a shell including a complex crossed-lamellar structure, had various combinations of cardinal and lateral teeth. Heteroconch diversifications were mainly in the Mesozoic and Cenozoic, but one Ordovician group, the Glyptarcoidea, is a good ancestor for the Pteriomorphia.

The following new taxa are proposed: Cardiolarioidea superfam. nov., Eritropidae fam. nov. and Catamarcaidae fam. nov.

The phylogeny of the bivalves has become much better understood over the past 30 years and credit for this is due to the renewal of interest in early Palaeozoic bivalves following a period of many years during which interest in these fossils appears to have been minimal. In turn, this followed on from the frenzy of naming new species and genera in the middle years of the last century. The lately renewed interest was initiated by two important publications: Babin's (1966) description of the Ordovician faunas of Brittany and Pojeta's (1971) now classic review of Ordovician bivalves. Both of these works demonstrated the unexpected diversity of Ordovician bivalves, but at that time links between various groups of Ordovician bivalves were largely unknown and several major groups of bivalves were then unknown before the mid Ordovician. Since then important faunas have been described from the early and mid-Ordovician (e.g. Pojeta & Gilbert-Tomlinson 1977; Babin 1982*a*; Babin & Gutiérrez-Marco 1991; Cope 1996, 1999), which have extended the known ranges of many bivalve groups and yielded intermediate fossils crucial to the understanding of bivalve evolution.

The earliest bivalves

Knowledge of Cambrian bivalves has also increased dramatically over past 30 years. Pojeta (1971) examined the various putative Cambrian bivalves then recorded and concluded that none of them could be unequivocally regarded as a bivalve. Pojeta's (1971) suspicions about *Lamellodonta* were confirmed when Havlíček & Kříž (1978) demonstrated that the mid-Cambrian genus, thought at the time of publication of the bivalve *Treatise* (Cox *et al.* 1969) to be the earliest bivalve, was in fact a distorted obolellid brachiopod. Pojeta

From: Harper, E. M., Taylor, J. D. & Crame, J. A. (eds) *The Evolutionary Biology of the Bivalvia*. Geological Society, London, Special Publications, **177**, 81–95. 1-86239-076-2/00/$15.00

(1973, 1975) undertook a study of the early Cambrian genus *Fordilla* and established that it was indeed a bivalve. Jell (1980) described the early Cambrian genus *Pojetaia* that is now known to be widely distributed in early and mid-Cambrian rocks, and subsequently new forms have been described from the early and mid-Cambrian (Berg-Madsen 1987; Hinz-Schallreuter 1995; Geyer & Streng 1998). In addition to *Pojetaia*, the middle Cambrian bivalves include the genera *Tuarangia* and *Camya*, which are both of unknown affinities. The genus *Arhouriella*, described from the Middle Cambrian of Morocco (Geyer & Streng 1998), is clearly different from other Cambrian bivalve taxa described hitherto and the authors of the genus were unable to suggest a suitable higher level taxon for it. *Arhouriella* is based on two right valves (only one, the holotype, shows many features) but this identification depends upon the stated orientation. It bears two peg-like teeth and also seems to have a long posterior lateral tooth, not noted in the description but visible in Geyer & Streng's illustration (1998, pl. 7, fig. A; see also Fig. 1a). Geyer & Streng (1998, pp. 93, 94) suggested that *Arhouriella* had an amphidetic ligament; the authors suggest two ligamental sites, one anterior and one posterior to the umbones (Geyer & Streng 1998, fig. 8). If the site identified as a posterior ligamental site is such, then this valve is correctly identified as a right valve. If, however, the other site (Geyer & Streng 1998, fig. 8, af) was the sole ligamental site, then the specimen should be interpreted as a left valve. One peculiar feature of *Arhouriella* is that it apparently does not have properly differentiated dental sockets; the authors believed that this was due to the teeth growing later over the mid part of a juvenile elongate amphidetic ligament Although there are some superficial resemblances to nucinelloids (particularly if the holotype were in reality a left valve) too little is known about *Arhouriella* to draw any firm conclusion of its affinities.

There still remains only a single bivalve specimen known worldwide from rocks of late Cambrian age (Berg-Madsen 1987) and this persistent gap in the bivalve record is particularly unfortunate since it is now apparent that this fairly clearly coincides with the appearance of the filibranch gill (Cope 1995, 1997*b*). It is also very apparent that many of the problems that still remain in early bivalve evolution will only be resolved by the discovery and description of faunas from this late Cambrian lacuna.

Cambrian bivalves are certainly very small and these early forms need not have been shallow infaunal forms that comparisons with modern nuculoids have suggested. Tevesz & McCall (1976) suggested that bivalves of < 1 cm in length could have functioned equally well as epifaunal crawlers (as their tergomyan ancestors almost certainly were). Morton (1995) also viewed these early forms as surface dwellers that used their foot both for feeding and locomotion. Cope (1995) noted that very small bivalves could only have survived infaunally in very fine sediment. Thus, the life habits of these early forms are not well-established, but the advent of larger sizes in the earliest Ordovician would have made a greater variety of habitats possible.

Knowledge of the earliest Ordovician bivalve faunas is also scant. From the early Tremadoc there are only a handful of poorly preserved moulds described by Harrington (1938), whilst the late Tremadoc faunas described by Harrington (1938), Pojeta & Gilbert-Tomlinson (1977) and Babin (1982*a*), totalling perhaps ten species from the whole of the Tremadoc Series, merely hint at the variety of forms that are certain to be discovered. The discoveries of the past 25 years have taken back the origins of many groups to far earlier in the geological past than was conceived when the *Treatise* (Cox *et al.* 1969) was published, but there remains this major gap in knowledge.

By the early Ordovician, bivalves were apparently restricted to the peri-Gondwanan seas (Cope & Babin 1999), whereas in the early Cambrian they appear to have a cosmopolitan distribution. In the mid Cambrian they are known from Baltica and Gondwana, and the single late Cambrian specimen is also Baltican (Berg-Madsen 1987). Early bivalves were largely restricted to inshore silty muds (Cope 1996) but the late Cambrian margins of both Gondwana and Baltica were largely rifted, thus such close inshore facies are consequently rare and this may well explain the, so far, disjunct early bivalve fossil record (Cope & Babin 1999).

Bivalve taxonomy

In trying to produce a satisfactory classification for the bivalves, a large number of schemes have been, and continue to be, proposed. The principal divide in these schemes has been between the zoologists who naturally rely on soft-part anatomy in drawing distinctions and the palaeontological approach that has largely relied upon features such as dentition and musculature, together with the fossil record. More recently, studies of shell microstructure (Taylor *et al.* 1969, 1973; Carter 1990) have added further information to help elucidate phylogeny, whilst molecular phylogenies are now also being reconstructed, based on 18S ribosomal DNA (rDNA) (e.g. Adamkewicz *et al.* 1997). However, with the discovery of new faunas from the Cambrian and Ordovician rocks over the past 25

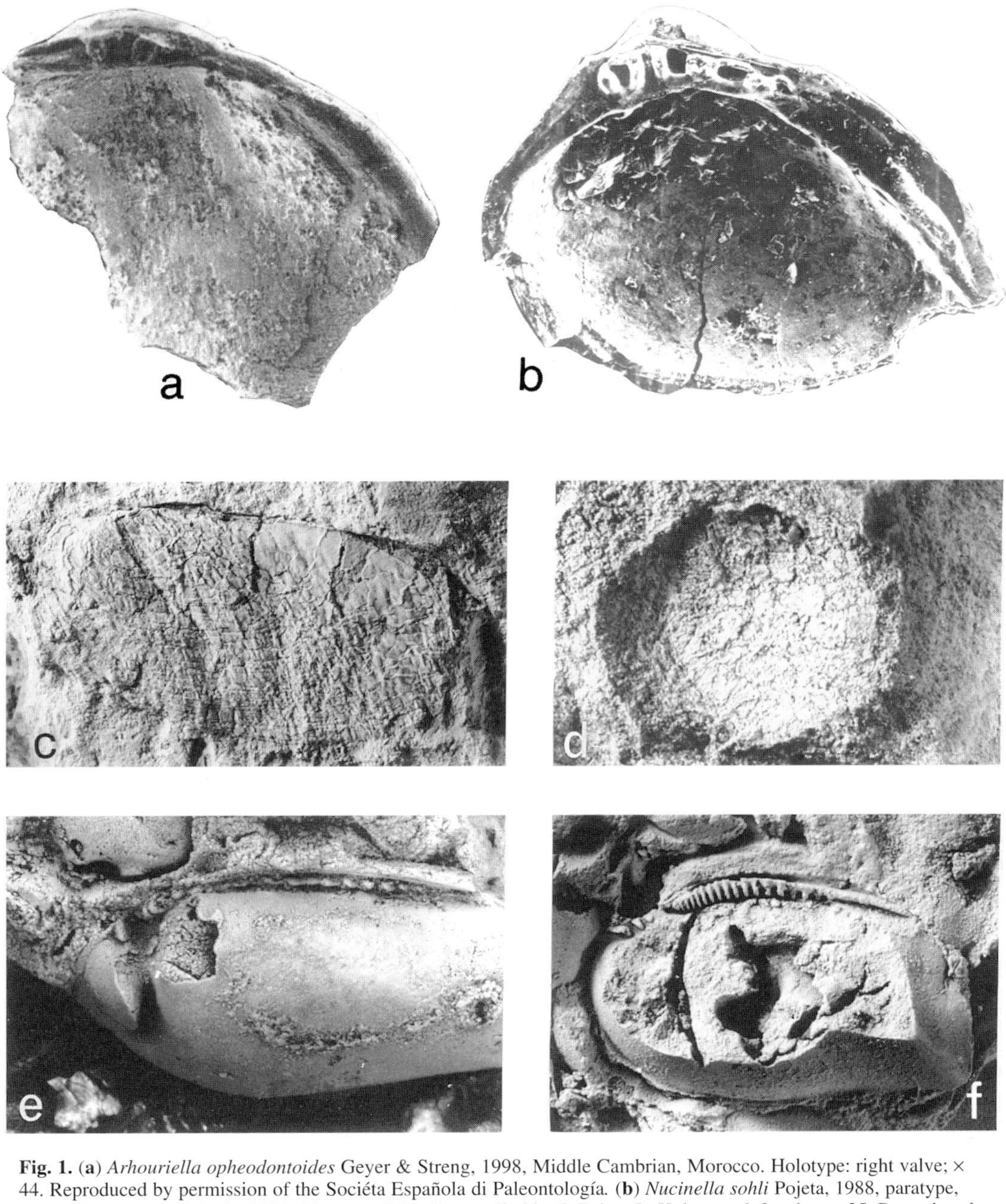

Fig. 1. (**a**) *Arhouriella opheodontoides* Geyer & Streng, 1998, Middle Cambrian, Morocco. Holotype: right valve; × 44. Reproduced by permission of the Sociéta Española di Paleontología. (**b**) *Nucinella sohli* Pojeta, 1988, paratype, Upper Cretaceous, Georgia, USA. Photograph kindly supplied by J. Pojeta Jr. Holotype: left valve; × 35. Reproduced by permission of the New Mexico Bureau of Mines and Mineral Resources. (**c**) and (**d**) *Ovatoconcha fragilis* Cope, 1996. Moridunian Stage, Arenig, Lower Ordovician, South Wales. (**c**) Holotype: right valve view of conjoined valves of a silicified composite mould, showing muscle scars; × 3. (**d**) External impression of the posterior portion of a left valve, to which the outer part of the silicified shell adheres, showing umbonal region and possible periostracal fringe beyond the mineralized shell margin. (**e**) *Noradonta redoniaeformis* (Thoral), Lower Arenig, Lower Ordovician, Montagne Noire. Holotype: left valve showing cardinal teeth (?two missing) lying beneath line of hinge along crenulate posterior tooth, posterior part of which appears to be of discrete teeth of palaeotaxodontid type. This species appears to be a link between *N. shergoldi* Pojeta & Gilbert-Tomlinson and the cardiolarioids; × 4. Negative kindly loaned by C. Babin. (**f**) *Eritropis peregrinata* Cope, Lower Llanvirn, Middle Ordovician, Mid-Wales. Holotype: left valve internal mould prepared to show cardiolarioid dentition; × 3.

years, the possibility of a taxonomic scheme based on phylogeny derived from the fossil record has become much more of a realizable objective.

One of the ways in which it has been possible to advance the understanding of the method(s) by which bivalves evolved is the recognition that the evolution of the filibranch gill from a protobranch gill was a single evolutionary step (Cope 1995). Too often in the past, palaeontologists have believed that it would not be possible to determine the gill grade of extinct bivalves and because of this there developed the feeling that all Ordovician bivalve groups must have been protobranch and that the filibranch gill evolved separately in different bivalve groups in the early Silurian. Such a view has also been expressed as recently as 1997: 'the gills [of most bivalve groups] probably crossed a functional threshold during the late Ordovician–early Silurian, operating as nutrient filters as well as respiratory organs' (Benton & Harper 1997, p. 177).

Recognizing that the hinges of some bivalves of palaeotaxodont grade were strongly asymmetrical, so that the hinge lay along the line of posterior taxodont teeth whilst the larger anterior teeth lay below the hinge, Cope (1995) suggested that these hinges were designed for wider valve opening; the larger anterior teeth ensuring that full dental articulation was preserved even with a wide shell opening. Cope (1995) correlated this with the evolution of the filibranch gill and the concomitant need to void pseudofaeces; later, Cope (1997*b*) erected the family Cardiolariidae to accommodate such forms. The Cardiolaridae were assigned to the subclass Palaeotaxodonta although, as argued by Cope (1997*b*), they were likely to have had filibranch gills. Contemporary nuculoids (members of the family Praenuculidae), on the other hand, show no such modifications and early, mid and late Ordovician praenuculid genera have hinge teeth that, with a wide valve opening, would readily have lost dental articulation as the valves were progressively opened. By the Silurian, however, nuculoid genera such as *Nuculoidea* show enlarged teeth at the anterior and posterior extremities of the hinge plate. These too are clearly designed to accommodate greater opening of the valves, but in this case the hinge is approximately symmetrical, there is an internal resilium and the augmented valve opening then possible seems more likely to be associated with a greatly enlarged foot, a feature characteristic of the Recent *Nucula*, and which can therefore be assumed to date back to no earlier than the Silurian when the earliest nuculoids with an internal resilium occur (Pojeta 1971).

It now seems clear that the position has been reached where the two subclasses recognized by neontologists can be separated with confidence, and thus the time is ripe to produce a unified taxonomic scheme that may be used by both zoologists and palaeontologists. In Cope's (1997*b*, p. 715) review of bivalve phylogeny, it is noted that 'it may be ... that a taxonomically more defensible position would be to regard ... major divisions of the class as superorders rather than subclasses'. Therefore, in this account of bivalve phylogeny, only two subclasses are recognized, the Protobranchia and the Autolamellibranchiata. Within each of these subclasses many of the highest category subdivisions are ordinal, but in some cases a superordinal category is also used for categories which had previously been regarded as subclasses by palaeontologists. This classification based on two subclasses was first employed by Grobben

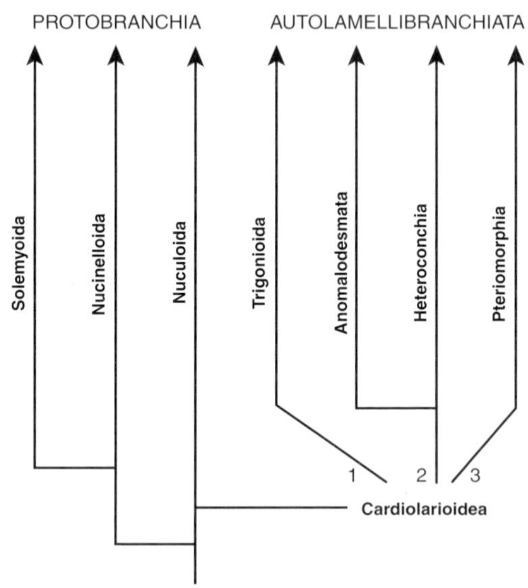

Fig. 2. Proposed phylogenetic relationships within the class Bivalvia. The separate derivation of the Trigonioida, Heteroconchia and Pteriomorphia from a cardiolarioid ancestor is supported by bivalves showing intermediate characters as follows: (1) cardiolarioid dentition links directly to the early Ordovician *Noradonta redoniaeformis* (Thoral) – see Fig. 1e; in turn, this can be linked to *N. shergoldi* (Pojeta & Gilbert-Tomlinson), *Tromelinodonta* and thence to *Lyrodesma* (see Babin 1982*b*, fig. 1); (2) Waller (1998, fig. 5) has shown how the dentition of a 'malletiid' or a cardiolarioid (Waller 1998, p. 36) links to other cardiolarioids like *Inaequidens*, thence via *Ekaterodonta* to a cycloconchid such as *Copidens* (or indeed to an earlier form such as *Carminodonta*); (3) Cope (1995, fig. 30.1) showed how the dentition of *Cardiolaria* linked it to *Glyptarca*, whilst Ratter & Cope (1998, fig. 7) showed how *Glyptarca* could have given rise to a form that was ancestral to both arcoids and cyrtodonts at the base of the pteriomorphian stocks.

(1894) but has subsequently been used by many other workers, including Allen (1985), Cox (1959), Waller (1978, 1990, 1998) and Yonge (1959).

Subclass Protobranchia Pelseneer, 1889

This group is unified by the protobranch gill that typifies the modern nuculoids and solemyoids. The protobranch ctenidium is also similar to the gill of primitive archaeogastropods, but differs in possessing latero-frontal cilia (Waller 1998) and is widely accepted as being close to the primitive molluscan gill. The earliest Cambrian bivalves, *Fordilla* and *Pojetaia*, have both been assigned to the palaeotaxodonts (Runnegar & Pojeta 1992). These authors recorded that the shell microstructure of these two genera was similar and resembled that of the outer shell layer of a Devonian nuculoid described by Carter (1990), which clearly played a major role in making that assignment. Runnegar & Pojeta (1992), however, also left open the possibilty that these two genera were stem-group bivalves. The Palaeotaxodonta must now be regarded as a paraphyletic unit, since the demonstration by Cope (1995, 1997*b*) that at least one group of them was likely to have had filibranch gills, whereas the majority of the palaeotaxodonts can be assigned to the subclass Protobranchia. Thus, palaeotaxodonts are here regarded as an informal grade.

Nuculoids

Many authors have considered that the genus *Nucula* is the most primitive living bivalve and have modelled the anatomy of early Palaeozoic palaeotaxodont bivalves on it. More recently, however, has come the realization that some of the features of *Nucula*, such as the enlarged foot and the large food-gathering palps, are specializations that need not have been present in the primitive bivalve (Morton 1995; Waller 1998). Allen (1978), in a study of deep-sea protobranchs, argued that both protobranchs and autolamellibranchs could have evolved from an ancestor like *Nucinella* (see below). He later (Allen 1985) concluded that nucinellids were more primitive than nuculoids.

It is at least possible that some of the Cambrian bivalves may be nuculoids. The strongest candidate here is *Pojetaia*, which has palaeotaxodont dentition and a shell structure having elements in common with a Devonian nuculoid (Runnegar & Pojeta 1992).

Most of the early Ordovician forms can be included in the family Praenuculidae, but many of these are so poorly known (or ill-preserved) that they tell us little. The three species figured by Harrington (1938), from the Tremadoc of Argentina, as species of *Ctenodonta* are probably praenuculids, as is at least one specimen of the types of the species figured by him as *Cosmogoniophorina tenuicostata*, as noted by Pojeta & Gilbert-Tomlinson (1977, p. 11). The only other possible Tremadoc praenuculid was a form figured by Pojeta & Gilbert-Tomlinson (1977, pl. 3, fig. 8) as *Deceptrix*? sp. A. Although almost certainly not a *Deceptrix* [see discussion by Cope (1997*b*, pp. 736–738)] this specimen, that has no preserved dentition, is of praenuculid shape. Cope (1996) figured two genera of praenuculids from the early Arenig of South Wales (*Paulinea* and *Pensarnia*) in which details of musculature and dentition were well preserved and these remain probably the best known early Ordovician nuculoids. It appears likely that the earliest praenuculids gave rise to the asymmetrically hinged cardiolariids; that group may well show the connection between the protobranch and autolamellibranch bivalves (Cope 1995, 1997*b*). In the mid Ordovician, praenuculids became more diverse and include forms like *Similodonta* and *Arcodonta* (see Cope 1999); the family continued on into the Devonian.

During the early Ordovician nuculoids diversified. The ctenodontids, which may have been derived from a praenuculid ancestor (see below), became anteriorly enlarged and their umbones migrated rearwards. The nuculanids are represented, probably from the mid Ordovician onwards, by the family Malletiidae, whose shell was posteriorly elongate.

The major innovation in the nuculoid stocks appeared in the early Silurian when the nuculids developed an internal ligament housed in a central resilifer on the hinge plate. Some of these forms developed larger teeth at each end of the hinge plate, suggesting that they had already evolved the larger foot that characterizes many modern nuculids.

Solemyoids

Solemyoids are a long-ranging conservative group of protobranch bivalves that appear to have changed little since the early Ordovician; they are always dimyarian, always edentulous and they often have a periostracum which is unusually thick. Harper (1997, p. 76) recorded periostracal thicknesses of 10–> 100 μm, whilst Beedham & Owen (1965) recorded up to 140 μm. The periostracum frequently projects beyond the calcified portion of the valves; the valves gape anteriorly and posteriorly (Pojeta 1988; Cope 1997*b*). The valves frequently have parallel dorsal and ventral margins and there are well-developed radial mantle muscles. In addition, Taylor *et al.* (1969) reported that the calcareous shell microstructure of *Solemya* was unique. It consists of an outer prismatic layer with a unique prism form and an inner layer (that is irregular in thickness and distribution) of finely laminated homogenous aragonite. Since then, Carter (1990) has examined further solemyids and found, for instance, that the Upper Carboniferous species *Acharax (Nacrosolemya) trapezoides* was partially nacreous.

The reduced gut of solemyoids was a cause of much speculation on their feeding habits that was resolved when Cavanaugh (1983) was able to show that they lived symbiotically with sulphur-oxidizing chemoautotrophic bacteria. It is now known that both solemyoids and nucinelloids are chemoautotrophic, as noted by Waller (1998); he therefore suggested that chemoautotrophic symbiotic bacteria were present 'at the base of the Solemyoidea' (in which he included the nucinelloids) and suggested that this was in the earlier part of the Ordovician. Cope (1996) described the solemyoid *Ovatoconcha* that already had anterior and posterior shell gapes, small posteriorly placed umbones, parallel ventral and dorsal shell margins, strongly anisomyarian musculature (with the anterior adductor the larger) and radial mantle muscles. In addition, Cope (1996) presented evidence to show that the shell of *Ovatoconcha* could have had a thick periostracum. The recent discovery of a specimen of this genus showing what appears to be a periostracal fringe extending beyond the calcified margins of the shell (but at an angle to them) further confirms the solemyoid nature of *Ovatoconcha*, and that early Arenig solemyoids had many of the characters of Recent forms. Figure 1d shows this new specimen (NMW 78. 17G. 1270b) [compare too with the *Solemya parkinsoni* figured by Pojeta (1988, pl. 1, fig. 9), in which the angle between the periostracal fringe and the calcified portion of the shell is more acute].

The origin of the solemyoids must thus go back further than the early Arenig. In many respects, the fact that *Ovatoconcha* is so like Recent solemyids reinforces Waller's (1998) opinion on the early origin of the chemoautotrophism, suggesting that the divergence into the praenuculoid type of nuculoid and the chemoautotrophic group of bivalves must have occurred very early in bivalve history.

The Solemyidae are thus a group that is edentulous through its known stratigraphical range and the present author has no problem in including the edentulous mid–late Ordovician genus *Psiloconcha* in the Solemyoidea, as was suggested by Pojeta (1988).

Nucinelloids

One bivalve group clearly related zoologically to the solemyids is the Nucinelloidea. The shell morphology of nucinelloids contrasts markedly with that of the solemyoids. Nucinelloids appear: always to be small (1–5 mm in length); always to be rare organisms (either as fossils or as extant bivalves); to be dentate; strongly anisomyarian or monomyarian; and their shell has no gapes. Allen & Sanders (1969) described a Recent species of the Jurassic–Recent genus *Nucinella*, suggesting that it was a living actinodont. It is now clear that any such similarity between that dentition and that of the Lower Palaeozoic heteroconch actinodonts is merely convergence, and they are in no way related [see also discussion by Cope (1996, pp. 990–991)].

Nucinellid shell microstructure was investigated by Taylor *et al.* (1973, p. 287), who reported that *Nucinella* had a homogenous shell similar to that of *Nuculana* but unlike that of the solemyoids or the nuculoids. Carter (1990, pp. 175–176), however, showed that the shell microstructure of nucinelloids is not always entirely homogenous; for the Cretaceous species *N. sohli* he recorded an irregular simple prismatic to fibrous prismatic outer layer, and a crossed acicular to homogenous middle shell layer, whilst Recent *N. walvis* has an irregular simple prismatic to homogemous outer shell layer, a fine complex crossed-lamellar to homogenous middle layer and an homogenous inner layer.

Nucinella has a range of Jurassic–Recent and can be related to the Permian family Manzanellidae, as was suggested by Keen & Newell (*in* Cox *et al.* 1969) who placed the two groups in the Limopsoidea; they noted that there was uncertainty with the orientation of these forms and suggested the possibility that they might be prosodetic. Following anatomical work by (amongst others) Allen & Sanders (1969), it was found that *Nucinella* was of solemyoid affinities. Pojeta (1988) suggested that the superfamily Nucinelloidea should contain the two families Nucinellidae and Manzanellidae; this arrangement was followed by Cope (1997*b*). Pojeta (1988, fig. 3) also suggested that the Upper Palaeozoic genus *Clinopistha* and the middle Ordovician genus *Dystactella* were directly ancestral to the Nucinelloidea, and Waller (1990, p. 61) gave assent to this interpretation. However, there must be major reservations about including in a lineage of small dentate bivalves that is unknown before the Permian, a group of bivalves that are, as far as is known, totally edentulous (and much larger); the only real point of similarity between *Clinopistha* and *Nucinella* is that they have similar shell shapes (Pojeta 1988, figs 2 and 4). The present author would prefer to regard both *Dystactella* and

Clinopistha as early offshoots of the solemyoid stock.

The present author believes that logic dictates that a dentate bivalve, lacking shell gapes and with the calcified margin of the valve coincident with the periostracal margin, is likely to be derived from an ancestor sharing these characters. Therefore, it is believed that it is likely that the nucinelloids were derived from an early dentate stock and are consequently likely to be more primitive than the solemyoids. Carter (1990, p. 178) concluded that the common ancestor of nucinellids and solemyids had shells that were at least partially nacreous. Allen's (1985) zoological investigations recognized *Nucinella* as a solemyoid more primitive than the nuculoids and suggested its simple peg-like teeth were more primitive than nuculoid teeth. These dentate forms constitute the superfamily Nucinelloidea, whilst the edentulous anteriorly elongate forms (presumably) developed from the same ancestor, often with the periostracum extending beyond the calcified margins of the valves and with well-developed radial mantle muscles, and anterior and posterior shell gapes. These constitute the superfamily Solemyoidea; the earliest member of this family known so far is the early Arenig genus *Ovatoconcha* Cope, 1996, also including the later Ordovician genus *Psiloconcha* as well as more recent genera.

Origin of the Solemyoida

Pojeta (1988) derived the solemyoids from the palaeotaxodont *Ctenodonta*. The present author does not agree that this genus is related in any way to the origin of the solemyoids as was suggested by Pojeta (1988) and followed by Waller (1990, 1998) – *Ctenodonta* is a perfectly normal 'mainstream' palaeotaxodont grade form that is anteriorly enlarged (like several other nuculoid taxa). The supposed radial mantle muscles shown by Pojeta (1988, e.g. pl. 13, fig. 10) are interpreted here as no more than a discontinuous pallial line, as noted by Pojeta when he previously figured the same specimen (1971, pl. 4, fig. 10).

The lectotype of the type species of *Ctenodonta*, *C. nasuta* (Hall), 1847, was designated by McAlester (1968, p. 22, and refigured 1968, pl. 1, fig. 1). The lectotype shows no feature of the dentition, but Pojeta (1971, pl. 4) refigured it, together with the paratype (*recte* paralectotype) and specimens of the same species showing a well-developed palaeotaxodont dentition. The name *Ctenodonta* has been much misused throughout the history of Lower Palaeozoic bivalve palaeontology and, as pointed out by Pojeta (1971, p. 15), '*Ctenodonta* ... has been used so widely for so many different nuculoids that the only information it now conveys is that a nuculoid is being described'. This is certainly the case for the nuculoids described by Harrington (1938) from the Tremadoc of Argentina as three new species of *Ctenodonta* (*C. famatinensis, C. laevigata* and *C. miniscularia*); none of these remotely approaches the ctenodontid form and they all resemble praenuculid shape. On the other hand, Pojeta & Gilbert-Tomlinson (1977, p. 9) placed *Palaeoneilo iruyensis* Harrington, 1938 tentatively in *Ctenodonta*. This species is represented by three specimens of which Harrington figured one (presumably the best preserved). Even this specimen (Harrington 1938, pl. 3, fig. 6) is very indifferently preserved, with no dentition and an outline that has been restored on the photograph. Its umbo is one-third of the length from the anterior margin (Harrington 1938, p. 132) and, although valve height is greater anteriorly than posteriorly, the present author would be loathe to give this specimen any generic identity. This appears to be the sole basis for Pojeta's claim (1988, p. 211) that *Ctenodonta* occurs in the early Ordovician and the reported mid Ordovician North American occurrences of *Ctenodonta* would now all be regarded as from the late Ordovician. The present author disagrees with Pojeta's (1988) placing of the late Ordovician species *Ctenodonta logani* Salter, 1859 within that genus, believing that that species belongs far more readily to *Tancrediopsis*, although the hinge-angle is a little more obtuse than in the type species *T. contracta* (Salter). Middle Ordovician *Ctenodonta* species are typified by those figured from Australia by Pojeta & Gilbert-Tomlinson (1977); these (*C. macalesteri* and *C. youngi*), together with *C. jonesii*, previously described by Johnston (1888), all have their umbones well into the anterior end of the shell. This is important in view of Pojeta's (1988) claimed derivation of the solemyids from the ctenodontids; Pojeta's conclusions were accepted by Waller (1990, 1998). However, analysis of the record of ctenodontids shows that the earlier species have their umbones well to the anterior of the shell, which is contrary to what would be expected if ctenodontids were closely related to solemyoids (which have their umbones at the posterior end of the shell).

Waller (1990, 1998) believed that the principal apomorphy linking ctenodontids and solemyoids was that they both had ligamental nymphs. However, the only ctenodontids to show such structures are a few of the late Ordovician forms figured by Pojeta (1988). No ligamental nymphs are known from the Gondwanan mid-Ordovician material (although it includes well-preserved silicified replicas). According to Morris (1979, p. 392), his advanced parinvincular ligament type (in which the ligament is inserted in nymphae) is present, amongst others, in the Ctenodontacea [*sic*] and, in a modified version, in the Solemyidae. Thus, in his opinion there are differences between the ligamental insertions in the ctenodontids and solemyoids. Because Waller (1990, 1998) believed these ligamental nymphs linked ctenodontids and solemyoids he was forced to conclude (1990, p. 61) that 'the anatomical features characteristic of the ... Nucinellidae + Ctenodontidae + Solemyidae originated *before* [his italics] the nymphae that are present in the Ctenodontidae and Solemyidae'. Now it can be shown that the solemyoids almost certainly predate the ctenodontids and that pre-late Ordovician ctenodontids apparently did not possess ligamental nymphs. It may be that nymphs were perforce developed in the ctenodontids as the rearward migration of the umbones shortened the effective length of the ligament, thus making stronger insertion essential. Thus, it follows that the nymphae are not apomorphic to the two groups but evolved independently.

The origin of the ctenodontids does not appear to have been previously considered, but it seems probable that they were derived from early Ordovician praenuculids; a relatively equilateral praenuculid genus perhaps similar to *Paulinea* (Cope 1996) could form a suitable ancestor. Ctenodontids could have evolved from such an ancestor by anterior enlargement and gradual rearward movement of the umbones. However, the origin of the solemyids was probably an earlier event of which, at present, there is no knowledge.

The very clear differences between the nucinelloids and the solemyoids from their first appearances in the Permian and early Ordovician, respectively, suggest a long separate history. Pending further information they are both retained in the order Solemyoida Dall, 1889. This comprises two superfamilies: the Solemyoidea Adams & Adams, 1857, which contains the single family Solemyidae Adams & Adams 1857; and the Nucinelloidea Vokes, 1956, in which two families are recognized, the Nucinellidae Vokes 1956, and the Manzanellidae Chronic, 1952.

Subclass Autolamellibranchiata Grobben, 1894

Grobben (1894, p. 73) introduced this subclass to include all bivalves with gills other than of protobranch grade. The subclass has been largely neglected by palaeontologists, but was reintroduced by Nevesskaya *et al.* (1971) and followed by other workers, including Starobogatov (1992) and Waller (1978, 1998). However, all of these authors have altered the original term Autolamellibranchiata of Grobben (1894) to Autobranchia. Although the International Code of Zoological Nomenclature (ICZN) rules, strictly speaking do not apply at subclass level, the present author feels that as all of these authors have credited Grobben (1894) with its authorship, it should be used in the original spelling and not the shortened term now used merely as a convenience.

The subclass is characterized by filibranch, eulamellibranch or septibranch gills; later gill types were independently derived from filibranch ancestors in separate bivalve groups as first demonstrated by Ridewood (1903). Cope (1995) suggested that the evolution of the filibranch gill was a single morphological change that occurred in one group of palaeotaxodont bivalves, later named the family Cardiolariidae (Cope 1997*b*). The characters of this group are a fairly straight line of posterior teeth along which the hinge line lay (Cope 1995, fig. 30.1.b) and a group of larger anterior teeth that lie beneath the hinge, constituting a hinge designed to allow full valve articulation with wider opening of the valves, such as would have been needed to void pseudofaeces concomitant with the evolution of the filibranch gill. Cope (1995) correlated this with the sudden increase in size, abundance and morphological diversity of bivalves in the earliest Ordovician. As more bivalves with the cardiolariid hinge are recognized as such, it has become clear that their morphological diversity is greater than can be comfortably included in a single family. Thus, proposed here is the superfamily Cardiolarioidea superfam. nov., to include the family Cardiolariidae Cope, 1997*b*, and the family Eritropidae fam. nov., to include such genera as *Eritropis* Pojeta & Gilbert-Tomlinson, 1977, which have a derived cardiolarioid dentition in which the posterior teeth have become elongate at right angles to the hinge and partially fused [see Cope (1999, pl. 2, figs 11 and 12; Fig. 1f)]. As noted by Cope (1999, p. 480), there is evidence that the juvenile dentition of *Eritropis* had a dentition which lacked a subumbonal lacuna. The dental elongation of the posterior teeth, that characterizes *Eritropis*, can also be seen in *Inaequidens* Pojeta & Gilbert-Tomlinson, 1977, which Cope (1997*b*) had already transferred to the Cariolariidae.

The primitive cardiolarioid can now be envisaged as having a row of subequal posterior teeth and a group of larger teeth lying below this hinge line. Such fossils may be expected from the late Cambrian or earliest Ordovician. It is noteworthy that amongst the genera of praenuculoids known thus far from the early Ordovician, some show a degree of this type of asymmetry, including *Pensarnia* Cope, 1996, but the posterior teeth tend to be on an arched plate rather than the straight posterior hinge plate of the cardiolarioids. The cardiolarioid asymmetry provides a basic hinge plan from which it is possible to readily derive the various autolamellibranchiate bivalve groups. Waller (1998, fig. 5) depicted a series of dentitions of early–mid-Ordovician bivalves showing how cycloconchid types could be derived from a 'malletiid' or cardiolariid type (Waller 1998, p. 33). The Cardiolarioidea are not assigned to an order for the present but form the rootstock of the superorder Heteroconchia Hertwig, 1895.

Whether the cardiolarioids will be confirmed in this fundamental position in autolamellibranch evolution depends upon the discovery of late Cambrian to early Ordovician representatives of that group. Their morphology strongly suggests such a position but the possibility that they represent a nuculoid evolutionary line convergent on the heteroconchs cannot be excluded at present.

One feature that occurs in many groups of the Autolamellibranchia is their retention of the byssus. This post-larval structure made possible the development of the epifaunal mode of life that is characteristic of many autolamellibranchiate bivalves.

Order Trigonioida Dall, 1889

Many authors treat the trigonioids as being derived from a cycloconchid ancestor. One of the criteria usually listed as occurring commonly in the trigonioids, and one that for many workers was the cause of the linkage of the Lower Palaeozoic lyrodesmatids to the Mesozoic trigonioids, was the presence in both of denticulate teeth. Johnston (*in* Johnston & Goodbody 1988, p. 340) showed that many families of Lower Palaeozoic bivalves have crenulations of the teeth and that in some families only a single species is crenulate. In fact, the present author believes that the position is yet more variable than that, e.g. in his description of *Glyptarca serrata* Cope (1996, p. 994) noted that out of a total of 484 specimens two had crenulations on a single tooth (a different one in each specimen). Thus, as pointed out by Johnston (*in* Johnston & Goodbody 1988), there seems little reason to accept crenulation of the teeth as a reliable character in familial diagnoses. Johnston (*in* Johnston & Goodbody 1988, p. 341), however, then goes on to point out that *Lyrodesma majus* (Ulrich) appears to possess ligamental nymphs and later (Johnston 1996) confirmed that the genus did indeed possess a short parinvincular ligament with well-developed nymphae resembling the schizodid ligament. He thus transferred the family Lyrodesmatidae to the order Trigonioida, confirming the classification of Newell & LaRocque (*in* Cox *et al.* 1969).

The Trigonioida are represented today by the single genus *Neotrigonia*, but their origins can be plausibly traced back to the early–middle Ordovician genus *Noradonta*, first described from the middle Ordovician of Australia by Pojeta & Gilbert-Tomlinson (1977) but subsequently found in the early Arenig of the Montagne Noire (Babin 1982*a*). The dentition of *Noradonta redoniaeformis* (Thoral) is very like that of a cardiolarioid in that there is a straightish posterior dentition lying along the hinge line, together with a group of larger subumbonal teeth lying below the hinge (Fig. 1e). The dentition of *Noradonta redoniaeformis* differs from that of a cardiolarioid such as *Deceptrix* in that the cardinal teeth are subparallel and the posterior teeth have become fused to form a single denticulate posterior tooth. The hinge of *Noradonta* also differs fundamentally from the hinges of cycloconchid genera such as *Copidens* Pojeta & Gilbert-Tomlinson or *Carminodonta* Cope in that the opisthodetic ligament in *Noradonta* is parallel to the posterior tooth (as it is also in *Cardiolaria*) whilst in the cycloconchids, with a sheaf of teeth, the teeth are terminated by the ligament and the teeth had to be resorbed to allow ligamental growth, as noted by Johnston & Zhang (1998, p. 372). From *Noradonta redoniaeformis* it is readily possible to derive the genus *Tromelinodonta*, from the upper Arenig of the Massif Armoricain of France (Babin 1982*b*), and in turn from *Tromelinodonta* to derive the genus *Lyrodesma*, which is widely reported from the Ordovician of Europe and North America, and is also known from the Lower Silurian of North America (Harrison & Harrison 1975). The earliest *Lyrodesma* recorded so far is from the early Llanvirn of Wales (Cope 1999) and the latest is from the Wenlock Series of Wales (Ratter, pers. comm.).

One feature that may be noted in this evolutionary sequence from the cardiolarioid ancestor through to the lyrodesmatids is the gradual diminution and loss of the strong asymmetry of the hinge that characterized early members of the lineage.

The relationship between the lyrodesmatids and the later trigonioids has been the subject of considerable debate, but the basic problem is a distinct stratigraphical gap between the latest lyrodesmatids and the earliest trigoniids. This gap is filled by the schizodids, which have a very similar ligamental insertion to the lyrodesmatids, but they lack the denticulate teeth of the other two groups. In view of the arguments of Johnston (*in* Johnston & Goodbody 1988, p. 340) this does not appear to be a significant phylogenetic obstacle. Newell & Boyd (1975) noted that while late Palaeozoic trigonioids lack the denticulate teeth of lyrodesmatids, they shared the same musculature with both them and the Mesozoic trigonioids. As suggested by Harrison (*in* Pojeta *et al.* 1986), the schizodid dental arrangement of the family Schizodidae probably arose through paedomorphic retention of the juvenile lyrodesmatid dentition. Pojeta (1978) drew attention to the similarities of the ventrally flared denticulate teeth, the dorsal placement of all the musculature and the positions of the pedal retractors at the extremities of the hinge plate in both lyrodesmatids and trigonioids.

The *Treatise* (Cox *et al.* 1969) included the Trigonioida in the subclass Palaeoheterodonta Newell, 1965, whilst Pojeta (1971) included the trigonioids within the order Actinodontoida of that subclass. However, its ligament, borne within strong nymphae, is unlike that of the other actinodontoids, as noted by Johnston (1996) and Johnston & Zhang (1998); the latter authors recommended the retention of the order Trigonioida. Here, the order Trigonioida is considered as being an entirely separate division of the subclass Autolamellibranchiata from the heteroconchs, with which they have frequently been placed hitherto (e.g. Cox *et al.* 1969; Pojeta 1971, 1978; Babin 1982*b*; Cope 1997*b*, 1999). The group is characterized throughout by a prismato-nacreous shell; it shares its columnar aragonite prismatic shell microstructure with that of the Unionioida, but this was probably derived from the simpler prismato-nacreous shell of its ancestors.

Anomalodesmata

The superorder Anomalodesmata constitute a group of essentially edentulous burrowing bivalves that are characterized by a prismato-nacreous shell similar to that of nuculoids (Taylor *et al.* 1973). Most modern forms have eulamellibranch-grade gills but some are septibranch. Another characteristic feature of anomalodesmatan shells is that they frequently display a fine-scale tuberculation. The earliest anomalodesmatan hitherto recorded is *Arenigomya* from the early Arenig of South Wales (Cope 1996), for which, like later Ordovician anomalodesmatans, there is as yet no evidence of a strongly parinvincular ligament, although such may have been present. Indeed, as these bivalves largely lacked dentition, the presence of a well-located ligament could well have been necessary for accurate valve articulation. *Arenigomya* is one of several anomalodesmatans that possess internal calcareous structures that may have been concerned with valve articulation (Cope 1996, fig. 7).

There has hitherto been much confusion over the affinities of the family Modiomorphoidea as defined in the *Treatise* (Cox *et al.* 1969). Such confusion is typified by Pojeta (1971, p. 20), 'Pelecypods which range in age from early Ordovician to late Permian have been placed in the Modiomorphidae, but the concept summed up by the name is vague and uncertain, and probably more than one family-level taxon is included in the Modiomorphidae as presently recognized'. It was many years later before this taxonomic problem was resolved. Fang & Morris (1997) demonstrated that there was a real difference in the hinge structures of Ordovician *Modiolopsis* and Devonian *Modiomorpha*. In modiomorphoids the hinge itself lies along the line of an elongate opisthidetic ligament housed within usually well-developed nymphae; the modiolopsoids lack the ligamental nymphs that the former possess. Fang & Morris (1997) also showed that the family Permophoridae was a junior subjective synonym of Modiomorphidae and that the *Treatise* (Cox *et al.* 1969) superfamily Modiomorphoidea should be replaced by the Modiolopsoidea Fischer, 1887 (*nom. trans.* Fang & Morris 1997) in which they included the families Modiolopsidae, Colpomyidae and the new family Modiolodontidae. They also suggested abandonment of the order Modiomorphoida Newell, 1969 (*in* Cox *et al.* 1969).

Carter & Aller (1975) showed that the periostracum of certain bivalves was embedded with calcified spicules. They recognized two types of spicules; the one in which the periostracal calcification took the form of spicules or granules cemented within the outer prismatic shell layer they reported in the Anomalodesmata, Permophoridae [= Modiomorphidae] and most of the Carboniferous–Recent order Myoida. The latter order has been included in the Anomalodesmata by some authors (e.g. Purchon 1987; Waller 1990).

Fang & Morris (1997) concluded that the Modiomorphidae should be assigned to the Anomalodesmata because of the characters that they shared. Cope (1997*b*) suggested that the similarities between mytilids and modiomorphoids were most likely due to convergence and cited evidence produced by Carter (1990) showing a lack of intermediate forms, despite an excellent fossil record of Late Palaeozoic modiomorphoids and mytilids. Johnston (1993) followed Waller's (1990) suggestion that the modiomorphoids belonged to the Anomalodesmata 'without prejudice'.

The origins of the anomalodesmatans are unknown. They differ from most possible ancestors in lacking teeth. However, the anomalodesmatans have a well-located ligament which clearly acted as an efficient substitute for the posterior teeth. This change may have been effected in one or more stages – either by the simple substitution of the ligament for the teeth, or by an initial fusion or reduction of the posterior teeth, followed by their loss. As the ligament is housed within well-developed nymphae, this clearly provided sufficient protection against any shearing motion along the hinge.

Modiolopsoids are essentially edentulous (Modiolopsidae), have a variable number of anterior teeth (Modiolodontidae) situated beneath the hinge line, or may have a blunt subumbonal articulating device (Colpomyidae). The former two are clearly not related to the anomalodesmatans and although the latter were included in the Isofilibranchia by Pojeta & Runnegar (1985) they noted that some post-Ordovician pholadomyoids had amorphous bulbous teeth comparable with those of colpomyids. Fang & Morris's (1997) placement is followed here, pending further information. It had previously been suspected that modiolopsoids were related to the heteroconchs, being derived from them by reduction in dentition. However, Carter & Seed (1998) demonstrated that some modiolopsids had multiple insertions of non-parinvincular ligaments, suggesting that they had affinities with the Pteriomorphia.

Superorder Heteroconchia

This term is used here to link the palaeoheterodonts (from which the trigonioids are excluded) and the heterodonts following Pojeta (1971, 1978). Waller (1998) used the term in the wider sense, as employed by Cox (1960), in which the paleo-heterodonts, heterodonts and anomalodesmatans are included. Waller (1998) claimed that a 3a/2/3b dental arrangement was a core homology linking these groups. There has been a great deal of debate about the various bivalve groups possessing the heterodont type of hinges that display combinations of cardinal and lateral teeth. The *Treatise* (Cox *et al*. 1969) divided these forms into two subclasses: the Palaeoheterodonta Newell, 1965 and the Heterodonta Neumayr, 1884. This scheme has subsequently been adopted by many workers (e.g. Carter 1971; Bailey 1983; Cope 1996, 1997*a*, *b*). Others have suggested that the two groups are best combined in the superorder Heteroconchia Hertwig, 1895 as they share similar dentitions and have other features in common (e.g. Pojeta 1971, 1978; Babin & Gutiérrez-Marco 1991)

Hitherto, the present author has subscribed to the view that it was possible to separate the two groups and it was suggested (Cope 1997*b*) that shell microstructural differences could distinguish between prismato-nacreous Palaeoheterodonta and a crossed-lamellar/complex crossed-lamellar shell of the Heterodonta. This view was based very much from an Ordovician perspective and pre-supposes that the shell microstructures of the many Lower Palaeozoic forms that are hitherto unknown would fit into this pattern. These views have changed largely as a result of examining large collections of Silurian bivalves made by Dr V. A. Ratter in Cardiff. His collections include topotype material of *Actinodonta cuneata* from the Llandovery Series of Pembrokeshire and a new genus and species that occurs in slightly younger rocks. *Actinodonta* has what can be regarded as typical palaeoheterodont ventrally divergent multidentate dentition, similar to that shown by Ordovician genera such as *Carminodonta* Cope, 1996 or *Copidens* Pojeta & Gilbert-Tomlinson, 1977, whilst the new genus has a dentition close to that of a crassatelloid heterodont. Other details of the shell differ in no marked way and the similarity of the two has allowed them to be incorporated in the same family. Thus, separation at a higher taxonomic level is untenable and so the palaeoheterodonts and heterodonts are united in the superorder Heteroconchia.

Interestingly, by separating out from this group the modiomorphoids [now included in the Anomalodesmata following Fang & Morris (1997)] and the trigonioids, the two principal groups of bivalves characterized throughout by prismato-nacreous shells have been removed. However, it is probable that some of the early heteroconchs retained this basic prismato-nacreous shell micro-structure. However, for later heteroconchs the outer shell layer is normally of prismatic calcite, the middle layer is of a crossed-lamellar structure and there is an inner complex crossed-lamellar layer. It may be that since bivalve groups quite clearly descended from the glyptarcoids are all charac-terized by a crossed-lamellar shell microstructure, then the superfamily Glyptarcoidea also possessed that type of shell microstructure.

The Glyptarcoidea have a dentition that radiates dorsally from a point within the centre of the valves (see Babin & Gutiérrez-Marco 1991, fig. 7; Cope 1996, fig. 5) rather than the subumbonal ventrally divergent dentition of the majority of heteroconchs. The anterior dentition of *Glyptarca*, originally figured from the Arenig of Ramsey Island by Hicks (1873) and redescribed by Carter (1971), with its strongly curved anterior teeth, led many authors (e.g. Pojeta 1971; Babin 1966; Morris 1978) to suggest that *Glyptarca* was the ideal ancestor for the cyrtodontids. However, when these authors made this suggestion, the dentition of *Glyptarca* was very imperfectly known as the genus was based on inadequate type material. In order to stabilize the nomenclature, Cope (1996) figured material with well-preserved dentition that showed that *Glyptarca* lacked the subumbonal dental lacuna that is characteristic of the cyrtodonts and indeed had a complex subumbonal dentition. *Glyptarca* could be quite readily derived from a cardiolariid ancestor, as demonstrated by Cope (1995), as could the actinodonts (Waller 1998, fig. 5 and text pp. 33–34).

It was the Mesozoic and Cenozoic evolution of the heteroconchs that was responsible for the pre-eminent position this superorder has in Recent bivalve faunas. It therefore lies outside of the scope of this review of early bivalve phylogeny.

Pteriomorphia

Sánchez & Babin (1993) erected the new bivalve genus *Catamarcaia* for a bivalve from the mid-Arenig of Argentina, which remains the earliest bivalve yet known with a duplivincular ligament. As shown by Cope (1997*a*) and Ratter & Cope (1998), the dentition of *Catamarcaia* could be derived quite readily from an ancestor like the heteroconch *Glyptarca* (see Ratter & Cope 1998, fig. 7). With the latter genus it shares strong anterior teeth and a complex subumbonal dentition, and differs in the posterior dentition and the lack of a duplivincular ligament. The superfamily Glyptarcoidea, however, is seen as the most likely origin for this stock. Thus, it seems that the Arcoidea could have been derived directly from early glyptarcoid heteroconchs. Cope (1995, 1997*a, b*) separated the order Arcoida (from which he excluded the cyrtodontids) from the remainder of the pteriomorphs on the grounds that they were the only group with a chevron-shaped duplivincular ligament and that they lacked the calcitic outer layer that characterizes most pteriomorphian shells. He suggested that Korobkov's (1954) term Neotaxodonta was the most appropriate available name for this group. It is now clear that this was an oversimplistic view and that there are many synapomorphies uniting the Arcoida with the other pteriomorphs (Waller 1998), so that they may be recombined in the superorder Pteriomorphia.

Following the discovery of *Catamarcaia*, other early arcoids have been discovered, including *Alytodonta* (Cope 1997*b*), *Trecanolia* and *Uskardita* (Ratter & Cope 1998), which link the group to the mid-Silurian genus *Freja*. Ratter & Cope (1998) assigned the last four genera to the new family Frejidae, leaving *Catamarcaia* without familial assignment. It is clear that *Catamarcaia* has dentition that differs from that of the Frejidae and also from that of the Parallelodontidae, as noted by Sánchez (1995), but the ligament of the Parallelodontidae is similar to that of the Frejidae, leading Ratter & Cope (1998, fig. 7) to suggest that both families were derived from a common post-*Catamarcaia* ancestor. It is proposed here that the monotypic family Catamarcaidae accommodates the genus *Catamarcaia*. The characteristics of this family include an asymmetric dentition, which includes lamellar anterior and posterior lateral teeth combined with a series of largely chevron-shaped cardinal teeth, together with a half chevron-shaped ligament.

There are still major gaps in our knowledge of the early arcoids, particularly since there are no records between *Catamarcaia* in the Arenig and *Alytodonta* in the early Llandovery. Ratter & Cope (1998) noted that Lower Palaeozoic arcoids are particularly characteristic of very close-inshore facies and this may explain their extreme rarity.

The earliest pteriomorphians are recorded from the late Tremadoc of Australia (Pojeta & Gilbert-Tomlinson 1977). Both of the recorded genera are cyrtodontids, but since they are preserved as sandstone moulds there is no knowledge of their ligament. However, because they are like later cyrtodontids that have multiple ligamental insertions, it is probably safe to assume that they also had this ligament type. As they predate *Catamarcaia* it is possible that they and *Catamarcaia* evolved from a common ancestor, such as that proposed by Cope (1997*a*).

In turn, from the cyrtodonts, ambonychiids could be readily derived by reduction of the anterior of the shell; the earliest ambonychiid known hitherto is the *Cleionychia* recorded from the middle Arenig of South Wales by Cope (1996). In addition, the derivation of pterioids implies the development of asymmetry between the valves and the earliest pterioid recorded hitherto was from the early Arenig of South Wales (Cope 1996). Thus, all these early pteriomorphian groups had evolved within the early Ordovician.

The presence of a multiple non-parinvincular ligament in the modiolopsoids, noted by Carter & Seed (1998), suggests that they belong within the Pteriomorphia. The earliest form that can be included in this superfamily is a colpomyid from the Tremadoc of Australia (Pojeta & Gilbert-Tomlinson 1977), but, as pointed out above, the colpomyids may be anomalodesmatans. Insofar as is known, no multiple ligamental insertions have been seen in a colpomyid. The earliest modiolopsids and modiolodontids appear to be from the early Arenig of South Wales (Cope 1996). The Tremadoc *Cosmogoniophorina* described by Harrington (1938) is probably a nuculoid (Pojeta & Gilbert-Tomlinson 1977). Modiolopsoids could have been derived from cyrtodonts, or they could have shared a common ancestor with reduced pteriomorphian dentition.

The other principal pteriomorphian group to evolve during the Ordovician was the limids. Tunnicliff (1987) described the new genus *Myodakyrotus* from the Upper Ordovician of North Wales. This group is dimyarian, as opposed to the normal monomyarian musculature in most limoids, and in many respects is intermediate between the cyrtodontids and the limids.

Conclusions

Bivalves first diversified in the Cambrian, perhaps from initial surface dwellers, to produce a limited variety of forms that included shallow infaunal burrowers that, like the majority of bivalves, were able to capitalize on the substrates of the Lower Palaeozoic seas. Probably still within the Cambrian, some of these forms migrated into partially anaerobic substrates where they lived symbiotically with chemoautotrophic sulphur-oxidizing bacteria, producing a rapid diversification amongst the protobranch bivalves. Towards the end of the Cambrian, or within the very earliest Ordovician, one group of palaeotaxodont bivalves developed the filibranch gill; this produced a major evolutionary burst in the autolamellibranchs that allowed bivalves to take advantage of the increased nutrient supplies in the Ordovician seas (Cope & Babin 1999). From the rootstock of the newly evolved heteroconchs a plethora of new bivalve groups diversified rapidly and colonized semi-infaunal and epifaunal habitats. By the end of the early Ordovician all main groups of bivalves had evolved and links between many of these groups can be established with varying degrees of confidence as the fossil record of the Cambrian and Ordovician bivalves becomes better known.

I am grateful for the assistance of Mrs D. G. Evans, of the Geology Department, National Museum of Wales, in the drafting of Fig. 2.

References

ADAMKEWICZ, S. L., HARASEWYCH, M. G., BLAKE, J., SAUDEK, D. & BULT, C. J. 1987. A molecular phylogeny of the bivalve mollusks. *Molecular Biology and Evolution*, **14**, 619–629.

ADAMS, H. & ADAMS, A. 1854–1858. *The Genera of Recent Mollusca Arranged According to their Organisation,* Vol. 2. J. van Voorst, London.

ALLEN, J. A. 1978. Evolution of the deep-sea protobranch bivalves. *Philosophical Transactions of the Royal Society of London, Series B,* **284**, 387–401.

—— 1985. The Recent Bivalvia: their form and evolution. *In:* TRUEMAN, E. R. & CLARKE, M. R. (eds) *The Mollusca. Vol. 10, Evolution.* Academic Press, New York, 337–403.

—— & SANDERS, H. L. 1969. *Nucinella serrei* Lamy (Bivalvia: Protobranchia), a monomyarian solemyoid and possible living actinodont. *Malacologia*, **7**, 381–396.

BABIN, C. 1966. *Mollusques bivalves et céphalopodes du Paléozoïque armoricain.* Imprimerie Commerciale et Administrative, Brest.

—— 1982*a*. Mollusqes bivalves et rostroconches. *In:* BABIN, C., COURTESSOLE, R., MELOU, M., PILLET, J., VIZCAINO, D. & YOCHELSON, E. L. *Brachiopodes (articulés) et mollusques (bivalves, rostroconches, monoplacophores, gastropodes) de l'Ordovicien inférieur (Trémadocien–Arenigien) de la Montagne Noire (France méridionale).* Mémoire de la Société des Études Scientifiques de l'Aude, Sival, Carcassonne, 37–49, pls 8–12.

—— 1982*b*. *Tromelinodonta* nov. gen., Bivalve lyrodesmatide (Mollusca) de l' Arenigien (Ordovicien inférieur). *Geobios*, **15**, 423–427.

—— & GUTTIÉRREZ-MARCO, J. C. 1991. Middle Ordovician bivalves from Spain and their phyletic and palaeogeographic significance. *Palaeontology*, **34**, 109–147.

BAILEY, J. B. 1983. Middle Devonian Bivalvia from the Solsville Member (Marcellus Formation), central New York State. *Bulletin of the American Museum of Natural History,* **174**, 193–326.

BEEDHAM, G. E. & OWEN, G. 1965. The mantle and shell of *Solemya parkinsoni* (Protobranchia: Bivalvia). *Proceedings of the Zoological Society of London*, **145**, 405–430.

BENTON. M. J. & HARPER, D. A. T. 1997. *Basic Palaeontology.* Addison Wesley Longman, Harlow.

BERG-MADSEN, V. 1987. *Tuarangia* from Bornholm (Denmark) and similarities in Baltoscandian and Australasian late Middle Cambrian faunas. *Alcheringa*, **11**, 245–259.

CARTER, J. G. (ed.) 1990. *Skeletal Biomineralization: Patterns, Processes and Evolutionary Trends.* Vols 1. and 2. Van Nostrand Reinhold, New York.

—— & ALLER, R. C. 1975. Calcification in the bivalve periostracum. *Lethaia*, **8**, 315–320.

—— & SEED, R. 1998. Thermal potentiation and mineralogical evolution in *Mytilus* (Mollusca; Bivalvia). *In:* JOHNSTON, P. A. & HAGGART, J. W. (eds) *Bivalves: An Eon of Evolution — Paleobiological Studies Honoring Norman D. Newell.* University of Calgary Press, Calgary, 87–117.

CARTER, R. M. 1971. Revision of Arenig Bivalvia from Ramsey Island, Pembrokeshire. *Palaeontology*, **14**, 250–261, pls 38–39.

CAVANAUGH, C. M. 1983. Symbiotic chemoautotrophic bacteria in marine invertebrates from sulphide-rich habitat. *Nature*, **302**, 58–61.

CHRONIC, H. 1952. Molluscan faunas from the Permian Kaibab Formation, Walnut Canyon, Arizona. *Bulletin of the Geological Society of America*, **63**, 95–166, pls 1–11.

COPE, J. C. W. 1995. The early evolution of the Bivalvia. *In:* TAYLOR, J. D. (ed.) *Origin and Evolutionary Radiation of the Mollusca.* Oxford University Press, Oxford, 361–370

—— 1996. Early Ordovician (Arenig) bivalves from the Llangynog Inlier, South Wales. *Palaeontology*, **39**, 979–1025.

—— 1997*a*. Affinities of the early Ordovician bivalve *Catamarcaia* Sánchez and Babin, 1993 and its role in bivalve evolution. *Geobios Mémoire Spécial*, **20**, 126–131.

—— 1997*b*. The early phylogeny of the class Bivalvia. *Palaeontology*, **40**, 713–746.

—— 1999. Middle Ordovician bivalves from mid-Wales and the Welsh Borderland. *Palaeontology*, **42**, 467–499 .

—— & BABIN, C. 1999. Diversification of bivalves in the Ordovician. *Geobios*, **32**, 175–185.

COX, L. R. 1959. The geological history of the Protobranchia and the dual origin of the taxodont Lamellibranchia. *Proceedings of the Malacological Society, London,* **33**, 200–209.

—— 1960. Thoughts on the classification of the Bivalvia. *Proceedings of the Malacological Society*, **34**, 60–88.

—— & 24 others 1969–1971. *In:* MOORE, R. C. (ed.) *Treatise on Invertebrate Paleontology. Part N. Mollusca 6, Bivalvia.* Geological Society of America, Boulder, CO, and University of Kansas Press, Lawrence, KS.

DALL, W. H. 1889. On the hinge of pelecypods and its development, with an attempt toward a better subdivision of the group. *American Journal of Science*, **38**, 445–462.

FANG, ZONG-ZIE & MORRIS, N. J. 1997. The genus *Pseudosanguinolites* and some modioliform bivalves (mainly Palaeozoic). *Palaeoworld*, **7**, 50–74.

GEYER, G. & STRENG, M. 1998. Middle Cambrian pelecypods from the Anti-Atlas, Morocco. *Revista Española di Paleontología*, **no. extra Homenajae al Prof. Gonzalo Vidal**, 83–96.

GROBBEN, K. 1894. Zur Kenntniss der Morphologie, der Verwandschaftsverhältnisse und des Systems der Mollusken. *Sitzungsberichte der Kaiserlichen Akademie der Wissenschaften, Mathematisch-Naturwissenschaftlichen Classe*, **103**, 61–86.

HARPER, E. M. 1997. The molluscan periostracum: an important constraint in bivalve evolution. *Palaeontology*, **40**, 71–97.

HARRINGTON, H. J. 1938. Sobre las faunas del Ordoviciano inferior del Norte Argentino. *Revista del Museo de La Plata, Nueva Serie*, **1**, 109–289.

HARRISON, W. B. & HARRISON, L. K. 1975. A Maquoketa-like molluscan community in the Brassfield Formation (Early Silurian) of Adams County, Ohio. *Bulletins of American Paleontology*, **67**, 197–231.

HAVLÍČEK, V. & KŘÍŽ, J. 1978. Middle Cambrian *Lamellodonta simplex* Vogel: 'bivalve' turned brachiopod *Trematobolus simplex* (Vogel). *Journal of Paleontology*, **52**, 972–975.

HERTWIG, C. W. T. R. 1895. *Lehrbuch der Zoologie.* Gustav Fischer, Jena.

HICKS, H. 1873. On the Tremadoc rocks in the neighbourhood of St. David's, South Wales and their fossil contents. *Quarterly Journal of the Geological Society of London*, **29**, 39–52, pls 3–5.

HINZ-SCHALLREUTER, I. 1995. Muscheln (Pelecypoda) aus dem Mittelkambrium von Bornholm. *Geschiebekunde aktuell*, **11**, 71–84.

IREDALE, T. 1939. Great Barrier Reef expedition 1928–1929. *Scientific Reports, British Museum (Natural History)*, **5**, [Mollusca, Part 1], 209–425.

JELL, P. A. 1980. Earliest known pelecypod on Earth. A new Early Cambrian genus from South Australia. *Alcheringa*, **4**, 233–239.

JOHNSTON, P. A. 1993. Lower Devonian Pelecypoda from south-eastern Australia. *Memoirs of the Association of Australasian Palaeontologists*, **14**, 1–134.

—— 1996. Middle Ordovician (Caradoc) pelecypods from the Advance Formation, British Columbia. *Bulletin of the Geological Survey of Canada*, **491**, 113–129.

—— & GOODBODY, Q. H. 1988. Middle Devonian bivalves from Melville Island, Arctic Canada. *In:* MCMILLAN, N. J., EMBRY, A. F. & GLASS, D. J. (eds) *Devonian of the World. 3. Paleontology, Paleoecology and Biostratigraphy.* Canadian Society of Petroleum Geologists, Memoir 14, 337–346.

—— & ZHANG R. 1998. Re-interpretation of *Sinodora* Pojeta and Zhang, 1984, an unusual bivalve (Trigonioida) from the Devonian of China. *In:* JOHNSTON, P. A. & HAGGART, J. W. (eds) *Bivalves: An Eon of Evolution – Paleobiological Studies Honoring Norman D. Newell.* University of Calgary Press, Calgary, 361–375.

JOHNSTON, R. M. 1888. *Systematic Account of the Geology of Tasmania.* J. Walch & Sons, Hobart.

KOROBKOV, I. A. 1954. *Sprvochnik i metodicheskoe rukovodstro po tretichnum Mollyuskam Plastinchatozhabernye.* Nauchno-tecnoï Isledov. Leningradskoi Otdelenie [in Russian].

MCALESTER, A. L. 1968. Type species of Paleozoic nuculoid bivalve genera. *Memoir of the Geological Society of America*, **105**.

MORRIS, N. J. 1978. The infaunal descendants of the Cycloconchidae: an outline of the evolutionary history of the Heteroconchia, Superfamilies Cycloconchacea to Chamacea. *Philosophical Transactions of the Royal Society of London, Series B*, **284**, 259–275.

—— 1979. On the origin of the Bivalvia. *In:* HOUSE, M. R. (ed.) *The Origin of Major Invertebrate Groups.* Systematics Association Special Volume No. 12. Academic Press, London, 381–413.

MORTON, B. 1995. The evolutionary history of the Bivalvia. *In:* TAYLOR, J. D. (ed.) *Origin and Evolutionary Radiation of the Mollusca.* Oxford University Press, Oxford, 337–359.

NEUMAYR, M. 1884. Zur Morphologie der Bivalvenschlosses. *Kaiserlich-Königlichen Akademie der Wissenschaften, Wien. Mathematische, Naturwissenschaftliche Klasse, Sitzungberichte, Abteilung 1*, **88**, 385–418.

NEVESSKAYA, L. A., SCARLATO, O. A., STAROBOGATOV, YA. L. & EBERZIN, A. G. 1971. New ideas on bivalve systematics. *Paleontological Journal*, **2**, 3–20. [English translation by American Geological Institute, published 1972].

NEWELL, N. D. 1965. Classification of the Bivalvia. *American Museum Novitates*, **2206**, 1–25.

—— & BOYD, D. W. 1975. Parallel evolution in early trigoniacean bivalves. *Bulletin of the American Museum of Natural History*, **154**, 53–162.

PELSENEER, P. 1889. Sur la classification phylogénétique des pélécypodes. *Bulletin Scientifique de la France et de la Belgique*, **20**, 27–52.

POJETA, J. 1971. Review of Ordovician pelecypods. *Professional Paper of the United States Geological Survey* , **695**.

—— 1973. *Fordilla troyensis* Barrande: the oldest known pelecypod. *Science*, **180**, 866–868.

—— 1975. *Fordilla troyensis* Barrande and early pelecypod phylogeny. *Bulletins of American Paleontology*, **67** (287), 363–384.

—— 1978. The origin and early taxonomic diversification of pelecypods. *Philosophical Transactions of the Royal Society of London, Series B*, **284**, 225–246, pls 1–15.

—— 1988. The origin and Paleozoic diversification of solemyoid bivalves. *Memoir of the New Mexico Bureau of Mines and Mineral Resources*, **44**, 201–271, pls 1–24.

—— & GILBERT-TOMLINSON, J. 1977. Australian Ordovician pelecypod molluscs. *Bulletin of the Bureau of Mineral Resources, Geology and Geophysics*, **174**, 1–64, pls 1–29.

—— & RUNNEGAR , B. 1976. The paleontology of rostroconch mollusks and the early history of the Phylum Mollusca. *Professional Paper of the United States Geological Survey*, **968**.

—— & —— 1985. The early evolution of diasome molluscs. *In:* TRUEMAN, E. R. & CLARKE, M. R. (eds) *The Mollusca, 10, Evolution*. Academic Press, Orlando, 295–336.

——, ZHANG, R. & YANG, Z. 1986. Systematic paleontology of Devonian pelecypods of Guangxi and Michigan. *Professional Paper of the United States Geological Survey*, **1394**, 57–108.

PURCHON, R. D. 1987. Classification and evolution of the Bivalvia: an analytical study. *Philosophical Transactions of the Royal Society of London, Series B*, **316**, 277–302.

RATTER, V. A. & COPE, J. C. W. 1998. New neotaxodont bivalves from the Silurian of South Wales and their phylogenetic significance. *Palaeontology*, **41**, 975–991.

RIDEWOOD, W. G. 1903. On the structure of the gills of lamellibranchs. *Philosophical Transactions of the Royal Society of London, Series B*, **194**, 147–284.

RUNNEGAR, B. & POJETA, J. 1992. The earliest bivalves and their Ordovician descendants. *American Malacological Bulletin*, **9**, 117–122.

SÁNCHEZ, T. M. 1995. Comments on the genus *Catamarcaia* Sánchez and Babin and the origin of the Arcoida. *Geobios*, **28**, 343–346.

—— & BABIN, C. 1993. Un insolite bivalve, *Catamarcaia* n.g., de l'Arenig (Ordovicien inférieur) d'Argentine. *Comptes rendus de l' Académie des Sciences, Paris, Série II*, **316**, 265–271.

STAROBOGATOV, YA. L. 1992. Morphological basis for phylogeny and classification of the Bivalvia. *Ruthenica*, **2**, 1–25.

TAYLOR, J. D., KENNEDY, W. J. & HALL, A. 1969. The shell structure and mineralogy of the Bivalvia. Introduction. Nuculacea–Trigonacea. *Bulletin of the British Museum (Natural History), Zoology Series, Supplement 3*.

——, —— & —— 1973. The shell structure and mineralogy of the Bivalvia. II. Lucinacea–Clavagellacea, Conclusions. *Bulletin of the British Museum (Natural History), Zoology Series*, **22**, 256–294, pls 1–15.

TEVESZ, M. J. S. & MCCALL, P. L. 1976. Primitive life habits and adaptive significance of the pelecypod form. *Paleobiology*, **2**, 183–190.

TUNNICLIFF, S. P. 1987. Caradocian bivalve molluscs from Wales. *Palaeontology*, **30**, 677–690, pls 76 and 77.

VOKES, H. E. 1956. Notes on the Nucinellidae (Pelecypoda) with a description of a new species from the Eocene of Oregon. *Journal of Paleontology*, **30**, 652–671.

WALLER, T. R. 1978. Morphology, morphoclines and a new classification of the Pteriomorphia (Mollusca: Bivalvia). *Philosophical Transactions of the Royal Society of London, Series B*, **284**, 345–365.

—— 1990. The evolution of ligament systems in the Bivalvia. *In:* MORTON, B. (ed.) *The Bivalvia – Proceedings of a Symposium in Honour of Sir Charles Maurice Yonge*. Hong Kong University Press, Hong Kong, 49–71.

—— 1998. Origin of the molluscan class Bivalvia and a phylogeny of major groups. *In:* JOHNSTON, P. A. & HAGGART, J. W. (eds) *Bivalves: An Eon of Evolution – Paleobiological Studies Honoring Norman D. Newell*. University of Calgary Press, Calgary, 1–45.

YONGE, C. M. 1959. The status of the Protobranchia in the bivalve Mollusca. *Proceedings of the Malacological Society, London*, **33**, 210–214.

A preliminary phylogeny for rudist bivalves: sifting clades from grades

PETER W. SKELTON[1] & ANDREW B. SMITH[2]

[1] *Department of Earth Sciences, Open University, Milton Keynes MK7 6AA, UK (e-mail: P.W.Skelton@open.ac.uk)*

[2] *Department of Palaeontology, Natural History Museum, Cromwell Road, London SW7 5BD, UK*

Abstract: The taxonomic history of the late Jurassic–Cretaceous rudist bivalves (Superfamily Hippuritoidea) is reviewed and a new phylogeny based on cladistic analysis of 32 skeletal characters is proposed. Informative characters include: relative thickness and structures of the outer (calcitic) shell layer; valve asymmetry and attachment to the substrate by either the left or right valve; form of the ligament; and, in the aragonitic inner shell, the dentition and shelly supports (myophores) for the adductor muscles, as well as accessory cavities and blind-ending (pallial) canals. The myocardinal complex (teeth and myophores) is especially important for discriminating clades previously lumped in paraphyletic or even polyphyletic taxa (e.g. Caprinidae, *sensu lato*, in the bivalve *Treatise* (Moore, R. C. (ed.) 1969. *Treatise on Invertebrate Paleontology, Part N. Mollusca 6, Bivalvia*, Geological Society of America and University of Kansas).

As outgroup, a megalodontid bivalve was used, as this shares three derived traits (massive dentition, modified parivincular ligament and posterior myophores) in common with rudists. The clade of all rudists is united by possession of an outer shell layer of fibrillar prismatic calcite (albeit greatly reduced in a few taxa). Two subclades are distinguished according to the attachment of the shell – either by the left valve (all 'diceratids', except *Diceras* and *Valletia*, plus the monophyletic requieniids), or by the right valve (*Diceras*, *Valletia* and all other rudists). Monophyly of some previously established families is substantiated. These include Caprinidae, *sensu stricto* (Skelton, P. W. 1978. *Philosophical Transactions of the Royal Society of London, Series B*, **284**, 305–318), Radiolitidae (with *Agriopleura* as sister taxon) and Hippuritidae (with *Tepeyacia* emerging as a possible sister taxon), while others are resolved into distinct clades. There is some evidence for the Polyconitidae Mac Gillavry (1937), though with negligible bootstrap support.

The rudists are an extinct group of sessile epifaunal bivalves that flourished in low-latitude shallow seas from late Jurassic to Cretaceous times. An expansive adaptive radiation spread them across the broad carbonate platforms that developed in the equatorial Tethys Ocean and on scattered seamounts in the Pacific (Masse & Philip 1986; Ross & Skelton 1993; Philip *et al.* 1995; Philip 1998). The shells and resulting debris from vast 'meadows' of rudists indeed contributed substantially to the growth of the platforms, especially in the late Cretaceous (Gili *et al.* 1995; Carannante *et al.* 1999).

The name (originally *'les rudistes'*) was first given by Lamarck (1819) to an assortment of inequivalve fossil shells with prominent beaks, comprising only two currently recognized rudist genera (*Radiolites* and *Sphaerulites*) plus a radiolitid internal mould, together with two inarticulate brachiopods (*Crania* and *Discina*) and a lidded solitary rugose coral (*Calceola*). Lamarck placed the group alongside the oysters in his *'Conchifères monomyaires'*. Other taxa now regarded as rudists were put elsewhere by Lamarck (1819) – *Diceras* among chamacean bivalves and *Hippurites* with the cephalopods. Deshayes (1825) later united the hippuritids with the radiolitids to comprise a more restricted grouping of *'rudistes'*, which he placed in the bivalves.

Many new descriptions followed, with much speculative phylogenetic pigeon-holing. The taxa now recognized as rudists were variously classified as brachiopods, corals, cirripedes, annelid tubes, or members of an independent group of molluscs, as well as bivalves, until they were finally assembled – together with *Chama* and a few other

From: Harper, E. M., Taylor, J. D. & Crame, J. A. (eds) *The Evolutionary Biology of the Bivalvia*. Geological Society, London, Special Publications, **177**, 97–127. 1-86239-076-2/00/$15.00

bivalve genera – within the Bivalvia (*'Familie Chamaceen'*), by Quenstedt (1852) [see Woodward (1855) for historical details]. Munier Chalmas (1873) was the first to extend the name *'rudistes'* to cover the expanded grouping of taxa from *Diceras* onwards. Newell (1965) renamed the group as 'Superfamily Hippuritacea Gray 1848', in keeping with modern nomenclatural rules, hence their coverage under that name in the *Treatise on Invertebrate Paleontology* (Dechaseaux *et al.* 1969). Following Ride *et al.* (1999) this should now be corrected to Hippuritoidea.

The diverse and elaborate (some would say, 'baroque') morphologies of the rudists present a good array of characters for cladistic analysis. Informative characters include: the relative thickness and mesostructural modifications of the outer (calcitic) shell layer; valve asymmetry and attachment to the substrate by either the left valve (LV) or the right valve (RV); the condition of the ligament; the configuration of the massive (pachyodont) dentition; and the arrangement of the internal shelly supports (myophores) for the adductor muscles, as well as various associated accessory cavities and other structures within the aragonitic inner shell.

So far, however, this character-rich group has not been extensively subjected to phylogenetic analysis using cladistics. Hand-drawn cladograms have been proposed for caprinids by Skelton (1991) and, more rigorously, by Skelton & Masse (1998) for those of the Lower Cretaceous. Smith (1994) and Steuber (1999*a*) presented hand-drawn cladograms for a wider selection of rudists, though based on limited data derived from the literature, while Stone & Telford (1999) constructed a computer-derived cladogram, though again based on rather few characters and taxa. The most rigorous and detailed analyses to date are computer-based cladograms constructed for various caprinid rudists, with selected outgroup taxa, by Chartrousse (1998*a*).

History of ideas on rudist systematics and phylogeny

Following the consensus on rudists as bivalves, authors at first tended to ally them with chamaceans, following Quenstedt (1852) (e.g. Bayle 1855; Woodward 1855; Zittel 1885; Fischer 1887). Although Douvillé (1886, 1887) also initially favoured such a link, he subsequently (Douvillé 1935) preferred a direct derivation from a byssate ancestor, such as *Pterocardia*. The chamid hypothesis was finally rejected by Kennedy *et al.* (1970), who convincingly showed chamids to be a later, independent offshoot from the carditids. However, *Pterocardia* (probably a specialized Jurassic cockle) is also an unlikely candidate for ancestry, possessing some derived features not shared with rudists, such as pronounced lateral teeth (a character typical of cardiids). Instead, the hypothesis most widely favoured today is for the derivation of the rudists from megalodontids [following Boehm (1882), and Dechaseaux (1952)]. Megalodontids share with the earliest rudists projecting spirogyrate umbones associated with a modified parivincular ligament, massive pachyodont dentition and ledge-like myophores supporting the posterior adductors (Fig. 1a; Skelton 1978).

The golden age for the systematic subdivision of rudists was in the late nineteenth and early twentieth centuries. The foundations for the current higher level classification of the group were laid down largely by Douvillé (1886, 1887, 1888, 1889, and many subsequent works, culminating in his remarkable synthesis of 1935 – published in his 89th year). The most important subsequent revision was that of Mac Gillavry (1937), whose investigations of Caribbean rudists led to several major conclusions concerning rudist phylogeny in general.

Following Munier Chalmas (1882), Douvillé (1887) recognized a fundamental distinction between a *'série normale'* of shells with one main tooth in the LV and two in the RV (Fig. 1b and c), and a *'série inverse'* showing two teeth in the LV (Fig. 1d) and one in the RV. To the *'série normale'* he assigned only the primitive *Tribu de Dicératinés* together with a minor offshoot, the *Bayleinés* (later included in the Family Requieniidae Douvillé, 1915), while six other *tribus* were included in the *'série inverse'*. Douvillé further noted that attachment to the substratum was by the LV in the 'normal' group, with the exception of *Diceras* itself, in which either valve might be attached, while the 'inverse' forms invariably attached by the RV. Douvillé (1887) initially accepted Munier Chalmas' [Munier Chalmas (1882)] explanation of dental inversion for the difference between his two *séries*, by analogy with *Chama*. However, he later argued (Douvillé 1896) that the change in dentition was due, instead, to suppression of the anterior tooth in the RV and enlargement of an incipient posterior tooth in the LV (e.g. Fig. 1b) in the first 'inverse' form, *Valletia*.

From specimens prepared by Bayle (1873), Douvillé (1935) further recognized that attachment by the LV, or by the RV, had been consistent within different *'Diceras'* species from their first appearances. Accordingly, he proposed a new genus, *Epidiceras*, for those species – the majority – that attached by the LV (Fig. 1b), in contrast to the few remaining species of *Diceras* (*sensu stricto*) in which attachment was by the RV, as in the type

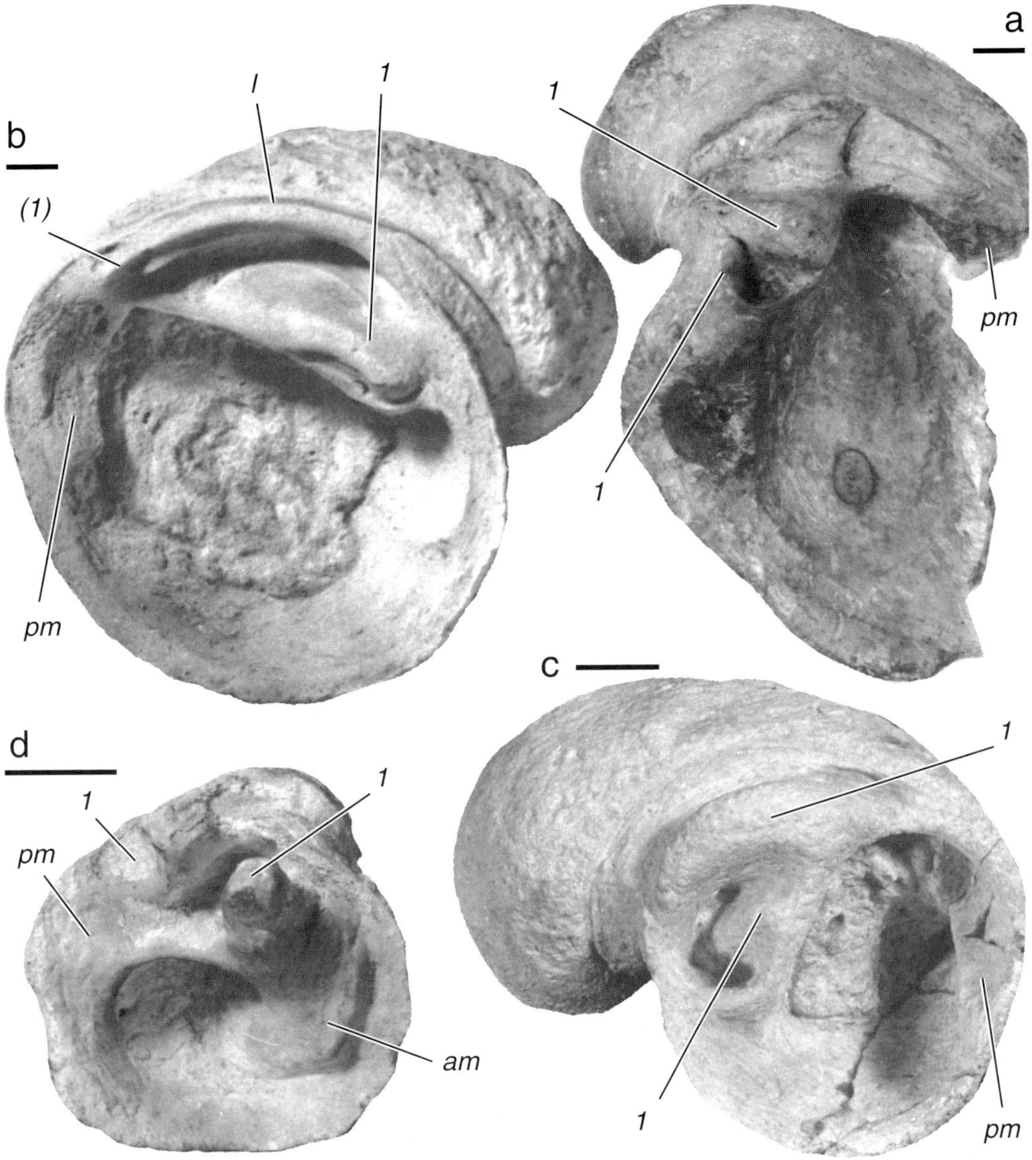

Fig. 1. (**a**) *Pachyrisma grande* Morris & Lycett, interior of right valve; Bathonian, Minchinhampton, Glos, UK (NHM, 49964). (**b**) *Epidiceras speciosum* (Deshayes), interior of left (attached) valve; Oxfordian, Dompcevrin, Meuse, France (NHM, LL 31920). (**c**) *Diceras arietinum* Lamarck, interior of right (attached) valve; as in (b), PWS collection. (**d**) *Monopleura varians* Matheron, interior of left (free) valve; Barremian, Brouzet-les-Alès, Gard, France (UCBL, EM 15681). Scale bars, 10 mm. *1*, tooth; *(1)*, incipient tooth; *am*, anterior myophore; *l*, ligamentary insertion site; *pm*, posterior myophore.

species *Diceras arietinum* Lamarck (Fig. 1c). Thus, two distinct evolutionary branches could be recognized, one commencing with *Epidiceras*, retaining 'normal' dentition, and the other starting with *Diceras*, giving rise (via *Valletia*) to the 'inverse' taxa.

Douvillé (1887) divided the 'inverse' forms among six *tribus*: (1) *Monopleurinés*, characterized by simple myophoral ledges adjoining the hinge plates (Fig. 1d), and interpreted by Douvillé as a basal group for each of the next three *tribus*; (2) *Caprotininés*, in which a projecting, inwardly facing posterior myophoral plate in the LV is separated from the valve wall by accessory cavities (Fig. 2a); (3) *Caprininés*, possessing pallial canals (longitudinal, blind-ending tubes within the inner shell) in one (Fig. 2c and d) or both valves; (4) *Hippuritinés*, containing radial canals (connecting to the outside via pores) within the operculiform LV (Fig. 3a and b), as well as a pair of radial infoldings of the postero-ventral valves walls, forming internal longitudinal pillars in the RV (Fig. 3c) and matching oscules in the LV (Fig. 3b); supposedly derived either from the *Caprotininés* or the *Caprininés*; (5) *Radiolitinés*, with elongate teeth and outwardly facing myophores projecting from the LV (Fig. 4a) and a pair of distinctly ornamented radial bands on the postero-ventral flanks of the shell (Fig. 4b); (6) *Ichthyosarcolithinés* (*sic*), with fine pallial canals, somewhat reminiscent of those in caprinids (Fig. 4d), but also with a myocardinal arcade (Fig. 4c) like that in radiolitids.

Soon after, Douvillé (1888) revised his earlier grouping of *Caprininés*, separating off the *Plagioptychinés* as a distinct *tribu*. Among the differences he noted was a contrast in myophoral orientation: the RV posterior myophore is more or less normal to the commissural plane in the *Caprininés* (*sensu stricto*) (Fig. 2e; see also Fig. 9b), but may be subparallel with it in the *Plagioptychinés* (Fig. 5b and c). Hence, he proposed that the latter were probably independently derived from the monopleurids and not from the caprinids. Thus, it was already evident that pallial canals had evolved convergently in a number of unrelated groups.

In a further revision of his phylogenetic scheme, Douvillé (1889) drew attention to what he now regarded as a fundamental difference in myophoral arrangement between *Gyropleura* and *Monopleura* [both previously included in the *Tribu de Monopleurinés* [Douvillé (1887)]. In *Gyropleura*, the RV posterior myophore forms a salient (backwardly sloping) shelf, while in *Monopleura* it is somewhat depressed (so inwardly inclined). The two genera were now interpreted as forming the rootstocks of two evolutionary branches of 'inverse' rudists (Fig. 6). The first branch (originating with the *Gyropleurinés*) was held to show an increasingly backward tilting of the RV posterior myophore, eventually giving rise to a deep accessory cavity between the erect myophore and the valve wall in both the caprotininids (e.g. Fig. 2b) and caprinids. The second branch (commencing with the *Monopleurinés*) was seen as showing the opposite trend, for greater inward depression of the RV posterior myophore, eventually leading to the arrangement seen in the radiolitids (Fig. 4a). In this new scheme, *Plagioptychus* was derived from a gyropleurinid ancestor, but again independently of the caprinids (Fig. 6).

In an important description of new or previously poorly understood taxa of Lower Cretaceous ('Urgonian') rudists, Paquier (1905) endorsed Douvillé's two-branch scheme for the 'inverse' rudists, according them familial rank under the names *Gyropleuridés* and *Monopleuridés*. Among the taxa that he assigned to the former were *Pachytraga* and *Retha*, which he regarded as the most primitive members of the *Caprotininés*. Despite their myocardinal similarities with an associated primitive caprinid, *Praecaprina*, he interpreted them as independently derived from a common gyropleurid root. However, this interpretation was rejected by Mac Gillavry (1937) and Skelton & Masse (1998), who recognized *Pachytraga* and *Retha* as the most primitive caprinids on the basis of their myocardinal arrangements.

Two further genera that Douvillé (1889) discussed in relation to his new scheme were *Polyconites* and *Horiopleura*. While acknowledging their external similarity, and even their shared possession of a distinctive accessory cavity behind the pedunculate posterior myophore of the LV (Fig. 7; 'ectomyophoral cavity' herein), he nevertheless separated them in his two main 'inverse' branches. He regarded *Horiopleura*, which has a prominent posterior myophoral ledge in the RV, as a gyropleurinid, but *Polyconites*, with its depressed RV myophore, as a monopleurinid [despite having previously allied it with *Caprotina* (Douvillé 1887)]. The weakness of this argument was exposed by Mac Gillavry (1937), who reunited the two genera in a newly proposed subfamily, Polyconitinae, interpreted by him as of monopleurinid

Fig. 2. (a) and (b) *Caprotina striata* d'Orbigny, interiors of left (free) valve (**a**) and right (attached) valve (**b**); Cenomanian, Le Mans, France (NHM, L96200). (c) and (d) *Caprina adversa* d'Orbigny, articulated shell, in posterior view, with outer layer of left valve worn (**c**), and broken/worn section across left valve, in adapical (commissural) view (**d**); Cenomanian, Charente, France. [(c), UCBL, EM 15684; (d), NHM, 30378]. (**e**) *Caprina douvillei* Paquier, adapical section of right valve, with part of posterior wall missing; note thin calcitic outer shell layer preserved around anterior and ventral margins; Lower Aptian, subsurface of Duval County, southern Texas, USA (Sligo Formation) (University of Texas at Austin collections, UT 48224). Scale bars, 10 mm for (a)–(d), 1 mm for (e). See Fig. 1 for key; *bc*, body cavity; *ecc*, ectomyophoral cavity; *enc*, endomyphoral cavity; *LV*, left valve; *pc*, pallial canals; *RV*, right valve.

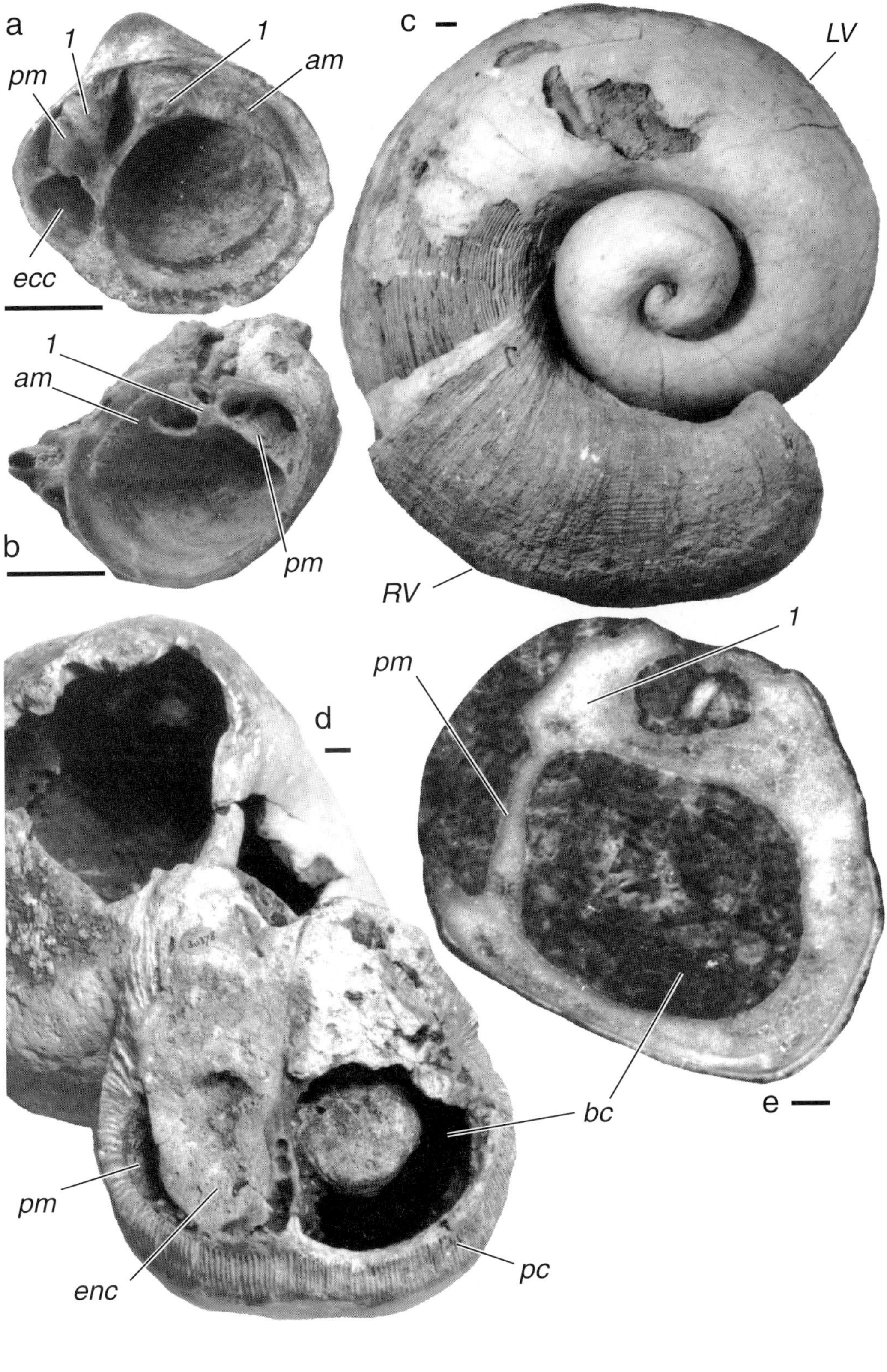
a
1
pm
1
am
ecc
b
1
am
pm
c
LV
RV
d
pm
enc
bc
pc
e
1
pm
bc

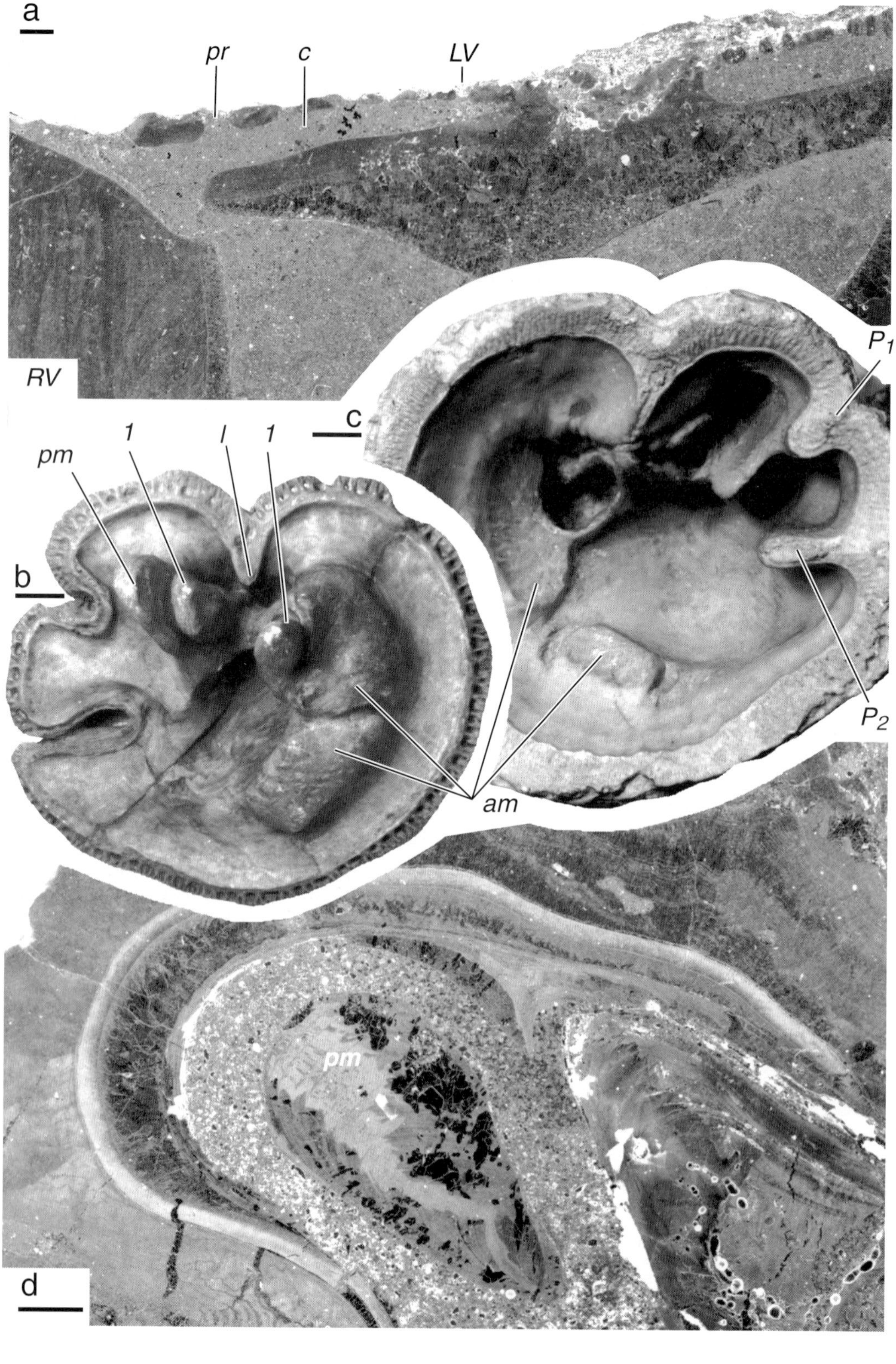
a
pr
c
LV
RV
b
pm
1
l
1
c
P1
P2
am
pm
d

ancestry. Skelton & Masse (1998) supported Mac Gillavry's conclusion and reaffirmed the independent origins of the polyconitids and the caprinids.

Mac Gillavry (1937) otherwise concurred with much of Douvillé's hypothesizing concerning morphological change in myocardinal systems as an account of general trends. However, he convincingly argued that such patterns had been repeated independently in many different taxa. Thus, he envisaged a somewhat bush-like phylogeny for the descendants of *Valletia*, with several distinct lineages clustering in three main types of evolutionary trend (Fig. 8). The first trend ('I type') culminated in the radiolitid-type configuration, with both LV myophores projecting so as to face outwards onto the inner wall of the RV. The second ('II type') involved (among other aspects) the projection of a tooth-like LV posterior myophore into a socket within the RV, as seen, for example, in the hippuritids (Fig. 3b and c). The third ('III type') was equivalent to Douvillé's (1889) postulated gyropleurinid trend, resulting in the formation of a posterior accessory cavity in the RV, though also involving the strong development of the RV tooth (e.g. Fig. 2e), in contrast to its reduction in representatives of the other two trends.

Mac Gillavry's (1937) analysis particularly emphasized the polyphyletic origins of canaliculate rudists. Thus, besides the separate origins of the ichthyosarcolitids, plagioptychids and caprinids, already signalled by Douvillé (1887, 1888), Mac Gillavry also argued for independent derivations, from monopleurinids, of the antillocaprinids, 'trechmannellids' (= dictyoptychids) and *Rousselia*, as well as for placement of *'Sabinia'* [= *Pseudosabinia*] *klinghardti* Boehm among the radiolitids. All these disparate taxa possess pallial canals. Mac Gillavry was thus the first to recognize the caprinids (*sensu stricto*) as an exclusively early–middle Cretaceous clade. Furthermore, he showed it to comprise two subclades – the largely Old World taxa allied with *Caprina*, itself (= Caprininae d'Orbigny), and the New World forms allied with *Caprinuloidea* (= Caprinuloidinae Mac Gillavry, 1970; syn. Coalcomaninae Coogan, 1973) (Chartrousse 1998*b*; Skelton & Masse 1998). He also argued that the tilting of the RV posterior myophore and associated formation of an accessory cavity, which Douvillé (1889) had regarded as characteristic of the gyropleurinids, had occurred independently in the caprinids, in the derived plagioptychid *Mitrocaprina*, in *Caprotina* and in species of *Horiopleura*.

The remarkable insights of Douvillé and Mac Gillavry were largely ignored in the grossly oversimplified classifications of the *Traité de Paléontologie* (Dechaseaux 1952) and the *Treatise on Invertebrate Paleontology* (Dechaseaux *et al.* 1969). Though useful as a compendium of genera, the latter (the *'Treatise'*) set out a retrograde classification (Table 1) that offered little beyond Douvillé's (1887) preliminary scheme and even overlooked aspects of it. In particular, the Family Caprinidae was presented as an ill-assorted grouping of all rudists with pallial canals and the Family Caprotinidae was likewise used as a generalized receptacle for rudists with all manner of accessory cavities, including such unlike taxa as *Horiopleura* and *Pachytraga*. The Polyconitidae remained unmentioned. Despite these shortcomings, the *Treatise* classification has been uncritically adopted by many authors, so undermining the potential value of rudists for elucidating large-scale patterns of taxonomic turnover and palaeobiogeography. Accurate analysis of such aspects requires the identification of monophyletic taxa (Smith 1994).

Skelton (1978) noted the likely polyphyletic nature of the *Treatise* 'caprinids' and 'caprotinids', and formally reinstated the findings of Douvillé and Mac Gillavry. He divided the *Treatise* 'caprinids' into four families, deemed to represent distinct clades of canaliculate rudists (although formal cladistics was not used to distinguish them). He recognized the Ichthyosarcolitidae Douvillé, Caprinidae d'Orbigny (*sensu stricto*), Plagioptychidae Douvillé and Antillocaprinidae Mac Gillavry, and later separated off a fifth family, the Dictyoptychidae Skelton [in Skelton & Benton 1993; see also Morris & Skelton (1995)].

Chartrousse (1998*b*) went on to demonstrate a significant additional difference in posterior

Fig. 3. (**a**) Negative print from peel of radial section through upper part of shell of *Vaccinites galloprovincialis* (Matheron); Santonian, Figuières, Bouches-du Rhône, France (PWS collection). (b) and (c) *Hippurites radiosus* Des Moulins, interiors of left valve (**b**) and right valve (**c**); Campanian of Barbezieux, Charente, France [NHM, (b), L 62201; (c), L 18965]. (d) *Hippurites socialis* Douvillé, negative print of peel of section across posterior part of right valve, seen in abapical view (i.e. towards commissure), with left valve posterior myophore at centre, showing myostracal prismatic structure (dark) on both sides, and with pillar P_1 of right valve at lower left; Santonian of Piolenc, Vaucluse, France (PWS collection). Scale bars, 10 mm for (b) and (c), 1 mm for (a) and (d). See Figs 1 and 2 for key; *c*, radial canal; P_1, and P_2, pillars.

Fig. 4. (**a**) *Radiolites sauvagesi* d'Hombres-Firmas(?), dorsal aspect of left (free) valve; Turonian(?) of Abeih, Lebanon (NHM, L 18719). (**b**) *Durania* cf. *apula* (Parona), postero-ventral aspect of right valve; Campanian, Khashm Hajajah, Riyadh, central Saudi Arabia (King Saud University, College of Science, El Asa'ad and Skelton collection, HN2.2). (c) and (d) *Ichthyosarcolites triangularis* Desmarest, interior of left valve, with broken myocardinal arcade (**c**), and worn surface of right valve, showing cement-filled pallial canals (**d**); Cenomanian, Charente, France [NHM, (c), 30382; (d), L63180]. Scale bars, 10 mm. See Fig. 1 for key; *mc*, myocardinal arcade; *rb*, radial band(s).

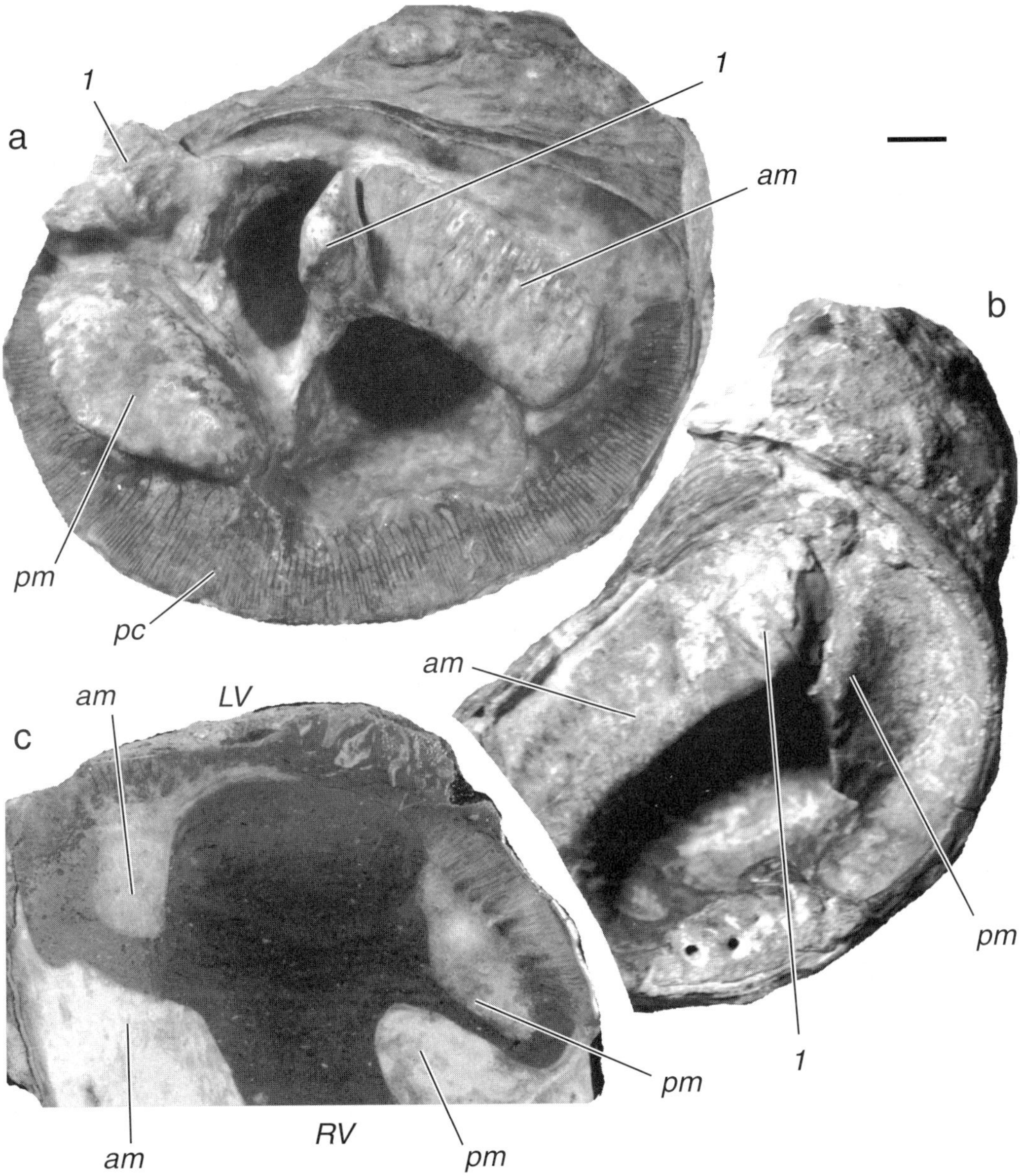

Fig. 5. (a) to (c) *Plagioptychus toucasi* Matheron, interiors of left valve (**a**) and right valve (**b**); Santonian, Le Beausset, Var, France (UCBL, EM 15685, EM 15686). (**c**) Antero-posterior radial section across both valves; same provenance (PWS collection). Scale bar, 10 mm. See Figs 1 and 2 for key.

myophoral organization between the two caprinid (*sensu stricto*) subfamilies (Fig. 9). The caprininids are characterized by an erect myophoral plate in the RV, which projects into the LV and faces back onto the inner side of the LV myophore. The 'coalcomaninids' (= Caprinuloidinae Mac Gillavry) show the opposite arrangement, with the LV myophore projecting into the posterior accessory ('myophoral') cavity in the RV and facing back onto an adductor insertion site on the posterior wall of the cavity.

A contrasting approach to the *Treatise* 'caprotinids' was adopted by Skelton (1978). In view of uncertainties at the time over how best to

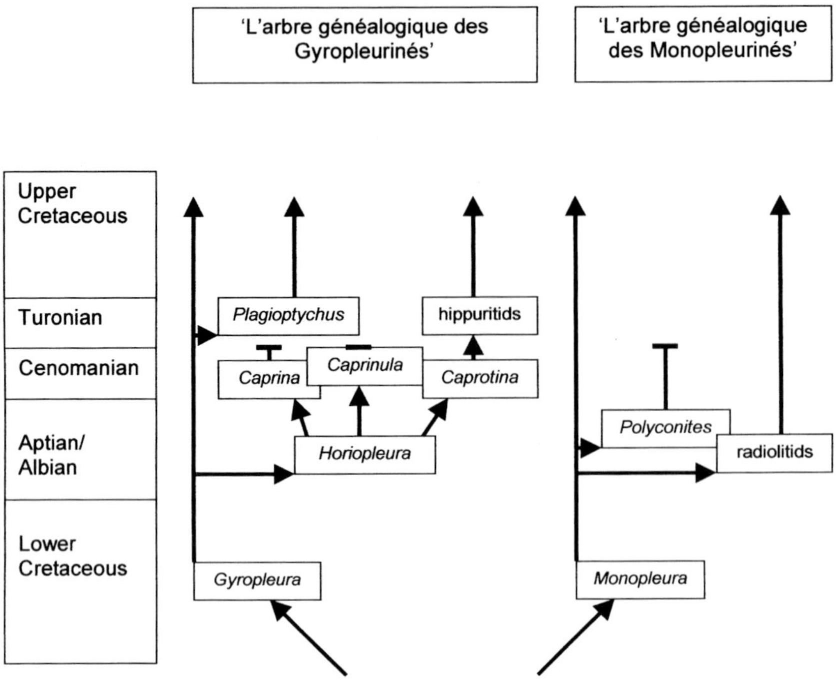

Fig. 6. Douvillé's (1889) proposed phylogenetic scheme for the 'inverse' rudists.

split up primitive uncoiled rudists, he expanded the latter family to include the monopleurids, so as to form a coherent basal group of uncoiled rudists. The paraphyletic nature of this last grouping was highlighted by Smith (1994) as symptomatic of the need for fuller cladistic analysis of the rudists – the main objective of the present paper.

Data and methods

Taxa

Thirty-five rudist taxa were selected for cladistic analysis, together with one megalodontid to serve as outgroup (Table 2). Most are single species, though in some instances it was felt appropriate to select whole genera for character coding. The taxa chosen provide a representative sampling across the various higher taxa recognized in earlier work. In almost all cases, characters have been checked directly from specimens (in many instances including type material), in addition to published figures and descriptions (Appendix 1). Where possible, some relatively plesiomorphic representatives from presumed monophyletic higher taxa were included, to optimize recovery of more basal branching patterns. In addition, however, a number of taxa of controversial or uncertain affinities (some with several derived characters) were included, in the hope of either clarifying their relationships or of detecting hitherto unexpected links between established higher taxa.

Dechaseaux *et al.* (1969) list 115 rudist genera, while a more recent, though still incomplete, website data base (Steuber 1999*b*) lists 142 genera from families so far covered, to which should be added another 12 or so genera from families still to be recorded (giving a total of around 154 genera). Future taxonomic revision will probably dispense with some of these genera through synonymy, though this loss is likely to be more than offset by the erection of new genera from regions where much primary taxonomic description remains to be done, such as Mexico and the Middle East. Hence, the present analysis is based on only a limited sampling of the group and the results must be regarded as preliminary. Future work will be directed to including more taxa and further refinement of character coding, in order to test and build upon the present conclusions.

The Middle Jurassic (Bathonian) megalodontid *Pachyrisma grande* Morris & Lycett (Fig. 1a) was chosen as the outgroup. It is one of the stratigraphically youngest species of described

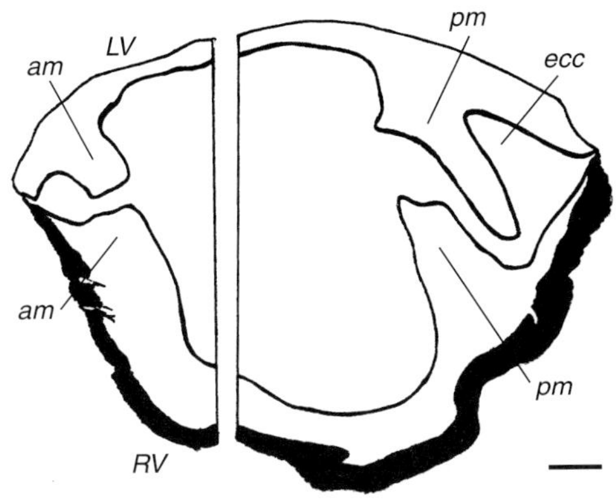

Fig. 7. *Horiopleura lamberti* Douvillé, traced drawing of radial sections across antero-dorsal (left) and posterior (right) parts of articulated shell, showing ectomyophoral cavity inserted behind pedunculate posterior myophoral plate of left valve. Albian of Santander, Spain (PWS collection). Scale bar, 10 mm. See Figs 1 and 2 for key.

megalodontids, predating by only two stages the oldest known (Oxfordian) rudists with which it already shares several derived characters with respect to other bivalves.

Spirogyrate rudists, sharing the plesiomorphic external condition of the ligament with the megalodontids (Skelton 1978, 1985), are represented by five species. These include the Oxfordian–Lower Kimmeridgian type species of Douvillé's (1935) two postulated ancestral diceratid genera, *Epidiceras sinistrum* (Deshayes) (Fig. 1b), in which attachment was by the LV, and *Diceras arietinum* Lamarck (Fig. 1c), attached by the RV. Two requieniids (attached by the LV) are also included – the earliest known (Upper Kimmeridgian–Tithonian) species, *Matheronia salevensis* Joukowsky & Favre, and the widespread Lower Cretaceous species, *Toucasia carinata* (Matheron). *Valletia*, attached by the RV, and the supposed progenitor of Douvillé's *'série inverse'* (Douvillé 1896), is represented by its earliest known species, the Upper Kimmeridgian *V. auris* Favre.

Caprinids (*sensu stricto*) are at present probably the best understood of the rudist higher taxa, having recently been revised (with cladistic analysis) by Chartrousse (1998*a*, *b*) and Skelton & Masse (1998). Six genera were thus deemed sufficient to represent the two subfamilies into which the family has been divided (see previous section). The earliest known members of each subfamily are *Pachytraga tubiconcha* Astre (Hauterivian; Caprininae) and *Retha tulae* (Felix) (?Hauterivian–Barremian; Caprinuloidinae). In these primitive forms, pallial canals are sparse (*Pachytraga*) or absent (*Retha*). Canaliculate caprinids are represented by the caprininids *Praecaprina varians* Paquier (Lower Aptian) and *Caprina adversa* d'Orbigny (Cenomanian) (Fig. 2c and d), and the caprinuloidinids *Amphitriscoelus waringi* Harris & Hodson (Lower Aptian) and *Coalcomana ramosa* (Boehm) (Lower Albian).

Monopleurids are represented by three species. One of the earliest species assigned to the genus *Monopleura* is *M. taurica* Pchelintsev, from the Berriasian of Crimea (Yanin 1975; Skelton 1985). Many quite disparate species have been assigned to this genus, which is in need of revision; the Barremian *M. varians* Matheron (Fig. 1d) is the type species. Coding for the genus *Agriopleura*, widely interpreted as the source of the radiolitids (Masse & Philip 1974), is based on the descriptions by Douvillé (1918) of two Barremian species [*A. blumenbachi* (Studer) and *A. marticensis* (Matheron)], which were regarded as synonymous by Toucas (1907). The question of synonymy in this case does not affect the characters coded for in the present analysis.

The radiolitids (e.g. Fig. 4a and b) are regarded as probably monophyletic (Skelton 1978), so just six representative taxa were selected for the purposes of this broad analysis. However, it is also by far the most diverse of the rudist families [Steuber (1999*b*) records 71 genera], so analysis of other taxa is desirable in the future, in order to test the assumption of monophyly. The earliest examples yet documented – from the Upper Aptian – have been assigned to *Eoradiolites plicatus* (Conrad) (Masse & Gallo Maresca 1997). These show limited development of the celluloprismatic mesostructure of the calcitic outer shell layer in the RV, characteristic of most radiolitids (Fig. 10a), but are otherwise similar to Lower Aptian *Agriopleura* sp. from Arabia (Skelton & Masse 2000). The cosmopolitan genus *Sauvagesia* Douvillé shows a more derived, polygonal cell pattern throughout its RV outer shell layer. Moreover, certain species have pallial canals in the LV [e.g. the Maastrichtian *S. macroplicata* (Whitfield) from Jamaica]. The Turonian *Durania cornupastoris* (Des Moulins) is similar to the more primitive species of *Sauvagesia*, though lacking a ligamentary invagination. The Campanian–Maastrichtian Old World genus *Osculigera* Kühn, however, shows more derived characters, with infolding of the radial bands to form pseudopillars, in addition to loss of the ligament. *Pseudosabinia klinghardti* (Boehm) is another Old World form with many derived characters. It is a large canaliculate recumbent rudist, locally abundant in Campanian platform deposits from the eastern Mediterranean to Arabia (Morris & Skelton 1995).

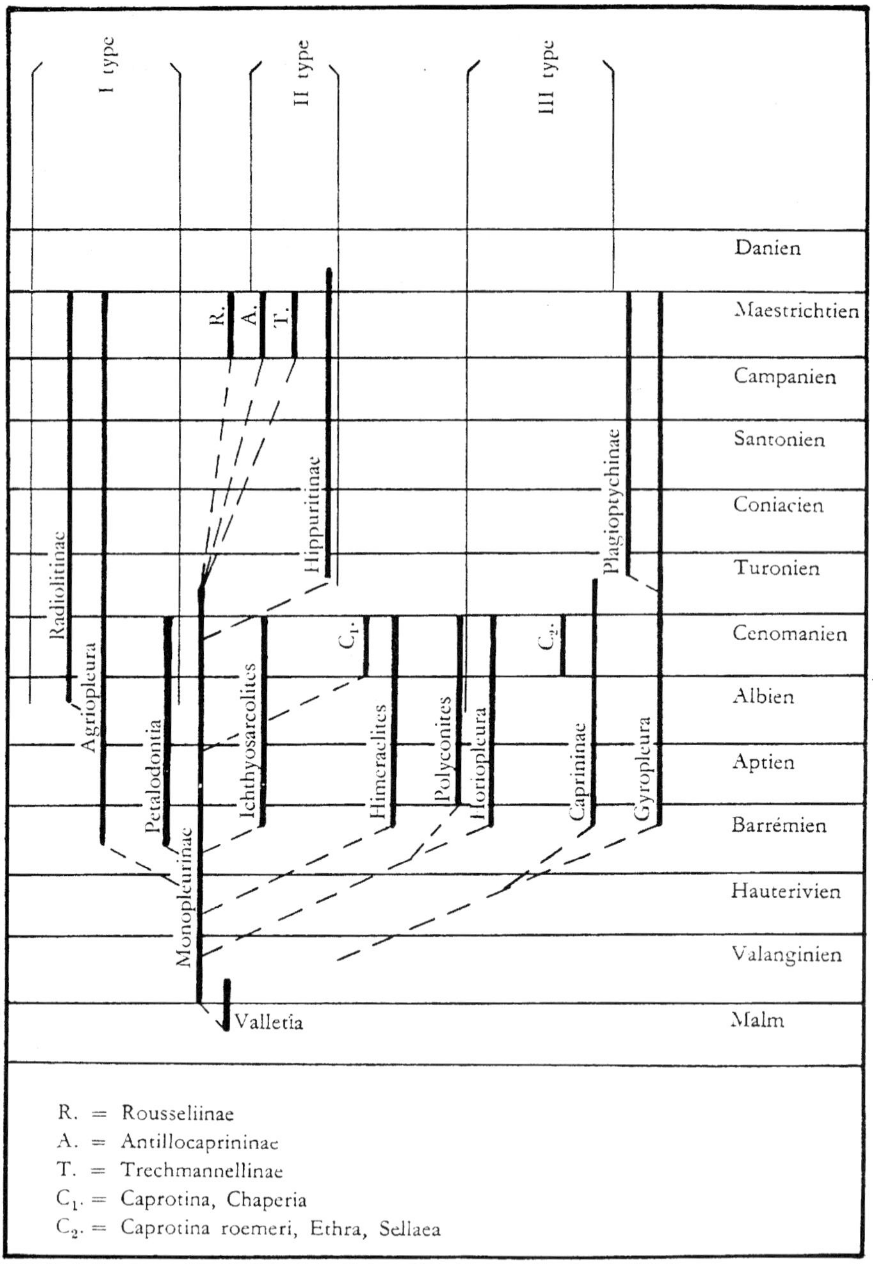

Fig. 8. Mac Gillavry's (1937) phylogenetic scheme for the 'inverse' rudists, comprising three main types of trend followed independently by several different lineages (as discussed in text). 'Trechmannellinae' is Dictyoptychidae Skelton; '*Ethra*' is *Retha* Cox. [Copied from Mac Gillavry (1937) with the author's permission].

Its possession of a fine celluloprismatic mesostructure in the outer shell layer of the RV (Fig. 10b) was noted by Skelton (1978), in support of Mac Gillavry's (1937) assertion of its sauvagesiine affinity. Philip (1986) accordingly removed the Genus *Sabinia* (to which *Pseudosabinia* was originally assigned) from the Caprinidae, where Dechaseaux *et al.* (1969) had placed it, to the

Table 1. *Rudist families as listed in* The Treatise on Invertebrate Paleontology (*Dechaseaux* et al. *1969*)

1	Diceratidae Dall (*Pa*)
2	Requieniidae Douvillé (*M*)
3	Monopleuridae Munier-Chalmas (*Pa*)
4	Caprotinidae Gray (*Po*)
5	Caprinidae d'Orbigny (*Po*)
6	Hippuritidae Gray (*M*)
7	Radiolitidae Gray (*M*)

Status of families according to Skelton (1978). *M*, monophyletic; *Pa*, paraphyletic; *Po*, polyphyletic.

Radiolitidae. However, apparent differences in myocardinal arrangement from the type species of *Sabinia*, *S. anienis* Parona, led Morris & Skelton (1995) to propose the new generic name *Pseudosabinia* for '*S.*' *klinghardti*. Reinspection of the type material of *S. anienis* will be necessary, however, before it can be included in this analysis as well. Another canaliculate radiolitid, this time from the Maastrichtian of the Caribbean region, is *Chiapasella radiolitiformis* (Trechmann). An interesting feature of this taxon is its possession of numerous tightly pleated infoldings in the outer shell layer of the RV, somewhat similar to those seen in *Torreites* (see below).

The Polyconitidae are represented by four species, from three genera. *Horiopleura* (Fig. 7) was selected as a typical genus, in preference to the relatively more derived *Polyconites*. Both the earliest known member of the family, the Upper Barremian–Lower Aptian species *H. dumortieri* (Matheron) (Masse 1996) and the Albian type species, *H. lamberti* Douvillé, were coded. The other two species included in the analysis are forms whose polyconitid affinities have recently been proposed. Although the Upper Aptian–Albian Japanese species *Praecaprotina yaegashii* (Yehara) (Fig. 11a) had originally been referred to *Horiopleura*, it was given its new generic name by Yabe & Nagao (1926), who suggested that it might be allied instead with *Pachytraga*. Skelton & Masse (1998) cast doubt on the latter interpretation, and re-emphasized its similarities with *Horiopleura*. The position of *Tepeyacia corrugata* Palmer (Fig. 11b–d), from Mexico, was formerly even less evident, as Palmer (1928) lacked information on its myocardinal features. He suggested that it was possibly a monopleurid, while Chubb (1971) referred a new Jamaican species (*T. multicostata* Chubb) to the Radiolitidae. Study of similar, but better preserved, specimens from the Upper Albian–Lower Cenomanian of SW Mexico

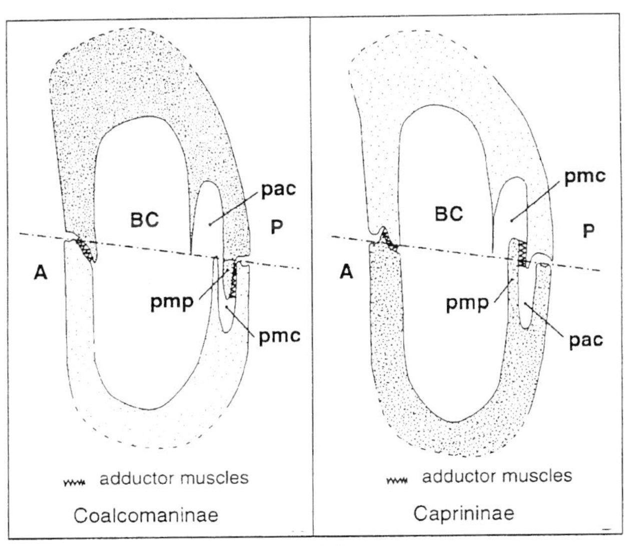

Fig. 9. Posterior myophoral configurations in the two subfamilies of the Caprinidae (*sensu stricto*), shown in diagrammatic antero-posterior sections through both valves, with left valve above and right valve below: (**a**) 'Coalcomaninae' Coogan (Caprinuloidinae MacGillavry); (**b**) Caprininae d'Orbigny. [Copied from Chartrousse (1998*b*) with the author's permission]. A, anterior; BC, body cavity; P, posterior; pac, 'posterior accessory cavity' (= LV posterior endomyophoral cavity in Caprinuloidinae, and RV posterior ectomyophoral cavity in Caprininae, herein); pmc, 'posterior myophoral cavity' (= RV posterior endomyophoral cavity in Caprinuloidinae, and LV posterior endomyophoral cavity in Caprininae, herein); pmp, posterior myophore.

led Pantoja-Alor & Skelton (1999) to transfer it to the polyconitids, based on its myocardinal arrangement (Fig. 11b and c). They also reassigned the type locality for the species, in eastern central Mexico, to the Albian, instead of the Turonian, as Palmer (1928) had erroneously supposed.

The systematic positions of the next three species are currently uncertain. The Lower Aptian species *Glossomyophorus costatus* Masse, Skelton & Sliskovic is a distinctive endemic of the eastern

Table 2. *Taxa used in this analysis (see Appendix 1 for sources)*

Outgroup (megalodontid)	
PAC	*Pachyrisma grande* Morris & Lycett
Spirogyrate rudists	
EPI	*Epidiceras sinistrum* (Deshayes)
MAT	*Matheronia salevensis* Joukowsky & Favre
TOU	*Toucasia carinata* (Matheron)
DIC	*Diceras arietinum* Lamarck
VAL	*Valletia auris* Favre
Caprinids (*sensu stricto*)	
RET	*Retha tulae* (Felix)
AMP	*Amphitriscoelus waringi* Harris & Hodson
COA	*Coalcomana ramosa* (Boehm)
PACH	*Pachytraga tubiconcha* Astre
PRA	*Praecaprina varians* Paquier
CAP	*Caprina adversa* d'Orbigny
Monopleurids	
MONt	*Monopleura taurica* Pchelintsev
MONv	*Monopleura varians* Matheron
AGR	*Agriopleura* Kühn
Radiolitids	
EOR	*Eoradiolites plicatus* (Conrad)
SAU	*Sauvagesia* Douvillé
DUR	*Durania cornupastoris* (Des Moulins)
OSC	*Osculigera* Kühn
PSAB	*Pseudosabinia klinghardti* (Boehm)
CHI	*Chiapasella radiolitiformis* (Trechmann)
Polyconitids	
HORd	*Horiopleura dumortieri* (Matheron)
HORl	*Horiopleura lamberti* Douvillé
PRAE	*Praecaprotina yaegashii* (Yehara)
TEP	*Tepeyacia corrugata* Palmer
Uncertain (plus caprotinid)	
GLO	*Glossomyophorus costatus* Masse, Skelton & Sliskovic
HIM	*Himeraelites* Di Stefano
CAPR	*Caprotina striata* d'Orbigny
Plagioptychid	
PLA	*Plagioptychus* Matheron
Other canaliculate rudists	
DICT	*Dictyoptychus morgani* (Douvillé)
ICH	*Ichthyosarcolites triangularis* Desmarest
ANT	*Antillocaprina occidentalis* (Whitfield)
Hippuritids	
HIP	*Hippurites* Lamarck
BAR	*Barrettia monilifera* Woodward
Torreitinids	
PTOR	*Praetorreites omanensis* Philip & Platel
TOR	*Torreites sanchezi* (Douvillé)

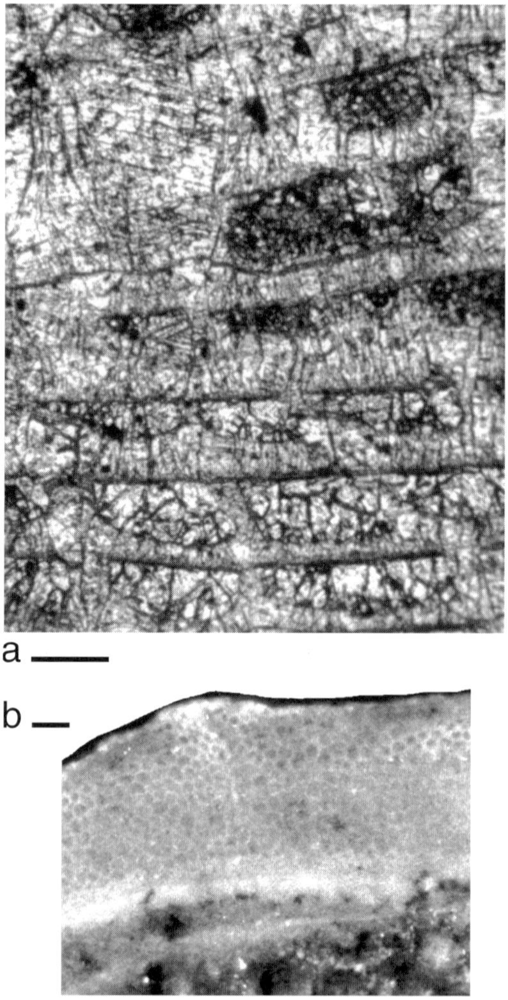

Fig. 10. (**a**) *Biradiolites angulosissimus* Toucas, radial section of part of celluloprismatic outer shell layer (acetate peel); vertical walls formed by localized ridges in the growth layers of fibrillar prismatic calcite, periodically capped off by the flat bases of successive growth layers, with cells infilled either by dark micrite (upper right) or by microspar (lower part); Santonian, Plan d'Aups, Var, France (PWS collection). (**b**) *Pseudosabinia klinghardti* (Boehm), commissural section across outer shell layer of right valve, showing fine polygonal celluloprismatic mesostructure; Campanian, NW Turkey, paratype specimen (NHM, L 49455). Scale bars, 1 mm.

Mediterranean/Arabian region. It displays a curious combination of monopleurid (e.g. subequal LV teeth) and 'caprotinid' features (e.g. projecting LV posterior myophore, facing in onto an erect myophoral wall in the RV) (Masse *et al.* 1984). *Himeraelites* Di Stefano has long been problematical (and still is), both from stratigraphical and taxonomic points of view (Masse *et al.* 1998*b*). Di Stefano (1888) created several species which are now in need of revision; most (if not all) may be synonymous with the (?Aptian–Albian)-type species, *H. vultur* Di Stefano. *Sellaea* and other species assigned to *'Caprotina'* by Di Stefano (1888) appear to be closely related to *Himeraelites*, on the basis of their myocardinal arrangements (Masse *et al.* 1998*b*), but distinct from *Caprotina* (*sensu stricto*; Douvillé 1887). The Cenomanian *Caprotina striata* d'Orbigny (Fig. 2a and b), by contrast, is securely placed as the type species of the genus and thus for the Family Caprotinidae. However, it is open to question how inclusive that family should be, given the establishment of the Polyconitidae by Mac Gillavry (1937), the placement of *Pachytraga* and *Retha* in the Caprinidae by Skelton & Masse (1998) and the distinct aspect of the *Himeraelites* group, discussed above. The broad paraphyletic concept of the caprotinids proposed by Skelton (1978) is no longer tenable in the context of the present analysis.

The Plagioptychidae is here exemplified by one species, the Santonian *P. toucasi* Matheron (Fig. 5), which shows particularly clearly the myocardinal features that led Douvillé (1888) to separate *Plagioptychus* from the caprinids. Only two other genera were assigned to the family by Mac Gillavry (1937) – *Mitrocaprina* and *Coralliochama* – both possessing more derived character states than *Plagioptychus*.

The next three species considered here are the type species and genera for each of the remaining three canaliculate families recognized by Skelton (1978) [and by Skelton & Benton (1993)] following Mac Gillavry (1937). *Dictyoptychus morgani* (Douvillé) (Dictyoptychidae) is a Maastrichtian endemic of the Middle East. *Antillocaprina occidentalis* (Whitfield) (Antillocaprinidae) is likewise a Maastrichtian endemic, though of the New World. *Ichthyosarcolites triangularis* Desmarest (Ichthyosarcolitidae; Fig. 4c and d) is widespread in the Cenomanian of the Old World Tethyan Realm.

As with the radiolitids, the monophyletic status of the Hippuritidae is regarded as secure (Skelton 1978), except in the case of *Torreites* [type species, *T. sanchezi* (Douvillé)], which lacks the characteristic system of radial canals and pores in the outer shell layer of the LV. Skelton & Wright (1987) followed earlier authors (e.g. Mac Gillavry 1935; Van Dommelen 1971) in interpreting it as an aberrant hippuritid. However, Philip & Platel (1994) proposed an independent origin, perhaps from plagioptychids, via their newly described genus *Praetorreites* (type species, *P. omanensis* Philip & Platel), in which pillars are only incipiently developed.

Although a large number of species have been assigned to the genus *Hippurites*, differences between them are often subtle. Moreover, many have been shown to represent successive chronospecies within lineages (e.g. Vicens 1992; Simonpiétri 1999). Therefore, in this analysis the genus has been treated as whole. *Hippurites radiosus* des Moulins was used as a standard for checking most characters, being well represented by many prepared whole specimens (Fig. 3b and c; e.g. Bayle 1855). Although this species had lost its ligament (Vicens 1992), others still retain a vestige of it. *Barrettia monilifera* Woodward, a Campanian New World endemic, is a more elaborate form, showing multiple radial infoldings (secondary pillars) of the outer shell layer (Fig. 12). Van Dommelen (1971) also noted the presence of pallial canals in the LV, besides the normal system of radial canals in the thin outer shell layer, an observation confirmed in the specimen illustrated here.

Definition of characters

The present analysis is based on 32 morphological characters, one of which (#01) has ordered multiple states, the rest being unordered. The full list of characters is given in Appendix 2, but a synopsis of how they are related to different components of the shell is given in Table 3.

The definition and coding of characters was the most problematical aspect of the present analysis. Groups with multicomponent skeletons, such as vertebrates and echinoids, lend themselves to cladistic analysis of fossil material because discrete homologous elements may be self-evident from the pattern of assembly of the skeleton. For example, it is possible to unambiguously identify the humerus in tetrapods as the bone in the upper forelimb. Bivalve shells, by contrast, must be metaphorically dissected into an arbitrary number of characters. An immediate problem in this respect is the non-independence of characters selected for coding. If a given aspect of the shell can be trivially resolved into many, potentially additive, characters, it may then dominate the parsimony analysis, swamping the signals from other aspects. In rudists, for example, it is easy to define several characters relating to pallial canals. Canals may be present in the LV and/or the RV, and around the anterior, posterior, dorsal and/or ventral margins, or within

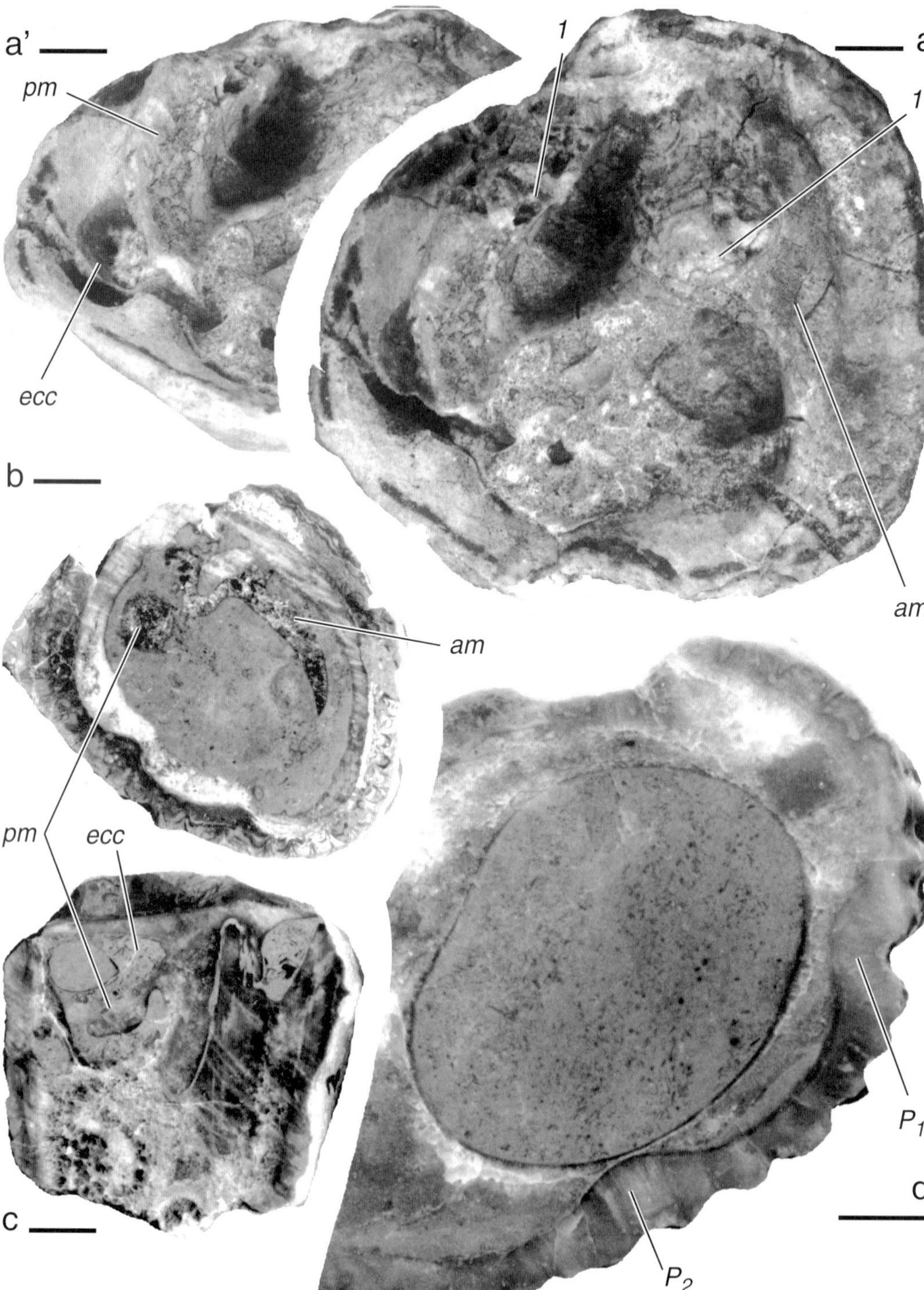
a'
pm
ecc
1
a
1
am
b
am
pm
ecc
c
P1
d
P2

the myophores and teeth, and so on. All of these characters are potentially linked, however, and a given taxon might have any or all of them. Hence, their contribution to the analysis is additive. In the first attempt at tree-building, pallial canals were inadvertently overemphasized in this way. The result (unsurprising, in retrospect) was a tree that assembled all rudists with pallial canals into a single clade, thus matching the *Treatise* grouping of 'caprinids' (Dechaseaux *et al.* 1969), which was criticized earlier. So, for the present analysis, pallial canals are limited to a single character (#32, with three unordered states) to avoid the effect of linkage. At first glance, it might appear from Table 3 that a similar error has now been committed by overemphasizing the posterior myophores (#22–28). However, #23–28 are mutually exclusive configurations (see Appendix 2) that are not susceptible to additive bias. While it would certainly be desirable to recover a more detailed coding of pallial canals in future analyses [as done by Chartrousse (1998*a*) for the caprinids *sensu stricto*], this will need to be balanced by a resolution of other components into comparable numbers of potentially additively linked characters.

Another problem is that many of the characters are of 'low burden', in the sense of Fortey & Jefferies (1982). As mere topological modifications of the shell's surface (e.g. dentition and myophores), they provide few structural clues from which to infer homology and frequent homoplasy can be expected. Myophoral configuration is again a case in point. There is no question as to the homology of the adductor scars. However, the only direct evidence for whether superficially similar myophores supporting them might be homologous or convergently evolved (analogous) resides in the details of their location and orientation. The earlier discussion of the history of ideas on rudist systematics noted the emphasis given by Douvillé and others to variations in myocardinal arrangements. Accordingly, in seeking to avoid unwarranted implications of homology, the several different types of myocardinal configuration mentioned above have been recognized as separate characters

Cavities that variously appear within the shell as a constructional consequence of myophoral reorientation pose a similar problem: to lump them all together just as 'accessory cavities' would again risk implying unwarranted homology. Yet the conception of the Caprotinidae in the *Treatise* (Dechaseaux *et al.* 1969) was essentially founded on this basis alone (Skelton 1978). To bring precision to the description of different cavities, it has been necessary to introduce two new terms which relate the position of a cavity to its neighbouring myophore in the same valve. A cavity situated on the outer side of a myophore, separating the latter from the valve margin, is termed an *ectomyophoral* cavity (like that seen, for example, behind the LV posterior myophore of *Horiopleura*; Fig. 7). A cavity on the inner side of a myophore, and delimited from the body cavity by a lamina or wall, is termed an *endomyophoral* cavity (like that to the inside of the LV posterior myophore in caprinids; Fig. 2d). This new terminology for accessory cavities is purely descriptive and unambiguous. In his studies of the caprinids, Chartrousse (1998*a*, *b*) distinguished between 'myophoral cavities', which receive a protruding myophore from the opposite valve, and 'veritable' accessory cavities, which do not (Fig. 9). However, the myophores rarely fit closely into such cavities – unlike teeth and sockets – and may indeed occupy but a small fraction of the available space (as in Fig. 9). Hence, the implied functional contrast between the two kinds of cavity (either housing a protruding myophore or not) is tenuous, with scope for uncertainty of classification. Chartrousse (1998*a*) also used the term 'perimyophoral cavity' for what is referred to here as the LV posterior endomyophoral cavity of caprinuloidinids. Again, Chartrousse's term is potentially ambiguous [the root 'peri-' (around) might suggest a position outside, rather than inside, the myophore]. By avoiding such ambiguities, it is hoped that the new terms presented here will offer further refinement in the quest for homologies.

In the end, the final arbiter for determining homology is the test of congruence in the most parsimonious cladograms – assuming a balanced selection of characters in the first place, as discussed above. In order to reduce the confounding effects of homoplasious 'noise', certain characters were eliminated during earlier iterations

Fig. 11. (a and a′) *Praecaprotina yaegashii* (Yehara), interior of left valve (**a**), with oblique ventral view (**a′**), showing ectomyophoral cavity with conical annexe inserting beneath ventral part of pedunculate posterior myophoral plate. See Appendix 1. (b)–(d) *Tepeyacia corrugata* Palmer, section across right valve, with myocardinal arcade of left valve, in abapical view (towards commissure) (**b**), antero-posterior radial section through both valves, showing typically polyconitid posterior myophore in left valve (note ectomyophoral cavity) (**c**), detail of section across right valve, in abapical view, showing incipient pillars and tightly pleated infoldings of outer shell layer (**d**); Upper Albian, Chumbítaro, Michoacán, Mexico [(b) and (c), Pantoja-Alor collection, Universidad Nacional Autónoma de Mexico; (d) PWS collection]. Scale bars, 10 mm for (a)–(c), 5 mm for (d). See Figs 1–3 for key.

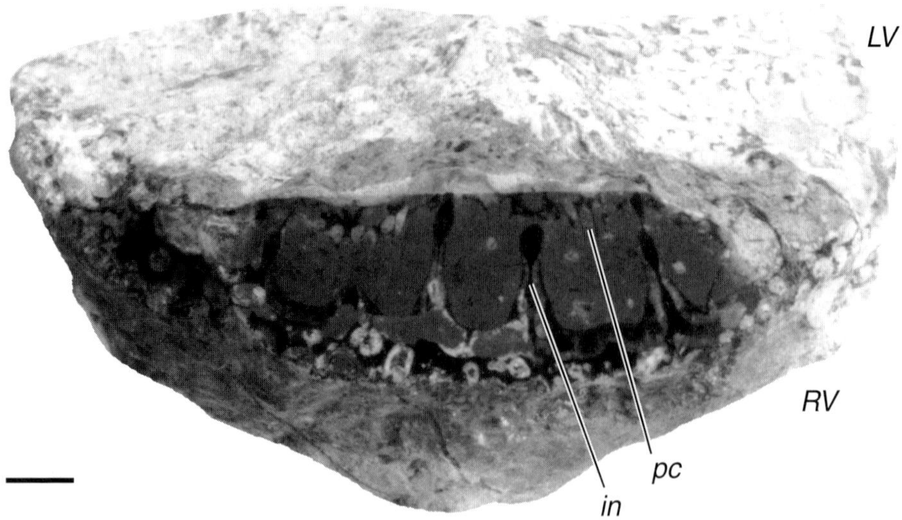

Fig. 12. *Barrettia monilifera* Woodward, tangential section across ventral margins of both valves, showing pallial canals in the left valve (within the inner shell) and moniliform infoldings of the outer shell layer of the right valve; Campanian, Puerto Rico (PWS collection). Scale bar, 5 mm. See Fig. 2 for key ; *in*, secondary infoldings of RV wall (in some hippuritids).

of the analysis, as they were found to be introducing considerable instability to branching patterns (a common symptom of homoplasy). Tabulae, for example, appear in several quite disparate taxa, including caprinids, hippuritids and radiolitids, and there can be little doubt that they have evolved independently in these (and other) groups. Accordingly, they have been eliminated from the present analysis.

Character coding

An outer shell layer of low magnesian calcite, with a fibrillar prismatic microstructure, seems to have been ubiquitous among rudists [see Skelton (1976, fig. 6) for an illustration of this in an hippuritid; and also Al-Aasm & Veizer (1986)]. Though notably thin in some taxa (e.g. caprinids; Fig. 2e), in no case has complete absence of the layer been verified. Characters #01–08 concern the relative thickness, and structural modifications of this layer. Primitively, it rarely exceeds 1 mm in thickness and may be but a fraction of this [e.g *Pachytraga*; illustrated in Skelton & Masse (1998, fig. 11)]. In some taxa, however, it is significantly thickened in parts of the shell, often up to several millimetres [e.g. *Agriopleura*; see Masse & Philip (1974)], and in some instances considerably more (e.g. hippuritids, Fig. 3a; radiolitids, Fig. 10a).

Distinctive structural modifications of the outer shell layer are characteristic of certain taxa. The outer shell layer of the RV may be thickened in such a way that its inner margin is exposed well beyond that of the LV (#02; e.g. *Torreites* and *Dictyoptychus*). Claims that this character is also present in *Durania* have been shown to be erroneous (Gili *et al.* 1995) and among the radiolitids considered here only the aberrant *Pseudosabinia* shows it.

The radial canals and external pores in the LV of hippuritids (#04; Fig. 3a and b) are interpreted as a highly modified suspension feeding system (Skelton 1976). It is unique to that family but absent in *Torreites* and *Praetorreites*. The outer shell layer of the latter genera do, however, show numerous tight longitudinal pleats in the RV, corresponding to salient ridges on the inner margin of that valve (#03; Skelton & Wright 1987).

Table 3. *Synopsis of characters coded in this analysis (see Appendix 2 for full list)*

8 (#01–08)	for calcitic outer shell layer
2 (#09–10)	for attachment to substrate
2 (#11–12)	for condition of ligament
3 (#13–15)	for shell shape
4 (#16–19)	for dentition
2 (#20–21)	for anterior myophores
7 (#22–28)	for posterior myophores
3 (#29–31)	for other myocardinal characters
1 (#32)	for pallial canals

Though no other hippuritid shows pleating of exactly this form, the multiple-fold hippuritids, of which *Barrettia* is an example, do contain numerous radially extended infoldings of the RV outer shell layer (Fig. 12; Van Dommelen 1971). This genus, too, has accordingly been coded for possession of character #03. Pleating similar to that in *Torreites* is nevertheless present in the RV outer shell layer of the polyconitid *Tepeyacia* [Fig. 11d; see also Pantoja-Alor & Skelton, (1999, fig. 1b)], as well as the radiolitid *Chiapasella*. Celluloprismatic mesostructure (#05 and 06; Fig. 10) is known only in radiolitids, as discussed in the previous section.

Both radiolitids and hippuritids, as well as some other rudists, possess a pair of modified radial zones of the outer layer around the posterior flanks of the shell. In most radiolitids, they are vertical deflections of the valve margins, marked externally by longitudinal bands (#07; Fig. 4b) which are usually distinctly ornamented. In a few radiolitids (such as *Osculigera*) they are infolded to form 'pseudopillars' [see also Steuber (1999*a*, fig. 8)]. In hippuritids, by contrast, these zones are invariably infoldings of the valve margins, creating internally projecting pillars in the RV, overlain by oscules in the LV (#08; Fig. 3b and c). Douvillé (1886) interpreted radial bands and pillars/oscules as siphonal structures, though Skelton (1976, 1979) doubted the presence of siphons in rudists, and suggested instead that the paired structures might represent the sites of faecal and pseudofaecal rejection. Regardless of the roles of such structures, the differences in construction between radial bands and pillars/oscules suggest that they are analogous rather than homologous. Hence, they are coded here as distinct characters. Nevertheless, pseudopillars have been coded along with pillars, since they are structurally equivalent. *Tepeyacia* shows incipient pillars with weak inflections of the valve margins (Fig. 11d; Palmer 1928).

Characters #09 and 10 concern attachment to the substrate by one or the other valve, as discussed in the earlier section on the history of ideas. The condition of the ligament (#11 and 12) was regarded by Skelton (1978, 1979, 1985) as a crucial factor in the evolutionary history of the rudists. The primitive condition, shared with the megalodontids, was an external, modified parivincular ligament (Fig. 1b; Skelton 1978). Though shortened, the ligament was still external in *Valletia*. In all subsequent 'inverse' rudists, by contrast, it was invaginated and greatly reduced to a residual connecting string between the valves (Skelton 1979), or entirely lost (#12). Invagination of the ligament permitted the growth of 'uncoiled' tubular shells (Skelton 1978, 1985)

The next three characters (#13–15) code for the external symmetry of the shell. Asymmetry is usually only moderately developed in the diceratids. In later forms, it is more pronounced, sometimes with a tendency to reduction of the free valve (RV, #14; or LV, #15).

The 'normal' and 'inverse' patterns of dentition (#16 and 17) were discussed in an earlier section (History of ideas ...). A distinctive dorsal geniculation of the central tooth in the RV (#18) is exclusive to the caprinuloidinids [see figures in Skelton & Masse (1998)], while complete loss of this tooth (#19) accompanies atrophy of the ligamentary invagination in certain radiolitids.

Anterior myophoral configuration is covered by characters #20 and 21. In more primitive rudists, the anterior adductor inserted directly on the valve walls (Fig. 1b and c). Most rudist taxa developed myophores, however, either as simple ledge-like extensions of the hinge plates (Fig. 1d) or as asymmetrical structures. In extreme cases, the LV projects across into the RV (Fig. 4a) and faces directly onto the inner wall of the latter (#21).

Posterior myophores are present in all rudists, as well as the megalodontid outgroup. Characters #22–28 are an attempt to systematize a large but disparate body of descriptive work on their different configurations. *Diceras* (Fig. 1c) and *Epidiceras* (Fig. 1b) share with *Pachyrisma* (Fig. 1a) the primitive condition of having myophoral ledges that pass beneath the hinge plates into the umbonal cavities. However, some specimens from the two diceratid genera may show myophoral union with the hinge plate in one valve, especially in later species (e.g. Skelton 1999, pl. 1, fig. 1 and pl. 2, fig. 1). The primitive condition is retained in the requieniids but in other rudist taxa the posterior myophores are invariably joined to the hinge plates (#22).

Various further discrete modifications of the posterior myophores are seen among uncoiled rudists. The LV myophore may be tilted steeply outwards and project across the commissural plane so as to face directly onto the posterior wall of the general cavity in the RV (#23), most notably in the radiolitids (Fig. 4a). By contrast, the LV myophore may be raised up as a flat or inwardly tilted plate, rooted by a stalk on the inner part of the myocardinal platform, and separated from the posterior valve wall by a deep ectomyophoral cavity (#24; e.g. Figs 2a, 7 and 11a). This configuration is typical of the polyconitids. Alternatively, in *Himeraelites*, the LV myophore is raised as a broad, inwardly sloping buttress, facing onto a prominent shelf in the RV (#25).

A more or less projecting LV posterior myophore may be rooted on the posterior valve wall, close to the valve margin. There, it may be inclined steeply inwards so as to face an erect, outward-facing myophoral plate in the RV, which is separated from the posterior valve wall by an ectomyophoral cavity (#26), as in caprininids (Figs 2d and e and 9b). Alternatively, the LV myophore may be inclined outwards so as to face back onto a low myophoral shelf on the posterior wall of the RV within an endomyophoral cavity in that valve (#27), as in caprinuloidinids (Fig. 9a).

Finally, the LV myophore may project, as a compressed tooth-like stucture, into a myocardinal socket in the RV (#28). This appears to be especially characteristic of the hippuritids (Fig. 3b and c) and specimens in which aragonitic shell microstructures have been preserved show myostraca (indicating muscle insertion) on both sides of the myophore (Fig. 3d; Skelton 1976).

Three further posterior myocardinal characters may be noted among uncoiled rudists. The first (#29) concerns the position of the inner (ventral) margin of the LV myocardinal platform. In some taxa this margin closely skirts the ventral sides of both teeth (e.g. *Agriopleura*). In others, the myocardinal margin is more or less ventrally extended and runs from the base of the anterior tooth to some way along the posterior myophore. Part, if not all, of the posterior myophore, as well as the posterior tooth, thus become separated from the general cavity [e.g. *Caprotina* (Fig. 2a) and hippuritids (Fig. 3b)]. Where this separation is strongly developed, a deep endomyophoral cavity (#30) may appear in the LV between the posterior myophore and the inner myocardinal margin, which now forms a thin erect lamina connecting the anterior tooth with the postero-ventral valve margin (Fig. 2d). This cavity is typical of caprinids (Skelton & Masse 1998), though also present to a variable extent in *Plagioptychus* (Fig. 5a). The third character (#31) is typical of certain polyconitids (Douvillé 1889) and comprises a marked conical annexe to the LV posterior ectomyophoral cavity, passing beneath the ventral part of the posterior myophore (Fig. 11a). The last character (#32) concerns the presence of pallial canals, as discussed earlier.

Cladistic analysis

Cladistic analysis of the 36 taxa listed in Table 2 was carried out using the software program *PAUP 4.0* (Swofford 1998; beta-test version), based on the 32 characters discussed above and listed in Appendix 2. The data matrix for the analysis is shown in Table 4. One character was coded as ordered (#01) and the rest as unordered; gaps were treated as 'missing'. *Pachyrisma* was given as the outgroup. Technical discussion of the principles involved in the analytical procedure can be found in Smith (1994, ch. 3).

Table 4. *Data matrix for the cladistic analysis*

	00000000011111111112222222222333
	12345678901234567890123456789012
PAC	00000000000000010000000000000000
EPI	10000000100010010000000000000000
MAT	10000000100011010000000000000000
TOU	20000000100011010000000000000000
DIC	10000000010010010000000000000000
VAL	10000000010010001001010000000000
RET	10000000011010001101010000100100
AMP	10000000011010001101010000100102
COA	10000000011010001101010000100102
PACH	10000000011010001001010001000100
PRA	10000000011010001001010001000101
CAP	10000000011010001001010001000101
MONt	10000000011010102001010000000000
MONv	10000000011010102001010000000000
AGR	20000010011010102000111000001000
EOR	20001010011010102000111000001000
SAU	2000111001101010200011100000100?
DUR	20001110011110100010111000001000
OSC	20001101011110100010111000001000
PSAB	21001110011010002000111000001002
CHI	201011?001111100000010111000001001
HORd	10000000011010102001010100000000
HORl	200000?0011010102001010100000010
PRAE	100000000110101020010101000000?10
TEP	201000010110101020010101000?0?0
GLO	20000010011010002001010200000000
HIM	100000000110101020010100100000?00
CAPR	10000000011010?02001010200000010
PLA	2?0000000110100020010101000001?1
DICT	21000000011110101001010000001002
ICH	10000000011110002000111000001002
ANT	10000000011010002001010000001002
HIP	20010001011?10102001010000010000
BAR	20110001011110102001010000010001
PTOR	21100001011110102001010000??001
TOR	21100001011110102001010000010001

Taxa represented by abbreviations in left-hand column as in Table 2; characters are numbered along the top. See Appendices for sources of information and character coding.

Results

An heuristic search (based on repeated branch swapping), with characters equally weighted, found 3190 most parsimonious trees, involving 67 steps. A majority rule tree for these, with nodes supported by < 50% of the most parsimonious trees collapsed together, produced mixed results. It showed satisfactory resolution of the more basal phylogenetic nodes, up to the advent of uncoiling and, thereafter, grouped the caprinids (*sensu stricto*) as a distinct clade. The remaining uncoiled taxa, however, were gathered in a largely unresolved polytomy, implying a considerable extent of homoplasy. The characters were therefore reweighted according to their rescaled consistency indices (to optimize homology, inferred on the basis of congruence) and an heuristic search repeated. This yielded 105 most parsimonious trees, for which the majority rule tree (with nodes supported by > 50% trees) is shown in Fig. 13.

Figure 13 resolves most of the clades expected from the classic works of Douvillé, Mac Gillavry and others, discussed earlier. In particular, all the deeper branching nodes were supported by 100% of the most parsimonious trees. Thus, the dichotomy between rudists attaching by the LV (with *Epidiceras* as sister group to the requieniids) and those attaching by the RV (with *Diceras* as sister group to the clade of all rudists with 'inverse' dentition) is fully supported. Likewise, *Valletia* consistently appears as sister group to the clade of all uncoiled rudists.

The expected clades of uncoiled rudists receiving 100% support include the caprinids (*sensu stricto*, with the two recognized subclades), the radiolitids (including *Pseudosabinia*), with *Agriopleura* as a sister group, and the hippuritids (with *Torreites* and *Praetorreites* forming a sister clade to *Hippurites* and *Barrettia*). The polyconitids also appear as a clade, though with the inclusion of *Caprotina*, *Glossomyophorus*, *Plagioptychus* and the hippuritids.

There are also some less conventional groupings. *Tepeyacia* is shown as the sister group to the hippuritids and *Glossomyophorus* is paired with *Plagioptychus*. *Ichthyosarcolites* emerges as a sister group to the *Agriopleura*/radiolitid clade and this grouping is placed in a polytomy alongside the other two canaliculate taxa, *Dictyoptychus* and *Antillocaprina*.

A bootstrap test (involving repeated random resampling from the data matrix) was also run, using 1000 replicates. This procedure effectively tests the breadth of support for nodes in the phylogeny; those supported by many characters scored high bootstrap percentage values and those relying on a few characters scored low values. The resulting bootstrap values are shown in Fig. 14. Most of the expected nodes remain moderately to well supported by the bootstrap test, especially that indicating monophyly of all rudists with 'inverse' dentition (*Valletia* onwards) with a bootstrap score of 98%. Monophyly of taxa attaching by the LV, of those attaching by the RV and of the uncoiled rudists (*Retha* onwards) are all given moderate support, with scores > 60%. Monophyly of the caprinids (*sensu stricto*) is weaker (55%), though its two subclades (especially the caprinuloidinids) are better supported, each scoring > 60%. Monophyly of the radiolitids (together with *Agriopleura*) is well supported, as is that of the hippuritids (including the torreitines), with scores of > 80% in each case.

The sister grouping of *Ichthyosarcolites* with the *Agriopleura*/radiolitid clade is also strongly supported (83%), though the association with the other two canaliculate forms is weaker. The pairing of *Tepeyacia* with the hippuritids gets only relatively weak support (40%), while the negligible values for monophyly of the polyconitids (including the other taxa cited previously) indicate that this grouping is reliant on very few characters and is therefore not robust.

One of the 105 most parsimonious trees, in which the polyconitid clade has nevertheless been retained (Fig. 15), has been selected here as a working phylogenetic hypothesis to illustrate the distribution of the major possible synapomorphies discussed earlier. It must be stressed that this should by no means be taken as the last word: future iterations, involving more taxa and characters, will be needed to test further the groupings shown and to resolve outstanding problems.

Discussion

In most respects, the results of the present analysis vindicate the main phylogenetic hypotheses emerging from the work of Douvillé, Mac Gillavry and others, as discussed earlier (History of ideas ...). Thus, little doubt can remain concerning the fundamental division of rudists into two clades, according to the valve of attachment (Douvillé 1935). The strength of support for *Valletia* as the sister group of all other rudists with 'inverse' dentition make it (or at least an early species of the genus) a likely progenitor for that clade, as postulated by Douvillé (1896). The monophyly – and prolific cladogenesis – of the uncoiled rudists (Skelton 1978, 1985) and the polyphyly of canaliculate rudists (Douvillé 1887, 1888; Mac Gillavry 1937; Skelton 1978) are also borne out. The support for these conclusions, both from the majority rule tree and the bootstrap test, render further discussion unnecessary.

The moderate bootstrap score for the monophyly of the caprinids (*sensu stricto*) suggests that further testing of this node is desirable, though the two subclades recognized by Mac Gillavry (1937), Coogan (1973), Chartrousse (1998*a*, *b*) and Skelton & Masse (1998) seem more secure. Evidently, more needs to be known about the basal caprinids, especially *Retha*, a point already noted by Chartrousse (1998*a*).

Although there is good support for the monophyly of the radiolitids (plus *Agriopleura*) and the hippuritids, only a few taxa were analysed for each. Further testing, with more taxa and more characters, would again be desirable. A particular point that calls for more probing analysis is the status of *Torreites* and *Praetorreites* [contrast

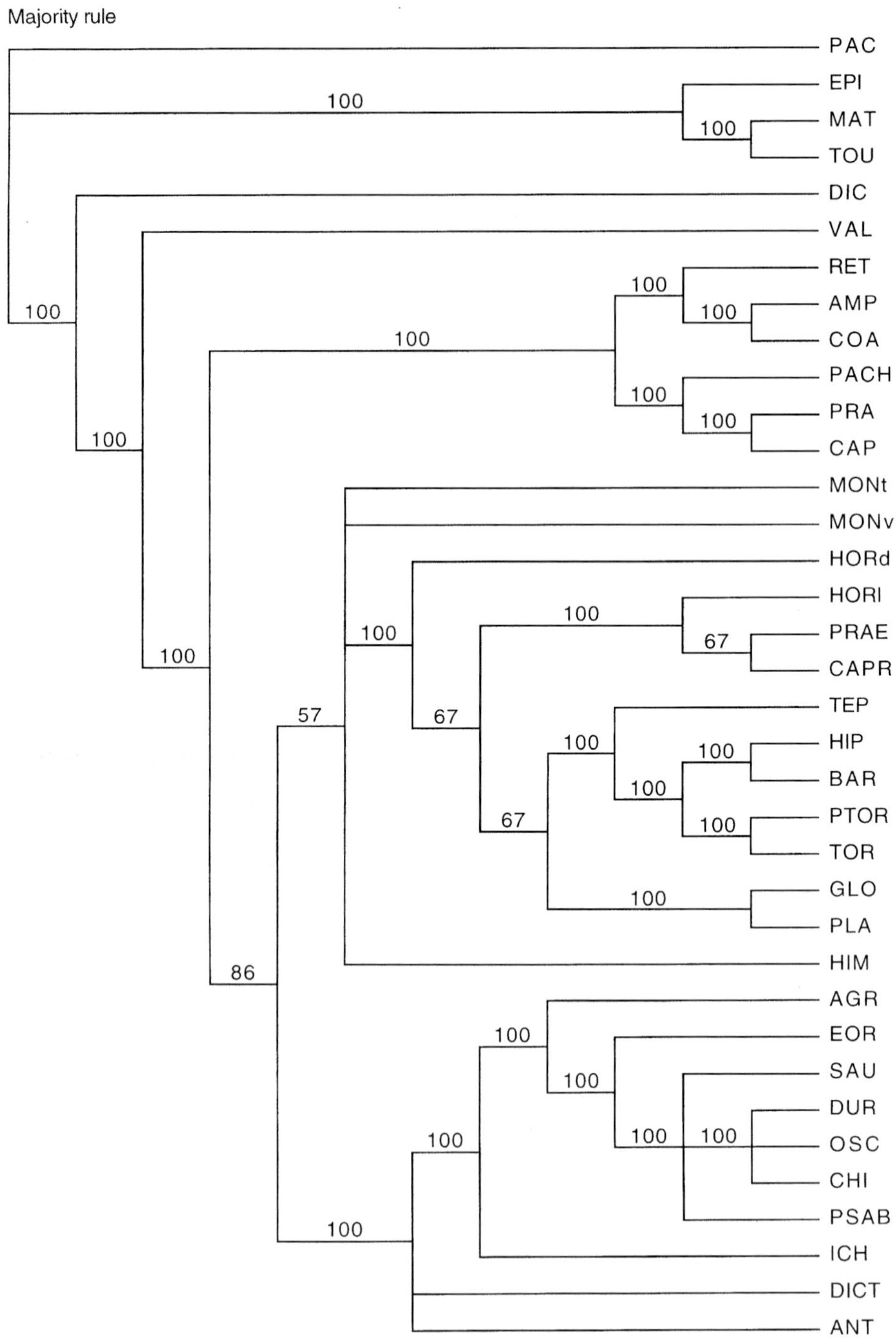

Fig. 13. Majority rule consensus tree for the 105 most parsimonious trees found in the present analysis (with simple reweighting of characters by maximum value of rescaled consistency indices). Abbreviations for taxa as in Table 2. Numbers are percentages (> 50%) of the set of most parsimonious trees that support each node; polytomy results from < 50% support for incorporated nodes. Analysis using *PAUP 4.0* (Swofford 1998).

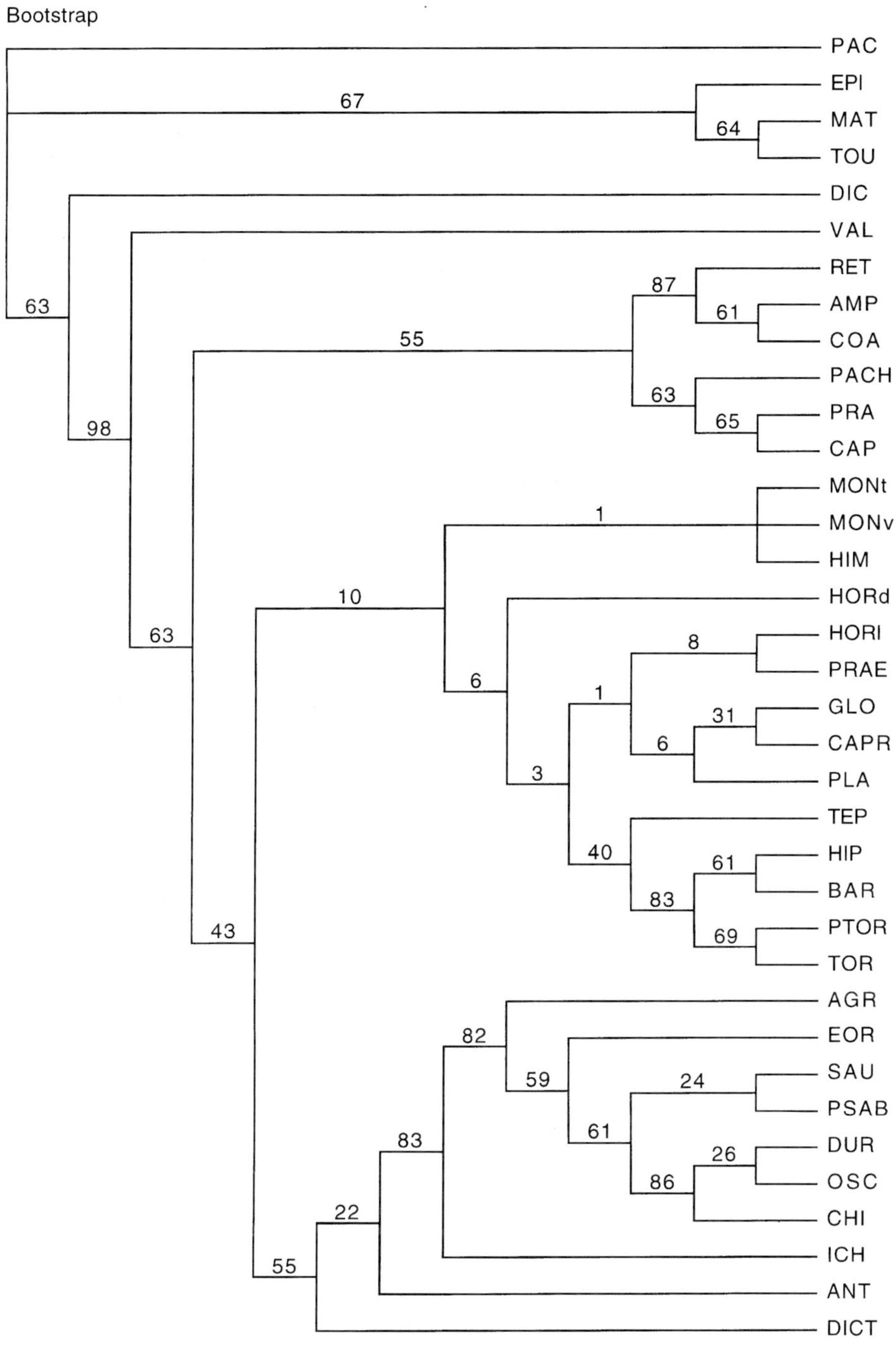

Fig. 14. Tree resulting from 1000 bootstrap replicates randomly resampled from the character matrix (with simple weighting). Abbreviations for taxa as in Table 2. Numbers are percentages of bootstrap trees supporting each node (giving an indication of the breadth of character support). Analysis using *PAUP 4.0* (Swofford 1998).

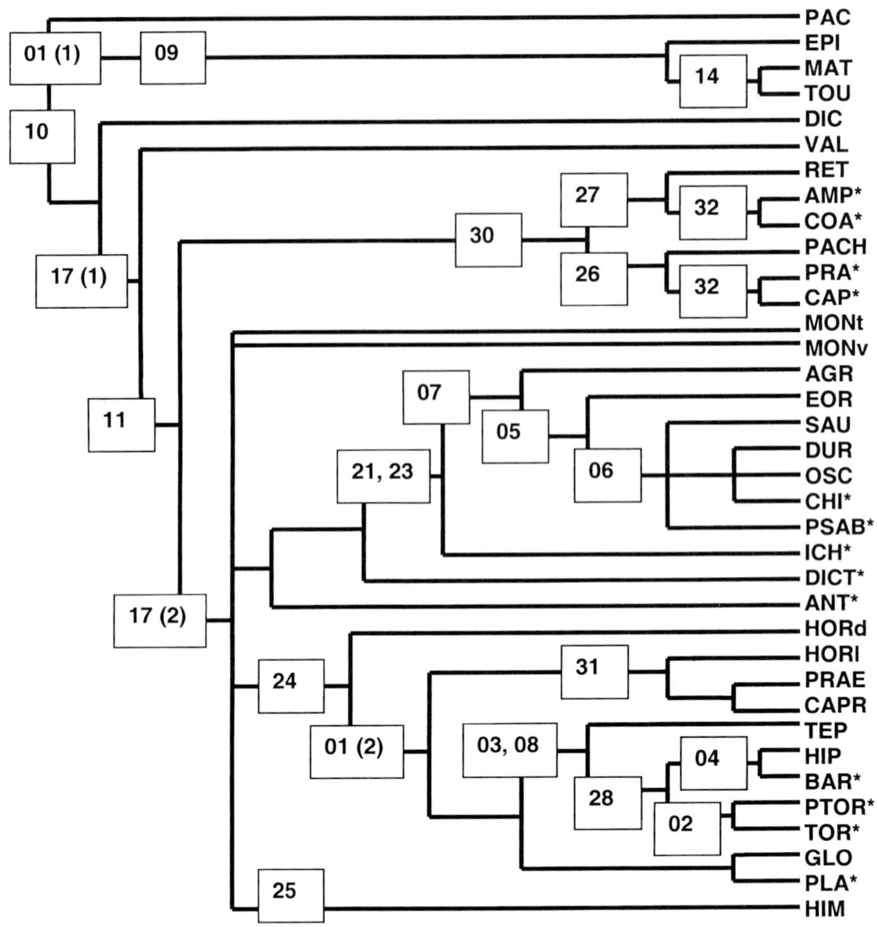

Fig. 15. One of the 105 most parsimonious trees found in the present analysis (based on simple reweighting of characters), with significant postulated synapomorphies indicated. Abbreviations for taxa as in Table 2 (*, possession of pallial canals) and the characters are numbered as in Appendix 2. Based on analysis using *PAUP 4.0* (Swofford 1998).

Skelton & Wright (1987), Philip & Platel (1994) and Morris & Skelton (1995)]. The present analysis retains the pair as a sister group to the other hippuritids. However, this leaves open the question as to whether some of their distinctive features are relatively derived, or plesiomorphic, and either further modified or lost in the other hippuritids (e.g. the pleated infoldings of the outer shell layer).

One of the most intriguing surprises of the present analysis is the sister grouping of *Tepeyacia* and the hippuritids, albeit with only weak bootstrap support. The suite of highly derived characters, already present in even the earliest known hippuritids (Simonpiétri 1999), has long left rudistologists puzzled as to their origins. Douvillé (1889) supposed them to have evolved from *Caprotina*, regarding their myocardinal arrangements as similar. Later (Douvillé 1891), he confusingly stated that the outer shell layer of hippuritids was homologous to the middle shell layer of *Plagioptychus* and the caprinids – supposing the radial canals in the hippuritid LV to correspond to the pallial canals in the latter taxa. This is clearly a false homology, however, as the outer shell layer in hippuritids, which contains the radial canals, has the same fibrillar prismatic calcite constitution as that of other rudists (Skelton 1976). The pallial canals of *Plagioptychus*, as well as the caprinids, in contrast, are in the aragonitic inner shell. They differ markedly in form, moreover, from the hippuritid canals with their externally communicating pores. Without commenting on this erroneous supposition of homology, Bilotte (1985) interpreted Douvillé's (1891) statement to imply a caprinid ancestry for the hippuritids and hypothesized a punctuational transition around the Cenomanian–Turonian boundary. With the evidence presented here for the sister grouping of *Tepeyacia* and the hippuritids, Bilotte's hypothesis now seems unlikely. Although *Plagioptychus* appears in the next sister group to the *Tepeyacia*/hippuritid clade in the majority rule tree (Fig. 13), albeit with negligible bootstrap support (Fig. 14), its pallial canals evidently evolved independently of the radial canal system in hippuritids.

Tepeyacia presents some tempting possibilities as a possible model for the origin of the hippuritids. It is not difficult to envisage a simple enhancement of the paired postero-ventral inflexions of the outer shell layer in the former (Fig. 11d) to yield the pillars and overlying oscules of the latter (Fig. 3b and c), nor do the differences in posterior myophoral configuration present much of a problem. If the polyconitid-style posterior myophore in the LV of *Tepeyacia* (Fig. 11b and c) were extended to project in tooth-like fashion into a socket in the RV, the muscle scar could then have become folded up on either side of the myophore, exactly as seen in hippuritids (Fig. 3d). The socket in the RV is partially created in hippuritids, in any case, as an embayment between the ligamentary infolding and the first pillar.

The origin of the complex canal and pore system of the hippuritids, however, is harder to interpret. One possible clue is offered by the sharp infoldings within the outer shell layer of *Tepeyacia* (Fig. 11d), corresponding to salient radial ridges on the RV inner margin. The spacing of these is similar to that of the radial canal mouths opening around the LV inner margin of hippuritids (Fig. 3b), hinting at a possible link between them. The retention of similar infoldings of the outer shell layer in *Torreites*, and perhaps, in modified form, in *Barrettia*, could then be plesiomorphic for the clade, in contrast to the species of *Hippurites* considered by the present authors, in which they appear to be lacking. In this respect, *Hippurites* ('*Batolites*') *organisans* Montfort, which does have tight infoldings in its outer shell layer (Dechaseaux 1952), merits further investigation. Another feature linking *Torreites* with *Barrettia* – though derived in this instance – is the possession of internal pallial canals within the aragonitic shell of the LV (Fig. 12; Van Dommelen 1971). There is evidently still much work to be done on the older (pre-Campanian) hippuritids of the New World, in particular, to clarify relationships within the group.

Finally, in relation to the affinities of *Tepeyacia*, it is also worth drawing attention to *Horiopleura*? *juxi* Steuber, recently described from the Aptian of Greece (Steuber 1999*a*), in which similar tight infoldings are also present around parts of the RV.

The location of *Glossomyophorus* in the polyconitid clade has little bootstrap support. This implies reliance upon a few characters, indicating a need for further testing with more (and better defined) characters. However, the similarity of its myocardinal system to those seen in other constituents (especially *Caprotina*), make the proposal quite plausible.

The polyconitid clade as a whole (*Horiopleura dumortieri* onwards; see Fig. 13), including the other taxa discussed above, received negligible bootstrap support. Hence, it would seem that it too is dependent on rather few characters – at least in the present analysis. Nevertheless, the consistency of their myophoral arrangements (albeit possibly modified in the hippuritids) weighs in favour of a close relationship between them. Further testing for the monophyly of this grouping is clearly needed.

The remaining grouping with weak bootstrap support is that of the three canaliculate taxa, *Dictyoptychus*, *Antillocaprina* and *Ichthyosarcolites*. These have previously been allocated among separate families (see earlier – History of ideas ...) and their markedly different myocardinal systems (Mac Gillavry 1937) would seem to cast doubt on close relationship between them. They have probably only become associated in the present analysis because of their relative lack of plesiomorphic character states which might tie them in with other taxa. Again, further analysis is called for.

Although the monophyletic status, or otherwise, of previously established families has been substantiated by the present analysis, several problems evidently remain. It would thus be premature to propose a formal reclassification here. This must await further analysis along the lines discussed above.

Conclusions

The following points emerge from the analysis.

- Diagnostic synapomorphies for the clade of all rudists are possession of an outer shell layer of fibrillar prismatic calcite and external asymmetry of the valves.
- Within the group, two clades are distinguished according to the attachment of the shell – either by the LV (all 'diceratids', except *Diceras* and *Valletia*, together with the monophyletic requieniids) or by the RV (*Diceras*, *Valletia* and all other rudists); 'Diceratidae' are thus paraphyletic.
- Monophyly of the rudists possessing 'inverse' dentition (two teeth in the LV and one in the RV)

is strongly supported, with *Valletia* as sister taxon, and plausible ancestor, to the uncoiled rudists (possessing an invaginated ligament).

- The Caprinidae (*sensu stricto*), Radiolitidae and Hippuritidae are confirmed as monophyletic groupings, as are the two caprinid subfamilies Caprininae and Caprinuloidinae (= Coalcomaninae).
- The identification of *Agriopleura* as the sister taxon for the radiolitids is not a surprise, as it has long been considered a likely ancestor. However, the emergence of *Tepeyacia* (an Albian Mexican form) as the sister taxon for hippuritids (albeit with weak bootstrap support) is intriguing and may offer some insight into the origins of this family of highly modified rudists.
- The rampantly polyphyletic nature of the *Treatise* 'Caprinidae' (*sensu lato*), asserted in earlier studies, is confirmed.
- The 'Cinderella' taxon of the Polyconitidae Mac Gillavry, which suffered undue neglect in the widely used *Treatise* classification promulgated in Dechaseaux *et al.* (1969), emerges from the majority rule tree based on reweighted characters, also including the hippuritids as a subclade. However, it has negligible bootstrap support and therefore needs more detailed testing.
- This analysis must be considered only as preliminary, as many relationships still have to be resolved and others remain only weakly supported; future iterations will require the inclusion of more taxa, and more, improved, definition and coding of characters.

Collaboration and discussion with many colleagues, and their generous sharing of specimens, have helped to sharpen this synthesis, and to them we express our gratitude, while reserving the blame for any remaining shortcomings for ourselves. We are particularly grateful to the referees, Alexandre Chartrousse and Thomas Steuber, who went through our initial submission with commendable care (including all those 0s and 1s in the data matrix!) and offered many valuable constructive criticisms. It was also an enormous pleasure to receive comments from Professor Henry Mac Gillavry, whose magisterial thesis on Cuba and rudists, published in 1937, is still an inspiration for rudistologists. Thanks also to Paul Jeffery (NHM) and Abel Prieur (UCBL) for access to specimens in their care and John Taylor and Andrew Lloyd at the Open University, who skillfully carried out the electronic composition of the photographic figures.

Appendix 1: sources for character coding of taxa

Abbreviations for Collections: NHM, Natural History Museum, London; UCBL, Université Claude-Bernard, Lyon, France; UPM, Université de Provence, Marseille, France.

PAC *Pachyrisma grande* Morris & Lycett: prepared topotype specimens from the Bathonian of SW England (NHM, nos 49964 and L6896).

EPI *Epidiceras sinistrum* (Deshayes): prepared specimens from the Middle Oxfordian of Dompcevrin (Meuse), NE France (Skelton 1978; NHM , LL 31920-1), and Bayle's (1873) collection of prepared specimens from around the Paris Basin (numerous specimens in UCBL).

MAT *Matheronia salevensis* Joukowsky & Favre: Joukowsky & Favre (1913), and specimens collected by C. Gourrat and D. Orbette (Société des naturalistes d'Oyonnax) from the Upper Kimmeridgian of the southern French Jura (Skelton 1999).

TOU *Toucasia carinata* (Matheron): Paquier (1903) and specimens from the Barremian of Orgon (Bouches-du-Rhône), SE France (NHM, L 10330 and unregistered).

DIC *Diceras arietinum* Lamarck: as for *Epidiceras*, also NHM specimens no. 33915 (*'D. originale' = D. arietinum*).

VAL *Valletia auris* Favre: Favre & Richard (1927) and topotype specimens from the Upper Kimmeridgian of Pierre Châtel (Ain), SE France (NHM , L 62220-8).

RET *Retha tulae* (Felix): Skelton & Masse (1998).

AMP *Amphitriscoelus waringi* Harris & Hodson: Chartrousse (1998*a*, *b*), also paralectotype specimen figured by Chartrousse (1998*b*, figs 2–6; Paleontological Research Institute, Ithaca, USA, no. 1539).

COA *Coalcomana ramosa* (Boehm): Chartrousse (1998*a*, *b*), also specimen figured by Douvillé (1900) (UCBL, EM 15687).

PACH *Pachytraga tubiconcha* (Astre): Skelton & Masse (1998).

PRA *Praecaprina varians* Paquier: Paquier (1905), also type specimens (UCBL).

CAP *Caprina adversa* d'Orbigny: Douvillé (1887, 1888), and specimens in NHM (no. 30378) and UCBL.

MONt *Monopleura taurica* Pchelintsev: Yanin (1975), also specimens donated to PWS by Professor Yanin.

MONv *Monopleura varians* Matheron: Douvillé (1918), also his figured specimens (UCBL, EM 15681-3), as well as topotype specimens from the Upper Barremian of Orgon (Bouches-du-Rhône), SE France (NHM , L 18597, unregistered '37'; and UPM).

AGR *Agriopleura* Kühn: *A. blumenbachi* (Studer) and *A. marticensis* (Matheron): Douvillé (1918) and specimens from the Barremian of SE France (NHM and UPM).

EOR *Eoradiolites plicatus* (Conrad): in Masse & Gallo Maresca (1997).

SAU *Sauvagesia* Douvillé: *S. sharpei* (Bayle), in

Toucas (1909), and *S. macroplicata* (Whitfield), in Chubb (1971).

DUR *Durania cornupastoris* (Des Moulins): Toucas (1909).

OSC *Osculigera* Kühn: *O.* cf. *vautrinioides* Vogel (Morris & Skelton 1995).

PSAB *Pseudosabinia klinghardti* (Boehm): Morris & Skelton (1995), as well as the type specimens (NHM, L 49454-6).

CHI *Chiapasella radiolitiformis* (Trechmann): Chubb (1971).

HORd *Horiopleura dumortieri* (Matheron): Barremian topotype specimens figured by Masse (1996) and Masse *et al.* (1998*a*), as well as weathered-out specimens, showing internal features, from the Lower Aptian of the Lusitanian Basin, Portugal, figured by Skelton & Masse (1998).

HORl *Horiopleura lamberti* Douvillé: Douvillé (1889), also RV figured by him (UCBL, EM 15688), and articulated specimen from Albian of Santander, NW Spain (NHM, no. 21253).

PRAE *Praecaprotina yaegashii* (Yehara): Yabe & Nagao (1926), also prepared topotype specimens of both valves, loaned to PWS from the Tohoku Imperial University, Sendai, Japan (nos 35442 and 66521).

TEP *Tepeyacia corrugata* Palmer: Palmer (1928), together with specimens from the Upper Albian–Lower Cenomanian upper part of the Mal Paso Formation, Guerrero, SW Mexico (Pantoja-Alor & Skelton 1999).

GLO *Glossomyophorus costatus* Masse, Skelton & Sliskovic: Masse *et al.* (1984), and numerous specimens observed by PWS in the Shu'aiba Formation of Arabia.

HIM *Himeraelites* Di Stefano: *H. vultur* Di Stefano (Di Stefano 1888) and congeneric specimens collected by J.-P. Masse from the ?Albian of Monte d'Ocre, central Italy [illustrated by Masse *et al.* (1998*b*)], also *'H. ugdulenae* Di Stefano*'* from Termini Imerese, Sicily, in the NHM (L 63120).

CAPR *Caprotina striata* d'Orbigny: specimens from the Cenomanian of Le Mans, France (NHM , L 96200, registered as '*C. semistriata* (d'Orbigny)').

PLA *Plagioptychus* Matheron: *P. toucasi* Matheron, prepared specimens of both valves figured by Chaper [(1873); copied in Dechaseaux (1952)] and Douvillé (1888) (UCBL, EM 15685, 6), and *P. aguilloni* (d'Orbigny) from the Santonian of SE France (La Cadière) (NHM, unregistered, Trechmann collection) and Austria (Gosau) (NHM, no. L 30151).

DICT *Dictyoptychus morgani* (Douvillé): Douvillé (1904), and numerous specimens from the Maastrichtian of the United Arab Emirates and Oman (NHM), listed in Morris & Skelton (1995), as well as sectioned specimens of '*D. persica* Cox' (syn.) from the Maastrichtian of western Iran (Pul-I-Karah) (NHM, L 58424, L 58420).

ICH *Ichthyosarcolites triangularis* Desmarest: Douvillé (1887) and specimens from the Cenomanian of western France (Charente), in the NHM (nos L 63180 and 30382).

ANT *Antillocaprina occidentalis* (Whitfield): Chubb (1971), and specimens from the Maastrichtian of Jamaica (NHM, L 63211) and of Cuba (Museo Nacional de Historia Natural, La Habana, Cuba).

HIP *Hippurites* Lamarck: especially *H. radiosus* des Moulins, prepared specimens from the uppermost Campanian/basal Maastrichtian of Charente, SW France in the NHM (nos L 18965, L 62199 and L 62201), as well as specimens of *H. socialis* Douvillé, with partial preservation of aragonite, from the Santonian of Piolenc, SE France, collected by PWS; with supplementary information on the genus from Douvillé (1891).

BAR *Barrettia monilifera* Woodward: Van Dommelen (1971), and specimens from the Campanian of Jamaica (NHM), Cuba (Museo Nacional de Historia Natural, La Habana, Cuba) and Puerto Rico (PWS collection).

PTOR *Praetorreites omanensis* Philip & Platel: Philip & Platel (1994), together with a sectioned, articulated topotype specimen from the collection of J. Philip.

TOR *Torreites sanchezi* (Douvillé): Douvillé's (1927) holotype (UCBL, EM 15689) and Palmer's (1933) specimens (Museo Nacional de Historia Natural, La Habana, Cuba); also Jung (1970), Van Dommelen (1971) and specimens of *T. s. milovanovici* Grubic, from Oman, described by Skelton & Wright (1987).

Appendix 2: Characters

CHARACTER	STATES
Calcitic outer shell layer	
#01 Outer shell layer of fibrillar prismatic calcite	0, absent; 1, thin (mostly < 1 mm); 2, thickened (>> 1 mm; often > 2 mm) in large parts of one or both valves *[ordered]*
#02 RV inner margin largely exposed beyond LV rim	0, no; 1, yes
#03 Numerous tightly infolded longitudinal pleats	0, no; 1, yes
#04 Radial canals in LV, connecting to outside via external pores and opening around inner valve margin	0, no; 1, yes
#05 Celluloprismatic mesostructure present in RV	0, no; 1, yes
#06 Cells of celluloprismatic mesostructure largely of polygonal plan	0, no; 1, yes
#07 Sharply demarcated pair of radial bands on postero-ventral flanks of shell	0, no; 1, yes
#08 Internal longitudinal swellings or pillars on postero-ventral RV wall, with matching indentations or oscules in LV	0, no; 1, yes

Appendix 2 *Continued*

Attachment of shell	
#09 Attachment to substrate by LV	0, no; 1, yes
#10 Attachment to substrate by RV	0, no; 1, yes
Ligament	
#11 Ligament insertion site invaginated	0, no; 1, yes
#12 Ligament lost	0, no; 1, yes
Shell shape	
#13 Valves externally asymmetrical	0, no; 1, yes
#14 RV relatively much reduced, hence low cap-shaped to operculiform	0, no; 1, yes
#15 LV relatively much reduced, hence low cap-shaped to operculiform	0, no; 1, yes
Pachyodont dentition	
#16 Two teeth in RV and one (plus one incipient) in LV	0, no; 1, yes
#17 One tooth in RV and two in LV	0, no; 1, yes, teeth very unequal in LV (*pt* < half *at* area); 2, yes, teeth equal to subequal in LV (*pt* > half *at* area).
#18 Section across projecting central tooth of RV shows sharp anteriorward geniculation of its dorsal part	0, no; 1, yes
#19 Two teeth in LV, only, slotting into grooves inside dorsal wall of RV	0, no; 1, yes
Anterior adductor myophores	
#20 Myophoral shelves adjoining hinge plates in both valves	0, no; 1, yes
#21 LV myophore projecting into RV, with adductor insertion rotated out to face anterior valve wall of RV	0, no; 1, yes
Posterior adductor myophores	
#22 Myophore(s) always attached to hinge plates	0, no; 1, yes
#23 LV myophore projecting into RV, with adductor insertion rotated out to face posterior body cavity wall of RV	0, no; 1, yes
#24 LV myophore a pedunculate plate, rooted on myocardinal platform, with flaring posterior rim widely separated from posterior valve margin by deep ectomyophoral recess and with adductor insertion on its inner surface opposing myophoral shelf or ridge in RV	0, no; 1, RV myophore a flat to gently sloping shelf; 2, RV myophore a postero-dorsally facing ridge
#25 LV myophore a massive buttress, rooted on myocardinal platform, with raised posterior margin separated from posterior valve rim by broad gutter, and with inwardly sloping adductor insertion on its inner face opposing broad postero-dorsally inclined shelf in RV	0, no; 1, yes
#26 LV myophore rooted on posterior valve wall at rear of myocardinal platform, with raised posterior margin separated from posterior valve rim by narrow gutter, and with adductor insertion rotated steeply inwards to face erect posterior myophoral plate in RV, which projects across into LV	0, no; 1, yes
#27 LV myophore rooted on posterior valve wall at rear of myocardinal platform, separated from posterior valve rim by narrow gutter, and projecting into RV, with adductor insertion rotated outwards to face low myophoral shelf on posterior RV wall, within posterior endomyophoral cavity	0, no; 1, yes
#28 LV myophore tooth-like and compressed, projecting into posterior myophoral socket in RV, with adductor insertions on both sides	0, no; 1, yes
Other myocardinal characters	
#29 Interior myocardinal margin between myophores in LV directly skirting ventral bases of both teeth	0, no; 1, yes
#30 Large posterior endomyophoral cavity in LV, bounded internally by lamina, and separating posterior tooth and posterior myophore from general cavity	0, no; 1, yes
#31 Conical annexe to ectomyophoral recess penetrating under ventral part of LV posterior myophore	0, no; 1, yes
Pallial canals	
#32 Pallial canals regularly developed in aragonitic inner shell layers	0, no; 1, yes, in LV only; 2, yes, in both valves

References

AL-AASM, I. S. & VEIZER, J. 1986. Diagenetic stabilization of aragonite and low-Mg calcite, I. Trace elements in rudists. *Journal of Sedimentary Petrology*, **56**, 138–152.

BAYLE, E. 1855. Observations sur la structure des coquilles des *Hippurites*, suivies de quelques remarques sur les *Radiolites*. *Bulletin de la Société géologique de France*, **12**(2), 772–807.

—— 1873. Observations sur quelques espèces du genre *Diceras*. *In*: BAYAN, J. F. F. (ed.) *Etudes faites dans la Collection de l'Ecole des Mines sur des fossiles nouveaux ou mal connus*. Unpublished report deposited at the Ecole de Mines (now at UCBL), Paris.

BILOTTE, M. 1985. Le Crétacé supérieur des plates-formes est-pyrénéenes. *Strata*, Series 2, **5**.

BOEHM, G. 1882. Über die Beziehungen von *Pachyrisma*, *Megalodon*, *Diceras* und *Caprina*. *Zeitschrift der deutschen geologischen Gesellschaft*, **34**, 602–617.

CARANNANTE, G., GRAZIANO, R., PAPPONE, G., RUBERTI, D. & SIMONE, L. 1999. Depositional system and response to sea level oscillations of the Senonian rudist-bearing carbonate shelves. Examples from central Mediterranean areas. *Facies*, **40**, 1–24.

CHAPER, M. 1873. Observations sur une espèce du genre *Plagioptychus*. *Etudes faites dans la Collection de l'Ecole des Mines*, **2**, 82–90.

CHARTROUSSE, A. 1998*a*. *Les Caprinidae (Rudistes) du Crétacé inférieur*. Thèse de Doctorat, Université de Provence – Centre de Sédimentologie et Paléontologie, Marseille.

—— 1998*b*. The myocardinal organization of coalcomaninid rudists revisited. *In*: MASSE, J.-P. & SKELTON, P. W. (eds) *Quatrième Congrès international sur les Rudistes*. Geobios, Mémoire spécial, **22**, 75–85.

CHUBB, L. J. 1971. Rudists of Jamaica. *Palaeontographica americana*, **7/45**, 157–257.

COOGAN, A. H. 1973. New rudists from the Albian and Cenomanian of Mexico and south Texas. *Revista del Instituto mexicano del Petróleo*, **5**, 51–83.

DECHASEAUX, C. 1952. Classe des lamellibranches. *In*: PIVETEAU, J. (ed.) *Traité de paléontologie, Vol. 2*. Masson, Paris, 220–364.

——, COX, L. R., COOGAN, A. H. & PERKINS, B. F. 1969. Superfamily Hippuritacea Gray, 1848. *In*: MOORE, R. C. (ed.) *Treatise on Invertebrate Paleontology, Part N, Mollusca 6, Bivalvia, 2*. Geological Society of America, Boulder, CO, and University of Kansas, Lawrence, KS, N749–N817.

DESHAYES, G. P. 1825. Quelques observations sur les genres Hippurite et Radiolite. *Annales des Sciences naturelles*, **5**(1), 205–211.

DOUVILLÉ, H. 1886. Essai sur la morphologie des rudistes. *Bulletin de la Société géologique de France*, **14**(3), 389–404.

—— 1887. Sur quelques formes peu connues de la famille des chamidés. *Bulletin de la Société géologique de France*, **15**(3), 756–802.

—— 1888. Etudes sur les caprines. *Bulletin de la Société géologique de France*, **16**(3), 699-730.

—— 1889. Sur quelques rudistes du terrain crétacé inférieur des Pyrénées. *Bulletin de la Société géologique de France*, **17**(3), 627–635.

—— 1891. Etudes sur les rudistes. Révision des principales espèces d'Hippurites. *Mémoires de la Société géologique de France, Paléontologie*, **6**(1), 1–31.

—— 1896. Observations sur la charnière des lamellibranches hétérodontes. *Bulletin de la Société géologique de France*, **24**(3), 26–28.

—— 1900. Sur quelques rudistes américains. *Bulletin de la Société géologique de France*, **28**(3), 205–221.

—— 1904. Sur quelques rudistes à canaux. *Bulletin de la Société géologique de France*, **4**(4), 519–538.

—— 1915. Les réquiénidés et leur évolution. *Bulletin de la Société géologique de France*, **14**(4), 383–389.

—— 1918. Le Barrémien supérieur de Brouzet. 3e partie, les rudistes. *Mémoires de la Société géologique de France, Paléontologie*, **52**.

—— 1927. Nouveaux rudistes du Crétacé de Cuba. *Bulletin de la Société géologique de France*, **7**(4), 49–56.

—— 1935. Les rudistes et leur évolution. *Bulletin de la Société géologique de France*, **5**(5), 319–358.

FAVRE, J. & RICHARD, A. 1927. Etude du Jurassique supérieur de Pierre-Châtel et de la cluse de la Balme (Jura méridional). *Mémoires de la Société paléontologique suisse*, **46**, 1–39.

FISCHER, P. 1887. *Manuel de conchyliologie et de paléontologie conchyliologique ou histoire naturelle des mollusques vivants et fossiles*. F. Savy, Paris.

FORTEY, R. A. & JEFFERIES, R. P. S. 1982. Fossils and phylogeny – a compromise approach. *In*: JOYSEY, K. A. & FRIDAY, A. E. (eds) *Problems of Phylogenetic Reconstruction*. Systematics Association Special Volume No. 21, Academic Press, London, 197–234.

GILI, E., MASSE, J.-P. & SKELTON, P. W. 1995. Rudists as gregarious sediment-dwellers, not reef-builders, on Cretaceous carbonate platforms. *Palaeogeography, Palaeoclimatology, Palaeoecology*, **118**, 245–267.

GRAY, J. E. 1848. On the arrangement of the Brachiopoda. *Annals and Magazine of Natural History*, **2**(2), 435–440.

JOUKOWSKY, E. & FAVRE, J. 1913. Monographie géologique et paléontologique du Salève (Haute Savoie, France). *Mémoires de la Société de Physique et d'Histoire naturelle de Genève*, **37**, 295–519.

JUNG, P. 1970. *Torreites sanchezi* (Douvillé) from Jamaica. *Palaeontographica americana*, **7/42**, 1–15.

KENNEDY, W. J., MORRIS, N. J. & TAYLOR, J. D. 1970. The shell structure, mineralogy and relationships of the Chamacea (Bivalvia). *Palaeontology*, **13**, 379–413.

LAMARCK, J. B. DE 1819. *Histoire naturelle des animaux sans vertèbres. Volume 6 (1)*. Verdière, Paris, 230–240.

MAC GILLAVRY, H. J. 1935. Remarks on rudists. *Koninklijke Akademie van Wetenschappen te Amsterdam, Proceedings of the Section of Sciences*, **38**, 558–565.

—— 1937. Geology of the province of Camagüey, Cuba,

with revisional studies in rudist paleontology. *Geographische en Geologische Mededeelingen, Physiographisch-Geologische reeks*, **14**.

—— 1970. Geological history of the Caribbean, 1. *Koninkl. Nederl. Akademie van Wetenschappen, Proceedings* **B73**(1), 64–83.

MASSE, J.-P. 1994. L'évolution des Requieniidae (rudistes) du Crétacé inférieur: caractères, signification fonctionelle adaptive et relations avec les modifications des paléoenvironnements. *Geobios*, **27**, 321–333.

—— 1996. Lower Cretaceous rudist biostratigraphy of southern France – a reference for Mesogean correlations. *In*: ALENCÁSTER, G. & BUITRÓN-SÁNCHEZ, B. E. (eds) *Número dedicado a la Tercera Conferencia Internacional sobre rudistas*. Revista Mexicana de Ciencias Geológicas, **12**, 236–256.

—— & GALLO MARESCA, M. 1997. Late Aptian Radiolitidae (rudist bivalves) from the Mediterranean and southwest Asiatic regions: taxonomic, biostratigraphic and palaeobiogeographic aspects. *Palaeogeography, Palaeoclimatology, Palaeoecology*, **128**, 101–110.

—— & PHILIP, J. 1974. Définition, position systématique, répartition stratigraphique et évolution du genre *Agriopleura* Kühn (rudiste). *Géologie méditerranéenne*, **1**, 53–62.

—— & —— 1986. L'évolution des rudistes au regard des principaux événements géologiques du Crétacé. *Bulletin des Centres de Recherches Exploration–Production Elf-Aquitaine*, **10**, 437–456.

——, ARIAS, C. & VILAS, L. 1998*a*. Lower Cretaceous rudist faunas of southeast Spain: an overview. *In*: MASSE, J.-P. & SKELTON, P. W. (eds) *Quatrième Congrès international sur les Rudistes*. Geobios, Mémoire spécial, **22**, 193–210.

——, GALLO MARESCA, M. & LUPERTO SINNI, E. 1998*b*. Albian rudist faunas from southern Italy: taxonomic, biostratigraphic and palaeobiogeographic aspects. *Geobios*, **31**, 47–59.

——, SKELTON, P. W. & SLISKOVIC, T. 1984. *Glossomyophorus costatus* nov. gen. nov. sp., rudiste (Caprotinidae) nouveau de l'Aptien du domaine méditerranéen central et oriental. *Geobios*, **17**, 723–732.

MORRIS, N. J. & SKELTON, P. W. 1995. Late Campanian–Maastrichtian rudists from the United Arab Emirates–Oman border region. *Bulletin of the British Museum (Natural History), Geology Series*, **51**, 277–305.

MUNIER CHALMAS, H. 1873. Prodrôme d'une classification des rudistes. *Journal de Conchyliologie*, **13**(3), 71–75.

—— 1882. I. Etudes critiques sur les rudistes. *Bulletin de la Société géologique de France*, **10**(3), 472–482.

NEWELL, N. D. 1965. Classification of the Bivalvia. *American Museum of Natural History, Novitates*, **2206**.

PALMER, R. H. 1928. The rudistids of southern Mexico. *Occasional Papers of the California Academy of Sciences*, **14**.

—— 1933. Nuevos rudistas de Cuba. *Revista de Agricultura, Comercio y Trabajo*, **14**, 95–125.

PANTOJA-ALOR, J. & SKELTON, P. W. 1999. Polyconitid rudists from the Mal Paso Formation (Albian–Lower Cenomanian) around Chumbítaro, Guerrero, SW Mexico. *In*: HÖFLING, R. & STEUBER, T. (eds) Fifth International Congress on Rudists – Abstracts and Field Trip Guides. *Erlanger Geologische Abhandlungen*, Sonderaband, **3**, 45–47.

PAQUIER, V. 1903. Les rudistes urgoniens. Première partie. *Mémoires de la Société géologique de France, Paléontologie*, **29**(11), 1–46.

—— 1905. Les rudistes urgoniens. II. Série inverse. *Mémoires de la Société géologique de France, Paléontologie*, **29**(13), 47–102.

PHILIP, J. 1986. Etude paléontologique du genre *Sabinia* (rudiste à canaux) des récifs du Campanien de Tunisie. *Geobios*, **19**, 247–251.

—— 1998. Biostratigraphie et paléobiogéographie des rudistes: évolution des concepts et progrès récents. *Bulletin de la Société géologique de France*, **169**, 689-708.

—— & PLATEL, J.-P. 1994. *Praetorreites*, nouveau genre de rudiste du Campanien d'Oman. *Geobios*, **27**, 303–319.

——, MASSE, J.-P. & CAMOIN, G. 1995. Tethyan carbonate platforms. *In*: NAIRN, A. E. M. *ET AL.* (eds) *The Ocean Basins and Margins, Vol. 8: The Tethys Ocean*. Plenum Press, New York, 239–265.

QUENSTEDT, F. A. 1852. *Handbuch der Petrefaktenkunde*. Laupp & Siebeck, Tübingen.

RIDE, W. D. L., *ET AL.* (eds). 1999. *International Trust for Zoological Nomenclature,* Fourth edition. International Trust for Zoological Nomenclature, C/O The Natural History Museum, London, UK.

ROSS, D. J. & SKELTON, P. W. 1993. Rudist formations of the Cretaceous: a palaeoecological, sedimentological and stratigraphical review. *In*: WRIGHT, P. (ed.) *Sedimentology Review*, **1**. Blackwell Scientific, Oxford, 73–91.

SIMONPIÉTRI, G. 1999. *Systématique, phylogenèse, ontogenèse chez les Hippuritidae (rudistes de Crétacé supérieur)*. Thèse de Doctorat, Université de Provence – Centre de Sédimentologie et Paléontologie, Marseille.

SKELTON, P. W. 1976. Functional morphology of the Hippuritidae. *Lethaia*, **9**, 83–100.

—— 1978. The evolution of functional design in rudists (Hippuritacea) and its taxonomic implications. *Philosophical Transactions of the Royal Society of London, Series B*, **284**, 305–318.

—— 1979. Preserved ligament in a radiolitid rudist bivalve and its implication of mantle margin feeding in the group. *Paleobiology*, **5**, 90–106.

—— 1985. Preadaptation and evolutionary innovation in rudist bivalves. *In*: COPE, J. C. W. & SKELTON, P. W. (eds) *Evolutionary Case Histories from the Fossil Record*. Special Papers in Palaeontology, **33**, 159–173.

—— 1991. Morphogenetic versus environmental cues for adaptive radiations. *In*: SCHMIDT-KITTLER, N. & VOGEL, K. (eds) *Constructional Morphology and Evolution*. Springer, Berlin, 375–388.

—— 1999. Synoptic guide to Kimmeridgian rudists for the Kelheim field visit. *In*: HÖFLING, R. & STEUBER, T. (eds) *Fifth International Congress on Rudists – Abstracts and Field Trip Guides*. Erlanger

Geologische Abhandlungen, Sonderaband, **3**, 83–89.

—— & BENTON, M. J. 1993. Mollusca: Rostroconchia, Scaphopoda and Bivalvia. *In:* BENTON, M. J. (ed.) *The Fossil Record 2*. Chapman & Hall, London, 237–263.

—— & MASSE, J.-P. 1998. Revision of the Lower Cretaceous rudist genera *Pachytraga* Paquier and *Retha* Cox (Bivalvia: Hippuritacea), and the origins of the Caprinidae. *In*: MASSE, J.-P. & SKELTON, P. W. (eds) *Quatrième Congrès International sur les Rudistes*. Geobios, Mémoire spécial, **22**, 331–370.

—— & —— 2000. Synoptic guide to Lower Cretaceous rudist bivalves of Arabia. *In*: SCOTT, R. W. & ALSHARHAN, A. (eds) Middle East Models of Jurassic/Cretaceous Carbonate Systems. *SEPM Special Volume*, **69**, 85–95.

—— & WRIGHT, V. P. 1987. A Caribbean rudist bivalve in Oman: island hopping across the Pacific in the late Cretaceous. *Palaeontology*, **30**, 505–529.

SMITH, A. B. 1994. *Systematics and the Fossil Record – Documenting Evolutionary Patterns*. Blackwell Scientific, Oxford.

STEFANO, G. D. 1888. Studi stratigrafici e paleontologici sul sistema Cretaceo della Sicilia. 1. Gli strati con *Caprotina* di Termini-Imerese. Atti della Reale Accademia di Scienze, Lettere e Belle Arti, (n.s.) **10**, 1–44.

STEUBER, T. 1999*a*. Cretaceous rudists of Boeotia, central Greece. *Special Papers in Palaeontology*, **61**.

—— 1999*b*. Database on rudist palaeontology. Website: http://www.geol.uni-erlangen.de/pal/Steuber/index.htm

STONE, J. R. & TELFORD, M. 1999. Using critical path method to analyse the radiation of rudist bivalves. *Palaeontology*, **42**, 231–242.

SWOFFORD, D. L. 1998. *PAUP**. *Phylogenetic Analysis Using Parsimony (*and other methods). Version 4.* Software diskette. Sinauer Associates, Sunderland, MA.

TOUCAS, A. 1907. Etudes sur la classification et l'évolution des Radiolitidés: *Agria* & *Praeradiolites*. *Mémoires de la Société géologique de France, Paléontologie*, **14**(36), 1–46.

—— 1909. Etudes sur la classification et l'évolution des Radiolitidés: *Sauvagesia* & *Biradiolites*. *Mémoires de la Société géologique de France, Paléontologie*, **17**(36), 79–132.

VAN DOMMELEN, H. 1971. *Ontogenetic, phylogenetic and taxonomic studies of the American species of* Pseudovaccinites *and of* Torreites *and the multiple-fold hippuritids*. Thesis, University of Amsterdam.

VICENS, E. 1992. Intraspecific variability in Hippuritidae in the southern Pyrenees, Spain: Taxonomic implications. *Geologica romana*, **28**, 119–161.

WOODWARD, S. P. 1855. On the structure and affinities of the Hippuritidae. *Quarterly Journal of the Geological Society of London*, **11**, 40–61.

YABE, H. & NAGAO, T. 1926. *Praecaprotina*, nov.gen., from the Lower Cretaceous of Japan. *Science Reports of the Tohoku University*, **9**(2), 21–24.

YANIN, B. T. 1975. Epibiosis and immuration among the rudists (*Monopleura*). *Paleontologicheskij Sbornik*, **12**, 72-76.

ZITTEL, K. A. 1885. *Handbuch der Palaeozoologie. II. Mollusca und Arthropoda*. Oldenbourg, München.

Relationships between the extant Anomalodesmata: a cladistic test

ELIZABETH M. HARPER[1], ELIZABETH A. HIDE[2] & BRIAN MORTON[3]

[1] *Department of Earth Sciences, Downing Street, Cambridge CB2 3EQ, UK (e-mail: emh21@cus.cam.ac.uk)*

[2] *Department of Geology and Zoology, National Museums of Scotland, Edinburgh EH1 1JF, UK*

Present address: Department of Earth Sciences, Downing Street, Cambridge CB2 3EQ, UK

[3] *The Swire Institute of Marine Science and Department of Ecology and Biodiversity, The University of Hong Kong, Hong Kong*

Abstract: Although ancient anomalodesmatans were apparently abundant shallow and deep burrowers, in Recent seas the subclass comprises some of the most specialized and rarest of all bivalves. The morphological adaptations associated with diverse life habits has persistently frustrated attempts to achieve a widely accepted scheme for the relationships between extant families.

A cladistic analysis, using 43 informative anatomical and shell characters for each of the extant anomalodesmatan families has produced a single, reasonably robust tree which is in broad, albeit imperfect, agreement with the known fossil record of the subclass. This total evidence tree places the Pandoridae, Lyonsiidae, Cleidothaeridae and Myochamidae, and also the Thraciidae, Periplomatidae and Laternulidae in monophyletic groups. Carnivory appears diphyletic, with the Parilimyidae separated from the 'septibranch' families (Cuspidariidae, Verticordiidae, Lyonsiellidae and Poromyidae) which form a monophyletic group. The enigmatic tube-dwelling Clavagellidae appear as a sister group to the 'septibranchs'.

Re-analysis of the data matrix using only those 18 characters which could be scored from shell characters alone, produced a tree which contradicted the total evidence tree rather than producing a poorly resolved version. The degree of convergence shown by shell characters make it, at least at present, difficult to include the extinct anomalodesmatan families in a cladistic analysis.

Nearly one-sixth of all bivalve families which ever lived are placed in the Anomalodesmata Dall, 1899. The earliest recognized anomalodesmatan, *Arenigomya carinata* Cope, is of Lower Ordovician (Lower Arenig) age (Cope 1996*a*), and from that time onwards members of the subclass are abundant in Palaeozoic and Mesozoic shallow-marine facies the world over. Then, they occupied predominantly shallow and deep-burrowing life habits, but also had endobyssate representatives (Bambach 1971; Runnegar 1974). During the Cenozoic, there was a marked proliferation in both the number of recognized anomalodesmatan families and the number of life habits exploited (Skelton *et al.* 1990, fig. 5.2) but there was also a concomitant decline in their abundance, many taxa being highly restricted in their distributions. Morton (1985*a*) attributed this decline to a failure to compete with more generalist heterodonts.

There are currently 14 recognized extant families of anomalodesmatans (Table 1), most of which comprise few genera and many occur within restricted geographic ranges. Although they occur in a wide range of habitats from shallow coastal waters to the deep sea, the niches of many are so highly specialized, e.g. pholadomyids, parilimyids and clavagellids, that they are among the rarest of all known bivalves. Virtually every life habit exploited by any bivalve has been evolved by anomalodesmatans: shallow and deep burrowing in soft sediments (Pholadomyidae, Pandoridae, Thraciidae, Laternulidae and Periplomatidae), and attachment to hard surfaces either by byssal threads (Lyonsiidae) or by permanent cementation (Cleidothaeridae and some Myochamidae). They also include deep-sea predators and scavengers (Parilimyidae, Verticordiidae, Cuspidariidae and Poromyidae) and the enigmatic tube-dwelling Clavagellidae, some of which have the ability to bore into hard substrata. The various adaptations that attend this range of life habits has produced within the anomalodesmatans such different morphologies that they have confounded previous attempts to analyse their phylogenetic

From: HARPER, E. M., TAYLOR, J. D. & CRAME, J. A. (eds) *The Evolutionary Biology of the Bivalvia*. Geological Society, London, Special Publications, **177**, 129–143. 1-86239-076-2/00/$15.00

Table 1. *List of the extant families of Anomalodesmata and the representative taxa scored in this cladistic analysis along with the principal study utilized*

Family	Representative taxa	Prime references
Clavagellidae	*Clavagella australis* Sowerby, 1830* *Brechites vaginiferus* (Lamarck, 1818)	Morton (1984*a*) Morton (1984*b*)
Cleidothaeridae	*Cleidothaerus maorianus* Finlay, 1827*†	Morton (1974)
Cuspidariidae	*Cuspidaria cuspidata* (Olivi, 1792)*†	Yonge & Morton (1980), Morton (1987)
Laternulidae	*Laternula truncata* (Lamarck, 1818)*	Morton (1973)
Lyonsiellidae	*Lyonsiella formosa* (Jeffreys, 1881)*	Morton (1984*c*)
Lyonsiidae	*Entodesma saxicola* Baird, 1836	Yonge (1952), Morgan & Allen (1976)
Myochamidae	*Myadora striata* (Quoy & Gaimard, 1835) *Myochama anomioides* Stutchbury, 1830*†	Morton (1977) Harper & Morton (2000)
Pandoridae	*Pandora inaequivalvis* (Linnaeus, 1789)*† *Frenamya ceylanica* Sowerby, 1835	Allen (1954) Morton (1984*d*)
Parilimyidae	*Parilimya fragilis* (Grieg, 1920)*	Morton 1982
Periplomatidae	*Periploma angasi* Crosse & Fischer, 1864*	Morton (1981*b*)
Pholadomyidae	*Pholadomya candida* Sowerby, 1823*†	Morton (1980)
Poromyidae	*Poromya granulata* (Nyst & Westendorp, 1839)*†	Yonge (1928), Yonge &Morton (1980)
Thraciidae	*Trigonothracia jinxingae* Xu, 1980	Morton (1995)
Verticordiidae	*Verticordia triangularis* Locard, 1898*	Allen & Turner (1974), Morton (1987)

Character 37 was also scored with reference to Morton (1985*b*). *, Representatives of the type genus; †, type species; ‡, Morton (1974) considers that *Cleidothaerus maorianus* may be a synonym of the type species *C. albidus* (Lamarck, 1819).

interrelationships. It is likely that these problems are exacerbated by convergent and parallel evolution, where lifestyles such as deep burrowing, cementing and even predation *appear* to have arisen independently in a number of different anomalodesmatan taxa. As a result, there are a plethora of different models to describe evolution within the subclass. Difficulties in relating the extant anomalodesmatan families to each other and to their Palaeozoic and Mesozoic precursors led Morton (1981*a*, p. 53) to regard them as 'the widely spaced outermost twigs of a tree, the roots of which have long since perished ...'. A more reliable phylogeny of the Anomalodesmata is crucial to enable better understanding of the nature of morphological evolution in the subclass, and would form a necessary phylogenetic basis for a classification of both fossil and modern representatives of the group.

There are two avenues of research which may illuminate the relationships between the extant anomalodesmatans: a molecular phylogeny and a cladistic study of their morphological characters. Many of the key anomalodesmatan taxa, however, are extremely rare, for example, the anatomical studies on *Parilimya fragilis* (Morton 1982) were carried out on a specimen collected in 1910, whilst the specimen of *Pholadomya candida* (the sole species of the ancient Pholadomyidae), studied by both Runnegar (1979) and Morton (1980), was collected in 1838. Although there have been detailed anatomical studies of at least one member of each of the families, this scarcity of material precludes any realistic chance of surveying the molecular phylogeny of the subclass comprehensively, at least at present. In any case, any molecular study should not be considered in isolation, but in conjunction with a comprehensive morphological study. In this paper the first cladistic analyses of the subclass is presented and consideration is given to the possibility of extending it to cover extinct families.

Use of cladistics in reconstructing bivalve phylogenies

Cladistic methods of phylogenetic reconstruction use the recognition of shared, derived characters (i.e. synapomorphies) to construct trees using these evolutionary innovations to support clades. Approaches of this type will, by definition, work perfectly when all characters imply the same tree, although experience shows that this is rarely the case and evolutionary innovations can and do occur more than once during the evolution of a clade, with the result that different characters may imply different trees. Because of these character incompatibilities, it is necessary to invoke the criterion of parsimony to choose between competing hypotheses of relationships. A data set in which there is either little or no convergent character state change will produce a well-supported cladogram; conversely, one in which many evolutionary reversals and parallel changes occur will be poorly supported.

Underlying the cladistic method, therefore, is the assumption that the majority of evolutionary change is divergent. It is unlikely that this is true for either the bivalves as a whole or the Anomalodesmata in particular. Yonge (1962), for example, suggested that the adult byssus has been evolved repeatedly among the Bivalvia. Nevertheless, the use of cladistic methodology allows a re-examination of characters of all types in a way which is consistent across their range. Although it is recognized that a cladistic data set can never be wholly objective, there are definite advantages in working with a wider ranging morphological data set compared to those studies which dwell on either one or a few characters, such as gills (Atkins 1936), stomach (Purchon 1987*a*) and hinge structures (Dall 1889; Douvillé 1913) or a combination of these (Nevesskaya *et al.* 1971). Use of either single or restricted characters in phylogenetic reconstructions makes the a priori assumption that the evolution of that character has shown no convergence. Purchon (1978, 1987*b*) did recognize the importance of considering multiple characters in establishing a classification for the Bivalvia, however, he adopted a phenetic approach which lacks an evolutionary basis.

Cladistic studies have been attempted for both the higher classification of the Mollusca (Salvini-Plawen & Steiner 1996) and for intraclass phylogenies; e.g. they have been widely used within the gastropods (Taylor *et al.* 1993; Hickman 1996) and have been used for the scaphopods (Steiner 1998). However, there are few rigorous computer-driven cladistic studies for the Bivalvia. Most such studies concentrated on relatively small taxonomic units, e.g. Cardioidea (Schneider 1992, 1995, 1998) and the Galeommatidae (Bieler & Mikkelsen 1992). An early cladistic analysis of the anomalodesmatan Pandoroidea was undertaken by Boss (1978) but it was not supported by a character matrix, nor was it possible at that time to use computer-driven methods to search for the most parsimonious tree. Thus far, the only analyses which have tackled higher bivalve taxa have been confined to the Pteriomorpha (Waller 1978, 1998) and these also have lacked supporting information. At least part of this reluctance to embrace cladistic methodology has been summed up by Seilacher (1984, p. 235), who suggested that convergence and parallelism are so common amongst the Bivalvia as to make the technique of limited use. Schneider (1995), however, has countered this by pointing out that instead of masking relationships a proper multicharacter analysis might reveal hitherto unnoticed polyphyletic traits, a point with which the present authors concur. Certainly, bivalves seem to be promising candidates for cladistic analysis, having a wide range of definable shell and tissue characters and an excellent fossil record, which exceeds the level of completeness suggested by Fortey & Jefferies (1982), which would be useful in the recognition of ancestral states and in discriminating between various trees (Harper 1998).

The present authors' decision to undertake a cladistic analysis of the Anomalodesmata forms part of a wider study of the evolution of this enigmatic subclass. It has the advantage that one of us (BM) has undertaken anatomical studies of members of every living anomalodesmatan family, thus providing a wealth of potentially useful phylogenetic information and also a vital consistency of approach to the recognition and scoring of characters. Morton's original observations have been extended to include characters which relate to gross shell morphology and detailed microstructure.

Although this study concentrates on the extant anomalodesmatans, it is hoped that eventually extinct families can be included in an overall phylogeny. To this end, the phylogenetic information of a data set which contains only shell characters is specifically compared with a more comprehensive morphological data set. Three particular areas of anomalodesmatan relationships have been identified to focus on, which are outlined below.

Relationships of the Pandoroidea

In the *Treatise,* Keen (*in* Moore 1969) recognized the Pandoroidea as encompassing the extant Pandoridae, Cleidothaeridae, Laternulidae, Lyonsiidae, Myochamidae, Periplomatidae and Thraciidae [as well as the extinct Margaritidae, although Runnegar (1974) suggested this family belonged to the Pholadomyoidea]. Runnegar (1974) also included the Verticordiidae within the Pandoroidea, and suggested that the Cleidothaeridae were derived from them. Boss (1978) and Yonge & Morton (1980), however, suggested, on the basis of hinge detail, that these families could be split into two superfamilies; the Pandoroidea (Pandoridae, Cleidothaeridae, Lyonsiidae and Myochamidae) and the Thracioidea (Laternulidae, Periplomatidae and Thraciidae). Furthermore, they pointed out that such a division was supported by the fossil record. One of the specific aims of this study, therefore, was to test the separation of these two superfamilies and to consider the position of the Verticordiidae in relation to both of them. Nevertheless, the *Treatise* (Moore 1969) classification of these taxa has been retained by some recent classifications (Amler 1999).

Relationships of the Clavagellidae to other anomalodesmatans

Representatives of the tube-dwelling Clavagellidae are some of the most enigmatic bivalves. Two lineages have been defined (Savazzi 1982) based on two extant genera, i.e. *Clavagella* (Cretaceous–Recent), which are facultative and obligate borers with their left valves attached to the adventitious tube, and the geologically younger *Brechites* (Oligocene–Recent), which has both valves incorporated into its tube and lacks the ability to bore. The relationships between these two taxa and other anomalodesmatans are open to considerable debate, having been placed consistently in their own superfamily, the Clavagelloidea within the Pholadomyoida (Keen & Smith *in* Moore 1969; Runnegar 1974). Some authors suggest an origin amongst the deep-burrowing Pholadomyoidea, e.g. Carter (1978) points to the Pandoroidea and Morton (1984*a*, *b*), more specifically, to the Laternulidae, whilst Morton (1981*a*) and Pojeta & Sohl (1987) consider the Pholadomyidae to be putative ancestors. In contrast, Savazzi (1982) has suggested an endolithic origin, arguing that the boring *Clavagella* is the more primitive of the two extant genera. Most recently, Savazzi (1999) has introduced the possibility that *Clavagella* and *Brechites* might exhibit parallel evolution and need not necessarily be closely related.

Relationships of the carnivorous anomalodesmatans

Carnivory has been recognized for some time in the deep-water Cuspidariidae, Poromyidae and Verticordiidae which are known to ingest small arthropods and polychaetes (Pelseneer 1891; Yonge 1928; Knudsen 1970; Morton 1987). The actual mechanism of prey capture has only been observed in the cuspidariids by Reid & Reid (1974) and Reid & Crosby (1980). The carnivorous taxa share a number of morphological adaptations to this habit, including possession of an eversible raptorial inhalant siphon, a modified stomach and reduction of the gill, and have often been grouped accordingly with various degrees of formality as the septibranchs (Knudsen 1970; Bernard 1974, 1979). More recently, Morton (1982) has suggested that species of the Parilimyidae may also be carnivores, because of their possession of taenoid muscles, a type-II stomach and a modified siphon, and argued that the family may have been the ancestral group of the other better known septibranchs.

Of all the problems with anomalodesmatan relationships, it is those of the carnivorous taxa which have been debated most. Numerous conflicting models have been put forward to explain the relationships between these taxa and other anomalodesmatan taxa. Figure 1 depicts these various hypotheses as a series of trees. In essence, the hypotheses can be divided into two groups: (1) those which view the carnivorous taxa as a monophyletic group, albeit with different relationships within it (hypotheses A–D); (2) those which propose at least a diphyletic arrangement, even deriving some of the carnivorous groups from outside the Anomalodesmata (hypotheses E–H). It should be noted that some authors (Bernard 1979; Morton 1987) regard the verticordiid genera *Lyonsiella*, *Laevicardia* and *Policordia* as members of a separate family, the Lyonsiellidae, which is, necessarily, the sister group of the Verticordiidae.

Materials and methods

Sixteen taxa have been included in the present analysis (Table 1), with one species from each of the currently recognized families of extant Anomalodesmata. There has been some debate about the desirability of using such an exemplar approach as opposed to using the ground-plan approach (summarizing the conditions found in all supraspecific units in a higher taxon) (Yeates 1995). However, the former is the only option available in this study because in the vast majority of cases the taxon used is the only member of the family ever to have been studied in adequate detail for this analysis. As indicated in Table 1, the taxon used in the cladistic analysis was a representative of the type genus and, most often, the type species itself. Two species (*Brechites* and *Clavagella)* from the Clavagellidae were included and these were scored separately so as to test the relationships between them. Similarly, for the Myochamidae the two genera *Myochama* and *Myadora* were treated separately so as to include two taxa displaying different life habits, i.e. cementing and burrowing. Only the Pandoridae was scored as a supraspecific taxon (combining information from *Pandora* and

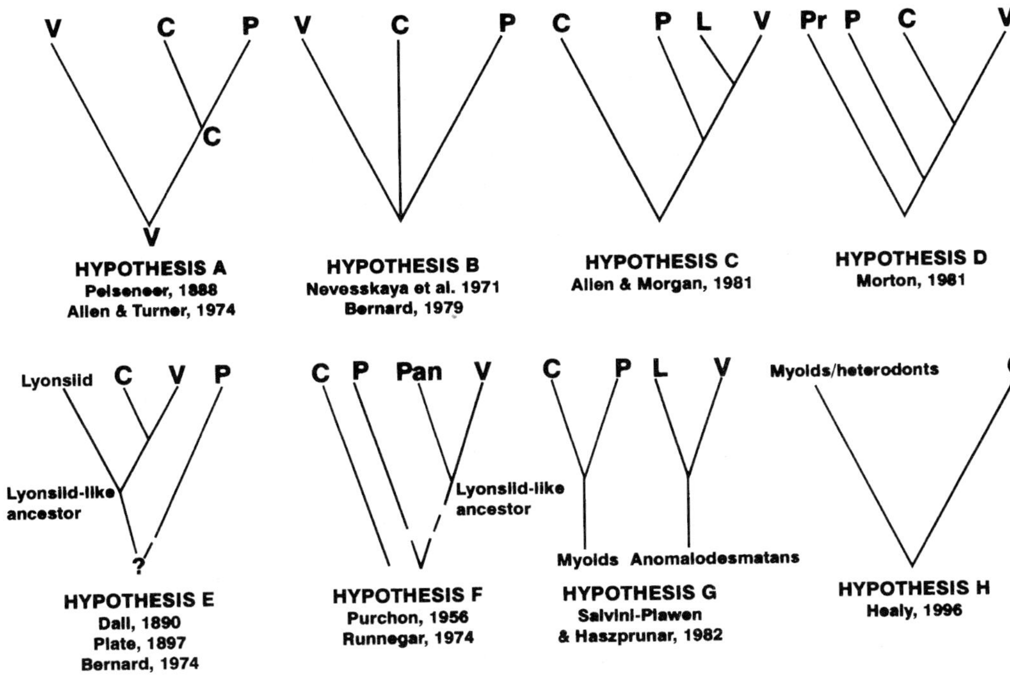

Fig. 1. Summary of the different hypotheses which have been suggested for the relationships of the carnivorous families of the Anomalodesmata. P, Poromyidae; C, Cuspidariidae; V, Verticordiidae; L, Lyonsiellidae; Pr, Parilimyidae; Pan, Pandoridae.

Frenamya) because of the lack of a very detailed study of either taxa.

Forty-three characters were selected as being potentially cladistically informative and all of these have discrete, non-continuous character states, in contrast to those employed by Schneider (1992) of which several were continuous. The characters and character states used in this analysis are defined in Appendix 1: the majority of these characters have been discussed by Morton (1985*a*). Characters were scored from either previously published descriptions (see Table 1) or, where the data were either missing or ambiguous, were investigated specifically for this study. The resultant data matrix is given in Appendix 2. It was deliberately chosen not to define any Dollo, i.e. non-reversible characters, because of potential unreliability of this, e.g. the fourth pallial aperture seems, at first sight, to be a distinctive anomalodesmatan character but is known to also occur in the Mactridae (Yonge 1948) and Solenidae (Yonge 1952).

The choice of a meaningful outgroup was complicated by a lack of consensus concerning the relationships between the Anomalodesmata and other bivalve taxa; Morton (1996) places the Myoida as a sister group to the Anomalodesmata, whilst Cope (1996*b*) has suggested that they may have evolved from a palaeoheterodont ancestor. At this stage in the understanding of anomalodesmatan relationships, it is not possible to include a non-anomalodesmatan as an outgroup without favouring one or other of the hypotheses as the origin of the group. As a result, the Pholadomyidae has been chosen as the outgroup, on the basis of the known longevity of the family [first recognized from the Hastarian by Skelton & Benton (1993)] and the general acceptance of the Pholadomyoidea as the stem group of the Anomalodesmata (Pojeta 1971; Runnegar 1974). In both trees obtained during this analysis, the Parilimyidae emerged as the closest taxon to *Pholadomya*, unsurprisingly as they were formerly considered to belong to the same family until Morton (1982) removed *Parilimya* and associated genera to erect the family. As a result, here the trees have been rooted using *Pholadomya* + *Parilimya* as the outgroup.

The data were analysed using PAUP version 4.0 b2 for Macintosh (Swofford 1998) with all characters treated as unordered and unweighted. The accelerated transformation (ACCTRAN) was used. Multistate characters were treated as polymorphic. Two different analyses were performed. The first used the entire data matrix of 43 characters, while the second used only the 18 characters which could be scored from the shells alone, i.e. the maximum number of characters which could be scored for extinct taxa (identified in Appendix 1). The second analysis was undertaken in an effort to assess the reliability of constructing trees based on characters which may be scored from fossil material and thus the possibility of adding extinct taxa to any later analysis. Branch and bound analyses were used to find the shortest tree(s), and where an analysis produced more than one most parsimonious tree, strict consensus trees were produced. Percentage bootstrap figures were obtained to indicate support for the branches (Felstenstein 1985).

In order to compare the results of the cladistic analyses with the known fossil record of the families, a database was established for the first occurrence of each anomalodesmatan family using information derived from Sepkoski (1992) and Skelton & Benton (1993), along with Sepkoski's unpublished compilation of generic longevities. Completeness of the fossil record may be assessed in a number of different ways, as suggested by Paul (1982). Of these, two are most applicable to investigating the incompleteness of records of whole extant families on a global basis and these were used to assess the incompleteness of the fossil record for each of the extant anomalodesmatan families: (1) recognition of families with no recognized fossil records; (2) those for which there are gaps during the known stratigraphic range of the family when no representative genera have been found.

Results

Total evidence analysis

The analysis of the complete character matrix produced a single most parsimonious tree of length 147 (Fig. 3) with a consistency index (CI) of 0.605 and a retention index (RI) of 0.570. As both CI and RI are intended as measures of the extent of homoplasy within the data set (with a value of 1 indicating that the data are entirely consistent), they support the contention that non-divergent change occurs frequently within the Anomalodesmata. Percentage bootstrap support values, shown on the tree, vary from close to 100% down to 58% for more poorly supported branches.

The total evidence tree supports the division of the Pandoroidea (*sensu* Keen *in* Moore 1969) into two monophyletic groups, the Thracioidea and Pandoroidea, as recommended by Boss (1978) and Yonge & Morton (1980). Indeed, it should be noted that Keen's Pandoroidea was found to be non-monophyletic. The Thracioidea form a deep branching group, the sister group to the rest of anomalodesmatans, implying that they arose early from the primitive anomalodesmatan stock, and the tree shows the Pandoroidea to be the more derived of the two superfamilies. Within the Thracioidea, the Laternulidae and Periplomatidae are sister groups and are more derived than the Thraciidae. Within the Pandoroidea, the Cleidothaeridae, Pandoridae and *Myochama* and *Myadora* form the sister group to the Lyonsiidae. The precise topology of that part of the tree relating to the Pandoroidea fits neither of the two models suggested by Boss (1978) and there is no support for the contention that the Verticordiidae should be allied to the Pandoroidea as the ancestors to the Cleidothaeridae (Runnegar 1974); indeed, the verticordiids appear to be highly derived.

The tube-dwelling *Clavagella* and *Brechites* appear together, implying that they are at least not distantly related, as a sister group of the septibranch taxa. This appears a slightly anomalous placement which supports none of the previous hypotheses for their relationships and deserves further investigation.

The total evidence tree provides some robust conclusions on the relationships of the carnivorous anomalodesmatans. The five families which are considered to be predatory are clearly differentiated into two clades; the Parilimyidae group allied closely with the Pholadomyidae [in which they were once included until Morton (1982) separated them], whilst the Lyonsiellidae, Verticordiidae, Poromyidae and Cuspidariidae form a highly derived monophyletic sister group to the clavagellids. The Lyonsiellidae and Verticordiidae are sister groups, and are themselves a sister group to the Cuspidariidae and Poromyidae. This tree allows rejection of hypothesis D, which suggests that all carnivorous taxa were derived from a parilimyid ancestor (Morton 1981*a*), and hypotheses E and F, which separate the taxa amongst the other anomalodesmatans. The fact that they form a derived clade also implies that models of relationships such as hypothesis F (Runnegar 1974) and model H (Healy 1996), which place the cuspidariids outside the Anomalodesmata, should be rejected. Of all the hypotheses presented in Fig. 1, the prsent tree supports hypothesis A, as proposed by Pelseneer (1888) and Allen & Turner (1974).

The completeness of the fossil record of the Anomalodesmata

The geological ranges of each of the extant families of anomalodesmatans are shown in Table 2. All families, with the exception of the Parilimyidae, have a fossil record. Indeed, only nine extant genera in the entire subclass have no recognized

Table 2. *Stratigraphic data for the first recognized appearance in the fossil record of each of the extant anomalodesmatan families*

Families	First appearances
Clavagellidae	*Clavagella* – Cretaceous (Cenomanian)
	Brechites – Oligocene
Cleidothaeridae	Lower Miocene
Cuspidariidae	Triassic (Ladinian)
Laternulidae	Triassic (Rhaetian)
Lyonsiellidae	Pliocene
Lyonsiidae	Paleocene Thanetian)
Myochamidae	*Myochama* – Upper Oligocene
	Myador – Upper Eocene
Pandoridae	Eocene (Ypresian)
Parilimyidae	Recent only [?Jurassic (Callovian) – Cretaceous (Maastrichtian) Sepkoski unpublished)
Periplomatidae	Jurassic (Tithonian)
Pholadomyidae	Carboniferous (Hastarian)
Poromyidae	Cretaceous (Aptian)
Thraciidae	Triassic (Rhaetian)
Verticordiidae	Cretaceous (Albian)

Data modified from Sepkoski (1992) and Skelton & Benton (1993).

fossil record. Apart from the Parilimyidae, all families have constituent genera with a continuous stratigraphic record. Such a level of completeness is high compared with some other higher non-molluscan taxa, e.g. echinoids (Raup 1979), and rather better than some other major bivalve clades, e.g. lucinoids (Harper 1998). Only the Parilimyidae has no widely recognized fossil record. The three genera, *Parilimya, Panacca* and *Nippopanacca*, placed within the family by Morton (1982) are recognized from Recent material only. This may, however, be a taxonomic artefact produced by the relatively recent erection of the family and it may prove that subsequent revision of the Pholadomyoidea will remove other fossil genera from the Pholadomyidae to the Parilimyidae. Indeed, Runnegar (1974) suggested that the Mesozoic genus *Procardia* might be closely allied to *Panacca* and, if this is so, it may be a parilimyid. Of all the anomalodesmatans the parilimyids have perhaps the lowest preservation potential being extremely thin shelled and living in deep water. The specimen of *Parilimya fragilis* examined by Morton (1982), for example, was collected from 1100 m in the North Atlantic.

A third method of assessing incompleteness in the fossil record of a taxon is the recognition of monotypic taxa, on the assumption that a very incomplete fossil will result in recognized taxa being so distinctive as to be given anomolously high taxonomic rank (Paul 1982). Of the extant anomalodesmatan families only the Cleidothaeridae are and have been monogeneric throughout their known duration and their coiled, highly inequivalve morphology is highly distinctive.

From the above, it appears that the Anomalodesmata have a fossil record complete enough to be useful in discriminating between cladistic hypotheses (Fortey & Jefferies 1982). The total evidence tree (Fig. 2) fits the recognized fossil record reasonably well without having to invoke many long range extensions unsupported by the fossil record (Smith 1994). In particular, as noted by Boss (1978), the division of the Thracioidea from the Pandoroidea is well supported by the appearance of the former in the Mesozoic whilst the latter are first recorded in sediments of Cenozoic age. The major discrepancy between the fossil record and the tree relates to the relationships of the 'septibranchs'. The recognition of the cuspidariids in rocks of Triassic and Jurassic age requires an extension of the fossil records of the three other septibranch families by at least 100 Ma. It may be possible to explain away this large gap in the fossil record by the fact that these bivalves often tend to live in deep water and are, like all anomalodesmatans, composed of diagenetically unstable aragonite. However, it is noted that Yin & Fürsich (1991, p. 159) remark that Jurassic cuspidariids have been identified on gross external morphology alone and thus their diagnosis may not be unequivocal, in fact, they point out these specimens resemble corbulids. If these records are

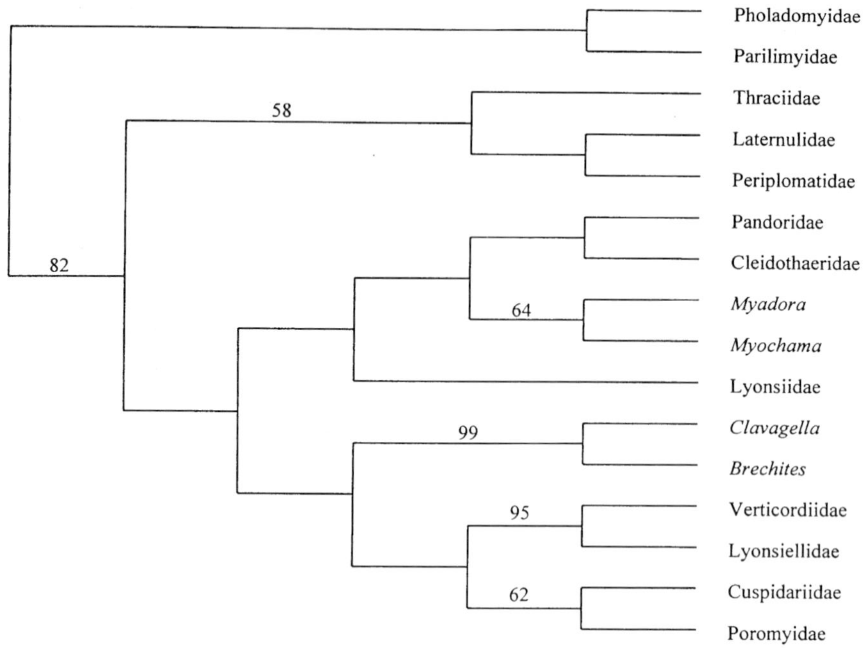

Fig. 2. Total evidence tree: Single most parsimonious tree produced from the analysis of the entire data set. Bootstrap values > 50% given on various branching points.

rejected as suspect, then the stratigraphic range of the Cuspidariidae only extends back as far as the Upper Cretaceous (Keen *in* Moore 1969), which is more in keeping with the total evidence tree. Future work should attempt to verify whether or not the Triassic and Jurassic taxa are correctly recognized as cuspidariids.

Shell character only tree

Only 18 (at most) of the 43 characters used in the above analysis can be scored from shells alone. Reanalysis of the data using only these produced 16 most parsimonious trees of length 49. The strict consensus tree of these is shown in Fig. 3 (CI = 0.714 and RI = 0.750) and is markedly different from the total evidence tree (Fig. 2). The branching pattern is much more pectinate and because it is a strict consensus tree, i.e. recognizes only those nodes present in all 16 most parsimonious trees, it is unable to resolve many of the relationships. Nevertheless, the separation of the Thracioidea from the Pandoroidea is still supported, but the order in which the two groups branch is reversed and thus in direct opposition to the order they first appear in the fossil record, implying that the pandoroids have a long unrecognized fossil record. The reduced data set is unable to resolve the relationships within the Pandoroidea. The tube-dwelling *Clavagella* and *Brechites* appear as a very deep branching group, again in opposition to their first appearance in the fossil record which suggests that they may have a substantial history which is not preserved in the fossil record.

The carnivorous taxa do not form a monophyletic group and *Poromya* branches extremely deeply, a position which is not supported by any accepted hypotheses of relationship (Fig. 2) nor the fossil record. The tree also implies that the Thracioidea arose from carnivorous ancestors, although none of the thracioid taxa are known to show that trait and it seems most unlikely that the very specialized multiple adaptations for carnivory (e.g. modified siphons, ctenidia, palps and stomach) could ever reverse to suspension feeding.

The Partition-Homogeneity Test (Farris *et al.* 1995) can be used to assess the extent of apparent incongruence between trees produced from two partitions of a complete data set. However, in this case, the two character sets are not mutually exclusive as there are a number of characters which can be scored from either the anatomical or the shell characters alone, e.g. details of the musculature and valve inequality. The test is therefore not appropriate in this case.

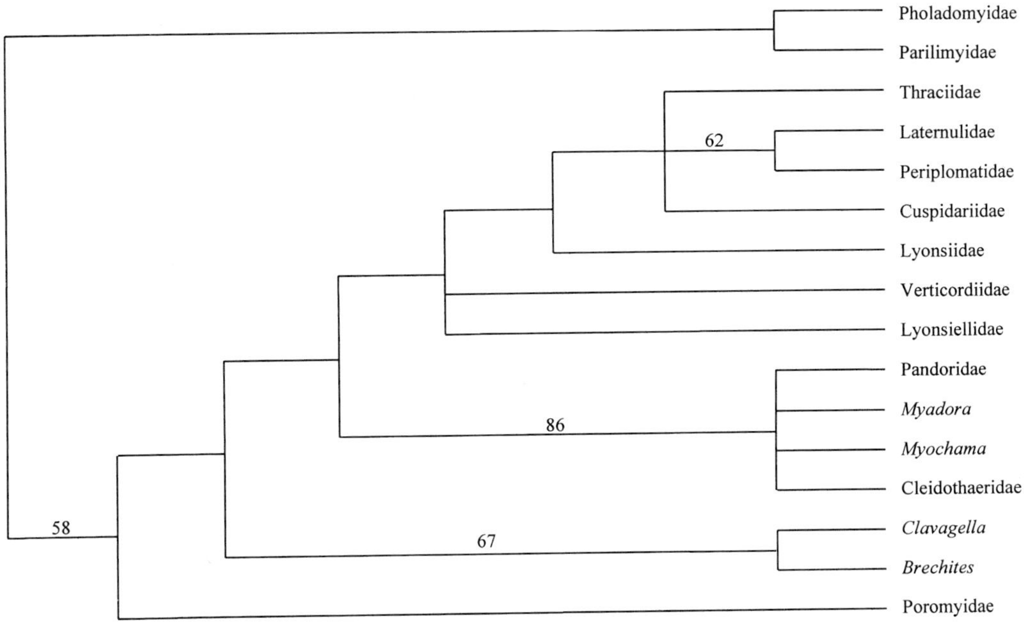

Fig. 3. Strict consensus tree of the 16 most parsimonious trees produced by the analysis of shell characters alone (these characters are identified in Appendix 1). Bootstrap values > 50% given on various branching points.

Discussion

Cladistic analysis of a wide array of anatomical and shell characters has produced a single reasonably robust phylogenetic tree. Although the CI and RI are not high, the total evidence tree is at least as robust as most of the molluscan cladistic trees published to date. Furthermore, the total evidence tree is broadly supported by the order of first appearance in the known stratigraphic record. The analysis has resolved a number of points: (1) it clearly supports the division of the Pandoroidea and Thracioidea as two monophyletic clades and has further resolved the relationships within these superfamilies; (2) it has shown that carnivory amongst the anomalodesmatans is at least diphyletic and supports hypotheses A–C (Fig. 1), in which the 'septibranch' taxa form a single clade rather than the suggestion that this group is polyphyletic, as suggested in hypotheses E and F, or even that the cuspidariids fall outside the Anomalodesmata (hypotheses G and H). Since they form the terminal group of the more derived anomalodesmatans there is scope to place the septibranchs into a separate higher taxon, as proposed by Knudsen (1970), although this leaves the Anomalodesmata as a paraphyletic clade. The present analysis also implies that the enigmatic tube-dwelling clavagellids are the sister group of the septibranchs, a conclusion which does not sit comfortably with any existing theories of clavagellid origins and which deserves further investigation.

Despite the fact that this analysis produced a single, acceptably robust tree it is apparent that the subclass has been dogged by non-divergent evolution. This is particularly evident if the occurrence of certain key characters are mapped, e.g. the presence or absence of hinge teeth (character 8) on to the total evidence tree (see Fig. 4). Similarly, the fourth pallial aperture, a character shared by many Anomalodesmata (Morton 1985*a*), appears to have been lost three times (but never regained).

Future refinement of characters might also lead to a more robust tree. In the current analysis character(s) for periostracal spicules (Carter & Aller 1975) or shell spinules (Prezant 1981) have not been included because of uncertainly, in the present authors' minds at least, about homology between these. Other characters, such as the apparent similarity of odour shared by some of the pandoroid families [noted by Boss (1978)] and which may be phylogenetically useful are unlikely to be scorable for every family.

The lack of congruence between the total evidence tree and that based solely on shell characters is startling, despite the proportion of

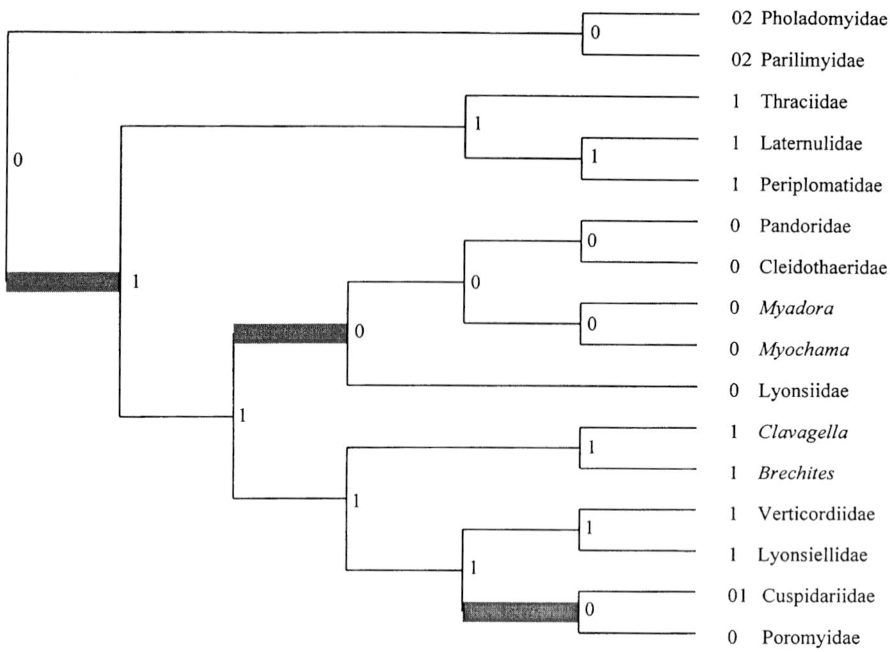

Fig. 4. Reversals in character state for the presence/absence of hinge teeth (character 8) on the total evidence tree.

characters that could be scored entirely on shell knowledge and the fact that the tree produced has RI and CI values which indicate it is acceptably robust to most investigators. This failure may have two explanations. Firstly, the number of characters may be too low and the distribution of character states within them too limited to be sufficient to resolve the phylogeny of such a large number of taxa. Alternatively, it may be that shell characters are influenced to such an extent by environmental and life habit that shell form masks phylogenetic information. These two explanations have the same net effect and only the first is potentially rectifiable by further investigations that might yield new cladistically informative characters. Although this may prove to be the case, the fact that the shell character tree is not just a poorly resolved version of the total data tree but clearly contradicts it, argues for at least part of the problem being due to the second explanation. On the total evidence tree it is possible to see that certain shell morphological traits (e.g. loss and gain of hinge teeth), which have been pivotal in traditional classifications, are polyphyletic (Fig. 4). It is thus unlikely that shell characters are suitable for use as the sole basis of phylogenetic reconstruction in the Anomalodesmata. This has obvious implications for any future studies that seek to incorporate extinct taxa in any cladistic analysis of the Anomalodesmata, especially since the unstable nature of their aragonitic shells over geological timescales means that it may not be possible to score even the reduced character set of Palaeozoic taxa where original shell microstructure and detailed information on muscle attachment sites have been lost during diagenesis. Full scoring of the shell character matrix for extinct families must await the discovery of specimens showing exceptionally well-preserved shell material. However, on the basis of this study, a full cladistic analysis of anomalodesmatan families (extant and extinct) should proceed using the total evidence character matrix, even though for the extinct taxa at least 50% of the character states will be encoded '?'. This finding has obvious relevance to cladistic studies of other bivalve groups. Reliance on hard-part characters alone is not necessarily a peculiarity of looking at fossil taxa; as it happens, the anatomy of all the extant families of the Anomalodesmata have been investigated in great detail but, as noted by Paul (1982) and Harper (1998), many molluscan taxa are known only from conchological detail. Nevertheless, the bivalve fossil record is excellent and therefore stratigraphic ranges are invaluable in assessing and 'callibrating' the trees, and provide a means of determining the polarity of some characters.

EMH holds a Royal Society University Research Fellowship for which she is grateful. We are also grateful to the Royal Society of London for funding BM's participation in this project. Sharon Capon redrafted Fig. 1. We also thank Gonzalo Giribet and Jay Schneider for their constructive reviews. This is Cambridge Earth Science Publication No. 5896.

Appendix 1: List of character and character states used in the cladistic analyses

All characters were used for the total evidence analysis, whilst those used in thr shell character only analysis are marked with an asterisk.

A. Shell microstructure and gross shell morphology

1. Shell microstructure*: (0) prismatic + nacre microstructures; (1) fine homogenous + nacre microstructures; (2) homogenous + homogenous microstructures; (3) coarse (> μm) homogenous + nacre microstructures
2. Adhesion of foreign material to shell exterior*: (0) absent; (1) present
3. Valve equality*: (0) equivalvy; (1) slight inequivalvy; (2) gross inequivalvy
4. Dorsal crack in shell*: (0) absent; (1) present
5. Permanent posterior gape*: (0) absent; (1) present
6. Permanent anterior gape*: (0) absent; (1) present
7. Valve concordance*: (0) concordant; (1) discordant

B. Hinge details

8. Hinge teeth*: (0) present, strong; (1) absent; (2) present, weak
9. Primary ligament*: (0) external; (1) internal
10. Ligament and chondrophore*: (0) not sunken; (1) simple sunken; (2) sunken between chondrophore; (3) sunken between coiled chondrophores
11. Secondary ligament: (0) thin; (1) thick
12. Lithodesma*: (0) absent; (1) present

C. General mantle characters

13. Mantle fusion type (Yonge 1982): (0) type A; (1) type B; (2) type C
14. Arenophilic glands: (0) absent; (1) present
15. Fourth pallial aperature: (0) absent; (1) present

D. Siphon characters

16. Siphon length*: (0) pallial sinus absent; (1) pallial sinus present but does not extend beyond the DV axis; (2) pallial sinus present and extends beyond the DV axis
17. Siphon separation: (0) separated; (1) fused
18. Siphon symmetry: (0) asymmetric; (1) symmetric
19. Siphon fusion type (Yonge 1982): (0) type A; (1) type B; (2) type C
20. Cilliary sense organs: (0) absent; (1) present
21. Siphonal tentacles: (0) absent on both; (1) present on exhalant only; (2) present on inhalant only; (3) present on both
22. Inhalant siphon: (0) typical bivalve form; (1) extensible 'hood'
23. Mucus external lining: (0) absent; (1) present

E. Musculature

24. Adductor musculature*: (0) isomyarian; (1) anterior larger; (2) posterior larger
25. Pedal gape muscles: (0) normal; (1) *Pholadomya* type (see Morton 1980)
26. Pedal musculature*: (0) normal; (1) reduced; (2) absent
27. Taenoid muscles: (0) absent; (1) present

F. General viscera

28. Palp type: (0) normal bivalve type; (1) funnel-shaped; (2) short, no cilia
29. Gills: (0) plicate; (1) non-plicate
30. Gills: (0) type E; (1) 'septibranch'
31. Gills, inner demibranch attached to the visceral mass: (0) present; (1) absent
32. Lips: (0) simple, unfused; (1) laterally fused; (2) medially fused
33. Stomach type II (Purchon 1956): (0) absent; (1) present
34. Sorting surfaces on stomach: (0) absent; (1) present
35. Rectum: (0) above heart; (1) below heart; (2) penetrates heart
36. Rectum: (0) passes over kidneys; (1) penetrates kidneys
37. Statocyst type (see Morton 1985*b*): (0) A; (1) B1; (2) B2; (3) B3; (4) C
38. Myadorid ray*: (0) absent; (1) present

G. Reproductive characters

39. Reproduction: (0) simultaneous hermaphrodites: (1) dioecious
40. Gonadial aperatures: (0) united; (1) separate; (2) united with urinary ducts

H. Other characters

41. Ability to cement*: (0) absent; (1) present
42. Ability to build calcareous tubes*: (0) absent; (1) present
43. Post-larval byssus*: (0) absent; (1) present

Appendix 2: Data matrix for the cladistic analyses

Representative taxa given in Table 1.

	1	2	3	4	5	6	7	8	9	10	11	12	13	14	15
Brechites	0	0	0	0	?	1	?	1	0	0	1	1	2	1	1
Clavagella	0	0	0	0	?	1	?	1	{01}	0	1	{01}	2	1	1
Cleidothaeridae	0	0	2	0	1	1	1	0	1	3	1	1	0	0	1
Cuspidariidae	2	0	{01}	0	0	1	0	{01}	1	2	1	1	1	0	0
Laternulidae	0	1	1	1	0	0	0	1	{01}	2	1	1	1	0	0
Lyonsiellidae	0	1	0	0	1	1	0	1	1	1	?	1	1	1	0
Lyonsiidae	{03}	1	1	0	{01}	1	0	0	1	1	1	1	1	{01}	1
Myadora	0	0	2	0	1	1	1	0	1	1	0	1	2	?	1
Myochama	0	0	2	0	1	1	1	0	1	1	0	1	2	0	0
Pandoridae	0	0	2	0	1	1	1	0	1	1	1	{01}	1	0	0
Parilimyidae	0	0	0	0	0	0	0	{02}	0	0	0	0	0	1	1
Periplomatidae	0	1	2	1	0	0	0	1	1	2	1	1	{01}	1	{01}
Pholadomyidae	0	?	0	0	0	0	0	{02}	0	0	0	0	1	0	1
Poromyidae	{12}	0	0	0	1	1	0	0	0	0	0	0	0	0	{01}
Thraciidae	2	1	1	0	0	0	0	1	1	2	1	1	1	?	0
Verticordiidae	0	1	0	0	1	1	0	1	1	1	{01}	1	1	{01}	0

	16	17	18	19	20	21	22	23	24	25	26	27	28	29	30
Brechites	0	0	0	2	1	3	0	0	2	0	1	0	0	0	0
Clavagella	{01}	0	0	2	1	3	0	0	2	0	1	0	0	0	0
Cleidothaeridae	0	0	0	0	0	2	0	0	1	0	2	0	0	0	0
Cuspidariidae	1	?	1	1	1	3	1	0	0	0	1	0	2	1	1
Laternulidae	0	1	0	2	1	3	0	0	2	0	1	0	0	0	0
Lyonsiellidae	0	0	?	0	1	?	1	0	0	?	1	1	1	0	?
Lyonsiidae	1	0	?	1	0	2	0	0	0	0	1	0	0	0	0
Myadora	0	1	0	0	0	3	0	0	0	0	1	0	0	0	0
Myochama	1	0	?	0	0	0	0	0	0	0	1	0	0	0	0
Pandoridae	0	1	0	1	0	3	0	0	0	0	1	0	0	0	0
Parilimyidae	1	1	1	1	0	0	1	0	0	1	2	1	1	0	0
Periplomatidae	1	0	0	0	0	3	0	1	0	0	1	0	0	0	0
Pholadomyidae	1	1	0	1	0	0	0	0	0	1	0	1	0	0	0
Poromyidae	0	?	1	0	1	3	1	0	0	0	1	0	1	1	1
Thraciidae	2	0	0	0	0	3	0	1	0	0	2	0	0	0	0
Verticordiidae	0	0	1	0	1	3	0	0	0	0	1	0	1	0	0

	31	32	33	34	35	36	37	38	39	40	41	42	43
Brechites	0	0	0	?	2	0	3	0	0	1	0	1	0
Clavagella	0	0	0	?	2	0	3	0	0	1	0	1	0
Cleidothaeridae	0	0	0	1	0	0	2	0	0	1	1	0	0
Cuspidariidae	1	0	1	0	2	0	4	0	1	1	0	0	0
Laternulidae	0	0	0	1	2	1	{12}	0	0	2	0	0	0
Lyonsiellidae	?	2	1	?	2	1	1	0	?	?	0	0	1
Lyonsiidae	0	0	0	1	2	0	1	0	0	1	0	0	1
Myadora	0	0	0	1	2	0	{23}	1	0	1	0	0	0
Myochama	0	0	0	1	2	0	3	1	0	1	1	0	0
Pandoridae	0	0	0	1	0	0	1	0	0	0	0	0	0
Parilimyidae	0	?	1	?	2	0	{12}	0	0	0	0	0	0
Periplomatidae	0	1	0	1	2	1	2	0	0	0	0	0	0
Pholadomyidae	0	1	0	0	1	0	0	0	0	0	0	0	0
Poromyidae	1	0	1	?	2	0	1	0	0	0	0	0	0
Thraciidae	0	0	0	0	2	1	{12}	0	1	0	0	0	0
Verticordiidae	0	2	1	0	2	1	1	0	0	1	0	0	1

References

ALLEN, J. A. 1954. On the structure and adaptations of *Pandora inaequivalvis* and *P. pinna*. *Quarterly Journal of Microscopical Science*, **95**, 473–482.

—— & MORGAN, R. E. 1981. The functional morphology of Atlantic deep water species of the families Cuspidariidae and Poromyidae (Bivalvia): an analysis of the evolution of the septibranch condition. *Philosophical Transactions of the Royal Society of London, Series B*, **294,** 413–546.

—— & TURNER, J. F. 1974. On the functional morphology of the family Verticordiidae (Bivalvia) with descriptions of new species from the abyssal Atlantic. *Philosophical Transactions of the Royal Society of London, Series B*, **268,** 401–520.

ALLER, R. C. 1974. Prefabrication of shell ornamentation in the bivalve *Laternula*. *Lethaia*, **7,** 43–56.

AMLER, M. R. 1999. Synoptical classification of fossil and Recent Bivalvia. *Geologica et Palaeontologica*, **33,** 237–248.

ATKINS, D. 1936. On the ciliary mechanisms and interrelationships of lamellibranchs. *Quarterly Journal of Microscopical Research*, **79,** 181–308.

BAMBACH, R. K. 1971. Adaptations in *Grammysia obliqua*. *Lethaia*, **4,** 169–183.

BERNARD, F. R. 1974. Septibranchs of the eastern Pacific (Bivalvia: Anomalodesmata). *Allan Hancock Monographs in Marine Biology*, **8,** 1–279.

—— 1979. New species of *Cuspidaria* from the north-eastern Pacific (Bivalvia: Anomalodesmata) with a proposed classification of the septibranchs. *Venus*, **38,** 14–24.

BIELER, R. & MIKKELSEN, P. M. 1992. Preliminary phylogenetic analysis of the bivalve family Galeommatidae. *American Malacological Bulletin*, **9**, 157–164.

BOSS, K. J. 1978. Taxonomic concepts and superfluity in bivalve nomenclature. *Philosophical Transactions of the Royal Society of London, Series B*, **284**, 417–424.

CARTER, J. G. 1978. Ecology and evolution of the Gastrochaenacea (Mollusca, Bivalvia) with notes on the evolution of the endolithic habitat. *Bulletin of the Peabody Museum of Natural History*, **41,** 1–92.

CARTER, J. G. & ALLER, R. C. 1975. Calcification of the bivalve periostracum. *Lethaia*, **8**, 315–320.

COPE, J. C. W. 1996*a*. Early Ordovician (Arenig) bivalves from the Llangynog inlier, South Wales. *Palaeontology*, **39,** 979–1025.

—— 1996*b*. The early evolution of the Bivalvia. *In:* TAYLOR, J. D. (ed.) *Origin and Evolutionary Radiation of the Mollusca*. Oxford University Press, Oxford, 361–370.

DALL, W. H. 1889. On the hinge of pelecypods and its development with an attempt toward a better subdivision of the group. *American Journal of Science*, **38,** 445–462.

DOUVILLÉ, H. 1913. Classification des lamellibranches. *Bulletin de la Société géologique de France 4eme Série*, **12,** 419–467.

FARRIS, J. S., KÄLLERSJÖ, M., KLUGE, A. G. & BULT, C. 1995. Testing significance of incongruence. *Cladistics*, **10,** 315–319.

FELSENSTEIN, J. 1985. Confidence limits on phylogenies: an approach using the bootstrap. *Evolution*, **39,** 783–791.

FORTEY, R. A. & JEFFERIES, R. P. S. 1982. Fossils and phylogeny – a compromise approach. *In:* JOYSEY, K. A. & FRIDAY, A. E. (eds) *Problems of Phylogenetic Reconstruction*. Systematics Association Special Volume, **21,** 197–234.

HARPER, E. M. 1998. The fossil record of bivalve molluscs. *In:* Donovan, S. K. & PAUL, C. R. C. (eds) *The Adequacy of the Fossil Record.* John Wiley & Sons, Chichester, 243–267.

—— & MORTON, B. 2000. Cementation in *Myochama anomioides* Stutchbury, 1830 (Bivalvia: Anomalodesmata: Pandoroidea) with reference to cementation. *Journal of Molluscan Studies*, **66**, 403–416.

HEALY, J. M. 1996. Molluscan sperm ultrastructure: correlation with taxonomic units within the Gastropoda, Cephalopoda and Bivalvia. *In:* TAYLOR, J. D. (ed.) *Origin and Evolutionary Radiation of the Mollusca*. Oxford University Press, Oxford, 99–113.

HICKMAN, C. S. 1996. Phylogeny and patterns of evolutionary radiation in trochoidean gastropods. *In:* TAYLOR, J. D. (ed.) *Origin and Evolutionary Radiation of the Mollusca*. Oxford University Press, Oxford, 177–198.

KNUDSEN, J. 1970. The systematics and biology of abyssal and hadal Bivalvia. *Galathea Report*, **11,** 7–236.

MOORE, R. C. 1969. *Treatise of Invertebrate Paleontology, Part* N(1–2). The Geological Society of America, Boulder, CO, and The University of Kansas, Lawrence, KS.

MORGAN, R. E. & ALLEN, J. A. 1976. On the functional morphology and adaptations of *Entodesma saxicola* (Bivalvia: Anomalodesmacea). *Malacologia*, **15,** 233–240.

MORTON, B. 1973. The biology and functional morphology of *Laternula truncata* (Lamarck, 1818) (Bivalvia: Anomalodesmata: Pandoracea). *Biological Bulletin*, **145,** 509–531.

—— 1974. Some aspects of the biology and functional morphology of *Cleidothaerus maorianus* Finlay (Bivalvia: Anomalodesmata: Pandoracea). *Proceedings of the Malacological Society of London*, **41,** 201–222.

—— 1977. The biology and functional morphology of *Myadora striata* (Quoy & Gaimard) (Bivalvia: Anomalodesmata: Pandoracea). *Journal of Molluscan Studies*, **43,** 141–154.

—— 1980. The anatomy of the 'living fossil' *Pholadomya candida* Sowerby, 1823 (Mollusca: Bivalvia: Anomalodesmata). *Videnskabelige Meddelelser fra Dansk naturhistorik Forening i Kjøbenhavn*, **142,** 7–102.

—— 1981*a*. The Anomalodesmata. *Malacologia*, **21,** 35–60.

—— 1981*b*. The biology and functional morphology of *Periploma (Offadesma) angasai* (Bivalvia: Anomalodesmata: Periplomatidae). *Journal of Zoology, London*, **193,** 39–70.

—— 1982. The functional morphology of *Parilimya fragilis (*Grieg, 1920) (Bivalvia: Parilimyidae fam. nov.) with a discussion of the origin and evolution of the carnivorous septibranchs and a reclassification of the Anomalodesmata. *Transactions of the Zoological Society of London*, **36,** 1–76.

—— 1984*a*. The biology and functional morphology of *Clavagella australis* (Bivalvia: Anomalodesmata). *Journal of Zoology, London*, **202,** 489–511.

—— 1984*b*. Adventitious tube construction in *Brechites vaginiferus* (Bivalvia: Anomalodesmata: Clavagellacea) with an investigation of the juvenile of '*Humphreysia strangei*'. *Journal of Zoology, London*, **203,** 461–484.

—— 1984*c*. Prey capture in *Lyonsiella formosa* (Bivalvia: Anomalodesmata: Verticordiacea). *Pacific Science*, **38,** 283–297.

—— 1984*d*. Aspects of the biology and functional morphology of *Frenamya ceylanica* (Bivalvia: Anomalodesmata: Pandoracea). *Journal of Conchology*, **31,** 359–371.

—— 1985*a*. Adaptive radiation in the Anomalodesmata. *In:* TRUMAN, R. R. & CLARK, M. R. (eds) *The Mollusca, Volume 10, Evolution.* Academic Press, Sydney, 405–459.

—— 1985*b*. Statocyst structure in the Anomalodesmata (Bivalvia). *Journal of Zoology, London*, **206,** 23–34.

—— 1987. Siphon structure and prey capture as a guide to affinities in the abyssal septibranch Anomalodesmata. *Sarsia*, **72,** 49–69.

—— 1995. The ecology and functional morphology of *Trigonothracia jinxingae* (Bivalvia: Anomalodesmata: Thracoidea) from Xiamen, China. *Journal of Zoology, London*, **237**, 445–468.

—— 1996. The evolutionary history of the Bivalvia. *In:* TAYLOR, J. D. (ed.) *Origin and evolutionary radiation of the Mollusca*. Oxford University Press, Oxford, 337–356.

NEVESSKAYA, L. A., SCARLATTO, O. A., STAROBOGATOV, Y. A. & EBERZIN, A. G. 1971. New ideas on bivalve systematics. *Paleontological Journal*, **2,** 141–155.

PAUL, C. R. C. 1982. The adequacy of the fossil record. *In:* JOYSEY, K. A. & FRIDAY, A. E. (eds) *Problems of Phylogenetic Reconstruction*. Systematics Association Special Volume*,* **21,** 75–117.

PELSENEER, P. 1888. Report on the anatomy of the deep-sea Mollusca. *Report on the Scientific Results of the Voyage of H.M.S. Challenger During the Years 1873–76, Zoology,* **27**(74), 8–40.

—— 1891. Contribution a l'étude des Lamellibranches. *Archives de Biologie, Paris*, **11**, 147–312.

PLATE, L. 1897. Giebt es septibranchiate muscheln? *Sitzungsberichteder Gesellschaff naturforschender Freunde zu Berlin,* **1897,** 24–28.

POJETA, J. 1971. Review of Ordovician pelecypods. *US Geological Survey, Professional Papers*, **695**, 1–46.

—— & SOHL, N. F. 1987. *Ascaulocardium armatum* (Morton, 1833). New genus (Late Cretaceous): the ultimate variation on the bivalve paradigm. *The Paleontological Society Memoir*, **24,** 1–77.

PREZANT, R. S. 1981. Comparative shell ultrastructure of lyonsiid bivalves. *Veliger*, **23,** 289–299.

PURCHON, R. D. 1956. The stomach in the Protobranchia and Septibranchia (Lamellibranchia). *Proceedings of the Zoological Society of London*, **127,** 511–525.

—— 1978. An analytical approach to a classification of the Bivalvia. *Philosophical Transactions of the Royal Society of London, Series B*, **284,** 425–436.

—— 1987*a*. The stomach in the Bivalvia. *Philosophical Transactions of the Royal Society of London, Series B*, **316,** 183–276.

—— 1987*b*. Classification and evolution of the Bivalvia: an analytical study. *Philosophical Transactions of the Royal Society of London, Series B*, **316,** 277–302.

RAUP, D. M. 1979. Biases in the fossil record of species and genera. *Bulletin of the Carnegie Museum of Natural History*, **13,** 85–91.

REID, R. C. & CROSBY, S. P. 1980. The raptorial siphon of the carnivorous septibranch *Cardiomya planetica* Dall (Mollusca: Bivalvia), with notes on feeding and digestion. *Canadian Journal of Zoology*, **58,** 670–679.

REID, R. G. B. & REID, A. M. 1974. The carnivorous habit of members of the septibranch genus *Cuspidaria* (Mollusca: Bivalvia). *Sarsia*, **56,** 47–56.

RUNNEGAR, B. 1974. Evolutionary history of the bivalve subclass Anomalodesmata. *Journal of Paleontology*, **48,** 904–939.

—— 1979. *Pholadomya candida* Sowerby: The last cadaver unearthed. *Veliger*, **22**, 171–172.

SALVINI-PLAWEN, L. V. & HASZPRUNAR, G. 1982. On the affinities of Septibranchia (Bivalvia). *Veliger*, **25**, 83–85.

—— & STEINER, G. 1996. Synapomorphies and plesiomorphies in higher classification of the Mollusca. *In:* TAYLOR, J. D. (ed.) *Origin and Evolutionary Radiation of the Mollusca*. Oxford University Press, Oxford, 29–51.

SAVAZZI, E. 1982. Adaptations to tube dwelling in the Bivalvia. *Lethaia*, **15,** 275–297.

—— 1999. Boring, nestling and tube–dwelling bivalves. *In:* SAVAZZI, E. (ed.) *Functional Morphology of the Invertebrate Skeleton*. Wiley & Sons, Chichester, 205–237.

SCHNEIDER, J. A. 1992. Preliminary cladistic analysis of the bivalve family Cardiidae. *American Malacological Bulletin*, **9**, 145–155.

—— 1995. Phylogeny of the Cardiidae (Mollusca, Bivalvia): Protocardiinae, Laevicardiinae, Lahilliinae, Tulogocardiinae subfam. n. and Pleuriocardiinae subfam. n. *Zoologica Scripta*, **24,** 321–346.

—— 1998. Phylogeny of the Cardiidae (Bivalvia): phylogenetic relationships and morphological evolution within the subfamilies Clinocardiinae, Lymnocardiinae, Fraginae and Tridacninae. *Malacologia*, **40,** 321–373.

SEILACHER, A. 1984. Constructional morphology of bivalves: evolutionary pathways in primary versus secondary soft-bottom dwellers. *Palaeontology*, **27,** 207–237.

SEPKOSKI, J. J. 1992. A compendium of fossil marine animal families. *Milwaukee Public Museum Contributions in Biology and Geology*, **83,** 1–156.

SKELTON, P. W. & BENTON, M. J. 1993. 'Mollusca: Rostroconchia, Scaphopoda and Bivalvia', *In:* BENTON, M. J. (ed.) *The Fossil Record 2*. Chapman & Hall, London, 237–263.

——, CRAME, J. A., MORRIS, N. J. & HARPER, E. M. 1990. Adaptive divergence and taxonomic radiation in post-Palaeozoic bivalves. *In:* TAYLOR, P. D. & LARWOOD, G. (eds) *Major Evolutionary Radiations*. The Systematics Association Special Volume 42. Clarendon Press, Oxford, 91–117.

SMITH, A. B. 1994. *Systematics and the Fossil Record*. Blackwell Scientific Publications, Oxford

STEINER, G. 1998. Phylogeny of Scaphopoda (Mollusca) in the light of new anatomical data on the Gadilinidae and some Problematica, and a reply to Reynolds. *Zoologica Scripta*, **27**, 73–82.

SWOFFORD, D. L. 1998. *PAUP**. *Phylogenetic Analysis Using Parsimony (* and other methods). Version 4.* Sinauer Associates Inc., Sundereland, MA.

TAYLOR, J. D., KANTOR, Y. I. & SYSOEV, A. V. 1993. Foregut anatomy, feeding mechanisms, relationships and classification of the Conoidea (= Toxoglossa) (Gastropoda). *Bulletin of the Natural History Museum, London (Zoology)*, **59,** 125–170.

WALLER, T. R. 1978. Morphology, morphoclines and a new classification of the Pteriomorphia (Mollusca: Bivalvia). *Philosophical Transactions of the Royal Society of London, Series B,* **284**, 345–365.

—— 1998. Origin of the molluscan class Bivalvia and a phylogeny of the major groups. *In:* JOHNSTON, P. A. & HAGGART, J. W. (eds) *Bivalves: An Eon of Evolution – Paleobiological Studies Honoring Norman D. Newell*. Calgary University Press, Calgary, 1 –45.

YEATES, D. K. 1995. Groundplans and exemplars: paths to the tree of life. *Cladistics*, **11**, 343–357.

YIN, J. & FÜRSICH, F. T. 1991. Middle and Upper Jurassic bivalves of the Tanggula Mountains, W. China. *Beringeria,* **4**, 127–192.

YONGE, C. M. 1928. Structure and function of the organs of feeding and digestion in the septibranchs, *Cuspidaria* and *Poromya*. *Philosophical Transactions of the Royal Society of London, Series B*, **216,** 221–263.

—— 1948. Cleansing mechanisms and the function of the fourth pallial aperture in *Spisula subtruncata* (da Costa) and *Lutraria lutraria* (L.). *Journal of the Marine Biological Association of the United Kingdom*, **27**, 585–596.

—— 1952. Structure and adaptation in *Entodesma saxicola* (Baird) and *Mytilimeria nuttallii* Conrad. *University of California Publications in Zoölogy*, **55**, 439–450.

—— 1962. On the primitive significance of the byssus in the Bivalvia and its effects in evolution. *Journal of the Marine Biological Association of the United Kingdom*, **42**, 112–125.

—— 1982. Mantle margins with a revision of siphonal types in the Bivalvia. *Journal of Molluscan Studies*, **48**, 102–103.

—— & MORTON, B. 1980. Ligament and lithodesma in the Pandoracea and Poromyacea with a discussion on the evolutionary history of the Anomalodesmata (Mollusca: Bivalvia). *Journal of Zoology, London*, **191**, 263–292.

Prodigious polyphyly in imperilled freshwater pearly-mussels (Bivalvia: Unionidae): a phylogenetic test of species and generic designations

CHARLES LYDEARD[1], RUSSELL L. MINTON[1] & JAMES D. WILLIAMS[2]

[1] *University of Alabama, Biodiversity and Systematics, Department of Biological Sciences, Box 870345, Tuscaloosa, Alabama 35487, USA (e-mail: clydeard@biology.as.ua.edu)*

[2] *US Geological Survey, Biological Resources Division, 7920 NW 71st Street, Gainesville, FL 32653, USA*

Abstract: Unionid bivalves or freshwater pearly-mussels (Unionoidea: Unionidae) serve as an exemplary system for examining many of the problems facing systematists and conservation biologists today. Most of the species and genera were described in the late 1800s and early 1900s, but few phylogenetic studies have been conducted to test conventional views of species and classification. Pearly-mussels of Gulf Coastal drainages of the southeastern United States from the Escambia (southern Alabama to Florida) to the Suwannee Rivers (Florida) are a unique fauna comprised of approximately 100 species, with about 30 endemic to the region. In this study, mitochondrial cytochrome c oxidase subunit I and 16S rRNA gene sequences were used to test the monophyly and to estimate evolutionary relationships of five unionid species representing three different genera. The molecular phylogenies depict all three genera as polyphyletic. The prodigious polyphyly exhibited within unionids is due to incorrect notions of homology and false assumptions about missing anatomical data. In contrast, the molecular phylogeny provides evidence to support the recognition of all five unionid species as distinct evolutionary entities. Furthermore, molecular genealogical evidence supports the elevation of *Quincuncina infucata* (Conrad) of the Suwannee River to species level, for which *Q. kleiniana* (Lea) is available.

The Systematics Agenda 2000's global objective is to discover, describe and classify the world's species on the basis of evolutionary relationships before organisms are lost to extinction (Systematics Agenda 2000 1994; Eshbaugh 1995). Unionid bivalves or pearly-mussels (Unionoidea: Unionidae) serve as an exemplary model for many of the problems facing systematics and conservation biologists today. Unionids are distributed in freshwaters of Eurasia and North America, but reach their greatest diversity in North America with nearly 300 recognized species (Burch 1975; Bogan 1993; Lydeard & Mayden 1995; Neves *et al.* 1997; Turgeon *et al.* 1998). Despite their ecological significance as dominant macroinvertebrates in many streams and rivers (McMahon 1991) of the southeastern United States and their economic significance (i.e. a yearly 3 billion dollar cultured pearl industry; Jenkinson & Todd 1997), relatively little is known about the life history, ecology and phylogeny of the majority of taxa. Moreover, unionids are recognized as one of the most imperilled groups of animals around the world with 70% of the recognized species in North America considered either extinct, endangered, threatened or of special concern (Williams *et al.* 1993; Neves *et al.* 1997; Master *et al.* 1998).

Unionid bivalve species and genera are diagnosed on characteristics of the shell, soft anatomy and length of the brooding season (Heard & Guckert 1970; Burch 1975). Initially, over 1000 nominal unionid species were described. This trend was followed by a reduction in the number of species, largely due to the thought that much of the observed conchological variation represented population-level phenomena (Johnson 1970). Likewise, the number of North American unionid genera expanded from two (e.g. Lea 1834), to 27 (Simpson 1914), to 46 (Burch 1975) (systematists initially treated margaritiferid species as unionids). Currently, 286 species and 51 genera (including 17 monotypic genera) are recognized through consensus reached by a committee of malacologists (Turgeon *et al.* 1998). Although relative taxonomic

From: HARPER, E. M., TAYLOR, J. D. & CRAME, J. A. (eds) *The Evolutionary Biology of the Bivalvia*. Geological Society, London, Special Publications, **177**, 145–158. 1-86239-076-2/00/$15.00

stability has been achieved over the past decade, this has been done largely by subjective taxonomic decisions with the understanding that many species and generic hypotheses need to be tested further. A pre-Hennigian classification and species checklist gives the false impression that biologists really understand the phylogenetic relationships and species richness of a group. This information is subsequently used to locate biological hotspots for setting future conservation priorities (e.g. Dobson *et al.* 1997; Master *et al.* 1998). Following the initial spate of descriptions in the early 1900s, most systematic activity focused on what the appropriate name should be for a given taxon rather than determining whether the named taxon constitutes a real species or is a member of a natural group (i.e. clade) worthy of a name in the first place.

Recently, molecular data have been used to address various phylogenetic hypotheses within unionids. Lydeard *et al.* (1996) constructed a molecular phylogeny of 23 genera and 29 species of unionids and showed that the classification scheme currently employed (i.e. Heard & Guckert 1970) does not reflect evolutionary relationships. Mapping various conchological and soft-part anatomical characters that were the primary basis of the classification scheme (Heard & Guckert 1970) on the molecular phylogeny, and constructing a phylogeny based on DNA and morphological data, revealed many of the key characters traditionally used in higher unionid systematics to be homoplastic (e.g. number gills used as marsupium, shell texture). Regrettably, few of the genera and species examined by Lydeard *et al.* (1996) were tested for monophyly, so the question remains whether traditional views of species and genera reflect natural units.

Unionids of the Gulf Coastal drainages of the southeastern United States from the Escambia (southern Alabama and Florida) to the Suwannee Rivers (Florida) are a unique fauna comprised of approximately 100 species, of which about 30 are endemic to the region (Williams & Butler 1994). The region has figured prominently in aquatic biogeographic studies, particularly the ichthyofauna (e.g. Bermingham & Avise 1986). The most comprehensive look at the malacofauna of the region was by Clench & Turner (1956). Subsequent studies consisted primarily of additional notes on the distribution and taxonomy of the species (e.g. Johnson 1969; Athearn 1970; Butler 1989; Williams & Butler 1994). While many of the recognized species are threatened, endangered or of special concern, current taxonomy may not reflect natural evolutionary entities (Williams & Butler 1994).

In the present study, mitochondrial cytochrome c oxidase subunit I (COI) and 16S rRNA gene sequences were used to test monophyly and to estimate phylogenetic relationships of five Gulf Coastal unionid species (Fig. 1) representing three different genera (*Fusconaia succissa*, *F. escambia*, *Quincuncina infucata*, *Q. burkei* and *Obovaria rotulata*). These five species were chosen because: (1) very little is known of their life history, ecology and taxonomic status relative to species found in the Mississippi Basin; (2) they are all easternmost representatives of their respective genera and endemic to various drainages of the eastern Gulf Coast; and (3) phylogenetic data will provide useful data for wisely managing the five taxa, which are either endangered, threatened or of special concern.

Materials and methods

Terminal taxa and vouchers

Table 1 lists the taxa used in this analysis, localities and voucher information. Ingroup taxa were comprised of 23 unionid specimens representing 12 species and five genera. *Quadrula quadrula*, *Pleurobema decisum*, *Fusconaia flava*, *F. ebena*, *Lampsilis teres*, *Obovaria unicolor* and *O. olivaria* were included to place the relationships of the five Gulf Coastal species in a broader evolutionary context. It was attempted to include type species of each genus whenever possible, however, *Obovaria retusa* and *Pleurobema clava* are both federally listed as endangered and specimens could not be secured for this study. The chosen species represent the morpho type exhibited by the type species. As outgroups, one anodontine (*Anodonta cygnea*) and one margaritiferid species (*Cumberlandia monodonta*), based on higher level molecular and morphological phylogenetic analyses (Lydeard *et al.* 1996), were used.

DNA processing sequence procurement, alignment and analysis

Genomic DNA was isolated from frozen or 95% ethanol-preserved specimens using the QIAamp Tissue Kit (Qiagen™). Mitochondrial DNA (mtDNA) sequences were obtained for an amplified segment (*c.* 550 nucleotides) of the 16S rRNA gene using primers 16Sar-L-myt and 16Sbr-H-myt (Lydeard *et al.* 1996), and the COI gene (*c.* 650 nt) using primers LCO1490 and HCO2198 (Folmer *et al.* 1994). Details of the amplification protocols have been reported by Lydeard *et al.* (1996) for 16S rRNA and Roe & Lydeard (1998) for the COI genes. Sequences were determined using either manual sequencing protocols [see Lydeard *et al.* (1996) for 16S rDNA and Roe & Lydeard (1998) for COI] or automated sequencing, using an Applied Biosystems 373A or 377 DNA sequencer using Taq DyeDeoxy terminator chemistry according to the manufacturer's protocol. To ensure accuracy and to resolve any ambiguities, both DNA strands were sequenced.

Sequences were entered in the software program ESEE (Cabot & Beckenbach 1989) and aligned by eye with the aid of bivalve (*Mytilus edulis* Linnaeus and *Pecten maximus* Linnaeus) 16S rRNA secondary structure models (Lydeard *et al.* 2000). Insertions and deletions (indels) were treated as

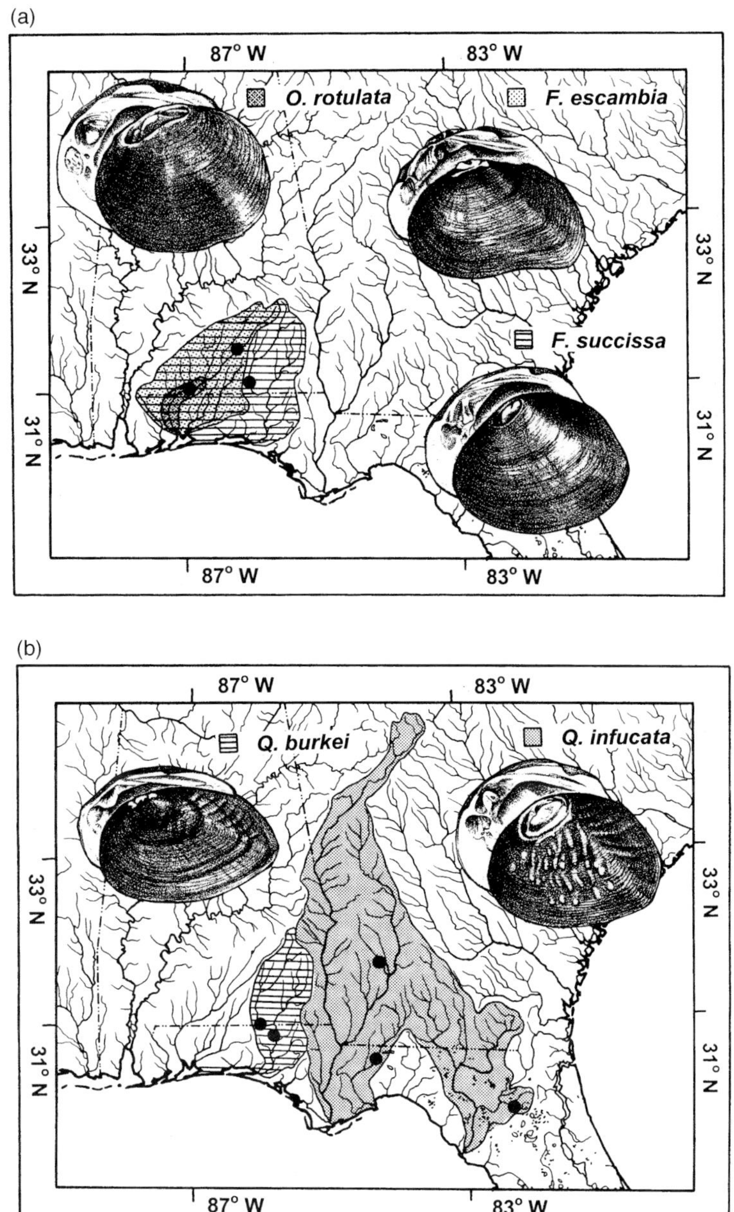

Fig. 1. Distribution and collection localities (i.e. dots on map) of: (**a**) *Fusconaia* and *Obovaria*; (**b**) *Quincuncina* species of eastern Gulf Coastal rivers of the southeastern United States. See Table 1 for names of rivers and voucher information.

Table 1. *Material used for phylogenetic analysis*

Quincuncina burkei Walker 118	Limestone Creek, Choctawhatchee Drainage, Florida UAUC575-577; AF232777, AF232802
Q. burkei 1	Pittman Creek, Choctawhatchee Drainage, Florida UAUC927-933; AF232779, AF232804
Q. burkei 2	Same as above UAUC927-933; AF232778, AF232803
Q. infucata (Conrad) 1	Ochlockonee River, Florida UAUC919-926; AF232780, AF232805
Q. infucata 45	Kinchafoonee Creek, Flint River, Georgia UAUC561; AF232782, AF232807
Q. infucata 46	Same as above UAUC6O5; AF232781, AF232806
Q. infucata 57	New River, Suwannee Drainage, Florida UAUC564; AF232783, AF232808
Q. infucata 58	Same as above UAUCS67; AF232784, AF232809
Lampsilis teres (Rafinesque)	Yellowleaf Creek, Coosa Drainage, Alabama UAUCOO6; AF232785, AF232810
Obovaria unicolor (I. Lea)	Sipsey River, Tombigbee Drainage, Alabama UAUC51; AF232786, AF232811
0. olivaria (Rafinesque)	Mississippi River, Missouri mantle clip only – Bernard Sietman; AF232787, AF232812
0. rotulata (Wright) 1	Conecuh River, Alabama UAUC5O8; AF232788, AF232813
0. rotulata 2	Same as above UAUC522; AF232789, AF232814
Fusconaia ebena (I. Lea)	Kentucky Reservoir, Tennessee Drainage, Tennessee UAUC71; AF232790, AF232815
F. escambia Clench and Turner 2	Conecuh River, Escambia Drainage, Alabama UAUC1448; AF232791, AF232816
F. escambia 11	Same as above UAUC1449; AF232792, AF232817
F. succissa (I. Lea) A	Conecuh River, Escambia Drainage, Alabama UAUC9I 1-918; AF232796, AF232821
F. succissa B	Same as above UAUC911-918; AF232793, AF232818
F. succissa 8	Same as above UAUC1456; AF232794, AF232819
F. succissa 119	Pea River, Choctawhatchee Drainage, Alabama UAUC525-528; AF232795, AF232820
F. flava (Rafinesque)	Ohio River, Indiana–Kentucky UAUC146; AF232797,AF232822
Pleurobema decisum (I. Lea)	Sipsey River, Tombigbee Drainage, Alabama AUC2O8; AF232776, AF232801
Quadrula quadrula (Rafinesque)	Ohio River Indiana–Kentucky UAUCI45; AF232798, AF232823
Anodonta cygnea (Linnaeus)	Lake Obertrurnersee Salzburgh, Austria; mantle clip only – R. A. Patzner source; AF232799, AF232824
Cumberlandia monodonta (Say)	Clinch River, Tennessee UAUCOO7; AF232800, AF232825

Locality, University of Alabama Unionid Collection (UAUC) numbers and GenBank accession numbers are provided.

missing data. Aligned sequences were analysed using maximum parsimony in *PAUP**, *version 4.0b2*a (Swofford 1998). A boostrap analysis (Felsenstein 1985) with 1000 iterations was conducted to estimate internal stability of the matrix and Bremer support index values were calculated using Autodecay 3.03 (Erriksson 1997); ten random-addition sequences were used to determine the decay value for each node of each tree. A skewness test statistic (g^1) was also calculated, based on the distribution of tree lengths of a random sample of 10 000 topologies (Hills & Huelsenbeck 1992). A partition homogeneity test was conducted to examine the extent of conflict between the two gene portions using the incongruence length difference (ILD) test (Farris *et al.* 1995) as implemented in *PAUP** (simple addition-sequence, tree bisection–reconnection, heuristic search option with 500 replicates analysed).

The Evolutionary Species Concept (Mayden 1997) was used as the primary model in the recognition of evolutionary lineages. The operational species definition employed is historically based (Baum & Donughue 1995) and includes the criterion of monophyly in the general sense (de Quieroz & Donoghue 1988) or 'exclusivity' (i.e. where an exclusive group of organisms is one whose members are more closely related to each other than they are to any organisms outside the group) to denote the recognition of species.

Results

Sequence data

The aligned data matrix is available from the authors. GenBank accession numbers are shown in Table 1. The region of 16S rDNA and COI that was sequenced resulted in an aligned data matrix of 838 base pairs (bp) (385 bp 16S rDNA and 453 bp COI). Individual 16S rDNA sequences ranged from 375 (*Lampsilis teres*) to 384 (*Cumberlandia monodonta*); no length variation was found in the COI sequences. The mitochondrial 16S rDNA sequences exhibited uncorrected p-distance values from 0 (*Q. burkei* and *Q. infucata*) to 2.1% (*Q. infucata* Suwannee River v. *Q. infucata* Apalachicola, Chattahoochee, Flint Basin) within species and 0 (*F. escambia* v. *Q. burkei*) to 16.2% (*Lampsilis teres* v. *Q. infucata*) among species within the ingroup. Outgroup and ingroup taxa differed by 13.3–26.9%. The mitochondrial COI gene exhibited uncorrected p-distance values from 0 (*Obovaria rotulata*) to 3.5% (*F. escambia* and *Q. burkei*) within species and 0.4 (*F. ebena* v. *O. rotulata*) to 18.9% (*Quadrula quadrula* v. *F. escambia*) among species within the ingroup. Outgroup and ingroup taxa differed by 13.2–20.5%. The mitochondrial 16S rDNA aligned data matrix had 153 variable and 97 parsimony informative sites, while the COI gene portion had 171 variable and 126 parsimony informative sites. The number of parsimony informative sites for each codon position of the mtDNA COI gene is 1st = 15, 2nd = 6 and 3rd = 105.

Base compositional bias is common in mitochondrial DNA sequences. The average nucleotide base frequences for the COI gene for all taxa was A = 0.16, C = 0.16, G = 0.26 and T = 0.42, and the 16S rRNA gene A = 0.35, C = 0.20, G = 0.22 and T = 0.23. There was no significant difference in base frequencies among taxa for either gene (X^2 = 17.86, P = 1.00, COI gene; X^2 = 14.85, P = 1.00, 16S rRNA gene).

Phylogenetic analysis

Maximum parsimony analysis of the COI gene was conducted by treating each character transformation as unordered and of equal weight, and weighting transversions twice transitions at the third codon position based on evidence of site saturation (see Roe & Lydeard 1998). Parsimony analysis of the COI data using equal weighting resulted in four equally parsimonious trees (consistency index (CI) = 0.531; homoplasy index (HI) = 0.469; retention index (RI) = 0.756; total length = 417). The strict consensus tree is shown in Fig. 2a. All four equally parsimonious trees support the monophyly of *Fusconaia escambia*, *F. succissa*, *Quincuncina infucata* and *Obovaria rotulata*. *Quincuncina burkei* was paraphyletic, but a monophyletic *Q. burkei* tree is only one step longer and not significantly different to the most parsimonious topology based on Templeton's (1983*a*, *b*) Wilcoxon sign-ranked test ($P = 0.835$). The genera *Obovaria*, *Fusconaia* and *Quincuncina* were found to be polyphyletic, with *Q. burkei* sister to *F. escambia*, *Q. infucata* sister to *F. succissa* and *F. ebena* sister to *O. rotulata*. A tree constraining the genera to be monophyletic is 100 steps longer, which is significantly longer than the most parsimonious trees ($P < 0.001$). Relationships among the deepest nodes are poorly supported and differ among the most parsimonious trees. Indeed, the outgroup species *Cumberlandia monodonta* is part of the ingroup for two of the four most-parsimonious trees. Parsimony analysis of the COI data weighting transversions twice transitions in the third codon position resulted in a single most parsimonious tree (CI = 0.533; HI = 0.467; RI = 0.76; total length = 550). A phylogram is shown in Fig. 2b. The topology is equivalent to one of the

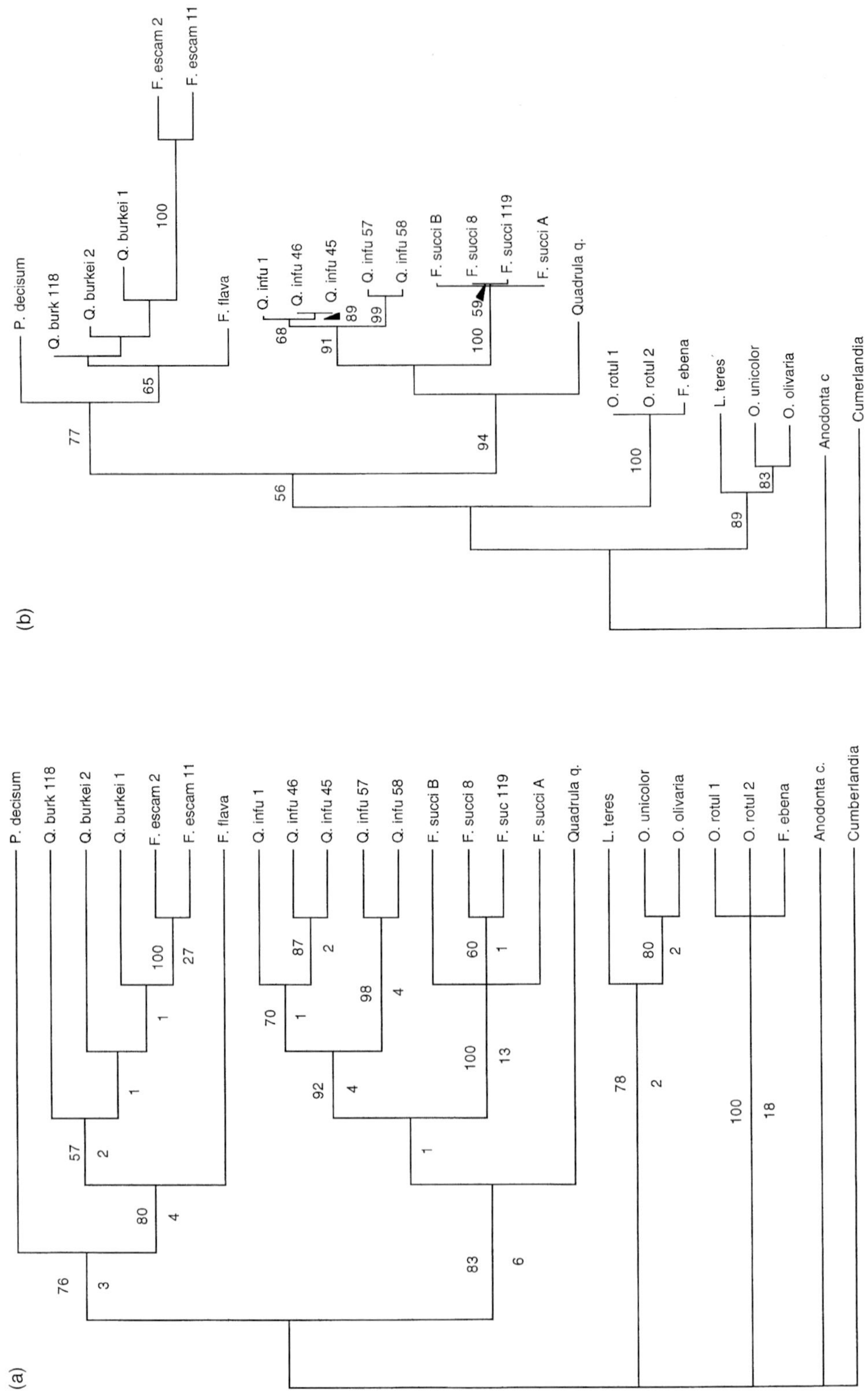
(b)
P. decisum
Q. burk 118
Q. burkei 2
Q. burkei 1
F. escam 2
F. escam 11
F. flava
Q. infu 1
Q. infu 46
Q. infu 45
Q. infu 57
Q. infu 58
F. succi B
F. succi 8
F. succi 119
F. succi A
Quadrula q.
O. rotul 1
O. rotul 2
F. ebena
L. teres
O. unicolor
O. olivaria
Anodonta c
Cumberlandia
(a)
P. decisum
Q. burk 118
Q. burkei 2
Q. burkei 1
F. escam 2
F. escam 11
F. flava
Q. infu 1
Q. infu 46
Q. infu 45
Q. infu 57
Q. infu 58
F. succi B
F. succi 8
F. suc 119
F. succi A
Quadrula q.
L. teres
O. unicolor
O. olivaria
O. rotul 1
O. rotul 2
F. ebena
Anodonta c.
Cumberlandia

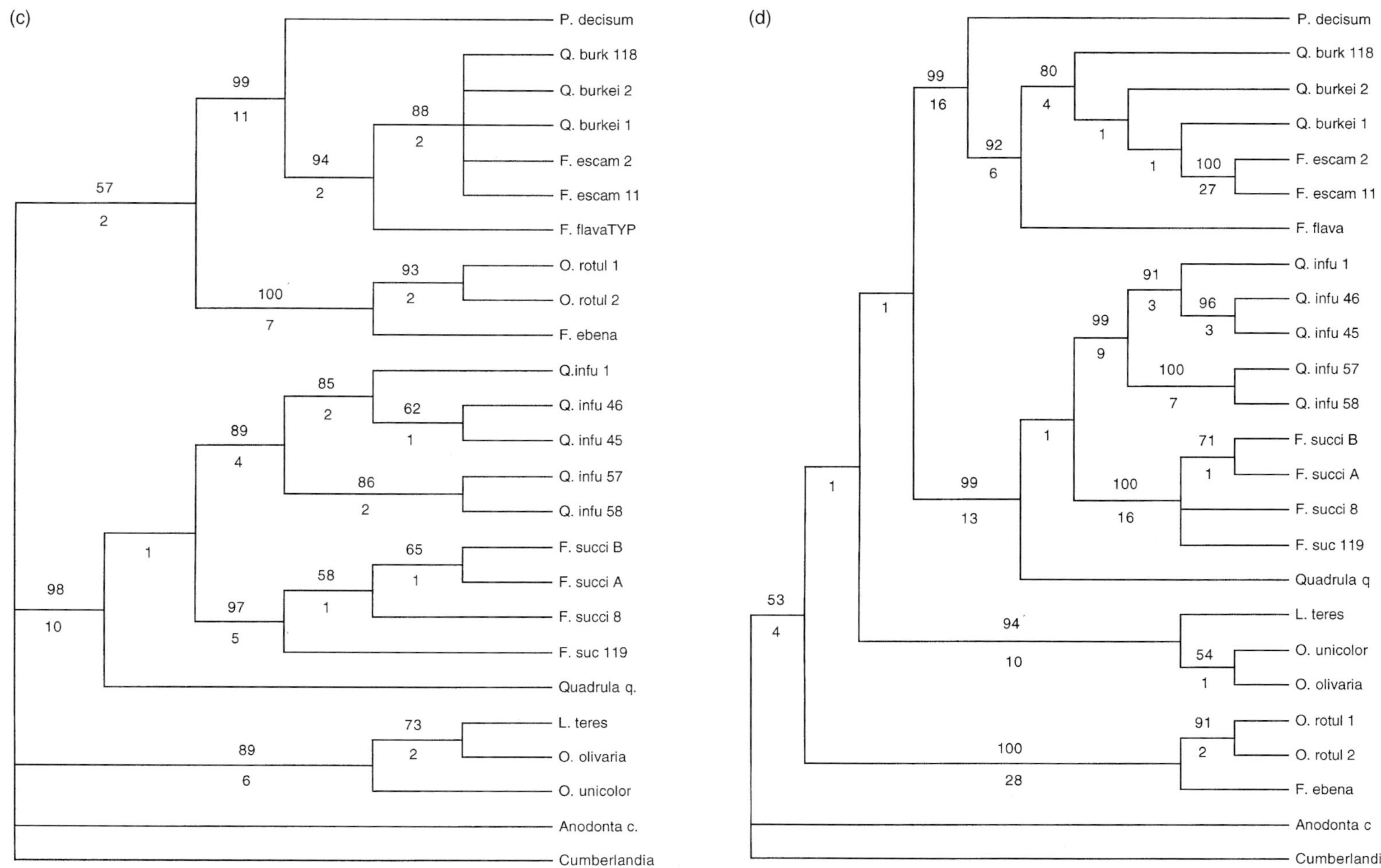

Fig. 2. Most parsimonious strict consensus trees or phylograms generated from: (**a**) mitochondrial cytochrome c oxidase sequences using equal weighting (strict consensus tree of four equally parsimonious trees); (**b**) mitochondrial cytochrome c oxidase sequences weighting transversions twice transitions in the third codon position (single most parsimonious phylogram; (**c**) mitochondrial 16S rDNA sequences using equal weighting (strict consensus tree of three equally parsimonious trees); (**d**) mitochondrial cytochrome c oxidase and 16S rDNA sequences combined (strict consensus tree of two equally parsimonious trees). Numbers above branches indicate the proportion of 1000 bootstrap replicates that supported the depicted clades. Numbers below branches are Bremer support index values. Taxon names and numbers correspond to those shown on Table 1.

four most parsimonious trees obtained using equal weighting with the ingroup depicted as being monophyletic. The COI data contains significant phylogenetic structure ($g^1 = -0.59887$ and -0.559, equal and unequal weighting, respectively).

Phylogenetic analysis of the mitochondrial 16S rDNA data was conducted treating all substitutions as unordered and of equal weight (Lydeard *et al.* 1996; Mulvey *et al.* 1997). Three equally parsimonious trees were obtained from the maximum parsimony analysis (CI = 0.643; HI = 0.356; RI = 0.617; total length = 302). A strict consensus tree of the three most parsimonious trees is shown in Fig. 2c. The topology is similar to the COI-based topologies with the monophyly of *Fusconaia escambia*, *F. succissa*, *Quincuncina infucata* and *Obovaria rotulata*. *Quincuncina burkei* is paraphyletic, but a monophyletic *Q. burkei* is an equally parsimonious solution (total length = 302). The genera *Obovaria*, *Fusconaia* and *Quincuncina* were found to be polyphyletic, with *Q. burkei* sister to *F. escambia*, *Q. infucata* sister to *F. succissa* and *F. ebena* sister to *O. rotulata*, which is identical to the topologies obtained using the COI gene. The primary difference between the phylogenies obtained from the two genes is that the *O. rotulata* + *F. ebena* clade is sister to *Pleurobema decisum* + *F. flava* + *F. escambia* + *Q. burkei* on the 16S rDNA-based phylogeny. This alternate placement of *O. rotulata* + *F. ebena* is not unusual given the generally weak support of the deeper nodes. The mitochondrial 16S rDNA sequences exhibit significant phylogenetic structure ($g^1 = -0.474$).

Prior to conducting a phylogenetic analysis combining the DNA sequences from the two mitochondrial genes, a partition homogeneity test was conducted. Results from the partition homogeneity test were not significant ($P = 0.16$), indicating that DNA sequences from the two genes could be combined. Maximum parsimony analysis of the entire DNA sequence data matrix, treating all substitutions as unordered and of equal weight, resulted in two equally parsimonious trees (CI = 0.566; HI = 0.434; RI = 0.798; total length = 729). A strict consensus tree of the two most parsimonious trees is shown in Fig. 2d. The topology is equivalent to one of the most parsimonious trees obtained from the COI phylogenetic analysis (i.e. unordered and of equal weighting). The only difference between the combined gene tree topology and the 16S rDNA-based topology is the more basal placement of the *O. rotulata* + *F. ebena* clade. Parsimony analysis of the entire DNA sequence data matrix weighting transversions twice transitions in the third codon position of the COI gene resulted in two equally parsimonious trees (CI = 0.561; HI = 0.439; RI = 0.794; total length = 863) that are identical to the two equally parsimonious trees obtained using equal weighting.

Discussion

Each of the five Gulf Coastal species were initially recognized and described based on a variety of conchological features that were implicitly thought to reflect the natural history and coherence of the species. It is evident that although no rigorous quantitative method existed at the time each of the Gulf Coastal species were described, the conchological 'bauplan' captured by the early naturalists seems to reflect the unique historical legacy of each species. The molecular phylogeny based on the two mitochondrial gene sequences combined depicted four of the five Gulf Coastal unionid species (*Fusconaia succissa*, *F. escambia*, *Obovaria rotulata* and *Quincuncina infucata*) as monophyletic. Although *Quincuncina burkei* was depicted as paraphyletic, constraining the monophyly of the species was not found to be statistically significantly different. Lack of reciprocal monophyly between *Q. burkei* and its sister taxon *F. escambia* is likely to be an indication of insufficient time for the two genes to coalesce (Avise & Ball 1990). Evidence for this supposition is found in the differences obtained between the two genes. The mitochondrial 16S rRNA gene is the most conservative of the two genes examined in this study and neither species was monophyletic. The more rapidly evolving COI gene, however, exhibited much greater variation and substantial genetic divergence of a monophyletic *F. escambia* from the paraphyletic *Q. burkei* lineages (a minimum p-distance value of 7.95% between *F. escambia* and *Q. burkei*). It is hypothesized that a more rapidly evolving gene would yield reciprocal monophyly between *F. escambia* and *Q. burkei* and its continued recognition as a distinct species is recommended. It is probably better to err in favour of recognition of a distinct evolutionary entity than to fail to recognize it (Daugherty *et al.* 1990), particularly given the numerous conchological differences between the two species (see below).

It has been debated whether the issue of monophyly should be extended from higher categorical levels to the species level (e.g. de Quieroz & Donoghue 1988; Nixon & Wheeler 1990). Some versions of the Phylogenetic Species Concept recognize species as minimal monophyletic groups (de Queiroz & Donoghue 1988, 1990; Mishler & Donoghue 1982; Donoghue 1985; Mishler & Brandon 1987; McKitrick & Zink 1988; Mayden 1997). This has proven useful in a variety of studies on unionids and has resulted in synonymizing some species and subspecies such as

Megalonaias boykiniana and *Amblema plicata perplicata*, and the discovery of cryptic species such as *Amblema elliottii* (Mulvey *et al.* 1997; Roe & Lydeard 1998; King *et al.* 1999). Molecular data often provides information that may not always be gleaned from morphology alone. Using an 'exclusivity' approach (Baum & Donoghue 1995) to recognizing species it is evident that the Suwannee River populations are geographically and genetically isolated from populations from the Apalachicola and Ochloconee Drainages (Figs 1 and 2). Additionally, this genetic isolation is reinforced by the absence of *Q. infucata* in the coastal drainages between the Ochlockonee and the Suwannee. Given the reciprocal monophyly of the populations and the degree of genetic differentiation, these two populations should be accorded specific status.

In stark contrast to the species-level findings, all three Gulf Coastal genera examined (*Fusconaia*, *Quincunina* and *Obovaria*) were found to be polyphyletic. A genus should reflect a historical group of species that share common ancestry. Ortmann & Walker (1922) placed *Q. burkei* and *Q. infucata* in *Quincuncina* (type species: *Q. burkei*) based on the presence of a complex zig-zag sculpture on the shell (*Q. mitchelli*, from Texas, was later described and is now presumed to be extinct; R. Howell pers. comm.). Similar zig-zag patterns are also found on other unionid species, including several species of *Quadrula* and *Tritogonia verrucosa* (Rafinesque). Given the topology of the molecular phylogeny, it is evident the zig-zag pattern is not homologous.

The only character described in the literature distinguishing the genus *Fusconaia* from other genera is a subcylindrical-shaped placentae (Ortmann 1912; Burch 1975). Soft parts from gravid females have been examined in *F. flava* (type species of *Fusconaia* Ortmann 1912) and *Quincuncina burkei*. The most characteristic anatomical feature of *Q. burkei* noted by Ortmann & Walker (1922) is that the gravid female has subcylindrical placentae like *Fusconaia*. No reports of gravid female soft parts have been published about *F. succissa* (Ortmann 1923) nor *F. ebena* (Ortmann 1912), which was placed in *Fusconaia* based on the similarity of the shell to other *Fusconaia* species. The fact that all taxa that are thought to possess a subcylindrical-shaped placentae form a clade, supports the hypothesis that the feature may be homologous. In contrast, Simpson (1900) placed both *F. succissa* and *Q. infucata* in the genus *Quadrula* because of the similarity of the shell. The molecular phylogeny depicts both *F. succissa* and *Q. infucata* as sister taxa and sister to *Quadrula quadrula* (type species of *Quadrula* Rafinesque, 1820). Davis (1984) conducted a phenetic analysis of 14 allozyme loci and found *Q. infucata* clustered with *Quadrula*, and that *F. flava* and *F. succissa* belong to different genera based on a high genetic distance value and their placement on an ordination diagram.

The molecular phylogeny depicts *Q. burkei* + *F. escambia* as sister to *F. flava*, which in turn is sister to *Pleurobema decisum*. The primary difference between *Fusconaia* and *Pleurobema*, which were placed in the same tribe (Pleurobemini) by Heard & Guckert (1970), is that all four gills are used as marsupia in *Fusconaia* and only the outer pair is used in *Pleurobema*. *Fusconaia ebena* is consistently placed outside the *Fusconaia flava* + *F. escambia* + *Q. burkei* + *Pleurobema* clade. Lydeard *et al.* (1996) showed that the number of gills used as a marsupium is homoplastic with use of only the outer gills being the derived condition. Previous investigators have debated the placement of *Fusconaia* (Heard & Guckert 1970; Davis & Fuller 1981). Part of the problem may be due to the assumption that *Fusconaia* was monophyletic. Other unionid species will have to be examined before this issue is resolved fully.

Obovaria is comprised of six species and is placed in the tribe Lampsilini (Burch 1975). Several reproductive and soft-part anatomical features are synapomorphies of the tribe (Lydeard *et al.* 1996) and four of five *Obovaria* species possess these lampsiline features. Simpson (1900*a*, *b*) examined *Unio rotulata* (Wright 1899) and placed it in the genus *Obovaria* based on general similarity of shell characteristics. Subsequent workers (Simpson 1914; Johnson 1967, 1969; Turgeon *et al.* 1998) continued to recognize *rotulata* as a species of *Obovaria* on the basis of shell characters, but Butler (1989) and Williams & Butler (1994) placed it in the genus *Fusconaia* also on the basis of shell characters (i.e. teeth and deep beak cavity). Examination of soft parts from a gravid female would easily remedy the situation, but the rarity of the species and difficulty in finding gravid females has precluded any investigation of soft parts. The species was completely overlooked by Clench & Turner (1956) and formerly only known by the holotype. The molecular phylogeny depicts the lampsiline species *O. unicolor*, *O. olivaria* and *Lampsilis teres* in a clade excluding *O. rotulata*, which is sister to *F. ebena*. This is strong evidence for the removal of *O. rotulata* from the genus *Obovaria*, however, what to name this clade awaits the inclusion of additional species, including type species.

Phylogenetic taxonomy is concerned with the representation of phylogenetic relationship, specifically, with issues related directly to the naming of taxa (de Queiroz & Gauthier 1990, 1992, 1994). Several case studies have been done advocating the use of phylogenetic taxonomy (e.g. Bryant 1996; Sereno 1999). Ideally, classification of unionids should reflect phylogeny (Lydeard & Roe 1998). It is evident from the molecular phylogenetic hypotheses of eastern Gulf unionid genera that the current classification scheme does not reflect phylogeny. A complete systematic re-evaluation of all North American and Old World unionids (e.g. Hoeh *et al.* 2000), using morphological and molecular data, is recommended here to strive for a phylogenetically, based classification.

Species accounts

The following is a description of the taxa examined in this study including recommended taxonomic changes to better reflect patterns of cladogenesis and genealogical affinities. Until higher order relationships are resolved, it is suggested that the polyphyletic genera be denoted with quotes.

'Quncuncina' infucata (Conrad, 1834) – Sculptured pigtoe

Unio infucatus Conrad 1834, p. 45, pl. 3, fig. 2.
Unio securiformis Conrad 1849, p. 300; Conrad 1850, p.275, pl. 37, fig. 1.

Clench & Turner (1956) limited the type locality to the Flint River, Albany, Dougherty County, Georgia, assuming that this was probably where Conrad collected. A map of Conrad's route through Georgia in 1833 (Wheeler 1935, p. 27) indicates that Conrad crossed the Flint River near Knoxville, located in Crawford County, Georgia, which is most likely to be where the type specimens were collected. The figured types of *Unio infucatus* and *U. securiformis* have not been found (Johnson & Baker 1973).

Description: Shell small, attaining a length of *c.* 50 mm; moderately thick, subcircular in outline and moderately inflated. Surface of shell varies from almost smooth to faintly sculptured, with small nodules or nodulous ridges arranged in a chevron-shaped pattern. Periostracum smooth on the disk, slightly roughened on the posterior slope; colour variable, ranging from greenish in juveniles to dark brown or almost black in adults, the majority of specimens being a dark brown. Beaks high (usually severely eroded), anterior to the centre, their sculpture consisting of prominent, irregular ridges, which curve up sharply behind. Posterior slope flat to slightly concave; the posterior ridge narrowly rounded, shell tapering to a blunt point on the base line posteriorly. Ligament short and thick, hinge plate wide and moderately thick. There are two pseudocardinal teeth in each valve, two laterals in the left valve and one, which is often somewhat double, in the right. The beak cavities are shallow, compressed. Anterior and posterior muscle scars clearly outlined. Nacre usually purplish to bluish white and iridescent posteriorly.

Soft anatomy: A brief description of soft anatomy of the sculptured pigtoe was presented by Lea (1863, pp. 40 and 43) based on specimens from the Chattahoochee River at Columbus, Uchee Bar below Columbus, and Roswell, Cobb County, Georgia. The gills are large, almost semicircular, and the inner ones are larger than the outer. The branchial uterus occupies all four gills. The branchial opening is large with numerous brown papillae. The anal opening is dark but lacks papillae. Supraanal opening large, coloured, slightly united below. Colour of the body is white.

Distribution: Occurs in the Apalachicola River drainage in Alabama, Florida, and Georgia, and the Ochlockonee System in Florida and Georgia. It is not known to occur in the coastal drainages between the Ochlockonee and Suwannee Rivers.

Conservation status: The sculptured pigtoe is considered of Special Concern (Williams *et al.* 1993; Lydeard *et al.* 1999). This status appears appropriate considering the disappearance of this species from the entire main channel of the Chattahoochee River and portions of the Apalachicola River. It has never been considered for endangered or threatened status by the US Fish and Wildlife Service.

'Quincuncina' kleiniana (Lea) – Suwannee pigtoe

Unio kleinianus Lea 1852*a*, p. 251; 1852*b*, p. 265, pl. 17, fig. 18.

Description: Shell small, attaining a length of 60 mm; moderately thick, subcircular in outline, moderately inflated. Surface generally covered with nine to 12 nodulous ridges which originate along the posterior ridge, extending anteriorly and ventrally across the disk, and posteriorly and ventrally across the posterior slope, forming chevron-shaped sculpture, often breaking up in larger (length > 35 mm) individuals. Periostracum smooth, dark brown to black, somewhat shiny on the disk. Beaks (usually eroded) moderately full and high, extending to or slightly above the margin of the shell in specimens with minimal beak erosion. Posterior ridge broadly rounded, ending at the base of the shell in a blunt point; anterior end round; ventral margin straight or slightly curved; posterior end obliquely truncated. Pseudocardinal teeth triangular, usually double in the left valve, single in the right. Beak cavities moderately deep.

Anterior scars small, the posterior scars about twice the diameter of the anterior. Nacre purplish to white. *Quadrula kleiniana* is usually larger and more sculptured than *Q. infucata*. Simpson (1914), in comparing *kleiniana* and *infucata*, observed 'there are intermediates that might almost as well be placed in one species as the other.'

Distribution: Endemic to the Suwannee River Drainage in northern Florida and southern Georgia. It occurs in the Withlacoochee and Alapaha Rivers, southcentral Georgia, and the Santa Fe River in northeast Florida, all tributaries of the Suwannee River.

Conservation status: The species should be considered threatened given that it has completely disappeared from large portions of the main channel of the Suwannee River.

'Fusconaia' succissa (Lea, 1852*b*) – Purple pigtoe

Unio succissus Lea 1852*b*, p. 275, pl. 21, fig. 32.
Unio cacao Lea, 1859, p. 154; Lea 1860, p. 26, pl. 56, fig. 169.
Quadrula wrightii Simpson, 1914, p. 868.

Description: Shell attains a length of 61 mm, subcircular in outline, moderately thick shell, smooth and moderately inflated. Olivaceous brown when young (*c.* 30 mm in length), becoming dark brownish black in larger individuals. Posterior ridge broadly rounded, poorly defined, the slope nearly flat. Shell outline rounded anteriorly and convex posteriorly, the beaks to the ventral margin and almost straight to slightly convex ventrally. Umbos (usually eroded) anterior to the centre, broad and extend to the margin of the shell. Ligament short and narrow. Periostracum smooth on the disk and somewhat roughened around the margin and on the posterior slope. Nacre white to purplish, most young specimens are entirely purple. Muscle scars, anterior and posterior, are well defined. Beak cavities rather deep. Hinge plate moderately broad and moderately arcuate. Right valve usually with one pseudocardinal and one lateral tooth, and the left valve with two pseudocardinal and two lateral teeth.

Distribution: *'Fusconaia' succissa* ranges from the Escambia to Choctawhatchee River Systems in Alabama and Florida. It occurs in three river drainages: the Escambia, Yellow and Choctawhatchee River systems in Alabama and Florida.

Conservation status: The species is considered of Special Concern (Williams *et al.* 1993; Lydeard *et al.* 1999).

'Quincuncina' burkei Walker – tapered pigtoe

Quincuncina burkei Walker 1922, p. 1–6, pls 1, 4.

Description: Shell small, attaining a length of 75 mm. Shell moderately inflated. Beaks only slightly elevated above the hinge line, their sculpture consisting of strong, subcircular ridges, stronger along the umbonal ridge and curved up sharply behind, fading out anteriorly and becoming nearly parallel with the growth lines. Anterior end regularly rounded; base line curved; posterior end somewhat produced, subtruncate, curving down rather abruptly and subangulated as it approaches the posterior point, which is below the median of the disk. Posterior ridge strong and angulated by the junction of the surface ridges. Posterior slope with strong ridges, curving upwards, extending from the posterior ridge to the posterior margin, these form a sharp angle on the posterior ridge with heavier ridges extending downwards and forward, which become more or less broken and tuberculous towards the margin and much weaker on the anterior end, where they assume a rather quincuncial arrangement. Epidermis in mature shells black or sometimes dark brown, in young shells brown or occasionally greenish-yellow, in which case obscure radial stripes of darker green are visible. Pseudocardinals double in both valves. In the right valve the anterior is low and oblique, the posterior strong and erect. In the left valve the anterior is rather long and projects obliquely forward, the posterior is larger, erect and more or less split up. The laterals, two in the left valve and (usually) one in the right, are only a little curved, that in the right valve is sometimes more or less inclined to be double. Beak cavities not very deep or compressed. Anterior muscle scars well marked, the superior one deep and extending under the base of the anterior pseudocardinal. Posterior muscle scars distinct but not deeply impressed. Nacre light purplish, deeper in the beak cavities and iridescent behind (Ortmann & Walker 1922).

Distribution: *Quincuncina burkei* is an endemic of the Choctawhatchee River System of Alabama and Florida.

Conservation status: Quincuncina burkei is considered threatened (Williams *et al.* 1993; Lydeard *et al.* 1999).

'Fusconaia' escambia Clench & Turner, 1956 – Narrow pigtoe

Fusconaia escambia Clench & Turner 1956, pp. 152–153, pl. 7, fig. 34.

Description: *Fusconaia escambia* is a small species, rarely exceeding 50 mm in length and subcircular in outline. It has a smooth, moderately heavy shell, somewhat inflated and with broad, full and high umbos. The posterior ridge is well developed and ends in a point (angle slightly > 90°) posteriorly. The posterior slope is slightly concave. Colour of *F. escambia* varies from dark reddish brown to blackish. Internally, the hinge plate is broad, heavy and slightly arcuate. There is one large pseudocardinal tooth in the right valve and two in the left valve. The nacre is highly iridescent and white to salmon coloured.

Distribution: The narrow pigtoe was first discovered in the Escambia River near Century, Escambia and Santa Rosa Counties, Florida, which was designated the type locality (Clench & Turner 1956). Subsequently, it was found to occur upstream in the Conecuh River below Gantt Lake, Covington County, Alabama. Johnson (1969, p. 35) reported the first record of *Fusconaia escambia* from the Yellow River, Okaloosa County, Florida.

Conservation status: The species is considered threatened (Williams *et al.* 1993; Lydeard *et al.* 1999).

'*Obovaria*' *rotulata* (Wright 1899) – round ebony shell

Unio rotulata B. H. Wright, 1899, pp. 22–23.
Obovaria rotulata Simpson, 1900*b*, p. 78, pl. IV, fig. 2.

Description: '*Obovaria*' *rotulata* is a medium-sized species that attains a length of *c.* 65 mm. It has an almost circular outline and a heavy, somewhat inflated, thick shell with a smooth black epidermis. The pseudocardinal teeth are low and usually double in the left valve and single in the right valve. The lateral teeth are long, slightly curved, and separated from the pseudocardinal teeth by a broad smooth interdentum. The nacre is typically white to silvery, and iridescent.

Distribution: '*Obovaria*' *rotulata* is endemic to the Escambia River Drainage in Escambia County, Alabama, and Escambia and Santa Rosa Counties, Florida (Johnson 1969).

Conservation status: The species is considered endangered (Williams *et al.* 1993; Lydeard *et al.* 1999).

Thanks to A. Bogan, P. Harris, R. Mayden, K. Roe, P. J. West, the Advanced Systematics Discussion Group at UA, and an anonymous reviewer for helpful comments on the manuscript. The illustrations of the shells were done by D. Neely. Thanks to L. Harper for the opportunity to contribute to the symposium volume. We appreciate the help Jayne Brim-Box, Christine O'Brien, Holly Blalock, Malcolm Pierson, Patty Morrison, Bernard Sietman for their assistance in obtaining specimens. This research was supported in part by a Research Grants Committee Award (2-67858) from the University of Alabama, U.S. Fish & Wildlife Service, the U.S. Geological Survey, and the National Science Foundation (DEB-9707623).

References

ATHEARN, H. 1970. Discussion of Dr. Heard's paper on Eastern freshwater mollusks, the south Atlantic, and gulf drainages. *Malacologia,* **10**, 1–56.

AVISE, J. C. & BALL, R. M., JR 1990. Principles of genealogical concordance in species concepts and biological taxonomy. *Oxford Surveys in Evolutionary Biology,* **7**, 45–67.

BAUM, D. A. & DONOGHUE, M. J. 1995. Choosing among alternative 'phylogenetic' species concepts. *Systematic Botany*, **20**, 560–573.

BERMINGHAM, E. & AVISE, J. C. 1986. Molecular zoogeography of freshwater fishes in the southeastern United States. *Genetics,* **113**, 939–965.

BOGAN, A. E. 1993. Freshwater bivalve extinctions (Mollusca: Unionoida): A search for causes. *American Zoologist,* **33**, 599–609.

BRYANT, H. N. 1996. Explicitness, stability, and universality in the phylogenetic definition and usage of taxon names: a case study of the phylogenetic taxonomy of the Carnivora. *Systematic Biology,* **45**, 174–189.

BURCH, J. B. 1975. *Freshwater unionacean clams (Mollusca: Pelecypoda) of North America.* Revised edition. Malacological Publications, Hamburg, MI.

BUTLER, R. S. 1989. Distributional records for freshwater mussels (Bivalvia: Unionidae) in Florida and south Alabama, with zoogeographic and taxonomic notes. *Walkeriana*, **3**, 239–261.

CABOT, E. L. & BECKENBECK, A. T. 1989. Simultaneous editing of multiple nucleic acid and protein sequences with ESEE. *Cabios*, **5**, 233–234.

CLENCH, W. J. & TURNER, R. D. 1956. Freshwater mollusks of Alabama, Georgia and Florida from the Escambia to the Suwannee River. *Bulletin of the Florida State Museum, Biological Sciences,* **1**, 95–239.

CONRAD, T. A. 1834. *New fresh water shells of the United States, with coloured illustrations and a monograph of the genus* Anculotus *of Say. Also a synopsis of the American naiades.* Judah Dodson, Philadelphia.

—— 1849. Descriptions of new freshwater shells. *Annals and Magazine of Natural History*, **4**, 2nd series, 300–303.

—— 1850. Descriptions of new freshwater and marine shells. *Journal of the Academy of Natural Sciences of Philadelphia*, New Series, **I,** 275–280, pls 37–39.

DAUGHTERTY, C. H., CREE, A., HAY, J. M. & THOMPSON, M. B. 1990. Neglected taxonomy and continuing extinctions of tuatara (*Sphenodon*). *Nature*, **347**, 177–179.

DAVIS, G. M. 1984. Genetic relationships among some North American Unionidae (Bivalvia): sibling species, convergence, and cladistic relationships. *Malacologia*, **25**, 629–648.

—— & FULLER, S. L. H. 1981. Genetic relationships among recent Unionacea (Bivalvia) of North America. *Malacologia*, **20**, 217–253.

DE QUEIROZ, K. & DONOGHUE, M. J. 1988. Phylogenetic systematics and the species problem. *Cladistics*, **4**, 317–338.

—— & —— 1990. Phylogenetic systematics and species revisited. *Cladistics*, **6**, 83–90.

—— & GAUTHIER, J. 1990. Phylogeny as a central principle in taxonomy: phylogenetic definitions of taxon names. *Systematic Zoology,* **39**, 307–322.

—— & —— 1992. Phylogenetic taxonomy. *Annual Review of Ecology and Systematics*, **23**, 449–480.

—— & —— 1994. Toward a phylogenetic system of biological nomenclature. *Trends in Ecology and Evolution*, **9**, 27–31.

DOBSON, A. P., RODRIGUEZ, J. P., ROBERTS, W. M. & WILKCOVE, D. S. 1997. Geographic distribution of endangered species in the United States. *Science*, **275**, 550–553.

DONOGHUE, M. J. 1985. A critique of the biological species concept and recommendations for a phylogenetic alternative. *Bryologist*, **88**, 172–181.

ERRIKSSON, T. 1997. *AutoDecay, Version 3.0.* Botaniska Institutionen, Stockholm University, Stockholm.

ESHBAUGH, W. H. 1995. Systematics Agenda 2000: an historical perspective. *Biodiversity and Conservation*, **4**, 455–462.

FARRIS, J. S., KÄLLERSJÖ, M., KLUGE, A. G. & BULT, C. 1995. Testing significance of incongruence. *Cladistics*, **10**, 315–319.

FELSENSTEIN, J. F. 1985. Confidence limits on phylogenies: an approach using the bootstrap. *Evolution*, **39**, 783–791.

FOLMER, O., HOEH, W. R., BLACK, M. B. & VRIJENHOEK, R. L. 1994. DNA primers for amplification of mitochondrial cytochrome C oxidase subunit I from metazoan invertebrates. *Molecular Marine Biology & Biotechnology*, **3**, 294–299.

HEARD, W. H. & GUCKERT, R. H. 1970. A re-evaluation of the Recent Unionacea (Pelecypoda) of North America. *Malacologia*, **10**, 333–355.

HILLIS, D. M. & HUELSENBACK, J. P. 1992. Signal, noise, and reliability in molecular phylogenetic analyses. *Journal of Heredity*, **83**, 189–195.

HOEH, W. R., BOGAN, A. E., CUMMINGS, K. S. & GUTTMAN, S. I. 2000. Evolutionary relationships among the higher taxa of freshwater mussels (Bivalvia: Unionoida): inferences on phylogeny and character evolution from analyses of DNA sequence data. *Malacological Review*, in press.

JENKINSON, J. J. & TODD, R. M. 1997. Management of native mollusk resources *In*: BENZ, G. W. & COLLINS, D. E. (eds) *Aquatic Fauna in Peril: the Southeastern Perspective.* Special Publication 1, Southeast Aquatic Research Institute, Lenz Design & Communications, Decatur, GA, 283–306.

JOHNSON, R. I. 1967. Additions to the unionid fauna of the Gulf drainage of Alabama, Georgia and Florida (Mollusca: Bivalvia). *Breviora Museum of Comparative Zoology*, **270**, 1–21, pls 1–3.

—— 1969. Further additions to the unionid fauna of the Gulf drainage of Alabama, Georgia and Florida. *The Nautilus*, **83**, 34–35.

—— 1970. The systematics and zoogeography of the Unionidae (Mollusca: Bivalvia) of the southern Atlantic Slope region. *Bulletin of the Museum of Comparative Zoology*, **140**, 263–450.

—— & BAKER, H. B. 1973. The types of Unionacea (Mollusca: Bivalvia) in the Academy of Natural Sciences of Philadelphia. *Proceedings of the Academy of Natural Sciences of Philadelphia,* **125**, 145–186, pls 1–10.

KING, T. L., EACKLES, M. S., GJETVAJ, B. & HOEH, W. R. 1999. Intraspecific phylogeography of the freshwater bivalve *Lasmigona subviridis* (Bivalvia: Unionidae): conservation implications of range discontinuity. *Molecular Ecology,* **8**, S65–S78.

LEA, I. 1834. Observations on the Naiades, and descriptions of new species of that and other families. *Transactions of the American Philosophical Society, New Series,* **4,** 63-121, pls 1–18.

—— 1852*a*. Descriptions of new species of the family Unionidae. *Proceedings of the American Philosophical Society*, **5**, 251–252.

—— 1852*b*. Descriptions of new species of the family Unionidae. *Transactions of the American Philosophical Society, New Series*, **10**, 253–302, pls 12–30.

—— 1859. Description of seven new species of Uniones from South Carolina, Florida, Alabama, and Texas. *Proceedings of the Academy of Natural Sciences of Philadelphia,* **11**, 154–155.

—— 1860. *Observations on the genus Unio, together with descriptions of new species, their soft parts, and embryonic forms in the family Unionidae, Volume 9.* Printed for the Author, Philadelphia.

—— 1863. *Observations on the genus Unio, together with descriptions of new species, their soft parts, and embryonic forms, in the family Unionidae. Volume 10.* Printed for the Author, Philadelphia.

LYDEARD, C. & MAYDEN, R. L. 1995. A diverse and endangered aquatic ecosystem of the Southeast United States. *Conservation Biology,* **9**, 800–805.

—— & ROE, K. J. 1998. Phylogenetic systematics: the missing ingredient in the conservation of freshwater unionid bivalves. *Fisheries*, **23**, 16–17.

——, MULVEY, M. & DAVIS, G. M. 1996. Molecular systematics and evolution of reproductive traits of North American freshwater unionacean mussels (Mollusca: Bivalvia) as inferred from 16S rRNA gene sequences. *Philosophical Transactions of the Royal Society of London, Series B*, **351**, 1593–1603.

——, GARNER, J. T., HARTFIELD, P. & WILLIAMS, J. D. 1999. Freshwater mussels in the Gulf region: Alabama. *Gulf of Mexico Science*, **1999**, 125–134.

——, HOLZNAGEL, W. E., SCHNARE, M. N. & GUTELL, R. R. 2000. Phylogenetic analysis of molluscan mitochondrial LSU rDNA sequences and secondary structures. *Molecular Phylogenetics and Evolution*, **15**, 83–102.

MCKITRICK, M. C. & ZINK, R. M. 1988. Species concepts in ornithology. *Condor*, **90**, 1–14.

MCMAHON, R. F. 1991. Mollusca: Bivalvia. *In*: THORP, J. H. & COVICH, A. P. (eds) *Ecology and classification of North American freshwater invertebrates*. Academic Press, San Diego, CA, 315–400.

MASTER, L. L., FLACK, S. R. & STEIN, B. A. (eds) 1998.

Rivers of Life: Critical Watersheds for Protecting Freshwater Biodiversity. The Nature Conservancy, Arlington, VA.

MAYDEN, R. L. 1997. A hierarchy of species concepts: the denouement in the saga of the species problem. *In*: CLARIDGE, M. F., DAWAH, H. A. & WILSON, M. R. (eds) *Species: the Units of Biodiversity*. Chapman & Hall, New York.

MISHLER, B. D. & BRANDON, R. N. 1987. Individuality, pluralism, and the phylogenetic species concept. *Biology and Philosophy*, **2**, 397–414.

—— & DONOGHUE, M. J. 1982. Species concepts: a case for pluralism. *Systematic Zoology*, **31**, 491–503.

MULVEY, M., LYDEARD, C., PYER, D. L., HICKS, K. M., BRIM-BOX, J., WILLIAMS, J. D. & BUTLER, R. S. 1997. Conservation genetics of North American freshwater mussels *Amblema* and *Megalonaias*. *Conservation Biology*, **11**, 868–878.

NEVES, R. J., BOGAN, A. E., WILLIAMS, J. D., AHLSTEDT, S. A. & HARTFIELD, P. W 1997. Status of aquatic mollusks in the southeastern United States: a downward spiral of diversity. *In*: BENZ, G. W. & COLLINS, D. E. (eds) *Aquatic Fauna in Peril: the Southeastern Perspective*. Special Publication 1, Southeast Aquatic Research Institute, Lenz Design & Communications, Decatur, GA, 43–85.

NIXON, K. C. & WHEELER, Q. D. 1990. An amplification of the phylogenetic species concept. *Cladistics* **6**, 211–223.

OORTMANN, A. E. 1912. Notes upon the families and genera of the najades. *Annals of the Carnegie Museum*, **8**, 222–365, pls 18–20.

—— 1923. The anatomy and taxonomy of certain Unioninae and Anondontinae from the Gulf drainage. *The Nautilus*, **36**, 73–85.

—— & WALKER, B. 1922. A new genus and species of American naiades. *The Nautilus*, **35**, 1–6.

ROE, K. J. & LYDEARD, C. 1998. Molecular systematics of the freshwater mussel genus *Potamilus* (Bivalvia: Unionidae). *Malacologia*, **39**, 195–205.

SERNO, P. C. 1999. Definitions in phylogenetic taxonomy: critique and rationale. *Systematic Biology*, **48**, 329–351.

SIMPSON, C. T. 1900*a*. Synopsis of the Naiades, or pearly fresh-water mussels. *Proceedings of the United States National Museum*, **22**, 501–1044.

—— 1900*b*. New and unfigured Unionidae. *Proceedings of the Academy of Natural Sciences of Philadelphia*, **52**, 74–86, pls 1–5.

—— 1914. *A Descriptive Catalogue of the Naiades, or Pearl Freshwater Mussels*. Ann Arbor Press, Ann Arbor, MI.

SWOFFORD, D. L. 1998. *PAUP*. Phylogenetic Analysis Using Parsimony (*and other methods). Version 4.0b1*. Sinauer Associates Inc., Sunderland, MA.

SYSTEMATICS AGENDA. 2000. 1994. *Systematics Agenda 2000: Charting the Biosphere. Technical Report*. Society of Systematic Biologists, American Society of Plant Taxonomists, Willi Hennig Society, Association of Systematics Collections, New York.

TEMPLETON, A. R. 1983*a*. Convergent evolution and non-parametric inferences from restriction fragment and DNA sequence data. *In*: WEIR, B. (ed.) *Statistical Analysis of DNA Sequence Data*. Marcel Dekker, New York, 151–179.

—— 1983*b*. Phylogenetic inference from restriction endonuclease cleavage site maps with particular reference to the evolution of humans and apes. *Evolution*, **37**, 221–244.

TURGEON, D. D., QUINN, JR, J. F., BOGAN, A. E. *ET AL*. 1998. *Common and Scientific Names of Aquatic Invertebrates from the United States and Canada: Mollusks*, 2nd edition. American Fisheries Society, Special Publication 26, Bethesda, MD.

WHEELER, H. E. 1935. *Timothy Abbott Conrad, with Particular Reference to his Work in Alabama One Hundred Years Ago*. Paleontological Research Institution, Ithaca, New York.

WILLIAMS, J. D. & BUTLER, R. S. 1994. Freshwater bivalves. *In*: DEYRUP, M. & FRANZ, R. (eds) *Rare and Endangered Biota of Florida. Volume IV*. University Press of Florida, Gainesville, 53–128.

——, WARREN, M. L. JR, CUMMINGS, K. S., HARRIS, J. L. & NEVES, R. J. 1993. Conservation status of freshwater mussels of the United States and Canada. *Fisheries*, **18**, 6–22.

WRIGHT, B. H. 1899. New southern unios. *The Nautilus*, **13**, 22–23.

On becoming cemented: evolutionary relationships among the genera in the freshwater bivalve family Etheriidae (Bivalvia: Unionoida)

ARTHUR E. BOGAN[1] & WALTER R. HOEH[2]

[1]*North Carolina State Museum of Natural Sciences, Research Laboratory, 4301 Reedy Creek Road, Raleigh, NC 27607, USA*
(e-mail: arthur.bogan@ncmail.net)

[2]*Department of Biological Sciences, Kent State University, Kent, OH 44242, USA*

Abstract: A robust phylogeny for the Unionoida is emerging and presumed relationships of some major clades are being questioned. The Etheriidae or freshwater oysters has been a distinct family for over 160 years and currently contains three cemented genera: *Acostaea* (Columbia, South America), *Pseudomulleria* (India) and *Etheria* (Africa and Madagascar). Starobogatov (1970, *Nauka*, 1–372), Mansur and da Silva (1990, *Amazoniana*, **11**(2), 147–166) and Bonetto (1997, *Biociências*, **5**, 113–142) present conflicting testable hypotheses regarding the evolution of these taxa. Using cytochrome c oxidase subunit I DNA sequences the evolutionary relationships of these three genera has been examined, by comparing them to representatives of 30 other unionoid taxa from around the world. These analyses place *Acostaea* and *Etheria* within the Mycetopodidae while *Pseudomulleria* falls within the Unionidae. A monophyletic Etheriidae, composed of cemented freshwater bivalves, is not supported by the present analyses. Furthermore, the analyses indicate that cementation in the Unionoida has evolved at least twice.

In a paper which discusses the evolution of cementing bivalves, Yonge (1979) identified the variety, origin and problems of convergence in them, finding that over 20 families of bivalves have become cemented. This paper focuses on a group of bivalves which has become cemented in freshwater, namely, the freshwater oysters (Unionoida: Etheriidae).

Bivalves found in freshwater environments represent families from most of the major bivalve subclasses (Bogan 1993). The greatest diversity in freshwater bivalves is found in the radiation of the Unionoida. Currently this order is divided into two superfamilies, six families and, it is estimated here, approximately 180 genera. The two major modern unionoid radiations occurred in the southeastern United States (Bogan 1993, 1998; Williams *et al.* 1993) and China (Liu 1979). Our understanding of evolutionary relationships within the Unionoida is based primarily on comparative anatomy (e.g. Simpson 1900, 1914; Ortmann 1912; Parodiz & Bonetto 1963; Heard & Guckert 1970).

The family Etheriidae has been recognized as a distinct taxon for well over 160 years (e.g. Deshayes 1830; Tryon 1884; Fischer 1886; Thiele 1934; Starobogatov 1970) but the evolutionary relationships of this family to other unionoid families, and among its constituent genera, have been debated (see Discussion). Most malacologists recognize three cemented genera in the Etheriidae: *Acostaea* (Columbia, South America), *Pseudomulleria* (India) and *Etheria* (Africa and Madagascar).

Prashad (1931) examined the convergence in the forms of the freshwater bivalve fauna of Southeast Asia and compared it with similar fauna from South America. At that time, he felt that the Unionoidea was polyphyletic and the Mutelidae with a taxodont hinge was derived from the Arcidae. He suggested that the Etheriidae relationships were undoubtedly with the Unionidae and that they were not related to the Mutelidae. Prashad also noted the similarities of *Etheria*, *Acostaea* and *Pseudomulleria* but felt they all belonged to distinct genera with independent origins and observed that 'the Etheriidae present the most noteworthy examples of the parallel evolution of similar forms from distinct ancestral types, living under identical conditions in widely separated countries.' Prashad's ideas represent the multiple origins hypothesis for the evolution of cementation in the Unionoida.

In contrast, Yonge (1978, p. 446), in discussing *Acostaea* and the evolution of the Etheriidae, posited, 'Certainly *Acostaea* and *Pseudomulleria* must have arisen from a common dimyarian stock; the mode of growth and of assumption of the

From: HARPER, E. M., TAYLOR, J. D. & CRAME, J. A. (eds) *The Evolutionary Biology of the Bivalvia*. Geological Society, London, Special Publications, **177**, 159–168. 1-86239-076-2/00/$15.00

monomyarian condition are too remarkable for convergence to be contemplated.' Yonge (1978) claimed that *Acostaea, Etheria* and *Pseudomulleria* were, in fact, each others closest relatives and so a monophyletic Etheriidae was supported. This is the single origin hypothesis for the genesis of cementation in the Unionoida.

Development of a unionoid phylogeny

Recent work on the evolutionary relationships among unionoid higher taxa has progressed from the efforts of Parodiz & Bonetto (1963) and Heard & Guckert (1970) to the immunoelectrophoretic analyses of Davis & Fuller (1981). The cladistic analysis of 16S DNA and morphology data sets presented by Lydeard *et al.* (1996) has clarified phylogenetic relationships among a number of North American unionoid genera, but did not include potentially closely related taxa from other continents or a suitable outgroup taxon.

Hoeh *et al.* (1998*b*) examined the higher level relationships of the Unionoida based on cytochrome c oxidase subunit I (COI) sequences and supported the hypothesis that *Neotrigonia* (Trigonioida: Trigoniidae) is the sister group to a monophyletic Unionoida. These conclusions support the monophyly of the Palaeoheterodonta (Trigonioida + Unionoida) as proposed by Waller (1990, 1998). Hoeh *et al.* (1998*a*) presented a COI sequence-based phylogeny for the Unionoida including 30 taxa representing five families (excluding representatives of the Etheriidae). Their analyses suggest that the hyriids, not margaritiferids, are a product of the most basal cladogenic event within the Unionoida and that the glochidial larvae is the ancestral larval type. The margaritiferids, mycetopodids, iridinids and hyriids were all depicted as monophyletic groups, with the Unionidae being paraphyletic. Hoeh *et al.* (2001) have expanded on the data presented in Hoeh *et al.* (1998*a*) by adding a morphological data set of 28 characters and produced a total evidence-based phylogeny for the Unionoida. The total evidence analysis supported the earlier phylogeny and character evolution hypotheses of Hoeh *et al.* (1998*a*). If it is assumed that the tree based on the total evidence analysis is a reasonable estimate of unionoid evolutionary history, then morphological character evolution within the Unionoida was very homoplasious.

Etheriid genera

The Etheriidae as used today contains three cemented genera: *Acostaea*, *Etheria* and *Pseudomulleria*. *Acostaea rivoli* (Deshayes 1830) was originally described in *Mulleria* Férussac, 1823 [non Leach 1814] and subsequently moved to *Acostaea* Orbigny, 1851. *Acostaea* is known from the Rio Magdalena in Columbia, South America. Arteaga Sogamoso (1994) discovered that the larvae of *Acosataea rivoli* is a lasidium which, from the diagnoses of the family Mycetopodidae, would argue that *Acostaea* belongs in this family.

The anatomy of *Etheria elliptica* Lamarck, 1807 has been reported, illustrated and discussed by Rang & Cailliaud (1834) and Anthony (1905, 1907). Heard & Vail (1976) examined the anatomy of *Etheria elliptica* and suggested that it belongs in the South American family Mycetopodidae. *Etheria elliptica* is widespread in Africa: the basins of the Nile, Lake Tanganyika and Lake Victoria; the basins of the Chad, Zaire, Niger and Senegal; part of the rivers in Angola and north Madagascar (Daget 1998); and it is known from the Miocene of northeast Zaire (Gautier 1965; Gautier & Van Damme 1973).

Specimens of *Pseudomulleria dalyi* (Smith 1898), from the Budra Drainage, Kadur district, State of Mysore, southern India, were initially placed in the genus *Mulleria* and in the family Etheriidae. Smith (1898) observed that *P. dalyi* was cemented by either the right or the left valves. Woodward (1898), in the same volume, carefully described the anatomy of *Pseudomulleria dalyi*, reported the monomyarian condition and noted that the rectum no longer passed through the heart. Woodward (1898) presented a cross-section diagram of the anatomy and figured the coiling of the intestine. The intestinal coiling, as figured by Woodward, appears very similar to the intestinal coiling of examined North American Unionidae (AEB, personal observations). Woodward (1898) concluded '... the details of the gills, the mantle lobes, and the kidney, *Mulleria* [= *Pseudomulleria*] approximates to the Unionidae.' However, Preston (1915) placed *Mulleria dalyi* in the Etheriidae and cited extensively from Woodward on the anatomy. The most recent coverage of the freshwater molluscan fauna of India by Subba Rao (1989) follows the taxonomy of the family put forward by Thiele (1934), recognizing *Pseudomulleria* as a subgenus of *Acostaea*.

Yonge (1953) recognized a single genus of monomyarian unionoid, *Acostaea*, with two subgenera, *Acostaea*, and *Pseudomullaria*. Both subgenera were monotypic and both species begin life as a young shell which is dimyarian and not cemented. Yonge (1953) noted no evidence for initial byssal attachment and observed that cementation takes place and growth continues at the anterior end of the shell with the subsequent loss of the anterior regions of the mantle, shell and the anterior adductor muscles.

Family level classification

Deshayes (1830) erected the family Etheriidae for *Etheria* Lamarck, 1807 and for many years it contained only *Etheria*. Lamarck (1819) placed the group close to the Chamidae and thought the group was marine, which confused the placement of this family (Fischer 1886). Swainson (1835) recognized the genera *Etheria* and *Mulleria* but placed them not within the Unionidae but rather within his Ostredea (*sic*), or oysters. Swainson (1840) recognized the family Etheriidae and included *Etheria* and *Mulleria*, and placed the family between the Unionidae and Ostreidae completing his circle of related families.

Tryon (1884) and Fischer (1886) placed the Etheriidae next to the Unionidae and included three genera: *Etheria*, *Mulleria* (+ *Acostaea*) and *Bartlettia*. Simpson (1896, 1900, 1914) did not mention the Etheriidae in any of his treatments of the unionoid bivalves.

Germain (1907) placed *Etheria* in the Etheriinae, which he considered a subfamily of the Unionidae because the juvenile *Etheria* looked like an *Anodonta*. Later, Dautzenberg & Germain (1914) recognized *Etheria* as belonging to a separate family, Etheriidae, but did not comment on this change of rank.

Thiele (1934) recognized the superfamily Unionacea and included Margaritiferidae, Unionidae, Mutelidae and Etheriidae [*Bartlettia*, *Etheria*, *Acostaea* (*Acostaea*) and *A. Pseudomulleria*]. Modell (1942, 1949) developed an alternative classification for the Unionoida. He placed *Acostaea* and *Bartlettia* in a subfamily Bartlettiinae, and *Etheria* and *Pseudomulleria* in the Etheriinae, both subfamilies in his inclusive Mutelidae. He later (Modell 1964) modified his ideas on the cemented bivalve placement, including only *Etheria* in the Etheriinae and including *Pseudomulleria* as a subgenus under *Acostaea*, which he moved to the Bartlettiinae.

Mandahl-Barth (1954) listed *Etheria* in the Etheriinae within the Mutelidae and subsequently (1988) decided that the group should have familial status. Pain & Woodward (1961) reviewed the family Etheriidae and included *Etheria*, *Bartlettia*, *Acostaea* and *Pseudomulleria*. They elevated *Pseudomulleria* from subgeneric to generic rank based on the disparate distribution of *Acostaea* in South America and *Pseudomulleria* in India.

Newell (1965) provided a complete classification of the Bivalvia and placed all of the living families of unionoid bivalves in the single superfamily Unionacea, recognizing four families; Unionidae, Mutelidae, Etheriidae and Margaritiferidae. Haas (1969*a*, *b*) recognized the families Etheriidae, Margaritiferidae, Mutelidae and Unionidae, all within a single superfamily, the Unionacea. He included *Etheria*, *Barlettia* and *Acostaea* with two subgenera, *Acostaea* and *Pseudomulleria*, in the Etheriidae.

Starobogatov (1970) split what others had lumped as the Etheriidae, placing the three genera into distinct families: *Acostaea* in the Mulleriidae, along with the Mycetopodidae, forming the Mullerioidea; *Etheria* remained in Etheriidae; and *Pseudomulleria* in Pseudomulleriidae, both families placed in the Etherioidea.

Van Damme (1984) summarized the freshwater molluscs of northern Africa and used the family Etheriidae. Kabat (1997) provided an examination of the dates and priority of the various family group names for the Etherioidea: Etheriidae Deshayes, 1830, Iridinidae Swainson, 1840(+ Mutelidae Gray 1847) and Mycetopodidae Gray, 1840, with these being the oldest available names.

Mansur & da Silva (1990) recently supported a monophyletic view of the Etheriidae, including *Bartlettia,* based on anatomical analyses. However, Bonetto (1997) has placed *Acostaea* in the Acostaeinae in the Mycetopodidae, *Etheria* in the Etheriinae and *Pseudomulleria* in the Pseudomulleriinae, with both subfamilies placed in the Mutelidae.

Recently, Daget (1998) has produced a complete catalogue of the freshwater bivalves of Africa and has recognized a single genus within the Etheriidae, *Etheria*, with a single species *Etheria elliptica* Lamarck, 1807. This volume contains a very detailed listing of four generic synonyms, 20 specific synonyms and a listing of citations using the various names and combinations.

Good (1998), in discussing the Late Triassic freshwater bivalve fauna of the North American southwest, placed the Etheriidae within the Unionoida but without further comment. He observed that the Etheriidae originated on Gondwanaland but erroneously included Australia as part of their modern range.

Other cemented freshwater bivalves

Recently, Bogan & Bouchet (1999) described *Posostrea*, a cemented corbiculid from Lake Poso, Sulawasi, Indonesia. This is the first record of cementation in the Corbiculidae and represents the only known cemented freshwater bivalve outside of the Unionoida.

Fossil record

The fossil record for the three etheriid genera examined here is restricted to *Etheria*, which occurs in the Miocene (Gautier 1965; Gautier & Van Damme 1973) of East Africa. Nothing is known of the fossil record regarding *Acostaea* and *Pseudomulleria*.

The works of Starobogatov (1970) and Bonetto (1997), as well as that of other authors, contradict the monophyletic nature of the Etheriidae, which has been supported most recently by Mansur & da Silva (1990), by suggesting that the Etheriidae is a polyphyletic assemblage. The conflicting views on the evolutionary relationships surrounding the Etheriidae hinder the development of a basic understanding of the circumstances involved in the evolution of the cemented habit in freshwater bivalves. Phylogenetic analyses of COI DNA sequences will be used herein to evaluate the taxonomic status of the Etheriidae, i.e. is the Etheriidae a monophyletic assemblage? These analyses will enable the following fundamental evolutionary questions to be addressed: did the cemented habit in unionoids evolve once (or multiple times); from which non-cemented ancestral taxon (or taxa) did it evolve?

Materials and methods

Organisms

The 34 bivalve species examined in this study are listed in Table 1 with their GenBank accession numbers for the COI sequences. The COI sequences representing the three etheriid genera are the only new sequences added in this paper. The other COI sequences have been analysed previously in Hoeh *et al.* (1996, 1998*a*, *b*, 2001). The names for the North American taxa follow Turgeon *et al.* (1998).

Methods

Total DNA was isolated from somatic (mantle) tissues from individuals representing *Acostaea*, *Etheria* and *Pseudomulleria*. Male gonadal tissues were specifically avoided to prevent comparisons of non-orthologous sequences due to the actual or potential presence of doubly uniparental inheritance of mitochodrial (mt)DNA in some bivalve taxa (e.g. see Skibinski *et al.* 1994; Zouros *et al.* 1994; Hoeh *et al.* 1996, 1997). Subsequently, a 710 base pair (bp) fragment of COI was polymerase chain reaction (PCR) amplified and cycle sequenced for each of the three taxa as described elsewhere (Folmer *et al.* 1994). Both strands of the COI fragment were sequenced from each of two individuals to guard against PCR-based contamination artifacts. The resulting sequences were readily aligned by eye, using MacClade (Maddison & Maddison 1997), with the one trigonioid and 30 unionoid COI sequences analysed previously (Hoeh *et al.* 1998*a*). Of the 34 total aligned COI sequences, 33 were of identical length (= 630bp) while that of *Acostaea* had a single inferred codon deletion (= 627bp). The autapomorphic nature of the inferred single codon deletion in *Acostaea* (i.e. it is not shared with any other taxon) excludes this mutation from playing a role in the subsequent phylogenetic analyses.

The suitability of the COI data set for

Table 1. *List of specimens used in this analysis and their associated GenBank accession numbers*

Taxa	GenBank numbers
In-group, **Order Unionoida**	
Superfamily Etherioidea, Family Etheriidae	
Etheria elliptica Lamarck, 1807	AF231742
Acostaea rivoli (Deshayes, 1827)	AF231739
Pseudomulleria dalyi (Smith, 1898)	AF231750
Family Iridinidae	
Mutela dubia (Gmelin, 1791)	AF231737
Mutela rostrata (Rang, 1835)	U56849
Family Mycetopodidae	
Anodontites guanarensis Marshall, 1927	AF231741
Anodontites trigonus (Spix, 1827)	AF231738
Monocondylaea minuana (d'Orbigny, 1835)	AF231745
Superfamily Unionoidea, Family Hyriidae	
Castalia stevensi (H. B. Baker, 1930)	AF231736
Diplodon deceptus (Simpson, 1914)	AF231744
Hyridella menziesi (Gray, 1843)	AF231747
Lortiella rugata (Sowerby, 1868)	AF231746
Velesunio angasi (Sowerby, 1867)	AF231743
Family Margaritiferidae	
Cumberlandia monodonta (Say, 1829)	AF231753
Margaritifera margaritifera (Linnaeus, 1758)	U56847
Family Unionidae	
Actinonaias ligamentina (Lamarck, 1819)	AF231730
Amblema plicata (Say, 1817)	U56841
Anodonta cygnea (Linnaeus, 1758)	U56842
Coelatura aegyptiaca (Cailliaud, 1827)	AF231735
Cyrtonaias tampicoensis (Lea, 1838)	AF231749
Elliptio dilatata (Rafinesque, 1820)	AF231751
Fusconaia flava (Rafinesque, 1820)	AF231733
Glebula rotundata (Lamarck, 1819)	AF231729
Gonidea angulata (Lea, 1838)	AF231755
Ligumia recta (Lamarck, 1819)	AF231748
Pleurobema clava (Lamarck, 1819)	AF231754
Potamilus alatus (Say, 1817)	AF231752
Pyganodon grandis (Say, 1829)	AF231734
Quadrula quadrula (Rafinesque, 1820)	AF231757
Strophitus undulatus (Say, 1817)	AF231740
Toxolasma lividus (Rafinesque, 1831)	AF231756
Unio pictorum (Linnaeus, 1758)	AF231731
Unio tumidus (Retzius, 1788; 2)	AF231732
Outgroup, **Order Trigonioida**	
Neotrigonia margaritacea (Lamarck, 1804).	U56850

phylogenetic analyses at this hierarchical level was evaluated by plotting the substitution pattern of transitions and transversions for each codon position (e.g. see Hoeh *et al.* 1998*b*). Furthermore, the degree of phylogenetic signal within the COI data set was evaluated using the g_1 statistic of a random tree distribution (from 10 000 000 random trees; e.g. see Hillis 1991; Hillis & Huelsenbeck 1992) and the permutation tail probability (PTP) test (Faith & Cranston 1991) as implemented in PAUP * (Swofford 1998). Phylogenetic analyses were carried out on the COI nucleotide sequences using the maximum parsimony (MP) algorithm contained in PAUP* (Swofford 1998). Based on previous morphological (e.g. see Atkins 1937; Taylor *et al.* 1969, 1973; Tevesz 1975; Popham 1979; Tevesz & Carter 1980; Smith 1986; Healy 1989; Waller 1990) and molecular (Hoeh *et al.* 1998*b*) systematic analyses that indicated the Trigonioida is the sister taxon to a monophyletic Unionoida, *Neotrigonia margaritacea* was used to root the resulting topologies. One thousand random terminal taxon addition order runs, combined with global branch rearrangement options, were employed to generate topologies from the MP analysis (all substitutions received equal weight). These options increased the probability of finding the best topology under the parsimony criterion (e.g. see Maddison 1991). The robustness of the resulting topologies was evaluated by bootstrap (10 000 replicates) and jackknife (50% deletion for 1000 replicates) analyses. In addition, character mapping, using MacClade (Maddison & Maddison 1997), was performed on the COI-based topologies to investigate their implications for the evolution of cementation within the Unionoida.

Results

Scatter plots of the relationship between the number of transitional and transversional substitutions, and the percentage of total uncorrected sequence divergence at each of the three codon positions for the COI sequences, revealed that only transitional substitutions at the third codon position had reached saturation (plots not shown). Since saturated categories of substitution can contribute to erroneous estimates of evolutionary history (e.g. see Swofford *et al.* 1996), all first and second position substitutions, together with only transversions at the third codon position, were included in the phylogenetic analyses. Of the 630 nucleotide positions in the transformed COI data matrix, 383 were constant while 194 were parsimony informative. Analysis of the tree length distribution of 10 000 000 randomly generated trees, using all 34 sequences, suggested that there is a significant amount of hierarchical structure within the transformed COI data set ($g_1 = -0.487$; with 247 variable sites, $p < 0.01$; Hillis & Huelsenbeck 1992). A PTP test on the transformed matrix also indicated significant hierarchical structure ($P = 0.001$). The findings from the plots of substitution pattern, g_1 statistics and the PTP test are consistent with the hypothesis that significant phylogenetic signal exists in the transformed COI nucleotide data matrix and validates its use in this particular phylogenetic context (e.g. see Swofford *et al.* 1996).

The strict consensus tree, derived from ten equally parsimonious trees (each of 879 steps; retention index = 0.5551) and produced by MP analysis of the transformed COI nucleotide matrix, is presented in Fig. 1, along with bootstrap (above branches, 10 000 replicates) and jackknife (below branches, 1000 replicates) percentages (only percentages > 50% are shown).

Discussion

Evolutionary relationships within the Unionoida

All ten equally parsimonious trees resulting from analyses of the transformed COI nucleotide matrix supported the monophyly of the Hyriidae, Margaritiferidae, Iridinidae, Mycetopodidae and Etherioidea (Mycetopodidae, Iridinidae and Etheriidae) (e.g. Fig. 1). However, the Unionoidea (Unionidae, Margaritiferidae, Hyriidae) was found paraphyletic since the etherioids are more closely related to the unionids than are the hyriids. These higher level phylogenetic relationships within the Unionoida are congruent with the results of Hoeh *et al.* (1998*a*, 2001).

The concept of the Etheriidae as a monophyletic bivalve family containing all of the cemented unionoid genera is rejected by the COI sequence analyses herein. Rather, the Etheriidae is shown to be a polyphyletic concept because two of the 'etheriid' genera are closely related to mycetopodids (*Acostaea* and *Etheria*) while another (*Pseudomulleria*) is a unionid. Constraining the parsimony analysis to produce a monophyletic Etheriidae produced three equally parsimonious trees of 900 steps each. These trees are 21 steps (2.4%) longer than the unconstrained trees. Therefore, the analysis of COI nucleotides contained herein rejects the notion of a monophyletic Etheriidae. Furthermore, these results suggest that cementation evolved at least twice (three times if Deltran character optimization is used) within the Unionoida, and that it arose from both etherioid and unionid ancestors (Fig. 1).

Figure 1 clearly shows that, as currently recognized, the Etheriidae is polyphyletic. However, the

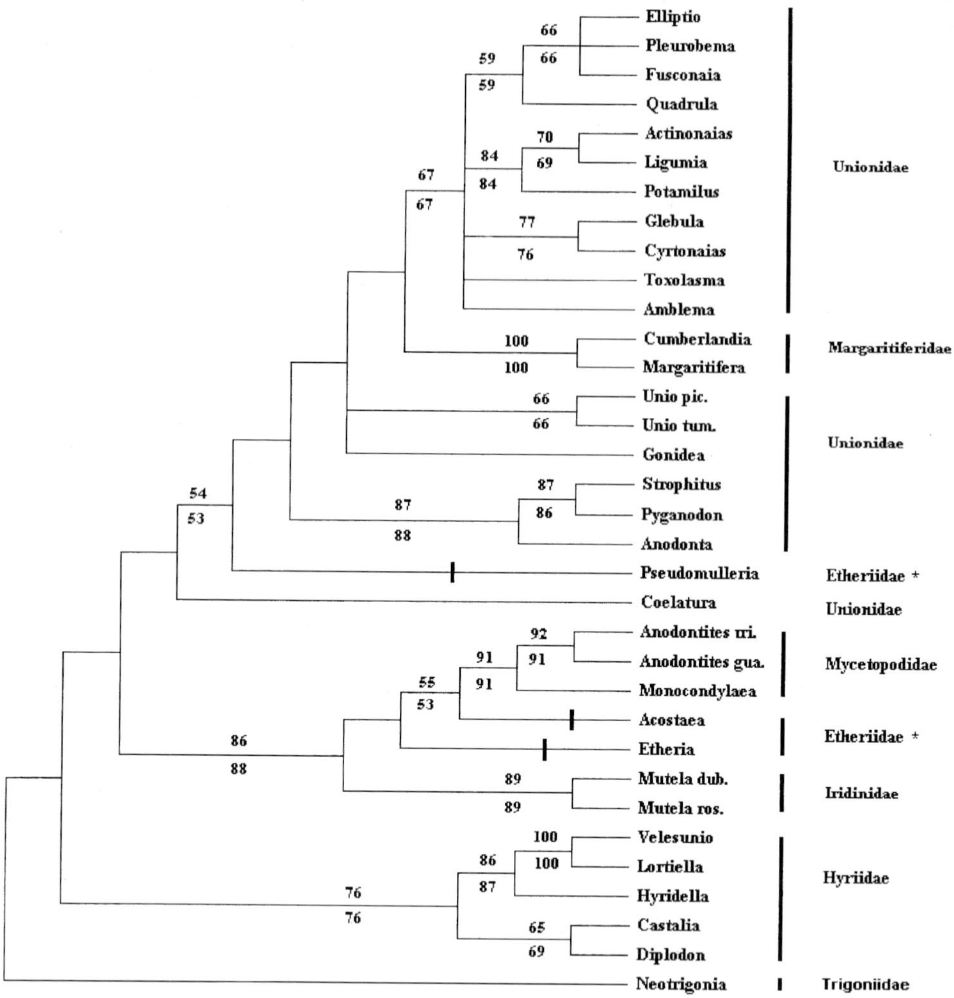

Fig. 1. Strict consensus tree of ten equally parsimonius trees. Tree length: 879 steps; retention index = 0.5551. Numbers above the branches are parsimony bootstrap percentages based on 10 000 replicates and those numbers below the line are parsimony jackknife percentages based on 1000 replicates. Current family names are listed on the right-hand side of the tree. *Neotrigonia* belongs to the Trigoniidae, Trigonioida, and is used as the outgroup (Hoeh *et al.* 1998*b*). The three independent origins of cementation, which are indicated by Deltran character optimization, are mapped onto the topology with hash marks.

present analyses used representatives of only 27 out of about 180 recognized unionoid genera. The topologies of the trees and evolutionary relationships of major clades may continue to change as more taxa are added. Furthermore, due to the incongruence between the trees generated from COI and anatomical data sets (Hoeh *et al.* 2001), and the relatively weak support for most basal nodes (Fig. 1), it is very premature to reassign higher taxa based on the clades produced herein.

The COI analysis implies that *Pseudomulleria* falls within the 'unionid' clade. The *Pseudomulleria* placement on the tree (Fig. 1) suggests that the larval structure, when discovered, will be a glochidium. Morrison (1973) separated this genus from *Acostaea* because the nacre was different. He felt that *Pseudomulleria* belonged within his Unionacea.

Acostaea appears closely related to the Mycetopodidae (Fig. 1). Morrison (1973) placed the Acostaeidae close to the Mycetopodidae in the Mutelacea as then recognized. Arteaga Sogamoso

(1994) confirmed that the *Acostaea* larval form was the lasidium.

Etheria also appears closely related to the Mycetopodidae but there is very weak support for this placement (Fig. 1). The placement of *Etheria* with the Mycetopodidae herein corroborates the hypothesis of Heard & Vail (1976). Heard & Vail (1976) suggested that the split of the South American and African 'Mutelids' into Mycetopodidae and Iridinidae based on geography needs to be re-examined, and that some South American genera such as *Leila* actually belong to the Iridinidae and not the Mycetopodidae.

Cementation in freshwater bivalves: hypotheses

Freshwater cemented bivalves have a tropical/subtropical distribution while the areas of highest unionoid diversity occur in more temperate zones. What has influenced the biogeography of cemented freshwater bivalves? Harper (1991) suggested a driving force in the evolution of cementation in marine bivalves is predation by such groups as crabs (Crustacea) and starfish (Echinoderms). Harper (1991) performed a test of the predation hypothesis on byssally attached v. cemented bivalves and found a significantly higher level of predation on those animals attached with a byssus. She suggests that cementation, as recorded in the fossil record, may be due to the concurrent development of crustacean and echinoderm bivalve predators and notes their co-occurrence in the fossil record.

Harper's (1991) idea of predation as a driving force in the origin of cementation of marine bivalves can be extended to freshwater. In this case, the freshwater crayfish and crabs may substitute for their marine relatives and the echinoderms. A problem with this hypothesis arises in North America with the world's greatest diversity of Unionoidea (Williams *et al.* 1993) and an equally diverse freshwater crayfish fauna but no freshwater crabs (Hobbs 1989; Taylor *et al.* 1996). There are no cemented freshwater bivalves in North America. The same holds for Europe and northern Asia (Liu 1979) where there are crayfish but no cemented bivalves or freshwater crabs. The tropics of Southeast Asia are home to *Modellnaia,* a crevasse-dwelling unionoid (Brandt 1974), and *Posostrea*, a cemented corbiculid. Both of these species apparently live with freshwater crabs. South America has a variety of freshwater bivalves, which includes the cemented *Acostaea* and the crevasse-dwelling unionoid *Bartlettia*, as well as a variety of freshwater crabs. The same picture holds for Africa with *Etheria*, and is assumed to be true for India with *Pseudomulleria*. Is the occurrence of cemented freshwater bivalves a reaction to predation by crabs? If this is the case, why is the number of cemented species so low?

An alternative hypothesis would be that the cemented freshwater bivalves represent a relatively recent opportunistic movement of species into high-energy environments and the evolution of local species into crevasse dwellers, such as *Bartlettia* in South America and *Modellnaia* in Thailand. This scenario of local species evolving to fill the cemented oyster niche in freshwater would explain why the different cemented bivalves are from different families. This might be the reason for the development of *Posostrea* in the absence of unionoids in ancient Lake Poso. In the case of *Etheria*, *Acostaea* and *Pseudomulleria*, they became fully cemented as opposed to crevasse dwellers. *Acostaea* and *Pseudomulleria* both begin life as a dimyarian shell and later became wedged in the substrate and developed into cemented freshwater oysters. These two species take the oyster habit to the extreme in becoming secondarily monomyarian as adults (see Yonge 1979). However, this ecological niche hypothesis fails to explain the lack of cemented bivalves in the highly diverse areas of the southeastern United States and China.

We would like to thank the following colleagues for kindly donating specimens for this analysis: Jay Cordiero, AMNH, New York, New York, for the specimens of *Etheria* from the Congo; Gamil Soliman, American University, Cairo, Egypt, for the *Coelatura* and *Mutela* specimens from Egypt; Dr Madyashtha for the specimens of *Pseudomulleria* from India, and Edgar Arteaga Sogamoso, Tolima, Columbia for the specimens of *Acostaea*. Paula Mikkelsen, AMNH, Kathie Way, BMNH and G. Thomas Watters are all gratefully acknowledged for their assistance with the literature. Gabriela M. Hogue very kindly provided a translation of the paper by Bonetto (1997).

References

ANTHONY, R. 1905. Influence de la fixation pleurothéque sur al morphologie des mollusques acéphales dimyaires. *Annales des Sciences Naturelles: Zoologie* 9e. Series **1**, 165–396, pls 7–9.

—— 1907. Étude monographique des Aetheriidae (Anatomie, morphogénie, systemématique). *Annales de la Société royale Zoologique et Malacologique de Belgique*, **41**, 322–430, pls 11 and 12.

ARTEAGA SOGAMOSO, E. 1994. El Lasidium de Acostaea rivoli Deshayes, 1827 (Mollusca: Bivalvia: Etheriidae) y su importancia para la ubicación taxonómica de esta especies. *Boletin Ecotropica: Ecosistemas Tropicales,* **27**,1–9.

ATKINS, D. 1937. On the ciliary mechanisms and

interrelationships of lamellibranchs. Part II: sorting devices on the gills. *Quarterly Journal of Microscopical Science*, **79**, 339–373.

BOGAN, A. E. 1993. Freshwater bivalve extinctions (Mollusca: Unionoida): A search for causes. *American Zoologist,* **33**(6), 599–609.

—— 1998. Freshwater molluscan conservation in North America: Problems and practices. *Journal of Conchology, Special Publication*, **2**, 223–230.

—— & BOUCHET, P. 1999. Cementation in the freshwater bivalve family Corbiculidae (Mollusca: Bivalvia): a new genus and species from Lake Poso, Indonesia. *Hydrobiologia,* **389**, 131–139.

BONETTO, A. A. 1997. Las 'ostras de agua dulce' (Muteloidea: Mutelidae). Su taxonomia y distribucion geografica en el conjunto de las Naiades del Mundo. *Biociências,* **5**, 113–142.

BRANDT, R. A. M. 1974. The non-marine aquatic Mollusca of Thailand. *Archiv für Molluskenkunde,* **105**(1–4), 1–423.

DAGET, J. 1998. *Catalogue raisonné des Mollusques bivalves d'eau douce africains*. Backhuys Publishers, Leiden, and OSTROM, Paris.

DAUTZENBERG, P. & GERMAIN, L. 1914. Récoltes Malacologiques du Dr. J. Bequeart dans le Congo Belge. *Revue Zoologique africaine,* **4**(1), 1–73.

DAVIS, G. M. & FULLER, S. L. H. 1981. Genetic relationships among recent Unionacea (Bivalvia) of North America. *Malacologia,* **20**(2), 217–253.

DESHAYES, G. P. 1830 [in 1792–1832]. Mollusques. Volume 2 *In*: *Encyclopédie méthodique: histoire naturelle des vers*. Panckoucke, Paris, 471–553.

FAITH, D. P. & CRANSTON, P. S. 1991. Could a cladogram this short have arisen by chance alone? On permutation tests for cladistic structure. *Cladistics,* **7**, 1–28.

FISCHER, P. 1880–1887. *Manuel de Conchyliologie et de Paléontologie conchyliologique ou histoire naturelle de Mollusques vivants et fossiles*. Librairie F. Savy, Paris.

FOLMER, O., BLACK, M., HOEH, W., LUTZ, R. & VRIJENHOEK, R. 1994. DNA primers for amplification of mitochondrial cytochrome *c* oxidase subunit I from diverse metazoan invertebrates. *Molecular Marine Biology and Biotechnology,* **3**, 294–299.

GAUTIER, A. 1965. *Geological investigation in the Sinda Mohari (Ituri, N.E. Congo), a monograph on the geological history of a region in the Lake Albert Rift*. Ganda-Congo Publications, Rijksuniversiteit Gent, 1–161.

—— & VAN DAMME, D. 1973. A revision of the Miocene freshwater molluscs of the Mohari Formation (Sinda–Mohari, Ituri, N. E. Zaire). *Annln. Kon. Mus. Midden Afrika*, 8, **45**(8), 43–62.

GERMAIN, L. 1907. Les Mollusques terrestres et fluviatiles de l'Afrique Centrale Française. *In*: CHEVALIER, A. (ed.) *l'Afrique Central Française. Mission Chari-Lac Tchad, 1902-1904.* A. Challamel, Paris, 475–617, figs 5–100.

GOOD, S. C. 1998. Freshwater bivalve fauna of the Late Triassic (Carnian–Norian) Chinle, Dockum, and Dolores Formation of the Southwest United States. *In*: JOHNSTON, P. A. & HAGGART, J. W. (eds) *Bivalves: An Eon of Evolution – Paleobiological Studies Honoring Norman D. Newell*. University of Calgary Press, Calgary, 223–249.

HAAS, F. 1969*a*. Superfamilia Unionacea. *Das Tierreich (Berlin)*, **88**.

—— 1969*b*. Superfamily Unionacea. *In*: MOORE, R. C. (ed.) *Treatise on Invertebrate Paleontology*. *Part N. Mollusca 1*. Geological Society of America and the University of Kansas, N411–N470.

HARPER, E. M. 1991. The role of predation in the evolution of cementation in bivalves. *Palaeontology* **34**(2), 455–460.

HEALY, J. M. 1989. Spermiogenesis and spermatozoa in the relict bivalve genus *Neotrigonia*: Relevance to trigonioid relationships, particularly Unionoidea. *Marine Biology*, **103**, 75–85.

HEARD, W. H. & GUCKERT, R. H. 1970. A re-evaluation of the recent Unionacea (Pelecypoda) of North America. *Malacologia,* **10**(2), 333–355.

—— & VAIL, V. A. 1976. Anatomical systematics of *Etheria elliptica* (Pelecypoda: Mycetopodidae. *Malacological Review,* **9**,15–24.

HILLIS, D. M. 1991. Discriminating between phylogenetic signal and random noise in DNA sequences. *In*: MIYAMOT, M. M. & CRACRAFT, J. (eds) *Phylogenetic Analysis of DNA Sequences*. Oxford University Press, New York, 278–294.

—— & HUELSENBECK, J. P. 1992. Signal, noise, and reliability in molecular phylogenetic analyses. *Journal of Heredity*, **83**, 189–195.

HOBBS, H. H., JR 1989. An illustrated checklist of the American crayfishes (Decapoda: Astacidae, Cambaridae, and Parastacidae). *Smithsonian Contributions to Zoology*, **480.**

HOEH, W. R., BOGAN, A. E. & HEARD, W. H. 2001. A phylogenetic perspective on the evolution of morphological and reproductive characteristics in the Unionoida. *In*: BAUER, G. & WACHTLER, G. (eds) *Ecology and Evolutionary Biology of Freshwater Mussels [Superfamily Unionoidea]*. Springer Verlag, Berlin, 257–280.

——, ——, CUMMINGS, K. S. & GUTTMAN, S. E. 1998*a*. Evolutionary relationships among the higher taxa of freshwater mussels (Bivalvia: Unionoida): inferences on phylogeny and character evolution from analyses of DNA sequence Data. *Malacological Review,* **31**, 121–140.

——, STEWART, D. T., SUTHERLAND, B. W. & ZOUROS, E. 1996. Multiple origins of gender-associated mitochondrial DNA lineages in bivalves (Mollusca: Bivalvia). *Evolution*, **50**(6), 2276–2286.

——, ——, SAAVEDRA, C., SUTHERLAND, B. W. & ZOUROS, E. 1997. Phylogenetic evidence for role-reversals of gender-associated mitochondrial DNA genomes in *Mytilus* (Bivalvia: Mytilidae). *Molecular Biology and Evolution,* **14**, 959–967.

——, BLACK, M. B., GUSTAFSON, R. G., BOGAN, A. E., LUTZ, R. A. & VRIJENHOEK, R. C. 1998*b*. Testing alternative hypotheses of *Neotrigonia* (Bivalvia: Trigonioida) phylogenetic relationships using cytochrome *c* oxidase subunit I DNA sequences. *Malacologia,* **40**, 267–278.

KABAT, A. 1997. Correct family names for the freshwater 'Muteloid' bivalves (Unionoida: Etheriodea).*Occasional Papers on Mollusks,* **5**(72), 379–392.

LAMARCK, J. B. 1819. *Histoire Naturelle des Animaux sans Vertèbres.* Tome Sixième. Paris.

LIU, Y.-Y. 1979 *Freshwater Mollusks. Economic Fauna of China.* Science Press, Beijing, People's Republic of China [in Chinese].

LYDEARD, C., MULVEY, M. & DAVIS, G. M. 1996. Molecular systematics and evolution of reproductive traits of North American freshwater unionacean mussels (Mollusca: Bivalvia) as inferred from 16S rRNA gene sequences. *Philosophical Transactions of the Royal Society of London, Series B,* **351**(1347), 1593–1603.

MADDISON, D. R. 1991. The discovery and importance of multiple islands of most-parsimonious trees. *Systematic Zoology,* **40**, 315–328.

MADDISON, W. P. & MADDISON, D. R. 1997. *MacClade: Analysis of Phylogeny and Character Evolution.* Sinauer Associates, Inc., Sunderland, MA.

MANDAHL-BARTH, G. 1954. The freshwater mollusks of Uganda and adjacent territories. *Annales du Musée Royal du Congo Belge, Tervuren (Beligique) Séries in 8°. Sciences Zoologiques,* **32**, 1–206.

—— 1988. *Studies on African Freshwater Bivalves.* Danish Bilharziasis Laboratory, Charlottenlund.

MANSUR, M. C. D. & DA SILVA, M. G. O. 1990. Morfologia e microanatomia comparada de *Bartlettia stefanensis* (Moricand, 1856) e *Anodontoides tenebricosus* (Lea, 1834) (Bivalvia, Unionoida, Muteloidea). *Amazoniana,* **11**(2), 147–166.

MODELL, H. 1942. Das natürliche system der najaden. *Archiv für Molluskenkunde,* **74**(5/6), 161–191.

—— 1949. Das natürliche system der najaden. 2. *Archiv für Molluskenkunde,* **78**(1-3), 29–48.

—— 1964. Das natürliche system der najaden. 3. *Archiv für Molluskenkunde,* **93**, 71–129.

MORRISON, J. P. E. 1973. The families of the pearly freshwater mussels. *Bulletin of the American Malacological Union, Inc.,* **1972**, 45–46.

NEWELL, N. D. 1965. Classification of the Bivalvia. *American Museum Novitates,* **2206**.

ORTMANN, A. E. 1912. Notes upon the families and genera of the najades. *Annals of the Carnegie Museum,* **8**(2), 222–365, pls 18–20.

—— 1921. South American naiades; a contribution to the knowledge of the fresh-water mussels of South America. *Memoirs of the Carnegie Museum,* **8**(3), 451–670, pls 34–48.

PAIN, T. & WOODWARD, F. R. 1961. A revision of freshwater mussels of the family Etheriidae. *Journal of Conchology,* **25**(1), 2–8.

PARODIZ, J. J. & BONETTO, A. A. 1963. Taxonomy and zoogeographic relationships of the South American naiades (Pelecypoda: Unionacea and Mutelacea). *Malacologia,* **1**(2), 179–213.

POPHAM, J. D. 1979. Comparative spermatozoon morphology and bivalve phylogeny. *Malacological Review,* **12**, 1–20.

PRASHAD, B. 1931. Some noteworthy examples of parallel evolution in the molluscan faunas of South-eastern Asia and South America. *Proceedings of the Royal Society of Edinburgh,* **51**(Part 1)(8), 42–53.

PRESTON, H. B. 1915. *The Fauna of British India, Including Ceylon and Burma. Mollusca. (Freshwater Gastropoda & Pelecypoda).* Taylor & Francis, London.

RANG, S. & CAILLAUD. 1834. Mémoires sur le genre Aethérie. *Nouv. Ann. Mus, series 3,* 128–144.

SIMPSON, C. T. 1896. The classification and geographical distribution of the pearly fresh-water mussels. *Proceedings of the United States National Museum,* **18**(1068), 295–343.

—— 1900. Synopsis of the naiades, or pearly fresh-water mussels. *Proceedings of the United States National Museum,* **22**(1205), 501–1044.

—— 1914. *A Descriptive Catalogue of the Naiades, or Pearly Fresh-water Mussels.* Parts I–III. Bryant Walker, Detroit, MI.

SKIBINSKI, D. O. F., GALLAGHER, C. & BEYNON, C. M. 1994. Sex-limited mitochondrial DNA transmission in the marine mussel *Mytilus edulis. Genetics,* **138,** 801–809.

SMITH, D. G.1986. The stomach anatomy of some eastern North American Margaritiferidae (Unionoida: Unionacea). *American Malacological Bulletin,* **4**, 13–19.

SMITH, E. A. 1898. Description of *Mulleria dalyi*, n. sp. from India. *Proceedings of the Malacological Society of London,* **3**, 14–16.

STAROBOGATOV, YA. I. 1970. Fauna of molluscs and zoogeographical regionalization of continental water bodies of the globe. *Nauka,* Leningrad [in Russian].

SUBBA RAO, N. V. 1989. *Handbook. Freshwater Molluscs of India.* Zoological Survey of India, Calcutta.

SWAINSON, W. 1835. *The Elements of Modern Conchology, Briefly and Plainly Stated, for the Use of Students and Ttravelers.* Baldwin and Craduce, London

—— 1840. *The Cabinet Cyclopaedia. Conducted by the Rev. Dionysius Lardner Assisted by Eminent Literary and Scientific Men. Natural History. A Treatise on Malacology or Shells and Shell-fish.* Longmans, London.

SWOFFORD, D. L. 1998. *PAUP*. Phylogenetic Analysis Using Parsimony (*and Other Methods).* Sinauer Associates Inc., Sunderland, MA.

——, OLSEN, G. J., WADDELL, P. J. & HILLIS, D. M. 1996. Phylogenetic inference. *In*: HILLIS, D. M., MORITZ, C. & MABLE, B. K. (eds) *Molecular Systematics* Sinauer Associates, Inc., Sunderland, MA, 407–514.

TAYLOR, C. A., WARREN, M. L. JR, FITZPATRICK, J. F. JR, HOBBS, J. J. III, JEZERINAC, R. F., PFLIEGER, W. L. & ROBSISON, H. W. 1996. Conservation status of crayfishes of the United States and Canada. *Fisheries,* **21**(4), 25–38.

TAYLOR, J. D., KENNEDY, W. J. & HALL, A. 1969. The shell structure and mineralogy of the Bivalvia. Introduction. Nuculacea–Trigonacea. *Bulletin of the British Museum (Natural History), Zoology, Supplement,* **3**, 1–125.

——, —— & —— 1973. The shell structure and mineralogy of the Bivalvia II. Lucinacea–Clavagellacea. Conclusions. *Bulletin of the British Museum (Natural History), Zoology,* **22**, 256–294.

TEVESZ, M. J. S. 1975. Structure and habits of the 'living fossil' pelecypod *Neotrigonia*. *Lethaia*, **8**, 321–327.

—— & CARTER, J. G. 1980. Environmental relationships of shell form and structure of unionacean bivalves. In: RHOADS, D. C. & LUTZ, R. A. (eds) *Skeletal Growth of Aquatic Organisms: Biological Records of Environmental Change*. Plenum Press, New York, 295–322.

THIELE, J. 1934. *Handbuch der systematischen weichtierkunde. Part* 2(3). Gustav Fischer, Jena, 779–1022.

TRYON, G. W., JR 1884. *Structural and Systematic Conchology: An Introduction to the Study of Mollusca. Volume 3.* Published by the author, Philadelphia.

TURGEON, D. D., QUINN, J. F., JR, BOGAN, A. E. *ET AL.* 1998. *Common and Scientific Names of Aquatic Invertebrates from the United States and Canada: Mollusks* 2nd Edition. American Fisheries Society, Special Publication **26**, ix–526.

VAN DAMME, D. 1984. *The Freshwater Mollusca of Northern Africa. Distribution, Biogeography and Palaeoecology. Developments in Hydrobiology, 25.* Dr W. Junk Publishers, Dordrecht.

WALLER, T. R. 1990. The evolution of ligament systems in the Bivalvia. *In*: MORTON, B. (ed.) *The Bivalvia – Proceedings of a Memorial Symposium in Honour of Sir Charles Maurice Yonge, Edinburgh.* Hong Kong University Press, Hong Kong, 49–71.

—— 1998. Origin of the molluscan Class Bivalvia and a phylogeny of major groups. *In*: JOHNSTON, P. A. & HAGGART, J. W. (eds) *Bivalves: an Eon of Evolution – Paleobiological Studies Honoring Norman D. Newell.* University of Calgary Press, Calgary, 1–45.

WILLIAMS, J. D., WARREN, M. L., JR, CUMMINGS, K. S., HARRIS, J. L. & NEVES, R. J. 1993. Conservation status of freshwater mussels of the United States and Canada. *Fisheries*, **18**(9), 6–22.

WOODWARD, M. F. 1898. On the anatomy of *Mulleria dalyi*, Smith. *Proceedings of the Malacological Society of London*, **3**, 87–91.

YONGE C. M. 1953. The monomyarian condition in the Lamellibranchia. *Transactions of the Royal Society of Edinburgh*, **62**(2), 443–478.

—— 1962. On *Etheria elliptica* Lam. And the course of evolution, including assumption of monomyarianism, in the Family Etheriidae (Bivalvia: Unionacea). *Philosophical transactions of the Royal Society of London, Series B*, **244**(715), 423–458.

—— 1978. On the monomyarian, *Acostaea rivoli* and evolution in the family Etheriidae (Bivalvia: Unionacea). *Journal of Zoology, London*, **184**, 429–448.

—— 1979. Cementation in bivalves. *In*: VAN DER SPOEL, S., VAN BRUGGEN, A. C. & LEVER, J. (eds) *Pathways in Malacology*. Dr W. Junk. Publishers, The Hague, 83–106.

ZOUROS, E., BALL, A. O., SAAVEDRA, C. & FREEMAN, K. R. 1994. An unusual type of mitochondrial DNA inheritance in the blue mussel *Mytilus*. *Proceedings of the National Academy of Science, USA*, **91**, 7463–7467.

Comparative sperm ultrastructure in pteriomorphian bivalves with special reference to phylogenetic and taxonomic implications

JOHN M. HEALY, JENNIFER L. KEYS & LINA Y. M. DADDOW

Department of Zoology and Entomology, University of Queensland, Australia, 4072

(e-mail: jhealy@zen.uq.edu.au)

Abstract: Pteriomorphian spermatozoa, like those of most other bivalves, are of the classic aquasperm type (conical acrosomal vesicle, short to rod-shaped nucleus, short midpiece composed of two centrioles and a ring of spherical mitochondria, a simple flagellum). Whereas most other bivalve subclasses show at least some defining acrosomal feature(s), this does not appear to be the case within the Pteriomorphia. While this could indicate non-monophyletic status, it also correlates with the fact that the Pteriomorphia are a very old and very successful group of bivalves. Acrosomal similarities suggest a close link between the Ostreoidea and Limoidea (acrosomal vesicle with wedge-shaped apical zone; radiating plates present but not well developed); and between the Pterioidea, Pinnoidea and Pectinoidea (dense anterior layer; very well developed radiating plates). For supposedly closely related taxa, the Arcoidea and Limopsoidea (both Arcoida) differ markedly from each other in acrosomal shape and substructure. The affinities of the Anomioidea and even more so the Mytiloida remain uncertain, the latter possibly connected with the Pterioida or, more likely, removed from the rest of the Pteriomorphia (mytiloid acrosomes show concentric lamellae). A very close relationship between the Pectinidae and Spondylidae of the Pectinoidea is demonstrated (dense anterior layer of acrosome recurved). Within the Mytilidae (Mytiloidea) there is substantial variation in sperm morphology between supraspecific taxa especially at the subfamial level.

The subclass Pteriomorphia, in its original, broad sense (see Beurlen 1944; Cox 1960) contains a large proportion of the world's most economically and ecologically important groups of Bivalvia [e.g. Mytiloidea (mussels and allies), Ostreoidea (rock oysters), Pterioidea (pearl oysters), Pectinoidea (scallops and allies)]. However, the composition and status of the Pteriomorphia and its relationship to other bivalve higher taxa remain controversial. Despite wide acceptance of the Pteriomorphia as a subclass-level division of the Bivalvia (Cox 1960, 1969; Newell 1965,1969; Boss 1982; Allen 1985; Vaught 1989; Wilson 1998), a number of workers have, on the basis of a reconsideration of anatomical and shell characters, suggested alternative classifications. Most notably the Russian school (Nevesskaya *et al.* 1971; Scarlato & Starobogatov 1978; Starobogatov 1992) does not recognize the division Pteriomorphia in any form, preferring instead to regard the constituent groups as subdivisions (orders or suborders) of the superorder Autobranchia Grobben 1894 [renamed Mytiliformii by Starobogatov (1992); the two other recognized divisions of the Bivalvia in this scheme being the Protobranchia and Septibranchia – renamed Nuculiformii and Conocardiiformii, respectively, by Starobogatov (1992)]. Purchon (1987) regarded the Pteriomorphia as no more than an order of lamellibranchs, and therefore unworthy of recognition as a discrete subclass. Waller (1978), using a cladistic approach, redefined the autobranch 'superorder' Pteriomorphia to exclude the Mytiloida and Arcoida – the latter two taxa being placed by him into the superorders Isofilibranchia Iredale 1939 and Prionodonta MacNeil 1937, respectively. More recently, Waller (1998) has returned to a 'traditional' concept of the Pteriomorphia and again included the Mytiloida and Arcoida. Nevertheless, the relationship of the Mytiloida to other bivalves, pteriomorphian or otherwise, has long posed a problem to students of bivalve evolution, partly because of the occurrence of the modioliform shell in several fossil lines of uncertain affinities (e.g. see Cox 1960; Waller 1978). Pojeta (1971, 1978) has however argued that the Ordovician modiomorphids 'are the likely ancestors of the Mytilidae' and that the Isofilibranchia should be considered as a distinct subclass from the Pteriomorphia *sensu lato* [but, for a critical discussion of these issues, see Waller (1998)].

From: Harper, E. M., Taylor, J. D. & Crame, J. A. (eds) *The Evolutionary Biology of the Bivalvia*. Geological Society, London, Special Publications, **177**, 169–190. 1-86239-076-2/00/$15.00

Comparative application of transmission electron microscopy (TEM), scanning electron microscopy (SEM; including freeze-fracture) and most recently molecular genetic analyses (allozyme and rDNA sequence data) have provided fresh and thought-provoking perspectives on bivalve phylogeny and higher classification [for recent discussions see Healy (1996), Salvini-Plawen & Steiner (1996), Steiner & Müller (1996), Adamkewicz *et al.* (1997), Campbell *et al.* (1998), Canapa *et al.* (1999), Giribet & Carranza (1999), Steiner (1999*a*, *b*)]. In most cases such 'new' sources of information have beautifully complimented what is known of comparative anatomy and shell morphology, and led to an improved understanding of bivalve relationships.

Although sperm ultrastructural studies of the Bivalvia were first attempted 40 years ago (e.g. Galtsoff & Philpott 1960) it is only in the last 25 years or so that this field of research has developed a truly comparative basis to match the earlier and more broad-ranging light microscopic accounts such as those of Retzius (1904, 1905), Franzén (1955) and Dan & Wada (1955). TEM has proved especially important because of its ability to reveal the finest internal details of spermatozoa, and it is work conducted using this technique that has contributed most to the growing understanding of bivalve sperm structure, biology and development [for class-wide reviews see Popham (1979) and Healy (1996)]. Spermatozoa of several groups of pteriomorphians have now been examined using TEM, including the Arcoidea, Limopsoidea, Mytiloidea, Pterioidea, Pinnoidea, Pectinoidea, Anomioidea and Ostreoidea (Table 1). The present study focuses on the taxonomic and potential phylogenetic utility of comparative sperm data in the Pteriomorphia *sensu lato*, based on available literature and new data for many unstudied or incompletely studied families including Noetiidae (Arcoidea), Glycymerididae (Limopsoidea), Ostreidae (Ostreoidea), Limidae (Limoidea), Spondylidae and Pectinidae (both Pectinoidea), Pteriidae and Pulvinitidae (both Pteriodea), and Pinnidae (Pinnoidea) (Table 1). Attention is especially focused on the comparative shape and substructure of the acrosomal complex (= acrosomal vesicle + subacrosomal material) and the adjoining nuclear apex.

Material and methods

Material of the following species was processed for TEM (systematic placement and collection data indicated in brackets): Arcoidea (Noetiidae) – *Striarca lactea* (Linnaeus, 1758), Adriatic Sea; Limopsoidea (Glycymerididae) – *Glycymeris holsericus* (Reeve, 1843), dredged 10 m, Banana Banks, Moreton Bay, southern Queensland; Limoidea (Limidae) – *Limaria fragilis* (Gmelin, 1791), from coral heads, Heron Island, Great Barrier Reef, off mid-Queensland coast; Ostreoidea (Ostreidae) – *Dendrostrea folium* (Linnaeus, 1758) attached to rock, 1 m depth, Orpheus Island, northern Queensland; Pectinoidea (Pectinidae) – *Gloripallium pallium* (Linnaeus, 1758), from under plate coral, 9 m depth, Heron Island; Pectinoidea (Spondylidae) – *Spondylus nicobaricus* Schreibers, 1793, Lady Musgrave Island reef, 8 m depth; *Pinctada* sp., *P. margaritifera* (Linnaeus, 1758), Orpheus Island, northern Queensland; Pinnoidea (Pinnidae) – *Atrina vexillum* (Born, 1778), Lady Musgrave Island, buried in sand in gap between coral bommies, outer reef slope, 8 m depth. Voucher specimen shells are deposited with the Queensland Museum, Brisbane (QMMO 66820–66826). Small (1–2 mm^3) pieces of mature testis or ovotestis were excised from the relaxed animal and placed directly into ice-cold 3% glutaraldehyde (prepared in 0.1 M sodium phosphate buffer containing w/v 10% sucrose) for 24 h, then rinsed in buffer (45 min) before being placed into a 1% solution of osmium tetroxide (buffer as for earlier steps) for 80 min. After a further rinse in buffer (45 min), tissues were dehydrated through a graded ethanol series (from 20% to absolute) before being gradually embedded in Spurrs epoxy resin. Semithin and ultrathin sections were cut using an LKB 2428 Ultrotome IV, collected on uncoated 200 μm mesh copper grids and stained with lead citrate and uranyl acetate according to the contrast-enhancing method of Daddow (1986). Stained sections were examined and photographed using a Hitachi A 300 transmission electron microscope operated at an accelerating voltage of 75 kV. All sperm organelle measurements are based on TEM observations.

Spermatozoan features of the Pteriomorphia

Mature spermatozoa of all examined pteriomorphian species are released directly into the ambient water and therefore can be termed 'aquasperm' (Jamieson 1987; Rouse & Jamieson 1987) or 'aquatic sperm' (Baccetti & Afzelius 1976). Rouse & Jamieson (1987) and Jamieson (1987, 1991) recognized two categories of aquasperm: (1) 'ect-aquasperm' for sperm which, at least potentially, fertilize eggs within the surrounding water; (2) 'ent-aquasperm' for sperm which are shed into the surrounding water but subsequently are swept by a feeding or inhalent current into the mantle cavity (in the case of molluscs) or tube (in the case of sedentary polychaetes) and fertilize eggs there.

Among the Pteriomorphia, most species appear to be free-spawners (i.e. with ect-aquasperm) but in some families, such as the Ostreidae, the occurrence of brooding in certain species indicates that at least some pteriomorphians fertilize within the mantle cavity (i.e. utilize ent-aquasperm). Aquasperm of pteriomorphians (and most other bivalves) exhibit, in anterio-posterior sequence, the following features: an acrosomal complex; a condensed nucleus (usually with scattered, irregularly shaped lacunae); a short midpiece (a pair of triplet microtubular centrioles surrounded by a ring of spherical mitochondria); a flagellum (tail), always of 9 + 2 microtubular configuration. The following section details the distinguishing sperm ultrastructural features of the various superfamilies of Pteriomorphia *sensu lato* (examined families indicated by *). Currently, no data exists for the Dimyoidea or Plicatuloidea. Primary emphasis is placed on the morphology and substructure of the acrosomal complex (acrosomal vesicle + subacrosomal material) and the shape of the nucleus. In addition to the micrographs of species examined, a diagrammatic summary of acrosomal ultrastructure is also presented, in order to facilitate comparisons between taxa (see Fig. 6).

Subclass Pteriomorphia – comparative spermatozoan structure

*Arcoidea (*Arcidae, *Noetiidae, Cucullaeidae) (Figs 1d–f, 3a–c and 6a and b)*

Among the Arcoidea, three species of the Arcidae have previously been examined for sperm ultrastructure: *Anadara trapezia* (Deshayes, 1840) (Popham 1979), *Barbatia obliquata* (Wood, 1828) and *B. foliata* (Forsskal, 1775) (Reunov & Hodgson 1994). Results presented herein for *Striarca lactea* (Linnaeus, 1758) (Noetiidae) are in general agreement with those obtained for arcids. Arcoidean spermatozoa consist of: (1) a conical, deeply invaginated acrosomal vesicle (length 0.7–1.0 µm in Arcidae and 0.5 µm in *S. lactea*), the contents of which are finely granular and homogeneous (with the exception of a thin, slightly less electron-dense peripheral layer); (2) an extensive deposit of coarsely granular subacrosomal material filling the vesicle invagination and the gap between the base of the vesicle and the nuclear apex (i.e. the acrosomal vesicle does not rest directly on the nuclear surface); (3) a short (2.0–3.0 µm) spheroidal nucleus (finely granular in arcids, coarsely granular in *S. lactea*); (4) four to five (rarely six) spherical mitochondria arranged around a pair of centrioles; (5) a satellite fibre complex connecting the distal centriole to the plasma membrane; (6) a 9 + 2 pattern axoneme within the flagellum. The studies by Popham (1979) and Reunov & Hodgson (1994) hint that sperm morphology does not vary profoundly between species or even between genera within the Arcidae. *Striarca lactea* differs from the three investigated arcids by having a shallow depression in the nuclear apex (v. apex flat or slightly convex in arcids) and in its more anterio-posteriorly compressed acrosomal vesicle. Sperm ultrastructural data for the Cucullaeidae, and for the many unstudied genera of Arcidae (*Arca*, *Trisidos*, *Bathyarca*, *Samacar*) and Noetiidae (*Sheldonella*, *Arcopsis*), are now required to test the generality of the sperm diagnosis outlined here.

*Limopsoidea (*Glycymerididae, Limopsidae, Philobryidae) (Figs 1a–c, 2d–f and 6c)*

Three species of the Limopsoidea have to date been examined for sperm ultrastructure, all belonging to the Glycymerididae: *Glycymeris yessoensis* Sowerby, 1886 (Pashchenko & Drozdov 1990, 1991), *Glycymeris* sp. (acrosome only; Sousa *et al.* 1998) and *Glycymeris holsericus* (Reeve, 1843) (this study). In *G. yessoensis* and *G. holosericus* the nucleus is slightly elongate (7.5–9.7 µm and widest basally in *G. yessoensis*; 4.8–5.0 µm and widest in posterior third in *G. holsericus)* and the midpiece is composed of spherical mitochondria (five in *G. yessoensis*; four in *G. holsericus)* surrounding a pair of triplet substructure centrioles. In *G. holosericus*, a short, periodically banded rootlet connects the proximal centriole to a shallow posterior nuclear fossa while the distal centriole is anchored to the plasma membrane by a radiating array of nine satellite fibres. The most profound difference between these species involves the morphology of the acrosomal complex. Pashchenko & Drozdov (1990, 1991) have reconstructed the acrosomal vesicle of *G. yessoensis* as being spherical and positioned on a short column (0.8 µm) of subacrosomal material (the latter attached to the nuclear apex). Unfortunately, their micrographs do not clearly support this interpretation, although they appear to have demonstrated a spherical acrosomal vesicle in the mid-spermatid stage of *G. yessoensis*. The acrosomal configuration described by Pashchenko & Drozdov (1990, 1991) is reminiscent of the acrosomal complex from heterobranch gastropod spermatozoa [see Healy (1996) for comparative figures and discussion] and also certain echinoid spermatozoa (e.g. see Jamieson 1991) and, if confirmed, would be unique within the Bivalvia. It seems likely that Pashchenko & Drozdov have either photographed tangentially sectioned acrosomes, or that the acrosomal vesicle in their material is either immature or (less likely) partially

Table 1. *Sperm ultrastructure in the Pteriomorphia: taxa examined to date and sources of data*

Classification	Species	Source
ARCOIDEA		
Arcidae	*Anadara trapezia* (Deshayes, 1840)	Popham, 1979
	Barbatia obliquata (Wood, 1828)	Reunov & Hodgson 1994
	Barbatia foliata (Forsskal, 1775)	Reunov & Hodgson 1994
Noetiidae	*Striarca lactea* (Linnaeus, 1758)	Present study
LIMOPSOIDEA		
Glycymerididae	*Glycymeris yessoensis* (Sowerby, 1866)	Paschenko & Drozdov 1990, 1991
	Glycymeris sp.	Sousa & Azevedo 1998
	Glycymeris holosericus (Reeve, 1843)	Present study
MYTILOIDEA		
Mytilidae		
	Modiolus kurilensis Bernard, 1983	Drozdov & Reunov 1986*a*, *b*
	Modiolus americanus (Leach, 1815)	Hylander & Summers 1977
	Aulacomya ater (Molina, 1782)	Hodgson & Bernard 1986*a*, *b*; Garrido & Gallardo 1996
	Arcuatula capensis (Krauss, 1848)	Reunov & Hodgson 1994
	Septifer keenae Nomura, 1936	Reunov & Drozdov 1986
	Brachiodontes semistriatus (Krauss, 1848)	Reunov & Hodgson 1994
	Bathymodiolus thermophilus Kenk & Wilson, 1985	Le Pennec & Beninger 1997
	Bathymodiolus puteoserpenyis von Cosel, Métivier & Hashimoto 1994	Le Pennec & Beninger 1997
	Bathymodiolus elongatus von Cosel, Métivier & Hashimoto, 1994	Le Pennec & Beninger 1997
	Bathymodiolus childressi Gustafson, 1998	Eckelbarger & Young 1999
	Adula falcatoides Habe, 1955	Reunov & Drozdov 1986
	Mytilus edulis Linnaeus, 1758	Niijima & Dan, 1965; Longo & Dornfeld 1967; Endo 1976; Drozdov & Reunov 1986*b*; Hodgson & Bernard 1986*b*; Tilney *et al.* 1987; Sousa & Azevedo 1988
	Mytilus coruscus Gould, 1861	Reunov & Drozdov 1987
	Mytilus galloprovincialis Lamarck, 1819	Hodgson & Bernard 1986*b*; Komaru *et al.* 1995
	Mytilus perna von Ihering, 1897	Bourcart *et al.* 1965
	Mytilus chilensis Hupé, 1854	Garrido & Gallardo 1996
	Choromytilus chorus (Molina, 1782)	Toledo *et al.* 1990; Garrido & Gallardo 1996
	Crenomytilus grayanus (Dunker, 1853)	Drozdov 1979; Drozdov & Mashansky 1979; Drozdov *et al.* 1981, 1982
	Semimytilus algosus (Gould, 1850)	Garrido & Gallardo 1996
	Trichomya hirsuta (Lamarck, 1819)	Present study
	Perumytilus purpuratus (Lamarck, 1819)	Garrido & Gallardo 1996
	Perna perna (Linnaeus, 1758)	Bernard & Hodgson 1985
	Perna viridis Linnaeus, 1758	Reunov *et al.* 1999
	Brachidontes virgiliae (Barnard, 1964)	Bernard *et al.* 1988
	Musculus discors (Linnaeus, 1767)	Franzén 1983

PTERIOIDEA		
Pteriidae	*Pinctada margaritifera* (Linnaeus, 1758)	Thielley *et al.* 1993; present study
	Pinctada sp.	Present study
Isognomonidae	*Isognomon isognomon* (Linnaeus, 1758)	Healy 1989, 1996
Malleidae	*Vulsella vulsella* (Linnaeus, 1758)	Lamprell & Healy 1998
PINNOIDEA		
Pinnidae	*Pinna nobilis* Linnaeus, 1758	De Gaulejac *et al.* 1995
	Atrina vexillum (Born, 1778)	Present study
PECTINOIDEA		
Pectinidae	*Pecten maximus* Linnaeus, 1758	Anderson & Personne 1970*a*, *b*,1976; Dorange & Le Pennec 1989
	Aequipecten irradians (Lamarck, 1819)	Linck 1973*a*, b
	Placopecten magellanicus (Gmelin, 1791)	Désilets *et al.* 1995
	Gloripallium pallium (Linnaeus, 1758)	Present study
Spondylidae	*Spondylus nicobaricus* Schreibers, 1793	Present study
ANOMIOIDEA		
Anomiidae	*Anomia trigonopsis* Hutton, 1877	Popham, 1979
LIMOIDEA		
Limidae	*Limaria fragilis* (Gmelin, 1791)	Present study
OSTREOIDEA		
Ostreidae	*Ostrea edulis* Linnaeus, 1758	O'Foighil 1989; Bozzo *et al.* 1993; Sousa & Oliveira 1994
	Dendrostrea folium (Linnaeus, 1758)	Present study
	Crassostrea virginica (Gmelin, 1791)	Galtsoff & Philpott 1960; Daniels *et al.* 1971; Eckelbarger & Davis 1996
	Crassostrea angulata (Lamarck, 1819)	Gutierrez *et al.* 1978; Sousa & Oliveira 1994
	Crassostrea gigas (Thunberg, 1793)	Osanai & Kyozula 1982; Kyozuka & Osanai 1985; Komaru *et al.* 1994; Gwo *et al.* 1996
	Saccostrea glomerata (Gould, 1850)	Healy & Lester 1991; Molinia & Swan 1991; Present study

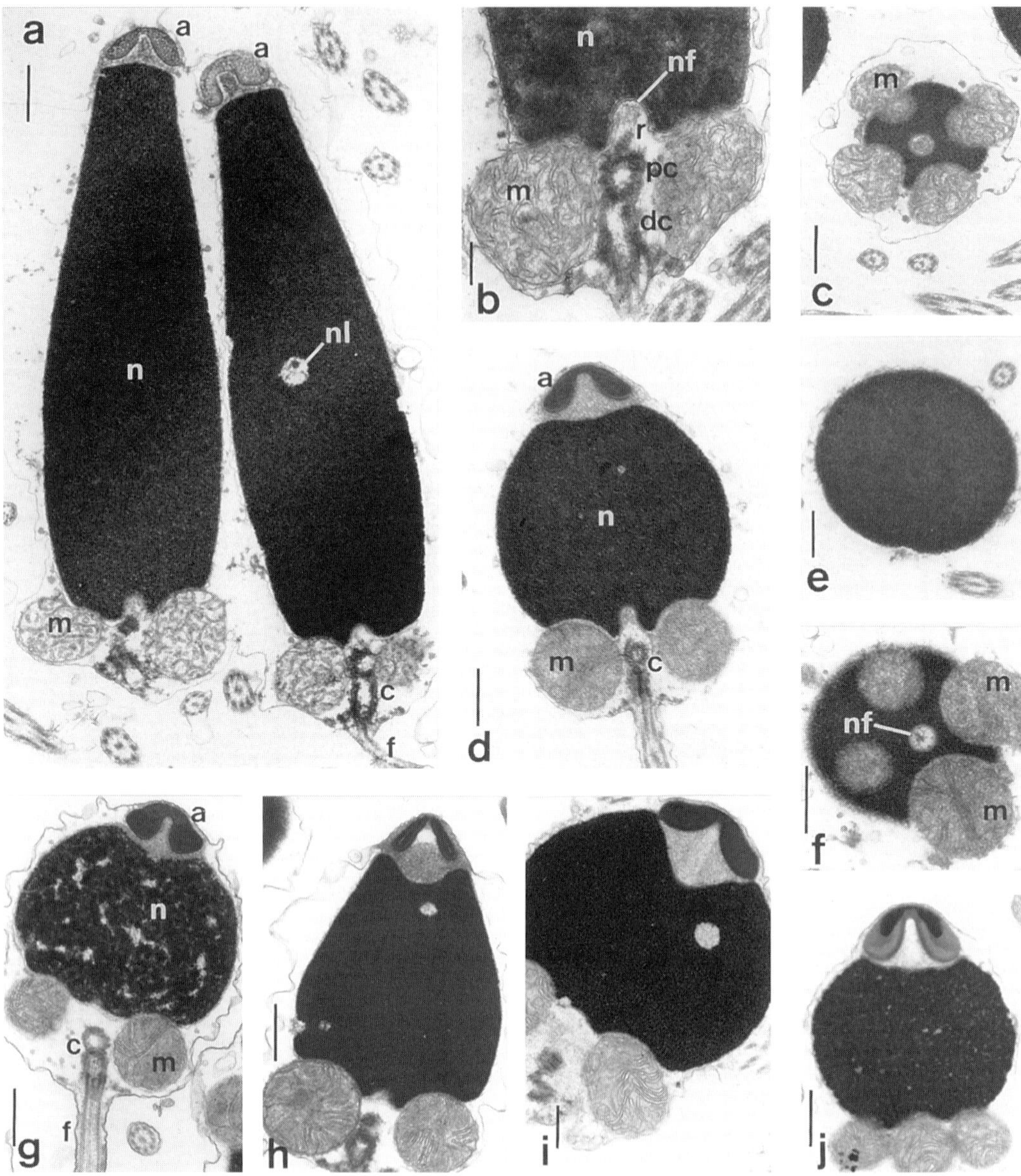

Fig. 1. Mature spermatozoa of various Pteriomorphia. (a)–(c) *Glycymeris holosericus* (Glycymerididae, Limopsoidea). (**a**) Longitudinal section (LS) through acrosomal complex (acrosomal vesicle + subacrosomal material), nucleus, midpiece and flagellum. (**b**) LS base of nucleus and midpiece, showing rootlet attached to proximal centriole. (**c**) Transverse section (TS) junction of nucleus and midpiece. (d)–(f) *Striarca lactea* (Noetiidae, Arcoidea). (**d**) LS acrosomal complex, nucleus, midpiece and portion of flagellum. (**e**) TS junction of nucleus and midpiece. (**f**) TS nucleus. (**g**) *Limaria fragilis* (Limidae, Limoidea), LS acrosomal complex, nucleus, midpiece. (**h**) *Gloripallium pallium* (Pectinidae, Pectinoidea). (**i**) *Dendrostrea folium* (Ostreidae, Ostreoidea), acrosomal complex, nucleus, midpiece. (**j**) *Pinctada* sp. (Pteriidae, Pterioidea), acrosomal complex, nucleus and midpiece. Scale bars are 0.5 μm, except (**b**) which is 0.25 μm. a, Acrosomal complex; c, centrioles; dc, distal centriole; f, flagellum; m, mitochondria; n, nucleus; nf, nuclear fossa; nl, nuclear lacuna; pc, proximal centriole, r, centriolar rootlet.

reacted. Clearly, additional observations on the mature spermatozoa of *G. yessoensis* are required to clarify what appears to be an anomalous result.

In *Glycymeris* sp., the acrosomal vesicle is tall-conical in longitudinal profile (length 1.2 μm), with subacrosomal material filling the deep posterior invagination (Sousa *et al.* 1998). Sousa *et al.* (1998) found a positive phosphotungstic acid reaction associated with the invagination portion of the acrosomal membrane and occasionally as a 'central cord like element' within the vesicle contents. The acrosomal vesicle of *G.* sp. slightly encloses the convex nuclear apex.

In *G. holsericus* the acrosomal vesicle has a wide, anteriorly truncate profile (length 0.3 μm, diameter 0.7 μm), with subacrosomal material occupying the extensive basal invagination of the vesicle (Figs 2d and 6c). This acrosomal morphology differs substantially from that of the Arcoidea, not only in its truncate longitudinal profile but also in the substructure of its contents (coarse granular with a dense peripheral layer, recurved posteriorly v. homogeneous contents in Arcoidea). In immature spermatozoa of *G. holosericus* the acrosomal vesicle is high-conical (Fig. 2f). However, during the final maturational phase, the vesicle assumes its final, compressed-conical form (Figs 2d and 6c). The substructure of the dense layer within the acrosomal vesicle could not be unequivocally established – some sections suggested periodic plates but others indicated a homogeneous texture. The resemblance of the mature acrosomal vesicle of *G. holosericus* to those of pandoroidean Anomalodesmata has previously been commented upon (Healy 1996) and seems worthy of further study, especially in relation to spermiogenic stages of acrosomal development. The recurved nature of the anterior dense layer in *G. holosericus* is also reminiscent of that seen in the Pectinoidea, but as yet no firm evidence of radiating plates has been observed in glycymeridid acrosomes. Comparative work on the three unstudied families of limopsoideans (Philobryidae and Limopsidae) will undoubtedly yield new clues concerning relationships within the Limopsoidea and maybe shed further light on the group's exact relationship to the Arcoidea and other bivalves.

*Limoidea (*Limidae) (Figs 1g, 2g–i and 6g)*

As members of this important superfamily have not as yet been examined for sperm ultrastructure, details are presented for an Indo-Pacific species, *Limaria fragilis* (Gmelin, 1791). The acrosomal vesicle (length 0.4 μm) has a low-conical shape with a deep basal invagination. Contents of the vesicle are moderately electron dense with the exception of a wedge-shaped, apical region which is conspicuously less electron dense and coarsely granular in appearance. Short, radiating plates are present within the peripheral region of the acrosomal vesicle (Fig. 2i). Subacrosomal material loosely fills the invagination of the acrosomal vesicle, as well as the broad, apical depression of the nucleus which extends posteriorly as a narrow canal (Figs 2g and h and 6g). The narrowness of this invagination almost certainly explains why it is usually not visible in longitudinal sections through the nucleus. The nucleus is rounded but slightly compressed anterio-posteriorly and its contents are coarsely granular, with scattered, electron-lucent lacunae. Four, five or occasionally six spherical mitochondria surround two triplet substructure centrioles. The proximal centriole bears a short rootlet for attachment to the nucleus while the distal centriole is fixed to the plasma membrane by an array of nine satellite fibres.

In general, the spermatozoa of *Limaria fragilis* resemble most closely those of the Ostreidae (Galtsoff 1960; Daniels *et al.* 1971; O'Foighil 1989; Healy & Lester 1991; Bozzo *et al.* 1993; Gwo *et al.* 1996), and to a lesser degree those of *Striarca lactea* (Arcoidea herein) and *Anomia trigonopsis* Hutton, 1877 (Anomioidea; Popham 1979). Especially significant is the apical zone of the *L. fragilis* acrosomal vesicle, which also occurs in the Ostreidae. In ostreids, the apical zone of the acrosomal vesicle often shows a layered or whorled internal substructure (Fig. 6k) but in certain species such an ordered substructure appears to be absent, as observed in *L. fragilis* (see following section – Ostreoidea). In all ostreid species examined to date, a well-developed axial rod forms a significant component of the subacrosomal material. No rod appears to be present in *L. fragilis* although the narrow and deep invagination of the nuclear apex may perhaps contain rod precursors. As Popham (1979) and Kafanov & Drozdov (1998) have shown, the occurrence of an axial rod among bivalves is often sporadic (e.g. constant in Ostreidae, occurring only in Mytilinae of the Mytilidae) and its presence among the Pteriomorphia appears to be closely linked with a very deep anterior invagination of the nucleus. Although the nucleus and midpiece morphology of Pectinoidea are similar to *L. fragilis* (and other eupteriomorphans), the pectinoidean acrosomal vesicle exhibits a dense reflected layer and no apical, wedge-shaped zone (Dorange & Le Pennec 1989; present study).

*Ostreoidea (*Ostreidae, Gryphaeidae) (Figs 1i, 3a–d and 6k)*

Sperm ultrastructure is known for five commercially important species of Ostreidae: *Ostrea*

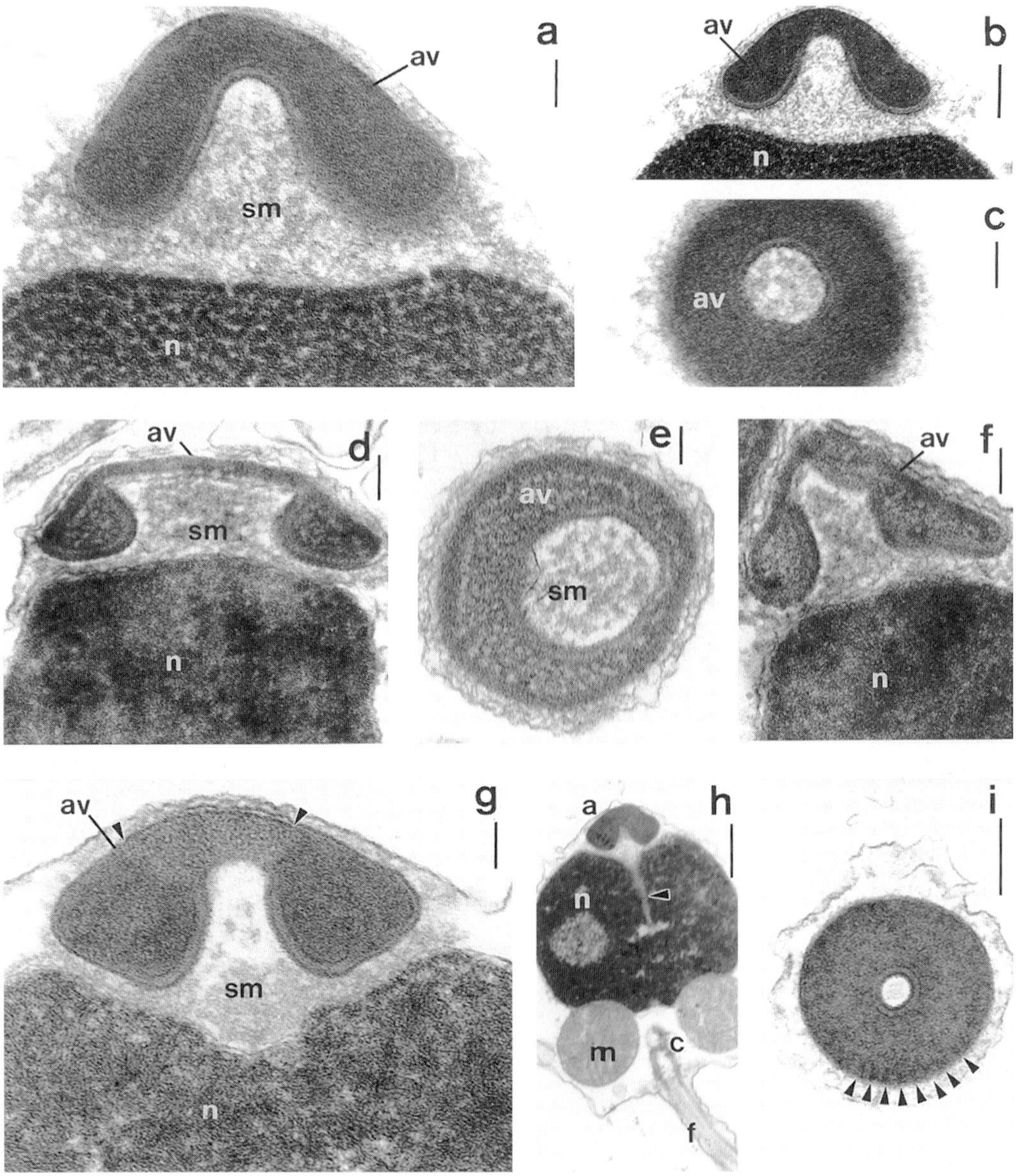

Fig. 2. Acrosomal morphology in Arcoidea, Limopsoidea and Limoidea
(a)–(c) *Striarca lactea* (Noetiidae, Arcoidea). (**a**) Longitudinal section (LS) through acrosomal complex and nuclear apex. (**b**) Transverse section (TS) through anterior region of acrosomal complex, showing radiating plates. (**c**) LS showing full extent of anterior depression of nucleus. (d)–(f) *Glycymeris holosericus* (Glycymerididae, Limopsoidea). (**d**) LS through acrosomal complex and nuclear apex. (**e**) TS through slightly immature acrosomal complex. (**f**) LS immature acrosomal complex showing conical shape of acrosomal vesicle. (g)–(i) *Limaria fragilis* (Limidae, Limoidea). (**g**) LS acrosomal complex and nuclear apex. Note wedge-shaped apical zone anteriorly (lateral extent indicated by arrowheads). (**h**) LS showing deep, narrow extension of anterior nuclear invagination (arrowhead). (**i**) TS showing radiating plates (some here indicated by arrowheads) associated with periphery of acrosomal vesicle. Scale bars are 0.1 μm, except (b) and (i) which are 0.25 μm, (h) which is 0.5 μm. a, Acrosomal complex; av, acrosomal vesicle; c, centrioles; f, flagellum; m, mitochondrion; n, nucleus; sm, subacrosomal material.

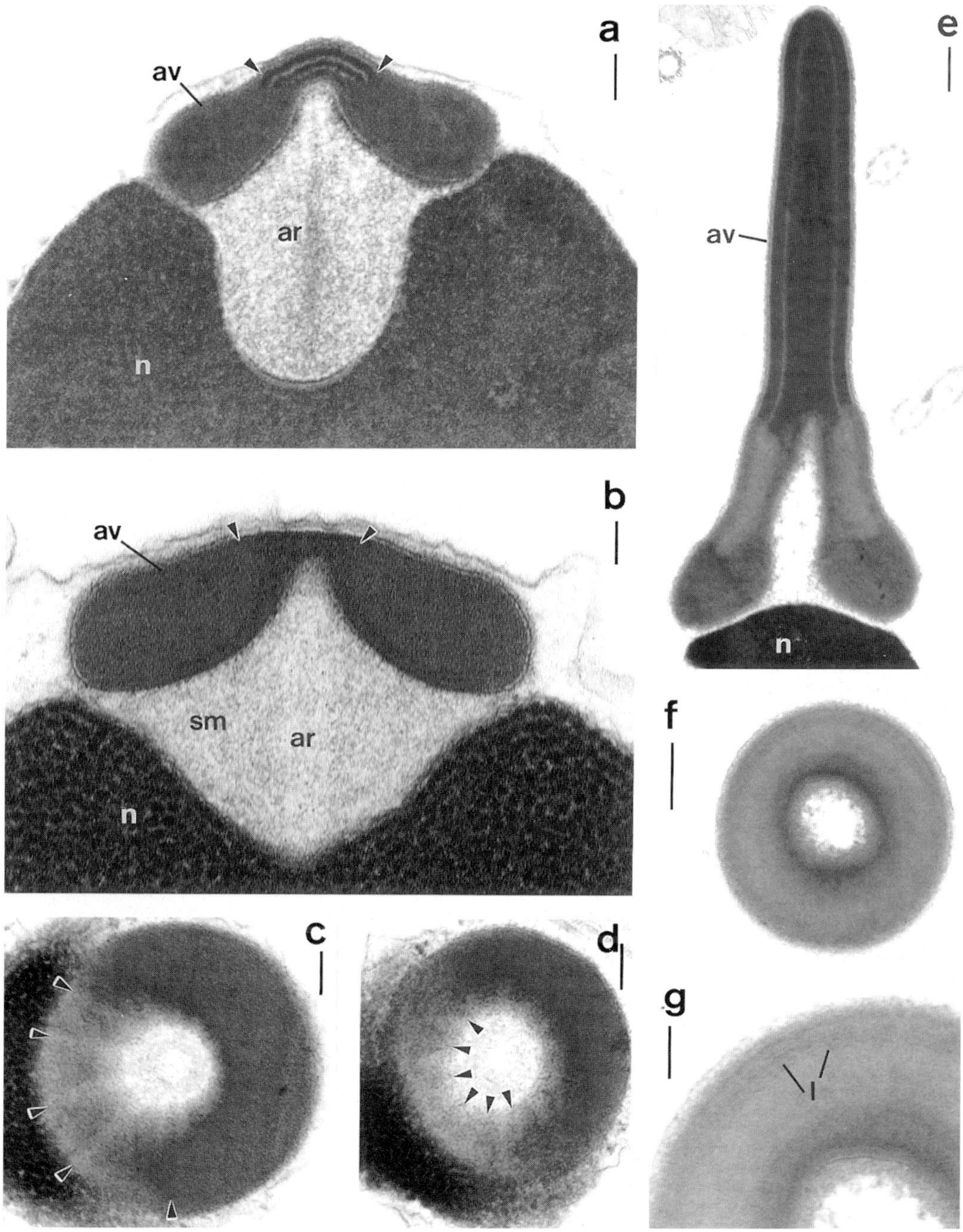

Fig. 3. Acrosomal morphology in Ostreoidea, Mytiloidea. (**a**) *Saccostrea glomerata* (Ostreidae, Ostreoidea). Longitudinal section (LS) through acrosomal complex and nuclear apex. Arrowheads indicating lateral extent of wedge-shaped apical zone. (b)–(d) *Dendrostrea folium* (Ostreidae, Ostreoidea). (**b**) LS through acrosomal complex and nuclear apex. Arrowheads indicating lateral extent of wedge-shaped apical zone. (c) and (d) Transverse section (TS) through basal region of acrosomal complex showing radiating plates (some indicated by arrowheads).(e)–(g) *Trichomya hirsuta* (Mytilidae, Mytiloidea). (**e**) LS through acrosomal complex (= acrosomal vesicle + subacrosomal material) and nuclear apex. (**f**) TS through basal region of acrosomal complex, showing concentric layers. (**g**) TS detail of fine, concentric lamellae at periphery of acrosomal vesicle. Scale bars are 0.1 μm, except (e) and (f) which are 0.25 μm. ar, Axial rod (differentiated component of subacrosomal material); av, acrosomal vesicle; l, fine, concentric lamellae of acrosomal vesicle contents; n, nucleus; sm, subacrosomal material surrounding axial rod.

edulis (Linnaeus 1758), *Crassostrea gigas* (Thunberg 1793), *C. virginica* Gmelin, 1791, *C. angulata* (Lamarck 1819) and *Saccostrea glomerata* (Gould 1850) (for a listing of studies see Table 1). In addition, the first TEM sperm data for a species of *Dendrostrea* [*D. folium* (Linnaeus 1758)] and additional micrographs of *S. glomerata* are presented. Ostreid spermatozoa exhibit a broad-conical acrosomal which is deeply invaginated and apically shows a wedge-shaped zone of varying substructure (Fig. 3a and b). In addition, radiating plates are associated with the basal region of the acrosomal vesicle in *D. folium* (Fig. 3c and d) but have not been detected (? or searched for) in other examined ostreids. Subacrosomal material, in the form of an axial rod surrounded by a granular matrix, fills not only the invagination of the acrosomal vesicle but also the deep and wide anterior invagination of the nucleus (Figs 1i and 3a and b). The nucleus is short and rounded, and shows a small fossa posteriorly to which a centriolar rootlet from the proximal centriole are attached. Nuclear contents show a coarse fibro-granular texture in addition to occasional electron-lucent lacunae. Four (rarely five) mitochondria surround the centrioles and a satellite fibre complex connects the distal centriole to the plasma membrane. In terms of sperm organelle shape, the five examined species of Ostreidae are remarkably uniform. However, there are interesting differences between species in relation to the apical, wedge-shaped zone of the acrosomal vesicle [homogeneous in *Crassostrea virginica*, *Ostrea edulis* and *Dendrostrea folium* (Fig. 3b), whorls in *C. gigas*, or three to five parallel layers in *C. angulata* and *Saccostrea glomerata* (Figs 3a and 6k)]. In *Dendrostrea folium*, the apical zone also extends posteriorly around the invagination of the acrosomal vesicle (Fig. 3b). Sousa & Oliveira (1994) and Sousa *et al.* (1998) have demonstrated, using phosphotungstic acid–chromic acid treatment of unosmicated spermatozoa (without heavy metal staining of ultrathin sections), that ostreids also differ somewhat in the cytochemistry of the acrosomal vesicle contents. The apical wedge-shaped zone always reacts positively to PTA, but in *Ostrea edulis* a layer associated with the posterior rim of the acrosomal vesicle is also revealed (Sousa *et al.* 1998). It seems likely that acrosomal substructure of the Ostreidae may be of considerable taxonomic significance once data are gathered for unstudied species and genera.

*Mytiloidea (*Mytilidae) (Figs 3e–g, 4 and 6d–f)*

Without doubt the Mytilidae have been the most thoroughly examined family of bivalves as regards comparative sperm ultrastructure (see Table 1). A significant proportion of the published work centres on widely cultivated species of the Mytilinae (e.g. *Mytilus edulis* Linnaeus, 1758, *M. galloprovincialis* Lamarck, 1819) and *Crenomytilus grayanus* (Dunker, 1853), focusing in particular on the acrosome reaction, acrosomal vesicle cytochemistry and sperm–egg interaction (for listing of studies see Table 1). However, in recent years all the remaining subfamilies (and most extant genera) have been studied for sperm ultrastructure from a comparative, often taxonomy oriented, viewpoint [see Table 1 and Kafanov & Drozdov (1998)]. This extensive body of research has shown that sperm structure varies profoundly within the Mytilidae – partly in connection with differing egg morphologies and fertilization environments [ect-aquatic v. ent-aquatic fertilization – for discussions of other bivalves see Popham (1974) and Franzén (1983)] but also reflecting apparently natural taxonomic subdivisions within the family. Recently, Kafanov & Drozdov (1998) have used sperm features to redefine three mytiliid subfamilies based primarily on nuclear morphology and presence/absence of an axial rod (Fig. 4). Mytiliid spermatozoa show varying degrees of acrosomal vesicle complexity ranging from simple (e.g. *Septifer* and *Bathymodiolus*) to very complex (e.g. *Mytilus*) (Figs 4 and 6d–f). Typically, an outer dense layer and an inner electron-lucent layer are present, but sometimes accompanied by one or more additional zones (Fig. 4). Endo (1976) and more recently Sousa & Azevedo (1995) have demonstrated the highly structured nature of the acrosome of *Mytilus edulis* using silver methenamine and phosphotungstic acid staining, and chemical and enzymatic pretreatments (differentiating up to five zones within the acrosomal vesicle contents; see Fig. 4). Fine, concentric lamellae appear to be characteristic of mytilid acrosomes [see Figs 3f and g and 4, also Garrido & Gallardo (1996)]. However, it is not known whether lamellae occur in the simplest mytilid acrosomes (e.g. for *Septifer* and *Bathymodiolus*). Typically, most mytiliid sperm possess five spherical mitochondria (surrounding the standard pair of centrioles) but usually some level of variation from four to five is observed (as with most bivalve spermatozoa) (Hodgson & Bernard 1986*a*, *b*; Kafanov & Drozdov 1998). Spermatozoa of *Modiolus* are exceptional among the Mytilidae and, in fact, among the Bivalvia in having a large number of midpiece mitochondria (12–14; Drozdov & Reunov 1986*a*; Kafanov & Drozdov 1998). Recent work by Reunov *et al.* (1999) suggests the possibility of sperm polymorphism in *Perna viridis* (Linnaeus, 1758) involving dual patterns of acrosomal and tail development.

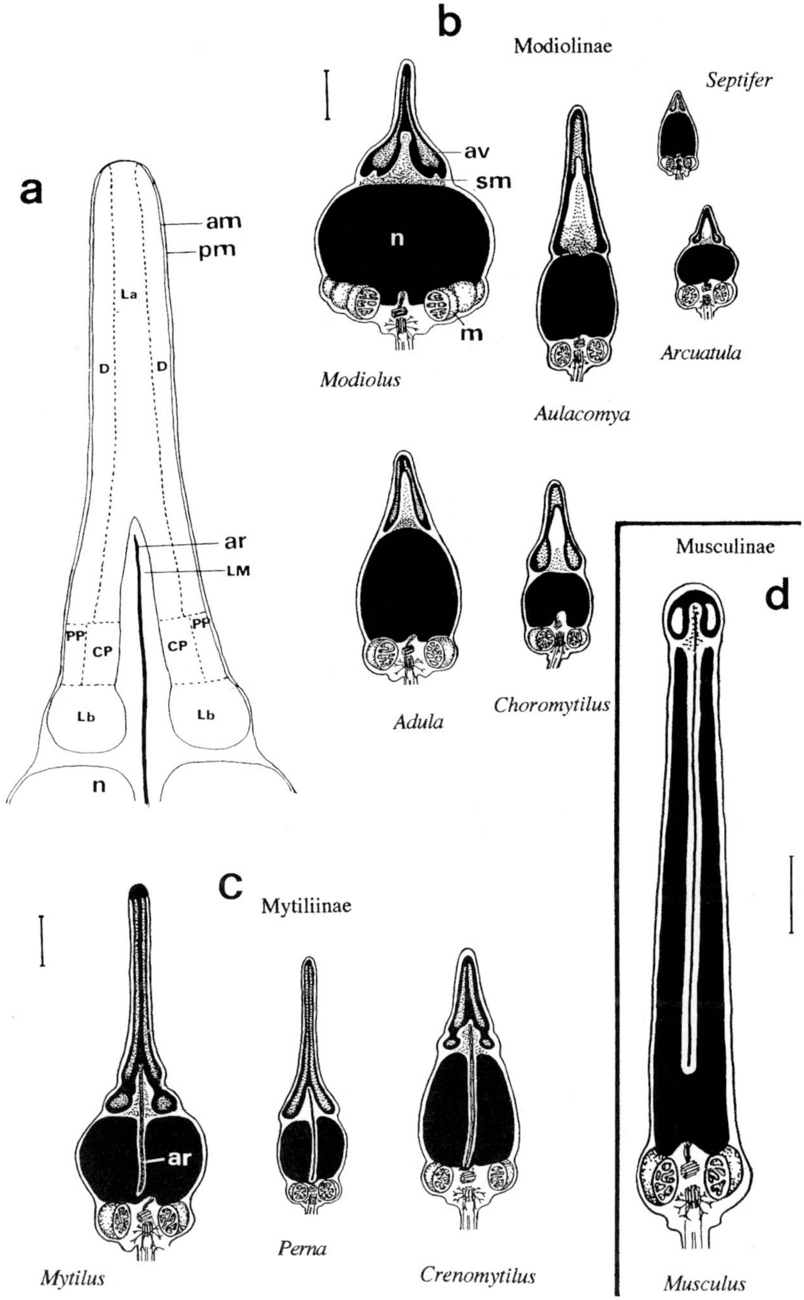

Fig. 4. (**a**) Acrosomal vesicle internal substructure in *Mytilus edulis* [Mytilidae, Mytiloidea – slightly modified from Endo (1976)]. (b)–(d) Comparative sperm structure in Mytiloidea [Mytilidae – slightly modified from Kafanov & Drozdov (1998)]. am, Acrosomal vesicle membrane; ar, axial rod (differentiated component of subacrosomal material); av, acrosomal vesicle; CP, central proximal portion of acrosomal vesicle; D, anterior portion of acrosomal vesicle; La, lysin in axially located strand; Lb, lysin in basal portion of acrosomal vesicle; lm, lumen formed by invagination of acrosomal membrane; m, mitochondria; n, nucleus; pm, plasma membrane; PP, peripheral proximal part of acrosomal vesicle; sm subacrosomal material. [Figures reproduced with permission of Academic Press and the Institute of Malacology (USA)]. Scale bars for (b)–(d) are 1.0 μm.

*Pinnoidea (*Pinnidae) (Figs 5a–c and 6j)*

Only a single species of Pinnidae has previously been examined for sperm ultrastructure – *Pinna nobilis* Linnaeus, 1758 (see De Gaulejac *et al.* 1995). Although the published information allows reconstruction of most features of the sperm of *P. nobilis*, details of the most important organelle, the acrosomal complex, require further elaboration. The present study of the spermatozoa of *Atrina vexillum* (Born, 1778) is in essential agreement with the finding of De Gaulejac *et al.* (1995) but has also shed additional light on the profile and substructure of the acrosomal vesicle. This has allowed the following characterization of pinnoidean sperm features: (1) the acrosomal vesicle is conical, *c.* 0.4 μm long and deeply invaginated basally; its contents are differentiated into a highly electron-dense anterior layer (exhibiting closely spaced radiating plates) and a less dense posterior layer; (2) very granular subacrosomal material fills the vesicle invagination and nuclear depression; (3) the nucleus is short (length 1.8–2.0 μm), squat and coarsely granular in texture, with a noticeable anterior depression; (4) five or, more uncommonly, four spherical mitochondria are associated with the paired centrioles (and associated satellite fibre complex and flagellum). The well-developed radiating plates, here associated with the dense anterior layer within the acrosomal vesicle, is essentially as seen in the Pterioidea and Pectinoidea (see previous and following sections), and spermatologically differentiates the Pinnoidea from the Anomioidea (acrosomal vesicle of simlar shape but evidently with homogeneous contents; Popham 1979).

*Pterioidea (*Pteriidae, *Isognomonidae, *Malleidae, Pulvinitidae) (Figs 1j, 5d–f and 6l)*

Despite their economic and ecological importance it is very surprising that only a few species of Pterioidea have been examined for sperm ultrastructure [*Pinctada margaritifera* (Linnaeus, 1758), *Isognomon isognomon* (Linnaeus, 1758) (Isognomonidae) and *Vulsella vulsella* (Linnaeus, 1758) (Malleidae); Table 1]. To this list is added observations for *Pinctada* sp. Spermatozoa of the Pterioidea can be characterized by the following features: (1) a conical (usually short, low-conical, 0.4–0.7 μm long) acrosomal vesicle, deeply invaginated basally and exhibiting marked differentiation of the acrosomal contents (a highly electron-dense anterior layer containing radiating plates; one or more areas of lower electron opacity); (2) a short (1.5–2.0 μm long) spheroidal nucleus, usually showing a broad anterior depression and a basal fossa; the apical depression is limited to a fine pit or absent in *Vulsella vulsella*; (3) usually five (sometimes four or six) spherical mitochondria; (4) the standard arrangement of centrioles and the satellite fibre complex. It is interesting to note that the differentiation of acrosomal vesicle contents is more complex within the Pteriidae than in either the Isognomonidae or Malleidae, perhaps mirroring the range of complexity occurring in the Mytilidae (Mytiloida).

*Pectinoidea (*Pectinidae, *Spondylidae, Propeamussiidae, Syncyclonemidae) (Figs 1h, 5g–j and 6m)*

Given their economic importance, it is surprising that sperm data are available for only three species of Pectinidae – *Pecten maxima* Linnaeus, 1758, *Aequipecten irradians* (Lamarck, 1819) and *Placopecten magellanicus* (Gmelin, 1791). Of these, only *Pecten maxima* has been adequately examined. In order to clearly establish characteristic sperm features of the Pectinidae details are presented for *Gloripallium pallium* (Linnaeus, 1758), a common and widespread Indo-Pacific species (Figs 1h, 5h–j and 6m). Although Dan & Wada (1955) studied the acrosome reaction

Fig. 5. Acrosomal morphology in Pterioidea, Pinnoidea, Pectinoidea. (a)–(c) *Atrina vexillum* (Pinnidae, Pinnoidea). (**a**) Longitudinal section (LS) through acrosomal complex (= acrosomal vesicle + subacrosomal material) and nuclear apex. (**b**) Transverse section (TS) through anterior region of acrosomal complex, showing radiating plates (arrowheads). (**c**) TS acrosomal complex below level of (b). (d)–(f) *Pinctada* sp. (Pteriidae, Pterioidea). (**d**) LS through acrosomal complex and nuclear apex showing dense anterior layer and electron-lucent zone. (**e**) TS through anterior region of acrosomal complex showing radiating plates (arrowheads). (**f**) TS acrosomal complex below level of (e). (**g**) *Spondylus nicobaricus* (Spondylidae, Pectinoidea) LS acrosomal complex and nuclear apex. Dense anterior layer recurved posteriorly. Inset TS acrosomal complex, faintly showing radiating plates. (h)–(j) *Gloripallium pallium* (Pectinidae, Pectinoidea). (**h**) LS through acrosomal complex and nuclear apex. Dense anterior layer recurved posteriorly. (**i**) TS through anterior region of acrosomal complex showing radiating plates (arrowheads). (**j**) TS acrosomal complex below level of (i) but still showing radiating plates (arrow heads). Scale bars are 0.1 μm. ar, Axial rod (differentiated component of subacrosomal material); av, acrosomal vesicle; n, nucleus; sm, subacrosomal material.

in *Spondylus cruentus* Lischke, 1868, no TEM level sperm data have been published for the Spondylidae, prompting inclusion of data for the Indo-Pacific species *S. nicobaricus* Schreibers, 1793 (Fig. 5g). Although pectinoids show variation between species in the longitudinal profile of the conical acrosomal vesicle (length 0.4 µm in *G. pallium*; 0.8–0.9 µm in *S. nicobaricus*, vesicle slightly ballooning anteriorly in *Pecten maxima* and in *S. nicobaricus*), the vesicle in all species exhibits a highly electron-dense internal layer which recurves posteriorly, giving a double-layered effect

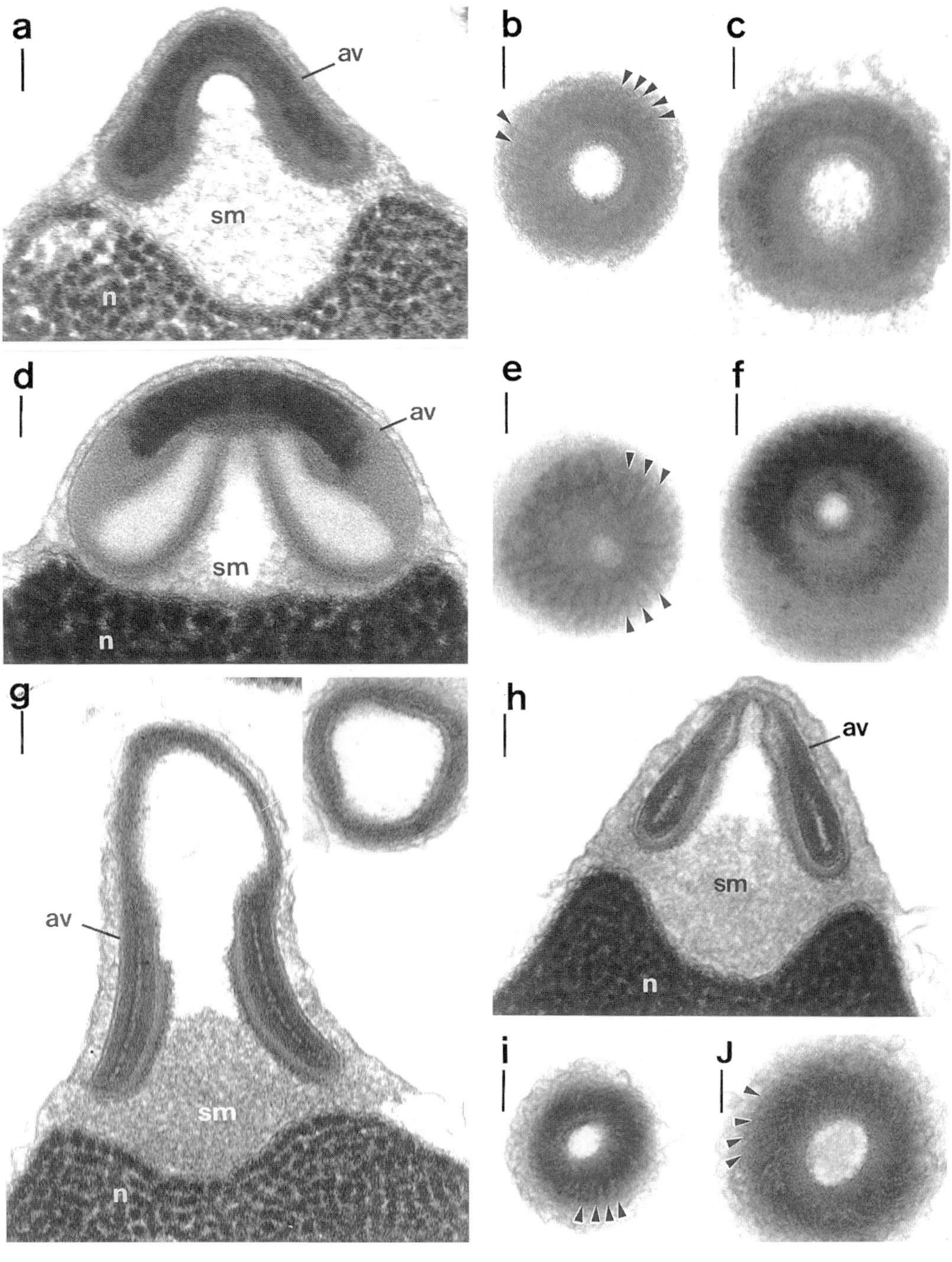

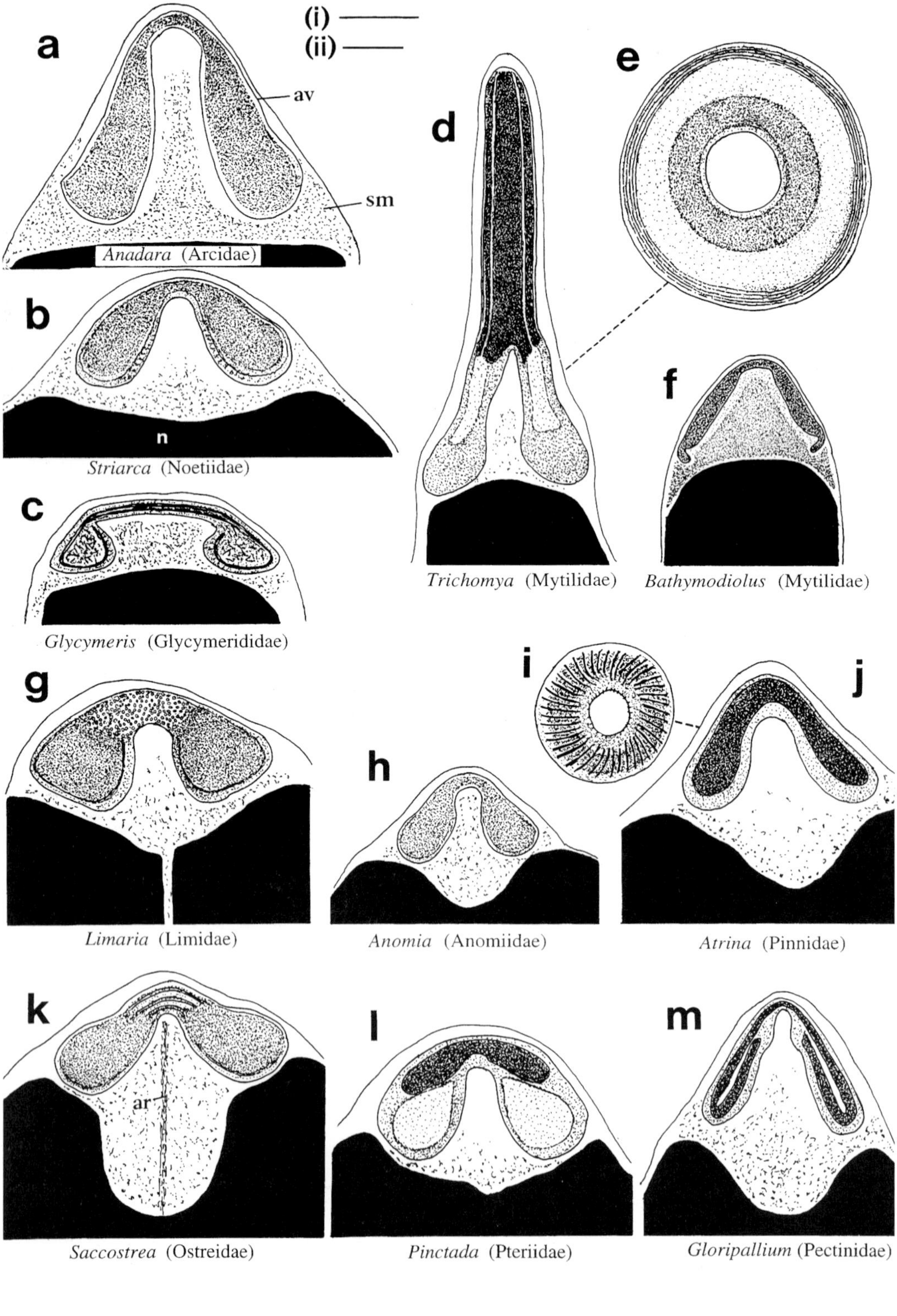
(i)
(ii)
a
av
sm
Anadara (Arcidae)
b
n
Striarca (Noetiidae)
c
Glycymeris (Glycymerididae)
d
Trichomya (Mytilidae)
e
f
Bathymodiolus (Mytilidae)
g
Limaria (Limidae)
h
Anomia (Anomiidae)
i
j
Atrina (Pinnidae)
k
ar
Saccostrea (Ostreidae)
l
Pinctada (Pteriidae)
m
Gloripallium (Pectinidae)

through most of its length. Of special interest is the presence of a very well-developed, radiating plate substructure within the dense layer – most closely resembling that seen in the Pinnoidea and Pterioidea (although in these two superfamilies, the dense anterior layer does not recurve). Pectinoidean sperm all have short, pyriform nuclei (with coarsely granular texture) and the midpiece exhibits five or sometimes four spherical mitochondria (associated with the standard paired centrioles, the satellite fibre complex and the flagellum).

*Anomioidea (*Anomiidae) (Fig. 6h)*

Since Popham's (1979) description of the spermatozoa of *Anomia trigonopsis* (as *A. descripta* Iredale, 1936) no additional data for the Anomioidea have been published. Fortunately, his TEM micrographs (only two, but informative longitudinal sections) and description are sufficiently detailed to allow reconstruction of almost the entire spermatozoon, which in its overall form resembles those of other eupteriomorphians. In *A. trigonopsis* the acrosomal vesicle is conical, deeply invaginated and its contents homogeneous and moderately electron dense. Subacrosomal material fills the basal invagination of the short (length 0.3 μm) acrosomal vesicle and an adjoining indentation of the nuclear apex. The nucleus is rounded and short (length 1.4 μm) but the presence of a basal fossa has yet to be established. Popham states that the midpiece contains four spherical mitochondria which surround two centrioles (the latter arranged at 90° to each other), suggesting a standard bivalve midpiece and flagellum arrangement.

Spermatozoa of the Pteriomorphia – taxonomic and phylogenetic implications

The results of the present study and of previous accounts on sperm ultrastructure in the Pteriomorphia *sensu lato* (and in bivalves in general) provide an opportunity for discussion of pteriomorphian classification and relationships. It is also timely that several molecular studies of the Bivalvia have been published in the last five years or so (e.g. Steiner & Müller 1996; Adamkewicz *et al.* 1997; Campbell *et al.* 1998; Canapa *et al.* 1999; Giribet & Carranza 1999; Steiner 1999*a*, *b*) because these, in addition to the extensive anatomical and fossil literature, provide a meaningful context for any discussion of sperm results.

One of the most important points to emerge from this study is that the Pteriomorphia either in the broad sense (i.e. including prionodonts and isofilibranchs; Beurlen 1944; Cox 1960) or the cladistically restricted sense of Waller (1978) [renamed Eupteriomorpha by Boss (1982)] cannot be defined on the basis of sperm synapomorphies. Whether this adds support to the view that the Pteriomorphia are not monophyletic (e.g. Starobogatov 1992) or simply reflects the fact that the group is both old and successful, is impossible to determine on the basis of sperm evidence alone. However, it is significant that several pteriomorphian (*sensu lato*) higher taxa can be recognized using the shape and internal substructure of the acrosomal vesicle and, occasionally also nuclear morphology. Unfortunately, complex spermatozoa such as those of caenogastropod and heterobranch gastropods [e.g. see Healy (1996) for review], do not occur in the Bivalvia because introspermic fertilization is unknown within the class. Hence the range of sperm characters available for taxonomic and/or phylogenetic scrutiny in the Bivalvia is limited. The most informative characters appear to be the shape and internal structure of the acrosomal vesicle and the spatial relationship of the vesicle to the nuclear apex. The least informative bivalve sperm characters include the structure of the flagellum (a constant 9 + 2 pattern) and mitochondrial shape [spheroidal in most, sometimes slightly elongate as

Fig. 6. Diagrammatic summary of acrosomal ultrastructure throughout the Pteriomorphia (based on all available literature and data presented in this study). (**a**) *Anadara trapezia* (Arcidae, Arcoidea). (**b**) *Striarca lactea* (Noetiidae, Arcoidea). (**c**) *Glycymeris holosericus* (Glycymerididae, Limopsoidea). (**d**) and (**e**) *Trichomya hirsuta* (Mytilidae, Mytiloidea; (**e**) is the transverse section (TS) showing peripheral lamellae which are probably characteristic of the Mytiloidea). (**f**) *Bathymodiolus childressi* (Mytilidae, Mytiloidea). (**g**) *Limaria fragilis* (Limidae, Limoidea). (**h**) *Anomia trapezoides* (Anomiidae, Anomioidea). (**i**) and (**j**) *Atrina vexillum* (Pinnidae, Pinnoidea – note dense anterior layer; (**j**) is TS showing radiating plates – such plates occur in all Pteriomorphia with the exception of the Arcoidea and Mytiloidea; the condition in the Anomioidea remains unknown). (**k**) *Saccostrea glomerata* (Ostreidae, Ostreoidea) note well-developed axial rod. (**l**) *Pinctada margaritifera* (Pteriidae, Pterioidea; note prominent dense anterior layer and electron-lucent zone). (**m**) *Gloripallium pallium* (Pectinidae, Pectinoidea; note recurved anterior dense layer). ar, Axial rod (differentiated component of subacrosomal material); av, acrosomal vesicle; n, nucleus; sm, subacrosomal material. (a) and (h) based on micrographs of Popham (1979); (f) based on micrographs of Eckelbarger & Young (1999); all other figures based on micrographs presented in this study. Scale bars are 0.3 μm, exept (d) and (f) which are 0.5 μm.

in some Veneroida; see Healy (1995)]. Acrosomal and nuclear length are probably tied to functional concerns such as the thickness of the egg jelly layer (acrosomal vesicle larger in mytiliids with thicker jelly coats than thin coats) and the environment of fertilization (ect-aquatic fertilizers with large acrosomes and rounded nuclei v. ent-aquatic fertilizers with smaller acrosomes and more elongate nuclei; Franzén 1983). Some groups such as the Ostreidae, with both ect-aquatic and ent-aquatic (brooding) species, appear to show no marked impact of altered fertilization biology on sperm organelle shape. What follows is a group-by-group discussion of taxonomic and phylogenetic implications of pteriomorphian sperm ultrastructure, based on all available sources of data. As a framework for this discussion the classification of Boss (1982) is adopted (but altered according to the International Code of Zoological Nomenclature recommended ending of '-oidea' for superfamilies).

Superorder Prionodonta [Order Arcoida (Arcoidea, Limopsoidea)]

The observed sperm ultrastructural differences between the Arcoidea and Limopsoidea are pronounced, and surprising given the general acceptance of a monophyletic Arcoida. The absence of marked differentiation of acrosomal vesicle contents in the Arcoidea (contents finely granular; no marked differentiation) contrasts with the differentiated acrosomal vesicle contents in the Limopsoidea (an anterior dense layer recurved posteriorly; other contents coarse granular). Indeed, the only notable sperm feature linking Arcoidea and Limopsoidea is the absence or poor development of an anterior depression of the nucleus, here judged as a likely symplesiomorphy (occurrence and strength of the nuclear depression being highly variable among pteriomorphians). The acrosomal vesicle of the Arcoidea (low-conical, contents uniformly fine granular) may represent a plesiomorphic condition from which the more complex vesicle of the Limopsoidea has arisen. This scenario would at least accord with the known later appearance of the Limopsoidea in the fossil record (Cox 1960, 1969; Newell 1969; Thomas 1978; Waller 1978). However, the similarity of immature acrosomes of *Glycymeris holsericus* to those of the Pectinoidea, Pterioidea and even certain Anomalodesmata (Healy 1996) hints perhaps at a more substantial evolutionary role for the Limopsoidea or their immediate ancestors. Clearly there exists a need for comparative work on other limopsoideans in order to investigate the relationship of this group to other pteriomorphians. In particular, it would be of interest to use sperm ultrastructure to test the suggestion of Thomas (1978), based on a consideration of hinge plate structure, that the Glycymerididae and Limopsidae may have arisen from the Cucullaeidae.

Superorder Isofilibranchia [Order Mytiloida (Mytiloidea)]

Kafanov & Drozdov (1998) have concluded that available sperm data support the recognition (or redefining) of three subfamilies within the Mytilidae based on differences in the form of subacrosomal material present and the shape of the nucleus (Fig 4): Modiolinae (e.g. *Modiolus*, *Brachidontes*, *Choromytilus*, *Aulacomya*, *Arcuatula*, *Adula* and *Septifer* – acrosomal vesicle conical, sometimes elongate; subacrosomal material diffuse, without axial rod; nucleus short, solid, without anterior fossa for axial rod); Mytilinae [e.g. *Mytilus*, *Crenomytilus* and *Perna* – acrosomal vesicle conical and usually elongate; subacrosomal material featuring well-developed axial rod of preformed (protein) fibres; nucleus short, almost completely penetrated by narrow anterior fossa for housing axial rod; see also of Tilney *et al.* 1987, fig. 2]; Musculinae (e.g. *Musculus* – acrosomal vesicle short, blunt-conical; nucleus elongate with elongate anterior fossa to house axial rod). This revised interpretation of mytiliid classification involves movement of certain genera from their more familiar subfamilial placements (e.g. *Musculus* from Crenellinae to Musculinae; *Adula* from Lithophaginae to Modiolinae) and undoubtedly provides a fresh, if somewhat radical, review of the subject. Further comparative studies of mytilid sperm ultrastructure will still be necessary to test any major changes to traditional (e.g. Soot-Ryen & Newell 1969) mytilid classification. The simple, cap-shaped acrosomal vesicle of *Bathymodiolus* spp. (Le Pennec & Beninger 1997; Eckelbarger & Young 1999) is almost certainly closer to the plesiomorphic mytiloid acrosome than the highly structured *Mytilus* type (Fig. 6f).

The relationship of the mytiloids to other pteriomorphians (*sensu lato*) is not greatly clarified by sperm ultrastructure. Waller (1978, 1998) has consistently argued that the Mytiloida form the sister taxon to all other pteriomorphians (Eupteriomorphia), whereas Campbell *et al.* (1998), on the basis of 18S ribosomal DNA (rDNA) studies, have suggested a close relationship between mytiloidans and the Ostreoida + Pterioidea (i.e. firmly among the Eupteriomorphia). The substructural complexity of the acrosomal vesicle and dominance of the highly electron-dense layer anteriorly suggests a possible connection between the Mytiloida and the Pterioidea (Pteriidae) but not with the Ostreoidea [or Ostreoida *sensu* Waller (1978)]. Certainly, the developing mytilid acrosomal vesicle is reminiscent of those of some Pterioidea and the Pinnoidea (Hodgson & Bernard 1986*a*, figs 12–14). However, the dense layer in the Pterioidea, as in the Pectinoidea and Pinnoidea, is associated with strongly developed radiating plates - structures which have never been observed in any of the many mytiloid species examined to date (mytiloids instead exhibiting fine, concentric lamellae; Garrido & Gallardo 1996; this study). Comparative studies of selective staining of acrosomal components (e.g. phosphotungstic acid and silver methenamine treatment; Endo 1976; Sousa *et al.* 1998) in mytiloids and pterioideans will be needed to address the question of homology in acrosomal features.

Superorder Eupteriomorphia [Order Pteroida (Suborder Pteriina (Pterioidea), Suborder Pinnina–Pinnoidea); Order Limoida (Limoidea); Order Ostreoida (Suborder Ostreina (Ostreoidea, Dimyoidea, Plicatuloidea), Suborder Pectinina (Pectinoidea, Anomioidea)]

Acrosomal substructure differs substantially among the Eupteriomorphia [= Pteriomorphia *sensu* Waller (1978)] but most appear to exhibit radiating plates indicating a possible synapomorphy. Three sub-divisions seem apparent based on sperm features: (1) the Ostreoidea and Limoidea (acrosomal vesicle contents showing wedge-shaped differentiated zone (granular or lamellate) apically); (2) the Pterioidea, Pinnoidea and Pectinoidea (acrosomal contents with a highly electron-dense layer associated with long radiating plates; sometimes with additional differentiation zones); (3) the Anomioidea (acrosomal vesicle with homogeneous contents; presence of radiating plates unknown). Perhaps the only uncontroversial aspect of these groupings is the implied close link between the Pterioidea and Pinnoidea. In his cladistic analysis of pteriomorphian relationships, Waller (1978) determined the Pterioida to be the sister group to the Limoida + Ostreoida and that the Limoida was the sister group to the Ostreoida (Suborders Ostreina and Pectinina), but in a recent reanalysis of living and fossil Pteriomorphia he (Waller 1998) actually redefines the Eupteriomorphia to exclude the Arcoida and Limoida. Allen (1985) found it difficult to accept Waller's (1978) conclusions and favoured the more traditional view (e.g. Thiele 1935) that the true relationship of the Limoida lay with the Pectinina rather than the Ostreina. Recent 18S rDNA work by Steiner (1999*b*) likewise favours this view. Morton (1996) considered the Ostreoidea to be most closely related to the Limoidea but, like Allen, Steiner (and possibly most authors), also recognized a firm bond between these taxa and the Pectinoidea. According to Waller (1978, 1998), similarities between limoidans and pectinids in the structure of the mantle tentacles, eyes and calcitic shell microstructure are due to convergence. The fact that the acrosomal vesicle of *Limaria fragilis* resembles more closely those of the Ostreidae than those of the Pectinina further suggests that any link between the Limoida and Pectinina is probably a distant one. Significantly, in trees derived from 18s rDNA analyses, ostreids are consistently positioned far from the pectinids (Steiner & Müller 1996; Adamkewicz *et al.* 1997; Campbell *et al.* 1998; Canapa *et al.* 1999; Girribet & Carranza 1999; Steiner 1999*a*, *b*). If Waller's (1978) cladogram is accepted, then the acrosomal similarities between the Ostreina and Limoida (notably the differentiated, apical wedge-zone) would have to be seen as plesiomorphic and the dense anterior layer in the Pectinoidea interpreted as convergent with that of the Pterioida. At present there is no evidence for either of these conclusions, but it will be vital to establish the homology of the apical wedge-shaped zone in the Limoidea and Ostreoidea using TEM and selective staining properties of mature spermatozoa (e.g. Sousa & Azevedo 1998) and detailed observations on acrosomal vesicle development. It will be particularly interesting, given the diversity of opinions among recent authors, to determine whether the acrosomes of the Dimyoidea and Plicatuloidea resemble those of the Limoidea, Ostreoidea, Anomioidea or the Pectinoidea. According to Steiner (1999*b*), the Plicatulidae, Anomiidae, Limidae and Pectinidae form a clade, based on 18 rDNA sequence data, with plicatulids and anomiids clustering together. The Dimyoidea + Plicatuloidea were regarded as a sister group to the Ostreoidea by Waller (1978) and as a sister group to the Pectinoidea by Morton (1996) [see also Watson (1930) and Yonge (1973, 1975)]. Also worthy of future study are the suggestions by Scarlato & Starobogatov (1975) and Starobogatov (1992) that the Ostreoidea and Gryphaeoidea are most closely related to the Arcoida but derived from different arcoid ancestors [although see Waller (1998) for refutation of the diphyletic Ostreoidea based on anatomical grounds]. Spermatologically, however, it would appear much easier to derive the Ostreoida and Limoida from an arcoid stock than from any other group of extant Pteriomorphia.

The precise nature of the relationship of the Pterioidea and Pinnoidea to the Pectinoidea (and, more generally, to other pteriomorphians) is uncertain. The present study and those of Dorange

& Le Pennec (1989) and Thielley *et al.* (1993) have shown that all three superfamilies exhibit a highly electron-dense portion of the acrosomal vesicle in addition to very well-developed radiating plates (this association, a possible synapomorphy of these taxa). This contrasts with acrosomes of the Mytiloida which, despite significant differentiation of the vesicle contents, appear to lack radiating plates in the electron-dense layer (e.g. Endo 1976; Hodgson & Bernard 1986*a*, *b*; Kafanov & Drozdov 1998; this study). For this reason, it is believed that derivation of the Pinnoidea from the Mytiloidea or immediate antecedents (e.g. Allen 1985) is unlikely unless the plateless condition is indeed plesiomorphic. Interestingly, the refolded configuration of the dense layer in Pectinoidea (Pectinidae and Spondylidae) could conceivably have been derived from the acrosomal configuration observed in prionodont *Glycymeris holsericus* (Limopsoidea, Arcoida), but radiating plates have yet to be demonstrated in acrosomes of the Arcoida. Although Waller (1978) considered the Pterioida to be not far removed from the Arcoida, this is not the case with his positioning of the Pectinoidea (furthest from the Arcoida in his cladogram). Molecular studies offer some support for a close connection between the Pectinoidea + Pterioidea and the Arcoida (Steiner & Müller 1996; Canapa *et al.* 1999) and, on the basis of comparative anatomy, Allen (1985) sees the Arcoidea as basal among all extant Pteriomorphia. As discussed above, the full extent of involvement of the Arcoida in pteriomorphian evolution can only be clarified by further investigation of living forms. Any suggestion that the Pterioidea may be diphyletic [e.g. Steiner 1999*b*, grouping Pteriids with Ostreidae rather than with Pinnidae; but see also Adamkewicz *et al.* (1997)] does not receive strong support from sperm ultrastructure. However, it is interesting to note that within the Pterioidea, acrosomal morphology in the Malleidae (Lamprell & Healy 1998) and Isognomonidae (Healy 1989, 1996) is less complex than that of the Pteriidae (Thielley *et al.* 1993; this study), but very similar to that observed in the Pinnoidea (De Gaulejac *et al.* 1995; this study). Although molecular studies have produced no strong consensus regarding the Pinnoidea–Pterioidea relationship (Campbell *et al.* 1998; Steiner 1999*b*), comparative anatomy (Waller 1978, 1998; Morton 1996;) shell microstructure (Carter 1990) and available sperm data clearly associate these taxa.

The relationship of the Anomioidea to other eupteriomorphians remains far from clear. Morton (1996) considered the group to be the sister taxon to the Pteriodea + Pinnoidea, whereas Waller (1978) concluded that anomioids belonged within the Ostreoida, specifically as the sister taxon to the Pectinoidea (the Ostreoidea being the sister taxon to the Anomioidea + Pectinoidea). Purchon (1987) affiliated the Anomioidea most closely with the Pectinoidea and Limoidea based on anatomical features, as has Steiner (1999*b*) based on molecular data. The apparent lack of differentiation of acrosomal contents in *Anomia trapezoides* [shown in micrographs by Popham (1979)] immediately distinguishes the Anomioidea from the Ostreoidea and Limoidea on the one hand and from the Pterioidea, Pinnoidea and Pectinoidea on the other. Has the anomioidean acrosome become secondarily simplified or is there a connection with the Arcoidea (which also lack differentiation of acrosomal contents)? At the present time it is impossible to offer any judgement on this issue until more detailed and comparative sperm data for the Anomiidae become available.

Sperm data strongly support the traditional view (e.g. Dakin 1928; Thiele 1935) of a close relationship between the Pectinidae and the Spondylidae, and in fact there do not appear to be any sperm features separating the two families. The acrosomal vesicle of *Gloripallium pallium* shows an apical constriction of the invagination not present in *P. maximus*, suggesting the possibility of generic level differences among the Pectinidae. Future studies of the Propeamussiidae and Syncyclonemidae, and further pectinid genera such as *Amusium*, may shed further light onto the relationships within the Pectinoidea, and between the Pectinoidea and other Pteriomorphia.

The authors thank T. Gorringe (Department of Zoology and Entomology) for his assistance with aspects of TEM and photography. Live or glutaraldehyde-fixed specimens of *Striarca lactea* and *Glycymeris holosericus* were generously supplied by G. Haszprunar (Zoologische Staatssammlung Mümchen) and K. Lamprell (Queensland Museum), respectively. This study was supported financially by a Senior Research Fellowship and a research grant from the Australian Research Council (JH) and by University of Queensland Research Grants (JH and JK). We also thank the Director and staff of the Queensland University Heron Island Research Station and the Orpheus Island Research Station for making available their facilities for this project. Academic Press and The Malacological Institute (USA) are thanked for granting permission to reproduce figures from Endo (1976) and Kafanov & Drozdov (1998). The manuscript has benefited from constructive comments by A. N. Hodgson (Rhodes University).

References

Adamkewicz, S. L., Harasewych, M. G., Blake, J., Saudek, D. & Bult, C. J. 1997. A molecular phylogeny of the bivalve mollusks. *Marine Biology and Evolution*, **14**, 619–629.

Allen, J. A. 1985. The Recent Bivalvia: their form and

evolution. *In*: TRUEMAN, E. R. & CLARKE, M. R. (eds) *The Mollusca, Volume 10: Evolution*. Academic Press, New York, 337–403.

ANDERSON, W. A. & PERSONNE, P. 1970*a*. The localization of glycogen in the spermatozoa of various invertebrate and vertebrate species. *Journal of Cell Biology*, **44**, 29–51.

—— & —— 1970*b*. Recent cytochemical studies on spermatozoa on some invertebrate and vertebrate species. *In*: BACCETTI, B. (ed.) *Comparative Spermatology*. Academic Press, New York, 431–449.

BACCETTI, B. & AFZELIUS, B. A. 1976. Biology of the Sperm Cell. *Monographs on Developmental Biology, Volume 10*. S. Karger, Basel.

BERNARD, R. T. F. & HODGSON, A. N. 1985. The fine structure of the sperm and spermatid differentiation in the brown mussel *Perna perna*. *South African Journal of Zoology*, **20**, 5–9.

——, DAVIES, B. R. & HODGSON, A. N. 1988. Reproduction in a brackish-water mytilid: gametogenesis and embryonic development. *The Veliger*, **30**, 278–290.

BEURLEN, K. 1944. Beiträge zur Stammesgeschichte der Muscheln. *Sitzungsberichte der Bayererische Akademie der Wissenschaften*, **1944**, 133–145.

BOSS, K. J. 1982. Mollusca. *In*: PARKER, S. P. (ed.) *Synopsis and Classification of Living Organisms Volume 2*. McGraw-Hill, New York, 945–1166.

BOURCART, C. LAVALLARD, R. & LUBET, P. 1965. Ultrastructure du spermatozoide de la moule (*Mytilus perna* von Ihering). *Compte Rendu Hebdomadaires des Séances de l'Academie des Sciences, Paris, D*, **260**, 5096–5099.

BOZZO, M. G., RIBES, E., SAGRISTA, E., POQUET, M. & DURFORT, M. 1993. Fine structure of the spermatozoa of *Crassostrea gigas* (Mollusca, Bivalvia). *Molecular Reproduction and Development*, **34**, 206–211.

CAMPBELL, D. C., HOEKSTRA, K. J. & CARTER, J. G. 1998. 18S ribosomal DNA and evolutionary relationships within the Bivalvia. *In*: JOHNSON, P. A. & HAGGART, J. (eds) *Bivalves: An Eon of Evolution – Paleobiological Studies Honoring Norman D. Newell*. University of Calgary Press, Calgary, 75–85.

CANAPA, A., MAROTA, I., ROLLO, F. & OLMO, E. 1999. The small-subunit rRNA gene sequences of venerids and the phylogeny of Bivalvia. *Journal of Molecular Evolution*, **48**, 463–468.

CARTER, J. G. 1990. Evolutionary significance of shell microstructure in the Palaeotaxodonta, Pteriomorphia, and Isofilibranchia (Bivalvia: Mollusca). *In*: CARTER, J. (ed.) *Skeletal Biomineralization: Patterns, Processes and Evolutionary Trends, Volume V*. Van Nostrand Reinhold, New York, 135–296.

COX, L. R. 1960. Thoughts on the classification of the Bivalvia. *Proceedings of the Malacological Society of London*, **34**, 60–88.

—— 1969. General features of Bivalvia. *In*: MOORE, R. C. (ed.) *Treatise on Invertebrate Paleontology. Part N. Volume 1. Mollusca 6, Bivalvia*. Geological Society of America, Boulder, CO, and the University of Kansas Press, Lawrence, KS, N2–N129.

DADDOW, L. Y. M. 1986. An abbreviated method of double lead stain technique. *Journal of Submicroscopic Cytology*, **18**, 221–224.

DAKIN, W. J. 1928. The anatomy and phylogeny of *Spondylus*, with a particular reference to the lamellibranch nervous system. *Proceedings of the Royal Society of London, Series B*, **103**, 337–354.

DAN, J. C. & WADA, S. K. 1955. Studies of the acrosome. IV. The acrosome reaction in some bivalve spermatozoa. *Biological Bulletin*, **109**, 40–55.

DANIELS, E. W., LONGWELL, A. C., MCNIFF, J. M. & WOLFGANG, R. W. 1971. Ultrastructure of spermatozoa from the American oyster *Crassostrea virginica*. *Transactions of the American Microscopical Society*, **90**, 275–282.

DÉSILETS, J., GICQUAUD, C. & DUBÉ, F. 1995. An ultrastructural analysis of early fertilization events in the giant scallop, *Placopecten magellanicus* (Mollusca, Pelecypoda). *Invertebrate Reproduction and Development*, **27**, 115–129.

DORANGE, G. & LE PENNEC, M. 1989. Ultrastructural characteristics of spermatogenesis in *Pecten maximus* (Mollusca, Bivalvia). *Invertebrate Reproduction and Development*, **15**, 109–117.

DROZDOV, A. L. 1979. Spermiogenesis and ultrastructure of spermatozoa of *Crenomytilus grayanus* (Dunker). *The Soviet Journal of Marine Biology*, **5**, 315–319.

—— & MASHANSKY, V. F. 1979. Acrosomal reaction of the spermatozoa of *Crenomytilus grayanus*. *Tsitologiya*, **21**, 657-661 [in Russian with English summary].

—— & REUNOV, A. A. 1986*a*. Spermatogenesis and the sperm ultrastructure in the mussel *Modiolus difficilis*. *Tsitologiya*, **28**, 1069-1074 [in Russian with English summary].

—— & —— 1986*b*. The morphology of the gametes of the blue mussel *Mytilus edulis* from the White and Japan Seas and Avacha Bay. *The Soviet Journal of Developmental Biology*, **12**, 236–239.

——, PODGORNAYA, O. I. & MASHANSKY, V. F. 1981. Effect of cytochalasin B on acrosomal reaction of sperms of grey sea urchins and of Gray's mussel. *Doklady Academii Nauk SSSR*, **257**, 1244-1246 [in Russian with English summary].

——, —— & —— 1982. Factor preventing of actin polymerization in the sperm's acrosome in grey sea urchin and Gray's mussel. *Dokladi Academii Nauk SSSR*, **267**, 479-481 [in Russian with English summary].

ECKELBARGER, K. J. & DAVIS, C. V. 1996. Ultrastructure of the gonad and gametogenesis in the eastern oyster, *Crassostrea virginica*. II. Testis and spermatogenesis. *Marine Biology*, **127**, 89–96.

—— & YOUNG, C. M. 1999. Ultrastructure of gametogenesis in a chemosynthetic mytilid bivalve (*Bathymodiolus childressi*) from a bathyal, methane seep environment (northern Gulf of Mexico). *Marine Biology*, **135**, 635–646.

ENDO, S. 1976. Silver methenamine and phosphotungstic acid staining of the acrosome of *Mytilus edulis*. *Experimental Cell Research*, **100**, 71–78.

FRANZÉN, Å. 1955. Comparative morphological investigations into the spermiogenesis among Mollusca. *Zoologiska Bidrag Fran Uppsala*, **30**, 399–456.

—— 1983. Ultrastructural studies of spermatozoa in three bivalve species with notes on evolution of elongated sperm nucleus in primitive spermatozoa. *Gamete Research*, **7**, 199–214.

GALTSOFF, P. S. & PHILPOTT, D. E. 1960. Ultrastructure of the spermatozoon of the oyster *Crassostrea virginica*. *Journal of Ultrastructural Research*, **3**, 241–253.

GARRIDO, O. & GALLARDO, C. S. 1996. Ultrastructure of sperms in bivalve molluscs of the Mytilidae family. *Invertebrate Reproduction and Development*, **29**, 95–102.

GAULEJAC, DE, B., HENRY, M. & VICENTE, N. 1995. An ultrastructural study of gametogenesis of the marine bivalve *Pinna nobilis* (Linnaeus, 1758). II. Spermatogenesis. *Journal of Molluscan Studies*, **61**, 393–403.

GIRIBET, G. & CARRANZA, S. 1999. Point Counter Point – What can 18S rDNA do for bivalve phylogeny? *Journal of Molecular Evolution*, **48**, 256–258.

GROBBEN, C. 1894. Zur Kenntniss der Morphologie, der Verwandtschaftsverhältniss und des Systems der Mollusken. *Sitzungsberichte der Akademie Wissenschaften Wien*, **103**, 61–86.

GUTIERREZ, M., PÉREZ CRESPO, J. & PASCUAL, E. 1978. Ultraestructura de ovocitos y espermatozoides del ostión, *Crassostrea angulata* Lmk. de la costa sudatlántica de España. *Investigacion Pesquera*, **42**, 167–178.

GWO, J.-C., LIOU, C.-H. & CHENG, C.-H. 1996. Ultrastructure of the spermatozoa of the Pacific oyster *Crassostrea gigas* (Mollusca, Bivalvia, Ostreidae). *Journal of Submicroscopic Cytology and Pathology*, **28**, 395–400.

HEALY, J. M. 1989. Spermiogenesis and spermatozoa in the relict bivalve genus *Neotrigonia*: relevance to trigonioid relationships, particularly with Unionoidea. *Marine Biology*, **103**, 75-85.

—— 1995. Comparative sperm ultrastructure in veneroid bivalves (Mollusca). *In*: JAMIESON, B. G. M., AUSIO, J. & JUSTINE, J.-L. (eds) *Advances in Spermatozoal Phylogeny and Taxonomy*. Memoires du Muséum National d' Histoire Naturelle, **166**, 155–166.

—— 1996. Molluscan sperm ultrastructure : correlation with taxonomic units within the Gastropoda, Cephalopoda and Bivalvia. *In*: TAYLOR, J. D. (ed.) *Origin and Evolutionary Radiation of the Mollusca*. Oxford University Press, Oxford, 99–113.

—— & LESTER, R. J. G. 1991. Sperm ultrastructure in the Australian oyster *Saccostrea commercialis* (Iredale & Roughley) (Bivalvia, Ostreidae). *Journal of Molluscan Studies*, **57**, 219–224.

HODGSON, A. N. & BERNARD, R. T. F. 1986*a*. Ultrastructure of the sperm and spermatogenesis of three species of Mytilidae (Mollusca, Bivalvia). *Gamete Research*, **15**, 123–135.

—— & —— 1986*b*. Observations on the ultrastructure of the spermatozoon of two mytilids from the south-west coast of England. *Journal of the Marine Biological Association of the United Kingdom*, **66**, 385–390.

HYLANDER, B. L. & SUMMERS, R. G. 1977. An ultrastructural analysis of the gametes and early fertilization in two bivalve molluscs, *Chama macerophylla* and *Spisula solidissima* with special reference to gamete binding. *Cell and Tissue Research*, **182**, 469–489.

IREDALE, T. 1939. Mollusca, Part 1. *British Museum (Natural History) Great Barrier Reef Expedition, Scientific Reports*, **5**, 209–425.

JAMIESON, B. G. M. 1987. A biological classification of sperm types, with special reference to annelids and molluscs, and an example of spermiocladistics. *In*: MOHRI, H. (ed.) *New Horizons in Sperm Cell Research*. Japan Scientific Societies Press and Gordon and Breach Science Publishers, New York, 311–332.

—— 1991. *Fish Evolution and Systematics: Evidence from Spermatozoa*. Cambridge University Press, Cambridge.

KAFANOV, A. I. & DROZDOV, A. L. 1998. Comparative sperm morphology and phylogenetic classification of Recent Mytiloidea (Bivalvia). *Malacologia*, **39**, 129–139.

KOMARU, A., KONISHI, K. & WADA, K. T. 1994. Ultrastructure of spermatozoa from induced triploid Pacific oyster. *Crassostrea gigas*. *Aquaculture*, **123**, 217–222.

——, SCARPA, J. & WADA, K. T. 1995. Ultrastructure of spermatozoa in induced tetraploid mussel *Mytilus galloprovincialis* (LMK). *Journal of Shellfish Research*, **14**, 405–410.

KYOZUKA, K. & OSANAI, K. 1985. Sperm entry in oyster (*Crassostrea gigas*) oocytes. *Bulletin of the Marine Biological Station, Asamushi, Tohoku University*, **18**, 25–34.

LAMPRELL, K. & HEALY, J. 1998. *Bivalves of Australia. Volume 2*. Backhuys Publishers, Leiden.

LE PENNEC, M. & BENINGER, P. G. 1997. Ultrastructural characteristics of spermatogenesis in three species of deep-sea hydrothermal vent mytilids. *Canadian Journal of Zoology*, **75**, 308–316.

LINCK, R. W. 1973*a*. Comparative isolation of cilia and flagella from the lamellibranch mollusc, *Aequipecten irradians*. *Journal of Cell Science*, **12**, 345–367.

—— 1973*b*. Chemical and structural differences between cilia and flagella from the lamellibranch mollusc, *Aequipecten irradians*. *Journal of Cell Science*, **12**, 951–981.

LONGO, F. J. & DORNFELD, E. J. 1967. The fine structure of spermatid differentiation in the mussel, *Mytilus edulis*. *Journal of Ultrastructure Research*, **20**, 462–480.

MACNEIL, F. S. 1937. The systematic position of the pelecypod genus *Trinacria*. *Journal of the Washington Academy of Science*, **27**, 452–458.

MABUCHI, Y. & MABUCHI, I. 1973. Acrosomal ATPase in starfish and bivalve mollusk spermatozoa. *Experimental Cell Research*, **82**, 271–279.

MOLINIA, F. C. & SWAN, M. A. 1991. Effect of ethanol and methanol on the motility of *Saccostrea commercialis* sperm and sperm models. *Molecular Reproduction and Development*, **30**, 241–249.

MORTON, B. 1996. The evolutionary history of the Bivalvia. *In*: TAYLOR, J. D. (ed.) *Origin and*

Evolutionary Radiation of the Mollusca. Oxford University Press, Oxford, 337–359.

NEVESSKAYA, L. A., SCARLATO, O. A., STAROBOGATOV, Y. I. & EBERZIN, A. G. 1971. New ideas on bivalve systematics. *Paleontological Journal*, **5**, 141–155.

NEWELL, N. D. 1965. Classification of the Bivalvia. *American Museum Novitiates*, **2206**. 1–25.

—— 1969. Classification of Bivalvia. *In*: MOORE, R. C. (ed.) *Treatise on Invertebrate Paleontology. Part N. Volume 1. Mollusca 6, Bivalvia*. Geological Society of America, Boulder, CO, and the University of Kansas Press, Lawrence, KS, N205–N224.

NIIJIMA, L. & DAN, J. 1965. The acrosome reaction in *Mytius edulis*. Fine structure of the intact acrosome. *Journal of Cell Biology*, **25**, 243–248.

O' FOIGHIL, D. 1989. Role of spermatozeugmata in the spawning ecology of the brooding oyster *Ostrea edulis*. *Gamete Research*, **24**, 219–228.

OSANAI, K. & KYOZUKA, K. 1982. Cross fertilization between sea urchin eggs and oyster spermatozoa. *Gamete Research*, **5**, 49–60.

PASCHENKO, S. V. & DROZDOV, A. L. 1990. The ultrastructural study of the spermatogenesis in the surf clam *Glycymeris yessoensis*. *Tsitologiya*, **32**, 325–330.

—— & —— 1991. The ultrastructure of gametes and the acrosome reaction of sperm in the bivalve mollusc *Glycymeris yessoensis*. *Tsitologiya*, **33**, 20–24.

POPHAM, J. D. 1974. Comparative morphometrics of the acrosomes of the sperms of 'externally' and 'internally' fertilizing sperms of the shipworms (Teredinidae, Bivalvia, Mollusca). *Cell and Tissue Research*, **150**, 291–297.

—— 1979. Comparative spermatozoon morphology and bivalve phylogeny. *Malacological Review*, **12**, 1–20.

—— & MARSHALL, B. 1977. The fine structure of the spermatozoon of the protobranch bivalve, *Nucula hartvigiana* Pfeiffer. *Veliger*, **19**, 431–433.

POJETA, J., JR 1971. Review of Ordovician pelecypods. *US Geological Survey Professional Paper*, **695**.

—— 1978. The origin and early taxonomic diversification of pelecypods. *Philosophical Transactions of the Royal Society of London, Series, B*, **284**, 225–246.

PURCHON, R. D. 1987. Classification and evolution of the Bivalvia : an analytical study. *Philosophical Transactions of the Royal Society of London, Series B*, **316**, 277–302.

REUNOV, A. A. & DROZDOV, A. L. 1986. Ultrastructure of spermatozoa in the bivalves *Adula falcatoides* and *Septifer keenae*. *Marine Biology (Vladivostok)*, **5**, 74–76 [in Russian with English summary].

—— & —— 1987. Spermatogenesis and ultrastructure of spermatozoa of the mussel *Mytilus cornuscus*. *Tsitologiya*, **29**, 260–266 [in Russian with English summary].

—— & HODGSON, A. N. 1994. Ultrastructure of the spermatozoa of five species of South African bivalves (Mollusca), and an examination of early spermatogenesis. *Journal of Morphology*, **219**, 275–283.

——, AU, D. W. T. & WU, R. S. S. 1999. Spermatogenesis of the green-lipped mussel *Perna viridis* with dual patterns of acrosome and tail development in spermatids. *Helgolander Marine Research*, **53**, 62–69.

RETZIUS, G. 1904. Zur Kenntnis der Spermien der Evertebraten. *Biologische Untersuchungen*, **11**, 1–32, Pls I–XIII.

—— 1905. Zur Kenntnis der Spermien der Evertebraten II. *Biologische Untersuchungen*, **12**, 1-32, Pls XI–XIX.

ROUSE, G. W. & JAMIESON, B. G. M. 1987. An ultrastructural study of the spermatozoa of the polychaetes *Eurythoe complanata* (Amphinomidae), *Clymenella laseroni* and *Micromaldane laseroni* (Maldanidae) with definition of sperm types in relation to reproductive biology. *Journal of Submicroscopic Cytology*, **19**, 573–584.

SALVINI-PLAWEN, L. V. & STEINER, G. 1996. Synapomorphies and plesiomorphies in higher classification of Mollusca. *In*: TAYLOR, J. D. (ed.) *Origin and Evolutionary Radiation of the Mollusca*. Oxford University Press, Oxford, 29–51.

SCARLATO, O. A. & STAROBOGATOV, Y. I. 1975. New data for constructing a system of classification for the Bivalvia. *Malacological Review*, **11**, 65–66.

—— & —— 1978. Phylogenetic relations and the early evolution of the class Bivalvia. *Philosophical Transactions of the Royal Society of London, Series B*, **284**, 217–224.

SOOT-RYEN, T. & NEWELL, N. D. 1969. Mytilidae. *In*: MOORE, R. C. (ed.) *Treatise on Invertebrate Paleontology. Part N. Volume 1. Mollusca 6, Bivlavia*. Geological Society of America, Boulder, CO, and the University of Kansas Press, Lawrence, KS, N271–N280.

SOUSA, M. & AZEVEDO, C. 1988. Comparative silver staining analysis on spermatozoa of various invertebrate species. *International Journal of Invertebrate Reproduction and Development*, **13**, 1–8.

—— & OLIVEIRA, E. 1994. An ultrastructural study of *Crassostrea angulata* (Mollusca, Bivalvia) spermatogenesis. *Marine Biology*, **120**, 545–551.

——, GUERRA, R., OLIVEIRA, E. & TORRES, A. 1998. Comparative PTA staining of molluscan spermatozoa. *Journal of Submicroscopic Cytology and Pathology*, **30**, 183–187.

STAROBOGATOV, Y. I. 1992. Morphological basis for phylogeny and classification of Bivalvia. *Ruthenica*, **2**, 1–25.

STEINER, G. 1999*a*. What can 18S rDNA do for bivalve phylogeny? Response [to Giribet and Carranza]. *Journal of Molecular Evolution*, **48**, 258–261.

—— 1999*b*. Molecular phylogeny of pteriomorph Bivalvia inferred from 18s rDNA sequences. *In*: *Biology and Evolution of the Bivalvia: Paper and Poster Abstracts*. Malacological Society of London, 47–48.

—— & MÜLLER, M. 1996. What can 18S rDNA do for bivalve phylogeny. *Journal of Molecular Evolution*, **43**, 58–70.

THIELE, J. 1935. *In*: BIELER, R. & MIKKELSEN, P. (eds) *Handbuch der Systematischen Weichtierkunde*.

[English translation by BHATTI, J. S.] Amerind Publishing, New Delhi.

THIELLEY, M., WEPPE, M. & HERBAUT, C. 1993. Ultrastructural study of gametogenesis in the French Polynesian black pearl oyster *Pinctada margaritifera* (Mollusca, Bivalvia), I – spermatogenesis. *Journal of Shellfish Research*, **12**, 41–47.

THOMAS, R. D. K. 1978. Limits to opportunism in the evolution of the Arcoida (Bivalvia). *Philosophical Transactions of the Royal Society of London, Series B*, **284**, 345–365.

TILNEY, L. G., FUKUI, Y. & DEROSIER, D. J. 1987. Movement of the actin filament bundle in *Mytilus* sperm: a new mechanism is proposed. *Journal of Cell Biology*, **104**, 981–993.

TOLEDO, P. H., BELLOLIO, G. & DUPRÉ, E. 1990. Fecundacion y desarrollo embrionario de *Choromytilus chorus* (Molina, 1782) (Bivalvia, Mytilidae). *Revista Biologia Marina, Valparaiso*, **25**, 47–57.

VAUGHT, K. C. 1989. *A Classification of Living Mollusca*. American Malacologists, Florida.

WADA, S., OKUNO, M., NAKAMURA, K.-I. & MOHRI, H. 1992. Dynein of sperm flagella of oyster belonging to Protostomia also has a two-headed structure. *Biology of the Cell*, **76**, 311–317.

WALLER, T. R. 1978. Morphology, morphoclines and a new classification of the Pteriomorphia (Mollusca: Bivalvia). *Philosophical Transactions of the Royal Society of London, Series B*, **284**, 345–365.

—— 1998. Origin of the molluscan class Bivalvia and a phylogeny of major groups. *In*: JOHNSON, P. A. & HAGGART, J. (eds) *Bivalves: An Eon of Evolution – Paleobiological Studies Honoring Norman D. Newell*. University of Calgary Press, Calgary, 1–45.

WATSON, H. 1930. On the anatomy and affinities of *Plicatula*. *Proceedings of the Malacological Society of London*, **19**, 25–31.

WILSON, B. R. 1998. Pteriomorphia. *In*: BEESLEY, P., ROSS, G. & WELLS, A. (eds) *Mollusca: the Southern Synthesis, Fauna of Australia, Volume 5, Part A*. CSIRO Publishing, Melbourne, xvi, 1–563, 858–862.

YONGE, C. M. 1973. Functional morphology with particular reference to hinge and ligament in *Spondylus* and *Plicatula* and a discussion on relations within the superfamily Pectinacea (Mollusca: Bivalvia). *Philosophical Transactions of the Royal Society of London, Series B*, **267**, 173–208.

—— 1975. The status of the Plicatulidae and the Dimyidae in relation to the superfamily Pectinacea (Mollusca: Bivalvia). *Journal of Zoology, London*, **176**, 545–553.

Relevance of sperm ultrastructure to the classification of giant clams (Mollusca, Cardioidea, Cardiidae, Tridacninae)

JENNIFER L. KEYS & JOHN M. HEALY

Department of Zoology and Entomology, The University of Queensland, Brisbane 4072 Qld, Australia (e-mail: jkeys@zen.uq.edu.au)

Abstract: Examination of sperm ultrastructure in six out of eight of the living species of giant clams ('Tridacnidae') supports reduction in status of the Tridacnidae to a subfamily of the Cardiidae (Tridacninae), as suggested by recent cladistic analyses based on shell, anatomical and molecular characters. Tridacninae spermatozoa are all of the aquasperm type, featuring, in anterior–posterior sequence: a conical acrosomal vesicle, an oblong to rod-shaped nucleus, a short midpiece region (proximal and distal centrioles surrounded by four round mitochondria); and a flagellum (9 + 2 pattern axoneme). Substantial differences occur between species with respect to the shape and length of the nucleus, and in the spatial relationship between the acrosomal complex and the nuclear apex. Although the two extant genera can be distinguished on sperm features – *Tridacna*, nuclear peg associated with acrosome, centriolar connective absent; *Hippopus*, nuclear peg absent, connective present – no defining feature of the Tridacninae can be detected. Within *Tridacna*, the species *T.* (*Chametrachea*) *maxima* and *T.* (*C.*) *crocea* are distinguished from other species of the genus by a much finer nuclear peg and a considerably smaller acrosome. In contrast, and against expectation, *T.* (*C.*) *squamosa* shows acrosomal and nuclear dimensions very close to those obtained for *T.* (*Tridacna*) *gigas*. Recently proposed phylogenies and classificatory changes for 'tridacnids' are discussed in the light of available sperm data.

The giant clams ('tridacnids') form a small but significant group of heterodont bivalves, noted for their importance to coral reef ecosystems in the Indo-Pacific and the fact that they include the largest living externally shelled molluscs (Yonge 1932, 1936; Rosewater 1965, 1982). They range in size from *c.* 15 cm long in *Tridacna crocea* Lamarck, 1819, > 1 m in *T. gigas* (Linnaeus, 1758).

Traditionally, giant clams have been regarded as a distinct family Tridacnidae Lamarck, 1819 within the Order Veneroida. There has been much discussion as to whether the family should be placed into its own superfamily, Tridacnoidea (Keen 1969; Kafanov & Popov 1977; Boss 1982; Purchon 1987*a*, *b*; Braley & Healy 1998) or included in the Cardioidea (Yonge 1975, 1980; Schneider 1992, 1995, 1998*b*; Maruyama *et al.* 1998). Recently, Schneider & O'Foighil (1999) have concluded, after cladistic analyses of shell, anatomical and molecular characters, that giant clams form a monophyletic subfamily within the Cardiidae and are therefore undeserving of either a separate family or superfamily.

There are only eight extant species of giant clams (with possibly a ninth, *T. rosewateri* Sirenko & Scarlato 1991) most of which are confined to the Indo-West Pacific and occur sympatrically. The widely accepted classification of the group follows that of Rosewater (1965), who recognized six genera only two of which, *Hippopus* and *Tridacna*, survive today. Since the appearance of Rosewater's system, an additional three species have been described: *T. rosewateri* Sirenko & Scarlati, 1991, *T. tevoroa* Lucas *et al.*, 1990 and *Hippopus porcellanus* Rosewater, 1982. The genus *Hippopus* Lamarck, 1799, consists of *H. hippopus* (Linnaeus, 1758) and *H. porcellanus* Rosewater, 1982. The genus *Tridacna* Bruguière, 1797 consists of three subgenera, *Tridacna sensu stricto*, *Persikima* and *Chametrachea*. *Tridacna sensu stricto* contains *T. gigas* Linnaeus 1758, *Persikima* Iredale, 1937 contains *T. derasa* (Röding, 1798), and *T. tevoroa* Lucas *et al.* 1991 and *Chametrachea* comprises *T. maxima* (Röding, 1798), *T. squamosa* Lamarck, 1819 and *T. crocea* Lamarck, 1819. *Tridacna rosewateri* from the western Indian Ocean (from the Saya de Malha Bank) was described by Sirenko & Scarlati (1991) on the basis of shells only and is regarded by Benzie & Williams (1998) as a likely junior synonym of *T. squamosa*.

Although the general reproductive biology of the giant clams is well known (e.g. see Braley 1988, 1992, 1994; Nash *et al.* 1988; Shelly & Southgate 1988; Norton & Jones 1992), very few

From: HARPER, E. M., TAYLOR, J. D. & CRAME, J. A. (eds) *The Evolutionary Biology of the Bivalvia*. Geological Society, London, Special Publications, **177**, 191–205. 1-86239-076-2/00/$15.00

ultrastructural studies of gamete morphology or development have been published, with the exception of a brief transmission electron microscopical (TEM) account of mature and developing oocytes of *T. crocea* (Kawaguti *et al.* 1982) and approximate dimensions of the sperm acrosome of *T. maxima* (Healy 1995*a*). Comparative accounts of sperm and spermatogenic morphology have previously provided valuable insights into the taxonomy and phylogeny of bivalves (e.g. see Gharagozlou-Van Ginnekin & Pochon-Masson 1971; Popham 1974, 1979; Franzén 1983; Hodgson & Bernard 1986*a*, *b*; Healy 1989, 1995*a*, *b*, 1996*b*; Hodgson *et al.* 1990). Recently, Keys & Healy (1999) have presented a detailed study of sperm ultrastructure of *T. maxima* (supplemented by comparative sperm data for the Cardiidae, both new and from the literature), the results of which suggest that Schneider (1992, 1998*b*) and Schneider & O'Foighil (1999) were probably correct in regarding the Tridacnidae as a subfamily of the Cardiidae. Keys & Healy (1999) suggested that more information was needed both for the Tridacnidae and Cardiidae in order to understand supraspecific relationships within the Cardioidea better. In the present paper, sperm ultrastructure for six of the eight living species of giant clams is described and then this information is used to assess views, both current and past, concerning the classification and phylogeny of the group.

Materials and methods

Sexually mature specimens of *Tridacna (Tridacna) gigas*, *T. (Chametrachea) crocea* and *Hippopus hippopus* were collected from the Orpheus Island fringe reef (18°37'S; 146°30'E), and *T. (C.) squamosa* from the Pelorus Island fringing reef (18°33'S; 146°30'E), Central Section of the Great Barrier Reef in November 1998. *T. (C.) maxima* was collected from the Lady Musgrave Island Reef (23°54'S; 152°23'E), Capricorn Bunker Group, Mackay/Capricorn Section of the Great Barrier Reef in January 1998. The individuals were collected by SCUBA diving at depths ranging from 2.5 to 9 m. Gonadal (ovotestis) material was removed and fixed for 6 h in cold (2–6°C) 1.5% glutaraldehyde prepared in 0.1 M sodium phosphate buffer containing 10% w/v sucrose. Museum material was used for *Hippopus porcellanus*. The material was taken from a specimen held in the 'wet' collection of the Australian Museum (Sydney). The specimen (C159293) was collected from the reef north of Fondeado Island, Hondo Bay, Palawan, Philippines, (9°57'N; 118°55'E) in May 1989. This specimen had been fixed and stored in 3% buffered formalin (prepared in seawater). Gonadal material was removed and further fixed in 1.5 % glutaraldehyde for 24 h. Museum material of *Tridacna (Persikima) derasa* and *T. (P.) tevoroa* was also dissected but proved to contain no spermiogenic tissues.

After initial fixation, all samples were rinsed thoroughly in 0.1 M sodium phosphate buffer containing 10% w/v sucrose, then transferred to a 1% osmium tetroxide solution (prepared in sucrose-adjusted sodium phosphate buffer) for 80 min (all stages at 0–6°C). Osmication was followed by further buffer rinses before being dehydrated through a graded ethanol series (20–100%) and infiltrated and embedded in Spurr's low-viscosity resin. Semi-thin survey sections and ultrathin sections were cut using a diamond knife mounted in an LKB Bromma, Nova Ultratome. Ultrathin sections (silver–gold interference colour) were collected on uncoated 200 mesh copper grids, stained according to the contrast-enhancing, lead citrate and uranyl acetate method of Daddow (1986) and examined using a Hitachi 300 transmission electron microscope operated at 75 kV.

Shells of the specimens used in this study have been retained as voucher material and are lodged with the Queensland Museum [QMMO 65406 *Hippopus hippopus* (Linnaeus, 1758), QMMO 65407 *Tridacna. (T.) gigas* Linnaeus, 1758, QMMO 62408 *T. (C.) maxima* (Röding, 1798), 65408 *T. (C.) squamosa* Lamarck, 1819 and QMMO 65409 *T. (C.) crocea* Lamarck, 1819].

Results

General description

Mature sperm from all species follow the same general construction, namely, a blunt-conical acrosome positioned at the apex of an elongate or oblong nucleus (acrosome + nucleus = collectively the 'head'), a pair of orthogonally arranged centrioles surrounded by a ring of four (rarely three) rounded mitochondria and a flagellum (Fig. 1). The acrosomal vesicle is membrane bound and deeply invaginated. The invagination is narrow posteriorly and expands into a wide cavity anteriorly. Acrosomal contents are differentiated into an extensive and highly electron-dense basal ring, surrounded by a thin layer of less electron-dense material which also fills the anterior region of the vesicle (Fig. 1a–g). The proximal centriole lies at 90° relative to the distal centriole (and spermatozoon's longitudinal axis) and is connected to a shallow, basal invagination of the nucleus by a thin layer of dense material. The distal centriole is continuous with the flagellum and is attached, via an array of nine, terminally forked, satellite fibres,

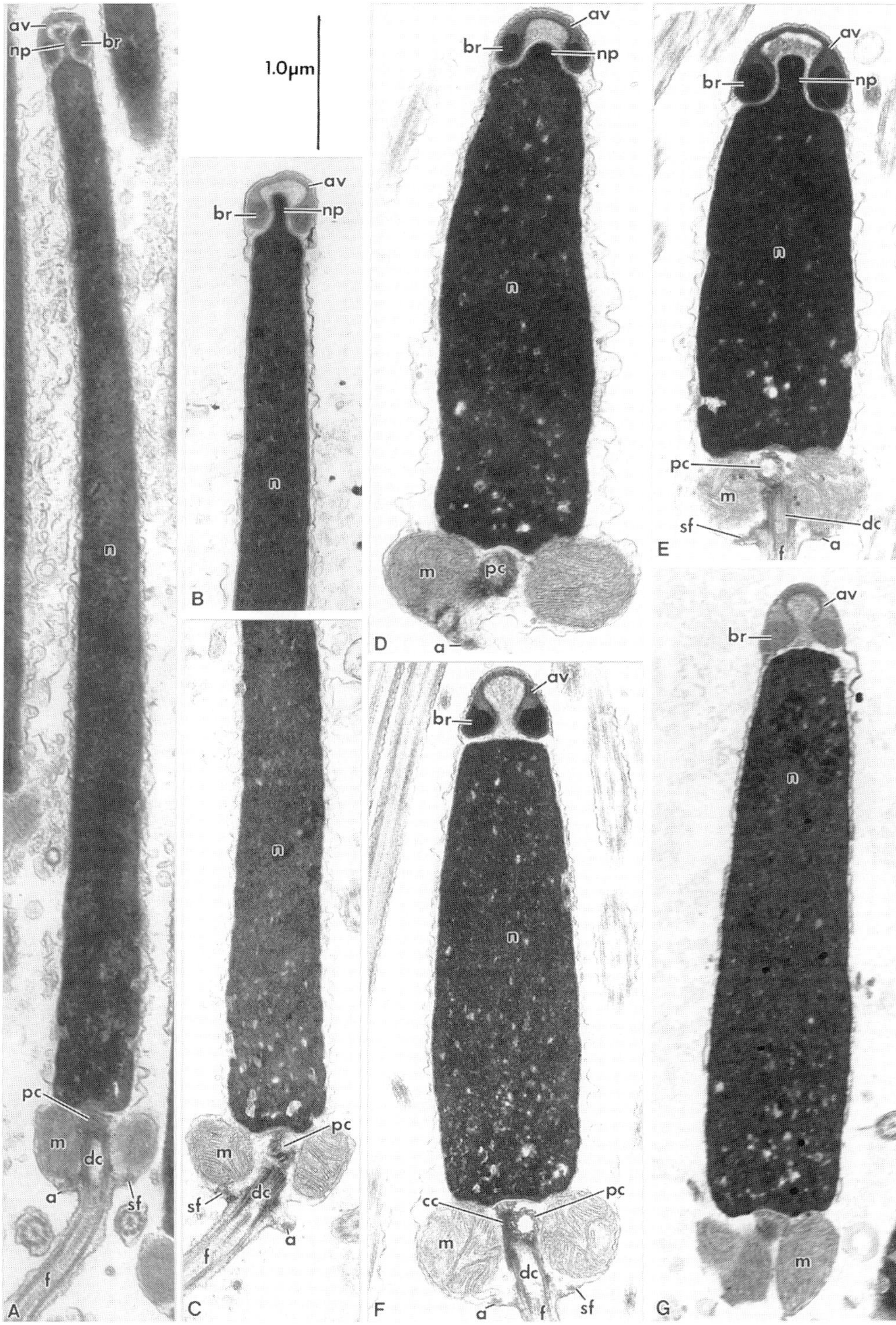

Fig. 1. Transmission electron micrographs (longitudinal sections) of mature spermatozoa of: (**a**)*Tridacna (Chametrachea) maxima*; (**b**) and (**c**) *Tridacna (C.) crocea*; (**d**) *Tridacna (C.) squamosa*; (**e**) *Tridacna (Tridacna) gigas*; (**f**) *Hippopus hippous*; (**g**) *Hippopus porcellanus*. a, Annulus; av, acrosomal vesicle; br, basal ring component of acrosomal vesicle contents; cc, centriole connective; dc, distal centriole; f, flagellum; m, mitochondria; n, nucleus; np, nuclear peg; pc, proximal centriole; sf, satellite fibres. All to same scale as indicated.

to a dense annulus lining the plasma membrane (Fig. 1a, c, e and f). At the base of the nucleus occur four (or three in some cases) ovate to spherical mitochondria with well-defined cristae (Fig. 1a–g). The flagellum consists of a classic 9 + 2 microtubular substructure axoneme sheathed by the plasma membrane. Although details of the flagellar termination were not examined in detail, it appears that there is substantial variation in the pattern of microtubular changes. The doublets may eventually be transformed into singlets and the central microtubules may or may not persist posteriorly. The number of doublets or singlets decreases as the terminal tip of the spermatozoon is approached.

In all *Tridacna* species the sperm nucleus exhibits an apical projection, termed the nuclear peg (Fig. 1a, b, d and e). The degree of development of the peg varies considerably between species within the genus. In contrast, the spermatozoa of both species of *Hippopus* lack a nuclear peg (Fig. 1f and g) but exhibit a dense structure (possible rootlet) linking the proximal centriole to the distal centriole and to the base of the nucleus (Figs 1f and 2i–l). The ultrastructural features of the mature spermatozoa of the six species of giant clams are summarized diagrammatically in Fig. 3. The relative sizes of the acrosomes and nuclear apex regions of the six species of giant clam are shown diagrammatically in Fig. 4. All the dimensions given in the results are from one specimen per species (all n values refer to the number of sperm measured). All dimensions are from TEM micrographs except for total lengths, which were obtained from light microscopy.

Tridacna (Chametrachea) maxima

A detailed description of the spermatozoa of this species has been presented by Keys & Healy (1999). However, for the purposes of this comparative account, the basic spermatozoan features of *Tridacna maxima* are here summarized. Mature spermatozoa are *c.* 50 μm (n = 7, $\bar{x}$ = 49.6 m ± SD = 1.6) in length (= head + flagellum). The acrosomal vesicle is *c.* 0.4 μm long (n = 9, $\bar{x}$ = 0.37 μm ± SD = 0.02), membrane bound and deeply invaginated. The nucleus is solid, elongate (length ≅ 7.7 μm; n = 4, $\bar{x}$ = 7.66 μm ± SD = 0.23) and rod shaped. The top of the nucleus projects deeply inside the invagination of the acrosome producing a nuclear peg, with the tip swelling noticeably (Fig. 1a). The peg is *c.* 0.3 μm long (n = 7, $\bar{x}$ = 0.28 μm ± SD = 0.02) and narrow throughout most of its length (*c.* 0.1 μm; n = 6, $\bar{x}$ = 0.05 μm ± SD = 0.01) but markedly expanded anteriorly (diameter ≅ 0.1 μm; n = 8, $\bar{x}$ = 0.11 μm ± SD = 0.01). Posterior to the peg, the diameter of the nucleus ranges from *c.* 0.4 μm (n = 10, $\bar{x}$ = 0.36 μm ± SD = 0.03) immediately below the acrosome to 0.7 μm (n = 6, $\bar{x}$ = 0.69 μm ± SD = 0.03) approaching the mitochondrial region (Fig. 1a). Irregular electron-lucent lacunae are present throughout the nucleus, including the peg region. The nucleus exhibits a short, shallow basal invagination housing a thin layer of dense material lining the anterior surface of the proximal centriole. Indentations of the nucleus also occur in regions of contact with the mitochondria (Fig. 1a). At the base of the nucleus occur four (or three in some cases) ovate to spherical mitochondria with well-defined cristae (Figs 1a, 2a and b). The flagellum is *c.* 40 μm long (n = 7, $\bar{x}$ = 39.8 μm ± SD = 1.3) and 0.2 μm in width (n = 7, $\bar{x}$ = 0.20 μm ± SD = 0.01).

Tridacna (Chametrachea) crocea

Mature spermatozoa of *Tridacna crocea* are the same shape as those of *T. maxima*, only differing slightly in size. They measure *c.* 56 μm (n = 8, $\bar{x}$ = 56.2 μm ± SD = 2.3) in length. The acrosomal vesicle is *c.* 0.4 μm long (n = 6, $\bar{x}$ = 0.39 μm ± SD = 0.04) and 0.5 μm wide (n = 8, $\bar{x}$ = 0.45 μm ± SD = 0.02). The nucleus is solid, elongate (length ≅ 7.2 μm; n = 4, $\bar{x}$ = 7.20 μm ± SD = 0.08) and rod shaped. The top of the nucleus narrows into a raised peg which projects inside the invagination of the acrosome producing a nuclear peg (Fig. 1b). The nuclear peg is *c.* 0.3 μm long (n = 4, $\bar{x}$ = 0.29 μm ± SD = 0.02) and narrow throughout most of its length (*c.* 0.1μm; n = 4, $\bar{x}$ = 0.08 μm ± SD = 0.02) but expanded anteriorly (diameter ≅ 0.1 μm; n = 4, $\bar{x}$ = 0.12 μm ± SD = 0.02). Posterior to the peg, the diameter of the nucleus ranges from *c.* 0.4 μm (n = 5, $\bar{x}$ = 0.37 μm ± SD = 0.05) immediately below the acrosome (Fig. 1b) to 0.7 μm (n = 5, $\bar{x}$ = 0.71 μm ± SD = 0.07) approaching the mitochondrial region (Fig. 1c). The flagellum is *c.* 45 μm long (n = 8, $\bar{x}$ = 44.5 μm ± SD = 2.2) and 0.2 μm in width (n = 8, $\bar{x}$ = 0.23 μm ± SD = 0.01).

Tridacna (Chametrachea) squamosa

Mature spermatozoa of *Tridacna squamosa* measure *c.* 54 μm (n = 10, $\bar{x}$ = 54.0 μm ± SD = 1.6) in length. The acrosomal vesicle is *c.* 0.5 μm long (n = 7, $\bar{x}$ = 0.45 μm ± SD = 0.03), and 0.6 μm wide (n = 8, $\bar{x}$ = 0.63 μm ± SD = 0.02). The nucleus is short, solid and rod shaped (length ≅ 3.7 μm; n = 5, $\bar{x}$ = 3.72 μm ± SD = 0.11). The top of the nucleus narrows into a raised peg which projects inside the invagination of the acrosome producing a nuclear peg (Fig. 1d). Posterior to the peg, the diameter of the nucleus ranges from *c.* 0.7 μm (n = 6, $\bar{x}$ = 0.68 μm ± SD = 0.05) immediately below the acrosome to 1.0 μm (n = 6, $\bar{x}$ = 1.00 μm ± SD = 0.06)

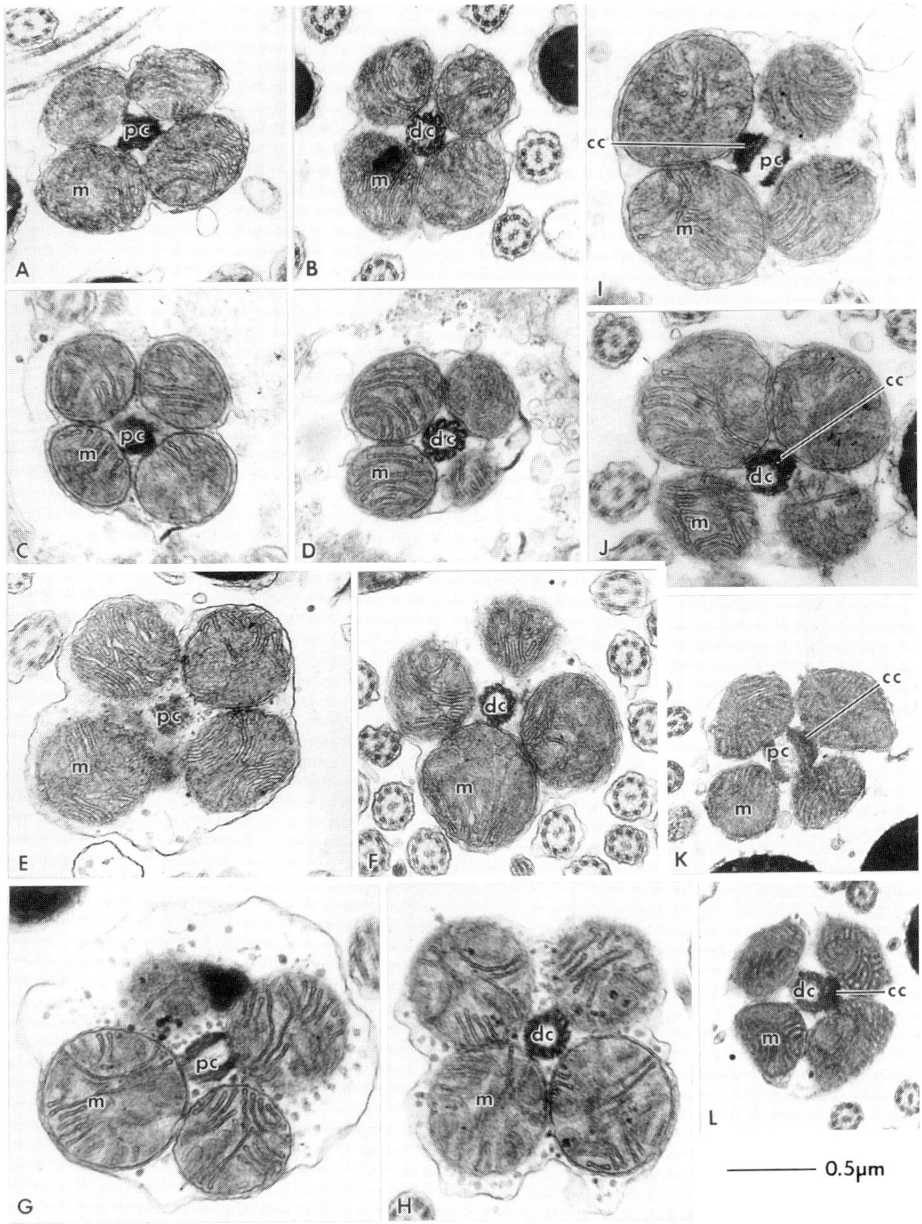

Fig. 2. Transmission electron micrographs (transverse sections) of the mitochondrial region showing proximal and distal centrioles and the presence or absence of a centriolar connective (possible rootlet) for: (**a**) and (**b**) *Tridacna (C.) maxima*; (**c**) and (**d**) *Tridacna (C.) crocea*; (**e**) and (**f**) *Tridacna (C.) squamosa*; (**g**) and (**h**) *Tridacna (T.) gigas*; (**i**) and (**j**) *Hippopus hippopus*; (**k**) and (**l**) *Hippopus porcellanus*. cc, Centriolar connective; dc, distal centriole; m, mitochondria; pc, proximal centriole. All to same scale as indicated.

approaching the mitochondrial region (Fig. 1d). The flagellum is *c*. 47 μm long ($n = 10$, $\bar{x} = 47.3$ μm ± SD = 1.6) and 0.3 μm in width ($n = 8$, $\bar{x} = 0.25$ μm ± SD = 0.01).

Tridacna (Tridacna) gigas

Mature spermatozoa of *Tridacna gigas* measure *c*. 56 μm ($n = 10$, $\bar{x} = 55.9$ μm ± SD = 2.2) in length.

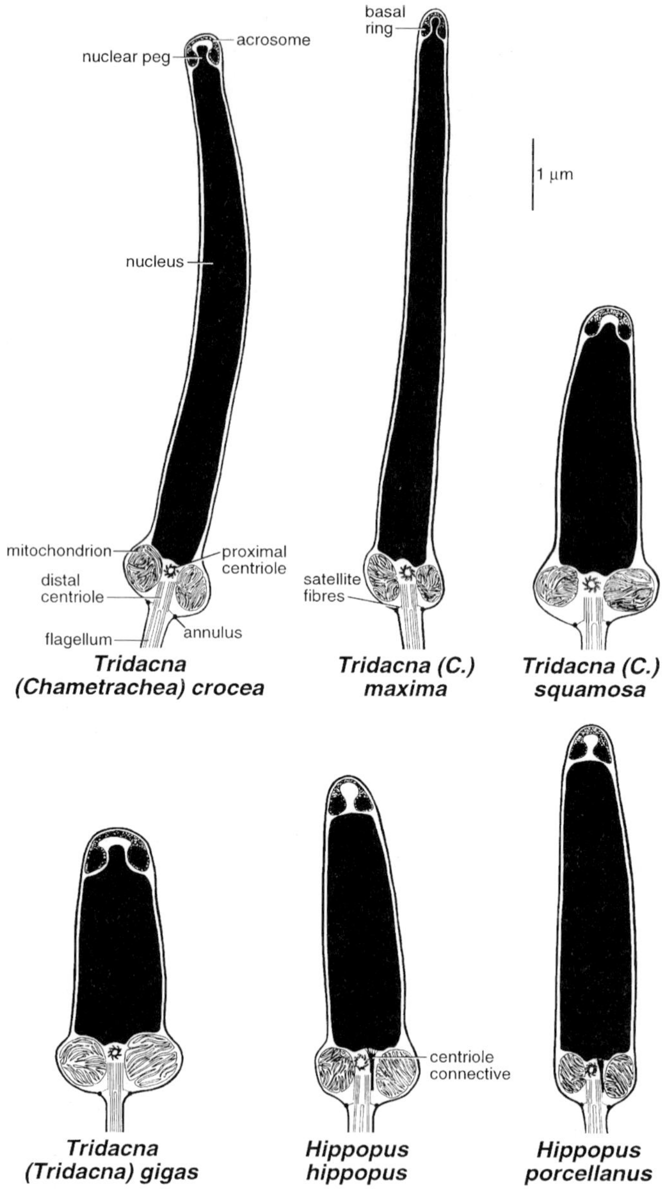

Fig. 3. Diagram summarizing comparative spermatozoa within the Tridacninae. Visual summary of longitudinal sections of the mature spermatozoa. All drawn to same scale as indicated. Outer mitochondrial membrane and nuclear membranes not shown.

The acrosomal vesicle is *c.* 0.6 μm long (n = 12, $\bar{x}$ = 0.59 μm ± SD = 0.02) and 0.8 μm wide (n = 12, $\bar{x}$ = 0.82 μm ± SD = 0.13) (Fig.1e). The nucleus is short, solid and rod shaped (length ≅ 3.0 μm; n = 7, $\bar{x}$ = 3.02 μm ± SD = 0.10). The top of the nucleus narrows into a raised peg which projects inside the invagination of the acrosome producing a nuclear peg (Fig. 1e). Posterior to the peg, the diameter of the nucleus ranges from *c.* 0.9 μm (n = 11, $\bar{x}$ = 0.87 μm ± SD = 0.05) immediately below the acrosome to 1.2 μm (n = 7, $\bar{x}$ = 1.16 μm ± SD = 0.05) approaching the mitochondrial region (Fig. 1e).

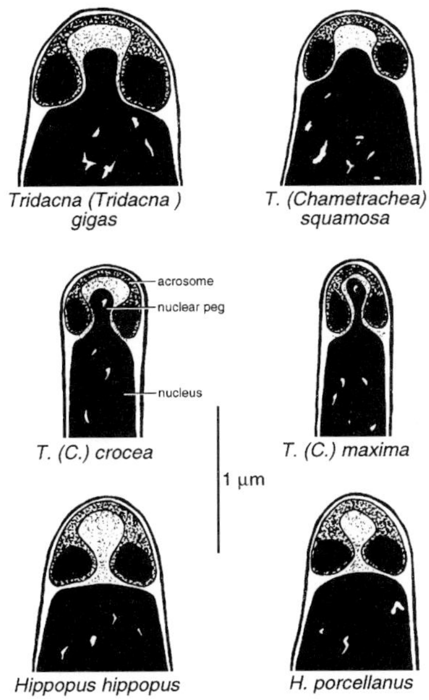

Fig. 4. Diagram comparing acrosomal and nuclear apex morphology within the Tridacninae. Note size difference between species and the presence of a nuclear peg in the *Tridacna* species and the absence of a nuclear peg in *Hippopus* species. All drawings to same scale as indicated. Nuclear membranes not shown.

The flagellum is *c.* 50 µm long (n = 10, $\bar{x}$ = 50.3 µm ± SD = 2.2) and 0.2 µm in width (n = 8, $\bar{x}$ = 0.23 µm ± SD = 0.01).

Hippopus hippopus

Mature spermatozoa of *Hippopus hippopus* measure *c.* 54 µm (n = 10, $\bar{x}$ = 53.9 µm ± SD = 2.6) in length. The acrosomal vesicle is *c.* 0.5 µm long (n = 10, $\bar{x}$ = 0.49 µm ± SD = 0.02) and 0.6 µm wide (n = 60, $\bar{x}$ = 0.63 µm ± SD = 0.03) (Fig.1f). The invagination is narrow posteriorly, expands into a wide cavity anteriorly and is occupied by sub-acrosomal material. The nucleus is solid (length ≅ 3.6 µm; n = 7, $\bar{x}$ = 3.57 µm ± SD = 0.09) and oblong-shaped with no evidence of a nuclear peg. The diameter of the nucleus ranges from *c.* 0.6 µm (n = 5, $\bar{x}$ = 0.62µm ± SD = 0.03) immediately below the acrosome to 1.1 µm (n = 5, $\bar{x}$ = 1.05 µm ± SD = 0.11) approaching the mitochondrial region (Fig. 1f). There is a well-developed centriolar connective (possible rootlet) lying lateral to, and in contact with, the centrioles (Fig. 1f). The flagellum is *c.* 47 µm long (n = 10, $\bar{x}$ = 46.9 µm ± SD = 2.6) and 0.3 µm in width (n = 8, $\bar{x}$ = 0.26 µm ± SD = 0.01).

Hippopus porcellanus

Mature spermatozoa of *Hippopus porcellanus* are *c.* 60.0 µm (n = 1) long. The acrosomal vesicle is *c.* 0.4 µm long (n = 7, $\bar{x}$ = 0.44 µm ± SD = 0.02) and 0.6 µm wide (n = 8, $\bar{x}$ = 0.60 µm ± SD = 0.02). The invagination is narrow posteriorly, expands into a wide cavity anteriorly and is occupied by subacrosomal material (Fig. 1g). The nucleus is solid (length ≅ 3.7 µm; n = 6, $\bar{x}$ = 3.73 µm ± SD = 0.21) and oblong shaped with no evidence of a nuclear peg. The diameter of the nucleus ranges from *c.* 0.6 µm (n = 8, $\bar{x}$ = 0.55 µm ± SD = 0.04) immediately below the acrosome to 0.9 µm (n = 6, $\bar{x}$ = 0.88 µm ± SD = 0.15) approaching the mitochondrial region (Fig. 1g). There is a well-developed centriolar connective (possible rootlet) lying lateral to, and in contact with, the centrioles (Fig. 2). The flagellum is *c.* 52.7µm long (n = 1) and 0.2 µm in width (n = 11, $\bar{x}$ = 0.21 µm ± SD = 0.01).

Discussion

Structural comparisons

Giant clam spermatozoa are all of the aquasperm type (*sensu* Jamieson 1987; Rouse & Jamieson 1987) and show several similarities to other bivalves [for a recent comparative review of bivalve sperm see Healy (1996*a*)]. In particular, tridacnid sperm resemble those of other heterodonts, especially the Cardiidae (Keys & Healy 1999; Fig. 5). This result is consistent with the traditional view of a close relationship between the Tridacnidae and Cardiidae, but also with recent moves to downgrade tridacnids to the status of a cardiid subfamily (Schneider 1992, 1995, 1998*a*, *b*; Schneider & O'Foighil 1999). Although the results of the present study support Schneider's decision, in general in the following discussion the vernacular names 'tridacnids' and 'cardiids' have been used in their traditional sense, partly for ease of reference and also to avoid ambiguity [the generic composition of various nominal subfamilies of the Cardiidae varies significantly between authors, as pointed out by Schneider (1992)].

The pronounced basal ring of the acrosomal vesicle in all examined tridacnids (Keys & Healy 1999; present study) (Fig. 4) occurs in other cardiids (Popham 1979; Sousa & Azevedo 1988; Sousa *et al.* 1995) (Fig. 5) and all other investigated heterodonts, and is particularly well-developed in the Veneroida (e.g. Longo & Anderson 1969; Gharagozlou-Van Ginnekin & Pochon-Masson 1971; Hylander & Summers 1977; Popham 1979; Hodgson *et al.* 1987, 1990; Healy 1995*a*, *b*, 1996*a*). Such a ring has also been demonstrated in

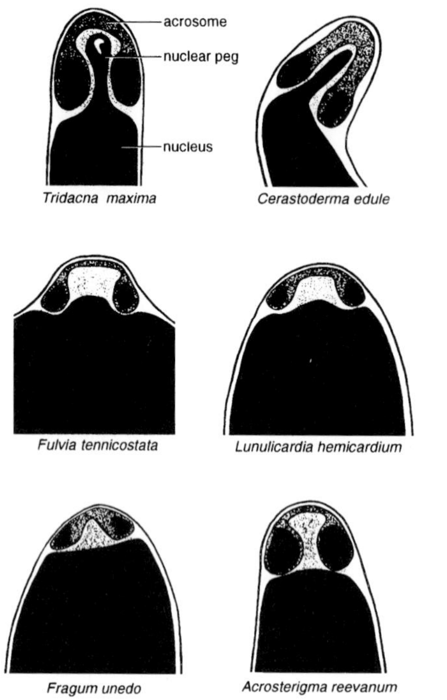

Fig. 5. Diagram comparing *T. maxima* acrosome and nuclear apex morphology with that of various species of Cardiidae. Note close similarity between *T. maxima* and *C. edule* (both species have a distinct nuclear peg projecting deeply into the acrosomal vesicle invagination). Nuclear membranes not shown. [(Copied from Keys & Healy 1999: sources of original data (micrographs) are: *Tridacna maxima* (Keys & Healy 1999), *C. edule* (Sousa & Azevedo 1988; Sousa *et al.* 1995), *F. tenuicostata* (Popham 1979), and *L. hemicardium*, *F. unedo* and *A. reeveanum* (Keys & Healy 1999)]

spermatozoa of certain pteriomorphians such as *Mytilus* spp. (Niijima & Dan 1965; Hodgson & Bernard 1986*a*, *b*) but, as demonstrated by Kafanov & Drozdov (1998), this feature is not even a consistent feature of the Mytiliidae (Mytiloidea). Although as yet uninvestigated, it is likely that the acrosomal complex of the Tridacnidae interacts with the egg in the manner described by Hylander & Summers (1977) for *Chama macerophylla* (Gmelin 1791) (Chamoidea) and *Spisula solidissima* (Dillwyn, 1817) (Mactroidea).

Within the Tridacnidae (Keys & Healy 1999; present study) and among other investigated members of the Cardiidae (Popham 1979; Azevedo 1988; Sousa *et al.* 1995; Keys & Healy 1999), there is substantial variation in the spatial relationship between the acrosomal complex and the nuclear apex (Figs 3 and 4). A pronounced nuclear peg occurs in all examined members of the genus *Tridacna* (see Fig. 4) and in at least one species of Cardiinae, *Cerastoderma edule* (Linnaeus, 1758) (Sousa & Azevedo 1988; Sousa *et al.* 1995) (Fig. 5). Interestingly, a broad, low nuclear peg is present in the cardiids *Fulvia tenuicostata* (Lamarck, 1819) and *Lunulicardia hemicardium* (Linnaeus, 1758) (poorly developed in the latter species) (Popham 1979; Keys & Healy 1999) (Fig. 5) but is absent in both species of *Hippopus* (Tridacnidae, present study) (Fig. 4) and in the cardiids *Fragum unedo* (Linnaeus, 1758) and *Acrosterigma reeveanum* (Dunker, 1852) (Keys & Healy 1999) (Fig. 5). The degree of similarity between the very fine nuclear peg of *Tridacna (Chametrachea) maxima* and *T. (C.) crocea* on the one hand and *Cerastoderma edule* on the other is difficult to judge because of the lack of detail in the micrograph presented by Sousa & Azevedo (1988) and Sousa *et al.* (1995). No apical swelling of the peg is evident in *C. edule* and unlike all tridacnids (but like certain other cardiids) the nucleus of this species shows evidence of helical coiling, at least anteriorly (Karpevich 1961). The reason as to why the peg should be attenuate, broad or even absent in different species of tridacnids and cardiids is unknown. A nuclear peg has not been reported in other bivalves, or even in other molluscs, although some species of the gastropod *Patella* show a narrowing of the nucleus which sits within the acrosomal invagination (Hodgson & Bernard 1988; Hodgson *et al.* 1996). A similar structure to the nuclear peg has been noted in the spermatozoa of hagfish (see Jespersen 1975; Morisawa 1999. Presumably the peg serves to support the acrosomal vesicle or, in the case of *T. (C.) maxima* and *T. (C.) crocea*, to support and anchor the acrosomal vesicle (in these two species the peg and acrosomal vesicle exhibit a ball-in-socket configuration). Clearly there is a need for comparative ultrastructural data on fertilization within the Cardioidea in order to shed light on the function of the nuclear peg.

Popham (1974) has found that externally fertilizing spermatozoa in the shipworms (Teredinidae) have a larger acrosomal vesicle and longer axial rod (= 'acrosomal rod', 'perforatorium') than those that fertilize 'internally' (i.e. within the mantle cavity after being drawn in by the siphonal currents). Both *T. (C.) maxima* and *T. (C.) crocea* not only have much smaller acrosomes than all other investigated tridacnids but also attenuate nuclei – a combination of features suggestive of mantle cavity fertilization (cf. with Galeommatoidea; O'Foighil 1985). Giant clams are simultaneous hermaphrodites and gonads contain both male and female follicles. In natural spawning situations clams will initially spawn sperm and later spawn eggs, if the eggs are mature (Lucas 1988; Nash *et al.* 1988). In aquaculture situations, fertilization has been shown to occur outside the mantle cavity (R. Braley, pers. comm.; J. Lucas, pers. comm.) and it is likely that this also occurs in the field. According to Franzén (1983) bivalves with an elongate sperm nucleus often have large eggs (diameter $> 90\ \mu m$) and often exhibit lecithotrophic or direct development (see also Drozdov & Kasyanov 1986). Within the Tridacnidae there does not appear to be any detectable correlation between nuclear length and egg size, although it is interesting to note that differences in the diameter of eggs of investigated

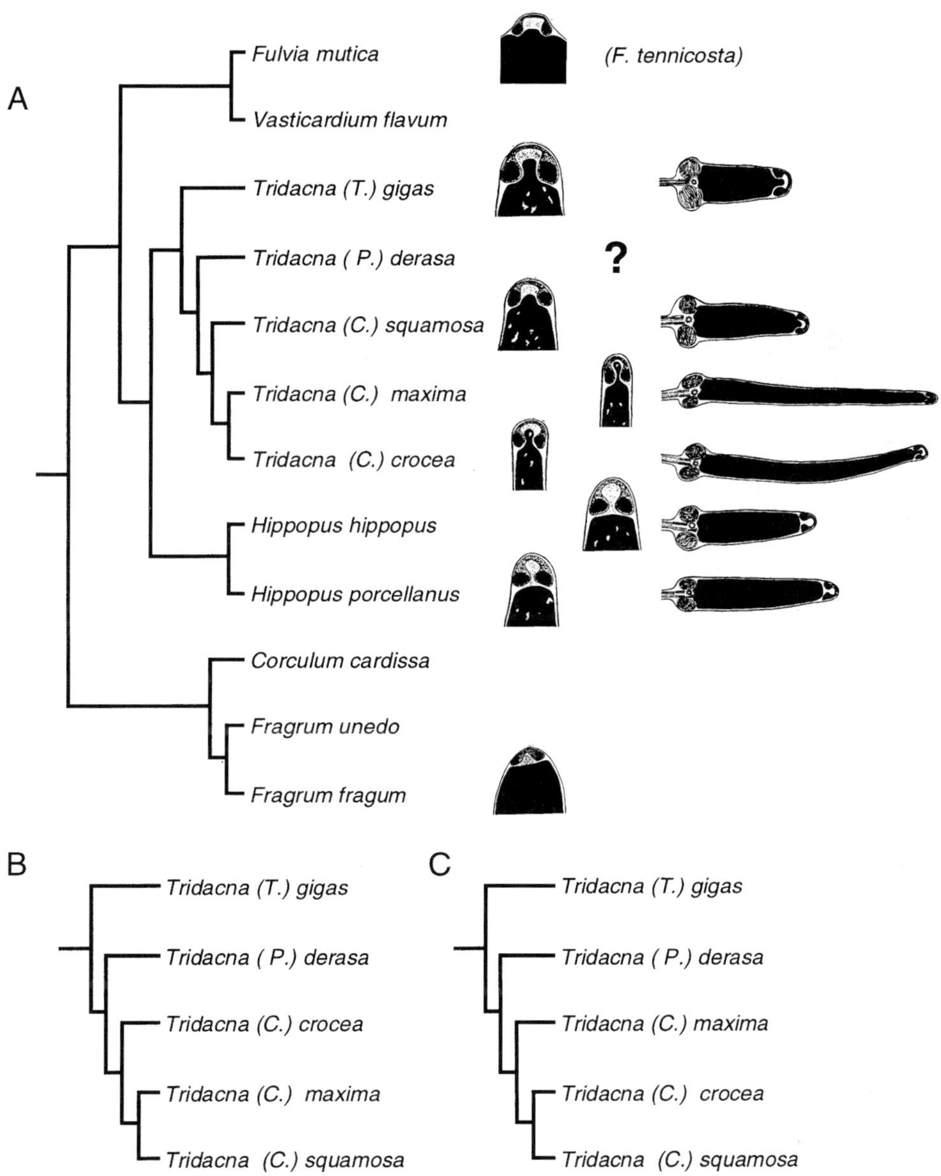

Fig. 6. (a) Results of the present study and those from Fig. 5 superimposed on consensus tree of Maruyama *et al.* (1998) derived from an analysis of 18S rDNA of zooxanthellate and azooxanthellate cardiodians (compared with other bivalves and a gastropod). (b), (c) Alternative topologies for *Tridacna* (also from Maruyama *et al.* 1998).

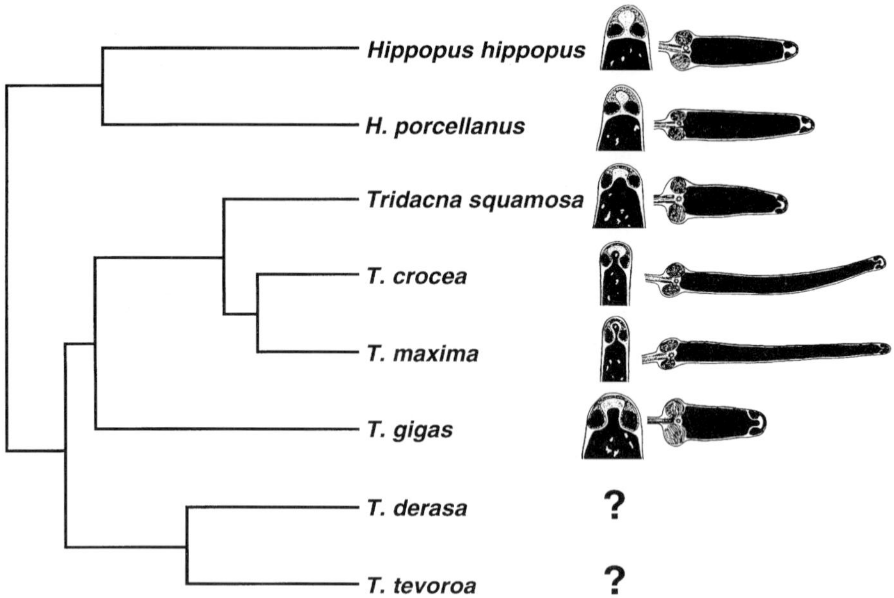

Fig. 7. Results of the present study superimposed on consensus tree of Benzie & Williams (1998) derived through cladistic study of protein allozymes.

Tridacna species (≅ 100 μm) and of both species of *Hippopus* (≅ 115 μm) (Kawaguti *et al.* 1982; Norton & Jones 1992) do correlate with the presence or absence of a nuclear peg.

In general, the morphology of the midpiece in all investigated tridacnids (including the internal substructure and arrangement of the mitochondria, positioning of centrioles, presence of a satellite fibre complex and annulus) and the morphology of the flagellum (with 9 + 2 microtubular pattern axoneme) is similar to that reported in most other heterodonts [for a summary see Healy (1996*a*, *b*)]. An interesting difference between the genera *Tridacna* and *Hippopus* is the presence, in the latter, of a centriolar connective (possible rootlet) lying lateral to the proximal and distal centrioles, and appearing to connect the two (see Figs 1–3). A centriolar rootlet has been observed in a number of heterodonts (Teredinidae, Veneridae) (Popham 1974; Popham *et el.* 1974; Reunov & Hodgson 1994) and pteriomorphians (Arcidae, Spondylidae, Ostreidae, Mytilidae) (Daniels *et al.* 1971; Healy & Lester 1991; Reunov & Hodgson 1994; Gwo *et al.* 1996), but in all of these cases the rootlet is attached to the adnuclear surface of the proximal centriole. No periodic banding was observed in the connective of *Hippopus* but it must be added that the presence or absence of such substructure varies considerably within the Bivalvia. In general, the function of the rootlet, like that of the satellite fibres complex, is believed to be for anchorage of the centrioles and the attached flagellum. It would be of great interest to determine why the spermatozoa of *Hippopus* have evolved a well-developed centriolar connective (possible rootlet) whereas those of *Tridacna* have not.

Taxonomic and phylogenetic considerations

In most recent literature, the giant clams are allocated to their own family, the Tridacnidae and allied with the Cardiidae, usually within a superfamily Cardioidea (Yonge 1936, 1975, 1980; Stasek 1962; Schneider 1992, 1995, 1998*b*). In view of their highly modified lifestyle they have, on occasions, also been deemed worthy of their own superfamily (Rosewater 1965; Keen 1969; Kafanov & Popov 1977; Boss 1982; Braley & Healy 1998). On the basis of a detailed cladistic analysis of shell and anatomical characters, Schneider (1992, 1998*b*) found that tridacnids formed a monophyletic group within the Cardiidae and should therefore be considered a subfamily of that family (thereby avoiding a paraphyletic Cardiidae). Questions about the relationship of the Tridacnidae to the Cardiidae and the possibility that tridacnids may constitute merely a subfamily of the Cardiidae were independently raised by Maruyama *et al.* (1998) based on their analysis of 18S rDNA of

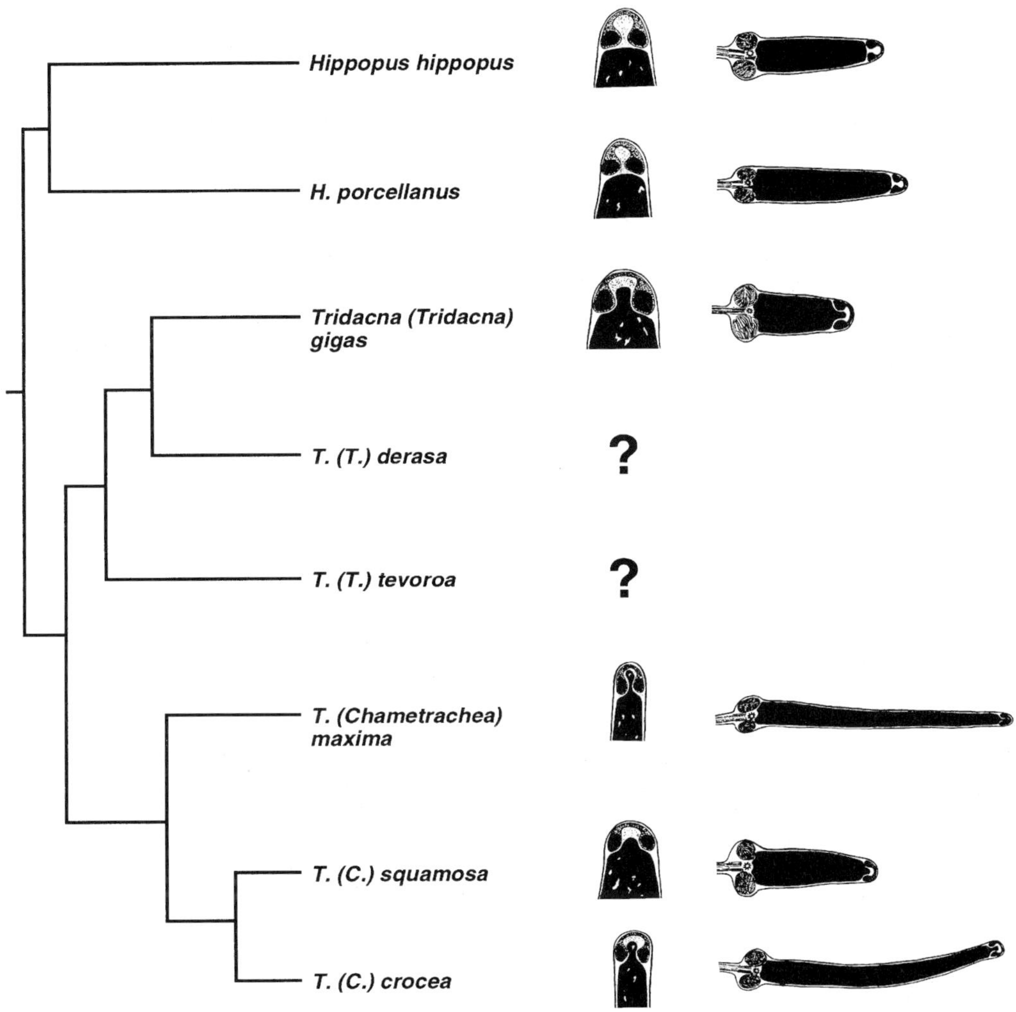

Fig. 8. Results of the present study superimposed on consensus tree of Schneider & O'Foighil (1999) based on molecular study (18S rDNA).

zooxanthellate and azooxanthellate cardioideans (compared with other bivalves and a gastropod).

Results of the present study and previous sperm ultrastructural accounts of the Cardioidea (Popham 1979; Sousa & Azevedo 1988; Sousa *et al.* 1995; Keys & Healy 1999) confirm the most widely held view that a very close relationship exists between tridacnids and cardiids. The view expressed by some authors that tridacnids are more closely allied to the Carditidae than the Cardiidae (Hedley 1921; McLean 1974; Allen 1985) is not supportable in the light of sperm ultrastructural evidence. Spermatozoa of the Carditidae exhibit a number of distinctive features not encountered in cardioideans, including an elongate acrosomal vesicle, a midpiece with seven to eight tightly packed mitochondria and transformation of the proximal centriole into a modified rootlet [features shared with the Crassatellidae; for comparative figures and further discussion see Healy (1995*b*, 1996*a*)].

Hippopus and *Tridacna* are considered by Stasek (1962) to have each evolved from a *Byssocardium*-like ancestor in the early Miocene. Of the two extant genera, *Hippopus* is usually accepted as the more primitive (Rosewater 1965; Yonge 1980), partly for anatomical reasons but also because its fossil record extends further back in time than any species of *Tridacna* (Stasek 1962; Schneider &

O'Foighil 1999). Molecular studies have unequivocally supported not only the division of giant clams into two genera (*Hippopus* and *Tridacna*) but also provided clues, albeit sometimes conflicting, about relationships within the group (Benzie & Williams 1998; Maruyama *et al.* 1998; Schneider & O'Foighil 1999) (see Figs 6–8).

The lack of a nuclear peg in *Hippopus* (present in *Tridacna*) combined with the presence of a centriolar rootlet (lacking in *Tridacna*) suggest either that these two genera arose from differing ancestral sources or that the sperm differences encountered are the result of a combination of secondary loss and/or reappearance of structures. The presence of a well-developed narrow nuclear peg in the cardiid *Cerastoderma edule* (Sousa & Azevedo 1988; Sousa *et al.* 1995), and of a more poorly developed peg or peg vestige in some other examined cardiids (*Fulvia tenuicostata* and *Lunulicardia hemicardium*) (Popham 1979; Keys & Healy 1999), indicate that the nuclear peg cannot be regarded as an autapomorphy of *Tridacna* (although the expanded-tip nuclear peg of *Chametrachea* may prove to be a autapomorphy of that subgenus). Indeed, the absence of a peg in *Hippopus* and in the cardiids *Fragum unedo* and *Acrosterigma reeveanum* (Keys & Healy 1999) makes it difficult to determine if 'presence of a peg' or 'absence of a peg' represents the plesiomorphic state for this character in the Tridacninae. If Yonge (1936) is correct in concluding that tridacnids arose from a *Cerastoderma*-like ancestor [a view endorsed by Schneider (1998*b*)], then the absence of a nuclear peg in *Hippopus* must be due to secondary loss (?genetic suppression or genuine loss). Clearly, sperm data for the many unstudied cardiid genera are required to establish basic defining features for each of the nominal subfamilies. Such information should then allow a more meaningful assessment of cardioidean classification at the subfamily level.

Most of the recently proposed phylogenies for the genus *Tridacna* agree with the traditional placement of species among three subgenera: *Tridacna sensu stricto, Persikima* and *Chametrachea* (Benzie & Williams 1998; Maruyama *et al.* 1998; Schneider 1998*b*) (see Figs 6 and 7). However, Schneider & O'Foighil have concluded, on the basis of 16S rDNA analyses, that *T. (Persikima) tevoroa* was the sister taxon to *T. (Tridacna) gigas* + *T. (Persikima) derasa,* thereby rendering *Persikima* paraphyletic (Fig. 8). They therefore abandoned the subgenus *Persikima* altogether and incorporated *T. tevoroa* and *T. derasa* within *Tridacna sensu stricto*. Previously Lucas *et al.* (1991) had made the suggestion, based on shell and anatomical details, that *T. tevoroa* may be transitional between *Hippopus* and *T. derasa*, which to some extent prompted Schneider & O'Foighil's re-evaluation of *Persikima*. Unfortunately, no sperm data are as yet available for either *T. (P.) tevoroa* or *T. (P.) derasa* to test the ideas expressed by Schneider & O'Foighil (1999) and Lucas *et al.* (1991).

Within the subgenus *Chametrachea*, *T. (Chametrachea) maxima* and *T. (C.) crocea* are distinguished from *T. (C.) squamosa* by their strongly attenuate nucleus (featuring a narrow, terminally expanded nuclear peg) and relatively small acrosome (Fig. 3). In contrast, and against expectation, *T. (C.) squamosa* shows acrosomal and nuclear dimensions very close to those obtained for *T. (Tridacna) gigas* (Fig. 3). Given the general agreement between authors regarding the truly monophyletic status of *Chametrachea* (Schneider 1992, 1998*b*; Benzie & Williams 1998; Maruyama *et al.* 1998; Schneider & O'Foighil 1999) (see Figs 6–8), the sperm differences between *T. (C.) squamosa* on the one hand and *T. (C.) maxima* + *T. (C.) crocea* on the other are very interesting, for they clearly support the results of Benzie & Williams (1998) (Fig. 7) and at least one of the alternative trees presented by Maruyama *et al.* (1998) (Fig. 6). Acceptance of Schneider & O'Foighil's (1999) tree (Fig. 8) and one of the alternative trees of Maruyama *et al.* (1998), in which *T.(C.) maxima* becomes the sister taxon to *T.(C.) squamosa* + *T.(C.) crocea*, would entail significant nuclear and acrosomal changes in *T.(C.) squamosa*.

We express our thanks to the directors and staff of the Heron Island and Orpheus Island Research Stations, and to Mr B. O'Kane, Ms N. Wilson and Mr D. Souter (Department of Zoology and Entomology, University of Queensland) for their assistance during the course of the fieldwork. In addition, we thank Mr I. Loch (Australian Museum, Sydney) for allowing us the chance to obtain formalin-fixed tissues from Hippopus porcellanus held in the museum collection. Mrs L. Daddow and Mr T. Gorringe (Department of Zoology and Entomology, University of Queensland) provided some assistance with aspects of electron microscopy and photography. We also thank Drs J. Schneider & D. O'Foighil for allowing us to see the manuscript of their paper on the molecular phylogeny of the Tridacniinae. The referees are also thanked for their constructive comments on the original text. This study was supported financially by a research grant and Senior Research Fellowship from the Australian Research Council (JH) and University of Queensland research grants (JH and JK).

References

ALLEN, J. A. 1985. The recent Bivalvia: their form and evolution. *In*: TRUEMAN, E. R. & CLARKE, M. R. (eds) *The Mollusca Volume 10: Evolution*. Academic Press, New York, 337–403.

BENZIE, J. A. H. & WILLIAMS, S. T. 1998. Phylogenetic

relationships among giant clam species (Mollusca: Tridacnidae) determined by protein electrophoresis. *Marine Biology,* **132**, 123–133.

BOSS, K. J. 1982. Mollusca. *In*: PARKER, S. P. (ed.) *Synopsis and Classification of Living Organisms, Volume 2.* McGraw-Hill, New York, 945–1166.

BRALEY, R. D. 1988. Reproductive condition and season of the giant clams *Tridacna gigas* and *T. derasa* utilising a gonad biopsy technique. *In*: COPLAND, J. W. & LUCAS, J. S. (eds) *Giant Clams in Asia and the Pacific.* Australian Centre for International Agricultural Research, Canberra, 98–104.

—— 1992. *The Giant Clam: Hatchery and Nursery Culture Manual, ACIAR Monograph, Volume 15.* Australian Centre for International Agricultural Research, Canberra, 144.

—— 1994. Reproduction in the Giant Clam *Tridacna gigas* and *T. derasa in situ* on the North-Central Great barrier Reef Australia, and Papua New Guinea. *Coral Reefs,* **3**, 221–227.

—— & HEALY, J. M. 1998. Superfamily Tridacnoidea. *In*: BEESLEY, P. L., ROSS, G. J. B. & WELLS, B. (eds) *Mollusca: The Southern Synthesis, Volume 1.* CSIRO Publishing, Melbourne, 332–336.

DADDOW, L. Y. M. 1986. An abbreviated method of double lead stain technique. *Journnal of Submicroscopic Cytology,* **18**, 221–224.

DANIELS, E., LONGWELL, A. C., MCNIFF, J. M. & WOLFGANG, R. W. 1971. Ultrastructure of the spermatozoa from the American oyster *Crassostrea virginica. Transactions of the American Microscopical Society,* **90,** 275–282.

DROZDOV, A. L. & KASYANOV, V. L. 1986. Size and form of gametes in marine bivalve mollusks. *The Soviet Journal of Marine Biology*, **11**, 204-212 [English translation of *Biologiya Morya*, 1985, **4**, 33–40].

FRANZÉN, Å. 1983. Ultrastructural studies of spermatozoa in three bivalve species with notes on evolution of elongated sperm nucleus in primitive spermatozoa. *Gamete Research,* **7**, 199–214.

GHARAGOZLOU-VAN GINNEKIN, I. D. & POCHON-MASSON, J. 1971. Étude comparative infrastructure du spermatozoïde chez les Palourdes de France. *Archives de Zoologie Expérimentale et Générale,* **112**, 805–817.

GWO, J.-C., LIOU, C.-H. & CHENG, C.-H. 1996. Ultrastructure of the spermatozoa of the Pacific oyster *Crassostrea gigas* (Mollusca, Bivalvia, Ostreidae). *Journal of Submicroscopic Cytology and Pathology,* **28**, 395–400.

HEALY, J. M. 1989. Spermiogenesis and spermatozoa in the relict bivalve genus *Neotrigonia*: relevance to trigonioid relationships, particularly Unionoidea. *Marine Biology,* **103**, 75–86.

—— 1995*a*. Comparative spermatozoal ultrastructure and its taxonomic and phylogenetic significance in the bivalve order Veneroida. *In*: JAMIESON, B. G. AUSIO, J. & JUSTINE, J.-L. (eds) *Advances in Spermatozoal Phylogeny and Taxonomy, Volume 166.* Mémoires du Muséum National D'Histoire Naturelle, Paris, 155–166.

—— 1995*b*. Sperm ultrastucture in the marine bivalve families Carditidae and Crassatellidae and its bearing on unification of the Crassatelloidea with the Carditoidea. *Zoologica Scripta,* **24**, 21–28.

—— 1996*a*. Molluscan sperm ultrastructure: correlation with taxonomic units within the Gastropoda, Cephalopoda and Bivalvia. *In*: TAYLOR, J. D. (ed.) *Origin and Evolutionary Radiation of the Mollusca* Oxford University Press, Oxford, 99–113.

—— 1996*b*. Spermatozoan ultrastructure in the trigonioid bivalve *Neotrigonia margaritacea* Lamarck (Mollusca): Comparison with other bivalves, especially Trigonioida and Unionoida. *Helgoländer Meeresuntersuchungen,* **50**, 259–264.

—— & LESTER, R. J. G. 1991. Sperm ultrastructure in the Australian oyster *Saccostrea commercialis* (Iredale and Roughley) (Bivalvia, Ostreidae). *Journal of Molluscan Studies,* **57**, 219–224.

HEDLEY, C. 1921. A revision of the Australian *Tridacna. Records of the Australian Museum,* **13,** 163–172.

HODGSON, A. N. & BERNARD, R. T. F. 1986*a*. Ultrastructure of the sperm and spermatogenesis of three species of Mytilidae (Mollusca, Bivalvia). *Gamete Research,* **15**, 123–135.

—— & —— 1986*b*. Observations on the ultrastructure of the spermatozoon of two mytilids from the south-west coast of England. *Journal of the Marine Biological Association of the United Kingdom,* **66**, 385–390.

—— & —— 1988. A comparison of the structure of the spermatozoa and spermatogenesis of 16 species of patellid limpet (Mollusca: Gastropoda: Archaeogastropoda). *Journal of Morphology,* **195**, 205–223.

——, —— & VAN DER HORST, G. 1990. Comparative spermatology of three species of *Donax* (Bivalvia) from South Africa. *Journal of Molluscan Studies,* **56**, 257–265.

——, DE VILLIERS, C. J. & BERNARD, R. T. F. 1987. Comparative spermatology of two morphologically similar species of *Solen* (Mollusca : Bivalvia). *South African Journal of Zoology,* **22**, 264–268.

——, RIDGWAY, S., BRANCH, G. M. & HAWKINS, S. J. 1996. Spermatozoan morphology of 19 species of prosobranch limpets (Patellogastropoda) with a discussion of patellid relationships. *Philosophical Transactions of the Royal Society of London, Series B*, **351**, 339–347.

HYLANDER, B. L. & SUMMERS, R. G. 1977. An ultrastructural analysis of the gametes and early fertilization in two bivalve molluscs *Chama macerophylla* and *Spisula solidissima* with special reference to gamete binding. *Cell and Tissue Research,* **182**, 469–489.

JAMIESON, B. G. M. 1987. A biological classification of sperm types, with special reference to annelids and molluscs, and an example of spermiocladistics. *In*: MOHRI, H. (ed.) *New Horizons of Sperm Cell Research,* Japanese Scientific Society Press, Tokyo, 311–332.

JESPERSEN, Å. 1975. Fine structure of spermiogenesis in eastern Pacific species of hagfish (Myxinidae). *Acta Zoologica (Stockholm),* **56**, 189–198.

KAFANOV, A. I. & DROZDOV, A. L. 1998. Comparative sperm morphology and phylogenetic classification of recent Mytiloidea (Bivalvia). *Malacologia,* **39**, 129–139.

—— & POPOV, S. V. 1977. On the system of the Cenozoic Cardioidea (Bivalvia). *Paleontological Journal,* **1977**, 307–314.

KARPEVICH, A. F. 1961. Adaptive character of spermatozoa and eggs' morphology in Bivalvia – Molluscs. *Zoologicheskii Zhurnal Akademiia Nauk SSSR,* **40**, 340–350.

KAWAGUTI, S., MASUDA, Y. & MURAKOSHI, M. 1982. Electron microscopic studies on the gonad of the boring clam, *Tridacna crocea. Bulletin of the Kawasaki College of Applied Health Professions,* **2**, 101–107.

KEEN, M. 1969. Superfamily Tridacnacea Lamarck, 1819. *In*: MOORE, R. C. (ed.) *Treatise on Invertebrate Paleontology. Part N. Mollusca 6. Volume 2. Bivalia.* Geological Society of America, Boulder, CO, and University of Kansas Press, Lawrence, KS, 594–595.

KEYS, J. L. & HEALY, J. M. 1999. Sperm ultrastructure in the giant clam *Tridacna maxima* (Tridacnidae: Bivalvia: Mollusca) from the Great Barrier Reef. *Marine Biology,* **135,** 41–46.

LONGO, F. J. & ANDERSON, E. 1969. Spermiogenesis in the surf clam *Spisula solidissima* with special reference to the formation of the acrosomal vesicle. *Journal of Ultrastructure Research,* **27**, 435–443.

LUCAS, J. S. 1988. Giant Clams: Description, distribution and life history. *In*: COPLAND, J. W. & LUCAS, J. S. (eds) *Giant Clams in Asia and the Pacific, Volume 9.* Australian Centre for International Agricultural Research, Canberra, 21–32.

——, LEDUA, E. & BRALEY, R. D. 1990. *A new species of gaint clam (Tridacnidae) from Fiji and Tonga.* Australian Centre for International Agricultural Research, Canberra.

——, —— & —— 1991. *Tridacna teveroa* Lucas, Ledua & Braley: A recently described Species of Giant Clam (Bivalvia; Tridacnidae) from Fiji and Tonga. *The Nautilus,* **105**, 93–103.

MCLEAN, R. A. 1974. A revision of the Pelecypod Family Tridacnidae. *Notulae Naturae of The Academy of Natural Sciences of Philadelphia,* **195**, 1–7.

MARUYAMA, T., ISHIKURA, M., YAMAZAKI, S. & KANAI, S. 1998. Molecular phylogeny of zooxanthellate bivalves. *Biological Bulletin,* **195**, 70–77.

MORISAWA, S. 1999. Acrosome reaction in spermatozoa of the hagfish *Eptatretus burgeri* (Agnatha). *Development, Growth and Differentiation,* **41**, 109–112.

NASH, W. J., PEARSON, R. G. & WESTMORE, S. P. 1988. A histological study of reproduction in the giant clam *Tridanca gigas* in the North-Central Great Barrier Reef. *In*: COPLAND, J. W. & LUCAS, J. S. (eds) *Giant Clams in Asia and the Pacific, Volume 9.* Australian Centre for International Agricultural Research, Canberra, 89–94.

NIIJIMA, L. & DAN, J. 1965. The acrosome reaction in *Mytilus edulis* I. Fine structure of the intact acrosome. *Journal of Cell Biology,* **25**, 243–248.

NORTON, J. H. & JONES, G. W. 1992. *The Giant Clam: an anatomical and histological atlas, ACIAR Monograph.* Australian Centre for international Agricultural Research, Canberra, 142.

O'FOIGHIL, D. 1985. Sperm transfer and storage in the brooding bivalve *Mysella tumida. Biological Bulletin,* **169**, 602–614.

POPHAM, J. D. 1974. Comparative morphometrics of the acrosomes of the sperms of 'externally' and 'internally' fertilizing sperms of the shipworms (Teredinidae, Bivalvia, Mollusca). *Cell and Tissue Research,* **150**, 291–297.

—— 1979. Comparative spermatozoon morphology and bivalve phylogeny. *Malacological Review,* **12**, 1–20.

——, DICKSON, M. R. & GODDARD, C. K. 1974. Ultrastructural study of the mature gametes of two species of *Bankia* (Mollusca, Teredinidae). *Australian Journal of Zoology,* **22**, 1–12.

PURCHON, R. D. 1987*a*. The stomach in the Bivalvia. *Philosophical Transactions of the Royal Society of London, Series B,* **316**, 183–276.

—— 1987*b*. Classification and evolution of the Bivalvia: An analytical study. *Philosophical Transactions of the Royal Society of London, Series B,* **316**, 277–302.

REUNOV, A. A. & HODGSON, A. N. 1994. Ultrastructure of the spermatozoa of five species of Southern African bivalves (Mollusca), and an examination of early spermatogenesis. *Journal of Morphology,* **219**, 275–283.

ROSEWATER, J. 1965. The family Tridacnidae in the Indo-Pacific. *Indo-Pacific Molluca,* **1**, 347–396.

—— 1982. A New Species of *Hippopus* (Bivalvia: Tridacnidae). *The Nautilus,* **96,** 3–6.

ROUSE, G. W. & JAMIESON, B. G. M. 1987. An ultrastructural study of the spermatozoa of the polychaetes *Eurythoe complanata* (Amphinomidae), *Clymenella laseroni* and *Micromaldane laseroni* (Maldanidae) with definition of sperm types in relation to reproductive biology. *Journal of Submicroscopic Cytology,* **19**, 573–584.

SCHNEIDER, J. A. 1992. Preliminary cladistic analysis of the bivalve family Cardiidae. *American Malacological Bulletin,* **9**, 145–155.

—— 1995. Phylogeny of the Cardiidae (Mollusca, Bivalvia): Protocardiinae, Laevicardiinae, Lahilliinae, Tulongocardiinae subfam. n. and Pleuriocardiinae subfam. n. *Zoologica Scripta,* **24**, 321–346.

—— 1998*a*. Phylogeny of stem-group eucardiids (Bivalvia: Cardiidae) and the significance of the transitional fossil *Perucardia. Malacologia,* **40**, 37–62.

—— 1998*b*. Phylogeny of the Cardiidae (Bivalvia): Phylogenetic relationships and morphological evolution within the subfamilies Clinocardiinae, Lymnocardiinae, Fraginae and Tridacninae. *Malacologia,* **40**, 321–373.

—— & O'FOIGHIL, D. 1999. Phylogeny of giant clams (Cardiidae: Tridacninae) based on partial mitochondrial 16S rDNA gene sequences. *Molecular Phylogenetics and Evolution,* **13**, 59–66.

SHELLY, C. C. & SOUTHGATE, P. C. 1988. Reproductive

periodicity and morphometry of *Hippopus hippopus* and *Tridacna crocea*. *In*: COPLAND, J. W. & LUCAS, J. S. (eds) *Giant Clams in Asia and the Pacific, Volume 9.* Australian Centre for International Agricultural Research, Canberra, 86–89.

SIRENKO, B. I. & SCARLATO, O. A. 1991. *Tridacna rosewateri* sp. n. A new species of giant clam from the Indian Ocean. *La Conchiglia,* **22**, 4–9.

SOUSA, M. & AZEVEDO, C. 1988. Comparative silver staining analysis on spermatozoa of various invertebrate species. *International Journal of Invertebrate Reproduction and Development,* **13**, 1–8.

——, OLIVEIRA, E. & OLIVEIRA, V. 1995. Comparative silver staining of molluscan spermatozoa. *Mémoires du Muséum National D'Histoire Naturelle,* **166**, 179–187.

STASEK, C. R. 1962. The form, growth and evolution of the Tridacnisae (giant clams). *Archives de Zoologie Expérimentale et Générale,* **101**, 1–40.

YONGE, C. M. 1932. Giant Clams and Burrowing Clams. *Natural History,* **32**, 244–251.

—— 1936. *Mode of life, feeding, digestion and symbiosis with zooxanthellae in the Tridacnidae.* Great Barrier Reef Expedition 1928–29 Scientific Reports, **1**(11), 283–321.

—— 1975. Giant Clams. *Scientific American,* **232**, 96–105.

—— 1980. Functional morphology and evolution in the Tridacnidae (Mollusca: Bivalvia: Cardiacea). *Records of the Australian Museum,* **33**, 735–777.

Functional anatomy, chemosymbiosis and evolution of the Lucinidae

JOHN D. TAYLOR & EMILY A. GLOVER

Department of Zoology, The Natural History Museum, London SW7 5BD, UK
(e-mail: j.taylor@nhm.ac.uk)

Abstract: All Lucinidae species studied so far possess sulphide-oxidizing, chemosymbiotic bacteria housed in bacteriocytes of gill filaments. The ecology, functional anatomy and evolution of the Lucinidae must be considered in relation to this symbiosis. The ctenidia have been extensively studied but other anatomical structures peculiar to lucinids have received much less attention. Reviewed are the morphological diversity of living lucinids, highlighting features of their anatomy including ctenidia, pallial apertures, anterior adductor muscle, pallial blood vessel and mantle gills. The latter are much more complex than previously understood and are here redescribed. They comprise folded structures located near the anterior adductor muscle in *Codakia*, *Phacoides* and *Lucina*, and on the septum of *Anodontia*. These are interpreted as secondary respiratory surfaces, their location enabling the separation of the anterior inflow of oxygenated water from sulphide-containing water. The latter is released from the sediment by the probing activities of the highly extensible foot and is pumped over the gill through the pedal gape and perhaps also via the exhalant tube. The shell features of *Ilionia* from the Silurian Period suggests that the lucinid chemosymbiosis is an ancient association.

The discovery of chemoautotrophic endosymbioses between bivalve molluscs and sulphur-oxidizing bacteria has resulted in a burgeoning interest in the biology of some previously rather neglected bivalve groups (Reid 1990; Distel 1998). Of the five extant families in which the chemosymbiosis has been demonstrated, the Lucinidae are by far the most diverse, are geographically the most widespread, live in the greatest variety of marine habitats, and have the longest and richest fossil history. Despite lucinids being capable of particle feeding, carbon isotope ($\delta^{13}C$) measurements from a few species demonstrate that a considerable proportion of the hosts' nutrition is derived from the intracellular, bacterial symbionts (Cary *et al.* 1989; Le Pennec *et al.* 1995). The ubiquity of the symbiosis in lucinids and their nutritional dependence upon it (Fisher 1990; Reid 1990) demands that any consideration of evolutionary history of the Lucinidae should take into account the importance of adaptations to this association.

In this paper, general aspects of the biology of Lucinidae are reviewed and it is then shown how major features of their anatomy and, in particular, mantle respiratory structures, are related to the symbiosis. Finally, it is considered how morphological evidence from the shells of fossil lucinids may provide evidence for estimating the age of the symbiosis. Although four families of living bivalves (Lucinidae, Fimbriidae, Thyasiridae and Ungulinidae) are assigned to the Lucinoidea, the phylogenetic relationships between them are rather uncertain, and in this review only the Lucinidae and Fimbriidae are considered. Despite the biological interest in the family there is no recent phylogenetic framework, using either morphological or molecular characters, on which to test hypotheses concerning evolution of symbiosis within the group, the latest being the phenetic study of North American taxa by Bretsky (1970, 1976). Additionally, the systematic revision of the whole family (Chavan 1969) is based on rather poorly defined shell characters and there is much confusion concerning the definition of subfamilies and genera. It is also clear that Recent lucinids are much more diverse than has been previously recognized and there are many undescribed taxa, particularly from the tropics (Glover & Taylor 1997; Taylor & Glover 1997*a*).

The chemosymbiosis

Bacteria [gamma subdivision of Proteobacteria (Durand *et al.* 1996; Distel 1998)] have now been reported from bacteriocytes in modified gill filaments of at least 30 species of Lucinidae, representing 18 different genera from several distinct clades (Table 1) and inhabiting a range of habitats from tropical to temperate latitudes. From this it may be a reasonable extrapolation that the

From: HARPER, E. M., TAYLOR, J. D. & CRAME, J. A. (eds) *The Evolutionary Biology of the Bivalvia*. Geological Society, London, Special Publications, **177**, 207–225. 1-86239-076-2/00/$15.00

Table 1. *Genera and species of Lucinidae and Fimbriidae confirmed with bacteria in bacteriocytes*

Anodontia alba (Link, 1807) – Giere (1985) (as *A. philippiana*)
A. edentula (Linnaeus, 1758) – Janssen (1992)
A. philippiana (Reeve, 1850) – Taylor & Glover unpublished
A. omissa (Iredale, 1930) – Taylor & Glover unpublished
Callucina radians (Conrad, 1841) – Schweimanns & Felbeck (1985), Giere (1985)
Cardiolucina semperiana (Issel, 1871) – Taylor & Glover (1997a)
C. australopilula Taylor & Glover 1997 – Taylor & Glover (1997*a*)
Codakia orbicularis (Linnaeus, 1758) – Gros *et al.* (1998*a*)
C. tigerina (Linnaeus, 1758) – Janssen (1992), Taylor & Glover unpublished
C. rugifera (Reeve, 1850) – Taylor & Glover unpublished
Ctena orbiculata (Montagu, 1808) – Berg *et al.* (1982)
Myrtea spinifera (Montagu, 1803) – Southward (1986), Dando *et al.* (1985)
Loripes lucinalis (Lamarck, 1818) – Southward (1986), Herry *et al.* (1988)
Lucina pensylvanica (Linnaeus, 1758) – Gros *et al.* (1996)
Lucinella divaricata (Linnaeus, 1758) – Herry & Le Pennec (1987)
Lucinisca nassula (Conrad, 1846) – Berg & Alatalo (1984)
Lucinoma annulata (Reeve, 1850) – Distel & Felbeck (1987)
L. aequizonata (Stearns, 1890) – Distel & Felbeck (1987), Cary *et al.* (1989)
L. atlantis (McLean, 1936) – Kennicutt *et al.* (1985)
L. borealis (Linnaeus, 1758) – Southward (1986)
Parvilucina multilineata (Toumey & Holmes, 1857) – Giere (1985)
P. tenuisculpta (Carpenter, 1864) – Reid & Brand (1986)
'Parvilucina' costata (d'Orbigny, 1842) – Giere (1985)
Phacoides pectinatus (Gmelin, 1791) – Frenkiel *et al.* (1996)
Pillucina fischeriana (Issel, 1871) – Taylor & Glover unpublished
P. pisidium (Dunker, 1860) – Kiiashko *et al.* (1989), Rodionov & Yushin (1991)
Rasta thiophila Taylor & Glover, 1997 – Taylor & Glover (1997b)
Stewartia floridana (Conrad, 1833) – Fisher & Hand (1984), Distel & Felbeck (1987)
Wallucina assimilis (Angas, 1867) —Barnes & Hickman (1999), Taylor & Glover unpublished
Fimbria fimbriata (Linnaeus, 1758) – Janssen (1992), Taylor & Glover unpublished

This list extends the previous compilations of Reid (1990), Fisher (1990) and Anderson (1995). Some names have been updated from the original references.

symbiosis is present in most, if not all, living taxa of Lucinidae. Indeed, the association has been regarded as obligate (Reid 1990; Anderson 1995). Bacteria have also been recorded from the gills of *Fimbria fimbriata* (Janssen 1992; pers. obs.). Additionally, some, but not all, species of Thyasiridae contain chemosymbiotic bacteria in gill filament bacteriocytes (Southward 1986; Reid 1990). Although only a few species have yet been studied, ribsomal DNA (rDNA) sequences show that the symbionts are not necessarily unique to each lucinid species. For example, *Codakia orbicularis* and *Lucina pensylvanica* host the same symbiont, but this differs from that associated with *Phacoides pectinatus* (Durand *et al.* 1996).

Experiments with *Codakia orbicularis* have shown that the bacterial symbionts are transmitted from the seagrass sediment to the newly settled juveniles of the host lucinid (Gros *et al.* 1998*a*). The bacteria enter by endocytosis at the apical poles of undifferentiated cells in the gill filaments which then differentiate into bacteriocytes. Juvenile *Codakia* with symbiotic bacteria are twice as large as aposymbiotic individuals grown in the same conditions. Molecular evidence also indicates environmental cyclical transmission of bacteria in other lucinid species, including *Lucinoma aequizonata* (Gros *et al.* 1998*b*, 1999). There is no evidence for vertical transmission of bacteria through gametes as found in *Solemya* (Kreuger *et al.* 1996) and *Calyptogena* (Cary & Giovannoni 1993).

Habitat and mode of life of Lucinidae

Lucinids, particularly species of *Codakia*, *Ctena*, *Lucina*, *Lucinisca*, *Loripes* and *Anodontia*, are common in shallow seagrass beds (*Thalassia* especially) of the tropical Atlantic and Indo-Pacific Oceans and are often the numerically dominant bivalves in these habitats (Allen 1958; Moore *et al.* 1968; Stanley 1970; Taylor & Lewis 1970; Jackson 1973). Lucinids are also present in seagrass at temperate latitudes, e.g. *Wallucina assimilis* around southern Australia (Barnes & Hickman 1999) and *Lucinella divaricata* in the northeastern Atlantic (Herry & Le Pennec 1987). *Ctena*, *Divaricella* and *Divalinga* species often occur in unvegetated sands of reef habitats (Allen 1958; pers. obs.). Some lucinids are abundant in organic-rich sediments around mangroves, such as *Phacoides pectinatus* (Frenkiel *et al.* 1996), *Anodontia philippiana*, *Austriella corrugata* (Deshayes, 1843) and *Pillucina* spp. (pers. obs.). Species from such habitats often show considerable anaerobic capabilities (Anderson 1995). *Lucinoma borealis* and *Loripes lucinalis* occur in intertidal mud and sands in the northeastern Atlantic (Dando *et al.* 1986; Southward 1986; Herry *et al.* 1988). Other *Lucinoma* species occur in offshore continental shelf and slope habitats (Cary *et al.* 1989; Okutani & Hashimoto 1997), around cold seeps (Callender & Powell 1997) and in oxygen-minimum zones (Bottjer *et al.* 1995; P. G. Oliver pers. comm.). Diverse assemblages of previously unknown species and genera of lucinids have been recovered from deep-water (*c.* 1000 m) sediments with a high content of decaying vegetation off Indonesia and the Philippines (Von Cosel & Bouchet pers. comm.). However, *Fimbria fimbriata* lives in clean, coralline sands (Morton 1979) and the present authors have found *Codakia rugifera* in New South Wales living on their sides exposed in permanent intertidal rock pools with decaying algae.

The symbiont bacteria utilize reduced sulphur as an energy source and require access to dissolved sulphide from interstitial water. Although some lucinids are associated with sediments with a high free-sulphide content (Fisher & Hand 1984), others live in habitats where the sulphide concentrations are very low (Dando *et al.* 1985; Cary *et al.* 1989), and various alternative pathways have been proposed for the sulphide oxidation. Cary *et al.* (1989) suggested that *Lucinoma aequizonata* used foot probing to exploit isolated pockets of sulphurous mud. Another mechanism proposed by Dando *et al.* (1994) for *Lucinoma borealis*, is the mining through oxidation of insoluble, sediment-bound, iron sulphides by oxygenated water drawn down the anterior inhalant tube of the bivalve. Details of sulphur acquisition strategies are available for very few species but lucinids exhibit such a diversity of morphologies and live in such a wide variety of habitats that it is likely that they utilize a number of different chemical pathways and behaviours to acquire reduced sulphides from the environment.

Many lucinids burrow deeply and live near the interface of aerobic and anaerobic zones, or even within the latter. Most live vertically, hinge uppermost, in the sediment and maintain an anterior, mucus-lined inhalant connection to the surface which is formed by the highly extensible cylindrical foot. Observations of *Lucinoma borealis* and *Myrtea spinifera* show that the position of the anterior tube may be shifted every few days (Dando *et al.* 1985, 1986). Lucinids also probe down with the foot into the reducing layers and Stanley (1970), using X-rays of *Codakia orbicularis* and *Phacoides pectinatus* in their life positions, demonstrated that in addition to the anterior inhalant tube there were long, narrow, tube-like probings radiating downwards from the ventral margin of shell (see also Turner 1985, fig. 3). Similar tunnels have been reported for *Lucinoma* and *Myrtea* (Dando *et al.* 1985) and Cary *et al.* (1989) recorded that *Lucinoma aequizonata* uses the vermiform foot to penetrate up to three times its body length down into the sediment. Four to five tunnels were formed each week which lasted for periods of ten to 15 days with the foot remaining extended in the probings for prolonged periods of time. Taylor & Glover (1997*b*) suggested that the long, periostracal pipes radiating from the ventral margin of the shell in *Rasta thiophila* were more permanent conduits for the channelling of interstitial water into the mantle cavity. By contrast, *Lucina pensylvanica* has a different life position and Stanley's (1970, pl. 18) X-rays show the anterior end of the shell uppermost, with a more or less vertical anterior tube to the sediment surface but with very long traces down into the sediment below the bivalve. Stanley (1970) thought the posterior tunnels were made by the posterior exhalant tube but they are more likely to have been produced by the foot. *Fimbria fimbriata* lives shallowly buried with the anterior end close to the surface and there is no anterior tube formed by the foot, the anterior inflow of water is directed from the surface around the mantle edge (Morton 1979).

Functional anatomy of Lucinidae

It has long been recognized that Lucinidae have many unusual anatomical features when compared with other bivalves but before the discovery of the chemosymbiosis these were largely unexplained or misunderstood. In an early account, Duvernoy (1854) clearly identified the unusual gill, the reduced labial palps and the mantle plications as rather peculiar features of lucinids. The most comprehensive anatomical description of Lucinidae is that of Allen (1958), who described the functional anatomy of nine species, along with species of Thyasiridae and Ungulinidae. Otherwise, a few species were briefly described by Pelseneer (1911) and more recently others been described in more detail, including Allen & Turner (1970) and Morton (1979) for *Fimbria fimbriata*, Narchi & Farani Assis (1980) for *Phacoides pectinatus*, Oliver (1986) for *Keletistes rhizoecus*, Reid & Brand (1986) for *Parvilucina tenuisculpta* and Taylor & Glover (1997*b*) for *Rasta thiophila*. Additionally, over the last 15 years there have been many

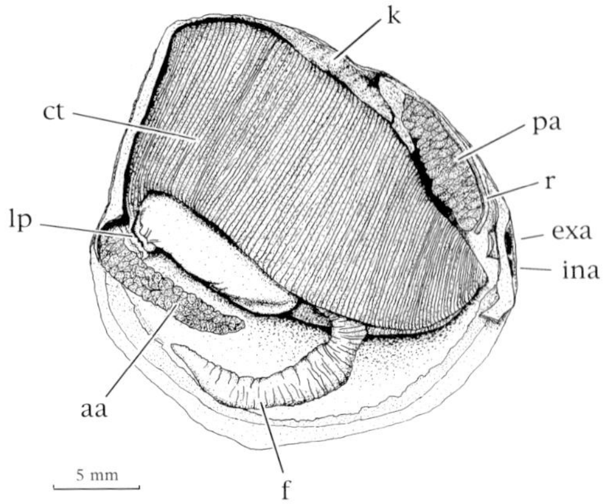

Fig. 1. *Rasta thiophila* Taylor & Glover, 1997. Houtman Abrolhos Islands, Western Australia. General anatomy with left mantle removed. aa, Anterior adductor muscle; ct, ctenidia; exa, exhalant aperture; f, foot; ina, inhalant aperture; k, kidney; lp, labial palps; pa, posterior adductor; r, rectum.

detailed studies of gill morphology and ultrastructure (see below).

Major features of the anatomy (Fig. 1) which are characteristic of Lucinidae are: an anteriorly extended body; an often elongate anterior adductor muscle with a degree of separation from the line of pallial attachment; an anterior inhalant tube formed by the foot; anterior–posterior water flow in the mantle cavity; a highly extensible vermiform foot; posterior exhalant siphon and usually a posterior 'inhalant' aperture; often a plicated inner mantle surface near the anterior adductor muscle – the mantle gills; thick and large inner demibranchs of ctenidia only, with the outer demibranchs absent; specialized bacteriocytes in gill filaments; very small labial palps; simple stomach and short gut; and a large pallial blood vessel which runs diagonally from near the ventral tip of the anterior adductor muscle to the heart.

The simplified, but thick and large, gills, the highly reduced labial palps, the simple stomach and the short gut are all features associated with the nutritional dependence of the bivalves on chemosymbiosis. Similar adaptations occur in other chemosymbiotic bivalves, e.g. *Solemya*, *Calyptogena* and *Bathymodiolus* (Fisher 1990; Reid 1990). However, the significance of some other lucinid features is less clear and these are reviewed in this paper; their function is then considered in relation to the chemosymbiotic lifestyle of the Lucinidae. Several of these distinctive anatomical features (elongate anterior adductor muscle, mantle gills and pallial blood vessel) leave traces on the insides of the shells (Fig. 2) and these can be recognized in fossil lucinids back to the Silurian Period.

Ctenidia and bacteriocytes

The structure of the gills housing the bacteria has been much studied and fine structural details are now available for species of several genera, for example: Gros *et al.* (1996) for *Lucina* (as *Linga*) *pensylvanica*; Reid & Brand (1986) for *Parvilucina tenuisculpta*; Frenkiel & Mouëza (1995) for *Codakia orbicularis*; Distel & Felbeck (1987) for *Lucinoma aequizonata;* Distel & Felbeck (1987) and Frankiel *et al.* 1996) for *Phacoides* (as *Lucina*) *pectinatus;* Herry & Le Pennec (1987) for *Lucinella divaricata;* Herry *et al.* (1988) and

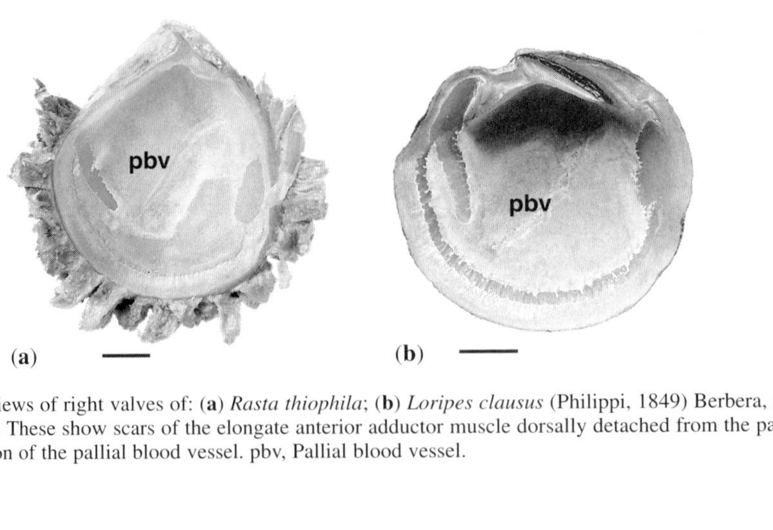

Fig. 2. Internal views of right valves of: (**a**) *Rasta thiophila*; (**b**) *Loripes clausus* (Philippi, 1849) Berbera, Somalia BMNH 1963341. These show scars of the elongate anterior adductor muscle dorsally detached from the pallial line and the impression of the pallial blood vessel. pbv, Pallial blood vessel.

Southward (1986) for *Loripes lucinalis* and other references cited in Table 1.

Although the detailed gill structure varies between genera, the basic plan is similar in all lucinids examined. The ctenidia consist of inner demibranchs only, which are large and thick compared with most other heterodont bivalves. The ctenidia of *Fimbria* are relatively less voluminous and thinner than those of lucinids. All lucinids possess a shallow, ventral food groove. In section (Figs 3 and 4), individual filaments comprise a narrow ciliated zone with frontal, eulatero-frontal cirri and lateral cilia similar to those of other bivalves. Inwards of this is an intermediary zone of several large cells [transition zone of Distel & Felbeck (1987); storage epithelium of Reid & Brand (1986)], followed by the thick bacteriocyte zone of subfilamental tissue consisting of sheets of

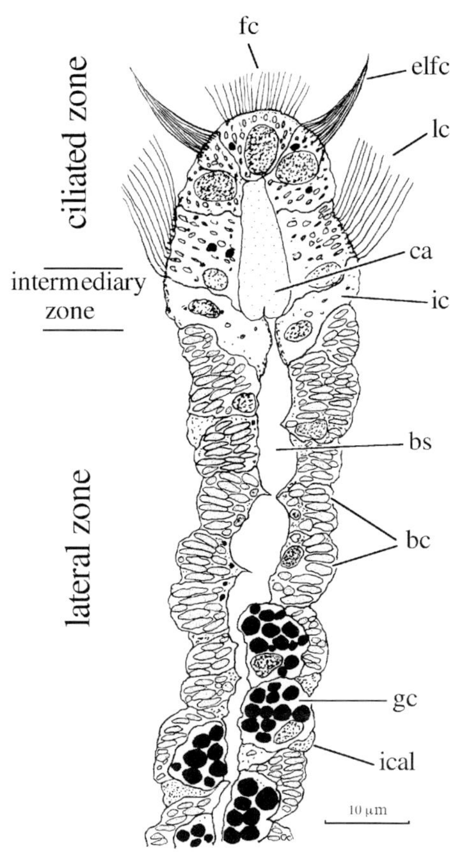

Fig. 3. Generalized diagram of part of a single gill filament of *Codakia orbicularis*. Based on TEM figure of Frenkiel & Mouëza (1995, fig. 4). bc, Bacteriocyte; bs, blood space; ca, collagen axis of gill filament; elfc, eulatero-frontal cilia; fc, frontal cilia; gc, granule cell; ic, intermediary cell; ical, intercalary cell; lc, lateral cilia.

domed cells separated by narrow, central, blood space. The bacteriocytes are packed with bacteria which are usually rod shaped and contained in membrane-bound vesicles. Often the bacteria are aligned with their long axes normal to the microvilli-covered, epithelial surface (Fig. 4b). Narrow intercalary cells separate the bacteriocytes and lateral extensions from them often overlap the outer surface of the bacteriocyte cells (Figs 3 and 4c). Occasional mucocytes occur between the bacteriocytes.

The function and significance of the intercalary cells is unknown but has been the subject of some speculation. The intercalary cells are usually narrow at the base and broaden distally into lateral extensions. These often overlap the apices of the bacteriocyte cells to form a partial 'cover' (Fig. 4c) which has led to suggestions that they may function to periodically restrict the contact of the bacteriocytes with seawater within the mantle cavity (Distel & Felbeck 1987; but see Frenkiel & Mouëza 1995).

Although the structure of the gill filaments has been well described there are many outstanding problems concerning the symbiosis. Particularly intriguing is the variety of granules seen in cells of the bacteriocyte zone of different lucinids (e. g. Reid & Brand 1986; Frenkiel & Mouëza 1995; Anderson 1995; Frenkiel *et al.* 1996) and the functional significance of these is, as yet, unresolved. Remarkable are the accumulations of rounded granules (2–5 µm in diameter) which fill cells within the bacteriocyte zone (Fig. 4d) of some genera (e.g. *Codakia*, *Lucinisca* and *Lucinoma*; Southward 1986; Gros *et al.* 1996). In *Codakia orbicularis* (Gros *et al.* 1996), where the granule cells are particularly abundant, they have been shown to be cystine rich and proteinic (Frenkiel & Mouëza 1995), and analytical scanning electron microscopy (SEM) shows them to be sulphur rich (pers. obs.). The granule cells become more abundant with age of the bivalve and in *C. orbicularis* the lateral zone of the gill filaments becomes divided into a superficial bacteriocyte zone and a deeper zone dominated by granule cells (Frenkiel & Mouëza 1995). Also contained within the bacteriocyte cells are a variety of other granules, some of which are elemental sulphur (Vetter 1985; Anderson 1995) whilst others are iron rich or contain a variety of metals (Reid & Brand 1986).

Variation between lucinid genera in the structure of the gill filaments concerns the number and types of cells in the intermediary zone, the presence and abundance of granule cells, the abundance of mucocytes and the presence of peroxisomes and cytoplasmic haemoglobin in bacteriocytes (Frenkiel & Mouëza 1995; Frenkiel *et al.* 1996; Gros *et al.* 1996).

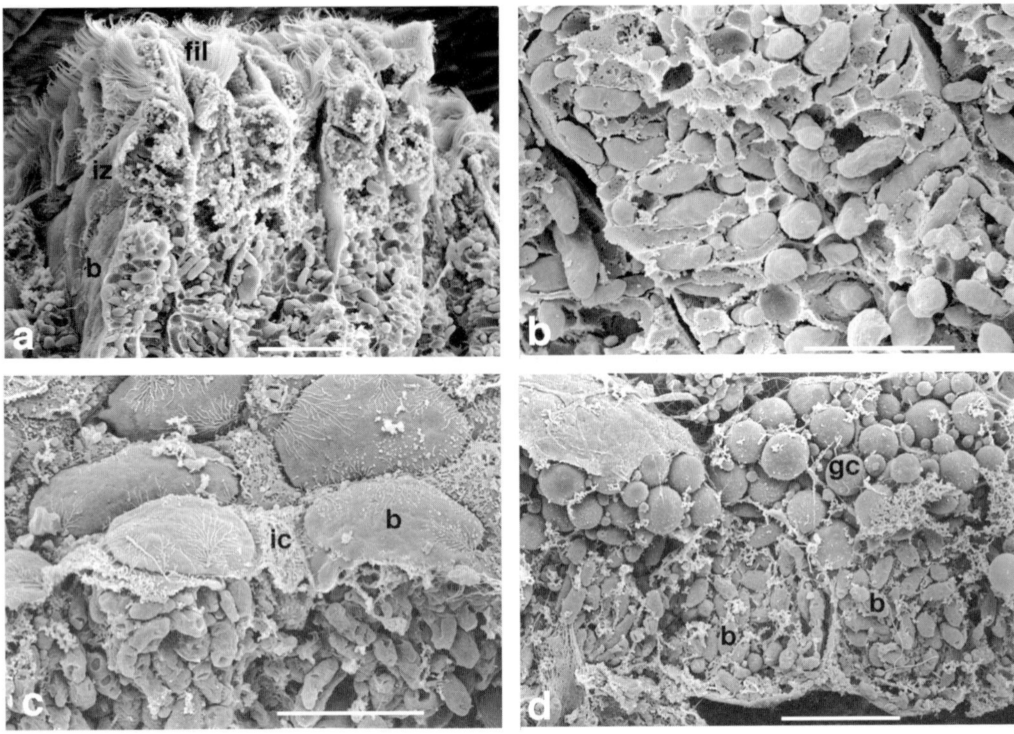

Fig. 4. (**a**) *Pillucina* cf. *fischeriana* (Issel, 1869), Port Douglas, Queensland, Australia. Section of several gill filaments showing ciliated zone, intermediary zone and lateral zone with bacteriocytes. Scale bar, 20 μm. (**b**) *Pillucina* cf. *fischeriana*. Section of gill filament with bacteria. Scale bar, 10 μm. (**c**) *Codakia tigerina* (Linnaeus, 1758). Bacteriocyte zone of gill filament showing bacteriocytes separated by intercalary cells. Scale bar, 10 μm. (**d**) *Codakia tigerina*. Section through gill filament showing (upper) cells packed with spherical granules with bacteriocytes below. Scale bar, 10 μm. All figures SEMs of critical-point dried tissue. b, Bacteria in bacteriocytes; fil, filament; gc, granule cell; ic, intercalary zone; iz, intermediary zone.

Mantle gills and septum

Mantle gills are structures located on the inner mantle surface between the anterior adductor muscle and the ventral mantle margin. They were first noticed in *Codakia tigerina* by Duvernoy (1854, p. 115, pl. 5, fig. 1) and later Semper (1880, fig. 48a) figured similar structures in *Lucina philippensis* (possibly *Anodontia philippiana*), calling them mantle gills 'Mantelkiemen'. Subsequently, Pelseneer (1911) illustrated mantle gills in *Lucina exasperata* Reeve, 1850 (= *Codakia tigerina*) and in *Lucina tumida* Reeve, 1850 (an *Anodontia* species), and suggested a possible respiratory function because of their similarity to the gills of opisthobranch gastropods. Later, Allen (1958) illustrated and described, with sections, the mantle gills of *Codakia orbicularis* and *Lucina pensylvanica*. Because of the extensive blood space associated with them he also interpreted them as respiratory surfaces. Additionally, Narchi & Farani Assis (1980) illustrated in *Phacoides pectinatus*, a row of mantle gills located in a channel alongside the elongate anterior adductor muscle.

Present observations show that mantle gills are more complex than previously illustrated and

Fig. 6. (**a**) *Codakia tigerina*. Mantle gills ventral to anterior adductor muscle. Scale bar, 1 mm. (**b**) *Anodontia philippiana* (Reeve, 1850). Septum and digitiform mantle gills, anterior of bivalve to the right. Scale bar, 1 mm. (**c**) *Codakia tigerina*. Section through mantle gills showing blood spaces and inner epithelial surface with ciliary tufts. Scale bar, 500 μm. (**d**) *Anodontia philippiana*. Microvilli-covered epithelium of mantle gills with ciliary tufts. Scale bar, 10 μm. All figures SEMs of critical-point dried tissue.

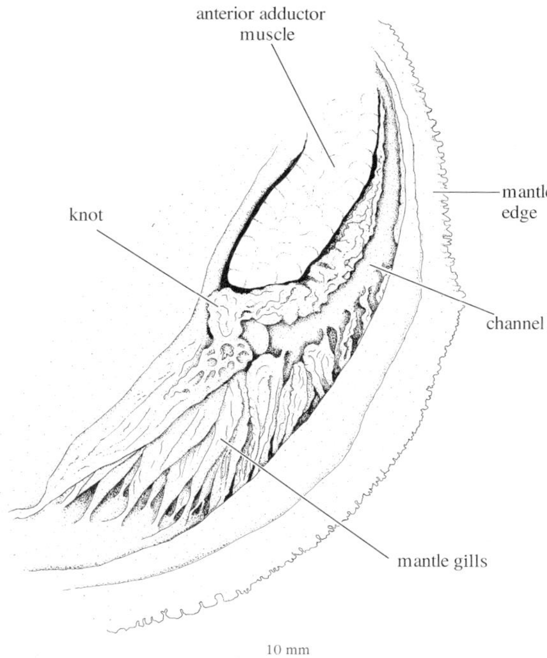

Fig. 5. *Codakia tigerina*. Inner surface of the left anterior mantle showing the highly plicate mantle gills located anterior and ventral to the anterior adductor muscle.

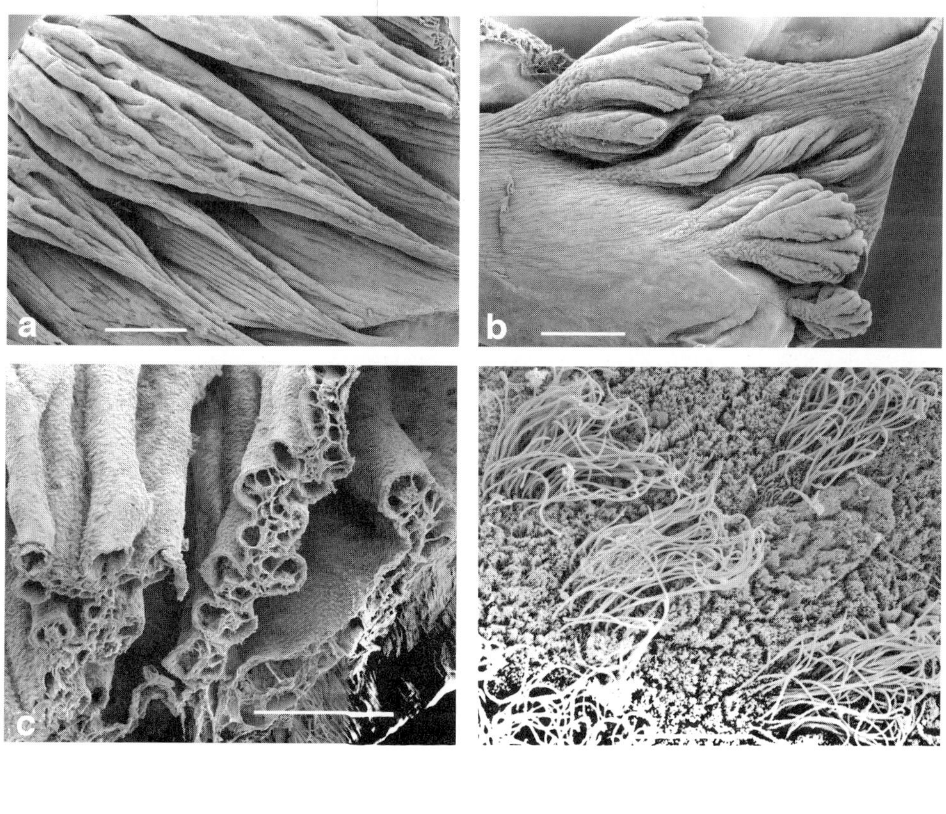

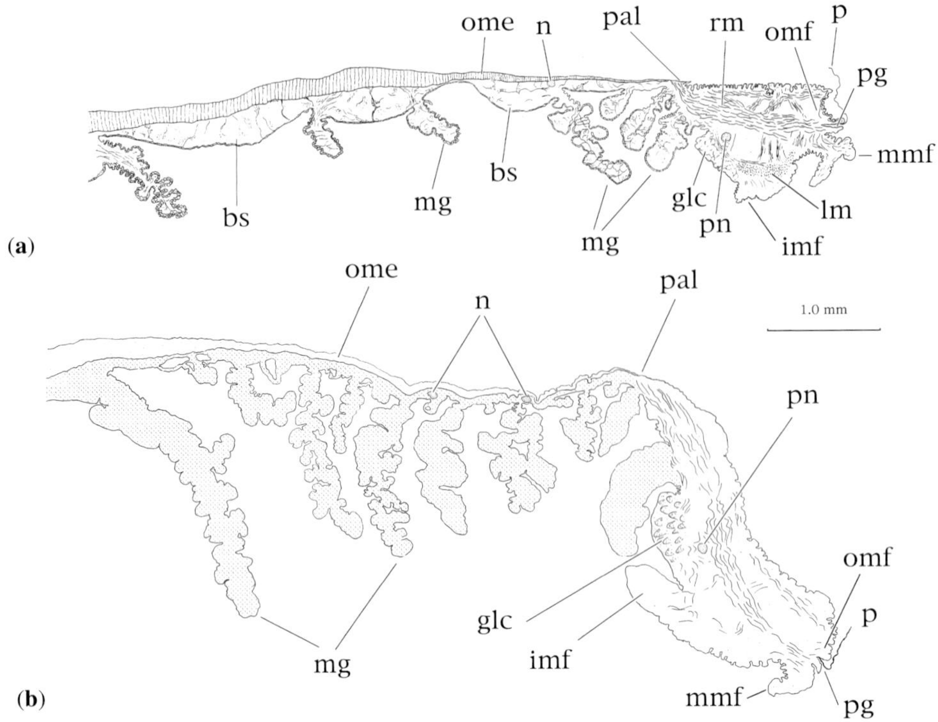

Fig. 7. (**a**) *Codakia orbicularis* (Linnaeus, 1758), Islamorada, Florida Keys. Transverse section through mantle edge and mantle gills ventral to the anterior adductor muscle showing the extensive blood spaces. (**b**) *Codakia tigerina*, Leyte Island, Philippines. Semidiagrammatic transverse section of mantle edge and mantle gills immediately ventral to the anterior adductor muscle. Mantle gills stippled. bs, Blood space; glc, gland cells; imf, inner mantle fold; lm, longitudinal mantle muscles; mg, mantle gills; mmf, middle mantle fold; n, nerve; ome, outer mantle epithelium; omf, outer mantle fold; p, periostracum; pal, pallial attachment; pg, periostracal groove; pn, pallial nerve; rm, radial mantle muscles.

described herein are the mantle gills of *Codakia orbicularis*, *C. tigerina, Lucina pensylvanica*, *Anodontia philippiana* and *A. omissa*.

In *Codakia tigerina* (Fig. 5) and *C. orbicularis* the mantle gills are located around the anterior and ventral sides of the anterior adductor muscles. They comprise a series of highly plicate ridges (Fig. 6a). Anterior to the adductor there is a grooved channel (Fig. 5) which is bounded on either side by ridges. The ridge alongside the adductor muscle being long, convoluted and continuous. After the ventral termination of the muscle the ridges extend posteriorly and ventrally into the pedal gape area. Coalescence of the ridges produces a prominent knot of tissue at the ventral end of the adductor muscle.

In section the mantle gills (Figs 6c and 7) form a series of projecting outgrowths of the epithelium of the inner mantle surface. The numbers of ridges seem variable but for the specimens sectioned there are 11 subparallel ridges in *Codakia tigerina* and seven in *C. orbicularis*. One or two ridges are located on the inner surface of thickened mantle edge. Each ridge has a thin epithelium of low cells and abundant ciliated tufts over the surface (Figs 6c, 7 and 8). The ridges are convoluted in profile and mainly occupied by interconnecting chambers of blood space which join up with the blood space of the inner mantle epithelium (Fig. 8).

Reinvestigation of *Lucina pensylvanica* revealed that Allen's (1958, p. 430, fig. 4) description and illustration of the mantle gill only hinted at the complexity of this structure. The mantle gills are large bipectinate structures (Fig. 9) comprising anastomosing folds of mantle epithelium which are aligned normal to and join the prominent pallial blood vessel. They extend postero-dorsally from close to the posterior tip of the anterior adductor muscle across the inner surface of the mantle, following the line of the pallial blood vessel. In section, as illustrated by Allen (1958, fig. 5a), the individual folds of the mantle gills have a large

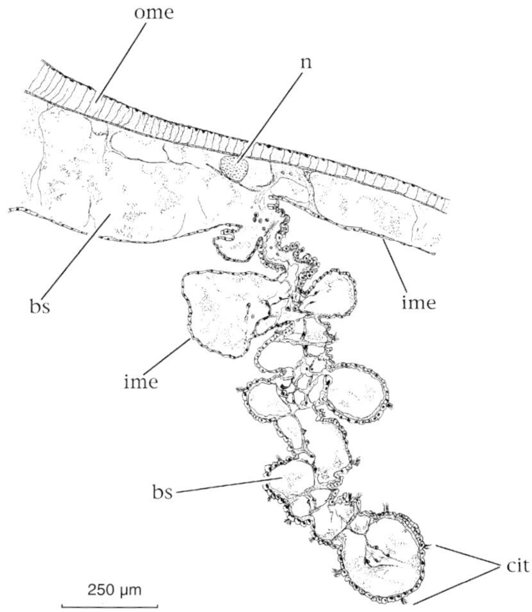

Fig. 8. *Codakia orbicularis*. Section of a single mantle gill. bs, Blood space; cit, ciliary tufts; ime, inner mantle epithelium; n, nerve; ome, outer mantle epithelium.

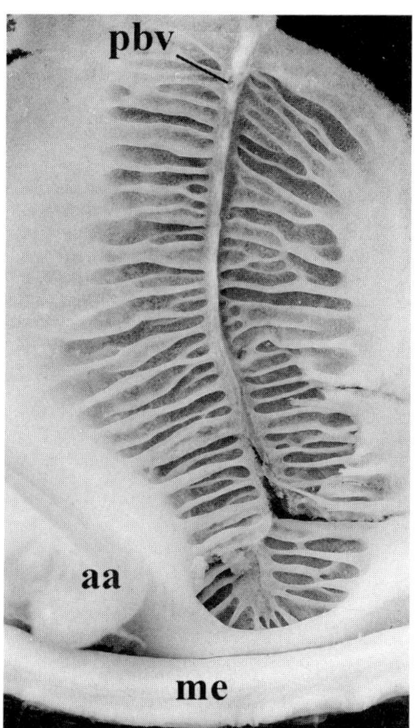

Fig. 9. *Lucina pensylvanica*. Upper Matecumbe Key, Florida. Inner surface of left mantle near the anterior adductor muscle showing the bipectinate mantle gill and the pallial blood vessel. Scale bar, 5 mm. aa, Anterior adductor muscle; me, mantle edge; pbv, pallial blood vessel.

blood space and ciliated epithelium. The histology is similar to that of the mantle gills of *Codakia orbicularis* but lower in profile. The ventral tip of the mantle gill is separated from the inner fold of the mantle edge by a thickened ridge of mantle epithelium which extends posteriorly from the anterior adductor muscle.

Structures possibly homologous to the mantle gills of *Codakia* occur in *Anodontia* species. These possess septa of thin, ciliated epithelial tissue across each side of the mantle cavity which originate from the ventral end of the anterior adductor muscle and run posteriorly to the site of posterior mantle fusion (Fig. 10). In section, each septum is seen to divide the antero-ventral part of the mantle cavity. In *Anodontia philippiana,* the anterior ends of the left and right septa bear a series of highly plicate structures which terminate anteriorly in finger-like projections (Fig. 6b). The innermost ridge of the septum is thickly ciliated, but most of the ventral flanks and plications carry ciliary tufts arising from a microvilli-covered surface (Fig. 6d).

Sections through the septum and digitiform structures of *A. philippiana* (Fig. 11) show that the septum comprises two epithelial layers enclosing a narrow, central blood space which is crossed by trabeculae. The epithelial layers consist of low cells with ciliary tufts. At the site of the digitiform structures, the septum is thrown into a series of

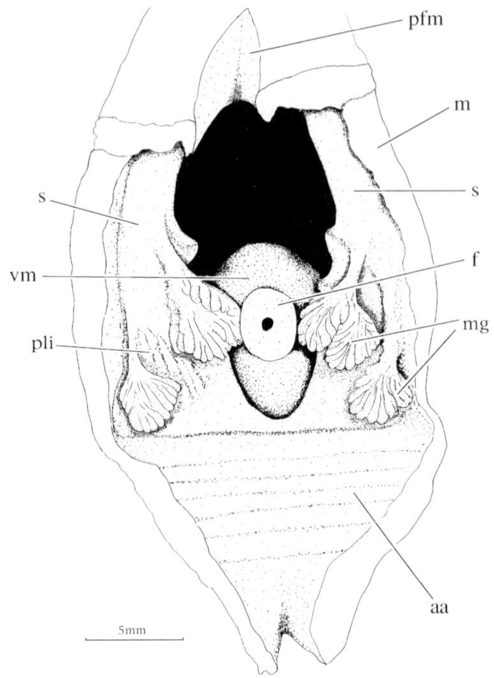

Fig. 10. *Anodontia philippiana* (Reeve, 1850). Ventral view of septum and digitiform mantle gills. Left and right mantle lobes and the foot removed (cut surface indicated by cross-hatching). aa, Anterior adductor muscle; f, foot; m, mantle; mg, mantle gills; pfm, posterior mantle fusion; pli, plications; s, septum; vm, visceral mass.

labyrinthine folds, the plicated surface being the summits of the folds. The blood spaces are broader and the trabeculae thicker on the outermost parts of the folds (Fig. 11b). Sections through the septum of *Anodontia omissa* reveal a similar histology but the labyrinthine folds are absent. Although few species have yet been examined, the septum may prove to be an autapomorphy of *Anodontia* species, whilst the plicated knots may be present only in the larger species.

The number of lucinid species bearing mantle gills is not yet known but likely homologous structures have been observed in species other than those described above. Narchi & Farani Assis (1980) illustrate in *Phacoides pectinata* a series of 16 small, separated, plicate ridges located in a channel between the elongate, anterior adductor muscle and the mantle edge. These are arranged more or less transversely to the anterior water current (Fig. 12) and have a similar histology to the mantle plications of *Codakia*. In *Austriella corrugata* (present authors' unpublished observations) there are thick, single ridges extending posteriorly from the ventral end of the anterior adductor muscle. Allen (1958, figs 17 and 18) also illustrates in *Divaricella quadrisulcata* Orbigny, 1846 and *Pegophysema alba* (as *Lucina chrysostoma*) a 'triangular fold' located in the same position as the mantle gills. Although the present authors did not report such structures in *Rasta thiophila* (Taylor & Glover 1997*b*) re-examination of thin sections shows that the inner mantle has small plications in the area close to the anterior adductor.

Similar mantle gill-like structures are found in another lucinoid family, the Fimbriidae. In *Fimbria fimbriata* there are elongate, wedge-shaped structures arising from the inner mantle epithelium which extend as ridges from the ventral edge of the anterior adductor muscle (Allen & Turner 1970; Morton 1979). A groove runs along the summit of the free margin which is edged by convoluted lips

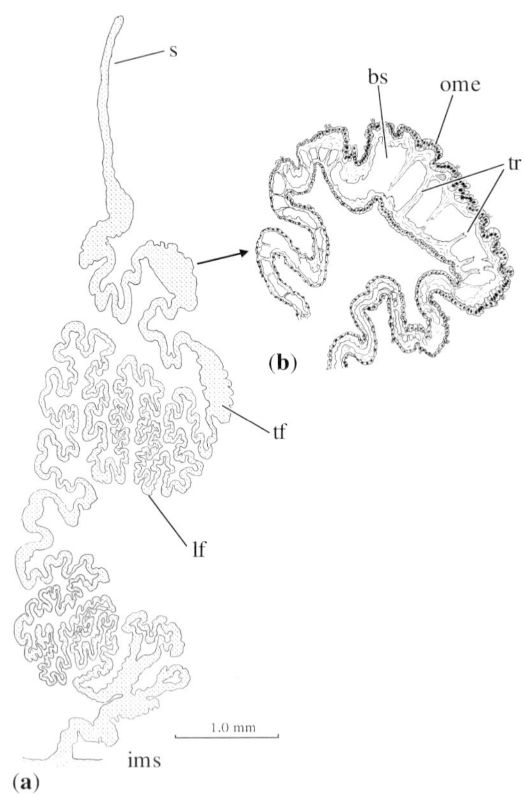

Fig. 11. *Anodontia philippiana*. (**a**) Transverse section through the mantle gills in the anterior part of the septum. Inner mantle surface at base. (**b**) Enlargement of part of the septum showing blood spaces and trabeculae. bs, Blood space; lf, labyrinthine fold; ims, inner mantle surface; ome, outer mantle epithelium; s, septum; tf, thickened fold; tr, trabeculae.

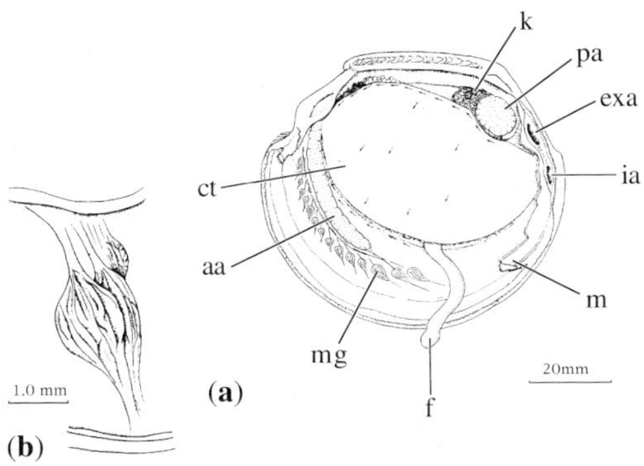

Fig. 12. *Phacoides pectinatus* (Gmelin, 1791). Brazil [modified from Narchi & Farani Assis (1980)]. (**a**) General anatomy with the left mantle removed to show the row of mantle gills lying alongside the anterior adductor muscle. (**b**) Detail of single mantle gill. aa, Anterior adductor muscle; ct, ctenidia; exa, exhalant aperture; f, foot; ia, inhalant aperture; k, kidney; m, mantle; mg, mantle gills; pa, posterior adductor muscle.

(Fig. 13). In section, the central area of the ridge is a spongy network of blood spaces with thin, cellular walls and covered by a ciliated epithelium. The distal groove is lined with gland cells opening through the epithelium. The blood spaces connect with the pallial blood vessel. A different interpretation was presented by Morton (1979) who called these same structures mantle palps. He suggested that they removed particulate material from the foot, trapping it in copious amounts of mucus

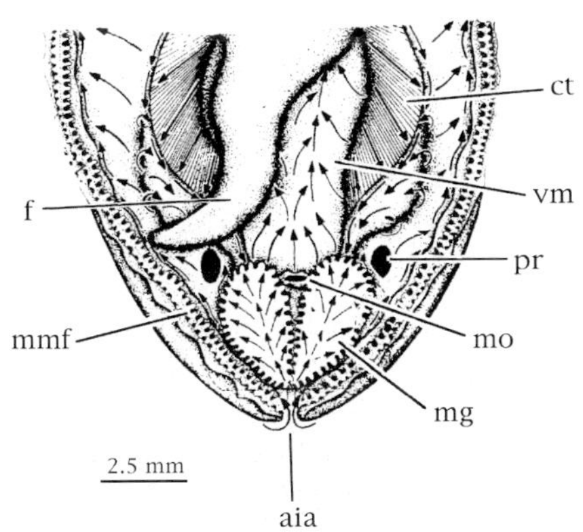

Fig. 13. *Fimbria fimbriata* (Linnaeus, 1758) [from Morton (1979, fig. 14)], Lizard Island, Queensland, Australia. Ventral view of anterior of the animal showing mantle gills positioned in the inhalant water current. aia, Anterior inhalant aperture; ct, ctenidia; f, foot; mg, mantle gills; mmf, middle mantle fold; mo, mouth; pr, pedal retractor muscle; vm, visceral mass.

produced by the epithelium in the glandular groove. The histology of the main part of these structures in *Fimbria* is similar to that of the ridges in *Codakia tigerina*, although the distal glandular groove and the lip-like edges are not found in the latter species.

From the present authors' observations it is also clear that many other lucinids, particularly the small species in genera such as *Cardiolucina*, *Pillucina*, *Wallucina* and *Parvilucina*, do not possess any obvious mantle gills. Nonetheless, in these species, as in those with mantle gills, an extensive, thickened blood space lies beneath the inner ciliated mantle epithelium.

Although there has been no direct physiological study, the histology, the abundant blood supply, the highly plicated surfaces and the position in the incoming anterior water current are all strong evidence that the mantle gills in these different lucinids are respiratory structures. Their histology and morphology are also very similar to the gills of opisthobranch gastropods such as *Acteon tornatilis* (Fretter & Graham 1954) and the pallial gills of basommatophoran pulmonates. Sections through the gill plications of *Cylindrobulla* (Morton 1988, fig. 6c and d) show a structure very similar to that of *Codakia orbicularis* and *C. tigerina* (Fig. 7), and the gills of *Akera bullata* are thrown into complex folds (Morton 1972) similar to those seen in the septum of *Anodontia philippiana* (Fig. 11).

Anterior adductor muscle

There are two features of the anterior adductor muscle which are peculiar to lucinids. Firstly, the muscle is enlarged and elongated ventrally and sometimes dorsally, in comparison with the posterior muscle which is reniform or subcircular in outline (Fig. 2). Secondly, the adductor muscle is detached from the line of pallial attachment for a proportion of its length (Fig. 2). These two features vary considerably between lucinid genera. Thus, *Miltha* and *Phacoides* (Fig. 12) have extremely long, narrow adductor muscles which are extended ventrally and detached from the pallial line for much of their length. At the other extreme, the adductor muscle of *Cardiolucina* species is small and detached only slightly from the pallial line at the ventral end (Taylor & Glover 1997*a*). However, in most lucinids there is often a considerable elongation of the anterior adductor muscle, while the detachment of the muscle from the pallial line has become an important distinguishing character of lucinids.

Given the widespread distribution of these two features of the anterior adductor muscle in lucinids, what, if any, is their functional significance? The present authors believe that they are functionally linked and associated with the anterior inhalant tube and the anterior-posterior water flow through the mantle cavity. In lucinids, the fusion of the anterior mantle is located much further antero-dorsally than in most heterodont bivalves and this exposes the ciliated outer surface of the adductor muscle. The detachment of the ventral end of the muscle from the pallial line forms an elongate channel or approximately triangular area of mantle which is also ciliated. The cilia on the face of the adductor muscle and in the anterior channel create the flow of water down the anterior inhalant tube and into the mantle cavity. In some species, *Phacoides pectinatus* and *Codakia* spp., the channel between the anterior adductor is occupied by mantle gills which also extend beyond the muscle posteriorly along the inner mantle surface. Only in *Lucina pensylvanica* have the mantle gills been observed extending dorsally from the anterior adductor muscle. Allen (1958) showed that the area between the adductor muscle and mantle edge was responsible for sorting most of the particulate material entering via the inhalant tube and that little material passed to the gills. This area of the anterior mantle is well endowed with blood space and the large pallial blood vessel which runs to near the ventral end of the anterior adductor muscle links up with these blood spaces. It is concluded that the mantle around the anterior adductor, with or without the mantle gills, has become the major respiratory surface of the lucinids. The elongate adductor muscle also acts as a partition, effectively separating the anterior part of the mantle cavity from the ctenidia. This is seen to an extreme extent in *Phacoides pectinatus* (Fig. 12) and *Miltha childreni* (Gray, 1825) where the muscle is very long. In *Anodontia* species, an even larger partition is created by the septum which joins the ventral end of the adductor muscle and extends nearly to the posterior apertures (Fig. 10).

In *Fimbria fimbriata,* the anterior adductor muscle is short and not detached from the pallial line. This species burrows with the antero-dorsal end close to the sediment surface and no anterior tube is formed (Morton 1979). The anterior water current is drawn down the antero-dorsal mantle edge into the mantle cavity and over the mantle gills which lie in its path (Fig. 13)

Pallial blood vessel

Another feature of lucinids not found in other bivalves is the large pallial blood vessel (Figs 2 and 9). In *Codakia, Lucina, Anodontia* and *Phacoides,* as in all lucinids and in *Fimbria,* this large vessel runs diagonally antero-ventrally across the mantle connecting the heart to a sinus near the ventral tip of the anterior adductor muscle. Here it branches, joining the thick blood space of inner mantle epithelium and, where present, the mantle gills and septum. The blood vessel often leaves an impressed diagonal groove within the inner shell layer of lucinids (Fig. 2). This groove is commonly visible on the inner shell of fossil species and is particularly deeply impressed in *Megaxinus rostratus* from the Pliocene Period of Italy (Glover & Taylor 1997, fig. 7i).

Posterior mantle apertures

In most lucinid species there are two posterior mantle apertures (Fig. 14) formed by fusion of the inner and, in a few species, middle mantle folds (Allen 1958) [because the middle mantle fold often comprises several lobes in lucinids (Fig. 7) the homologies of the mantle folds are difficult to recognize]. The ventral of these two apertures, corresponding to the inhalant aperture of most heterodont bivalves, is a simple orifice, fringed with papillae in some species. The extent of mantle fusion ventral to the inhalant aperture varies between lucinid genera, e.g. it is long in *Anodontia* and short in *Codakia*. In some lucinids, e.g. *Parvilucina tenuisculpta* (Reid & Brand 1986), *P. multilineata* (pers. obs.) and *Cardiolucina*

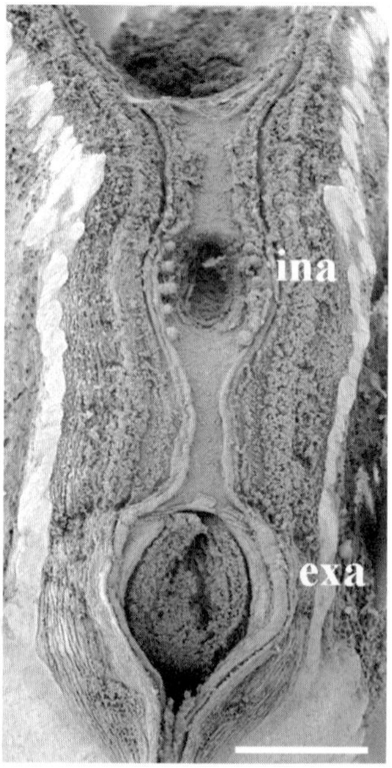

Fig. 14. *Pillucina* cf. *fischeriana* Port Douglas, Queensland, Australia. SEM of critical-point dried preparation of posterior mantle fusion and apertures. Scale bar, 0.5 mm. exa, Exhalant aperture, ina, inhalant aperture.

australopilula (Taylor & Glover 1997*a*), there is no fused inhalant aperture. The dorsal posterior aperture, the exhalant, is usually larger, sometimes fringed with papillae, and comprises an extensible, thin, but muscular, tube which is inverted, on retraction, into the posterior mantle cavity between the two gill demibranchs. In some species, the siphon can be extended to at least six times the length of the shell (Allen 1958). The inhalant aperture may function only for the rejection of pseudofaeces. However, the exhalant siphon is clearly more active and shows prominently in Stanley's (1970) X-rays of lucinid life positions. As well as directing the outward flow of water from the gills, Reid (1998) has suggested that water could also be drawn into the suprabranchial mantle cavity when the siphon is inverted.

Reid & Brand (1986) described in *Parvilucina tenuisculpta* a tissue connection between the posterior end of the ctenidia and the inner mantle close to the exhalant aperture. They suggested that this arrangement acted as a bellows, enabling the flushing of the suprabranchial chamber and thereby drawing in sulphide-rich water. Additionally, they speculated that this structure might be more widely distributed amongst lucinids but the present authors have seen it only in *Cardiolucina australopilula* (Taylor & Glover 1997*a*), *Parvilucina multilineata, Pillucina fischeriana and Wallucina assimilis.*

Foot

The form of lucinid foot (Fig. 1) is unusual for heterodont bivalves and its activities are an important part of the varied behaviour associated with chemoautotrophy. As well as burrowing, the foot forms the anterior, mucus-lined, inhalant tube and the ventral tunnels below the animal. Additionally, the foot may be responsible for drawing interstitial water up into the mantle cavity from the ventral tunnels either by piston action or through currents created by the ciliated epithelium. In most lucinids the main part of foot is vermiform, long and cylindrical and can be extended for at least four to six times the length of the animal. *Fimbria fimbriata* has a broader, less extensible foot (Morton 1979). Many lucinid species have a stubby, flattened posterior component – the heel (Allen 1958) – but this is absent in others (e.g. *Rasta*; Taylor & Glover 1997*b*). The long, cylindrical part of the foot has layers of circular and longitudinal muscle (Allen 1958; Taylor & Glover 1997*b*) surrounding a central blood space. The distal tip is bluntly pointed but can become bulbous. The outer epithelium consists of columnar cells which become highly ciliated near the tip. Also at the tip, there are two types of subepithelial gland cells with different staining properties (Allen 1958; Morton 1979). In *Rasta,* Taylor & Glover (1997*b*) showed that red staining gland cells extended further back along the foot than the blue staining cells which occupied only the very tip. Allen (1958) thought that these gland cells were responsible for producing the mucus which lines the anterior inhalant tube, however, the presence of two types of cell suggests that their role is more complex.

General functional summary

The location of sulphide-oxidizing, bacterial chemosymbionts in the gills of lucinids necessitates separation of the respiratory surfaces from the location of the chemosymbionts. Otherwise, sulphides in the water being drawn into the mantle cavity by the foot and by currents generated from the ciliated zone of the gills, and maybe also through the posterior apertures, would be oxidized before reaching the bacteriocytes. Thus, the relocation of major respiratory surfaces near the anterior tube, partly separated from the rest of the mantle cavity by the often elongated adductor muscle, effectively partitions the sites of respiration and chemosynthesis. In some larger lucinids the anterior inner mantle surface becomes highly plicated to form the mantle gills, whereas in other species the unfolded inner mantle is the probable respiratory surface. The septum of *Anodontia* species, extending posteriorly from the adductor muscle, further divides off the antero-ventral part of the mantle cavity. The highly plicate outer face of the septum forms the respiratory surface. The mantle gills of *Lucina pensylvanica* are positioned rather differently and lie alongside the pallial blood vessel across the mantle. However, X-rays show the life position of this species oriented with the anterior end uppermost and the hinge line vertical in the sediment, with posterior tunnels developed vertically below the bivalve (Stanley 1970). This probably allows an anterior–posterior separation of the oxygenated water from the interstitial water being drawn in from the posterior. The anterior part of the mantle of lucinids also has to become the main area of particle sorting, removing sediment particles along the mantle to be expelled as pseudofaeces through the posterior inhalant aperture. Because at least some lucinids are partially heterotrophic, food particles are passed along the surface of the adductor muscle to the mouth (Allen 1958; Morton 1979). Thus, amongst lucinids, the relatively enlarged anterior end of the body, the elongate anterior adductor muscle, the dorsal separation of the muscle from the line of pallial attachment and the mantle gills can all be considered as adaptations to the chemosymbiosis enabling the respiratory and sulphide–oxidizing functions to be spatially separated within the mantle cavity.

Information on the life habits of lucinids is summarized diagrammatically in Fig. 15 which is based on observations of *Codakia* species. This highlights the location of the respiratory area at the mantle gills in the path of the anterior current being drawn down the anterior tube. The ciliated foot is active in tunnelling in the sediment below the bivalve and in drawing interstitial water into the mantle cavity to the gills. It should be emphasized that lucinids are diverse in morphology and occupy a wide range of different habitats so that variations on this schematic model may be expected.

How old is the symbiosis?

The pervasiveness of the chemosymbiosis among living lucinids suggests that the association with chemosynthetic bacteria may be ancient and ancestral to the family (Reid & Brand 1986). It is also possible, as suggested by Distel (1998), that the symbiosis was acquired later and spread laterally, with non-symbiotic taxa becoming extinct.

There are several approaches to the interpretation of fossil lucinids and determining whether they possessed a chemoautotrophic symbiosis, none of which is direct. These methods include the examination of the association of lucinids with particular habitats and co-occurring faunas, and comparison with Recent analogues, measurement of the isotopic signatures in the shells (CoBabe 1991), study of symbiosis-associated morphological features on the fossils themselves and, finally, a phylogenetic approach whereby phylogenies of the bacteria might be tracked against those for the bivalves. For the latter method there is, as yet, no robust phylogenetic framework, either molecular or morphological, for the lucinids.

Species of several chemosymbiont-bearing bivalve families are associated with hydrothermal vents and cold-seep communities (Distel 1998). Recent lucinids are not usually associated with hot vents but they (usually *Lucinoma* species) are found commonly in cold-seep communities along with the other bivalve chemoautotrophs *Bathymodiolus* and *Calyptogena* (Callender & Powell 1997; Sibuet & Olu 1998). This association has been recognized in the fossil record from the Mesozoic onwards, e.g. in the Jurassic (Oxfordian) of France (Gaillard *et al.* 1992; Peckmann *et al.* 1999); Early Cretaceous (Barremian) of Greenland

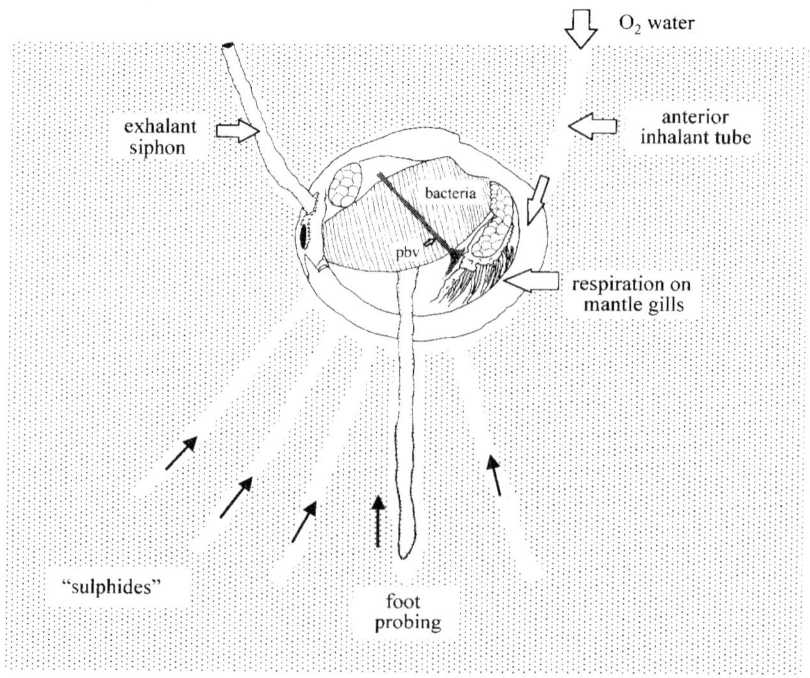

Fig. 15. Diagram summarizing the life position and major water currents of a typical lucinid based on *Codakia* spp. pbv, Pallial blood vessel.

(Kelly *et al.* 2000); Late Cretaceous (Campanian) Tepee Buttes of Colorado (Callender & Powell 1997; Kauffman *et al.* 1996); Eocene of California (Squires & Gring 1996; Saul *et al.* 1996) and the 'calcari a *Lucina*' from the Miocene of Italy (Taviani 1994).

Recent lucinids are also frequently associated with dysaerobic habitats (Bottjer *et al.* 1995; Reid 1998) and such associations are also seen in the fossil record (Hickman 1994). However, species of other bivalves, from families such as Nuculidae, Mytilidae, Tellinidae and Cardiidae, which do not possess chemosymbionts, are also often found in such habitats. The mere presence of bivalves in palaeohabitats interpreted as dysaerobic does not necessarily mean that they possessed chemosymbionts and some other corroborative evidence is needed.

The most convincing evidence for the presence of chemosymbiosis may come from the morphology of the shell and the preserved life positions of the fossil lucinids. Many Cenozoic and Mesozoic lucinids are similar to living species in shell morphology, with detached anterior muscle scars and the trace of the pallial blood vessel impressed into the shell, and the present authors believe that these features indicate that the bivalves were chemosymbiotic. An interesting example is '*Lucina*' *pandata* (Conrad, 1833) from the Eocene of Alabama (Fig. 16), in which there is a line of discrete, claw-like, impressions located between the elongate anterior adductor muscle and the pallial line. These impressions may represent the traces of a line of mantle gills similar to those seen in the living *Phacoides pectinatus* (Fig. 12).

Although lucinids are much more diverse and abundant from the late Mesozoic onwards, they have a long history into the Palaeozoic, possibly to the early Ordovician Period (Pojeta 1978). However, a major problem with the Palaeozoic fossils is the actual recognition of lucinids, since their history and systematics is rather confused and many taxa need reinvestigation. For example, Johnston (1993) provides fine illustrations of some putative lucinoideans from the Devonian of Australia. Some, such as *Paracyclas proavia* (Goldfuss 1840), have lucinid-like features with an elongate, but not detached, anterior adductor muscle scar. It is difficult to place these fossils and they were accommodated in a new family, Paracycladidae, interpreted by Johnston (1993) as close relatives of the Ungulinidae. Despite the Ungulinidae being classified within the Lucinoidea, their relationship to the other families is far from clear and furthermore no evidence for chemosymbiosis has yet been found. Also, Johnston & Goodbody (1988) illustrate some new genera from the Devonian of Arctic Canada similar to *Tanaodon* which they also claim have lucinoidean features, but again their affinities remain uncertain.

Fig. 16. '*Lucina*' *pandata* (Conrad, 1833), Eocene, Claiborne Formation, Alabama, USA. BMNH Palaeontology Dept LL26958. Interior of left valve showing the line of impressions of the possible mantle gills, cf. with *Phacoides pectinatus* in Fig. 12.

Although species of the Ordovician bivalve *Babinka* have been claimed as early lucinoids, or their direct ancestor (McAlester 1965; Pojeta 1978), the evidence is equivocal and other authors (Cope 1997) regard the genus as belonging to the Actinodontoida. McAlester (1965) highlighted several features of *Babinka* as having lucinoid affinities: an anteriorly expanded shell shape; the elongate, anterior adductor muscle; a simple, non-sinuate pallial line; the typical lucinoid hinge, dentition and ligament. Although the shell is slightly anteriorly extended and somewhat lucinoid in appearance, McAlester (1965) in his advocacy exaggerated the degree of elongation of the anterior adductor scar which is, in fact, only slightly larger than the posterior. It is felt that the evidence connecting *Babinka* with the lucinoids is still uncertain.

Apart from *Babinka,* the earliest bivalve which can be confidently assigned to the Lucinidae is *Ilionia prisca* (Hisinger, 1837) from the Silurian Period (Ludlow) of Gotland, Sweden (Liljedahl 1991). This is a relatively large bivalve, *c.* 65 mm in length. It is anteriorly extended with a large, elongate, anterior adductor muscle which is ventrally detached from the pallial line (Fig. 17). In shape, *Ilionia* is similar to the Recent lucinid *Eomiltha voorhoevei* (see Oliver 1995, fig. 1029). *Ilionia* are preserved in life position in a carbonate mud with the hinge line horizontal in the sediment. The only other common bivalve in the habitat is another likely chemosymbiont host, the solemyid *Janeia silurica*. Liljedahl (1991, fig. 6b) interpreted *Ilionia* as a lucinid with an anterior tube excavated by the foot and an anterior–posterior water current.

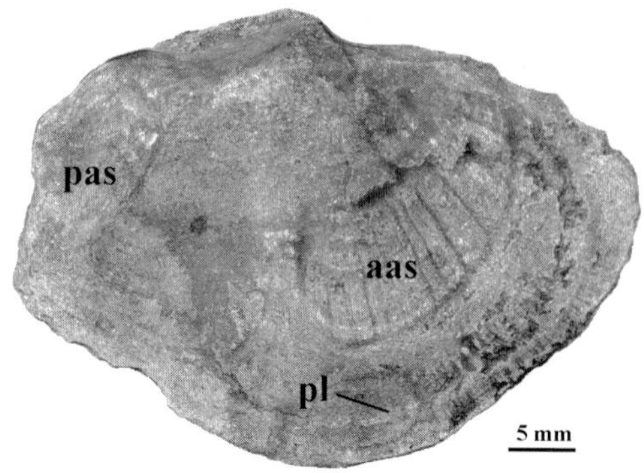

Fig. 17. *Ilionia prisca* (Hisinger, 1837). Silurian, Gotland, Sweden. BMNH Palaeontology Dept PL360. Internal cast. aas, Anterior adductor scar; pas, posterior adductor scar; pl, pallial line.

The large anterior area lying between the adductor muscle and the pallial line is interpreted, by comparison with Recent species, as the site of respiration. The ciliated outer surface of the adductor would generate the respiratory current and the elongate muscle projecting deeply into the mantle cavity would function to partition the oxygenated water from the sulphide-containing water being drawn into the mantle cavity to the gills by the activity of the foot. All the internal features of *Ilionia* support the interpretation of the animal possessing chemosymbionts.

The presence of posterior mantle fusion forming both inhalant and exhalant apertures, as in most other heterodont bivalves, suggests that lucinids may have originally been shallow burrowing, filter feeders. Reid & Brand (1986) proposed a scenario whereby acquisition of a loose symbiosis with sulphide-oxidizing bacteria may have encouraged the bivalves to burrow more deeply to access sulphides from anoxic sediment layers. It is known that some, perhaps all, lucinids acquire their symbionts by cyclical transmission from the sediment (Gros *et al.* 1998*a*) and that the initial association was most likely derived from this source. McFall-Ngai (1998) provides a model of an evolutionary progression of increasing dependence of the host and microbial cells. Bacteria might initially have occupied a superficial extracellular position on the gills but increasing intimacy with the host may have progressed to an intracellular location with coevolved specialized bacteriocytes. Increasing nutritional dependence on the chemosymbiosis necessitated a reorganization of the life position of the lucinids, with the anterior ventilatory connection to the surface being maintained by the elongate foot. The posterior inhalant aperture lost its function as the main conduit of water into the mantle cavity. The gills became simplified with the loss of the outer demibranch but thickened to accommodate more bacteriocytes in the subfilamental tissue. Loss of the outer demibranch of the gills, the reduced labial palps, the simple gut and the reversion to an anterior–posterior water flow are all likely paedomorphic features (Reid & Brand 1986; Reid 1990). Evidence suggests that chemosymbiosis has existed in Lucinidae since at least the Silurian Period, and accommodation and morphological adaptation to the symbiosis were responsible for major changes in body organization and behaviour. Despite the probable early acquisition of the chemosynthetic lifestyle, lucinids remained only minor faunal constituents until the late Mesozoic–Cenozoic when they diversified along with many other heterodont clades (Skelton *et al.* 1990).

We thank D. G. Reid and P. S. Rainbow for useful discussions and collection of specimens. This work is partially supported by a grant from the Australian Biological Resources Study (ABRS).

References

ALLEN. J. A. 1958. On the basic form and adaptations to habitat in the Lucinacea (Eulamellibranchia). *Philosophical Transactions of the Royal Society of London, Series B,* **241**, 421–484.

—— & TURNER, J. F. 1970. The morphology of *Fimbria fimbriata* (Linné) (Bivalvia: Lucinidae). *Pacific Science,* **24**, 147–154.

ANDERSON, A. E. 1995. Metabolic responses to sulfur in lucinid bivalves. *American Zoologist*, **35**, 121–131.

BARNES, P. A. G. & HICKMAN, C. S. 1999. Lucinid bivalves and marine angiosperms: a search for causal relationships. *In*: WALKER, D. I. & WELLS, F. E. (eds) *The Seagrass Flora and Fauna of Rottnest Island, Western Australia.* Western Australian Museum, Perth, 215–238.

BERG, C. J. & ALATALO, P. 1984. Potential of chemosynthesis in molluscan mariculture. *Aquaculture*, **39**, 165–179.

——, ——, CAVANAUGH, C. M., FELBECK, H., JANNASCH, H. W. & SOMERO, G. N. 1982. Possible chemoautotrophic nutrition in Bahamian bivalves. *In:* PRUDER, C. D., LANGDON, C. & CONCKLIN, D. (eds) *Proceedings of the 2nd International Conference on Aquaculture Nutrition.* Louisiana State University, Baton Rouge, 425.

BOTTJER, D. J., CAMPBELL, K. A., SCHUBERT, J. K. & DROSER, M. L. 1995. Palaeoecological models, non-uniformitarianism, and tracking the changing ecology of the past. *In*: BOSENCE, D. W. & ALLISON, P. A. (eds) *Marine Palaeoenvironmental Analysis from Fossils.* Geological Society, London, Special Publications, **83**, 7–26.

BRETSKY, S. S. 1970. Phenetic and phylogenetic classification of the Lucinidae (Mollusca: Bivalvia).

Bulletin of the Geological Institute of the University of Uppsala, **2**, 5–23.

—— 1976. Evolution and classification of the Lucinidae (Mollusca: Bivalvia). *Palaeontographica Americana*, **8**, 219–337.

CALLENDER, W. R. & POWELL, E. N. 1997. Autochthonous death assemblages from chemautotrophic communities at petroleum seeps: palaeoproduction, energy flow and implications from the fossil record. *Historical Biology*, **12**, 165–198.

CARY, S. C. & GIOVANNONI, S. J. 1993. Transovarial inheritance of endosymbiotic bacteria in clams inhabiting deep-sea hydrothermal vents and cold seeps. *Proceedings of the National Academy of Sciences USA,* **90**, 5695–5699.

——, FRY, B., FELBECK, H. & VETTER, R. D. 1989. Habitat characterization and nutritional strategies of the endosymbiont-bearing bivalve *Lucinoma aequizonata*. *Marine Ecology Progress Series*, **55**, 31–45.

CHAVAN, A. 1969. Superfamily Lucinacea Fleming 1828. *In:* MOORE, R. C. (ed.) *Treatise on Invertebrate Paleontology. Part N. Mollusca 6. Bivalvia Volume 2*. Geological Society of America and University of Kansas, Lawrence, KS, N491–N518.

COBABE, E. A. 1991. Detection of chemosymbiosis in the fossil record: the use of stable isotopy on the organic matrix of lucinid bivalves. *In:* DUDLEY, E. C. (ed.) *The Unity of Evolutionary Biology.* Proceedings of the Fourth International Congress of Systematic and Evolutionary Biology. Dioscoroides Press, Portland, OR, 669–673.

COPE, J. C. W. 1997. The early phylogeny of the Class Bivalvia. *Palaeontology*, **40**, 713–746.

DANDO, P. R., RIDGWAY, S. A. & SPIRO, B. 1994. Sulphide 'mining' by lucinid bivalve molluscs: demonstrated by stable sulphur isotope measurements and experimental models. *Marine Ecology Progress Series*, **107**, 169–175.

——, SOUTHWARD, A. J. & SOUTHWARD, E. C. 1986. Chemoautotrophic symbionts in the gills of the bivalve mollusc *Lucinoma borealis* and the sediment chemistry of its habitat. *Proceedings of the Royal Society of London, Series B*, **227**, 227–247.

——, ——, ——, TERWILLIGER, N. B. & TERWILLIGER, R. C. 1985. Sulphur-oxidising bacteria and haemoglobin in gills of the bivalve mollusc *Myrtea spinifera*. *Marine Ecology Progress Series,* **23**, 85–98.

DISTEL, D. L. 1998. Evolution of chemoautotrophic endosymbioses in bivalves. *Bioscience*, **48**, 277–286.

—— & FELBECK, H. 1987. Endosymbiosis in the lucinid clams *Lucinoma aequizonata*, *Lucinoma annulata* and *Lucina floridana*: a reexamination of the functional morphology of the gills as bacteria-bearing organs. *Marine Biology,* **96**, 79–86.

DURAND, P, GROS, O., FRENKIEL, L. & PRIEUR, D. 1996. Phylogenetic characterization of sulfur-oxidising bacterial endosymbionts in three tropical Lucinidae by 16S rDNA sequence analysis. *Molecular Marine Biology and Biotechnology*, **5**, 37–42.

DUVERNOY, M. 1854. Mémoires sur la système nerveux des mollusques acéphales. *Mémoires de l'Académie des Sciences de l'Institut de France*, **24,** 1–210

FISHER, C. R. 1990. Chemoautotrophic and methanotrophic symbioses in marine invertebrates. *Reviews in Aquatic Sciences*, **2**, 399–436.

FISHER, M. R. & HAND, S. C. 1984. Chemautotrophic symbionts in the bivalve *Lucina floridana* from seagrass beds. *Biological Bulletin, Woods Hole*, **167**, 445–459.

FRENKIEL, L. & MOUËZA, M. 1995. Gill ultrastructure and symbiotic bacteria in *Codakia orbicularis* (Bivalvia, Lucinidae). *Zoomorphology*, **115**, 51–61.

——, GROS, O. & MOUËZA, M. 1996. Gill structure in *Lucina pectinata* (Bivalvia: Lucinidae) with reference to hemoglobin in bivalves with symbiotic sulphur-oxidising bacteria. *Marine Biology*, **125**, 511–524.

FRETTER, V. & GRAHAM, A. 1954. Observatioins on the opisthobranch mollusc *Acteon tornatilis* (L.). *Journal of the Marine Biological Association, UK*, **33**, 565–585.

GAILLARD, C., RIOS, M., ROLIN, Y. & ROUX, M. 1992. Fossil chemosynthetic communities related to vents and seeps in sedimentary basins: the pseudobioherms of southeastern France compared to other world examples. *Palaios*, **7**, 451–465.

GIERE, O. 1985. Structure and position of bacterial endosymbionts in the gill filaments of Lucinidae from Bermuda (Mollusca, Bivalvia). *Zoomorphology*, **105**, 296–301.

GLOVER, E. A. & TAYLOR, J. D. 1997. New species and records of *Rastafaria* and *Megaxinus* (Bivalvia: Lucinidae) from the western Indian Ocean and Red Sea, with a reappraisal of *Megaxinus*. *Journal of Conchology,* **36**, 1–18.

GROS, O., DUPLESSIS, M. R. & FELBECK, H. 1999. Embryonic development and endosymbiont transmission mode in the symbiotic clam *Lucinoma aequizonata* (Bivalvia: Lucinidae). *Invertebrate Reproduction and Development*, **36**, 93–103.

——, FRENKIEL, L. & MOUËZA, M. 1996. Gill ultrastructure and symbiotic bacteria in the tropical lucinid, *Linga pensylvanica* (Linne). *Symbiosis*, **20**, 259–280.

——, —— & —— 1998*a*. Gill filament differentiation and experimental colonization by symbiotic bacteria in aposymbiotic juveniles of *Codakia orbicularis* (Bivalvia: Lucinidae). *Invertebrate Reproduction and Development*, **34**, 219–231.

——, DURAND, P., FRENKIEL, L. & MOUËZA, M. 1998*b*. Putative environmental transmission of sulfur-oxidising gill endosymbionts in four tropical lucinid bivalves, inhabiting various environments. *FEMS Microbiology Letters*, **160**, 257–262.

HERRY, A. & LE PENNEC, M. 1987. Endosymbiotic bacteria in the gills of the littoral bivalve molluscs *Thyasira flexuosa* (Thyasiridae) and *Lucinella divaricata* (Lucinidae). *Symbiosis*, **4**, 25–36.

——, DIOURIS, M. & LE PENNEC, M. 1988. Chemoautotrophic symbionts and translocation of fixed carbon from bacteria to host tissues in the littoral bivalve *Loripes lucinalis* (Lucinidae). *Marine Biology*, **101**, 305–312

HICKMAN, C. S. 1994. The genus *Parvilucina* in the

Eastern Pacific: making evolutionary sense of a chemosymbiotic species complex. *Veliger*, **37**, 43–61

JACKSON, J. B. C. 1973. The ecology of molluscs of *Thalassia* communities, Jamaica, West Indies. 1. Distribution, environmental physiology, and ecology of common shallow-water species. *Bulletin of Marine Science*, **23**, 313–350.

JANSSEN, H. H. 1992. Philippine bivalves and micro-organisms: past research. present progress and a perspective for aquaculture. *Philippine Scientist*, **29**, 5–32.

JOHNSTON, P. A. 1993. Lower Devonian Pelecypoda from southeastern Australia. *Memoirs of the Association of Australasian Palaeontologists*, **14**, 1–134.

—— & GOODBODY, Q. H. 1988. Middle Devonian bivalves from Melville Island, Arctic Canada. *Memoirs of the Canadian Society of Petroleum Geologists*, **14**(3), 337–345

KAUFFMAN, E. G., ARTHUR, M. A., HOWE, B. & SCHOLLE, P. A. 1996. Widespread venting of methane-rich fluids in Late Cretaceous (Campanian) submarine springs (Tepee Buttes), Western Interior seaway, U.S.A. *Geology*, **24**, 799–802.

KELLY, S. R. A., BLANC, E., PRICE, S. P. & WHITMAN, A. G. 2000. Early Cretaceous giant bivalves from seep-related limestone mounds, Wollaston Foreland, Northeast Greenland. *This volume*.

KENNICUTT, M. C., BROOKS, J. M., BIDIGARE, R. R., FAY, R. R., WADE, T. L. & MCDONALD, T. J. 1985. Vent type taxa in a hydrocarbon seep region on the Louisiana slope. *Nature*, **317**, 351.

KIIASHKO, S. I., RODIONOV, I. A. & IUSHIN, V. V. 1989. Contribution of symbiotic chemoautotrophic procaryota to the diet of shallow water lucinacean bivalves. *Doklady Akademii Nauk SSSR*, **308**(4), 1021–1023. [in Russian].

KREUGER, D. M., GUSTAFSON, R. G. & CAVANAUGH, C. M. 1996. Vertical transmission of chemoautotrophic symbionts in the bivalve *Solemya velum* (Bivalvia: Protobranchia). *Biological Bulletin*, **98**, 60–65.

LE PENNEC, M., BENINGER, P. & HERRY, A. 1995. Feeding and digestive adaptations of bivalve molluscs to sulphide-rich habitats. *Comparative Biochemistry and Physiology*, **111A**, 183–189.

LILJEDAHL, L. 1991. Contrasting feeding strategies in bivalves from the Silurian of Gotland. *Palaeontology*, **34**, 219–235.

MCALESTER, A. L. 1965. Systematics, affinities, and life habits of *Babinka*, a transitional Ordovician lucinoid bivalve. *Palaeontology*, **8**, 231–246.

MCFALL-NGAI, M. J. 1998. The development of cooperative associations between animals and bacteria: establishing détente among domains. *American Zoologist*, **38**, 593–608.

MOORE, H. B., DAVIES, L. T., FRASER, T. H., GOLRE, R. H. & LOPEZ, N. R. 1968. Some biomass figures from a tidal flat in Biscayne Bay, Florida. *Bulletin of Marine Science*, **18**, 261–279.

MORTON, B. 1979. The biology and functional morphology of the coral-sand bivalve *Fimbria fimbriata* (Linnaeus, 1758). *Records of the Australian Museum*, **32**, 389–420.

MORTON, J. E. 1972. The form and functioning of the pallial organs in the opisthobranch *Akera bullata* with a discussion of the of the nature of the gill in Notaspidea and other tectibranchs. *Veliger*, **14**, 337–349.

—— 1988. The pallial cavity. *In:* TRUEMAN, E. R. & CLARKE, M. R. (eds) *The Mollusca, Volume 11. Form & Function.* Academic Press, London, 253–286.

NARCHI, W. & FARANI ASSIS, R. C. 1980. Anatomia functional de *Lucina pectinata* (Gmelin, 1791) Lucinidae-Bivalvia. *Boletim de Zoologia Universidade de Sao Paulo*, **5**, 79–110.

OKUTANI, T. & HASHIMOTO, J. 1997. A new species of lucinid bivalve (Heterodonta: Lucinidae) from Kanesu-no-Se Bank near mouth of Surga Bay, with a review of the Recent species of the chemosynthetic genus *Lucinoma* from Japan. *Venus*, **56**, 271–280.

OLIVER, P. G. 1986. A new lucinid bivalve from the Niger Delta and an appraisal of the *Loripes* group (Bivalvia, Lucinacea). *Basteria*, **50**, 47–64.

—— 1995. Bivalvia. *In:* DANCE, S. P. (ed.) *Seashells of Eastern Arabia*. Motivate Publishing, London, 196–281.

PECKMANN, J., THIELE, V., MICHAELIS, W., CLARI, P., GAILLARD, C., MARTIRE, L. & REITNER, J. 1999. Cold seep deposits of Beauvoisin (Oxfordian: southeastern France) and Marmorio (Miocene: northern Italy): microbially induced authigenic carbonates. *International Journal of Earth Sciences*, **88**, 6–75.

PELSENEER, P. 1911. Les lamellibranches de l'expédition du Siboga, partie anatomique. *Siboga Expéditie., Siboga Expédition Monograph*, **53**a, 1–125.

POJETA, J. 1978. The origin and early taxonomic diversificatiopn of the pelecypods. *Philosophical Transactions of the Royal Society of London, Series B*, **284**, 225–246.

REID, R. G. B. 1990. Evolutionary implications of sulphide-oxidising symbioses in bivalves. *In:* MORTON, B. (ed.) *The Bivalvia – Proceedings of a Memorial Symposium in Honour of Sir Charles Maurice Yonge, Edinburgh, 1986*. Hong Kong University Press, 127–140.

—— 1998. Superfamily Lucinoidea. *In:* BEESLEY, P. L., ROSS G. J. B. & WELLS, A. (eds) *Mollusca: the Southern Synthesis. Fauna of Australia. Volume 5, Part A*. CSIRO Publishing, Melbourne, 309–315.

——& BRAND, D. G. 1986. Sulfide-oxidising symbiosis in lucinaceans: implications for bivalve evolution. *Veliger*, **29**, 3–24.

RODIONOV, I. A. & YUSHIN, V. V. 1991. Procaryotic symbionts in gill cells of the bivalve mollusc *Pillucina pisidium. Biologiya Morya* (Vladivostok), **1991**(1), 39–46 [in Russian].

SAUL, L. R., SQUIRES, R. L. & GOEDERT, J. L. 1996. A new genus of cryptic lucinid? bivalve from Eocene cold seeps and turbidite-influenced mudstone, western Washington. *Journal of Paleontology*, **70**, 788–794.

SCHWEIMANNS, M. & FELBECK, H. 1985. Significance of the occurrence of chemoautotrophic bacterial symbionts in lucinid clams from Bermuda. *Marine Ecology Progress Series*, **24**, 113–120.

SKELTON, P. W., CRAME, J. A., MORRIS, N. J. & HARPER, E. M. 1990 Adaptive divergence and taxonomic radiation in post-Palaeozoic bivalves. *In*: TAYLOR, P. D. & LARWOOD, G. P. (eds) *Major Evolutionary Radiations*. Systematic Association, Special Volume, **42**, 91–117.

SEMPER, K. 1880. *Die Natürlichen Existenzbedingungen der Thiere*. F.A. Brockhaus, Leipzig.

SIBUET, M. & OLU, K. 1998. Biogeography, biodiversity and fluid dependence of deep-sea cold-seep communities at active and passive margins. *Deep-Sea Research II*, **45**, 517–567.

SOUTHWARD, E. C. 1986. Gill symbionts in thyasirids and other bivalve molluscs. *Journal of the Marine Biological Association, UK*, **66**, 889–914.

SQUIRES, R. L. & GRING, M. P. 1996. Late Eocene chemosynthetic? bivalves from suspect cold seeps, Wagonwheeel Mountain, Central California. *Journal of Paleontology*, **70**, 62–73.

STANLEY, S. M. 1970. Relation of shell form to life habits of the Bivalvia (Mollusca). *Geological Society of America Memoir*, **125**, 1–296.

TAVIANI, M. 1994. The 'calcari a *Lucina*' macrofauna reconsidered: deep-sea faunal oases from Miocene-age cold vents in the Romagna Apennine, Italy. *Geo-Marine Letters*, **14**, 185–191.

TAYLOR, J. D. & GLOVER, E. A. 1997*a*. The lucinid bivalve genus *Cardiolucina* (Mollusca, Bivalvia, Lucinidae): systematics, anatomy and relationships. *Bulletin of the Natural History Museum, London (Zoology)*, **63**, 93–122.

—— & —— 1997*b*. A chemosymbiotic lucinid bivalve (Bivalvia: Lucinoidea) with periostracal pipes: functional morphology and description of a new genus and species. *In*: WELLS, F. E. (ed.) *The Marine Flora and Fauna of the Houtman Abrolhos Islands, Western Australia*. Western Australian Museum, Perth, 335–361.

—— & LEWIS, M. S. 1970. The flora, fauna and sediments of the marine grass beds of Mahé, Seychelles. *Journal of Natural History*, **4**, 199–220.

TURNER, R. D. 1985. Notes on mollusks of deep-sea vents and reducing sediments. *American Malacological Bulletin*, **1**, 23–34.

VETTER, R. D. 1985. Elemental sulphur in the gills of three species of clams containing chemautotrophic symbiotic bacteria: a possible inorganic storage compound. *Marine Biology* **88**, 33–42.

Early Cretaceous giant bivalves from seep-related limestone mounds, Wollaston Forland, Northeast Greenland

SIMON R. A. KELLY, ERIC BLANC, SIMON P. PRICE[1] & ANDREW G. WHITHAM

CASP, University of Cambridge, West Building, 181A Huntingdon Road, Cambridge CB3 0DH, UK

[1]Present address: Shell UK Exploration and Production, 1 Altens Farm Road, Nigg, Aberdeen AB12 3FY, UK

Abstract: Anomalous mound-forming limestones, here termed the Kuhnpasset Beds, occur within Late Barremian (Early Cretaceous) mudstones on Wollaston Forland, Northeast Greenland. The normal mudstones contain a sparse fauna of small nuculoids, arcoids and inoceramids; by contrast, the mounds contain an unusual faunal assembage, dominated by large bivalves. These include an abundant lucinid, *Cryptolucina kuhnpassetensis* sp. nov., and, less commonly, *Solemya*, both known seep-associated genera. Locally, a large modiomorphid, *Caspiconcha whithami* gen. et sp. nov., is common and reaches > 300 mm in length and has a shell up to 28 mm thick. Also, the wood-boring bivalve *Turnus* is abundant in driftwood. Gastropods are rare, but the associated cephalopod fauna includes ammonites, belemnites, nautiloids and a remarkable large orthoconic phragmocone. The form of the mounds with calcite-cemented tube systems, associated laminated calcite crusts and void fills, together with the fauna, is analogous to those of methane-based cold-seep complexes. However, preliminary studies indicate that much of the original aragonitic shell is now replaced by silica. This precluded conclusive geochemical studies based on the shells themselves. It is believed that the mounds formed on the seafloor in a mid- to outer shelf situation at the end of a period of extensional rifting on the eastern Greenland passive Atlantic margin. The vents occur near the footwall crest of a tilted fault block. The underlying faults may have provided routes or influenced direction of movement for nutrient migration. Source rocks were probably the Late Jurassic black shales from depths of < 600–1200 m. If methane was being generated, it was probably forming by shallow-depth organic breakdown rather than by thermogenic processes, which require greater burial.

Bivalves from deep-sea environments have been attracting increasing attention since the description of *Calyptogena* by Boss & Turner (1980) from hydrothermal vents on the Galapagos Rift. Faunas are known from both hot vents and cold seeps. Hot-vent faunas may be associated with various situations, including active generative plate margins at mid-ocean ridges (Van Dover 1985). Cold-seep faunas are usually associated with migration of fluids on continental shelves and are known from forearc basins and from passive margins, and are often related to faulting at depth (Peek *et al.* 1997). Methane-seep faunas may be associated with localized limestone cementation and even chimney formation (Kulm & Suess 1990; Campbell & Bottjer 1993; Jørgensen 1994). Carbonates may be also associated with sulphate reduction from hydrogen sulphide. Massive sulphide deposits are associated with their own specialized biota (Little *et al.* 1999). The geological history of vent-related bivalves is documented for much of the Phanerozoic back to the Silurian at least, but the succession is incomplete and is patchily known (Little *et al.* 1999). Vent- and seep-related faunas are not depth related and are known from environments ranging from deep sea (*c.* 4000 m depth; Mayer *et al.* 1988) to intertidal (Dando *et al.* 1994). They are not geographically restricted and the geological record demonstrates that they have occurred from the Arctic (Beauchamp & Savard 1992) to the Antarctic (Kelly *et al.* 1995). The limited sites available for their particular specialized physiology to function also controls their distribution. Ephemeral chemoautotrophic assemblages may even develop on foundered shipping cargoes (Dando *et al.* 1992) and on recently dead whale carcasses (Smith *et al.* 1989), e.g. on Oligocene whale carcasses (Squires *et al.* 1991). These sites are ephemeral and only exist with living faunas while the key nutrients, such as hydrogen sulphide or methane, are being generated. As soon as the source declines, the specialized biota die.

From: HARPER, E. M., TAYLOR, J. D. & CRAME, J. A. (eds) *The Evolutionary Biology of the Bivalvia*. Geological Society, London, Special Publications, **177**, 227–246. 1-86239-076-2/00/$15.00

Organisms require special lifestyles to inhabit seeps. Bivalves have adapted to living under such unusual conditions by chemosymbiotic association with bacteria which are cultivated on modified gills. In the absence, or decrease in the significance of, filter feeding the gut may become reduced in lucinids (Reid & Brand 1986; Taylor & Glover 2000). Some solemyids are completely gutless (Cavanaugh 1985). There are two principal chemautotrophic types: the sulphur oxidizers and the methanotrophs (Childress *et al.* 1986). Adult bivalves are relatively static and are life dependant on the activity of the seep. Special reproductive strategies must be adopted for survival of their species. Ideally, growth to sexual maturity must be rapid in case the seep ceases. Annual shell growth rates of 60 mm a^{-1} have been recorded (Lutz *et al.* 1985, 1994). The larvae must develop initially under seep conditions and then adapt to normal marine conditions during the pelagic dispersal

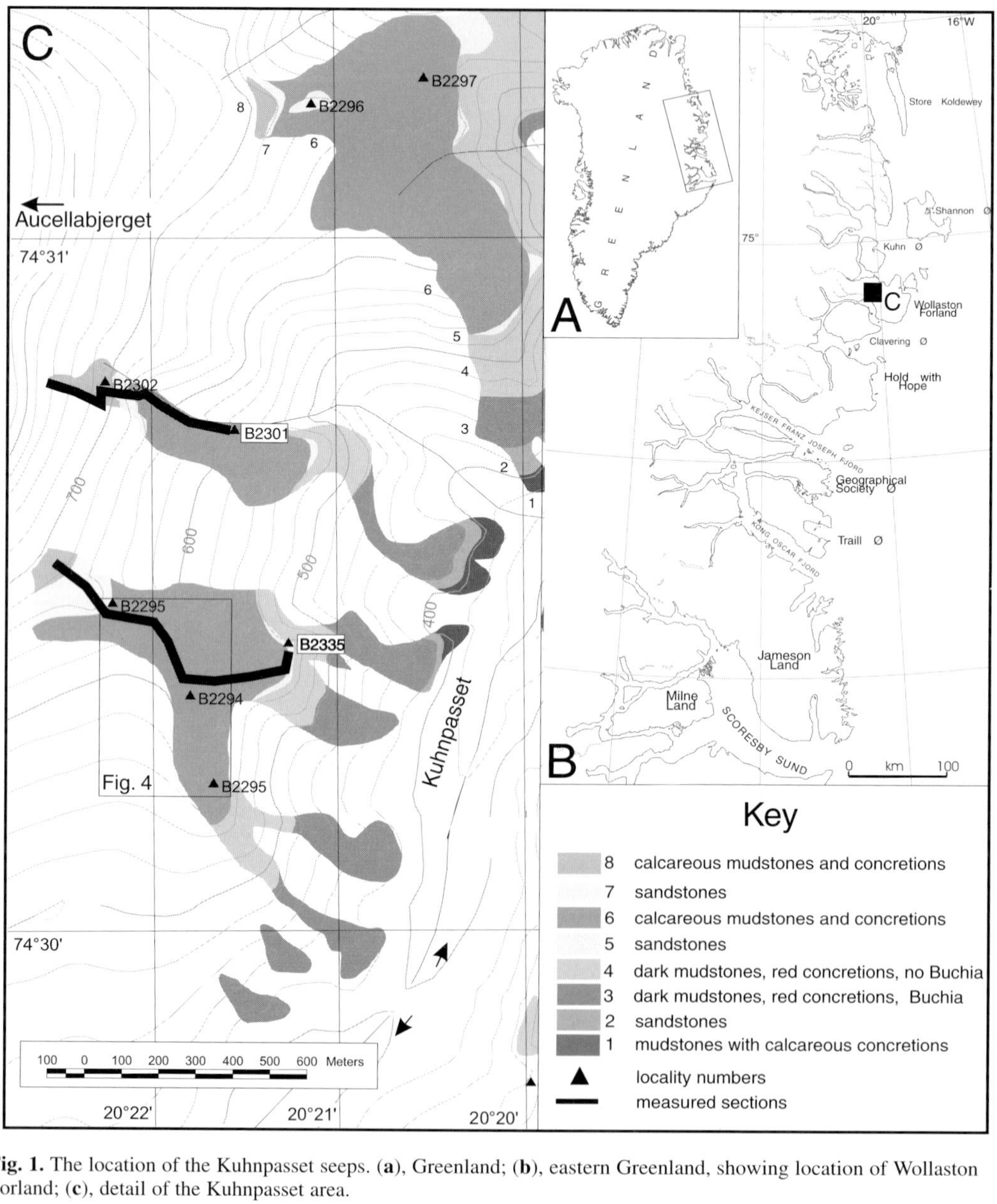

Fig. 1. The location of the Kuhnpasset seeps. (**a**), Greenland; (**b**), eastern Greenland, showing location of Wollaston Forland; (**c**), detail of the Kuhnpasset area.

phase before seeking a further seepage to colonize, and in which to grow to adulthood (Lutz *et al.* 1984; Berg 1985).

The initial field recognition of fossil methane seeps may be problematic, especially at low latitudes, where limestones are normal shallow-water sediments. However, at high latitudes limestones are atypical. At these latitudes the occurrence of anomalous localized limestone bodies in sediments which otherwise suggest deep-water conditions is usually the first indication that one is dealing with a seep. For example, the 3 m thick limestone in the Fossil Bluff Group of Antarctica described by Kelly *et al.* (1995) was the only carbonate bed in *c.* 7 km of forearc basin clastic deposits. Authigenic limestones have not been recorded previously from Cretaceous rocks in eastern Greenland. However, it is only by laboratory based geochemical studies that the methane source can be confirmed.

This article documents the discovery, by Cambridge Arctic Shelf Programme (CASP) geologists, of limestone mounds with an abundant bivalve fauna in eastern Greenland. The site at Kuhnpasset (Fig. 1c) represents a seep of Barremian (Early Cretaceous) age in a passive margin situation. The observations made are based upon field sedimentological descriptions and observations on the unusual bivalves themselves. Initial geochemical studies of the shells failed because of extensive silicification. However, future stable isotope studies of the adjacent carbonates should provide more conclusive evidence on the nature of the fluids supporting the vent fauna.

General setting

The newly discovered limestone mounds of the Kuhnpasset Beds (Kelly *et al.* 1999) occur in a mudstone-dominated succession on the west side of Kuhnpasset, which forms the eastern slopes of Aucellabjerget, in western Wollaston Forland (Figs 1–5). The mounds are distributed laterally for *c.* 2 km, at an altitude of *c.* 550–650 m, along the outcrop of near-horizontal strata. Gneissic Caledonian basement lies at depth (Fig. 3), probably at about sea level, beneath Kuhnpasset. Gneiss crops out in adjacent areas on the upthrown parts of tilted fault blocks. The basement is overlain unconformably by Middle–Late Jurassic sediments. These, in turn, are overlain unconformably by the marine Wollaston Forland Group (Surlyk 1978), which was deposited during Volgian–Hauterivian north-northeast–south-southwest extensional faulting, with lithologies ranging from conglomerates to mudstones with red beds (Surlyk 1977). This dating is based on modifications to the buchiid bivalve stratigraphy of Surlyk & Zakharov (1982). A mudstone-dominated succession of Hauterivian–Aptian age (Nøhr-Hansen 1993; this study) is overlain by Tertiary plateau basalts (Fig. 5a and b). Albian strata are known further east in Wollaston Forland (Maync 1949; Nøhr-Hansen 1993). There is no useful formal lithostratigraphy for the post-rift mudstone-dominated succession at Kuhnpasset, although Nøhr-Hansen (1993, fig. 3) used the generalized term 'mid-Cretaceous sandy shale sequence', which is not satisfactory. These sediments correlate with the finer grained parts of the Steensbybjerg Formation of Hold with Hope to the south (Kelly *et al.* 1998).

Strata here attributed to the Kuhnpasset Beds were first described by Maync (1949, p. 87), who recognized large extremely fossiliferous concretions at an altitude of 550–560 m on the southeastern slope of Aucellabjerget. He documented an 'Aptian' fauna, including *Lytoceras polare* Ravn etc., and large white-shelled bivalves (*?Aucellina* sp.), gastropods and fossil wood in great quantities. The distinctive macrofauna was not formally described and has not been studied until now. Maync's (1949) record of *?Aucellina* has not been confirmed.

The sections

The two principal stratigraphical sections containing well-exposed Kuhnpasset Beds with rich bivalve faunas are shown in Fig. 2. In section B2301 a complete section through the beds is steeply exposed in a gully, but only contains a single limestone mound in vertical section. Section B2335 shows mounds at different heights in the section. A single, poorly exposed, mound occurs near the base, but at least 26 mounds are concentrated along the blunt crest of a ridge (Fig. 4). Other poorer exposures at B2296 and B2297 (Fig. 1) reveal much fragmentary limestone and bivalve material weathering out, but no detailed section measurements were made.

Form and composition of mounds

Over 30 limestone mounds have been recorded in the Kuhnpasset area so far (Figs 1 and 4; Table 1) and they have received individual numbers for recording purposes. Where they are best exposed in section B2335, the mounds vary from 1 to 3 m in diameter and are up to 1.8 m in height. They are subcircular to suboval in plan view. The original profiles of most mounds have been modified by Recent erosion and weathering, as well as obscured by downwash. The present exposures show generally lenticular to subconical profiles with fairly flat bases (Fig. 5c–f). They owe their

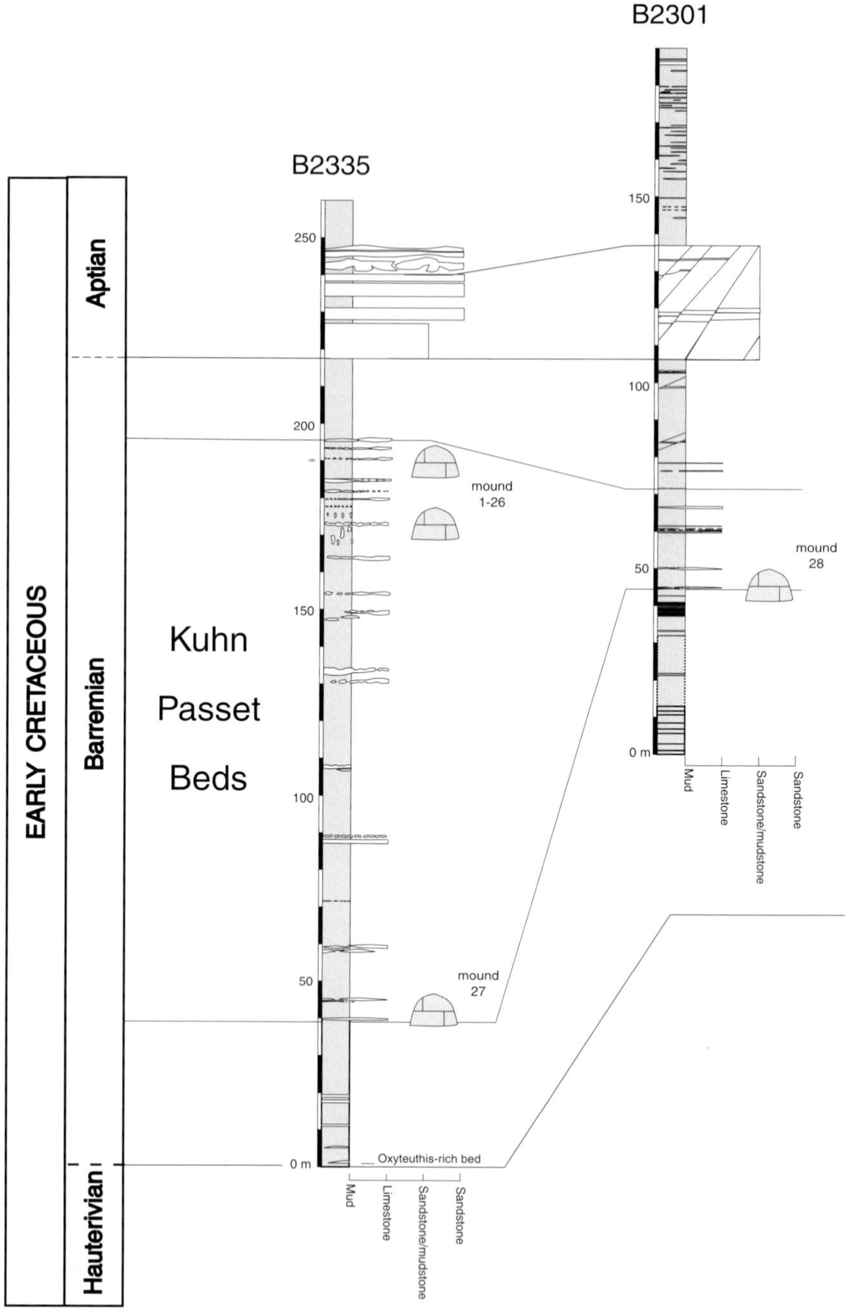

Fig. 2. Correlation of the principal logged sections at Kuhnpasset. For location see Fig. 1c.

prominence to the cementation of the mounds, in contrast to the poorly cemented adjacent, normal mudstones. In this respect they are similar to the weathered-out carbonate mounds of the Cretaceous Teepee Buttes of Colorado (Petta & Gerhard 1977; Kauffman *et al.* 1996).

The composition of the mounds is not uniform and their characteristic features are shown in Table

Table 1. *Distribution of lithological features and fossils in Kuhnpasset limestone mounds*

Locality No.	Limestone mound no.	Width (m)	Height (m)	Homogenized, bioturbated sediment	Vertical/oblique calcite cemented tubes, *c.* 50–100 mm diameter	Concentrically laminated carbonate crusts	Vugs	Septarian fracturing with sparry infill	*Caspiconcha whithami* gen. et sp. nov.	*Cryptolucina kuhnpassetensis* sp. nov.	*Solemya* sp.	*Propeamussium* sp.	Nuculacean	*Turnus* sp.	Wood	Plant frond	Gastropod	*Lytoceras* sp.	Aconeceratid	Heteromorph	*Cymatoceras* sp.	*Oxyteuthis* sp.	Giant phragmocone
B2335	1	2.5 × 2.5	1.8			p	p		c			1			p			p				p	
B2335	2	2.5 × 3.0	1.0				p			1					p			c	1	1			1
B2335	3a	inc	inc			p	p	p										p					
B2335	3b	3.0 × 3.0	1.0	p	p													p			1		1
B2335	3c	inc	inc				p																
B2335	4	inc	inc	p	p			p															
B2335	5	no records																					
B2335	6	no records																					
B2335	7	1.5 × 1.5	0.8	p	p		p			p	1				p		1						2
B2335	8	1.5 × 1.5	0.8							p	4												
B2335	9	3.0 × 3.0	inc	p	p			p							p								
B2335	10	1.5 × 1.5	0.3		p		p																
B2335	11	1.5 × 2.0	0.8	p	p																		
B2335	12	no records																					
B2335	13	no records																					
B2335	14	no records																					
B2335	15	1.0 × 1.0	inc	p																			
B2335	16	1.0 × 1.0	0.4	p	p		p																
B2335	17	1.0 × 1.0	*c.* 1.0		p																		
B2335	18	1.0 × 1.5	0.8	p	p			p															
B2335	19	1.0 × 1.5	0.8+	p	p			p										1		1			
B2335	20	no records																					
B2335	21	inc	inc							p				c	p	p		1					
B2335	22	inc	inc	p			p		9	p													
B2335	23	no records																					
B2335	24	2.0 × 3.0	1.0				p		p	p	c			c	p								
B2335	25	no records																					
B2335	26	1.5 × 2.5	1.0				p		c	p	p			c	c								
B2335	27					p		p	p									q					
B2296	28	inc	inc		p																		
B2296	29	inc	inc				p			p	p								1	1			
B2301	30	2.0	1.0				p		p	c								p	p				

For location of mounds see Figs 1 and 4. Numeral, number of individuals; c, common/abundant; fr, frond; inc, incomplete exposure; p, present; shaded, bivalve bearing.

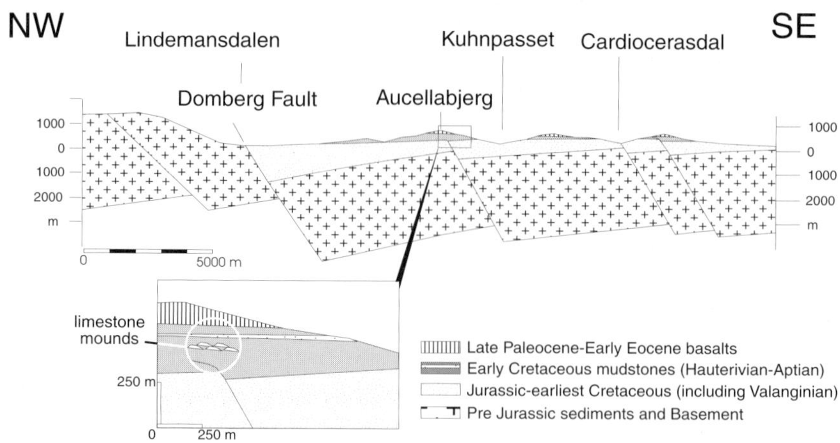

Fig. 3. Simplified geological cross-section of west Wollaston Forland. Inset shows detail of position of limestone mounds (circled).

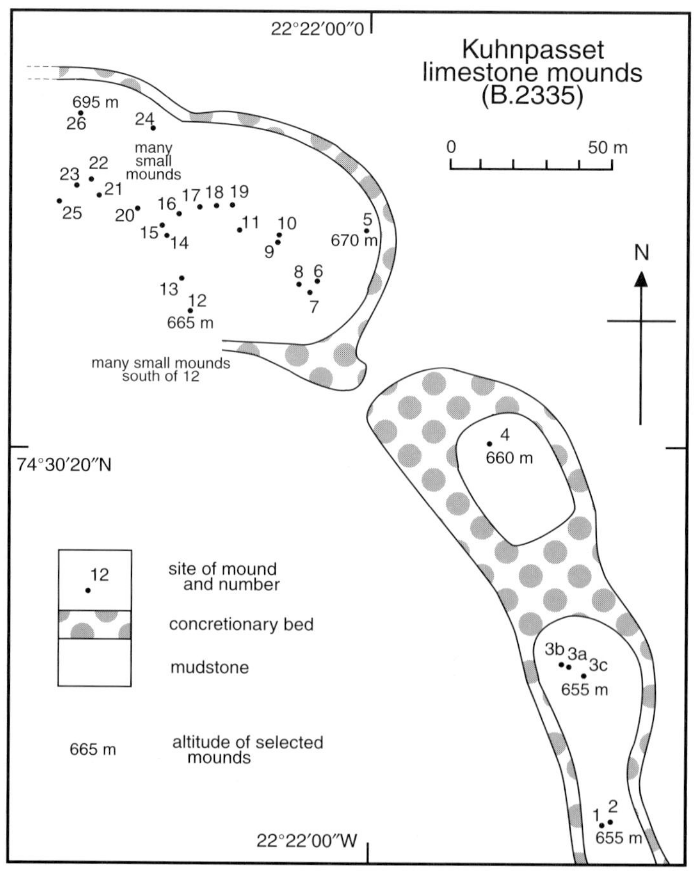

Fig. 4. Sketch map of the field occurrences of numbered limestone mounds at Kuhnpasset.

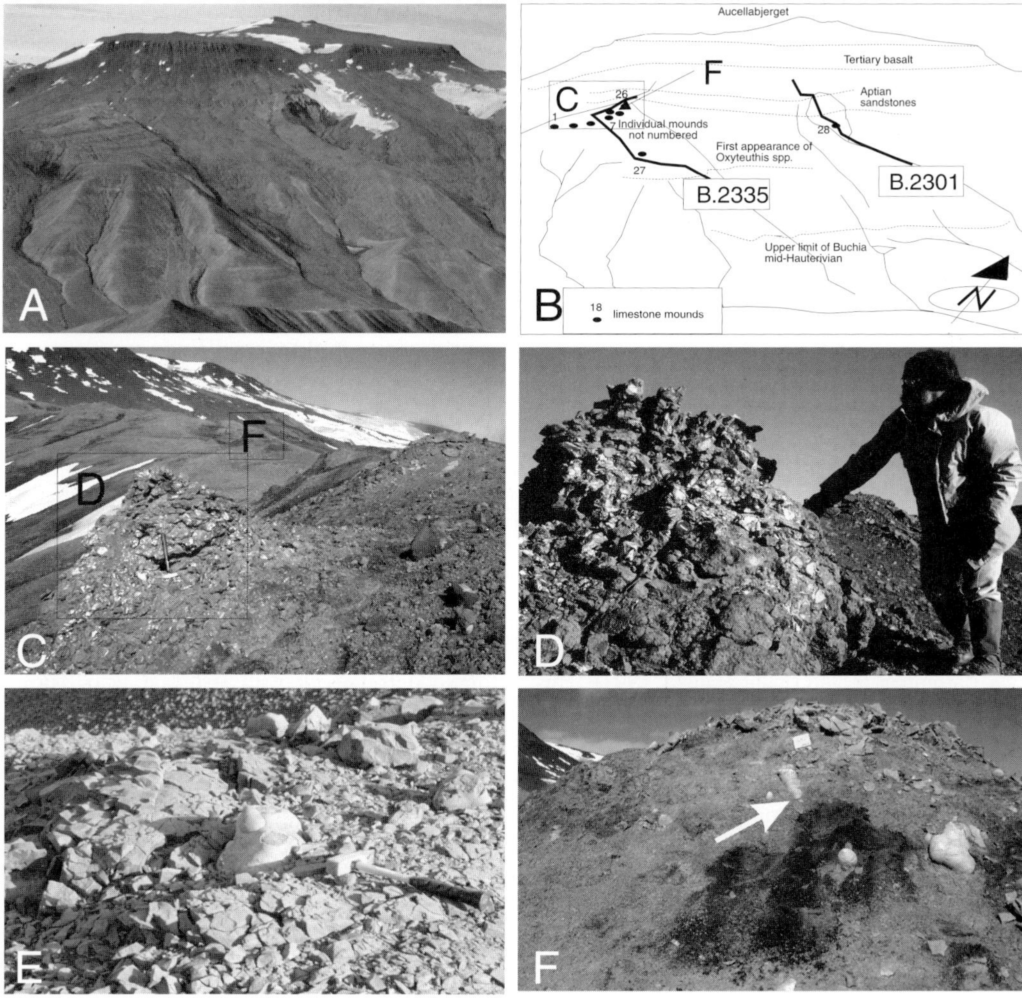

Fig. 5. Kuhnpasset limestone mounds. (**a**) General view of the west face of Kuhnpasset and Aucellabjerget and (**b**) sketch of geological interpretation; (**c**), view northward across locality B2335 with limestone mound 1 in foreground; (**d**), detail of shell-rich mound 1; (**e**), oblique view of top of mound 28 at locality B2296, showing cemented vertical tube at top of seep; (**f**), lateral surface view of mound 4, locality B2335, showing vertical cemented tube (arrowed) at base of seep.

1. The mounds themselves are constructed of grey, unbedded, muddy to sandy limestone. Bedding has been homogenized, probably by burrowing activity. Betweeen the mounds are dark grey, poorly consolidated silty to sandy mudstones. Within the mounds preferentially cemented tubular structures commonly occur. They are oblique to subvertical, usually *c.* 50–100 mm in diameter (Fig. 5e and f) and probably represent the seep plumbing system. Septarian fracturing may occur at the core of the mounds. Irregular to rounded, complex, laminated carbonate structures up to 300 mm in diameter occur in several seeps. They all have an almost blackened exterior and are partly pink on the interior. They show complex growth history with laminations usually in the < 1 mm scale, radiaxial cements and late stage sparry calcite. These compare to those illustrated by Campbell & Bottjer (1993, p. 332, fig. 5) and Kelly *et al.* (1995, p. 278, fig. 6A). Voids in some tubular structures and steinkerns of bivalves usually have a sparry calcite infilling. The fauna within the mounds is preserved uncrushed, while that of the surrounding mudstones is usually flattened. This demonstrates early diagenetic cementation within the mounds. The shells of most of the molluscs represented in the mounds

were originally aragonitic but are now largely silicified. This alteration precludes geochemical study of the isotopic compositions of these shells, which could have established more precisely the source of the carbonate.

Beneath the principal mound occurrence, in locality B2335, is a level of regular subspherical, sideritic(?) 100–150 mm concretions, which appear to be related to horizontal burrows of a *Thalassinoides* type, because they occasionally demonstrate partially preserved Y-shaped branching and vertical tubes typical of that genus. Some burrows show a diameter of *c.* 40 mm.

Age of the Kuhnpasset Beds

Although Maync (1949) ascribed an Aptian age to the strata here termed the Kuhnpasset Beds, the present authors believe that they are Late Barremian. Equivalent mudstones in the adjacent Cardiocerasdal were regarded as Aptian–Albian by Surlyk (1977, p. 47, fig. 37). However, the limestone mound levels occur within the dinoflagellate cyst *Pseudoceratium toveae* subzone of the *Batioladinium longicornutum* zone, which is Late Barremian in age (Nøhr-Hansen 1993). This dating is confirmed by CASP's independent studies on the dinocysts (A. M. Koraini pers. comm.), by the associated ammonites (lytoceratids, *Sanmartinoceras, Audouliceras* and *Epicheloniceras*) (R. Casey and P. F. Rawson pers. comm.) and by the belemnites (*Oxyteuthis*) (P. Doyle pers. comm.). The Late Barremian Kuhnpasset macrofaunas correlate with so-called Aptian faunas previously described from Kuhn Ø (Bøgvad & Rosenkrantz 1934) and Store Koldewey (Frebold 1935) [See also list of Donovan (1957, pp. 208–209), especially taxa labelled 'Upper Aptian']. Deshayesitid ammonites occur in the sandstones from above the mound-bearing levels of Kuhnpasset at locality B2303 (Fig. 5a and b) and indicate an Early Aptian age (Kelly & Whitham 1999).

Fauna of the Kuhnpasset Beds

The benthic fauna of the limestone mounds is anomalous, being very local in distribution, but very abundant, and of an unusually large size in comparison to the otherwise undistinguished benthos in adjacent sediments. Within the mounds, the fauna is dominated by bivalves, particularly by a large lucinid, *Cryptolucina kuhnpassetensis* sp. nov. (Figs 6–8), and by a modiomorphid, *Caspiconcha whithami* gen. et sp. nov. (Figs 9 and 10), which is the largest of the bivalves, reaching > 300 mm in length (Appendix 1). Common driftwood is present with abundant wood-boring *Turnus,* in the borings of *Teredolites clavatus* Leymerie, which may be fortuitously associated with the seeps (Fig. 11). All these bivalves occur usually in life position.

The *Cryptolucina* and *Caspiconcha* are crowded around the cores of some of the limestone mounds,

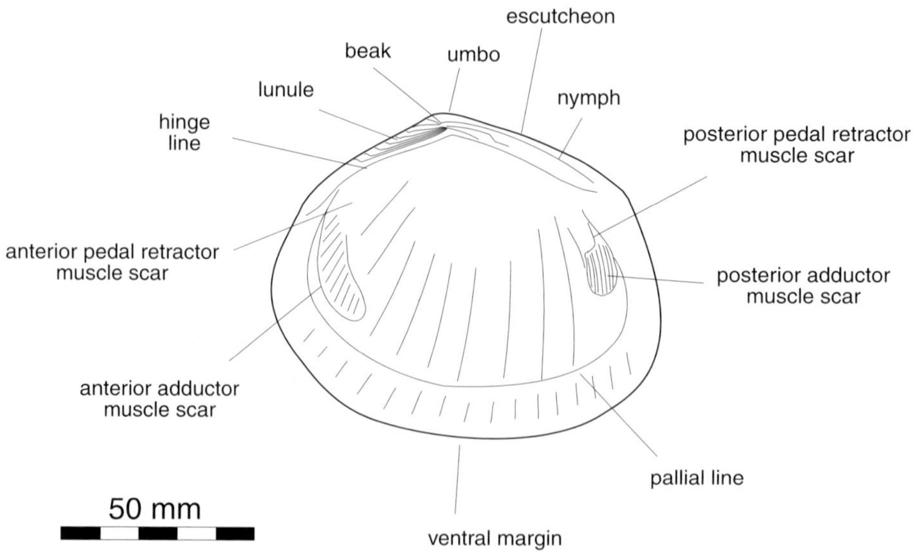

Fig. 6. *Cryptolucina kuhnpassetensis* sp. nov., reconstruction of right valve interior based on specimens from locality B2335.

with *Solemya* (Fig. 12) located less commonly in the outer parts of the mounds. The bivalves are locally rock forming (Figs 5c and d and 8) but were present in only 11 of the 32 mounds examined (Table 1). Other benthos in the mounds are rare but include a bathrotomariid and a limpet-like gastropod. In addition to the ammonites mentioned above, other cephalopods occur, including the nautiloid *Cymatoceras* and a remarkable unidentified orthoconic phragmocone, which reaches *c.* 300 mm in length and > 150 mm in diameter, but with no evidence for a rostrum. These cephalopods may represent active predators and scavengers, presumably attracted to the rich fauna of the mounds.

Lucinids (Figs 6–8)

Lucinids are well known for their chemosymbiotic association with sulphide-oxidizing bacteria located on modified gills (Reid & Brand 1986; Taylor & Glover 2000). The group has a history going back at least to the Silurian (Liljedahl 1991). *Cryptolucina kuhnpassetensis* occurs in 13 of the 30 mounds, usually as paired valves. Locally, it forms dense clusters in life position (Fig. 8) at the core of mounds with scattered isolated valves and fragments in the vicinity. It is referred to an edentulous genus, *Cryptolucina,* originally described from the carbonate mounds of the Humptulips Formation, Eocene of Washington, western USA (Saul *et al.* 1996).

Lucinids are characteristic of many methane-seep limestone mounds. They often reach a very large size but few are systematically well described. Mesozoic examples include: the Oxfordian Terres Noires of southeastern France (Rolin *et al.* 1990; Gaillard *et al.* 1992), up to 145 mm in length; the Tithonian Fossil Bluff Group of Antarctica (Kelly *et al.* 1995), reaching only 30 mm in length; possibly the Late Jurassic–Early Cretaceous Knoxville Beds of California (Saul *et al.* 1996); the Hauterivian Lentiles à Peregrinelles of southeastern France (Thieuloy 1972; Lemoine *et al.* 1982), up to 210 mm in length. *Nymphalucina* occurs in the Campanian Tepee Buttes of Colorado (Petta & Gerhard 1977; Howe & Kauffman 1986). Saul *et al.* (1996) record post-Mesozoic occurrences of *Cryptolucina*.

Modiomorphids (Figs 9 and 10)

Modiomorphids have a history of seep association going back to the Devonian (Little *et al.* 1999).

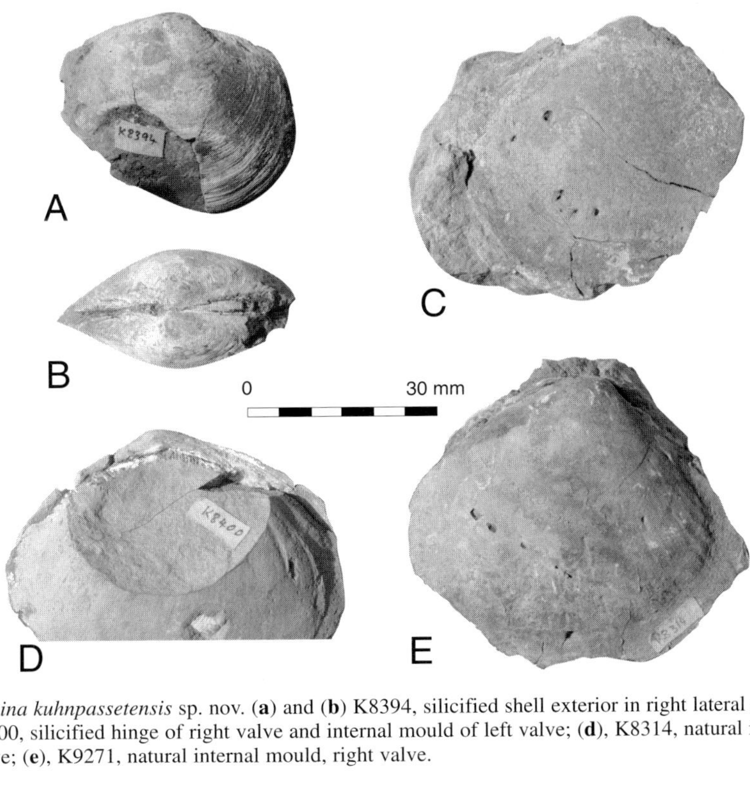

Fig. 7. *Cryptolucina kuhnpassetensis* sp. nov. (**a**) and (**b**) K8394, silicified shell exterior in right lateral and vertical aspects; (**c**), K8400, silicified hinge of right valve and internal mould of left valve; (**d**), K8314, natural internal mould, right valve; (**e**), K9271, natural internal mould, right valve.

Fig. 8. Loose block immediately below limestone mound 7, locality B2335, showing dense *in situ* cluster of *Cryptolucina kuhnpassetensis* sp. nov.

Caspiconcha whithami gen. et sp. nov. is the largest bivalve from the Kuhnpasset seeps, having a shell length of > 300 mm – all specimens seen were of this general size. The species occurred in six out of the 30 mounds. It was seen in close-packed clusters at the core of two mounds, with valves in occlusion and having the posterior directed obliquely upwards. The elongate shape and ventral margin inflexion suggest that it was byssally attached like a large mussel. *Caspiconcha* is more rectangular in profile than the large vesicomyid, *Calyptogena,* which occurs in a number of Tertiary as well as Recent vent sites. The maximum dimension of the fossil *Calyptogena* is also only *c.* 200 mm, although Recent examples have been collected up to 400 mm in length (Corliss & Ballard 1977).

Solemyids (Fig. 11)

Solemya is well known in sulphide-rich environments, occurring in seeps (Kreuger *et al.* 1990), and is known back to the Devonian (Cox in Cox *et al.* 1969). Modern *Solemya* is known to be a symbiont with sulphur-oxidizing chemautotrophic bacteria (Cavanaugh 1985). The genus is known from Late Jurassic seeps associated with limestone mounds in California (Campbell & Bottjer 1993). The Kuhnpassset specimens occur with valves in occlusion in life position, with the long axis in an inclined position. Specimens are usually *c.* 50 mm in length and occur isolated from the main clusters of *Cryptolucina* and *Caspiconcha* in the periphery of the mounds. The figured specimen demonstrates a gape in the antero-ventral margin. This specimen also shows an adherent limpet-type gastropod (Fig. 11).

Wood-boring bivalves (Fig. 12)

Many well-preserved pieces of silicified wood, showing abundant *Teredolites clavatus* Leymerie borings (Kelly & Bromley 1984), occur in association with the mounds. The borings contain *Turnus* sp., which is well known as a wood-boring bivalve in the Cretaceous (Kelly 1988) and is global in its distribution. The borings are up to 25 mm in length and 12 mm in diameter, which is normal for *Teredolites clavatus*. The *Turnus* shells occupy about one-third of the length of the borings, which suggests that they were predominantly filter feeding from the water column, although some wood eating and digestion may also have been taking place. Shells which occupy a smaller proportion of the boring generally have enlarged guts to house cellulose-digesting bacteria, typical of wood-eating bivalves. While *Turnus* is not known to be an obligatory vent-inhabiting bivalve, its preservation in great numbers here is possibly related to the early carbonate cementing at the seep sites. Away from the seeps, the wood would have undergone rapid breakdown, and the fragile and thin shells of *Turnus* would have been destroyed by a combination of premortal predation, pre- and post mortem breakdown of the wood, and post-mortal bioturbation and diagenesis. Driftwood commonly occurs in marine sediments, and may travel great distances before becoming waterlogged and finally reaching the seafloor. The association of common bivalve-bored wood with carbonate mounds is unusual. However, it is also known in the Late Cretaceous of Vancouver (P. A. Johnston pers. comm.) and in the Tertiary of Alaska (K. A. Campbell pers. comm.).

Normal benthos

The normal benthic fauna of the mudstones adjacent to the seeps contrasts with that of the seeps themselves. Bivalves are uncommon, and typically consist of small nuculaceans and arcaceans, with larger inoceramids. A comparable fauna has been described from the so-called Aptian of Store

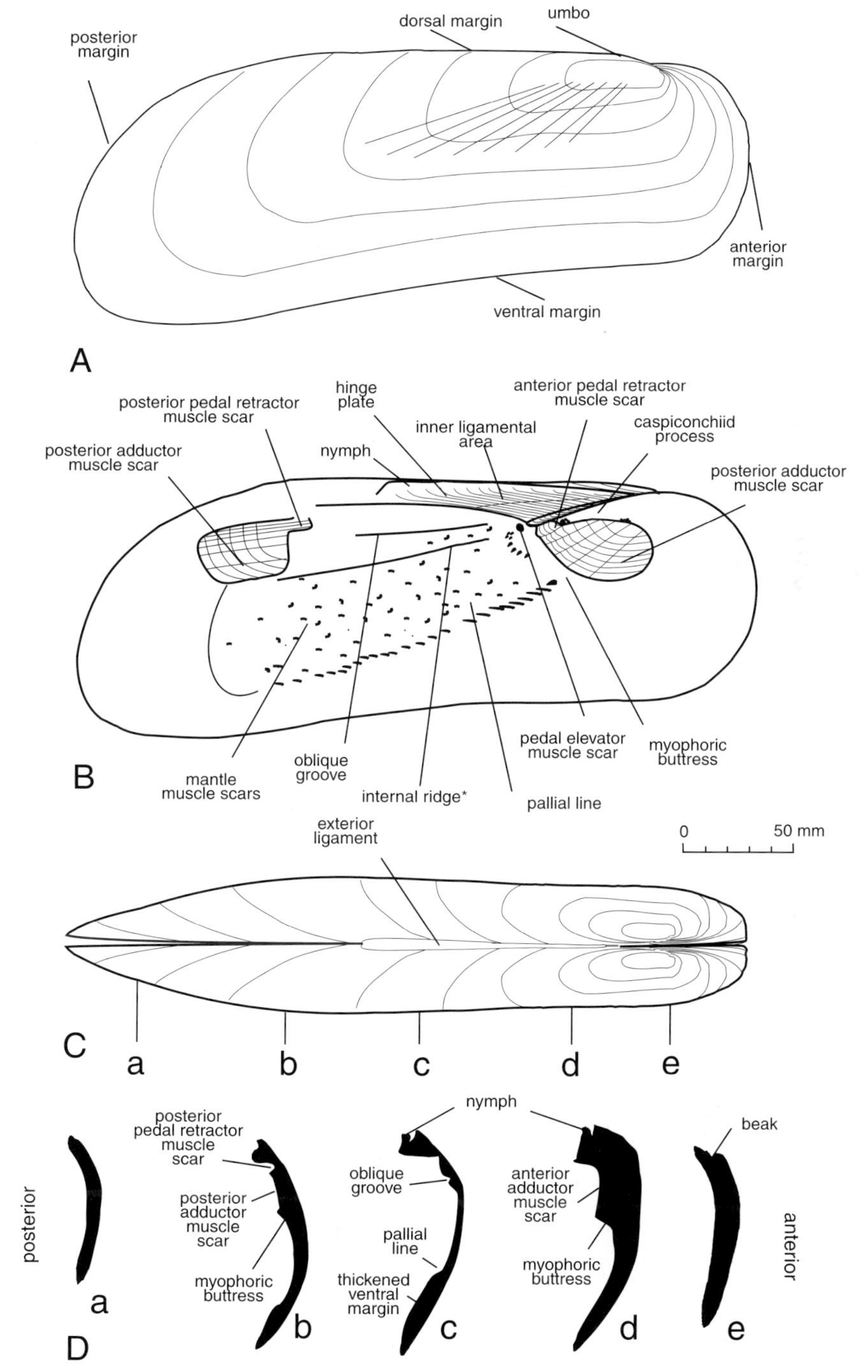

Fig. 9. Sketch reconstruction of *Caspiconcha whithami* gen. et sp. nov. (**a**), Right valve, exterior aspect; (**b**), left valve interior aspect; (**c**), valves in occlusion, dorsal aspect; (**d**), vertical sections showing shell thickness at sites a–e indicated on (C). Based on specimens from locality B2335.

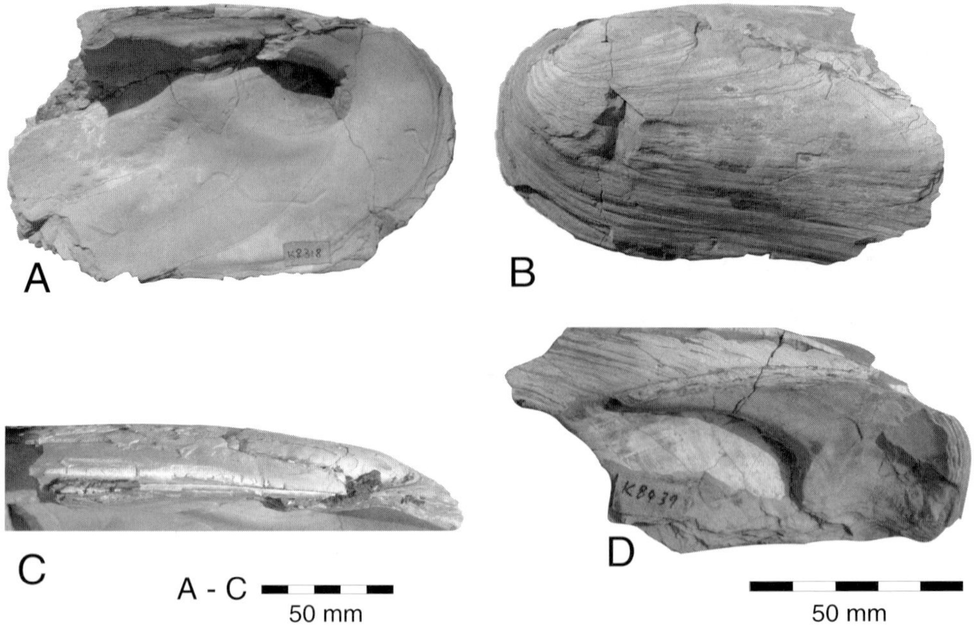

Fig. 10. Silicified shells of *Caspiconcha whithami* gen. et sp. nov. (**a**)–(**c**) K8318, left valve, in lateral, interior and dorsal aspects, respectively; (**d**), K8939, left valve, interior aspect. All specimens from limestone mound 1, locality B2335.

Fig. 11. *Solemya* sp. (**a**) and (**b**) K9269, natural internal mould in left lateral and dorsal aspects showing attached limpet-type gastropod, limestone mound 8, locality B2335.

Koldewey by Frebold (1935). The nuculaceans were deposit feeders, collecting organic material from the sediment; the arcaceans were likely byssate filter feeders. The inoceramid, *Neocomiceramus subneocomiensis* Glazunova, reached a diameter of *c.* 150 mm; it was a byssate, or reclining, filter feeder.

Discussion

Large size of bivalves

The large size of the bivalves deserves comment. *Cryptolucina* reaches 26 mm length and *Caspiconcha* over 300 mm, which is the largest known modiomorphid species. Normally thick and robust shells indicate a turbulent high-energy environment such as encountered in shallow-marine conditions, including rocky shores or reefs. However, the fine-grained nature of the Kuhnpasset Beds indicates a deeper, lower energy environment. The main reason for large size is probably the need to produce very large numbers of progeny to ensure successful and multiple colonization of any other newly formed seeps which may be very distant from the host site. Recent seep-related bivalves, such as vesicomyids, reach a length of 400 mm (Corliss & Ballard 1977). Berg (1985) examined

Fig. 12. Silicified wood with *Teredolites clavatus* Leymerie borings occupied by *Turnus* sp. K8408, limestone mound 7, locality B2335.

the reproductive strategies of molluscs from hydrothermal vents and observed an unusually high fecundity in modern *Calyptogena*. This genus has a particularly large gonad, producing copious lecithotropic larvae year round, as also recorded by Fiala-Medioni & Le Pennec (1989). The bivalve may simply reproduce continuously because its food source, chemautotrophic bacteria, is always available. With no seasonality in growth, the shells are free to grow to large sizes. This also appears to be true for Mesozoic brachiopods (Campbell & Bottjer 1995). Therefore, the large sizes of *Cryptolucina* and *Caspiconcha* may be to house a well-developed gonad. The thick shell in *Caspiconcha* could also have deterred powerful predators. One may also speculate that the thickness is a by-product of the chemosymbiotic processing of large quantities of methane.

Shape of the limestone bodies

The precise shape of limestone mounds at Kuhnpasset is difficult to assess because as soon as the mounds are exposed they are subject to erosion, and boundaries are commonly covered by downwash. However, their physical size and shape appears to be similar to the Late Eocene mounds at Menlo, Washington, described by Goedert & Squires (1990) and illustrated by Campbell & Bottjer (1993, fig. 9). The pyramidal shape of some Kuhnpasset mounds matches that of the slightly larger mound shown in cross-section (Campbell & Bottjer 1993, fig. 6) from the Oligocene Keasey Formation, Oregon. The absence of brecciated fragments of limestone in the Kuhnpasset mounds, and the compactness of each mound, indicates continuously active seepage rather than one in which ponding of gas, such as methane at shallow depth, erupts and causes pockmark structures, such as that which occurs today in the North Sea (Hovland & Judd 1988) or that which occurred in the Late Jurassic of Antarctica (Kelly *et al.* 1995). There does not appear to be any evidence for the mound forming a strong topographical feature on the seafloor. It is more probable that the upper surface of the mound was only in slight relief, as demonstrated in Fig. 13. In fact, the term 'mound' may be a misnomer for these structures, which only take their mound form, standing in relief as relict features, when the softer sediments surrounding them have been removed.

Faunal comparison with other seeps

Faunally, the Kuhnpasset seeps are characterized by large species of *Cryptolucina*, with very large *Caspiconcha* gen. nov. and normal sized *Solemya* occurring less commonly. *Cryptolucina* is particularly important in the Late Jurassic and Early Cretaceous seeps of southeastern France (Lemoine *et al.* 1982; Gaillard *et al.* 1992). There does seem to be a strong relationship between this genus and methane-seep limestones. The presence of

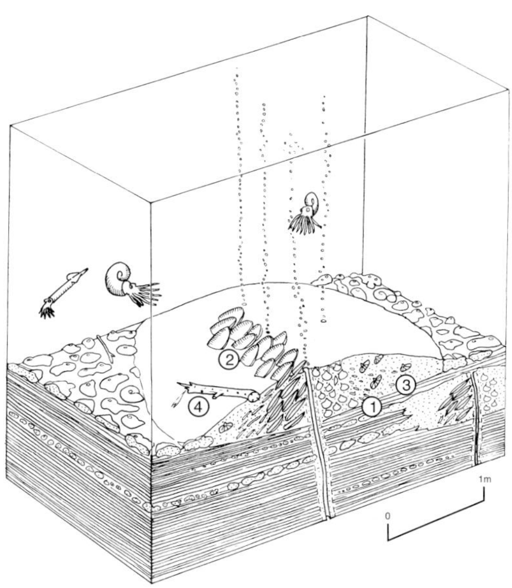

Fig. 13. Sketch reconstruction of appearance of a Kuhnpasset seep, cut away to show interior. 1, *Cryptolucina kuhnpassetensis* sp. nov.; 2, *Caspiconcha whithami* gen. et sp. nov.; 3, *Solemya* sp.; 4, *Turnus* sp. in waterlogged drift-wood.

Caspiconcha is difficult to compare because it is unique. However, as a probable mussel-type habit is envisaged, this may compare with the *Modiolus*-bearing seeps in the Eocene of Washington (Squires & Goedert 1991). Other Mesozoic seeps are dominated by brachiopods with subordinate bivalves such as *Grammatodon* and *Nucula,* as in the Albian of the Northwest Territories of the Canadian Arctic (Beauchamp *et al.* 1989). Alternatively, they are gastropod dominated, as in the Late Jurassic of the Antarctic (Kelly *et al.* 1995) and the Mesozoic of California (K. A. Campbell, pers. comm.). Brachiopods have not been found in the Kuhnpasset mounds and gastropods are uncommon.

Source rock

A total of *c.* 600 m of sedimentary rocks separate the Kuhnpasset Beds from the directly underlying Caledonian basement gneisses, and laterally this may extend to 1200 m because of fault-block rotation (Fig. 3). The Kuhnpasset Beds lie within mudstones that postdate the extensional faulting of the Wollaston Forland Group (Volgian–Hauterivian). This, in turn, overlies unconformably block-faulted Middle Jurassic deposits which rest on the basement. The vents occur in the footwall crest of a tilted fault block. The faulting probably influenced the direction of upward-migrating fluid and gas. Seepage at a low rate through the silty mudstones during the Late Barremian led to the development of the limestone mounds at the sediment–water interface. By Early Aptian time the seepage had ceased. The source of the seepage probably lies in the breakdown of organic matter in the underlying Middle Jurassic–Early Cretaceous (Hauterivian) strata, in which the Bjernberg Formation (Oxfordian–Volgian) dark shales are well-known as being organic rich, with total organic content levels of up to 9% (Price & Whitham 1997). For a further description and discussion of hydrocarbon sources and traps in eastern Greenland see also Christensen (1994). The source, if methane, is unlikely to be thermogenic because of the relatively shallow depth to basement of 600–1200 m.

Conclusions

Anomalous limestone mounds occur with an abundant, thick-shelled, bivalve-dominated benthic fauna in the Kuhnpasset Beds in a relatively low-energy depositional environment. The bivalves include especially abundant and large *Cryptolucina kuhnpassetensis* sp. nov., which is the most characteristic genus in the mounds. The giant modiomorphid, *Caspiconcha whithami* gen. et sp. nov., reaches > 300 mm in length and is one of the larger recorded vent-related bivalves. These taxa occur in dense clusters at the core of the mounds where many are preserved in life position. *Solemya* in life position and common waterlogged driftwood infested with the wood-boring bivalve *Turnus* are also associated with the mounds. The arrangement of the fauna around the vents is reconstructed in Fig. 13. The vents probably did not cause much topographic relief at the time of formation but are today exposed as mounds because they are more cemented than adjacent rocks. Because of diagenetic silicification of most of the shells, it has not yet been possible to conclusively prove a geochemical association with methane seepage at Kuhnpasset, but the lithological and faunal association make this possibility very likely. Future stable isotope geochemical work on the carbonate cements should indicate whether the limestones are derived from methane or hydrogen sulphide. It is believed that the source of the nutrient would have been from the breakdown of organic matter in the underlying Jurassic and Cretaceous rocks. If methane is present, a thermogenic source is unlikely because of the relatively shallow depth available to the source rock. The principal source would probably have been from the organic-rich shales of the underlying Bernjerg Formation (Oxfordian–Volgian). The nutrient probably migrated along faulted or fault-related surfaces which formed during Volgian–Hauterivian extensional rifting, reaching the surface of a tilted fault block by seepage in the Late Barremian.

The authors thank the following: the consortium of companies which supported the CASP East Greenland Project; D. Pedersen (Iceland) for assistance in the field; K. A. Campbell (University of Auckland, New Zealand), J. A. Crame (British Antarctic Survey, Cambridge) and L. R. Saul (Natural History Museum of Los Angeles County, California) for constructive criticism of early drafts of the article; T. Waller, J. Pojeta (USGS), N. J. Morris, J. Taylor (The Natural History Museum, London) and N. Malchus (Berlin) for discussion of the bivalves; R. Casey (The Natural History Museum, London) and P. F. Rawson (University College, London) for discussion of the ammonites; A. M. Koraini (Petronas, Malaysia) for dinoflagellate cyst dating; and P. Doyle (Greenwich University) for discussion of the belemnites.

Appendix 1: systematic descriptions of bivalves

Simon R. A. Kelly

All described specimens are from the 'mid-Cretaceous sandy shales', locality B2335 (Fig. 1) on the west side of Kuhnpasset, Wollaston Forland, Northeast Greenland.

The described specimens have been transferred from the CASP Collections to the Sedgwick Museum, University of Cambridge where X-prefixed numbers are allocated. CASP and the corresponding Sedgwick Museum numbers are listed in Appendix 2.

Family Lucinidae Fleming 1828
Subfamily Milthinae? Chavan in Cox *et al.* 1969
Genus *Cryptolucina* Saul *et al.* 1996

Type species. Cryptolucina megadyseides Saul *et al.* 1996, from the Humptulips Formation, middle–Late Eocene, of Grays Harbour County, Washington, western USA.

Diagnosis. Large thick-shelled lucinid lacking differentiated antero-dorsal and postero-dorsal areas, with commarginal ornament over whole flank and traces of radial ornament in the mid-flank. Interior usually smooth to pustulose, sometimes showing fine radial traces. Hinge strong but edentulous. Anterior adductor muscle scars large and elongate.

Remarks. Saul *et al.* (1996) placed *Cryptolucina* with some doubt in the Subfamily Milthinae (Chavan in Cox *et al.* 1969). However, the *Miltha* lineage itself is characterized by a well-developed hinge and dentition (Bretsky 1976, fig. 6). Several lucinid genera are edentulous, such as *Anodontia, Myrteopsis* and *Stewartia,* but they have more differentiated antero-dorsal and postero-dorsal regions. *Pseudomiltha* (Fischer 1887), from the Eocene of France, has a discoidal shell, obsolete dentition and shows strong fine radial ornament on the exterior and a distinct pallial blood vessel scar (Chavan in Cox *et al.* 1969). In the present study, the edentulous condition is seen in all medium to large size specimens but there are no small individuals that can be checked to show whether this condition exists throughout life. It is possible that the young specimens may demonstrate dentition which becomes obsolete in adults, but this cannot be verified at present. *Cryptolucina* is slightly elongate and its general external form is very close to *Nymphalucina* (Speden 1970), which has a well-developed dentition and occurs in seep limestones (Petta & Gerhard 1977). A number of elongate lucinids are recorded in the literature but few have proper descriptions including the shell interior and hinge. They include: *'Lucina' colusaensis* Stanton (1895, p. 60, pl. 11, figs 4 and 5), from the Late Jurassic and Early Cretaceous Knoxville Beds, California, USA; specimens identified as an undetermined lucinacean from the Oxfordian of the Terres Noires, Drôme, France (Rolin *et al.* 1990, fig. 6a); and an undescribed species from the Tithonian of the Antarctic Peninsula (Kelly *et al.* 1995). While some of these taxa may eventually prove to be *Cryptolucina*, the Greenland material described here represents the oldest known example of the genus.

Included taxa. Cryptolucina elassodyseides Saul *et al.* and *C. megadyseides* Saul *et al.* (1996), Eocene, Washington State, USA; *C. kuhnpassetensis* sp. nov., Barremian of northeast Greenland; possibly also unidentified lucinids from Oxfordian of France and Tithonian of Antarctica.

Cryptolucina kuhnpassetensis *sp. nov.* *(Figs 6–8)*

Type specimens. Holotype: K8400; paratypes: K8314, K8315, K8393, K8394, K8395, K8398, K8399, K8401, K9271. K9272, K9273.

Description. Medium–large sized lucinid, up to *c.* 130 mm in length, slightly more elongate than high, and of moderate inflation; sub-ovate in lateral aspect with weakly projecting prosogyrate umbones and small beaks; beak and umbo not coincident; evenly rounded anterior margin and slightly truncated posterior. Small lanceolate and shallow lunule, with beak slightly overhanging proximal end. Elongate, sharp-edged, lanceolate escutcheon. Flank ornament of fine commarginal growth lines; posterior area not differentiated, apart from slightly increased curvature of the growth lines at the postero-ventral margin; very weak radial ribbing visible between umbo and mid-flank on some specimens. Interior smooth with fine radial striae; larger specimens often coarsely pustulate. Adductor muscle scars lightly impressed, often showing radial striae similar to elsewhere on the interior; anterior adductor muscle scar large and elongate, and separated from the distinct integripalliate pallial line; posterior adductor muscle scar sub-pyriform. Pedal elevator muscle under hinge, immediately behind beak and below umbo. Pallial blood vessel scar not seen. Hinge plate well-developed (only seen clearly in holotype, left valve, Fig. 7d, and K8315) but edentulous; nymph elongate extending almost whole length of escutcheon. Ventral margin smooth, but slightly thickened between margin and pallial line.

Dimensions.

Specimen number	Height (mm)	Length (mm)	Length posterior (mm)	Width (mm)
K8314	98	115	74	51*†
K8392	60	97	60	40*
K8393	92	114	58	50*
K8394	71	–	–	46*
K9395	–	74	40	33*
K9398	67	86	50	40*†
K9399	85	102	55	36*†
K8400	86	112	78	26
K9271	*c.* 95	*c.* 129	92	30†
K9272	63	75	54	33*†

*, Two valves; †, internal mould.

Remarks. Cryptolucina kuhnpassetensis is not as large as the type species, *C. megadyseides*, which reaches 180 mm in length and is characterized by a more indented lunular margin. *C. elassodyseoides* (Saul *et al.* 1996) is more elongate, having a straight to slightly indented ventral margin. Internal moulds of *C. kuhnpassetensis* compare closely to specimens figured as 'Bivalvia gen. indet.' by Thieuloy (1972, pl. 1, fig. 10; pl. 2, fig. 6; pl. 3, fig. 1) and identified as *Pseudomiltha* aff. *germaini* (Pictet & Campiche 1864) by Lemoine *et al.* (1982), which reaches 210 mm in length but has a more rounded posterior. However, the dentition of the type and other specimens of *P. germaini* is unknown.

Subclass Anomalodesmata Dall 1889
Order Pholadomyoida Newell 1965
Superfamily Modiomorphoidea Miller 1877
Family Modiomorphidae Miller 1877
(= Permophoridae Van de Poel 1959, *pro* Pleurophoridae Dall 1895)
Subfamily Myoconchinae Newell 1957
Genus *Caspiconcha* nov.

Etymology. Derived from the Cambridge Arctic Shelf Programme (CASP) and *concha*, the latin for a shell (feminine).

Type species. Caspiconcha whithami sp. nov.

Remarks. The only Cretaceous bivalve that *Caspiconcha* gen. nov. resembles is *Myoconcha*, sharing the modioliform shape, extreme reduction of the shell anterior and the deep set adductor muscle scars, and differing from it in the lack of external radial ornament and the edentulous hinge. The Myoconchinae and Permophorinae were united in the Family Permophoridae which was placed in the Carditacea by Chavan (in Cox *et al.* 1969, p. 543). However, Morris (1978, p. 273) believed that the Permophoridae were not carditaceans because of the presence of homogeneous shell structure in permophorids, as opposed to crossed lamella and complex crossed lamella shell structure in carditoideans. Fang Zong-jie & Morris (1997) synonymised the Permophoridae in the Modiomorphidae and placed it in the Subclass Anomalodesmata. Despite the shells of *Caspiconcha* having been secondarily replaced by silica, the original structure would have been aragonite. N. Malchus (pers. comm.) believed that the shell of the type specimen showed relict nacreous texture, which supports the placement of *Caspiconcha* in the Anomalodesmata.

Little *et al.* (1999) described a Middle Devonian modiomorphid, *Sibaya*, from sulphide ore deposits of the Ural Mountains, Russia, thus demonstrating a long-term association of the modiomorphids with vent communities since the Palaeozoic. *Caspiconcha* is known only from the type species which occurs in limestone mounds of Kuhnpasset. However, preliminary observations on undescribed specimens from the Raukumara Peninsula, New Zealand, in the McKay Collection (Institute of Geological & Nuclear Sciences, Wellington, New Zealand; J. S. Crampton, pers. comm.), indicate that a similar or related large thick-shelled bivalve may have existed in the Late Cretaceous (?Campanian–Maastrichtian).

Caspiconcha contrasts with a number of other modioliform and other large vent-related bivalves. The mytilid, *Bathymodiolus* (Kenk & Wilson 1985) is calcitic and has a thin shell with a completely different hinge. The elongate aragonitic vesicomyid, *Calyptogena* (Dall 1891), has a heterodont dentition, a more rounded outline, and a thinner and less attenuated shell anterior. Another vesicomyid, *Pleurophopsis* (Van Winkle 1919), from the Oligocene of Trinidad, has deep-set adductor muscle scars with an anterior myophoric buttress, a well-developed pair of cardinal teeth in each valve and a weak, but integripalliate, pallial line. It was described by Keen (in Cox *et al.* 1969, p. 664) as having a 'flange above the anterior adductor scar'. This structure could be analogous to the caspiconchid process (Figs 9b and 10d). Although the beaks are located well to the anterior in *Pleurophopsis*, they do not overhang the anterior adductor scar as in *Caspiconcha*.

Range. Barremian of northeast Greenland.

Caspiconcha whithami *sp. nov.* *(Figs 9 and 10)*

Etymology. After A. G. Whitham, who discovered the first specimens and has been the leader of the CASP expeditions to eastern Greenland since 1990.

Type specimens. Holotype: K8318 (Fig. 10a–c); paratypes: K8319, K8433, K8434, K8435, K8439, K8440, K8442, K8443, K8447, K9276.

Description. Shell large and elongate, subtrapezoidal or cuneiform to modioliform in lateral aspect, with weak posterior carina and moderate inflation with slightly flattened flanks; *c.* 300 mm in length, *c.* 120 mm in height and *c.* 75 mm in width (valves in occlusion). Equivalve, with strongly inequilateral valves. Beaks close to anterior and situated in front of the anterior adductor muscle scar and below the hinge line in mature specimens. Umbo flush with dorsal margin. Shell thickness variable, up to 28 mm thick at myophoric buttress, tending to be thick around dorsal, anterior and ventral margins, and thinner at mid-flank; early formed shell becoming partially enveloped within later shell growth. Dorsal margin straight in lateral profile for the length of the nymph and feebly curved towards the posterior; ventral margin weakly indented, probably for byssal attachment; anterior margin short and rounded; posterior margin probably rounded – obliquely truncated. Shell only moderately inflated in dorsal aspect with bluntly rounded anterior and wedge-shaped posterior. Exterior ornament commarginal, blunt in mid-flank to slightly lamellose at periphery. A change in preservation from white inner shell to brown outer shell may reflect existence of a well-developed periostracum. Interior smooth, except that the mid-flank area is pitted with small mantle muscle scars. Adductor muscle scars very deeply inset (Fig. 10a and d); pallial line marked by oblique row of elongate pits from posterior adductor muscle scar in a postero-ventral direction, before swinging upwards in a more continuous line to the posterior adductor muscle scar. Anterior adductor scar oval, deeply set on dorsal side to shallow along ventral side over myophoric buttress; floor of scar showing very fine commarginal ornament and fine radial striae, in which up to four stronger radial ribs are seen towards the postero-dorsal part; postero-dorsal portion of scar abuts onto base of hinge plate, but dorsal to anterior border overgrown by a recurved pointed area, here termed the caspiconchiid process (Figs 9b and 10d); caspiconchiid process continuous with main interior of shell; the ventral part of the caspiconchiid process having a number of irregular pits, including a group of larger ones at the posterior (below the point) and a larger area of smaller pits to the anterior. Posterior adductor muscle scar less deeply inset than anterior one; subquadrate, having panhandle shape at the antero-dorsal corner at conjunction with posterior pedal retractor muscle scar. Hinge stout but edentulous (Figs 9b, 10a and d), with very long straight nymph and ligament groove (Fig. 10c) on dorsal side supporting the exterior ligament. Resilifer obliquely triangular and elongate with pointed anterior end below beak and greatest width immediately posterior to anterior adductor muscle scar, tapering gently posteriorly and terminating well before the nymph. The earlier formed part of the shell shows exfoliation; traces of fine radial striae present in mid-flank areas lacking outer surface.

Dimensions.

Specimen number	Length (mm)	Length (posterior) (mm)	Length (anterior) (mm)	Height (mm)	Width (mm)
K8318	*c.* 250	*c.* 220	30	*c.* 108	38
K8319	–	–	36	*c.* 116	30
K9276	*c.* 300	–	–	110	72*

*, Two valves.

Remarks. No complete valves were collected. The most complete valve is that of the holotype, K8318, a left valve missing substantial portions of the posterior and postero-ventral margins. A paratype, K9276, shows valves in occlusion but is missing parts of the shell, especially at the anterior and posterior ends. Precise information has not been obtained for the morphology of the shell posterior to the posterior adductor scar but has been pieced together from collected fragments (Fig. 9).

The relatively small size of the shell anterior may indicate that the size of the foot is relatively small (thus indicating a sedentary lifestyle). The presence of a weakly modioliform shape with ventral indentation may indicate the presence of a byssus in the adult. The very large posterior to the shell may have been for the housing of modified gills for bacterial farming. It is likely that the gut was reduced or absent, as in chemosymbiotic-related bivalves. The strong adductor musculature may have been for protection from crustacean and molluscan predators. The heavy degree of growth around the hinge of *Caspiconcha* may mean that the edentulous state is a secondary feature which developed from a juvenile with hinge dentition but which became obsolete in the adult; however, this idea remains unproven.

Irregularities in growth lines and evidence of recovery from shell damage, especially at the commissural margins, during life are present in several specimens. These structures may be related to shell crowding within the seep or to predation. In K9276, there is a large lateral displacement of the commissural margin affecting the symmetry of the valves. One may speculate that this is close to the position where a byssus may be expected to emerge, or the point at which two individual specimens came in contact within the seep but were unable to move apart, hence causing growth irregularity. The interior surface of the type specimen shows evidence of irregular vermiform grooves, *c.* 1.5 mm wide and up to 51 mm in length, and with a semi-lunular cross-section. Their position and cross-section suggests that these were caused by vermiform organisms, present between the shell interior and the mantle, and which perhaps were passive borers, inhibiting shell deposition (Fig. 10a). These structures are analogous to those formed during bioimmuration (Todd 1993), but may have been caused by a burrowing organism moving between the mantle and the shell. A single exterior boring on the same specimen, situated near the centre of the ventral margin is a hemispherical-based boring *c.* 1.5 mm deep into the shell, which is *c.* 11 mm in thickness.

Appendix 2: CASP and corresponding Sedgwick Museum catalogue numbers.

Taxon	CASP number	Sedgwick Museum number
Caspiconcha whithami gen. et sp. nov.	K8318	X30000
	K8319	X30001
	K8433	X30002
	K8835	X30003
	K8439	X30004
	K8440	X30005
	K8442	X30006
	K8443	X30007
	K8447	X30008
	K9276	X30009
Cryptolucina kuhnpassetensis sp. nov.	K8314	X30010
	K8315	X30011
	K8392	X30012
	K8393	X30013
	K8394	X30014
	K8395	X30015
	K8398	X30016
	K8399	X30017
	K8400	X30018
	K8401	X30019
	K9271	X30020
	K9272	X30021
	K9273	X30022
Solemya sp.	K9269	X30023
Turnus sp. in borings of *Teredolites clavatus* Leymerie	K8408	X30024

References

BEAUCHAMP, B. & SAVARD, M. 1992. Cretaceous chemosynthetic carbonate mounds in the Canadian Arctic. *Palaios*, **7**(4), 434–450.

——, KROUSE, H. R., HARRISON, J. C., NASSICHUK, W. W. & ELIUK, L. S. 1989. Cretaceous cold-seep communities and methane-derived carbonates in the Canadian Arctic. *Science*, **244**, 53-56.

BERG, C. 1985. Reproductive strategies of mollusks from abyssal hydrothermal vent communities. *Bulletin of the Biological Society of Washington*, **6**, 185–197.

BØGVAD, R. & ROSENKRANTZ, A. 1934. Beiträge zur Kenntnis der untern Kreide Ostgrönlands. *Meddelelser om Grønland*, **93**(1), 1–28.

BOSS, K. J. & TURNER, R. D. 1980. The giant white clam from the Galapagos Rift, *Calyptogena magnifica* species novum. *Malacologia*, **20**, 161–194.

BRETSKY, S. S. 1976. Evolution and classification of the

Lucinidae (Mollusca; Bivalvia). *Palaeontographica Americana*, **8**(50), 215–337.

CAMPBELL, K. A. & BOTTJER, D. J. 1993. Fossil cold seeps. *National Geographic Research and Exploration*, **9**(3), 326–343.

—— & —— 1995. Brachiopods and chemosymbiotic bivalves in Phanerozoic hydrothermal vent and cold seep environments. *Geology*, **23**(4), 321–324.

CAVANAUGH, C. M. 1985. Symbioses of chemoautotrophic bacteria and marine invertebrates from hydrothermal vents and reducing sediments. *In:* JONES, M. L. (ed.) The hydrothermal vents of the eastern Pacific: an overview. *Bulletin of the Biological Society of Washington*, **6**, 373–388.

CHILDRESS, J. J., FISHER, C. R., BROOKS, J. M., KENNICUTT II, M. C., BIDIGARE, R. R. & ANDERSON, A. E. 1986. A methanotrophic marine molluscan (Bivalvia: Mytilidae) symbiosis: mussels fueled by gas. *Science*, **233**, 1306–1308.

CHRISTENSEN, F. G. 1994. Seeps and other bitumen showings: a review of origin, nomenclature and occurences in Greenland. *Grønlands Geologiske Undersøgelse, Open File Series*, **94/7**, 1–21.

CORLISS, J. B. & BALLARD, R. D. 1977. Oases of life in the cold abyss. *National Geographic Magazine*, **152**(4), 440–453.

COX, L. R., NEWELL, N. D., BOYD, D. W. *ET AL.* 1969. *In*: MOORE, R. C. (ed.) *Treatise on Invertebrate Paleontology. Part N. Volume 1, Part 2. Mollusca 6, Bivalvia*. Geological Society of America and University of Kansas, Boulder, N491–N952.

DALL, W. H. 1889. On the hinge of pelecypods and its development, with an attempt toward a better subdivision of the group. *American Journal of Science*, **38**, 445–462.

—— 1891. On some new or interesting west American shells obtained from the dredgings of the U.S. Fish Commission steamer Albatross in 1888, and from other sources. *Proceedings of the United States National Museum*, **14**, 173–191.

—— 1895. Tertiary mollusks of Florida. Part 3. A new classification of the Pelecypoda. *Transaction of the Wagner Free Institute of Science, Philadelphia*, **3**, 483–565.

DANDO, P. R. & HOVLAND, M. 1992. Environmental effects of submarine seeping natural gas. *In:* DAVIES, A. M. (ed.) Methane in marine sediments. *Continental Shelf Research*, **12**(10), 1197–1207.

——, O'HARA, S. C. M., NIVEN, S. J., SCHMALJOHANN, R., SCHUSTER, U. & TAYLOR, L. J. 1994. The effects of methane seepage at an intertidal/shallow subtidal site on the shore of the Kattegat, Hedestrand, Vendsyssel, Denmark. *Bulletin of the Geological Society of Denmark*, **41**(1), 65–79.

——, SOUTHWARD, A. J., SOUTHWARD, E. C., DIXON, D. R., CRAWFORD, A. & CRAWFORD, M. 1992. Shipwrecked tube worms. *Nature*, **356**, 667.

DONOVAN, D. T. 1957. The Jurassic and Cretaceous systems in East Greenland. *Meddelelser om Grønland*, **155**(4), 1–214.

FANG ZONG-JIE & MORRIS, N. J. 1997. The genus *Pseudosanguinolites* and some modioliform bivalves (mainly Palaeozoic). *Palaeoworld*, **7**, 49–74.

FIALA-MEDIONI, A. & LE PENNEC, M. 1989. Adaptive features of bivalve molluscs associated with fluid venting in the subduction zones off Japan. *Palaeogeography, Palaeoclimatology and Palaeoecology*, **71**, 161–167.

FISCHER, P. H. 1887. *Manuel de conchyliologie et de paléontologie conchyliogique. Pt 3. Classe des pelécypodes*. F. Savy, Paris.

FLEMING, J. 1828. *A History of British Animals, Exhibiting the Descriptive Characters and Systematical Arrangement of the Genera and Species of Quadrupeds, Birds, Reptiles, Fishes, Mollusca and Radiata of the United Kingdom; Including the Indigenous, Extirpated, and Extinct Kinds; Together with Periodical and Occasional Visitants*. Bell & Bradfute, Edinburgh.

FREBOLD, H. 1932. Die Lagerungsverhältnisse der Unterkreide im nördlichen teil von Ostgrönland und die Frage der Prätertiären Fjordanlage. *Meddelelser om Grønland*, **84**(6), 1–40.

—— 1935. Marines Aptien von Koldewey Insel (nördliches Ostgrönland). *Meddelelser om Grønland*, **95**(4), 1–112.

GAILLARD, C., RIO, M. & ROLIN, Y. 1992. Fossil chemosynthetic communities related to vents or seeps in sedimentary basins: the pseudobioherms of southeastern France compared to other world examples. *Palaios*, **7**(4), 451–465.

GOEDERT, J. L. & SQUIRES, R. L. 1990. Eocene deep-sea communities in localised limestones formed by subduction-related methane seeps, southwestern Washington. *Geology*, **18**, 1182–1185.

HOVLAND, M. & JUDD, A. G. 1988. *Seabed Pockmarks and Seepages. Impact on Geology, Biology and the Marine Environment*. Graham & Trotman, London.

HOWE, B. & KAUFFMAN, E. G. 1986. The lithofacies, biofacies and depositional setting of tepee buttes, Cretaceous submarine springs, between Colorado Springs and Boone, Colorado. *In:* KAUFFMAN, E. G. (ed.) *Cretaceous Biofacies of the Central Part of the Western Interior Seaway: A Field Guidebook*. The 4th North American Paleontological Convention, Boulder, 155–175.

JØRGENSEN, N. O. (ed.) 1994. *Gas in Marine Sediments, 2nd Conference*. Bulletin of the Geological Society of Denmark, **41**(1), 1–109.

KAUFFMAN, E. G., ARTHUR, M. A., HOWE, B. & SCHOLLE, P.A. 1996. Widespread venting of methane-rich fluids in Late Cretaceous (Campanian) submarine springs (Teepee Buttes), Western Interior seaway. *Geology*, **24**, 799–802.

KELLY, S. R. A. 1988. Cretaceous wood-boring bivalves from western Antarctica with a review of the Mesozoic Pholadidae. *Palaeontology*, **31**(2), 341–372, pls 26 and 27.

—— & BROMLEY, R. G. 1984. Ichnological nomenclature of clavate borings. *Palaeontology*, **27**(4), 793–807.

—— & WHITHAM, A. G. 1999. Deshayesitid ammonites from the lower Aptian (Lower Cretaceous) of North-East Greenland. *Scripta Geologica, Special Issue*, **3**, 83–95.

——, BLANC, E., PRICE, S. P. & WHITHAM, A. G. 1999. Giant bivalves from a Barremian (Early Cretaceous) seep system in Wollaston Forland. *Palaeontological Association Newsletter*, **42,** 24.

——, DITCHFIELD, P. W., DOUBLEDAY, P. A. & MARSHALL, J. D. 1995. An Upper Jurassic methane-seep limestone from the Fossil Bluff Group fore-arc basin of Alexander Island, Antarctica. *Journal of Sedimentary Research*, **A65**, 274–282.

——, WHITHAM, A. G., KORAINI, A. M. & PRICE, S. P. 1998. Lithostratigraphy of the Cretaceous (Barremian-Santonian) Hold with Hope Group, NE Greenland. *Journal of the Geological Society, London*, **155**(6), 993–1008.

KENK, V. C. & WILSON, B. R. 1985. A new mussel (Bivalvia, Mytilidae) from hydrothermal vents in the Galapagos Rift zone. *Malacologia*, **26**, 253–271.

KREUGER, D. M., GUSTAFSON, R. G. & CAVANAUGH, C. M. 1990. Vertical transmission of chemautotrophic symbionts in the bivalve *Solemya velum. Biological Bulletin*, **98**, 60–65.

KULM, L. D. & SUESS, E. 1990. Relationship between carbonate deposits and fluid venting: Oregon accretionery prism. *Journal of Geophysical Research*, **95(B6)**, 8899–8915.

LEMOINE, M., ARNAUD-VANNAU, A., ARNAUD, H., LÉTOLLE, R., MEVEL, C. & THIEULOY, J.-P. 1982. Indices possibles de paléo-hydrothermalisme marin dans le Jurassique et le Crétacé des Alpes occidentales (Océan Téthysien et sa marge continentale européene): essai d'inventaire. *Bulletin de la Société Géologique de France*, **24**(3), 641–647.

LILJEDAHL, L. 1991. Contrasting feeding strategies in bivalves from the Silurian of Gotland. *Palaeontology*, **34**, 219–235.

LITTLE, C. T. S., MASLENNIKOV, V. V., MORRIS, N. J. & GUBANOV, A. P. 1999. Two Palaeozoic hydrothermal vent communities from the southern Ural Mountains, Russia. *Palaeontology*, **42**, 1043–1078.

LUTZ, R. A., FRITZ, L. W. & RHOADS, D. C. 1985. Molluscan growth at deep-sea hydrothermal vents. *In:* JONES, M. L. (ed.) The hydrothermal vents of the eastern Pacific: an overview. *Bulletin of the Biological Society of Washington*, **6**, 199–210.

——, JABLONSKI, D. & TURNER, R. D. 1984. Larval development and dispersal at deep-sea hydrothermal vents. *Science*, **226**, 1451–1454.

——, SHANK, T. M., FORNARI, D. J., HAYMON, R. A., LILLEY, M. D., VON DAM, K. L. & DESBRUYÈRES, D. 1994. Rapid growth at deep-sea vents. *Nature*, **371**, 663–664.

MAYER, L. A., SHOR, A. N., CLARKE, J. H. & PIPER, D. J. W. 1988. Dense biological communities at 3850 m on the Laurentian Fan and their relationship to the deposits of the 1929 Grand Banks earthquake. *Deep-Sea Research*, **35**(8), 1235–1246.

MAYNC, W. 1949. The Cretaceous beds between Kuhn Island and Cape Franklin (Gauss Peninsula), northern East Greenland. *Meddelelser om Grønland*, **133**(3), 1–291.

MILLER, S. A. 1877. *The American Palaeozoic Fossils, a Catalogue of the Genera and Species*. S. A. Miller, Cincinnatti.

MORRIS, N. J. 1978. The infaunal descendants of the Cycloconchidae: an outline of the evolutionary history and taxonomy of the Heteroconchia, superfamilies Cycloconchacea to Chamacea. *Philosophical Transactions of the Royal Society of London*, **B284**, 259–275.

NEWELL, N. D. 1957. Notes on certain primitive heterodont pelecypods. *American Museum Novitates*, **1857** 1–14.

—— 1965. Classification of the Bivalvia. *American Museum Novitates*, **2206**, 1–25.

NØHR-HANSEN, H. 1993. Dinoflagellate cyst stratigraphy of the Barremian to Albian, Lower Cretaceous, northeast Greenland. *Grønlands Geologiske Undersøgelse, Bulletin*, **166**, 1–171.

PEEK, A. S., GUSTAFSON, R. G., LUTZ, R. A. & VRIJENHOEK, R.C. 1997. Evolutionary relationships of deep-sea hydrothermal vent and cold-seep clams (Bivalvia: Vesicomyidae): results from mitochondrial cytochrome oxidase subunit I. *Marine Biology*, **130**, 151–161.

PETTA, T. J. & GERHARD, L. C. 1977. Marine grass banks – possible explanation for carbonate lenses, Tepee Zone, Pierre Shale (Cretaceous), Colorado. *Journal of Sedimentary Petrology*, **47**(3), 1018–1026.

PICTET, F. J. & CAMPICHE, G. 1864. Description des fossiles du terrain Crétacé des environs de Ste Croix. *Materiaux pour la Paléontologie Suisse*, **2**, 701–752.

POEL, L. VAN DE. 1959. Faune malacologique du Hervien. *Bulletin de l'Institut Royal des Sciences Naturelles de Belgique*, **35** (15), 1–26; **35**(16), 1–28.

PRICE, S. P. & WHITHAM, A. G. 1997. Exhumed hydrocarbon traps in East Greenland: analogs for the Lower–Middle Jurassic play of Northwest Europe. *American Association of Petroleum Geologists, Bulletin*, **81**, 196–221.

REID, R. G. B. & BRAND, D. G. 1986. Sulfide-oxidising symbiosis in Lucinaceans: implications for bivalve evolution. *The Veliger*, **29**, 3–24.

ROLIN, Y., GAILLARD, C. & ROUX, M. 1990. Ecologie des pseudobiohermes des terres noires jurassiques liés à des paléo-sous-marines. Le site Oxfordien de Beauvoisin (Drôme, bassin du sud-est, France). *Palaeogeography, Palaeoclimatology and Palaeoecology*, **80**(2), 79–105.

SAUL, L. R., SQUIRES, R. L. & GOEDERT, J. L. 1996. A new genus of cryptic lucinid? bivalve from Eocene cold seeps and turbidite-influenced mudstone, western Washington. *Journal of Paleontology*, **70**(5), 788–794.

SMITH, C. R., KUKERT, H., WHEATCROFT, P. A., JUMARS, P. A. & DEMING, J. W. 1989. Vent fauna on whale remains. *Nature*, **341**, 27–28.

SPEDEN, I. G. 1970. The Type Fox Hills Formation, Cretaceous (Maestrichtian), South Dakota. Part 2. Systematics of the Bivalvia. *Peabody Museum of Natural History, Yale University, Bulletin*, **33**, 1–222.

SQUIRES, R. L. & GOEDERT, J. L. 1991. New late Eocene mollusks from localized limestone deposits formed by subduction-related methane seeps, southwestern Washington. *Journal of Paleontology*, **65**(3), 412–416.

——, —— & BARNES, L. G. 1991. Whale carcasses. *Nature*, **349**, 574.

STANTON, T. A. 1895. Contribution to the Knoxville Beds. *United States Geological Survey Bulletin*, **133**, 1–132.

SURLYK, F. 1977. Stratigraphy, tectonics and palaeogeography of the Jurassic sediments of the areas north of Kong Oscar Fjord, East Greenland. *Grønlands Geologiske Undersøgelse, Bulletin*, **123**, 1–56.

—— 1978. Submarine fan sedimentation along fault scarps on tilted fault blocks (Jurassic–Cretaceous boundary), East Greenland. *Grønlands Geologiske Undersøgelse, Bulletin*, **128**, 1–108.

—— & ZAKHAROV, V. A. 1982. Buchiid bivalves from the Upper Jurassic and Lower Cretaceous of East Greenland. *Palaeontology*, **25**(4), 727–753.

TAYLOR, J. & GLOVER, E. 2000. Functional anatomy, chemosymbiosis, and evolution of the Lucinidae. *This volume*.

THIEULOY, J.-P. 1972. Biostratigraphie des lentilles à peregrinelles (brachiopodes) de l'Hauterivien de Rottier (Drôme, France). *Geobios*, **5**(1), 5–53.

TODD, J. A. 1993. The bivalve shell as a preservation trap, as illustrated by the Late Jurassic gryphaeid, *Deltoideum delta* (Smith). *Scripta Geologica*, **2**, 417–433.

VAN DOVER, C. L. 1985. Ecology of mid-Atlantic ridge hydrothermal vents. *In*: PARSON, L. M. & WALKER, C. L. (eds) *Hydrothermal Venta and Processes*. Geological Society, London, Special Publications, **87**, 257–294.

VAN WINKLE, K. 1919. Remarks on some new species from Trinidad. *Bulletins of American Palaeontology*, **8**(33), 19–33.

The function of pallial eyes within the Pectinidae, with a description of those present in *Patinopecten yessoensis*

BRIAN MORTON

The Swire Institute of Marine Science and Department of Ecology and Biodiversity, The University of Hong Kong, Hong Kong, China

Abstract: The structure of the pallial, ectopic eye of *Patinopecten yessoensis* is described and shown to be of the typical pectinid form, located on the middle mantle fold. The cornea is, however, a tall epithelium and, with the lens, forms a Cartesian oval, unlike the lens alone in other pectinids which functions to counter spherical aberration. The basal cell layer beneath the retina and argentea is poorly pigmented, although this may be countered by the fact that the optic tentacle epithelium itself is apically heavily pigmented.

It is generally assumed that pectinid pallial eyes, being able to perceive a moving image, are used to warn of vicinal predators, resulting in swimming. However, this does not seem to be the case; although crabs and starfish, in particular, are known predators of scallops, which respond by swimming to the latter, they do so (usually) on receipt of mechanical and chemical stimuli, not visual. Some scallop species live in seagrass beds and there is evidence that these may afford protection. Scallops are visually attracted to the waving fronds.

Scallops may also make relocation movements and, in more specialized taxa, e.g. species of *Amusium*, can swim for several metres. The visual sense of scallops is poorly understood and although pallial eyes may have developed early in the ancestry of the Pectinidae, as an antipredation sense organ, they may, in the descendants, now have a different, but perhaps related, function, although what this is, is unknown. Pectinid pallial eyes may improve the efficiency of photon capture in low light intensity subtidal habitats but for what purpose is unclear, since even if they are functioning as optical 'burglar alarms', what could they see in such a situation, especially as there is no brain to formulate an image?

Pelseneer (1911) first attempted to classify eye structure in the Bivalvia, and Morton (2001) has reviewed cephalic and pallial eye structure in the class. The most well-understood bivalve pallial eyes belong to representatives of the Pectinidae, a family in which their development has been linked to the evolution of swimming (Yonge 1936). The first comprehensive study of a pectinid eye (*Pecten maximus*) was by Dakin (1910), who showed, importantly, that the retina comprised two cell layers, proximal and distal, with a crystalline argentea and a cup of pigmented cells below it, and a Cartesian oval-shaped lens above. Such an eye is clearly sophisticated and Wenrich (1916) first suggested that it might be able to form an image. Dakin (1928*a*) subsequently went on to describe the eyes of *Spondylus gaederopus* and *Amusium pleuronectes*, while Barber *et al.* (1967) described the fine structure of the eye of *Pecten maximus*. Dakin (1928*b*) also showed that the pallial optic nerves connect up with the visceral ganglia which develop inequilateral optic lobes related to the numbers of eyes on each mantle lobe. Even so, however, such lobes are small in comparison with those of, say, an octopus, which has well-developed visual acuity (Wells 1966). Despite a sophisticated eye structure, therefore, it is doubtful if the bivalve's visceral ganglia are capable of recreating any image.

The eye is of a similar construction in all pectinids, and Morton (1980, 1993, 1996) has described their general organization and relationships with the mantle folds in *Amusium pleuronectes*, *Leptopecten latiauratus* and *Minnivola pyxidatus*. The pallial eyes of the boreal *Patinopecten yessoensis* (Jay, 1857) have not been described and this study remedies this deficiency. The principal objective of this study is, however, a review and an analysis of the function of such, as will be shown, superficially complex eyes.

Materials and methods

Patinopecten yessoensis is a boreal, free-living scallop from the Bohai Sea (China), Korea and the Sea of Japan. It is imported into Hong Kong as a

From: HARPER, E. M., TAYLOR, J. D. & CRAME, J. A. (eds) *The Evolutionary Biology of the Bivalvia*. Geological Society, London, Special Publications, **177**, 247–255. 1-86239-076-2/00/$15.00

seafood where it is kept in chilled seawater aquaria. Specimens were dissected and pieces of the mantle margin removed and fixed in 5% neutral formalin. Following routine histological procedures, such tissues were sectioned at 6 µm and alternate slides stained in either Ehrlich's haematoxylin and eosin or Masson's trichrome.

Results

As in all scallops, the pallial eyes of *Patinopecten yessoensis* are developed on the middle mantle fold, the inner forming the velum that facilitates swimming. In adult specimens, there are always more eyes on the lower right ($n = c.$ 27), than on the upper left ($n = c.$ 45) mantle margins.

Sections of the eyes reveal the following structure (see Fig.1). The eye is situated on its own tentacle, the epithelial surface of which is darkly pigmented (POE) and which may be > 40 µm tall. In *Patinopecten yessoensis*, as in *Pecten* (= *Chlamys*) *pusio* (Patten 1886), the cornea is enlarged and comprises cells which can be up to 40 µm tall, each with, as in the general epithelium, a distal nucleus. Immediately under the cornea (C) is a lens (L), housed in its own chamber. In *P. yessoensis* and *P. pusio*, the enlarged cornea and the lens form a Cartesian oval, whereas in *P. maximus* the lens alone has this shape. The lens of *P. yessoensis* is *c.* 160 µm in diameter and has a core of darkly staining cells, surrounded by lighter ones. Beneath the lens is the distal retina (DR), from the surface of which arise ciliary appendages. These eventually unite to form the distal retinal nerve (DRN) that emerges from the eyes laterally at one point. The cells of the distal retina have a proximal nucleus. Beneath this is the proximal retina (PR), also with proximal nuclei and from the distal surface of which also arise ciliary appendages, which eventually form the proximal retinal nerve (PRN). The membranous appearance and ciliary character of the proximal retinal appendages suggest that they too, like those of the distal retina, are light receptive. The proximal retinal nerve emerges from the opposite side of the eye to the distal retinal nerve, eventually uniting their bases.

In life, forming a union with the proximal retina is the argentea (A), or tapetum. Such a structure is one of the few examples of an invertebrate tapetum outside the Arthropoda and gives the pectinid eye

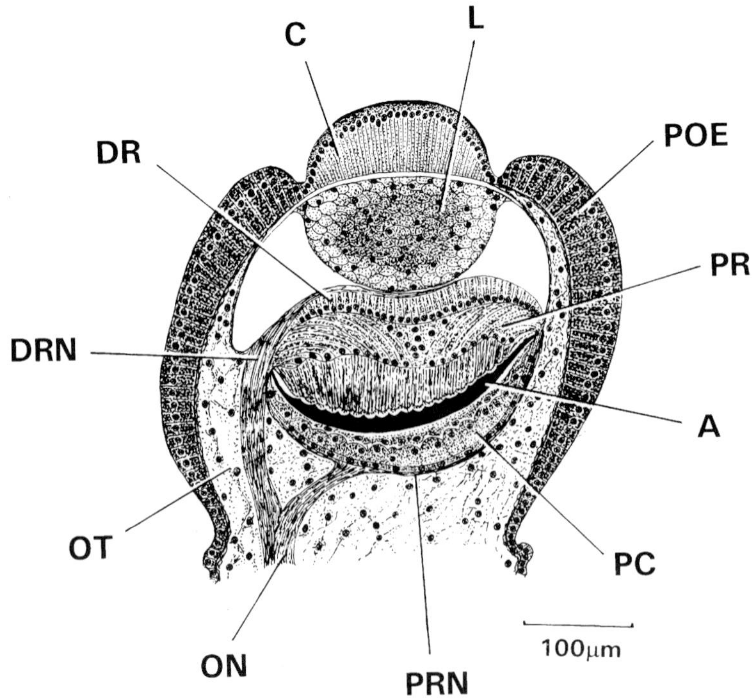

Fig. 1. *Patinopecten yessoensis* – a section through a pallial eye. A, Argentea; C, cornea; DR, distal retina; DRN, distal retinal nerve; L, lens; ON, optic nerve; OT, optic tentacle; PC, pigment cell; POE, pigmented outer epithelium; PR, proximal retina; PRN, proximal retinal nerve.

its metallic appearance. Beneath the argentea is a row of cells called the pigment cup (PC); in *Patinopecten yessoensis* these cells are only lightly pigmented but in *P. maximus* they are more heavily pigmented (Dakin 1910, 1928*a*). The distal (DRN) and proximal retinal nerves (PRN) unite outside the eye to become the optic nerve (ON). The pectinid eye does not have an accessory organ (Morton 2001), although it is possible that one or more of the complex array of tactile tentacles which adorn the middle mantle fold, and surround each eye, can be considered such a structure. No one has offered an opinion on the function of the accessory organ, except for *Laternula truncata*, which Adal & Morton (1973) suggested might protect the eye when the tentacle on which it is located bends back, with the others, to flick sand grains over the siphonal region for camouflage (i.e. it acts like a statocyst).

A review of the function of pectinid pallial eyes

The most basic response to light in the Bivalvia is the shadow 'off' reflex leading to either adduction, siphonal withdrawal or digging (or a combination of all three), first described by Sharp (1883). The pallial eyes of pectinids are the only ones in which a well-resolved image is formed by reflexion (Land 1968, 1981). Land (1965, 1966*a*) showed that the lens of *Pecten maximus* had a focal length which was at least twice as great as the depth of the eye, so that any image must be formed at the spherical argentea on the distal retina. Hartline (1938) showed that the distal retina responds to the offset of light (the primitive 'off' shadow reflex), whereas the proximal retina, which does not receive a focused image, responds to the onset of light, the 'on' reflex. The overall function of such an eye, therefore, is to permit responses to be made to moving objects which do not cast direct shadows (Land 1966*b*). Possibly, scallops see a simple image of an object as it goes past their eyes in sequence. Thus, Rees (1957, p. 23) records 'They [*Pecten maximus*] can also perceive movement whether or not it involves a change in brightness; they will react by *shutting their valves* [my italics] when, say, a small white card only three-fifths inches square is moved against a black background more than a foot away'. Similarly, Hartline (1925, p. 211) says of *Pecten maximus*: 'On cutting off the light the animal is stimulated and reacts by *closing its valves* [my italics]'. If, however, the visceral ganglia initiate a response, based on information received from the pallial eyes, to nothing more than gross movement, why do such sophisticated eyes develop when the same result could be achieved from the stimulus received from eyespots of a much simpler construction? All bivalves (except probably bathyal species), whether they possess pallial eyespots or not, have a shadow reflex which results in valve adduction. Such shadows are detected by photosensitive nerves (Charles 1966).

Of all the bivalves, representatives of the Pectinidae possess the most anatomically advanced eyes (Morton 2001). Buddenbrock & Moller-Racke (1953) showed that scallops (*Pecten maximus*) close their shell valves in response to small movements in the environment. But how valuable is this ability in the natural environment? In a laboratory study of *Chlamys bifrons* in South Australia, Pitcher & Butler (1987, p. 237) showed that control individuals mechanically stimulated at the mantle margin did not exhibit an escape response even after 2 min of contact and that 'even severe disturbance of the scallop did not always cause a response'. Moreover, when isolated with the starfish *Coscinasterias calamaria*, 'the majority of *C. bifrons* responded immediately *to contact* [my italics] with starfish, many responded before contact and very few took > 1 s to respond'.

The scallop *Minnivola pyxidatus* is a subtidal species in Hong Kong and possesses the usual array of advanced pallial eyes (Morton 1996). A known predator of this scallop is the muricid *Rapana bezoar* and in choice experiments with this species, and a range of potential prey with which the gastropod was familiar, *M. pyxidatus* was always the first to be attacked (Morton 1994). Despite its complex eyes, *M. pyxidatus* could not detect the large, advancing, predator and it was not until *R. bezoar* actually touched the long pallial tentacles of the scallop that any escape response could be detected, by which time it was usually too late (the gastropod, literally jumping onto and immobilizing its victim). Gould (1971, p. 79) states that '*Placopecten* and *Amusium* respond to predators by flight (photographs in Rees 1957)'. However, Rees' photographs show no such thing and are, on the contrary, of *Chlamys opercularis,* a species which because of functional design cannot swim at all well (Chapman *et al.* 1979) and responds only to the touch of a potential predator, the asteroid *Asterias rubens*, by gaping widely and then only at the last minute attempting to swim (Rees 1957, pp. 27–31, figs I–VI). The photographs reproduced by Rees (1957) actually show some *C. opercularis* overlain by the starfish, but still not responding by flight.

The well-known swimming escape of scallops '*from contact with, or odour of, asteroids* [my italics]' (Sloan 1980, p. 109) is also characterized by differing reports of such responses. Hancock (1974) showed that *Crossaster papposus* would not stimulate escape behaviour in *Pecten maximus* and *Chlamys opercularis*. Christensen (pers. comm. in

Hancock 1974) observed swimming by *C. opercularis* to *C. papposus* and Mackie (pers. comm. in Hancock 1974) observed a weak reaction in *Pecten* to an extract of the same starfish. Dickie & Medcof (1963) showed that *Placopecten* will respond strongly to *Asterias vulgaris* but not to *C. papposus*. Finally, Fange (1963) and Thomas & Gruffydd (1971) stated that *Pecten* responds only to *Marthasterias glacialis*. Stephens (1978) used a crude asteroid extract to stimulate a swimming response in *Aequipecten irregularis*.

Morphologically less specialized scallops, e.g. *Chlamys opercularis*, can swim distances of between 3.5 and 6.6 m (Chapman *et al.* 1979). This same species will swim when either mechanically or chemically stimulated by extracts of predatory starfish (Stephens & Boyle 1978). The same escape response is also induced in *Pecten ziczac* by such a chemical stimulus (Wilkens 1981). Of particular interest to this study, the same response was evoked in *Patinopecten yessoensis* (Karpenko 1980). In this case the scallops responded using chemoreceptors in their pallial tentacles to *Distolasterias nipon* and *Patiria pectinifera* (Asteroidea) (Karpenko 1981). The same escape response to these starfishes was also seen in *Swiftopecten swifti* (Dautov & Karpenko 1984) in the Sea of Japan. Clearly, it is the array of pallial tentacles on the middle mantle fold that are the mechanical means by which scallops detect potential predators. However, pectinids, e.g. *Placopecten magellanicus*, also possess an osphradium and an abdominal sense organ to sample water flowing through the mantle cavity and both may act as mechanoreceptors, either to monitor flow rate and/or as chemoreceptors (Moir 1977), detecting predator exudates.

Starfish often occur on subtidal scallop beds and in a review of asteriod feeding (Sloan 1980) it has been suggested that this has led to them being characterized as predator and prey. However, this is not necessarily true. On a subtidal scallop bed in Venezuala, *c.* 87% of the diet of *Luidia barimae* was ophiuroids (Penchaszadeh & Molinet *in* Sloan 1980). On an Isle of Man scallop bed, 97.4% of the diet of *Luidia ciliaris* was echinoderms, again mostly ophiuroids (Brun 1972, p. 225), with only 'small *Pecten*' found in the guts of two out of 108 specimens studied by Hunt (1925) at Plymouth. In Venezuala, *Luidia clathrata* had a diet of 45% 'molluscs' and 36% ophiuroids, whereas on a North Carolina scallop bed this asteroid was a scallop predator (Schwartz & Porter 1977), as was *Astropecten articulatus*. *Luidia alternata* was, however, a specialist predator on these coexisting scallop predators.

Barbeau & Scheibling (1994) undertook an experiment wherein tethered and untethered *Placopecten magellanicus* were offered as potential prey to the crab *Cancer irroratus* and the starfish *Asterias vulgaris*. Tethering did not influence the predation rate by the crab but it did limit the scallop's response to the starfish, thereby increasing the predation rate, demonstrating that swimming is an effective defence against this species. Kim (1969) showed that *Asterias rubens* could open *Patinopecten yessoensis* more quickly than a variety of other cemented and slow-swimming species. This was, however, because the scallop was secured to its substratum and, in life, might have been able to escape the predator.

Other scallop prey–predator interactions include the crab *Dyspanopeus sayi* feeding on juvenile *Argopecten irradians* in Long Island Sound, New York (Strieb *et al.* 1995). In the Gulf of St Lawrence, smaller *Chlamys islandica* are the prey of the crabs *Hyas araneus* and *Cancer irroratus* (Arsenault & Himmelman 1996). In laboratory experiments, *Pecten maximus* was fed on by *Liocarcinus puber*, *Carcinus maenas* and *Cancer pagurus* (Lake & Jones 1987). *Placopecten magellanicus* is preyed upon by *Cancer irroratus* and the American lobster *Homarus americanus* (Elner & Jamieson 1979). In both the laboratory and in the field, in Rhode Island, USA, *Argopecten irradians* is fed upon by the muricid *Urosalpinx cinerea* (Ordzie & Garofalo 1980). In Nova Scotia, *P. magellanicus* is the prey of *Cancer irroratus* and two starfishes, *Asterias vulgaris* and *A. forbesi* (Barbeau *et al.* 1998). In Northern Chile, *Argopecten purpuratus* is preyed upon by *Cancer polydon*, two starfishes, *Meyenaster gelatinosus* and *Luidia magellanicus*, and two gastropods, *Xanthochorus* sp. and *Priene rude* (Wolff & Alarcon 1993). *A. irradians* can, however, differentiate predators from non-predators, but *Asterias forbesi* (Asteroidea) and the gastropods *Eupleura caudata* and *Urosalpinx cinerea*, and *Thais lapillus* and *Busycon canaliculatum* all provoked swimming (Ordzie & Garofalo 1980).

In the Gulf of St Lawrence, USA, normal movements by *Placopecten magellanicus* reduced the overall predation rate by *Cancer irroratus*. However, experimentally tethered individuals were easily captured (Stokesbury & Himmelman 1996), e.g. *Patinopecten yessoensis* and *Asterias amurensis* (Kim 1969). In Tasmania, *Equichlamys bifrons* is predated upon by the starfish *Coscinasterias muricata* and *Octopus maorum*, but the predation rate was less in seagrass beds (Wolf & White 1997). In seagrass beds of *Halodule wrightii* and *Zostera marina*, predation upon *Argopecten irradians* by the whelk *Busycon carica*, and even by the Ring-billed gull, *Larus delawarensis*, and the Herring gull, *Larus argentum*, was reduced in comparison with those scallops living on sandy areas. *Zostera marina* beds also provided a significant refuge for juvenile scallops (*Argopecten irradians*) of < 15 mm shell length from predation by the crab *Dyspanopeus sayi* and the pufferfish *Sphoeroides maculatus* (Bricelj *et al.* 1993). This appears to be because small scallops can climb up the seagrass leaves and, thus, at high tide times are removed from the vicinity of at least the crab. Caddy (1972) showed that juveniles of *Placopecten magellanicus* can remain bysally attached up to a shell height of *c.* 120 mm and such individuals show less swimming response than larger adults. Therefore, seagrass beds may, at different stages of the life cycles of some scallops, confer protection from predators, e.g. *Argopecten irradians* orients towards them, perhaps visually sensing the moving fronds, from a distance of 25 cm but no further (Hamilton & Koch 1996).

Yonge (1936) considered that the most active scallop swimmers, i.e. *P. (Chlamys) opercularis* of European waters and the American *P. irradians*, both of which live in shoals on sandy bottoms at moderate depths, make extensive migrations from time to time. Rolfe (1973) similarly argued that there was evidence to suggest that the sudden disappearance of *Chlamys opercularis* from areas where previously good catches were obtained could be due to mass migration, perhaps as the result of swimming activity assisted by tidal currents.

In the Clyde Sea, the spat of *Chlamys septemradiata* settle near the shore where they byssally attach for 2 years. They subsequently assume a swimming habit and migrate to deeper waters (Allen 1953). Thayer (1972) accepted that the (reportedly) enhanced swimming activity and adductor muscle obliquity in the young of other scallop genera might also be associated with relocation of the juveniles to suitable adult habitats. Such scallops might be responding to subtle changes in light intensity because, as has been shown by Gomez & Nasi (1997), the pallial eyes of *Pecten irradians* are capable of light adaptation.

Moore & Marshall (1967) believed that transposition of *Aequipecten irradians* in the Niantic Estuary may be accounted for by tidal currents acting upon the, albeit weakly, swimming scallop. On the sea bottom of Iles-de-la-Madeleine, Canada, Cliché *et al.* (1994) showed that, after 44 days, 49% of seeded *Placopecten magellanicus* had moved > 60 m preferentially downstream to the south. Baird (1966) has reviewed the literature on migration in scallops, particularly the researches of Gibson (1956), Baird & Gibson (1956), Mason (1957) on *Pecten maximus*, and Dickie (1955) and Posgay (1963) on *Placopecten magellanicus*, and concluded that migrations do not occur although local movements may. Baird (1966) believed that the swimming habit in *Pecten maximus* is employed to enable the animal to move away from a substratum which is unsuitable for recessing. However, the evidence against migration is not as unequivocal as Baird (1966) believed; the wide distribution of many scallops, notably species of *Amusium* (Knudsen 1967), and aspects of scallop design, again especially observed in species of *Amusium*, give them the potential to be efficient, long-distance swimmers. It is extremely doubtful that the recorded swimming distances, *c.* 20–30 m, of *Amusium pleuronectes* are used merely for escape (Morton 1980). In this species, a recognizable escape reaction comprises between one to three adductions and probably has the effect of removing the scallop from an immediate predator. Conversely, the extended swims of species of *Amusium* (Morton 1980; Joll 1989), like those of *Placopecten magellanicus* (Caddy 1968; Stanley 1970), cover many metres and, in terms of functional morphology (i.e. the laterally thin, smooth, light shell, hydrodynamically efficient with internal ribs), it does not seem logical that such modifications arose simply to remove the animal from the immediate proximity of a predator. On the contrary, the adaptations of *A. pleuronectes*, including a fish-like crypsis (Thayer 1971), all point to a highly mobile mode of life. Under natural conditions in Shark Bay, Western Australia, *A. balloti* demonstrated a maximum swimming distance of 23 m, although a maximum, cumulative distance, swim of 31 m was recorded, at a speed of *c.* 1.6 m s^{-1} (Joll 1989). This species does not, however, make seasonal migrations (Joll 1989) and it too had to be physically made to swim by tapping the shell.

Swimming has probably evolved independently in several pectinoid lineages with this, rather than shell robustness, being the preferred defense against predators (Hayami 1991). Species of *Amusium* and *Placopecten* have the usual complex pectinid pallial eyes and representatives of both genera are capable of dramatic bouts of swimming (Caddy 1968; Morton 1980). In the case of *A. pleuronectes*, however, they rarely do so naturally and have to be encouraged, vigorously, to swim (Morton 1980). Occurring, as they do, at depths down to *c.* 200 m on the continental shelf, what, however, could they see moving anyway? Finally, Morton (1980) points out that the distance *A. pleuronectes* can swim, i.e. > 10 m at a speed of 0.73 m s^{-1}, in combination with the hydrodynamic attributes identified earlier, are unlikely to be adaptations evolved to avoid predation but, instead, point to a highly mobile lifestyle, perhaps associated with life-cycle changes.

Discussion

It appears that scallops, in general, respond to the

presence of possible predators by detecting either tactile or chemical cues, but rarely visual ones. They may possibly also seek out refuges, such as seagrass beds, visually, but they can also swim, not just to escape predators or to find refuges, but for other reasons too, e.g. short- or long-distance movements to keep away from predators and as part of seasonal changes in the environment, which may be related to the life cycle, respectively. For example the Antarctic scallop, *Adamussium colbecki*, although a poor swimmer, is among the earliest immigrants to areas of seabed uncovered by the retreat of glacier tongues in the Ross Sea (Ansell *et al.* 1998), and population densities can decline from tens per square metre to zero within hours (Ralph & Maxwell 1978). It is this species too for which there is an, albeit anecdotal, record of swimming possibly being induced by light. *In situ* observations of *A. colbecki* using a remotely operated vehicle (ROV) showed that 'the scallops appeared to be stimulated (to swim) either by mechanical stimuli or turbulence generated by the approach of the ROV *or possibly by the increase in light intensity* [my italics]' (Ansell *et al.* 1998, p. 372). A second example is given by Dibden & Joll (1998, p. 5), who stated that 'some scallops (*Amusium balloti*) usually began to swim in response to [the lights of] the underwater camera being towed along the bottom in their immediate vicinity and the flight response of a few often triggered a domino-effect in the surrounding scallops, resulting in the camera view being filled with swimming scallops'.

Primitive members of the Pectinidae, e.g. *Eopecten*, were byssally attached, resting on their right valves (Harper *et al.* 1996). Subsequently, the swimming habit was probably developed in a number of lineages (Hayami 1991), possibly in relation to the early Mesozoic Marine Revolution (Vermeij 1977) and the evolution of new suites of predators which fed on, among other things, bivalves. There was possibly an increase in swimming taxa at the beginning of the Mesozoic which might reflect such a natural selection pressure (Skelton *et al.* 1990). It has also been argued that many suspension-feeding bivalves became extinct during middle Palaeozoic and post-Triassic turbidity increases, while deposit feeding and swimming (i.e. pectinid) taxa diversified (Larson & Rhoads 1983; Thayer 1983). However, there is little evidence to support this, and, on the contrary, Skelton *et al.* (1990) suggest that over the last 250 Ma there has been progressive declines in both the actual number of pteriomorph families, in which the incidence of pallial eyes is most diverse (Morton 2001), and their relative importance. The same is true of swimming families, e.g. the Pectinidae, so that, assuming there is a relationship between swimming and the development of pallial eyes, such sophistications have not led to enhanced success. Conversely, the more recently diversifying Heterodonta have achieved their modern success largely without eyes (Morton 2001).

So, as discussed in this paper, in modern pectinids there seems to be little correlation between the evolution of complex pallial eyes and swimming as an escape response to approaching predators detected visually, and to light-induced patterns of short- or long-term movement. Is it possible, therefore, that the pectinid eye evolved in an ancestor(s) as a result of natural selection imposed by predation, changes in turbidity or because of a need to have better visual acuity in deeper waters, but have been retained in their descendants and now constitute an example of overdesign. Morton (2001) has argued that simpler eyes, seen in other bivalve lineages, could achieve the same functional bivalve response to perceived predation, i.e. adduction, digging or even swimming, as those of the Pectinidae. As Land (1981, p. 423) points out, 'It is true that some eyes are capable of supplying more information than others but at a higher cost in terms of space they take up and the metabolic energy they require'. Such eyes cannot, however, be that energetically expensive because they have been retained in the more modern *Spondylus gaederopus* (Spondylidae) and *Leptopecten latiauratus* (Pectinidae), which are cemented and byssally attached, respectively (Dakin 1928*a*, *b*; Morton 1993). Both possess the typically complex pallial eyes of the Pectinidae but neither species is capable of an escape response other than to adduct the shell valves. Clearly, the complex pectinid eye has been retained in these immobile scallops to initiate only the most basic bivalve defensive response, i.e. closure of the shell valves.

Spondylus is, then, perhaps the clue to the significance of the pectinid eye. If spondylid and pectinid eyes are homologous, as they seem to be, then this suggests that they evolved before the two families diverged in the early Jurassic. Evolved in an ancestral pectinid as the result of unknown natural selection pressures, sophisticated pallial eyes have, today, been retained and may fulfil a number of poorly understood functions in the variously adapted descendants. It is, however, also possible that the complex pectinid eye has evolved from a simpler structure to increase the efficiency of photon capture in low light intensities, such as occurs in the typical habitat of mobile scallops (i.e. the continental shelves), but, again, what advantage this confers upon them, since predators still seem to be typically detected by mechanical and chemical means, is unknown. An ability to detect changes in light intensity at deeper depths may be important for scallops and would explain their more sophisticated eyes. But, if so, why are they not needed in the much wider array of other subtidal bivalves which must have similar needs to detect light (or changes in it), e.g. to detect predators or coordinate reproductive cycles. As Morton (2001) concludes for bivalve pallial eyes in general: those of the Pectinidae seem, at present, to defy explanation. Nilsson (1994) suggests that bivalve eyes are optical 'burglar alarms', detecting non-image-forming movement. This seems probable in species occupying shallow, clear waters, but not in those, like the majority of scallops, occupying turbid continental shelf waters: nor does it explain why pectinid eyes are of such a sophisticated design when those of even the most basic form achieve the same objective.

Perhaps natural selection has acted so that even the smallest degrees of anatomical refinement confer a survival advantage in the face of predation.

Dr E. M. Harper, University of Cambridge, is thanked for hosting the conference at which this paper was presented and for constructive criticisms of the first draft of the manuscript.

References

ADAL, M. N. & MORTON, B. 1973. The fine structure of the pallial eyes of *Laternula truncata* (Bivalvia: Anomalodesmata: Pandoracea). *Journal of Zoology, London*, **171**, 533–556.

ALLEN, J. A. 1953. Observations on the epifauna of the deep-water muds of the Clyde Sea area, with special reference to *Chylamys septemradiata* (Müller). *Journal of Animal Ecology*, **22**, 240–260.

ANSELL, A. D., CATTANEO-VIETTI, R. & CHIANTORE, M. 1998. Swimming in the Antarctic scallop *Adamussium colbecki*: analysis of *in situ* recordings. *Antarctic Science*, **10**, 369–375.

ARSENAULT, D. J. & HIMMELMAN, J. H. 1996. Size-related changes in vulnerability to predators and spatial refuge use by juvenile Iceland scallops *Chlamys islandica*. *Marine Ecology Progress Series*, **140**, 115–122.

BAIRD, R. H. 1966. Notes on an escallop (*Pecten maximus*) population in Holyhead Harbour. *Journal of the Marine Biological Association of the United Kingdom*, **46**, 33–47.

—— & GIBSON, F. A. 1956. Underwater observations on escallop (*Pecten maximus* L.) beds. *Journal of the Marine Biological Association of the United Kingdom*, **35**, 555–562.

BARBEAU, N. A. & SCHEIBLING, R. E. 1994. Procedural effects of prey tethering experiments: predation of juvenile scallops by crabs and sea stars. *Marine Ecology Progress Series*, **111**, 305–310.

——, —— & HATCHER, B. G. 1998. Behavioural responses of predatory crabs and sea stars to varying density of juvenile sea scallops. *Aquaculture*, **169**, 87–98.

BARBER, V. C., EVANS, E. M. & LAND, M. F. 1967. The fine structure of the eye of the mollusc *Pecten maximus*. *Zeitschrift für Zellforschung und Mikroscopische Anatrome*, **76**, 295–312.

BRICELJ, V. M., BAUER, S. & TANIKAWA-OGLESBY, S. 1993. Constrasting foraging tactics of two predators of juvenile Bay scallops, *Argopecten irradians*, in the eelgrass canopy. *Journal of Shellfisheries Research*, **12**, 130–131.

BRUN, E. 1972. Food and feeding habits of *Luidia ciliaris* Echinodermata: Asteroidea. *Journal of the Marine Biological Association of the United Kingdom*, **52**, 225–236.

BUDDENBROCK, W. VON & MOLLER-RACKE, I. 1953. Über den Lichtsinn von *Pecten*. *Pubblicazioni della Stazione Zoologica di Napoli*, **24**, 217–245.

CADDY, J. F. 1968. Underwater observations on scallop (*Placopecten magellanicus*) behavior and drag efficiency. *Journal of the Fisheries Research Board of Canada*, **25**, 2123–2141.

—— 1972. Progressive loss of byssus attachment with size in the sea scallop, *Placopecten magellanicus*. *Journal of Experimental Marine Biology and Ecology*, **9**, 179–190.

CHAPMAN, C. J., MAIN, J., HOWELL, T. & SANGSTER, G. I. 1979. The swimming speed and endurance of the Queen scallop *Chlamys opercularis* in relation to trawling. *Progress in Underwater Science*, **4**, 57–72.

CHARLES, G. H. 1966. Sense organs (less cephalopods). *In*: WILBUR, K. M. & YONGE, C. M. (eds) *Physiology of Mollusca, Volume II*. Academic Press, New York, 455–521.

CLICHÉ, G., GIGUERE, M. & VIGNEAU, S. 1994. Dispersal and mortality of sea scallops, *Placopecten magellanicus* (Gmelin, 1791), seeded on the sea bottom off Iles-de-la-Madeleine. *Journal of Shellfisheries Research*, **13**, 565–570.

DAKIN, W. J. 1910. The eye of *Pecten*. *Quarterly Journal of Microscopical Science*, **55**, 49–112, pls 6 and 7.

—— 1928*a*. The eyes of *Pecten*, *Spondylus*, *Amussium* and allied lamellibranchs with a short discussion on their evolution. *Proceedings of the Royal Society of London, Series B*, **103**, 355–366.

—— 1928*b*. The anatomy and phylogeny of *Spondylus*, with a particular reference to the lamellibranch nervous system. *Proceedings of the Royal Society of London, Series B*, **103**, 337–354.

DAUTOV, S. SH. & KARPENKO, A. A. 1984. Behavior and mechanisms of locomotion of two species of scallops from the Sea of Japan. *Soviet Journal of Marine Biology*, **9**, 271–275.

DIBDEN, C. J. & JOLL, L. M. 1998. A research vessel survey of bottom types in the area of the Abrolhos Islands and mid-west trawl fishery. *Fisheries Research Report, Fisheries Western Australia*, **110**, 1–21.

DICKIE, L. M. 1955. Fluctuations in abundance of the giant scallop, *Placopecten magellanicus* (Gmelin), in the sea of the Bay of Fundy. *Journal of the Fisheries Research Board of Canada*, **12**, 797–857.

—— & MEDCOF, J. C. 1963. Causes of mass mortalities of scallops (*Placopecten magellanicus*) in the southwestern Gulf of St. Lawrence. *Journal of the Fisheries Research Board of Canada*, **20**, 451–482.

ELNER, R. W. & JAMIESON, G. S. 1979. Predation of sea scallops, *Placopecten magellanicus*, by the rock crab, *Cancer irroratus*, and the American lobster, *Homarus americanus*. *Journal of the Fisheries Research Board of Canada*, **36**, 537–543.

FANGE, R. 1963. Toxic factors in starfishes. *Sarsia*, **10**, 19–21.

GIBSON, F. A. 1956. Escallops *Pecten maximus* L. in Irish waters. *Scientific Proceedings of the Royal Dublin Society*, **27**, 253–270.

GOMEZ, M. & NASI, E. 1997. Light adaptation in *Pecten* hyperpolarizing photoreceptors: insentivity to calcium manipulations. *Journal of General Physiology*, **109**, 371–384.

GOULD, S. J. 1971. Muscular mechanics and the ontogeny of swimming in scallops. *Palaentology*, **14**, 61–94.

HAMILTON, P. V. & KOCH, K. M. 1996. Orientation toward natural and artificial grassbeds by swimming Bay scallops, *Argopecten irradians* (Lamarck, 1819). *Journal of Experimental Marine Biology and Ecology*, **199**, 79–88.

HANCOCK, D. A. 1974. Some aspects of the biology of the sunstar *Crossaster papposus* (L.). *Ophelia*, **13**, 1–30.

HARPER, E. M., RADLEY, J. D. & PALMER, T. J. 1996. Early Cretaceous cementing pectinid bivalves. *Cretaceous Research*, **17**, 135–150.

HARTLINE, H. K. 1925. The protosensory mechanisms of *Pecten irradians* – preliminary note. *American Journal of Physiology*, **122**, 211–212.

—— 1938. The discharges of impulses in the optic nerve of *Pecten* in response to illumination of the eye. *Journal of Cellular and Comparative Physiology*, **11**, 465–477.

HAYAMI, I. 1991. Living and fossil scallop shells as airfoils: an experimental study. *Paleobiology*, **17**, 1–18.

HUNT, O. D. 1925. The food of the bottom fauna of the Plymouth fishing grounds. *Journal of the Marine Biological Association of the United Kingom*, **13**, 560–599.

JOLL, L. M. 1989. Swimming behaviour of the scallop *Amusium balloti* (Mollusca: Pectinidae). *Marine Biology*, **102**, 299–305.

KARPENKO, A. A. 1980. Starfish avoidance response in the scallop *Patinopecten yessoensis* (Mollusca: Bivalvia). *Zoologicheskii Zhurnal*, **59**, 146–148.

—— 1981. How a prey distinguishes its predator; scallop-starfish relationship. *In*: ZHIRMUNSKIJ, A. V. (ed.) *Physiology and Biochemistry of Adaptations in Marine Animals*. Vladivostock, USSR, 147–153.

KIM, Y. S. 1969. An observation on the opening bivalve mollusks by starfish *Asterias amurensis*. *Bulletin of the Faculty of Fisheries, Hokkaido University*, **20**, 60–63.

KNUDSEN, J. 1967. The deep sea Bivalvia. *Scientific Reports of the John Murray Expedition*, **11**, 237–343.

LAKE, N. C. H. & JONES, M. B. 1987. Crab predation on scallop (*Pecten maximus*) and its implications for scallop culture. *Journal of the Marine Biological Association of the United Kingdom*, **67**, 55–64.

LAND, M. F. 1965. Image formation by a concave reflector in the eye of the scallop, *Pecten maximus*. *Journal of Experimental Biology*, **179**, 138–153.

—— 1966*a*. A multilayer interference reflector in the eye of the scallop, *Pecten maximus*. *Journal of Experimental Biology*, **45**, 433–477.

—— 1966*b*. Activity in the optic nerve in *Pecten maximus* in response to changes in light intensity, and to pattern and movement in the optical environment. *Journal of Experimental Biology*, **45**, 83–99.

—— 1968. Functional aspects of the optical and retinal organization of the mollusc eye. *In*: CARTHY, J. D. & NEWELL, G. E. (eds) *Invertebrate Receptors*. Symposia of the Zoological Society of London, **23**, 75–96.

—— 1981. Optics and vision in invertebrates. *In*: AUTRUM, H. (ed.) *Handbook of Sensory Physiology. Volume VII/6H. Comparative Physiology and Evolution of Vision in Invertebrates. B. Invertebrate Visual Centers and Behaviour I.* Springer Verlag, Berlin. 471–593.

LARSON, D. W. & RHOADS, D. C. 1983. The evolution of infaunal communities and sedimentary fabrics. *In*: TEVESZ, M. J. S. & MCCALL, D. C. (eds) *Biotic Interactions in Recent and Fossil Benthic Communities*. Plenum Press, New York, 627–649.

MASON, J. 1957. The age and growth of the scallop, *Pecten maximus* L., in Manx waters. *Journal of the Marine Biological Association of the United Kingdom*, **36**, 473–492.

MOIR, A. J. G. 1977. On the ultrastructure of the abdominal sense organ of the giant-scallop, *Placopecten magellanicus* (Gmelin). *Cell and Tissue Research*, **184**, 359–366.

MOORE, J. K. & MARSHALL, N. 1967. An analysis of the movements of bay scallops *Aequipecten irradians* in a shallow estuary. *Proceedings of the National Shellfisheries Association*, **57**, 77–82.

MORTON, B. 1980. Swimming in *Amusium pleuronectes* (Bivalvia: Pectinidae). *Journal of Zoology, London*, **190**, 375–404.

—— 1993. The biology and functional morphology of *Leptopecten latiauratus* (Conrad, 1837): an opportunistic scallop. *Veliger*, **37**, 5–22.

—— 1994. Prey preference and method of attack by *Rapana bezoar* (Gastropoda: Muricidae) from Hong Kong. *In*: MORTON, B. (ed.) *Proceedings of the Third International Workshop on the Malacofauna of Hong Kong and southern China, Hong Kong 1992*. Hong Kong University Press, Hong Kong, 309–325.

—— 1996. The biology and functional morphology of *Minnivola pyxidatus* (Bivalvia: Pectinoidea) in Hong Kong. *Journal of Zoology, London*, **240**, 735–760.

—— 2001. The evolution of eyes in the Bivalvia. *Oceanography and Marine Biology: an Annual Review*, in press.

NILSSON, D. -E. 1994. Eyes as optical alarm systems in fan worms and ark clams. *Philosophical Transactions of the Royal Society of London, Series B*, **346**, 195–212.

ORDIZE, C. J. & GAROFALO, G. C. 1980. Behavioral

recognition of molluscan and echinoderm predators by the Bay scallop, *Argopecten irradians* (Lamarck) at two temperatures. *Journal of Experimental Marine Biology and Ecology*, **43**, 29–37.

PATTEN, W. 1886. Eyes of molluscs and arthropods. *Mitteil, Zoologische Station Neapel*, **6**, 542–756, pls 28–32.

PELSENEER, P. 1911. Les lamellibrancles de l'expédition du Siboga. Partie anatomique. *Siboga Expeditie*, **53a**, 1–125.

PITCHER, C. R. & BUTLER, A. J. 1987. Predation by asteroids, escape responses, and morphometrics of scallops with epizoic sponges. *Journal of Experimental Marine Biology and Ecology*, **112**, 233–249.

POSGAY, J. A. 1963. Tagging as a technique in population studies of the sea scallop. *Special Publications of the International Commission on Atlantic Fisheries*, **4**, 268–271.

RALPH, R. & MAXWELL, J. G. H. 1978. Growth of the Antarctic lamellibranchs *Adamussium colbecki* and *Laternula elliptica*. *Marine Biology*, **42**, 171–175.

REES, W. J. 1957. The living scallop. *In*: COX, I. (ed.) *The Scallop*. London, The 'Shell' Transport & Trading Co., Ltd, London, 15–32.

ROLFE, M. S. 1973. *Notes on Queen Scallops and How to Catch Them. Shellfish Information Leaflet.* Ministry of Agriculture Fisheries and Food, UK, **27**, 1–13.

SCHWARZ, F. J. & PORTER, H. J. 1977. *Fishery Bulletin NOAA*, **75**, 427–446.

SHARP, B. 1883. On visual organs in *Solen*. *Proceedings of the Academy of Natural Sciences, Philadelphia*, **58**, 248–249.

SKELTON, P. W., CRAME, J. A., MORRIS, N. J. & HARPER, E. M. 1990. Adaptive divergence and taxonomic radiation in post-Palaeozoic bivalves. *In*: TAYLOR, P. D. & LARWOOD, G. P. (eds) *Major Evolutionary Radiations*. Clarendon Press, Oxford, 91–117.

SLOAN, N. A. 1980. Aspects of the feeding biology of asteroids. *Oceanography and Marine Biology: an Annual Review*, **18**, 57–124.

STANLEY, S. M. 1970. Relation of shell form to life habits in the Bivalvia (Mollusca). *Memoirs of the Geological Society of America*, **125**, 1–296.

STEPHENS, P. J. 1978. The sensitivity and control of the scallop mantle edge. *Journal of Experimental Biology*, **75**, 203–221.

—— & BOYLE, P. R. 1978. Escape responses of the Queen scallop *Chlamys opercularis* (L.) (Mollusc: Bivalvia). *Marine Behaviour and Physiology*, **5**, 103–113.

STOKESBURY, K. D. E. & HIMMELMAN, J. H. 1996. Experimental examination of movement of the giant scallop, *Placopecten magellanicus*. *Marine Biology*, **124**, 651–660.

STRIEB, M. D., BRICELJ, V. M. & BAUER, S. I. 1995. Population biology of the mud crab, *Dyspanopeus sayi*, an important predator of juvenile bay scallops in Long Island (USA) eelgrass beds. *Journal of Shellfisheries Research*, **14**, 347–357.

THAYER, C. W. 1971. Fish-like crypsis in swimming Monomyaria. *Proceedings of the Malacological Society of London*, **39**, 371–375.

—— 1972. Adaptive features of swimming monomyarian bivalves (Mollusca). *Forma et Functio*, **5**, 1–32.

—— 1983. Sediment-mediated biological disturbance and the evolution of marine benthos. *In*: TEVESZ, M. J. S. & MCCALL, P. L. (eds) *Biotic Interactions in Recent and Fossil Benthic Communities*. Plenum Press, New York, 479–625.

THOMAS, G. E. & GRUFFYDD, L. D. 1971. The types of escape reactions elicited in the scallop *Pecten maximus* by selected sea-star species. *Marine Biology*, **10**, 87–93.

VERMEIJ, G. J. 1977. The Mesozoic marine revolution: evidence from snails, predators and grazers. *Paleobiology*, **3**, 245–258.

WELLS, M. J. 1966. The brain and behaviour of cephalopods. *In*: WILBUR, K. M. & YONGE, C. M. (eds) *Physiology of Mollusca*. Academic Press, New York, 547–590.

WENRICH, D. H. 1916. Notes on the reactions of bivalve mollusks to changes in light intensity: image formation in *Pecten*. *Journal of Animal Behaviour*, **6**, 297–318.

WILKENS, L. A. 1981. Neurobiology of the scallop (*Pecten ziczac*): 1. Starfish-mediated escape behaviours. *Proceedings of the Royal Society of London, Series B*, **211**, 341–372.

WOLF, B. M. & WHITE, R. W. G. 1997. Movements and habitat use of the Queen scallop, *Equichlamys bifrons*, in the D'Entrecasteaux Channel and Huron River estuary, Tasmania. *Journal of Shellfisheries Research*, **16**, 533–539.

WOLFF, M. & ALARCON, E. 1993. Structure of a scallop *Argopecten purpuratus* (Lamarck, 1819) dominated subtidal macro-invertebrate asemblage in northern Chile. *Journal of Shellfisheries Research*, **12**, 295–304.

YONGE, C. M. 1936. The evolution of the swimming habit in the Lamellibranchia. *Mémoires du Musée Royal d'Histoire Naturelle de Belgique*, **3**, 77–100.

Functional anatomy of the digestive system of *Neoteredo reynei* (Bartsch, 1920) and *Psiloteredo healdi* (Bartsch, 1931) (Bivalvia: Teredinidae)

SÔNIA G. B. C. LOPES, OSMAR DOMANESCHI, DANIELA T. DE MORAES, MARISA MORITA & GEORGEANA DE L. C. MESERANI

Departamento de Zoologia, Instituto de Biociências, Universidade de São Paulo, PO Box 11.461, CEP 05422-970, São Paulo, Brazil (e-mail: sonialop@uol.com.br)

Abstract: Studies on the digestive system of the Teredinidae are useful for a better understanding of the evolution of these bivalves in relation to the xylophagous habit. *Neoteredo reynei* and *Psiloteredo healdi*, two common species in Brazilian mangroves, have evolved differently in their methods to use wood as food, despite the similarities in the anatomy and functioning of their globular type II stomachs. *N. reynei* is predominantly xylophagous throughout its life, while *P. healdi*, despite its predominant suspension-feeding habit, uses wood more efficiently as the animal grows older. The outstanding differences that allow these conclusions are the large size of the appendix and anal canal in *N. reynei*, always conspicuous and packed with wood, and the small appendix of *P. healdi*, which increases in size with age. Based on anatomical data and revision of the literature, it is suggested that in both species the appendix, and also the anal canal in *N. reynei*, is of primary importance in the digestion of wood and absorption of nutrients, counterbalancing the reduced specialized digestive diverticula.

The Teredinidae are important organisms in marine and estuarine ecosystems, contributing to the reduction of wood in the sea. They first appear in the fossil record in the Lower Cretaceous, their radiation being related to the evolution of woody plants, which provide substrata in the form of driftwood, and the evolution of salt-tolerant plants (Turner 1966; Turner & Johnson 1971). In the course of their evolution, the wood was used primarily for protection and, later, as food (Turner & Johnson 1971; Hoagland & Turner 1981).

Mechanisms of wood digestion

The mechanisms of wood digestion, as well as the origin of necessary cellulolytic enzymes, have been revised by Nair & Saraswathy (1971), Rosenberg & Cutter (1973), Morton & McQuiston (1974), Morton (1978, 1983) and Mann (1984). Endosymbiotic bacteria have been demonstrated in the ctenidia of the Teredinidae by Popham & Dickson (1973) and were later shown to digest cellulose *in vitro*, fix nitrogen (Warterbury *et al.* 1983), and to produce proteolytic, xylanolytic and cellulolytic enzymes (Greene & Freer 1986; Greene *et al.* 1988, 1989). These bacterial endosymbiont populations, previously described as the gland of Deshayes (Sigerfoos 1908), are now thought to be essential in this intriguing process of evolution that allowed the Teredinidae to use wood as food. Similar bacteria were recently demonstrated in the ctenidia of two *Xylophaga* species (Pholadidae) which are now considered the ecological equivalent of the Teredinidae in the deep sea (Distel & Roberts 1997).

Although the symbiont activity are appropriate for a wood-based diet, no direct evidence exists for their participation in the wood digestion in shipworms, nor it is clear how the digestive enzymes produced by the endosymbiont in the ctenidia are transported to the lumen of the gut (Distel & Roberts 1997). In the Teredinidae, a well-defined duct, previously described by Sigerfoos (1908) as a duct of the Deshayes gland, can be traced inside and along the extension of each afferent branchial vessel. The only reference in the literature that describes these ducts opening into an organ (the oesophagus) of the digestive system, and discharging its contents to help in the digestive process, is that of Saraswathy & Nair (1971) for *Teredo furcifera* von Martens 1894.

Morphological adaptations

The evolution of the capability of wood digestion in the Teredinidae is also related to the specialization

From: Harper, E. M., Taylor, J. D. & Crame, J. A. (eds) *The Evolutionary Biology of the Bivalvia*. Geological Society, London, Special Publications, **177**, 257–271. 1-86239-076-2/00/$15.00

of the digestive diverticula and modifications of the stomach. They are unique among the lamellibranchs in possessing specialized digestive diverticula related to the digestion of wood in addition the normal ones responsible for the digestion of suspension foods (Potts 1923; Yonge 1926; Turner 1966; Morton & McQuiston 1974; Morton 1983).

Three different types of stomach occur in the Teredinidae (Turner 1966): globular type I; globular type II; elongate type III. Globular type I is exclusive to the Kuphinae, globular type II is most frequently encountered in the Teredininae and elongate type III in the Bankiinae. Morphological characteristics of the stomach, such as its elongation, reduction of the crystalline style, enlargement of the appendix and contained typhlosole, have been considered to reflect the increasing ability of the Teredinidae to feed mainly on wood (Turner 1966; Turner & Johnson 1971; Morton 1978; Hoagland & Turner 1981). Reduction of the labial palps, and length and height of the demibranchs, are indicative that the suspension foods are relatively less important than wood in the diet of the shipworm (Turner 1966).

The Teredininae species *Neoteredo reynei* and *Psiloteredo healdi*, two common inhabitants in the same Brazilian mangroves, both have globular type II stomachs. Comparing their ctenidia, palps and appendix, Turner (1966) pointed out differences indicative that the former species is more dependent on wood as a source of food than the latter. The comparative analyses of the internal anatomy and the functioning of the digestive system performed in this study provide new data which allow a better understanding of how these two Teredininae species evolved different methods of using wood as food, despite having the same globular type II stomach.

Materials and methods

Neoteredo reynei and *Psiloteredo healdi* were collected in the same mangrove area in São Paulo State, Brazil (22°30'S; 45°15'W). The first species is more abundant in regions of the mangrove forest less frequently covered by seawater, while the latter is restricted to submerged logs in a section of a tributary river where the salinity is very low (Lopes & Narchi 1993). Logs collected in these regions were kept in aquaria at a salinity of 20‰ and 10‰ for *N. reynei* and *P. healdi*, respectively. More than 50 living, as well as many fixed, specimens of each species were analysed in the present study. Some of the anatomical details were obtained from transverse sections (8–10 μm thick) stained with Mallory's triple stain, or Ehrlichs's hematoxylin and eosin, according to the methods described by Pantin (1948). The epithelium of the appendix and anal canal of *N. reynei* was prepared for examination under a scanning electron microscope (SEM), using the same techniques adopted by Leonel *et al.* (1998).

Ciliary currents on the internal walls of the stomach were observed using powdered sand grains and suspensions of carmine, Acquadag and very fine sawdust fragments. Sawdust was obtained by rubbing a mangrove log with a dentist's drill. Such materials were applied to the ciliary tracts either mixed or separately.

Variations in the diameter of the appendix of *P. healdi* were evaluated in relation to the shell length. The latter was adopted as a parameter to express the size of the specimens, since these bivalves shrink when removed from wood.

Results

General disposition of organs in the mantle cavity

The dispositions of the major organs in the mantle cavity of *N. reynei and P. healdi* are shown in Fig. 1a and b, respectively. The ctenidia (prc) of *N. reynei* comprise only the outer demibranchs and these extend from the base of the siphons to the posterior extremity of the visceral mass, and occupy *c.* 20% of the body length. The demibranchs anteriorly reduce abruptly in height, becoming restricted to the food groove (fg) which extends forward to meet the very reduced labial palps (l). The anterior portion of the ctenidia present in most Teredinidae is lacking in this species.

The ctenidia of *P. healdi* differ from those of *N. reynei* in the presence of an anterior portion (apc) made up of only 15–19 filaments of the external lamella. The posterior region of the organ occupies *c.* 68% of the body length and is higher than the corresponding region in *N. reynei*. Immersed in the tissue of the interlamellar junctions of the demibranchs of both species occur the so-called gland of Deshayes. Cellulolytic nitrogen-fixing bacterium was demonstrated in the demibranchs of *P. healdi* by Waterbury *et al.* (1983). Histological sections allowed the duct of the gland of Deshayes to be traced inside each afferent branchial vessel of both species, but no direct connection with any organ of the digestive system were detected in the present work.

The dorsal labial palps of *P. healdi* have an expanded free distal end, unlike those of *N. reynei*, which are represented by a slender and low

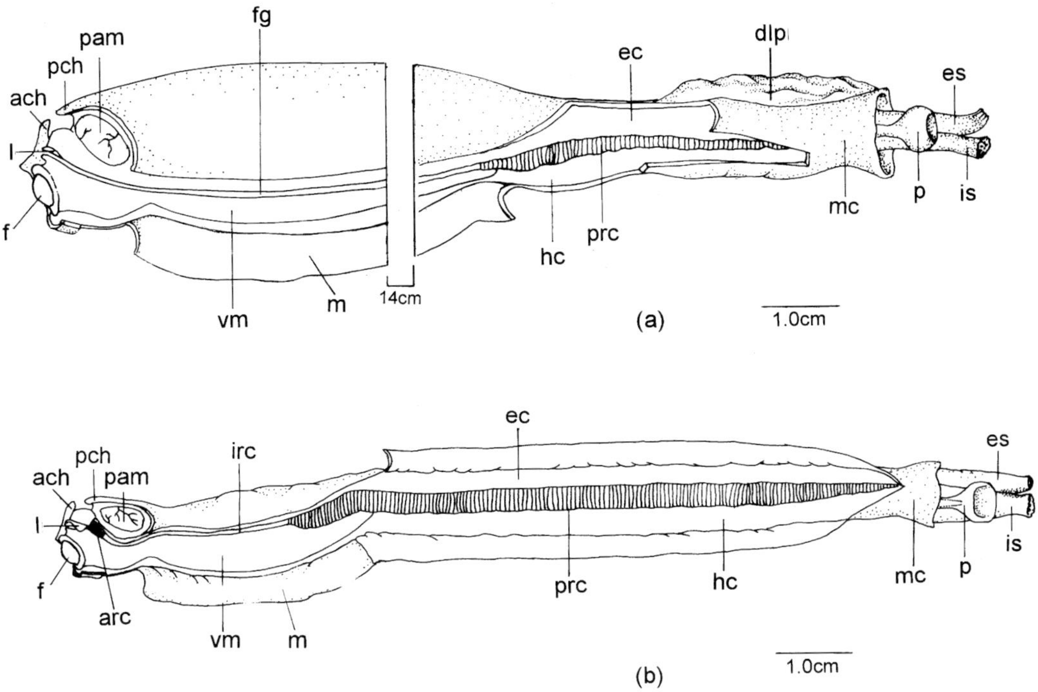

Fig. 1. (**a**) *Neoteredo reynei* and (**b**) *Psiloteredo healdi.* General topography of the organs and structures of the mantle cavity. ach, Anterior cephalic hood; arc, anterior region of the ctenidium; dlp, dorsal lappets; ec, epibranchial cavity; es, exhalant siphon; f, foot; fg, food groove; hc, hypobranchial cavity; is, inhalant siphon; irc, intermediate region of the ctenidium; l, labial palp; m, mantle; mc, mantle collar; p, pallet; prc, posterior region of the ctenidium; pam, posterior adductor muscle; pch, posterior cephalic hood; vm, visceral mass.

epithelial fold. The ventral labial palps of both species are equally reduced to slender and low folds.

General configuration of the digestive system

The general configuration of the digestive system of *N. reynei* and *P. healdi* are shown in Fig. 2a and b, respectively. The nomenclature adopted in the description of their structures is the same of Purchon (1960, 1987) and Lopes *et al.* (1998). A complete description is given for *N. reynei* and only significant differences in the digestive system of *P. healdi* will be pointed out.

N. reynei. The slit-like mouth is located dorsally to the foot and opens into a very short, dorso-ventrally flattened, oesophagus (o) that enters the stomach antero-dorsally. The stomach (st) has a globular region sheltered within the limits of the shell valves, from which the appendix (a) arises. This structure is an elongate and cylindrical pouch, corresponding to *c.* 30–44 % of the body length. Its diameter is relatively constant and roughly equal to that of the globular region of the stomach. The appendix opens into the right side of the posterior wall of the stomach via a narrow aperture, then enlarges abruptly in diameter and extends backwards horizontally through the visceral mass which lies outside the shell valves. The aperture of the appendix into the stomach is preceded by a corrugation of the stomach wall, which begins on the dorsal side of the organ and extends to the right. Here the wall bulges in a conical fashion as it turns ventrally and, at the same time, it spirals, embracing almost completely the proximal end of the appendix.

The globular region gives also rise to the style-sac (css), the dorsal hood (dh), the combined left pouch and left caecum (lp/lc) and the right caecum (rc). Well-developed normal digestive diverticula (NDD) surround the combined left caecum/left pouch and the right caecum, and hide the poorly developed specialized digestive diverticula (SDD). These two kinds of digestive diverticula open exclusively into these embayments of the stomach

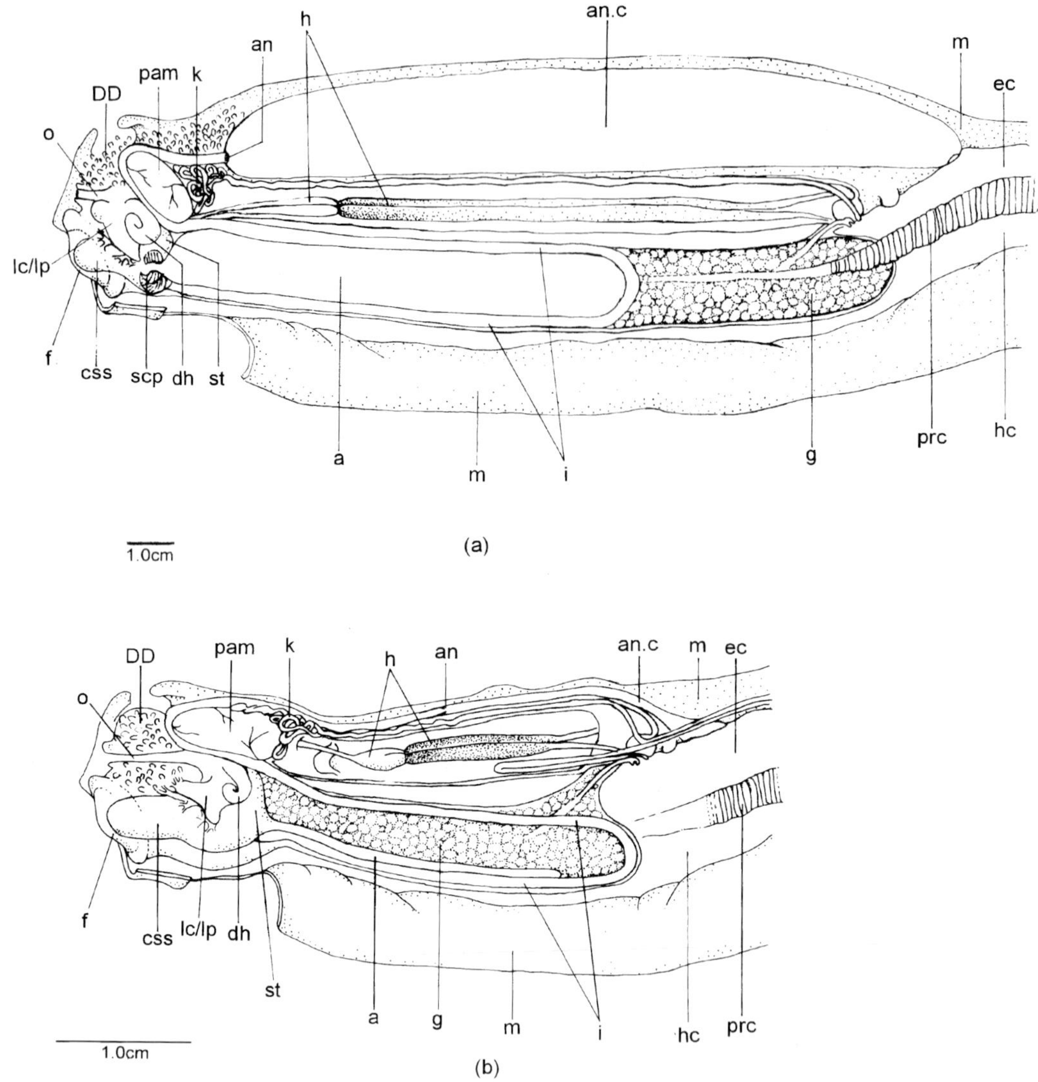

Fig. 2. (**a**) *Neoteredo reynei* and (**b**) *Psiloteredo healdi*. Diagrammatic representation of the anterior third of the body dissected to expose the digestive system. a, Appendix; an, anus; an.c, anal canal; css, crystalline style-sac; DD, digestive diverticula; dh, dorsal hood; ec, epibranchial cavity; f, foot; g, gonads; h, heart; hc, hypobranchial cavity; i, intestine; k, main portion of the kidney; lc/lp, combined left caecum/left pouch; m, mantle; o, oesophagus; pam, posterior adductor muscle; prc, posterior region of the ctenidium; st, stomach.

and keep the same relative development independently of the size of the specimen. Externally, the NDD and the SDD are similar in appearance but differ in their histological organization, which corresponds to that illustrated by Saraswathy & Nair (1971) for the species they have analysed.

A short and distally widened style-sac opens at the floor of the stomach, close to, but separated from, the opening of the intestine (i). From its opening in the stomach, the style-sac slants to the right and ventrally ending where it makes contact with the sole of the foot. The blind tip of the style-sac bears a small bag-like evagination, the appendix of the style-sac.

The dorsal hood is a large conical pocket that emerges from the left dorsal side of the stomach and narrows gradually as it extends forward and describes a complete spiral. This is in a clockwise direction when the stomach is observed dorsally.

The left caecum and the left pouch form a combined structure which opens into the stomach through a common and broad opening, with no gap between them. The left caecum lies anteriorly to the dorsal hood and can be defined by the presence of a small dorsal swelling and two long, slender, antero-ventral compartments, into which several ducts from the NDD open. A few ducts from the poorly developed SDD open into the left caecum. The left pouch is a flattened, leaf-like, evagination ventral to the dorsal hood. Some ducts from the NDD open near the blind end of this pocket. The relationship between these ducts with the left pouch or left caecum is difficult to establish.

The right caecum is a large evagination near the entrance of the oesophagus into the stomach. Just after its emergence from the stomach, it receives ducts from the SDD and gives rise to three compartments. The short anterior first compartment extends ventrally and receives ducts from the NDD and the SDD. A second compartment, longer in length and wider in diameter than the preceding one, extends ventro-posteriorly close to the stomach wall and receives ducts only from the NDD. The third compartment begins dorsally as a broad cylindrical structure that declines abruptly in diameter half-way along its distal end, which lies dorsally to the posterior adductor muscle. This compartment receives a large number of ducts from the NDD and is completely enveloped by these digestive glands.

The intestine leaves the floor of the globular region of the stomach near to and ventral to the opening of the appendix. Thence, it extends ventralward for a short distance, and then turns abruptly forward and immediately swells to form a large bulbous region (bu). As the mid-gut proceeds forward, it returns to its initial diameter, curves to the right and extends onto and embraces dorsally the style-sac, making a loose loop. From here the intestine extends backwards, running ventrally to the stomach and the long appendix. At the distal end of the appendix the intestine makes a loop to run forwards, dorso-laterally to the structure. Then, the intestine embraces the anterior and the dorsal sides of the posterior adductor muscle and finishes in the anus (a), immediately after the posterior limit of the shell valves. This terminal part of the intestine, dorsal to the posterior adductor muscle, is completely enveloped by the NDD leading from the right caecum.

The anus (a) opens into the anal canal (an.c), an enormous, dorsally located, cylindrical tube that extends backwards throughout the extension of the visceral mass. A strong muscular sphincter controls the aperture of the anal canal into the epibranchial chamber. The walls of the anal canal are richly provided with blood vessels, cilia and microvilli (Fig. 3a). The species eliminates powdery faeces and, eventually, small oval faecal pellets.

Psiloteredo healdi. The general configuration of the digestive system of this species is similar to that of *N. reynei*. Differences between them are mainly related to: the longer oesophagus; the less-developed right caecum, with only two semi-cylindrical compartments; the shorter appendix (*c.* 15–20% of the body length), which varies in diameter in different specimens; a smaller semi-spiral conical projection in *P. healdi*. This species also has smaller amounts of SDD and a very large style-sac and crystalline style.

In *N. reynei* the gonads do not surround the appendix, while in *P. healdi* they almost conceal this pocket in mature individuals. The development of the appendix and gonads of *P. healdi* show a tendency to be inversely related to the animals age (Fig. 4). Of the 38 specimens examined (Fig. 5), most of those with a shell length ≤ 7.5 mm ($n = 20$)

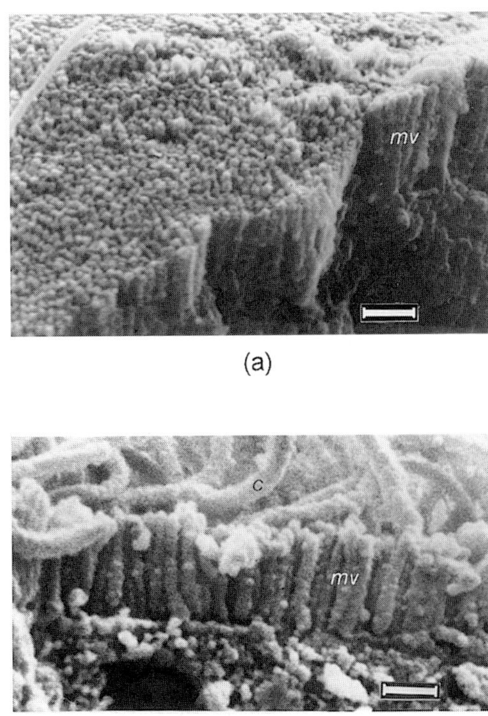

(a)

(b)

Fig. 3. *Neoteredo reynei.* (**a**) SEM of the internal surface of the anal canal and (**b**) SEM of the appendix. c, Cilia; mv, microvilli. Scale bars, 1.0 μm.

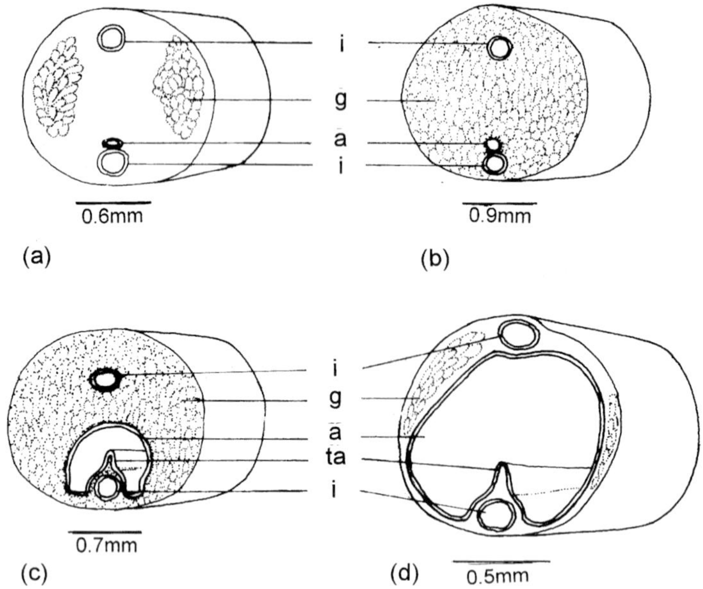

Fig. 4. *Psiloteredo healdi.* Transverse section through the visceral mass showing variations in the development of the appendix and gonads among 38 specimens from 3.0 to 17.0 mm in shell length. (**a**) Small appendix and poorly developed gonads; (**b**) small appendix and well-developed gonads; (**c**) large appendix and well-developed gonads; (**d**) large appendix and poorly developed gonads. a, Appendix; g, gonads; i, intestine; ta, typhlosole of the appendix.

had a small appendix frequently associated with a well-developed gonad. The reverse situation occurred in all animals with a shell length ≤ 2.5 mm ($n = 5$). Specimens of intermediate size (8.0–12.0 mm) ($n = 13$) exhibited a mosaic of the preceding situations.

The intestine of *P. healdi* differs from that of *N. reynei* mainly in its final portion, which extends well behind the posterior limit of the shell valves and finishes mid-way along the pericardial cavity. The very short anal canal begins at the anal opening, exhibiting a uniform diameter, similar to that of the intestine, throughout its length. Its wide opening into the epibranchial cavity lacks a muscular sphincter and faeces do not accumulate within this compartment of the digestive system. Faeces are always eliminated as long cylindrical rods.

Internal anatomy of the digestive system

Neoteredo reynei. The internal anatomy of the stomach of *N. reynei* and its ciliary currents are illustrated in Figs 6 and 7. The inner wall of the oesophagus (o) has longitudinal folds and two well-developed lateral grooves. Cilia impel particles entangled in mucus to the head of the crystalline style. Isolated particles are passed into the dorsal hood by cilia on a long fold (R) that extends from the left dorsal side of the aperture of the oesophagus to the distal end of the dorsal hood. The crystalline style (cs) is short and pear shaped, with the part that projects into the stomach being slender and frequently twisted. The inner dorsal surface of the style-sac (css) has a longitudinal groove.

The gastric shield (gs) is a saddle-shaped structure. It lines the floor and the posterior wall of the stomach from the opening of the style-sac and sends wing-shaped projections into the dorsal hood and left pouch. In several individuals, the gastric shield was brilliant yellow. The dorsal hood (dh) has folds on its inner wall (R, R1, R2 and R4) and a sorting area (SA3) flanked by R1 and R2. A fine longitudinal striated area occurs between folds R2 and R4.

Fold R1 begins as a smooth, low ridge in the distal end of the dorsal hood becoming higher; the side facing SA3 is slightly transversely plicate. It leaves the dorsal hood, passes dorsally to the opening of the oesophagus and finishes abruptly dorsal to the opening of the right caecum. The ciliary currents on both faces of fold R1 conduct particles obliquely onto the crest, where cilia throw them over fold R to be accumulated near the wing of the gastric shield.

Fold R4 is located posteriorly and begins as a

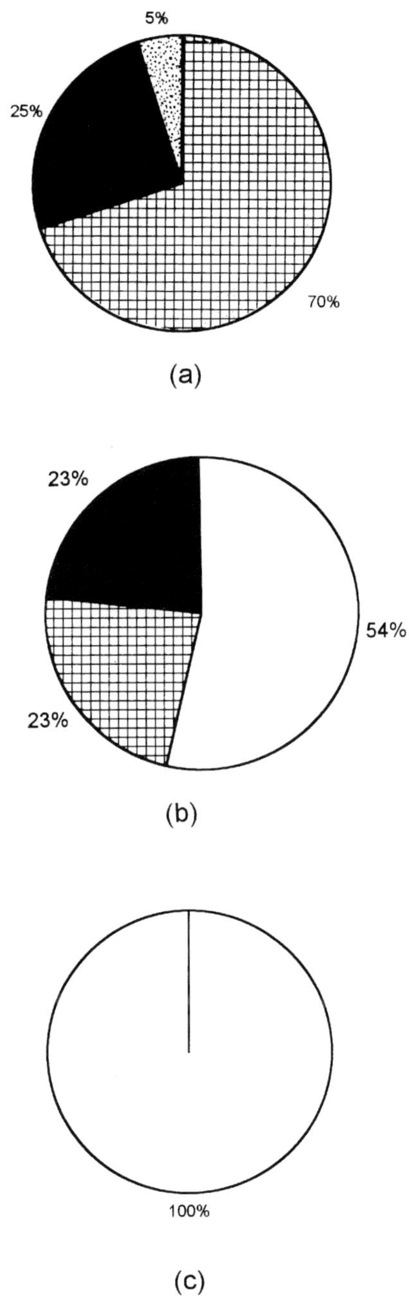

Fig. 5. *Psiloteredo healdi*. Relative percentages of the variations described in Fig. 4 among specimens of three classes of shell length: (**a**) 3.0–7.5 mm (n = 20); (**b**) 8.0–12.0 mm (n = 13); (**c**) 12.5–17.0 mm (n = 5). Black segments, small appendix and poorly developed gonads; cross-hatched segments, small appendix and well-developed gonads; stippled segments, large appendix and well-developed gonads; open segments, large appendix and poorly developed gonads.

smooth, low ridge in the distal end of the dorsal hood, immediately becoming higher with one face slightly transversely plicate. It leaves the dorsal hood, passes dorsally to the oesophagus, becoming smooth, and terminates in the dorsal side of the stomach. Ciliary currents on this fold conduct particles to its basal region where they are concentrated near the gastric shield.

Fold R2 and SA3 begin in the distal end of the dorsal hood and extend in a dorsal direction to the dorsal right side of the stomach, terminating in the distal end of the semi-spiral conical projection (scp).

Along the folds of SA3, ciliary currents conduct small particles from crest to crest in the direction of the dorsal hood. As the small particles are transported they are able to form large masses. These masses, and larger particles, are captured by ciliary activity in the grooves among the folds of SA3. In the right dorsal region of the stomach, these particles are conducted either to the intestinal groove or they pass over the typhlosoles and make contact with the strong ciliary currents on the floor of the stomach. In the region where folds R1 and R2 delimit SA3, the particles are conducted to R1.

The cilia on R2, and on the longitudinally striated area between R2 and R4, remove particles from the dorsal hood towards the semi-spiral conical projection. This structure has an extension of SA3 on its anterior wall and on the posterior wall there is another well-developed sorting area (SA11) which extends dorsalwards around, and anterior to, the opening of the appendix into the stomach. Fold R2 within the 'scp' separates SA3 from SA11. Deep troughs between the transverse folds of SA11 give the external wall of the 'scp' a corrugated appearance, which varies according to the contracted state of the stomach. Cilia on the folds of SA11 conduct particles obliquely from crest to crest onto the aperture of the appendix.

At the ventral edge of the mouth of the dorsal hood, another transversely folded sorting area (SA8) begins. It passes ventrally to the oesophageal opening and ends on the roof of the entrance to the right caecum. Isolated particles conducted by SA8 in the suboesophageal region are caught by a transverse ciliary tract near the opening of the oesophagus and are conducted onto fold R1. Those particles conducted by SA8 on the ventral edge of the dorsal hood are conducted into this structure and caught by a ciliary tract that accumulates them in the margin of the gastric shield.

On the anterior wall of the stomach, and ventral to SA8, there is another sorting area (SA7) consisting of large and well-spaced folds, which slightly enter the right caecum and the left one deeply. Within these caeca, SA7 becomes inconspicuous, being difficult to follow along its entire

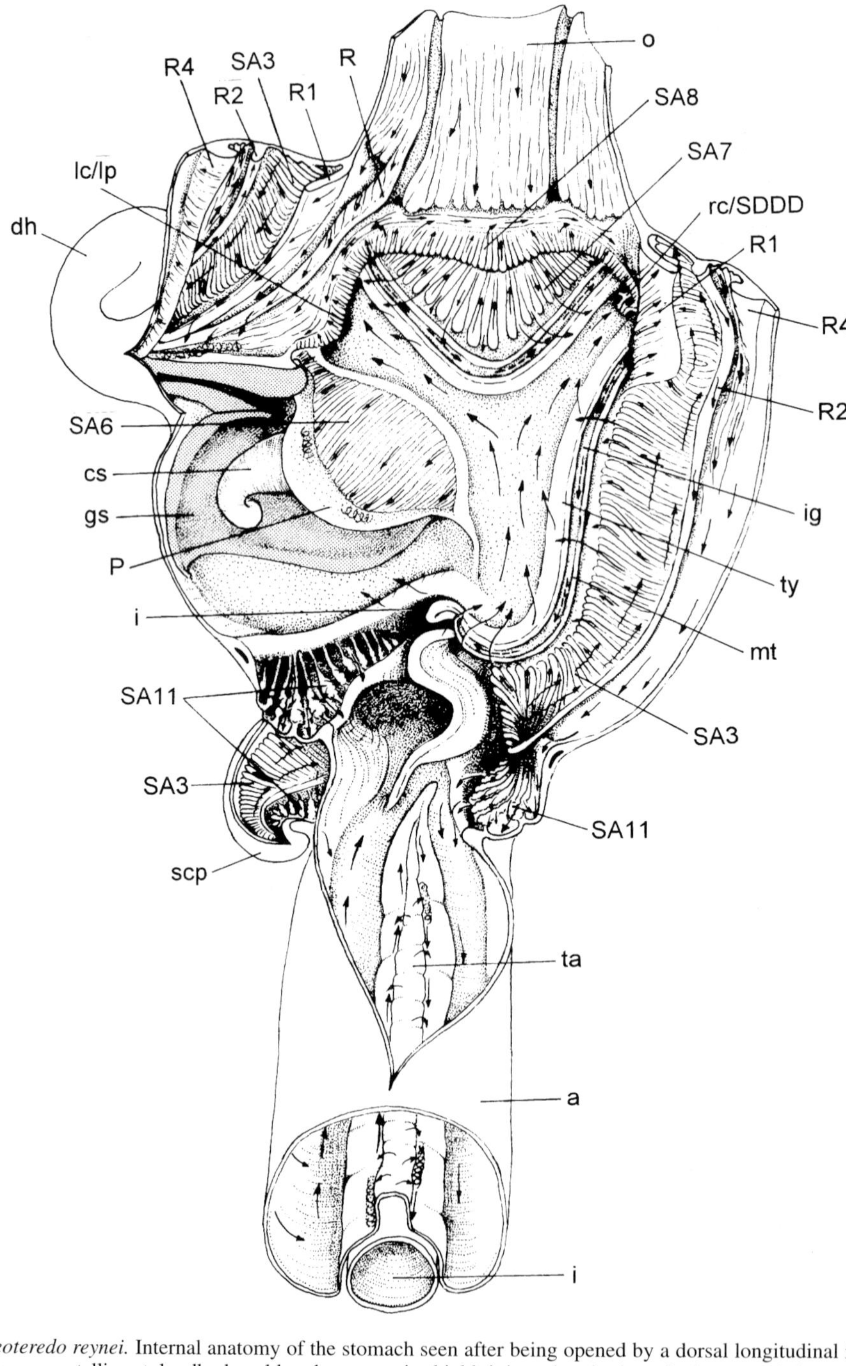

Fig. 6. *Neoteredo reynei.* Internal anatomy of the stomach seen after being opened by a dorsal longitudinal incision. a, Appendix; cs, crystalline style; dh, dorsal hood; gs, gastric shield; i, intestine; ig, intestinal groove; lc/lp, common opening of the combined left caecum/left pouch; mt, minor typhlosole; o, oesophagus; P, R, R1, R2 and R4, folds of the stomach wall; rc/SDDD, opening of the right caecum/openings of ducts from the specialized digestive diverticula at the entrance of the right caecum; SA3, SA6, SA7, SA8 and SA11, sorting areas; scp, semi-spiral conical projection; ta, typhlosole of the appendix; ty, major typhlosole. Arrows show the main ciliary currents.

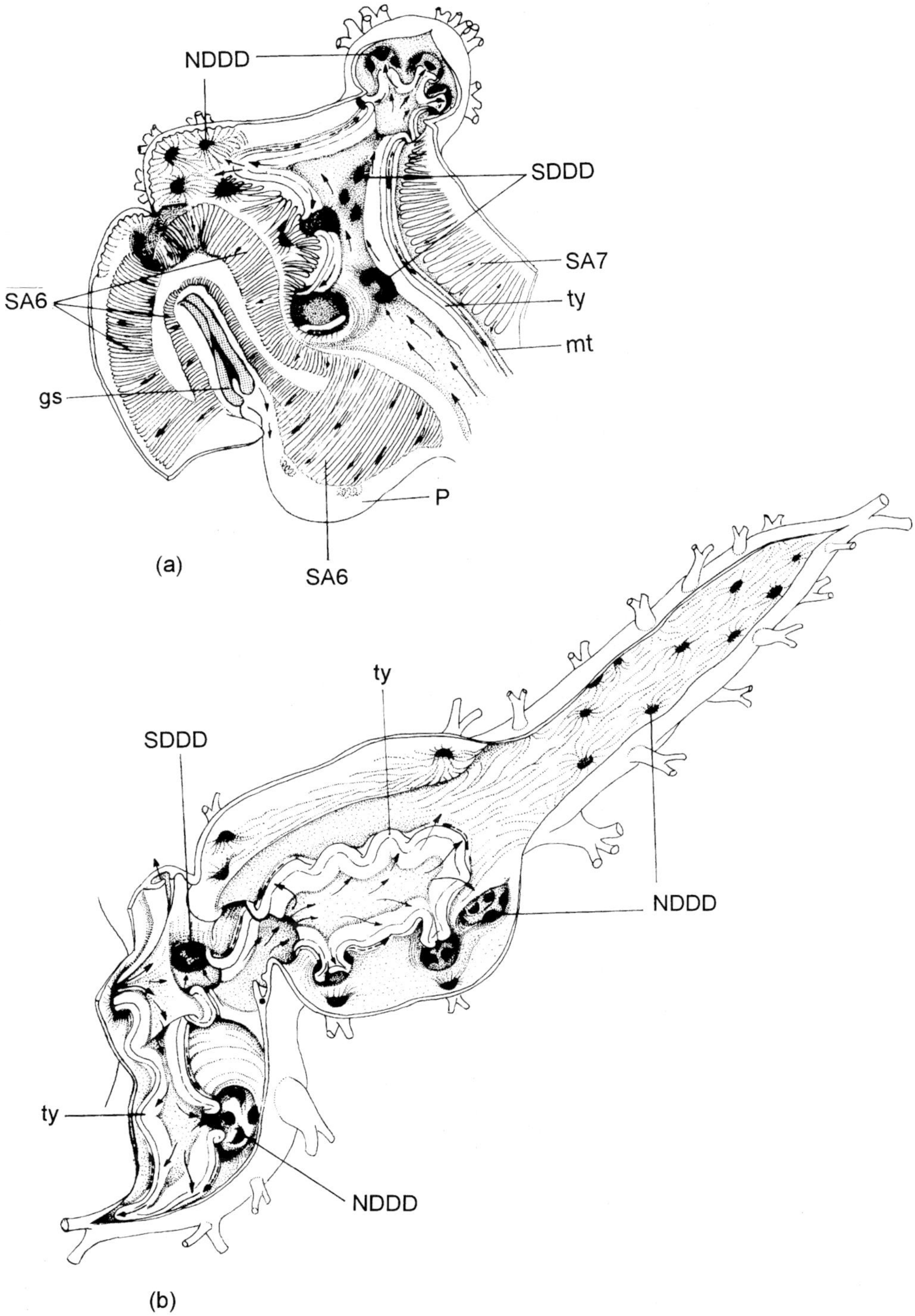

Fig. 7. *Neoteredo reynei*. Internal anatomies of: (**a**) the combined left caecum/left pouch; (**b**) the right caecum. gs, Wing of the gastric shield within the compartment corresponding to the left pouch; mt, minor typhlosole; NDDD and SDDD, openings of the ducts from the normal and specialized digestive diverticula, respectively; P, fold of the stomach wall; SA6 and SA7, sorting areas; ty, major typhlosole. Arrows show the main ciliary currents.

course. Where this sorting area is evident, cilia on the folds carry particles on to SA8 and those on the troughs in an opposite direction. Large and/or dense particles on the troughs of SA8 are conveyed to the intestinal groove (ig), which is protected by the major typhlosole (ty), and light material is thrown over this fold and returned to general circulation within the stomach.

The major typhlosole is accompanied along its entire course by the intestinal groove and begins as a low fold of the inner wall of the intestine, where it describes a loop over the dorsal surface of the style-sac. At the bulbous region of the intestine (bu), the major typhlosole becomes taller and spiralled but, before entering the stomach, it returns to its initial height.

The minor typhlosole (mt) arises within the bulbous region of the intestine and runs parallel to the major typhlosole, flanking the intestinal groove along its entire course inside the stomach. The minor typhlosole is not always evident but the major typhlosole is well developed, overlapping and protecting the intestinal groove and even the minor typhlosole.

The typhlosoles emerge from the intestine and run latero-ventrally on the right side of the stomach, where they penetrate deeply into the right caecum (rc). Leaving the right caecum, the typhlosoles cross the anterior floor of the stomach, ventrally to SA7 in the suboesophageal region, and penetrate the left caecum (lc), where they terminate.

The left caecum and the left pouch (lp) are not separated internally by a wall; these two pockets are herein regarded as the combined left caecum/left pouch (Fig. 7). The left caecum is defined by the course of the major typhlosole, which penetrates deeply within three main compartments, and sends flares into the mouth of some ducts from the NDD. The major typhlosole is always accompanied by the minor typhlosole and the intestinal groove, circumnavigating the left caecum in an anticlockwise direction when the stomach is observed dorsally, terminating near the mouth of the most ventral, long, tubular compartment. A few ducts from the SDD open into the left caecum within the area delimited by the major typhlosole. Such ducts are concentrated near the apertures of the two antero-ventral compartments. Strong ciliary currents along the major typhlosole and on the area circumscribed by this fold conduct particles onto the apertures of the ducts of the NDD and the SDD. Particles were not observed entering these ducts. Nevertheless, pressing on the NDD and the SDD frequently evoked the regurgitation of an amorphous mass of microscopic particles.

The left pouch is defined by the presence of a wide sorting area (SA6) which comprises fine transverse striations. This area extends along the roof and floor of the left pouch, and extends over the anterior face of a tall, shield-like, prominence (P) which protects that part of the crystalline style emerging into the stomach. Within the left pouch, a smooth area divides SA6 into two bands of different widths, the narrower band being contiguous with the gastric shield. The portion of SA6 on the floor of the left pouch conducts and concentrates particles near the wing of the gastric shield within this pocket, while that on the roof concentrates material on its anterior margin. Material entangled in mucus is probably captured by the rotating crystalline style. On fold 'P', sorting area SA6 conducts particles to the crystalline style.

The right caecum and its three compartments are invaded deeply by the typhlosoles and their accompanying intestinal groove (Fig. 7). The major typhlosole sends flares into some ducts of the NDD. Ducts from the SDD open into a shallow depression just at the entrance to the right caecum (Fig. 6) and into another one in the anterior compartment of this pocket. Vigorous ciliary currents on the major typhlosole and on the area circumscribed by this fold conduct particles toward the opening of the ducts of the digestive glands. Only large particles are passed into the intestinal groove and removed from the stomach.

The appendix has a well-developed typhlosole on its floor which extends from the mouth to the distal end of this pocket. The free dorsal margin of this fold forms a low irregularly swollen crest that shows slow movements. At the narrow entrance of the appendix, the proximal end of the typhlosole meets another elevated fold coming from the aperture of the intestine, which acts as a barrier against rejection of isolated particles within the stomach. All dissected animals had an appendix filled with wood particles. In a few specimens, weak ciliary activity was observed on the epithelium of the appendix. This epithelium is richly provided with microvilli (Fig. 3b).

Particles selected on SA11 enter the appendix at its right side. Those particles contacting the crest of the typholosole segregate laterally. Longitudinal currents along the base of the crest conduct a string of mucus with entangled material onto the distal end of the appendix on the right side, and in the opposite direction on the left side. A similar behaviour was detected on the floor of the appendix, where ciliary tracts conduct particles arriving from the lateral walls backwards towards the blind end of the appendix, and forwards and out of this pocket on the right side. Such currents seem to exert little influence on the wood packed within the appendix. Movements of the appendix and contractions of the body may be responsible for the discharge of particles from the appendix.

Particles on the right side of the proximal tip of

the typhlosole in the appendix were slowly removed onto a blind depression contiguous with and anterior to the aperture of the intestine. The major typhlosole is high at the entrance of the intestine and, along with the fold emerging from the intestine and extending onto the appendix aperture, creates an efficient barrier against excessive loss of material circulating within the stomach. Vigorous ciliary activity on the crest of both folds recapture isolated particles leaving the appendix and SA3 at the entrance of the spiral conical projection, returning them to the general circulation within the stomach. Only a few particles escape this barrier and are rejected via the intestine.

Ciliary mechanisms within the right side of the appendix, and on the crest of the major typhlosole, may be responsible for the return of wood particles stored in the appendix to the main cavity of the stomach. Vigorous ciliary tracts on the floor of this organ carry particles forward, segregating them into the right caecum and the combined left caecum/left pouch. Wood particles stored in the appendix can thus be resubmitted to the sorting devices on the stomach wall and re-presented to the digestive glands.

All material leaving the stomach via the intestinal groove and appendix enters the intestine and is accumulated in the enormous anal canal. This structure was always full, mainly with wood particles. Despite the presence of cilia on its epithelium, no ciliary currents were detected removing material into the epibranchial cavity. Discharge of material from the anal canal may be largely accomplished by muscular activity in its walls, and improved by the foreshortening of the animal. The efficient sphincter at the aperture of the anal canal controls the exit of material and contributes to its long residence in this compartment of the digestive system.

Psiloteredo healdi. The anatomy and functioning of the stomach of this species (Figs 8 and 9) is similar to that of *N. reynei*. Differences are mainly related to the anatomy of the organ, and refer to the greater extension of fold R4, the absence of a smooth area dividing SA6 within the left pouch, SA7 is not detected within the left caecum and SA11 is less developed.

Discussion

The functional anatomy of stomach type II of Turner (1966) is known only from the work of Purchon (1960) and from the present study. Lazier (1924), Morton (1970), Martinez (1987) and Lopes *et al.* (1998) studied type III stomachs. The few species analysed by these authors share the same structural type of stomach, which corresponds to type V of Purchon (1960), regardless of their globular or elongate form.

All of the folds described for the stomach of *N. reynei* and *P. healdi*, except for fold R4, are homologous to the corresponding folds described by Purchon (1960). Since Purchon's terminology was adopted in the present work, and he described only folds R, R1, R2 and R3 in the stomachs he studied, the newly described fold in *N. reynei* and *P. healdi* was identified as R4.

Elongation of the stomach was considered by Turner (1966) to be a specialization for increasing the efficiency of wood digestion. Stomach type III is morphologically more complex and provided with a larger amount of SDD than globular stomach type II. Intracellular digestion of wood fragments in these glands (Morton & McQuiston 1974), and the finding of a true cellulase within the SDD (Morton 1983), lead to the conclusion that in the species with large amount of SDD the wood is relatively more important than suspension in their diet.

N. reynei has a globular stomach type II and reduced SDD, conditions which, considered separately, could indicate a predominantly suspension-feeding mode of life. Other anatomical characters indicative of a preferred wood diet, such as the possession of extremely reduced labial palps, short ctenidia, a small crystalline style and, particularly, a large appendix always packed with wood particles, do not support this conclusion. The last condition indicates that the species keeps the wood stored for a long period of time. The identification of microvilli in the epithelial cells of this pocket strongly suggests that they play an important role in the absorption of nutrients, probably those derived from the digestion of its stored wood. The microvillar border identified by Bazylinski & Rosenberg (1983) in the epithelial cells of the appendix of several *Teredo* and *Bankia* species led these authors to attribute an absorptive function to this pocket. Mann (1984) subscribes to the hypothesis that the appendix acts as an actively mixed incubator for extracellular cellulase activity. Considering both hypothesis, the retention of wood fibres for a long period in the appendix of *N. reynei* could provide the necessary time for extracellular cellulase activity, following absorption, counterbalancing the scarcity of SDD and enlarging the capability of the species to feed on wood.

Wood particles leaving the appendix can return to the general circulation in the main cavity of the stomach and be resubmitted to the sorting devices on its walls. Those captured by the vigorous ciliary tracts on the major typhlosole and on the floor of the organ may be conducted onto the right caecum and onto the combined left caecum/left pouch, to be taken to and digested by the SDD.

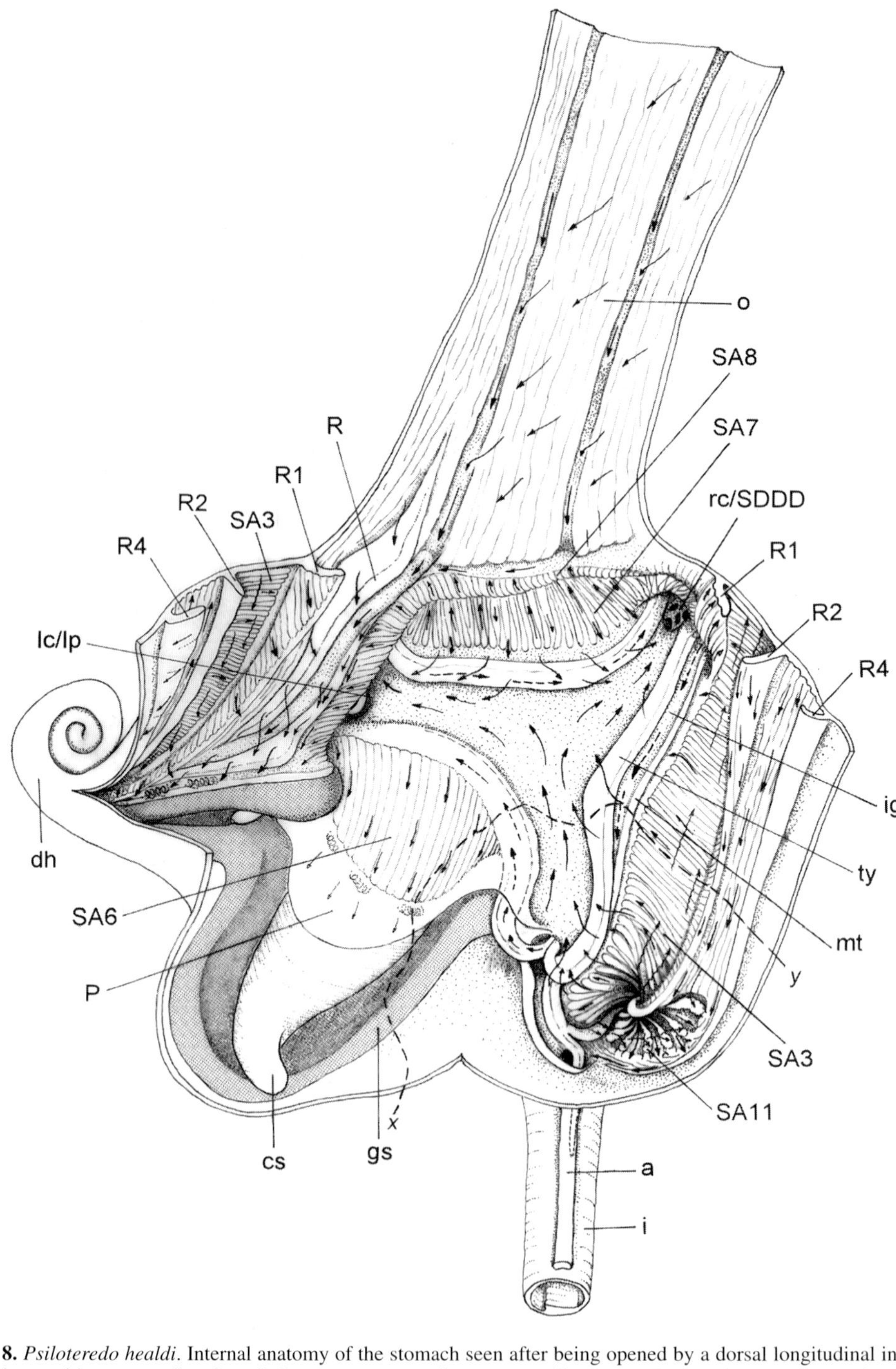

Fig. 8. *Psiloteredo healdi*. Internal anatomy of the stomach seen after being opened by a dorsal longitudinal incision. Specimen with an appendix of small diameter. a, Appendix; cs, crystalline style; dh, dorsal hood; gs, gastric shield; i, intestine; ig, intestinal groove; lc/lp, opening of the combined left caecum/left pouch; mt, minor typhlosole; o, oesophagus; P, R, R1, R2 and R4, folds of the stomach wall; rc/SDDD, opening of the right caecum/openings of ducts from the specialized digestive diverticula at the entrance of the right caecum; SA3, SA6, SA7, SA8 and SA11, sorting areas; ty, major typhlosole. Arrows show the main ciliary currents. *x–y*, boundary line that delimits the region of the stomach depicted with modification in Fig. 9c.

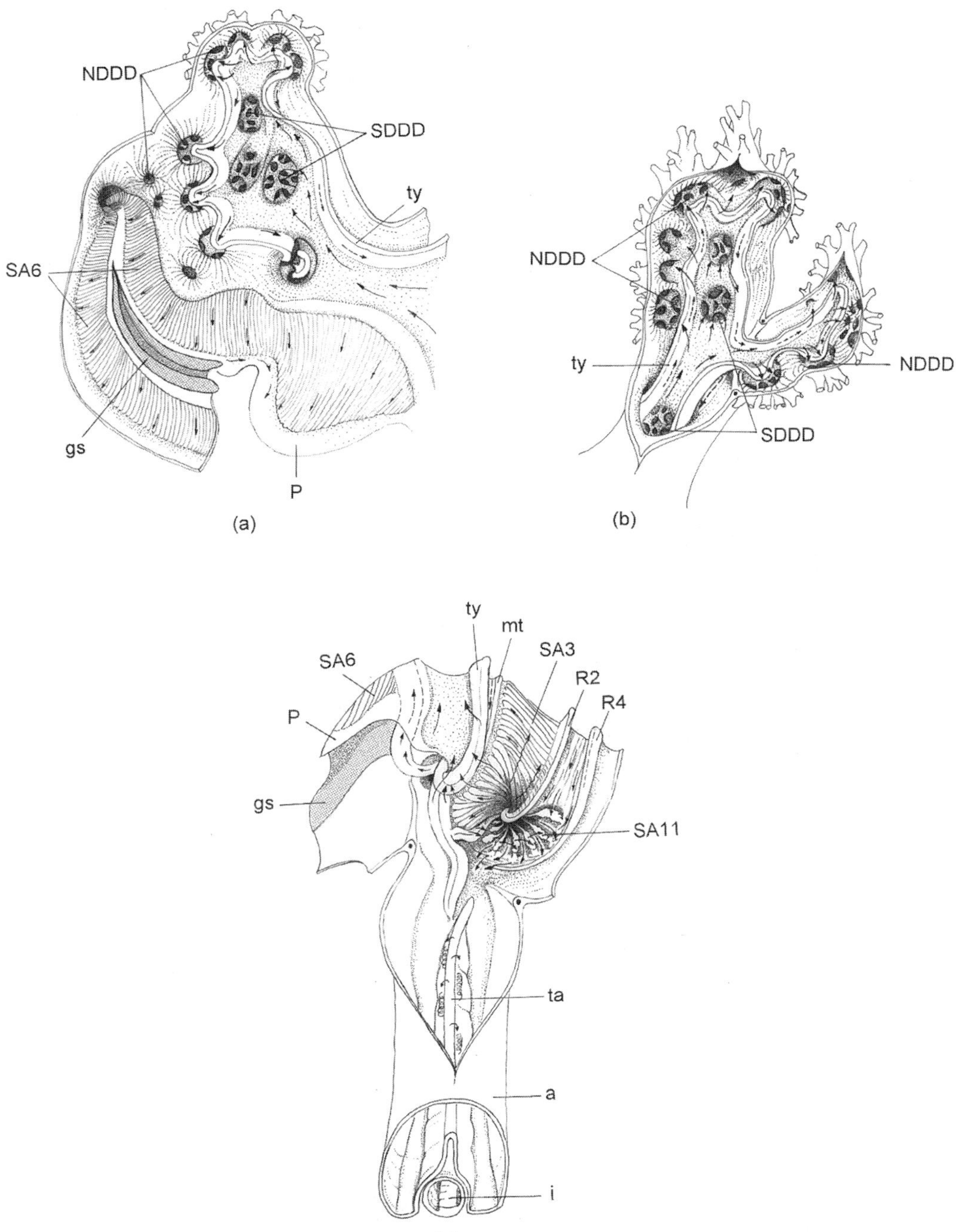

Fig. 9. *Psiloteredo healdi*. Internal anatomies of: (**a**) the combined left caecum/left pouch and (**b**) the right caecum. (**c**) Posterior region of a stomach with a well-developed appendix. Section equivalent to the area circumscribed by the broken line *x–y* in Fig. 8. a, Appendix; gs, gastric shield; i, intestine; mt, minor typhlosole; NDDD and SDDD, openings of ducts from the normal and specialized digestive diverticula, respectively; P, R2 and R4, folds of the stomach wall; SA3, SA6 and SA11, sorting areas; ta, typhlosole of the appendix; ty, major typhlosole. Arrows show the main ciliary currents.

Long residence of wood particles within the enormous anal canal of *N. reynei* may be another important device to allow enzymatic degradation of wood to be continued. The presence of epithelial cells richly supplied with microvilli, and the highly vascularized anal canal walls, suggest that absorption occurs in this pocket. These probable roles played by the anal canal greatly increases the capability of this species to use wood as food, despite the globular type II stomach and reduced SDD, as compared with stomach type III. These anatomical data can also help to explain why this species is more common in mangrove areas less frequently submerged by the tides as registered by Rancurel (1971) and Lopes & Narchi (1993). The specialization for great storage capacity for wood fragments in both the appendix and anal canal gives the species increased independence of suspended food, contributing to the successful occupation of wood substrata, even at high tide levels.

Anatomical characters suggestive of a suspension-feeding mode of life in *P. healdi* are: long ctenidia; large crystalline style; minute anal canal (always empty of faecal material); reduced appendix and SDD. The tendency of progressive development of the appendix as the animal grows older gives evidence of feeding flexibility in *P. healdi*. When the appendix is reduced, the animals rely mainly on suspended material. As the appendix enlarges, it can act as described for *N. reynei*, thereby allowing the animal to counterbalance its reduced SDD by using wood as food more efficiently. In *P. healdi*, the anal canal is always reduced and never retains faeces, which are eliminated immediately from the body as long faecal pellets. The digestion of wood apparently depends mainly on processes that occur inside the appendix, since the reduced SDD do not enlarge following the development of this pocket and the anal canal is extremely reduced.

Turner (1966) observed in *Psiloteredo*, and in other species with the gonads dorsal to the appendix, that the anal canal is always empty, flattened and inconspicuous when the gonads are swollen. Consequently, suspension food becomes more important than the wood in the reproductive period. Data obtained for *P. healdi* in the present work show that this relationship is not constant. Specimens with gonads, and an appendix and associated typhlosole which were well developed, as well as specimens with an inconspicuous appendix and poorly developed gonads, were present in the studied sample.

According to Hoagland & Turner (1981), the feeding flexibility in the Teredinidae can also be related to crowding when active wood boring may cease in favour of filter feeding. Experiments are required to test if crowding and/or other environmental circumstances induce switching from a wood- to a filter-feeding habit, or vice versa, in the species analysed in the present work. Feeding flexibility is here demonstrated only for *P. healdi*, where it is related to age. Once the hypothesis of digestion and absorption in the appendix and anal canal of *N. reynei* is confirmed, it can be stated unequivocally that the large amount of wood kept in these pockets is stored food, allowing survival under conditions of crowding or when growth is impossible. This may represent an adaptive device, since *N. reynei* does not have the same capability to filter food as other Teredinidae with large ctenidia, and it occurs mainly in mangrove areas less frequently submerged by the seawater.

The pathways whereby digestive enzymes from the intracellular symbiotic bacterial population present in the ctenidia of *P. healdi*, and probably in *N. reynei*, are ultimately transported to the lumen of the stomach to participate in cellulose degradation remain to be elucidated. Histological sections did not demonstrate connection of the so-called ducts of the gland of Deshayes present within the afferent branchial vessel of both species with the digestive system. According to Distel & Roberts (1997), the plausibility of these symbiotic mechanisms depend mainly on the demonstration of a complete route from the bacterial population in the ctenidia to the digestive system.

The authors thanks the Fundação de Amparo à Pesquisa do Estado de São Paulo (FAPESP), Coordenação de Aperfeiçoamento de Pessoal de Nível Superior (CAPES) and Conselho Nacional de Desenvolvimento Científico e Tecnológico (CNPq), Brazil, for financial support of this study. SEM facilities were made available through the Laboratório de Microscopia Eletrônica, Instituto de Biociências da Universidade de São Paulo (IB-USP), and field facilities through the Base Norte do Instituto Oceanográfico da USP. Thanks to Enio Mattos and Márcio Valentim Cruz for technical support in the SEM, and José Domingos Batista dos Reis for maintenance of the teredinids in aquaria. Special acknowledgements to Drs Brian Morton, Elizabeth B. Andrews and Liz Harper for their critical review of the text.

References

Bazylinski, D. A. & Rosenberg, F. A. 1983. Occurrence of a brush border in the caecum (appendix) of several *Teredo* and *Bankia* species. *The Veliger*, **25**, 251–254.

Distel, D. L. & Roberts, S. J. 1997. Bacterial endosymbionts in the gills of the deep-sea wood-boring bivalves *Xylophaga atlantica* and *Xylophaga washingtona*. *Biological Bulletin*, **192**, 253–261.

Greene, R. V. & Freer, S. N. 1986. Growth characteristics of a novel nitrogen-fixing cellulolytic bacterium. *Applied and Environmental Microbiology*, **52**, 982–986.

GREENE, R. V., COTTA, M. A. & GRIFFIN, H. L. 1989. A novel, symbiotic bacterium isolated from marine shipworm secretes proteolytic activity. *Current Microbiology*, **19**, 353–356.

——, GRIFFIN, H. L. & FREER, S. N. 1988. Purification and characterization of an extracellular endoglucanase from the marine shipworm bacterium. *Archives of Biochemistry and Biophysics*, **267,** 334–341.

HOAGLAND, K. E. & TURNER, R. D. 1981. Evolution and adaptation radiation of wood-boring bivalves (Pholadacea). *Malacologia,* **21**, 111–148.

LAZIER, E. L. 1924. Morphology of the digestive tract of *Teredo navalis*. *University of California Publications in Zoology,* **22**, 455–474.

LEONEL, R. M. V., LOPES, S. G. B. C. & ZAGO, D. 1998. Morphological basis of excretory function in *Nausitora fusticula* (Jeffreys, 1860) (Bivalvia, Teredinidae). *Journal of Molluscan Studies,* **64**, 223–237.

LOPES, S. G. B. C. & NARCHI, W. 1993. Levantamento e distribuição das espécies de Teredinidae (Mollusca: Bivalvia) no manguezal da Praia Dura, Ubatuba, São Paulo, Brasil. *Boletim do Instituto Oceanográfico*, **41**, 29–38.

——, —— & DOMANESCHI, O. 1998. Digestive tract and functional anatomy of the stomach of *Nausitora fusticula* (Jeffreys, 1860) (Bivalvia, Teredinidae). *The Veliger,* **41**, 351–365.

MANN, R. 1984. Nutrition in the Teredinidae. *In*: COSTLOW, J. D. & TIPPER, R. C. (eds) *Marine Biodeterioration: An Interdisciplinary Study. Proceedings of the Symposium on Marine Biodeterioration, Uniformed Services University of Health Science, 20–23 April 1981*. Naval Institute Press, Annapolis, MD, 24–29.

MARTINEZ, J. C. 1987. Structure et fonctionnement de l'appareil digestif de *Teredo navalis* L. (Teredinidae, Bivalvia). *Haliotis*, **16**, 197–207.

MORTON, B. 1970. The functional anatomy of the organs of feeding and digestion of *Teredo navalis* Linnaeus and *Lyrodus pedicellatus* (Quatrefages). *Proceedings of the Malacological Society of London,* **39**, 151–167.

—— 1978. Feeding and digestion in shipworms. *Oceanography and Marine Biology, Annual Review,* **16**, 107–144.

—— 1983. Feeding and digestion in Bivalvia. *In*: HOCHACHKA, P. W. (ed.) *The Mollusca, Volume 5*. Academic Press, New York, 65–147.

—— & MCQUISTON, R. W. 1974. The daily rhythm of activity in *Teredo navalis* Linnaeus correlated with the functioning of the digestive system. *Forma et Functio*, **7**, 59–80.

NAIR, N. B. & SARASWATHY, M. 1971. The biology of wood-boring teredinid molluscs. *Advances in Marine Biology*, **9**, 335–509.

PANTIN, C. F. A. 1948. *Notes on Microscopical Technique for Zoologists*. Cambridge University Press, Cambridge.

POPHAM, J. D. & DICKSON, M. R. 1973. Bacterial associations in the teredo *Bankia australis* (Lamellibranchia, Mollusca). *Marine Biology*, **19**, 338–340.

POTTS, F. A. 1923. The structure and function of the liver of *Teredo*, the shipworm. *Proceedings of the Cambridge Philosophical Society, Biological Sciences*, **1**, 1–17.

PURCHON, R. D. 1960. The stomach in the Eulamellibranchia: stomach types IV and V. *Proceedings of the Zoological Society of London*, **135**, 431–489.

—— 1987. The stomach in the Bivalvia. *Philosophical Transactions of the Royal Society of London,* **B316**, 183–276.

RANCUREL, P. 1971. Les Teredinidae (Mollusques Lamellibranches) dans les lagunes de Côte d'Ivoire. *Memoires Offices de la Recherche Scientifique et Technique Outre-Mer, Paris*, **47**, 1–235.

ROSENBERG, F. A. & CUTTER, J. 1973. The role of cellulolytic bacteria in the digestive processes of the shipworm. *In*: ACKER, R. F. *ET AL*. (eds) *Materials in the Sea*. Northwestern University Press, Evanston, 778–796.

SARASWATHY, M. & NAIR, N. B. 1971. Observations on the structure of the shipworms *Nausitora hedleyi*, *Teredo furcifera* and *Teredora princesae* (Bivalvia: Teredinidae). *Transactions of the Royal Society of Edinburgh*, **68**, 507–566.

SIGERFOOS, C. P. 1908. Natural history, organization and late development of the teredinidae or shipworms. *Bulletin of the Bureau of Fisheries, Washington*, **37**, 191–231.

TURNER, R. D. 1966. *A survey and illustrated catalogue of the Teredinidae (Mollusca: Bivalvia).* The Museum of Comparative Zoology, Cambridge.

—— & JOHNSON, A. C. 1971. Biology of the marine wood-boring molluscs. *In*: JONES, E. B. G. & ELTRINGHAM, S. K. (eds) *Marine Borers, Fungi and Fouling Organisms of Wood.* Organisation for Economic Co-operation and Development, Paris, 259–301.

WATERBURY, J. B., CALLOWAY, C. B. & TURNER, R. D. 1983. A cellulolytic nitrogen-fixing bacterium cultured from the gland of Deshayes in shipworms (Bivalvia: Teredinidae). *Science*, **221**, 1401–1403.

YONGE, C. M. 1926. The digestive diverticula in the Lamellibranchs. *Transactions of the Royal Society of Edinburgh*, **54**, 703–718.

Evolutionary trajectories of a redundant feature: lessons from bivalve gill abfrontal cilia and mucocyte distributions

PETER G. BENINGER[1] & SUZANNE C. DUFOUR[2]

[1]*Laboratoire de Biologie Marine, Faculté des Sciences, Université de Nantes, 44322 Nantes Cédex France (e-mail: Peter.Beninger@isomer.univ-nantes.fr)*

[2]*Scripps Institution of Oceanography, University of California, San Diego, La Jolla, CA 92093-0202, USA*

Abstract: Recent data on the distributions of cilia and mucocytes on the bivalve gill abfrontal surface are analysed with respect to evolutionary relationships of the principal autobranch gill types. From the primitive function as a mucociliary cleaning surface in the protobranchs, two evolutionary trajectories are evident: (1) progressive reduction of both cilia and mucocytes with resultant loss of surface function, seen in the homorhabdic filibranchs studied; (2) reduction of cilia but retention or increase in acid mucopolysaccharide-secreting (AMPS) mucocyte density in the eulamellibranchs, corresponding to the assumption of a new function, probably in the reduction of frictional resistance to flow in the water canals. Heterorhabdic gill abfrontal surfaces present a mixture of these characteristics: reduction of cilia and mucocytes on the ordinary filaments, and retention of both on the principal filaments. The retention of AMPS mucocytes on the abfrontal surface of the pseudolamellibranchs may be related to the degree of interlamellar fusion, reducing frictional resistance to water flow as in the eulamellibranchs. The gill abfrontal surface thus constitutes an excellent candidate for the study of the different evolutionary options and trajectories of a redundant feature.

Compared to the intensive anatomical and functional studies of the frontal surface of bivalve gills [see Winter (1978), Jørgensen (1990) and Beninger & St-Jean (1997) for reviews and references], the abfrontal surface has been virtually ignored, with only very cursory descriptions of surface and histological characteristics. This lack of interest is perhaps understandable from a functional point of view, since the immense majority of bivalves are suspension feeders and the frontal surface of the gill plays a key role in particle processing (Atkins 1938; Nelson 1960; Beninger & St-Jean 1997; Beninger *et al.* 1993, 1997*a*; Nielsen *et al.* 1993; Riisgård *et al.* 1996; Silverman *et al.* 1996; Ward *et al.* 1998). In contrast, the abfrontal surface is not involved in any stage of particle processing and, indeed, if any particles were available to it, there would be no route to the digestive tract. However, this surface presents a very interesting peculiarity in that its presumed original protobranch cleaning function has been obviated by the separation of the pallial cavity into more or less modified infra- and suprabranchial chambers (Yonge 1941; Morton 1996; Waller 1998), such that the risk of fouling is virtually absent in contemporary autobranchs (*sensu* Autobranchia, non-protobranch bivalves; Morton 1996; Salvini-Plawen & Steiner 1996). It therefore constitutes an excellent opportunity to study the evolution of a redundant feature throughout the Bivalvia. Two main evolutionary trajectories are available to such a structure: (1) retention or augmentation of the original functional characteristics, in response to either neutral or positive selection for a new function to which the original features were pre-adapted; (2) reduction or loss of the original functional characteristics in response to negative selection, i.e. the metabolic cost of maintaining these features.

The chief functional characteristics of the primitive bivalve gill abfrontal surface are cilia and mucocytes, present in the protobranch condition and variously reported in the autobranchs (Ridewood 1903; Atkins 1936, 1938; Nelson 1960; Jones *et al.* 1990; Richard *et al.* 1991; Beninger *et al.* 1997*a*; Dufour & Beninger in press) . The types and abundances of these two characteristics may thus serve as markers to trace the evolutionary trajectories of the abfrontal surfaces within the major bivalve taxa. Here such a study is presented,

From: HARPER, E. M., TAYLOR, J. D. & CRAME, J. A. (eds) *The Evolutionary Biology of the Bivalvia*. Geological Society, London, Special Publications, **177**, 273–278. 1-86239-076-2/00/$15.00

based largely on a recent detailed account of cilia and mucocyte densities on the abfrontal surfaces of eight bivalve species representing the four principal autobranch gill types, including primitive and advanced conditions (Dufour & Beninger in press). The elucidation of such heretofore poorly studied characteristics may also reinforce the significance of gill structure in taxonomy and phylogeny (Salvini-Plawen 1980; Salvini-Plawen & Steiner 1996).

Database

To date, reports on the distribution of bivalve abfrontal cilia and mucocytes have been confined to indications of their presence or absence, sometimes with subjective comments on their abundance (e.g. Atkins 1938; Nelson 1960; Eble & Scro 1996), and no systematic investigation of the entire abfrontal surface appears to have been carried out for any species. The data for the present work is therefore drawn from a systematic investigation of the types and distribution of cilia and mucocytes of the abfrontal surfaces of eight bivalve species, representing seven families and the four major autobranch gill types (Dufour & Beninger in press). Briefly, this study showed the following. (1) In the homorhabdic filibranchs, varying degrees of reduction of the abfrontal cilia and mucocytes were observed; *Mytilus edulis* presented the greatest density of cilia and mucocytes, and the most mucocyte secretion types; *Modiolus modiolus* presented a much smaller density of both cilia and mucocytes, and fewer mucocyte secretion types; *Arca zebra* displayed the greatest degree of reduction, with few cilia and low densities of mucocytes, and only one mucocyte secretion type. (2) In the homorhabdic eulamellibranchs, reduction of cilia was extreme and only one mucocyte secretion type was present – acid mucopolysaccharides (AMPS). However, the density of mucocytes was high, with an extraordinarily high density of *Spisula solidissima*. (3) In the heterorhabdic species, ciliation of the abfrontal surface of the ordinary filaments (OF) was greatly reduced, while that of the principal filaments (PF) was dense. The density of mucocytes was also greater on PF cf. OF, with a mixture of mucopolysaccharide types in the heterorhabdic filibranch *Placopecten magellanicus* and AMPS only in the pseudolamellibranch *Crassostrea virginica*. The observations of cilia and mucocyte densities may be summarized graphically, using scaleless axes in which each quadrant embodies a different functional outcome (Fig. 1). The relative positions of the gill types studied may thus be interpreted from a functional and evolutionary standpoint.

Discussion

The basic premise of this work is that functionally coupled organs key to the success of an organism respond more strongly to their respective selective pressures than other organs whose functioning is not significantly affected by those pressures. In the case of the Bivalvia, the ready availability of planktonic particles doubtless conferred a significant advantage to those individuals which presented gill modifications from the primitive protobranch type, enabling increasingly efficient capture and processing. Hence, the bivalve gill underwent rapid evolution within each of the major taxa, while other organs, such as the heart, retained their form, as witnessed in the uniformity of this organ throughout the class (Beninger & Le Pennec 1991; Eble 1996). It is thus possible to use the data on gill abfrontal ciliation to trace the evolutionary changes of this surface in the gill itself, and to relate these to changes in the overall form and function of the gill, independent of the larger phylogenetic trajectories.

There is little doubt that the filibranch condition evolved from the protobranch gill type (Yonge 1941; Morton 1979; Salvini-Plawen 1980; Morton 1996; Waller 1998). The gill of the family Nuculidae, which best represents the primitive protobranch condition (Yonge 1941), is characterized by a uniform, dense abfrontal ciliation which participates in particle transport, principally in cleaning (Orton 1912). Although no study of the mucocyte distribution on the nuculid gill has been made, particle transport, and especially cleaning a ciliated surface, involves mucociliary transport (see Beninger *et al.* 1997*b*), and thus a dense array of mucocytes. The abfrontal surface of the contemporary autobranch gill is most probably a vestigial mucociliary epithelium (Dufour & Beninger in press), whose original protobranch cleaning function was lost with the reflection of the gill filaments to form the homorhabdic filibranch gill, separating the pallial cavity into infra- and suprabranchial chambers. Indeed, the assumption of the filibranch condition may have been the most important factor in the diversification and proliferation of the Bivalves from the early Ordovician (Cope 1996). This is, however, the most primitive contemporary autobranch gill type, present in only 7% of extant families. Given the large species numbers and extensive habitats of the eulamellibranch heterodonts, the proportion of homorhabdic filibranch species is probably even smaller [data from Newell (1965)]. Within the taxa presenting this gill type, the data suggest an evolutionary gradation of the abfrontal surface. *Mytilus edulis* presents the most abundantly ciliated abfrontal epithelium, followed by *Modiolus*

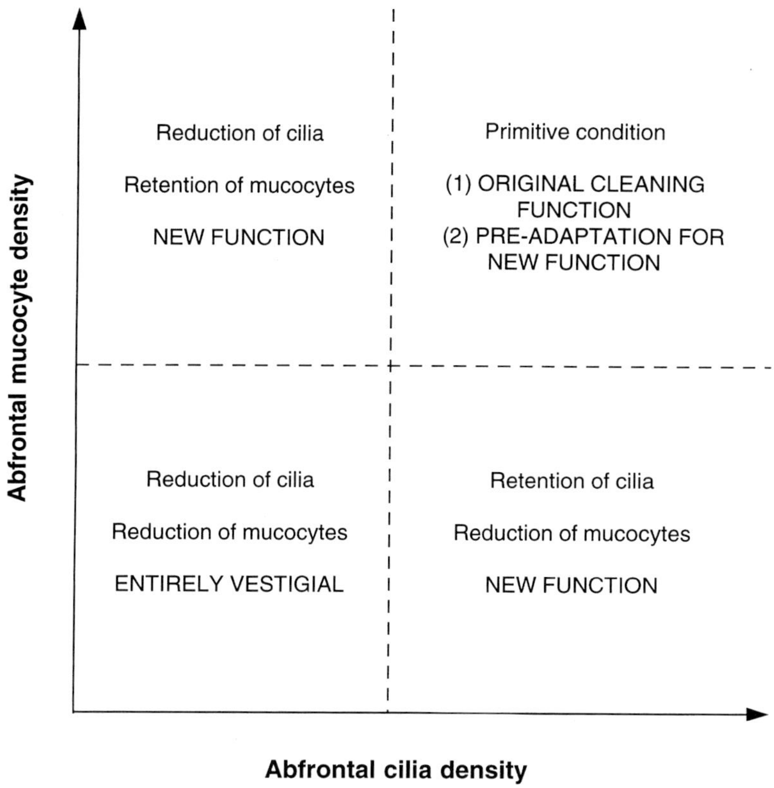

Fig. 1. Scaleless graph of relative abfrontal mucocyte and ciliary densities showing functional significance of positions within each of the quadrants. In the upper right quadrant, the original (protobranch) condition of high mucocyte and ciliary densities is situated, as well as eventual pre-adaptations of this condition for new functions. In the upper left quadrant, high mucocyte densities and low ciliary densities indicate the loss of the original cleaning function, with the assumption of a new function for this surface – possibly in reduction of drag and thus increase in efficiency of water flow. In the lower left quadrant, low mucocyte and ciliary densities indicate a loss of the original cleaning function, with no new surface function. Finally, in the lower right quadrant, low mucocyte densities and high ciliary densities indicate loss of the original cleaning function, and assumption of a new function accomplished by cilia alone (e.g. water pumping).

modiolus, with *Arca zebra* presenting only very sparse cilia (Fig. 2).

It is not known whether the more primitive *Mytilus edulis* homorhabdic filibranch gill has retained the original abfrontal cleaning function, but since this surface is normally exposed to moving, 1 µm filtered seawater (Møhlenberg & Riisgård 1978), any cleaning would be restricted to occasional removal of faeces or gametes not voided in the excurrent flow. Two more probable interpretations may be made: (1) the abfrontal cilia may be vestigial and largely non-functional in the primitive homorhabdic filibranch gill of *Mytilus edulis* [other vestigial mucociliary surfaces are known in this species, see Beninger *et al.* (1995)]; or (2) the dense abfrontal ciliation in *M. edulis* might assist in water pumping (Orton 1912; Jones *et al.* 1990, 1992; Jones & Richards 1993). It is currently impossible to visualize simple cilia *in vivo,* even using endoscopy (Beninger 2000), so this interesting hypothesis cannot be unequivocally confirmed. In this scenario, the abfrontal ciliation would be a pre-adaptation for more efficient water pumping, perhaps partly responsible for the numerical success of the Mytilidae. However, the abfrontal mucocytes would still be non-functional and vestigial.

The sparse ciliation and the few small mucocytes in *Modiolus modiolus* render cleaning or pumping functions impossible, as do the near-bare abfrontal surface and rare mucocytes of *Arca zebra* (Fig. 2). These three species thus present increasing degrees of reduction of the abfrontal mucociliary surface of the homorhabdic filibranch gill, from *Mytilus*

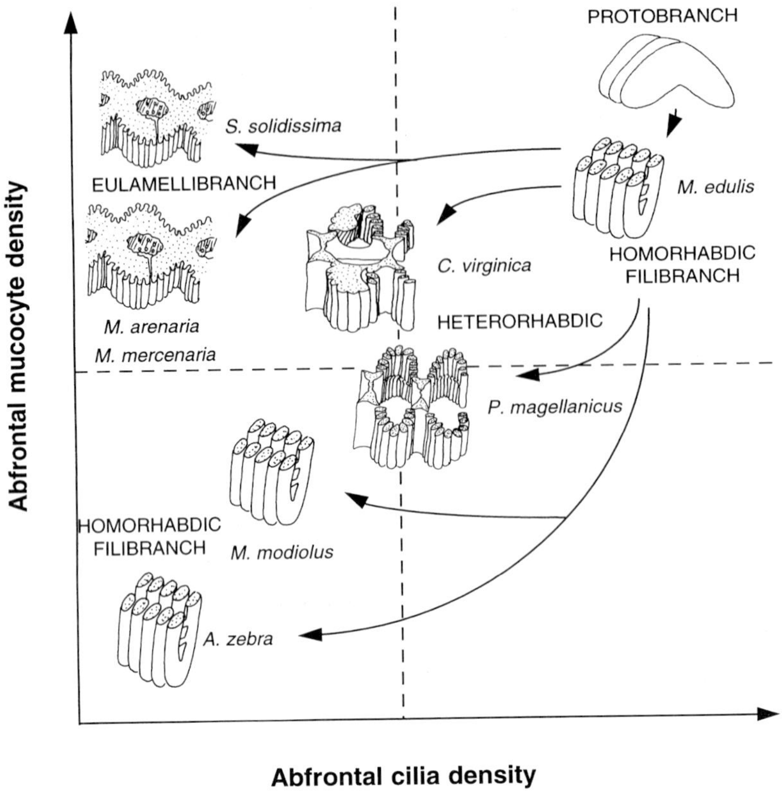

Fig 2. Relative positions of the gill types examined with respect to abfrontal mucocyte and cilia density. Note new function assumed by the abfrontal surface of eulamellibranch gills, loss of surface function in the advanced homorhabdic filibranchs *Modiolus modiolus* and *Arca zebra*. None of the representatives of the four principal gill types presents reduction of mucocytes with retention of cilia (lower right quadrant).

edulis to *Arca zebra*, paralleling the reduction and disappearance of the original cleaning function.

The great degree of reduction of the abfrontal cilia in the eulamellibranch gills studied clearly demonstrates the extension of the trend to reduction and loss of the primitive mucociliary function in the homorhabdic gill type (Fig. 2). However, the retention of a relatively high density of abfrontal mucocytes indicates that, in this gill type, there has either been neutral selection for this trait or that the mucocytes have been retained and redirected toward a new function by positive selection. The presence of only AMPS (i.e. viscous)-secreting mucocytes on the eulamellibranch abfrontal surface, and their extremely dense distribution in *Spisula solidissima*, argues strongly for the latter interpretation. The most probable new function of the mucocytes of the abfrontal surface in this gill type is the reduction of frictional resistance to water flow (Faillard & Schauer 1972; Hoyt 1975; Daniel 1981) across the epithelia of the suprabranchial chamber, which is highly modified to form water tubes. This would increase the efficiency of water flow in these species and parallels the pronounced morphological modifications of the gills, allowing enhanced flow as well as a consequently smaller gill:pallial cavity volume ratio.

The eulamellibranch gill is widely believed to be derived from the ancestral homorhabdic filibranch condition (Orton 1912; Yonge 1941; Morton 1979; Salvini-Plawen 1980; Allen 1985; Waller 1998). The modifications to the abfrontal surface described in the present study indicate that this gill type not only continues the trend to reduction or loss of the original cleaning function, but that this surface has assumed a new function commensurate with the increased efficiency of water flow in this gill type.

In the two heterorhabdic gill types studied, the total absence of cilia on the abfrontal surface of the OF plicae clearly demonstrates the loss of the primitive mucociliary cleaning function in these species. The presence of abundant ciliation and high densities of mucocytes on the PF may be related to the tardy evolutionary development of PF cf. OF. It is likely that PF are formed from modified OF, and both phylogenetic and ontogenetic studies show that PF arise well after OF (Le Pennec *et al.* 1988; Beninger *et al.* 1994). The developmental sequence of the PF may thus be quite different from that of the OF, with the notable retention of the primitive mucociliary characters. In *Placopecten magellanicus*, it has been suggested that the PF abfrontal mucocytes may provide the lubrication necessary for the retraction of the gill during rapid valve adduction, such as the swimming escape response common in juveniles or sudden valve closure in adults (Beninger *et al.* 1988). However, only about two-thirds of the abfrontal surface would actually be in contact with the apposing mantle surface under such conditions (the remaining third being the frontal surface of the ascending branch of the PF outer demibranch). The abfrontal secretions may also reduce friction between apposed lamellae following collapse of the ascending lamellae from the ciliary attachment to the mantle prior to retraction, assisting in the preservation of structural integrity during the clap response. In the Ostreidae, the relatively high degree of interlamellar and interfilamentar fusion has resulted in a suprabranchial cavity akin to the water canals in eulamellibranchs; the retention of AMPS mucocytes only on the PF may indicate that they play a similar role in the reduction of frictional resistance to water flow. In any event, the unique context of the evolutionary and developmental history of the heterorhabdic gill has resulted in a mixed condition (Fig. 2), with the phylogenetically older OF presenting a degree of reduction similar to the advanced homorhabdic filibranch or eulamellibranch condition, whereas the phylogenetically recent PF have retained the original mucociliary characters of the abfrontal surface.

It is noteworthy that no gill types appear in the lower right quadrant of Fig. 2. The abfrontal surface of a gill type in this region would present reduced mucocyte numbers and a high ciliary density; the fact that a high ciliary density is always accompanied by a moderate to high density of mucocytes reinforces the conclusion that this surface was originally mucociliary. The only conceivable new function for the abfrontal surface in this quadrant would be the propulsion of water; none of the gills examined appears to have derived from lines which selectively reduced mucocytes while retaining cilia for such a function.

While the present study proposes an evolutionary paradigm for the distribution of cilia and mucocytes on the abfrontal surface of bivalve gills, it clearly requires additional data from all gill types in order to evaluate the universality of this scheme. In particular, observations of the abfrontal surface of early developmental stages will be important in validating the evolutionary sequences proposed herein for the bivalve gill.

The authors wish to acknowledge funding from the Région Pays de la Loire through the Syndicat Mixte de Développement de l'Aquaculture aux Pays de la Loire during this work.

References

ALLEN, J. A. 1985. The recent Bivalvia: their form and evolution. *In:* TRUEMAN, E. R. & CLARKE, M. R. (eds.) *The Mollusca. Volume 10. Evolution.* Academic Press, Orlando, 337–403.

ATKINS, D. 1936. On the ciliary mechanisms and interrelationships of lamellibranchs. I. New observations on sorting mechanisms. *Quarterly Journal of Microscopical Science*, **79**, 181–308.

—— 1938. On the ciliary mechanisms and interrelationships of lamellibranchs. VII. Latero-frontal cilia of the gill filaments and their phylogenetic value. *Quarterly Journal of Microscopical Science,* **80**, 346–430.

BENINGER, P. G. 2000. Limits and constraints: a comment on premises and methods in recent studies of particle capture mechanisms in bivalves. *Limnology and Oceanography*, **45**, 1196–1199.

—— & LE PENNEC, M. 1991. Functional anatomy of scallops. *In*: SHUMWAY, S. E. (ed.) *Scallops: Biology, Ecology and Aquaculture.* Elsevier, Amsterdam, 133–223.

—— & ST-JEAN, S. D. 1997. The role of mucus in particle processing by suspension-feeding marine bivalves: unifying principles. *Marine Biology*, **129**, 389–397.

——, DUFOUR, S. C. & BOURQUE, J. 1997*a*. Particle processing mechanisms of the eulamellibranch bivalves *Spisula solidissima* and *Mya arenaria*. *Marine Ecology Progress Series*, **150**, 157–169.

——, DWIONO, S. A. P. & LE PENNEC, M. 1994. Early development of the gill and implications for feeding in *Pecten maximus* (Bivalvia: Pectinidae). *Marine Biology*, **119**, 405–412.

——, LE PENNEC, M. & SALAÜN, M. 1988. New observations of the gills of *Placopecten magellanicus* (Mollusca: Bivalvia), and implications for nutrition. I. General anatomy and surface micro-anatomy. *Marine Biology,* **98**, 61–70.

——, ST-JEAN, S. D. & POUSSART, Y. 1995. Labial palps of the blue mussel *Mytilus edulis* (Bivalvia: Mytilidae). *Marine Biology*, **123**, 293–303.

——, LYNN, J. W., DIETZ, T. H. & SILVERMAN, H. 1997*b*. Mucociliary transport in living tissue: the two-layer model confirmed in the mussel *Mytilus edulis* L. *Biological Bulletin*, **193**, 4–7.

——, ST-JEAN, S. D., POUSSART, Y. & WARD, J. E. 1993. Gill function and mucocyte distribution in *Placopecten magellanicus* and *Mytilus edulis* (Mollusca: Bivalvia): the role of mucus in particle transport. *Marine Ecology Progress Series*, **98**, 275–282.

COPE, J. C. W. 1996. The early evolution of the Bivalvia.

In: JOHNSTON, P. A. & HAGGART, J. W. (eds) *Bivalves: An Eon of Evolution – Paleobiological Studies Honouring Norman D. Newell*. University of Calgary Press, Calgary, 361–370.

DANIEL, T. L. 1981. Fish mucus: *in situ* measurements of polymer drag reduction. *Biological Bulletin of the Woods Hole Marine Laboratory*, **160**, 376–382.

DUFOUR, S. C. & BENINGER, P. G. In press. A functional interpretation of cilia and mucocyte distributions on the abfrontal surface of Bivalve gills. *Marine Biology*.

EBLE, A. F. 1996. The circulatory system. *In:* KENNEDY, V. S., NEWELL, R. I. E. & EBLE, A. F. (eds) *The Eastern Oyster* Crassostrea virginica. Maryland Sea Grant Program, College Park, 271–298.

—— & SCRO, R. 1996. General anatomy. *In:* KENNEDY, V. S., NEWELL, R. I. E. & EBLE, A. F. (eds) *The Eastern Oyster* Crassostrea virginica. Maryland Sea Grant Program, College Park, 19–73.

FAILLARD, H. & SCHAUER, R. 1972. Glycoproteins as lubricants, protective agents, carriers, structural proteins and as participants in other functions. *In:* GOTTSCHALK, A. (ed.) *Glycoproteins: Their Composition, Structure and Function. Part B.* 2nd Edition. Elsevier, Amsterdam, 1246–1267.

HOYT, J. W. 1975. Hydrodynamic drag reduction due to fish slimes. *In:* WU, T. V. T., BROKAW, C. J. & BRENNAN, C. (eds) *Swimming and Flying in Nature*. Plenum Press, New York, 653–673.

JONES, H. D., RICHARDS, O. G. & HUTCHINSON, S. 1990. The role of ctenidial abfrontal cilia in water pumping in *Mytilus edulis* L. *Journal of Experimental Marine Biology and Ecology,* **143**, 15–26.

——, ——, SOUTHERN, T. A. 1992. Gill dimensions, water pumping rate and body size in the mussel *Mytilus edulis* L. *Journal of Experimental Marine Biology and Ecology,* **155**, 213–237.

JONES, H. D. & RICHARDS, O. G. 1993. The effects of acetylcholine, dopamine, and 5-hydroxytryptamine on water pumping rate and pressure in the mussel *Mytilus edulis* L. *Journal of Experimental Marine Biology and Ecology*, **170**, 227–240.

JØRGENSEN, C. B. 1990. *Bivalve Filter-feeding: Hydrodynamics, Bioenergetics, Physiology and Ecology*. Olsen and Olsen, Fredensborg.

LE PENNEC, M., HERRY, A., LUTZ, R., FIALA-MÉDIONI, A. 1988. Premières observations ultrastructurales de la branchie d'un Bivalve Pectinidae hydrothermal profond. *Comptes rendus hebdomadaires de l'Académie des Sciences de Paris, Série III*, 627–633.

MØHLENBERG, F. & RIISGÅRD, H. U. 1978. Efficiency of particle retention in 13 species of suspension-feeding bivalves. *Ophelia,* **17**, 239–246.

MORTON, B. 1996. The evolutionary history of the Bivalvia. *In:* TAYLOR, J. D. (ed.) *Origin and Evolutionary Radiation of the Bivalvia.* Oxford University Press, Oxford, 337–359.

MORTON, J. E. 1979. *Molluscs* 5th Edition. Hutchinson & Co., London.

NEILSEN, N. F., LARSEN, P.S., RIISGÅRD, H. U. & JØRGENSEN, C. B. 1993. Fluid motion and particle retention in the gill of *Mytilus edulis*: video recordings and numerical modelling. *Marine Biology*, **116**, 61–71.

NELSON, T. C. 1960. The feeding mechanism of the oyster. II. On the gills and palps of *Ostrea edulis, Crassostrea virginica*, and *C. angulata*. *Journal of Morphology*, **107**, 163–191.

NEWELL, N. D. 1965. Classification of the Bivalvia. *American Museum Novitates,* **2206**, 1–25.

ORTON, J. H. 1912. The mode of feeding of *Crepidula*, with an account of the current-producing mechanism in the mantle cavity, and some remarks on the mode of feeding in Gastropods and Lamellibranchs. *Journal of the Marine Biological Association of the United Kingdom,* **9**, 444–478.

RICHARD, P. E., DIETZ, T. H. & SILVERMAN, H. 1991. Structure of the gill during reproduction in the unionids *Anodonta grandis*, *Ligumia subrostrata* and *Carunculina parva texasensis*. *Canadian Journal of Zoology*, **64**, 1744–1754.

RIDEWOOD, W. G. 1903. On the structure of the gills of Lamellibranchia. *Philosophical Transactions of the Royal Society of London, Series B*, **195**, 147–284.

RIISGÅRD, H. U., LARSEN, P. S. & NIELSEN, N. F. 1996. Particle capture in the mussel *Mytilus edulis*: the role of latero-frontal cirri. *Marine Biology*, **127**, 259–266.

SALVINI-PLAWEN, L. v. 1980. A reconsideration of systematics in the Mollusca (phylogeny and higher classification). *Malacologia,* **19**, 249–278.

—— & STEINER, G. 1996. Synaptomorphies and plesiomorphies in higher classification of Mollusca. *In:* TAYLOR, J. D. (ed.) *Origin and evolutionary radiation of the Bivalvia.* Oxford University Press, Oxford, 29–51.

SILVERMAN, H., LYNN, J. W. & DIETZ, T. H. 1996. Particle capture by the gills of *Dreissena polymorpha*: structure and function of latero-frontal cirri. *Biological Bulletin of the Woods Hole Marine Laboratory*, **191**, 42–54.

WALLER, T. R. 1998. Origin of the Molluscan class Bivalvia and a phylogeny of major groups. *In:* JOHNSTON, P. A. & HAGGART, J. W. (eds) *Bivalves: An Eon of Evolution*–Paleobiological Studies Honouring Norman D. Newell. University of Calgary Press, Calgary, 1–45.

WARD, J. E., LEVINTON, J. S., SHUMWAY, S. E. & CUCCI, T. 1998. Particle sorting in bivalves: *in vivo* determination of the pallial organs of selection. *Marine Biology*, **131**, 283–292.

WINTER, J. E. 1978. A critical review on some aspects of filter-feeding in lamellibranchiate bivalves. *Haliotis,* **7**, 71–87.

YONGE, C. M. 1941. The protobranchiate Mollusca; a functional interpretation of their structure and evolution. *Philosophical Transactions of the Royal Society of London, Series B*, **230**, 79–147.

Growth patterns of noetiid ligaments: implications of developmental models for the origin of an evolutionary novelty among arcoid bivalves

R. D. K. THOMAS[1], A. MADZVAMUSE[2], P. K. MAINI[3] & A. J. WATHEN[2]

[1]*Department of Geosciences, Franklin & Marshall College, Lancaster, Pennsylvania 17604-3003, USA (e-mail: r_thomas@acad.fandm.edu)*

[2]*Computing Laboratory, Wolfson Building, Parks Road, Oxford OX1 3QD, UK*

[3]*Centre for Mathematical Biology, Mathematical Institute, 24-29 St Giles', Oxford OX1 3LB, UK*

Abstract: The dorsal ligaments of arcoid bivalves typically consist of oblique, lamellar and fibrous sheets, alternating along the hinge so that their attachments form characteristic chevron patterns. New elements are added at or near the middle of the pattern, as the ligament grows ventrally and gets longer. Most Palaeozoic arcoids exhibit this growth pattern, which still predominates among their living descendants. Early in the Cretaceous, a novel pattern emerged, with vertical strips of lamellar ligament embedded in grooves in the sheet of fibrous ligament which is attached to each valve. In contrast with the chevron, duplivincular ligament, new elements are added to each end of the noetiid ligament, anteriorly and posteriorly. This distinctive growth pattern is the defining character of the family Noetiidae.

Remarkable variation among individuals within populations of a living limopsid arcoid includes forms with vertical strips of lamellar ligament. These variants suggest how the noetiid growth pattern could have been derived from the duplivincular pattern. Computer simulations show that such patterns can be generated by a reaction–diffusion mechanism of the sort first conceived by Turing (1952, *Philosophical Transactions of the Royal Society, London, Series B*, **237**, 37–72). Moreover, the noetiid growth pattern can simply be derived from the duplivincular pattern by a developmental switch based, for example, on a change in boundary conditions. These results indicate that striking differences in form may arise from modest changes in the developmental process. The evolution of the Noetiidae, members of which are quite disparate in overall shell form, should be reassessed. The derived character on which this family is based may not be uniquely shared, so the group could well be polyphyletic.

Debate arises in different contexts over rates of evolutionary change in form, the magnitude of change that can occur from one generation to the next, and the reliability of characters as indicators of common ancestry. However, these issues all turn on patterns of change that have a common basis. They all depend on degrees of genetically based change that are constrained by what is, or is not, permissible in the course of individual development. Thus, they are different aspects of a single problem.

The bivalve family Noetiidae is supposedly defined by a constellation of shell characters (Newell 1969), but only one of these appears in no other arcoids. Noetiid species vary considerably in size and shell sculpture. They live attached to rocks, nestle in crevices or burrow near the surface in soft sediments, over a fairly wide range of nearshore marine habitats. The family is united by the unique form of the ligament that links the two valves, dorsally (MacNeil 1937). It consists of vertical strips of elastic, lamellar material, embedded in fibrous ligament that is attached to the cardinal area of each valve (Fig. 1). The soft-part anatomy of the noetiids has not been systematically differentiated from that of other arcoids. Consequently, the group is defined by a single, rather striking character.

If the arcoids bearing this distinctive ligament are all descended from a single common ancestor, this trait is the synapomorphy that defines the clade, and the family Noetiidae is soundly based. On the other hand, if ligaments of this kind appeared independently, more than once, the taxon would be polyphyletic. The former alternative is more probable if the derived character represents a

From: HARPER, E. M., TAYLOR, J. D. & CRAME, J. A. (eds) *The Evolutionary Biology of the Bivalvia*. Geological Society, London, Special Publications, **177**, 279–289. 1-86239-076-2/00/$15.00

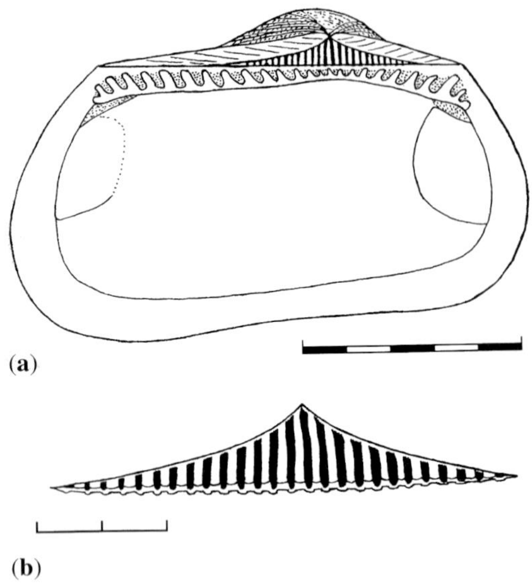

Fig. 1. (**a**) Left valve of *Striarca lactea* (Linné) showing allometric growth of the ligament, achieved by expansion from a small, triangular area to occupy a large proportion of the cardinal area in the adult shell. Scale bar, 5 mm. (**b**) Enlargement of (a) showing vertical strips of lamellar ligament embedded in fibrous ligament and not directly attached to the shell. Scale bar, 2 mm.

radical departure from the ancestral condition. Multiple origins are more likely if the development of the novel ligament arose as a simple modification of the preexisting mechanism of pattern formation.

In this paper, observations of unusual, naturally occurring ligaments are reported. These lead to a hypothesis for the means by which changes in ligament pattern formation have occurred, among the arcoids. The types of developmental processes by which such patterns can be formed are modelled to determine how novel patterns may have arisen. Finally, the forms of living and extinct noetiids are examined in relation to this analysis.

Growth and form of arcoid ligaments

The typical arcoid ligament consists of multiple sheets of lamellar material, embedded in a sheet of fibrous ligament. This is thinner, but possibly never entirely absent where the lamellar layers are attached, and thicker between them (Waller 1990). Dorsally, away from the region of ligament growth along the mantle isthmus, the brittle fibrous ligament splits as the growing valves diverge. The lamellar ligament is first stretched, exerting tension that causes the valves to spring open when the adductor muscles relax. Then, further dorsally, it tears as it exceeds its elastic limit (Owen 1959; Trueman 1969). In many burrowing bivalves, this unavoidable breakage of the lamellar ligament occurs at its anterior extremity, immediately below the umbones, where most divergence of earlier growth stages of the shell takes place. In most arcoids, the umbones are located about half-way back along the shell and they grow much more rapidly away from one another. In this case, breakage of the earlier formed dorsal parts of the ligament necessarily occurs along its whole length.

In the majority of living and extinct arcoids, the zones where lamellar ligament is secreted migrate anteriorly and posteriorly away from the umbones, along the mantle isthmus, as the shell grows. The tracks of these zones form two sets of oblique ridges and grooves on the ligamental areas where they are attached. The resulting chevron pattern (Fig. 2a) constitutes the duplivincular ligament that is characteristic of arcoids and many extinct Palaeozoic pteriomorphs (Newell 1937). This ligament has been considered primitive because its materials, with different physical properties, simply alternate rather than being segregated in positions appropriate to their functional roles. It is more appropriate to regard this serial construction as the simplest of several strategies that emerged early in the evolutionary radiation of the bivalves to offset the effects of breakage of the lamellar ligament in animals of increasingly larger size (Thomas 1978).

A number of variants on this general plan have evolved among the arcoids, some of them repeatedly (Fig. 3). The ligament may be more or less asymmetric, depending on the position of the umbones. Where they lie unusually far forward, or towards the rear, the anterior, or occasionally the posterior, series of lamellar sheets has been eliminated. In *Cucullaea*, which is relatively large but has a thin shell, there is generally only one pair of sheets of lamellar ligament, inserted in well-defined grooves at the anterior and posterior extremities of the cardinal area. *Limopsis*, which is small and paedomorphic in several shell characters (Tevesz 1977), likewise has a single pair of lamellar sheets, but here the entire ligament is inserted in a small, triangular pit, near the midpoint of the hinge line. In most arcoids with multiple sheets of lamellar ligament, their width and spacing is remarkably uniform, but in some forms, notably *Arca* itself, the pattern is quite irregular.

Growth and form of the noetiid ligament

The growth pattern of the noetiid ligament (Fig. 2b) departs from those of all other arcoids in two ways. Here, lamellar ligament is inserted in vertical strips, as opposed to forming oblique sheets. This is accomplished by keeping the site of secretion of

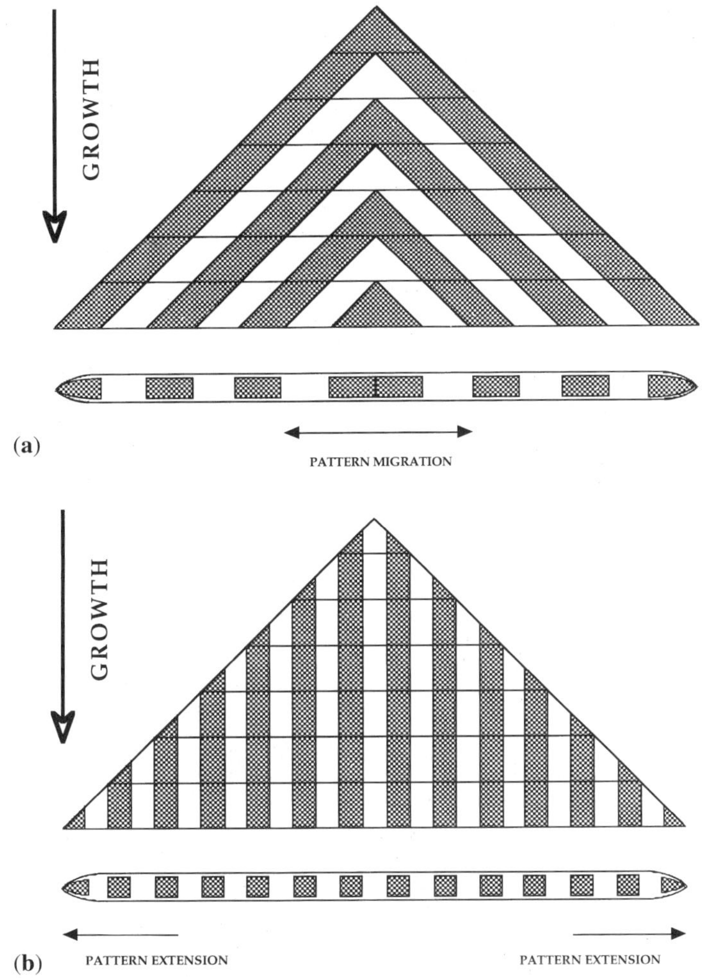

Fig. 2. Growth patterns of: (**a**) the duplivincular ligament typical of *Anadara*, *Barbatia* and *Glycymeris*; (**b**) the noetiid ligament. Each diagram shows a sagittal view of the ligament, as it is inserted on the attachment area of either valve, and a dorsal view of the mantle isthmus where new material is secreted.

each lamellar strip fixed, instead of migrating along the hinge axis. Consequently, new elements are added to the pattern at its anterior and posterior extremities, rather than medially beneath the umbones. This innovation evidently required the pattern-forming process responsible for formation of the ligament to be decoupled from that which controls growth of the hinge plate. In other arcoids, growth of the ligament and growth of the row of hinge teeth follow a common pattern. In both cases, new elements are inserted at or near the middle of the set, beneath the umbones.

A remarkable series of specimens of *Limopsis marionensis* Smith from the vicinity of the Falkland Islands suggests how this change in pattern formation may have been accomplished (Fig. 4). The shells are large for this genus, where a maximum dimension of 1–2 cm is usual. Their ligaments exhibit strong positively allometric growth. This is characteristic of many arcoids on account of the inherent weakness of their ligaments (Thomas 1976). The juvenile shells have typical limopsid ligaments with a single pair of anterior and posterior lamellar layers. The adults are quite

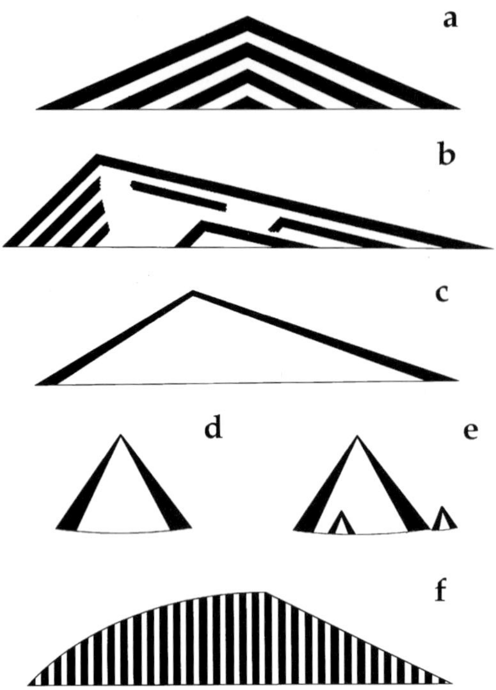

Fig. 3. Representative growth patterns of arcoid ligaments. (**a**) *Glycymeris*, *Anadara*, *Barbatia*, *Parallelodon*; (**b**) *Arca*; (**c**) *Cucullaea*; (**d**) *Limopsis*; (**e**) *Limopsis* variant, not uncommon in some species; (**f**) *Noetia (Eontia)*.

different in form, with varying numbers of more or less irregularly disposed, vertical strips of lamellar ligament. In the most extreme variants, the ligament extends to the extremities of the cardinal area and, apart from its irregularity, the growth pattern is that of a noetiid ligament.

This growth series shows that, after a phase of isometric expansion, the zones of secretion of the initial lamellar components of the ligament stopped migrating anteriorly and posteriorly. As a result, space for the insertion of additional strips of lamellar ligament became available at the anterior and posterior margins of the ligament, rather than medially as in other arcoids. The inferred relationship between this growth pattern and those of actual noetiids is confirmed by the ligaments of living and Cretaceous species of *Striarca*, where lamellar strips can be seen to grow first obliquely and then perpendicular to the hinge line (Fig. 5).

In combination with the regularly spaced chevron patterns of typical arcoid ligaments, these observations suggest a general model for development of the observed patterns. Along the mantle isthmus, alternating zones of activation and inhibition control the production of lamellar ligament that forms either oblique sheets or vertical strips, depending on the direction in which activation propagates as the shell grows. New elements are added wherever space available between existing zones of activation exceeds a limit that defines the repeat interval of the pattern. If this model is correct, only a single, simple change in instructions to the narrow field of cells constituting the mantle isthmus should be required to change the process from generating one pattern to produce another.

To assess the plausibility of this model, the present authors have developed computer simulations of arcoid and noetiid ligaments, based on growth programs that incorporate its assumptions.

Harmonic bifurcation model

A simple, one-dimensional model treats the mantle isthmus as a finite, expanding space that can accommodate an integral number of waves. This approach employs the kind of morphogenetic rules that were employed by Oster *et al.* (1988) to explain the patterns of cartilage condensation in the

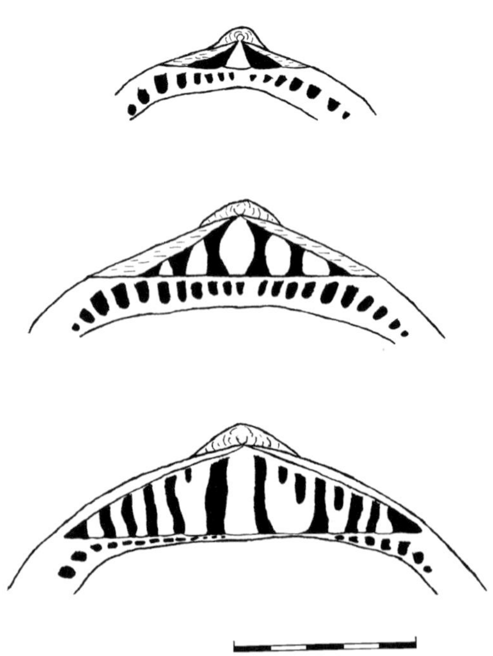

Fig. 4. Ontogeny of the ligament of *Limopsis marionensis* Smith from the South Atlantic, near the Falkland Islands. Scale bar, 1 cm. Specimens from BMNH lots 1964576 and 1964581, 'Discovery' stations 819 and 652, described by Dell (1964), but without reference to this aberrant growth pattern.

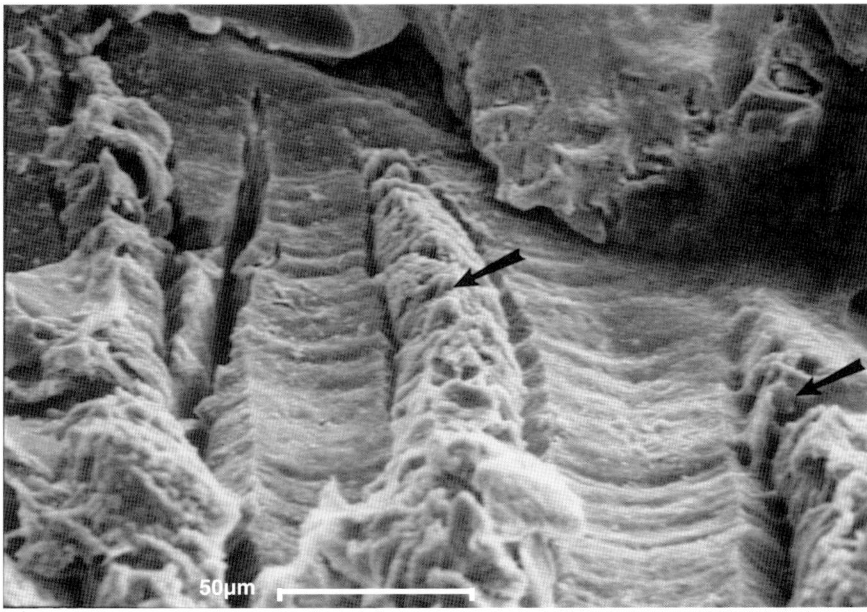

Fig. 5. Scanning electron micrograph of the ligament of *Striarca lactea* (Linné) from the Bay of Naples, Italy. Note strips of lamellar ligament (arrows) inserted in grooves in the fibrous ligament, turning away from the margin of the attachment area to assume a vertical orientation.

development of vertebrate limbs. Here, a sine function represents the net effect of activating and inhibiting processes that control the serial repetition of lamellar sheets as they grow to form a typical chevron ligament (Fig. 6). A threshold value of this function acts as a switch, calling for a change in the type of secretion each time it is passed. The level of the threshold determines the relative widths of the lamellar sheets and the fibrous intervals between them.

As growth increments, δh, are added to the ligament area of height h and length L, then:

$$h' = h + \delta h \text{ and } L' = L + f(\delta h) \quad (1)$$

where $f(\delta h)$ determines the shape of the attachment area (Fig. 7a). At growth stages in which the medial chevron is a new lamellar sheet, the relation between the length of the ligament and the number of alternations is given by:

$$L = 2ma + 2mb + r_{\text{lam}} \quad (2)$$

where m is the number of lamellar sheets, a and b are the lengths of the lamellar sheets and fibrous intervals in the plane of the hinge axis, respectively, and r_{lam} is the length of the newly developing lamellar sheet in the same plane (Fig. 7b). Likewise, when the medial chevron is a new fibrous interval, then:

$$L = 2(m + 1)a + 2mb + r_{\text{fib}} \quad (3)$$

where r_{fib} is the length of the newly developing fibrous interval. New elements are added to the pattern when:

$$r_{\text{lam}} = 2a \text{ or } r_{\text{fib}} = 2b \quad (4)$$

These relationships define the proportions of the triangular ligamental area as a whole and they maintain the constant widths and spacing of the elements of the growth pattern.

A computer program incorporating these principles, written by Luke Kiskaddon, produces simple graphic simulations of the growth patterns of arcoid ligaments. At each step in the process, the preexisting pattern serves as a template for the next growth increment. An instruction corresponding to $f(\delta h)$ controls the rate of anterior and posterior migration of the zones of secretion of lamellar and fibrous material. This variable determines the angle of the chevrons and hence the shape of the ligament's attachment area (see Equation 1 and Figs 3 and 7). The introduction of new medial increments is controlled by a sine function and by a limiting value that triggers the switch and sets the relative magnitudes of a and b (Fig. 8a).

The same model produces noetiid growth patterns if the zones of secretion of the two types of ligamental material remain fixed, producing strips that expand vertically downwards once they are established (Fig. 8b). Here, the sine function

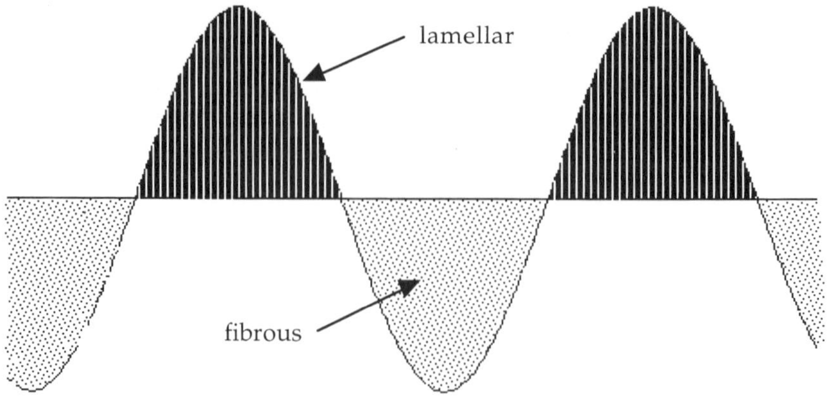

Fig. 6. Simple harmonic, sine wave model for control of secretion of lamellar and fibrous ligament along the mantle isthmus by arcoid bivalves.

governs the introduction of new elements at the anterior and posterior margins of the ligament, rather than medially. Thus, two simple changes in a program written to simulate typical arcoid ligaments are sufficient to convert it to generate noetiid growth patterns. Actually, one change would suffice if both programs were modified in such a way as to prompt the introduction of new elements wherever the available space exceeds a limiting value. This is what one would predict if a set of morphogens controls activation and inhibition uniformly along the length of the mantle isthmus.

These models confirm that quite simple changes in instructions are sufficient to derive a noetiid ligament from a developmental system originally programmed to produce the chevron patterns of typical arcoids. These simulations represent the outcomes of development in this system appropriately. However, they do not incorporate the dynamics of diffusing morphogens, or cell migration, that are generally inferred to control the process of pattern formation. A more realistic approach to this problem is to employ a Turing model (Turing 1952) that simulates the kinetics of the reactions that may be involved in development. Models of this kind have been used extensively by Meinhardt (1984, 1998) and Meinhardt & Klingler (1987) to simulate the development of colour patterns observed in molluscan shells. Some of these patterns are very similar, but not quite identical in form or regularity, to those that are involved here.

Reaction–diffusion model

The harmonic bifurcation model simulates interaction of a sine function with the growth of a bounded domain – the ligamental attachment area – as a mechanism that could generate the spatial patterns of arcoid ligaments. However, it does not account for the origin of the oscillating signal represented by the sine function. A model that has commonly been invoked to explain pattern formation of this sort involves the interaction of a short-range 'activator' and an 'inhibitor' with effects that extend over a greater distance. Turing (1952) first showed theoretically that such a system could spontaneously generate heterogeneous concentrations of reacting chemicals which could, in principle, lead to biological pattern formation. The stable, steady state patterns of morphogens predicted by Turing (1952) have yet to be documented in living systems, but they have been produced experimentally in chemical systems (Castets *et al.* 1990; Ouyang & Swinney 1991), confirming the feasibility of such models.

Turing models are based on the supposition that at least one of the reactants is migrating through the system by diffusion and that there is an autocatalytic step in a linked set of reactions. The concentration of one of the reactants triggers a specific developmental process when it exceeds a certain threshold. Fixed patterns, such as the segmentation of arthropod limbs (Meinhardt 1984), can be modelled by standing waves in such a system. Migrating colour patterns of molluscan shells (Meinhardt & Klingler 1987) and stripes of growing angelfish (Painter *et al.* 1999) are simulated by waves which move through time, due to growth of the domains where these patterns occur. The patterns generated by Turing models depend on diffusion rates, the kinetics of the reactions involved, the size and shape of their domain of activity, and whether or not the reactants can diffuse across the boundaries of the domain.

Models involving cell movement in response to mechanochemical cues have also been invoked to explain spatial pattern formation. Most of these models also involve a mechanism involving short-range activation and long-range inhibition (Murray 1993). The two sorts of models are sufficiently similar mathematically that many developmental patterns can equally well be represented by either one of them. Consequently, one of the simplest types of Turing models has been used here to simulate the growth patterns of arcoid ligaments.

The Schnakenberg (1979) [and see Murray (1993)] model is based on a hypothetical set of three reactions:

$$X \rightleftharpoons A, \quad 2X + Y \longrightarrow 3X, \quad B \longrightarrow Y \quad (5)$$

Applying the law of mass action simultaneously to these reactions, production of the activator, X, and the inhibitor, Y, occurs at rates $k_2a - k_1u + k_3u^2v$ and $k_4b - k_3u^2v$, respectively. The variables u, v, a and b are the concentrations of X, Y, A and B, respectively, and k_1, k_2, k_3 and k_4 are rate constants. The simplifying assumption that A and B occur in abundance is made, so that their concentrations may be approximated by constants. Then, the Schnakenberg model takes the form:

$$\frac{\delta u}{\delta t} = \gamma(a - u + u^2v) + \nabla^2 u \quad (6)$$

and

$$\frac{\delta v}{\delta t} = \gamma(b - u^2v) + d\nabla^2 v \quad (7)$$

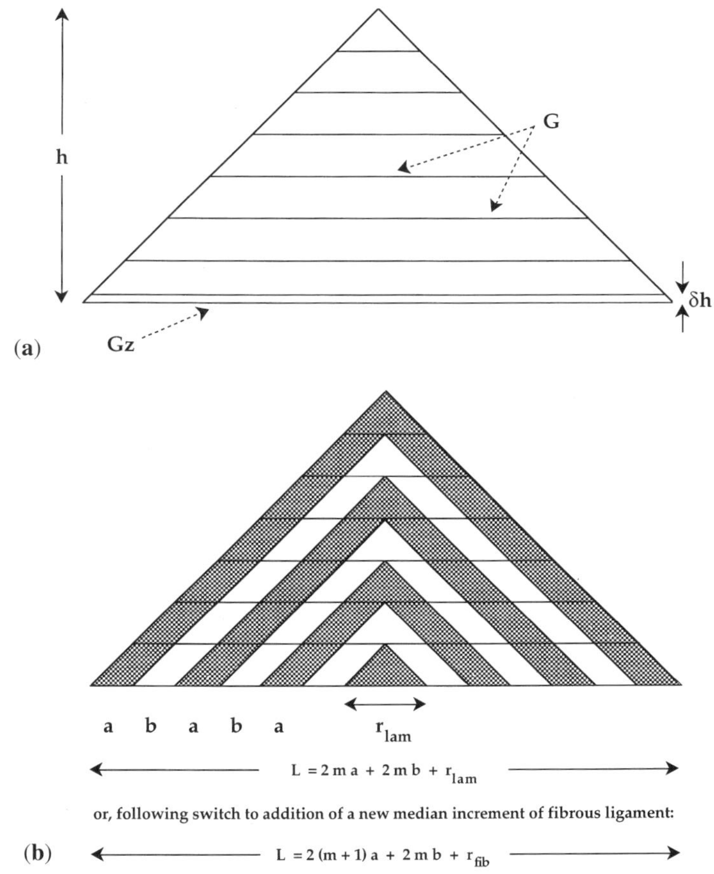

Fig. 7. Parameters of a growing duplivincular ligament. Gz, zone of secretion at the mantle isthmus; G, growth lines; h, height of ligamental attachment area; δh, height of a growth increment; L, length of ligamental attachment area; a and b, lengths of sheets of lamellar and fibrous material, respectively, in the plane of the mantle isthmus; r_{fib} and r_{lam}, length of the latest fibrous or lamellar element, respectively, being added to the pattern; m, number of pairs of ligament elements.

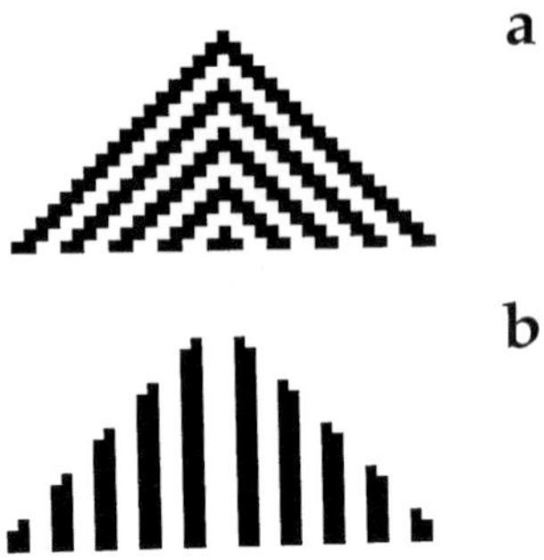

Fig. 8. Simulated growth patterns of: (**a**) the duplivincular or chevron ligament; (**b**) the noetiid ligament; (a) and (b) generated by two variants of a computer model written by Luke Kiskaddon. New elements are added to the pattern whenever the length of the latest median element (a) or distal elements (b) reaches a limiting value.

The last term in each equation models diffusion and the system has been rescaled so as to be nondimensional, reducing the number of constant parameters to four (a, b, γ and d). Here, γ is a function of the size of the ligament relative to the strength of the reaction terms (Murray 1993) and d is the ratio of the diffusion coefficients of the activator and inhibitor, X and Y, respectively.

The boundary conditions for this system take the form:

$$(1 - \theta_L)u_n + \theta_L(u - u_f) = 0$$
$$(1 - \theta_R)u_n + \theta_R(u - u_f) = 0$$
$$(1 - \theta_L)v_n + \theta_L(v - v_f) = 0$$
$$(1 - \theta_R)v_n + \theta_R(v - v_f) = 0$$

where u_n and v_n are the derivatives of u and v, respectively, normal to the domain boundary; they account for the flow of chemicals in and out of the domain of activity. θ_L, θ_R, u_f and v_f are constants. The subscripts R and L denote right- and left-hand boundaries of the model domain, corresponding to the anterior and posterior margins of the ligament. By choosing different values for these constants, different boundary conditions can be set. For example, when $\theta_L = \theta_R = 1$, then u and v take the fixed values u_f and v_f, respectively, on the boundaries. Alternatively, if $\theta_L = \theta_R = 0$, there is no chemical flux across the boundary, which is considered to be impermeable. In this model, growth is represented by increasing x along a one-dimensional interval equivalent to the length of the ligament, from its midpoint, at any time, t (Fig. 9). Initial conditions are given by:

$$u(\underline{x}, 0) = u_0(\underline{x}) \text{ and } v(\underline{x}, 0) = v_0(\underline{x}).$$

To simulate patterns of ligament growth on the expanding domain x_t, appropriate parameters must be set. With $a = 0.1$ and $b = 0.9$, the model yields the homogeneous steady state $u_s = 1.0$, $v_s = 0.9$. This is stable in the absence of diffusion. Standard linear stability analysis is then used to select values of d and γ at which this steady state gives way to linear instability, producing periodic concentrations of the reactants in the presence of diffusion. The number of sinusoidal peaks formed in the domain of ligament growth is determined by the values of the parameters [see Murray (1993) for full details]. In this analysis, the value of d is set at ten and γ is increased until the uniform steady state becomes unstable. At this point, γ is fixed and an exponentially increasing function is selected to define domain growth, representing the allometric growth of the ligamental area. Changes in the spatial pattern of chemical concentrations can now be simulated as this domain expands over time.

In this system, the growth pattern is determined by the boundary conditions and γ prescribes the number of cycles corresponding to alternations between the secretion of lamellar and fibrous ligament material. If $v = 0.7$ at the domain boundaries, and let $u_n = 0$ at these boundaries, the model simulates the symmetrical, chevron growth pattern of typical arcoid ligaments. With $\gamma = 29$, a growth pattern with one anterior and one posterior sheet of lamellar ligament is represented (Fig. 9a). This corresponds to the ligaments of most juvenile arcoids and to the adults of species with a single pair of lamellar sheets, such as *Cucullaea* and normal forms of the paedomorphic genus *Limopsis*. With $\gamma = 185$, the model simulates the characteristic growth pattern of a classic, symmetrical duplivincular ligament (Fig. 9b).

All that is necessary to switch this system to the developmental pattern of a noetiid ligament is a change in boundary conditions. With $v = 0.7$ at $x = 0$, instead of at the expanding margin of the domain, the extremities of the ligamental area are no longer fixed boundaries that anchor the growth pattern. Now, the pattern originates below the umbo and new lamellar elements are added at the distal margins of the ligament. Without any other change in the underlying system, the pattern simulating alternate lamellar and fibrous elements runs perpendicular to the hinge axis, like the ligaments of noetiids (Fig. 9c). Furthermore, by adjusting the concentration of the activator, X, to $u = 1.4$ at the expanding margin, on the right (Fig. 9d), it is possible to simulate the transition from one or more oblique lamellar chevrons, running parallel to the margin of the ligament, to vertical strips as seen in the ontogeny of *Limopsis marionensis* (Fig. 4) and in the early growth stages of the noetiid ligament in some species of *Striarca* (Fig. 5).

This interpretation of the models implies that the

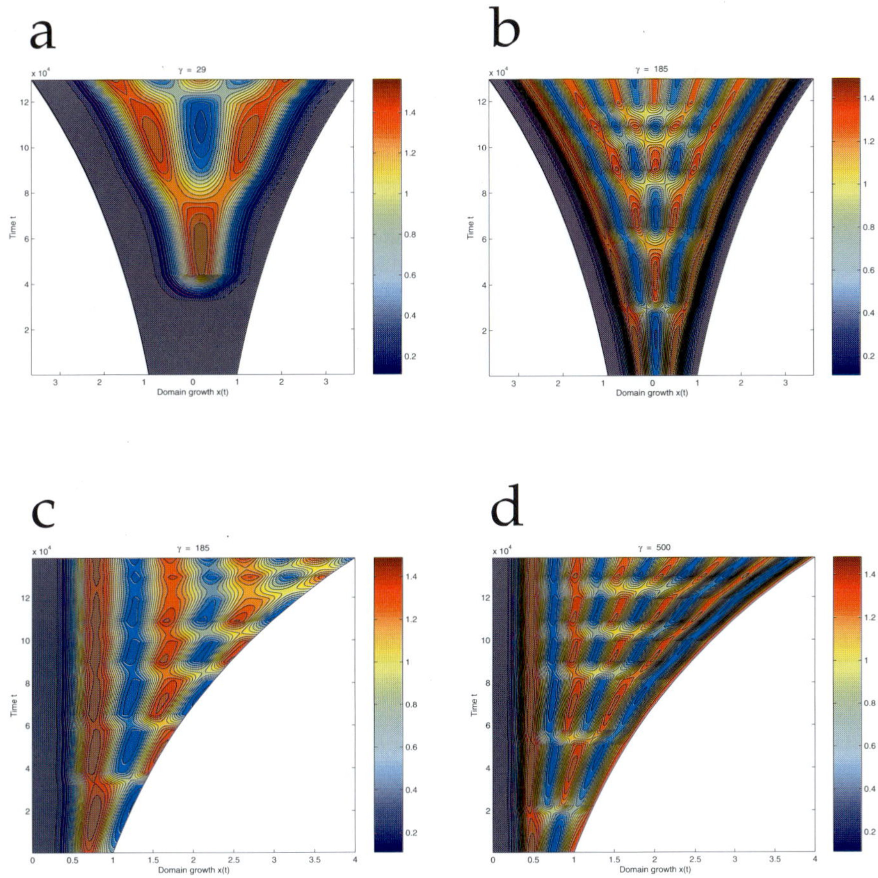

Fig. 9. Schnakenberg models simulating the growth patterns of arcoid ligaments. Colour spectrum represents the concentration, u, of the activator. Limiting values of u model control over the periodic insertion of sheets of lamellar ligament. (**a**) Juvenile of any arcoid with a duplivincular ligament, or adult *Limopsis*. (**b**) Duplivincular ligament of *Glycymeris*, *Anadara*, *Idonearca* or *Grammatodon*. These two simulations represent different stages of development under the same model, with $v = 0.7$ at the domain boundaries and with $u_n = 0$ at these boundaries. (**c**) One side of a noetiid ligament, simulated with $v = 0.7$ and $u_n = 0$ along $x = 0$, and with $v_n = 0$ at the expanding, distal margin of the ligament. (**d**) Model of an intermediate growth pattern. Strips of lamellar ligament first run parallel to the margin of the attachment area, then turn to run perpendicular to the hinge axis. Here, $v = 0.7$ along $x = 0$, and $u = 1.4$ along the distal margin, otherwise with zero flux.

concentration of the inhibitor, v, is fixed at the extremities of the ligament in typical arcoids, which is consistent with their growth patterns. It is less clear how the concentration of this morphogen might be fixed half-way along the mantle isthmus, as required to simulate the noetiid pattern. This may reflect the establishment of a pattern that itself serves as a template for the subsequent distribution of morphogens, except where growth permits the introduction of new elements. This calls for exploration of the possibility that a change in one of the rate-determining variables, rather than the boundary conditions, may serve as a developmental switch in this system (J. Hutchinson, pers. comm.).

Discussion

The models developed in this work show that what might have seemed to be rather fundamental differences in growth pattern, between the ligaments of typical arcoids and those of noetiids, may be controlled by little more than a relatively simple developmental switch. Shifting the site of addition of new elements to the pattern, from near the midpoint of the ligament to its extremities, requires only a change in boundary conditions of the expanding morphogenetic field where growth increments are added to the ligament. Consequently, the possibility that this evolutionary innovation emerged independently in more than one lineage must be taken seriously.

The noetiids are quite disparate in shell form, including even the disposition of the ligament. The genera assigned by MacNeil (1938) and Newell (1969) to the Noetiinae, including *Noetia* and *Eontia*, seem to constitute a cohesive group. They are generally characterized by quite strongly opisthogyrate umbones, broad primary ribs and crenulate interior margins. In this group, the ligament extends over much of the cardinal area, including all the space that is available anterior to the umbones (MacNeil 1938). Most of these animals are or were shallow burrowers, broadly comparable in shell form and mode of life with the arcid *Anadara*. Widely distributed today, from temperate to tropical latitudes, this group includes the earliest recorded noetiid, from the early Cretaceous (Aptian) of Lebanon.

The genera placed in the Striarcinae are more varied. This subfamily is defined primarily by the restriction of its ligament to a triangular area, directly below the umbones, which tend to be more nearly orthogyrate than those of other noetiids. In *Striarca*, a conservative genus that appeared very late in Cretaceous and is widely distributed today, the ligament is inserted in a shallow embayment on each cardinal area. Its growth is commonly allometric, so that it occupies an increasingly large proportion of the cardinal area in later ontogeny (Oliver & Cosel 1992). In contrast, the ligaments of other genera occupy small, deep resilifers on cardinal areas that may be broad, as in *Arcopsis*, or very much constricted, as in *Ovalarca*. In this subfamily, some genera have fine radial ribs while in others the exterior surface is marked only by concentric growth lines. The form, number and angular disposition of the hinge teeth are very varied. In short, it is by no means certain that the members of this group share a common ancestry with one another or with members of the Noetiinae.

A third subfamily, the Trinacriinae, consists of three extinct genera with a limited temporal and geographic distribution. This may well be a natural group, as MacNeil (1937) has shown. These are tiny forms, most of them shaped more like *Corbula* than an arcoid, but with taxodont hinges and vertically laminated ligaments. Interestingly, in different species and genera, these ligaments range from occupying a narrowly confined pit to extending over the entire anterior part of an expanded cardinal area (MacNeil 1937).

It is now apparent that none of the shell characters by which the family Noetiidae has been distinguished and subdivided is exempt from pervasive homoplasy. The phylogenetic relationships of these and other arcoids need to be reassessed. Arcoid shells are relatively simple structures, so evolutionary reversals (Stanley 1972) and convergence are not unexpected. Consequently, future attempts to infer true phylogenetic relationships within this group will require the careful integration of morphological, stratigraphic and biogeographical data.

Conclusions

The analysis developed here shows how the strikingly distinctive form of the ligament that characterizes the noetiids is related, in terms of plausible underlying developmental processes, to the duplivincular ligaments of more typical arcoids from which it evolved. Models indicate that the novel growth pattern of the noetiid ligament can be derived as a result of one or two very simple changes in the regulation of development of a typical arcoid ligament. The adaptive significance of this shift has not yet been determined. However, the strength and mechanical function of the noetiid ligament are comparable, in general terms, to those of more typical arcoid ligaments, such as that from which it evolved (Thomas 1978). Furthermore, extreme variants in a living species of *Limopsis* develop a ligament that is incipiently convergent in form with that of the noetiids.

Together, these observations suggest that bivalves with 'noetiid' ligaments are not necessarily monophyletic. The implications of this analysis for evolutionary systematics are quite provocative. The evolution of growth patterns which appear to be quite disparate, but are based on a common developmental process, may require no more genetic divergence than more subtle character shifts. What seem to be 'major' characters, assumed to be of great evolutionary significance, are not necessarily less labile or more constrained than characters like details of shell sculpture that are commonly supposed to be of more local (lower taxonomic level) significance. Consequently, growth processes must be taken into full account whenever a character is used to determine phylogenetic relationships.

The authors are grateful to L. Kiskaddon for his simulations, programmed when he was a student at Franklin & Marshall. His work established the viability of this project. We are grateful to K. L. Davis for excellent SEM photography (Fig. 5). We also thank J. A. Crame, J. Hutchinson, P. W. Skelton and T. R. Waller for detailed and constructive reviews that gave rise to significant improvements in the manuscript. We are grateful for financial support from Franklin & Marshall College

(RDKT) and the National University of Science and Technology of Zimbabwe (AM).

References

CASTETS, V., DULOS, E., BOISSONADE, J. & DE KEPPER, P. 1990. Experimental evidence of a sustained standing Turing-type nonequilibrium chemical pattern. *Physical Review Letters*, **64**, 2953–2956.

DELL, R. K. 1964. Antarctic and Subantarctic Mollusca: Amphineura, Scaphopoda and Bivalvia. *Discovery Reports*, **33**, 93–250.

MACNEIL, F. S. 1937. The systematic position of the pelecypod genus *Trinacria*. *Journal of the Washington Academy of Sciences*, **27**, 452–458.

—— 1938. Species and genera of Tertiary Noetinae. *United States Geological Survey, Professional Paper*, **189-A**, 1–50.

MEINHARDT, H. 1984. Models for positional signalling, the threefold subdivision of segments and the pigmentation patterns of molluscs. *Journal of Embryology and Experimental Morphology*, **83**, Suppl, 289–311.

—— 1998. *The Algorithmic Beauty of Sea Shells*, 2nd Edition. Springer, Berlin.

—— & KLINGLER, M. 1987. A model for pattern formation on the shells of molluscs. *Journal of Theoretical Biology*, **126**, 63–89.

MURRAY, J. D. 1993. *Mathematical Biology*, 2nd Edition. Springer, Berlin.

NEWELL, N. D. 1937. Late Paleozoic pelecypods: Pectinacea. *Kansas Geological Survey, Publications*, **10**(1), 1–123.

—— 1969. Family Noetiidae Stewart, 1930. *In*: MOORE, R. C. (ed.) *Treatise on Invertebrate Paleontology. Part N. Mollusca 6, Bivalvia.* Geological Society of America, Boulder, CO, and University of Kansas, Lawrence, KS, 261–264.

OLIVER, P. G. & COSEL, R. VON 1992. Taxonomy of tropical West African bivalves. V. Noetiidae. *Bulletin, Muséum national d'Histoire Naturelle, Paris*, 4e série, **14**, section A, 655–691.

OSTER, G. F., SHUBIN, N., MURRAY, J. D. & ALBERCH, P. 1988. Evolution and morphogenetic rules: the shape of the vertebrate limb in ontogeny and phylogeny. *Evolution*, **42**, 862–884.

OUYANG, Q. & SWINNEY, H. L. 1991. Transition from a uniform state to hexagonal and striped Turing patterns. *Nature*, **352**, 610–612.

OWEN, G. 1959. The ligament and the digestive system in the taxodont bivalves. *Proceedings of the Malacological Society of London*, **33**, 215–223.

PAINTER, K. J., MAINI, P. K. & OTHMER, H. G. 1999. Stripe formation in juvenile *Pomacanthus* explained by a generalized Turing mechanism with chemotaxis. *Proceedings of the National Academy of Sciences, USA*, **96**, 5549–5554.

SCHNAKENBERG, J. 1979. Simple chemical reaction systems with limit cycle behaviour. *Journal of Theoretical Biology*, **81**, 389–400.

STANLEY, S. M. 1972. Functional morphology and evolution of byssally attached bivalve mollusks. *Journal of Paleontology*, **46**, 165–212.

TEVESZ, M. J. S. 1977. Taxonomy and ecology of the Philobryidae and Limopsidae (Mollusca: Pelecypoda). *Peabody Museum, Yale University, Postilla*, **171**, 1–64.

THOMAS, R. D. K. 1976. Constraints of ligament growth, form and function on evolution in the Arcoida (Mollusca: Bivalvia). *Paleobiology*, **2**, 64–83.

—— 1978. Shell form and the ecological range of living and extinct Arcoida. *Paleobiology*, **4**, 181–194.

TRUEMAN, E. R. 1969. Ligament. *In*: MOORE, R. C. (ed.) *Treatise on Invertebrate Paleontology. Part N. Mollusca 6, Bivalvia.* Geological Society of America, Boulder, CO, and University of Kansas, Lawrence, KS, 58–64.

TURING, A. M. 1952. The chemical basis of morphogenesis. *Philosophical Transactions of the Royal Society, London, Series B*, **237**, 37–72.

WALLER, T. R. 1990. The evolution of ligament systems in the Bivalvia. *In*: MORTON, B. (ed.) *The Bivalvia – Proceedings of a Memorial Symposium in Honour of Sir Maurice Yonge, Edinburgh, 1986.* Hong Kong University Press, Hong Kong, 49–71.

Carboniferous praecardioid bivalves from the exceptional Buckhorn Asphalt biota of south-central Oklahoma, USA

T. E. YANCEY & M. J. HEANEY, III

Department of Geology and Geophysics, Texas A&M University, College Station, Texas 77843-3115, USA, (e-mail: tyancey@tamu.edu)

Abstract: Unusual Pennsylvanian bivalves recovered from the Buckhorn Asphalt Quarry of south-central Oklahoma are members of the Order Praecardioida, Family Lunulacardiidae. This is the youngest occurrence of praecardioids in North America. These praecardioids have an ontogeny characterized by a veneriform juvenile shell, changing abruptly to an elongate adult shell with sharp carinae, accompanied by strong rotation of the juvenile shell and change in ligament. These shells provide new evidence of dentition and ligament for lunulacardiids and the first documentation of praecardioid shell microstructure. The Buckhorn taxa possess stout hinge teeth, indicating that lunulacardiids are not edentulous. Preserved ligament in multiple growth stages indicates that beaks are prosogyrous and the flattened truncated portion of the shell is posterior. A revised diagnosis of the Lunulacardiidae is presented. New taxa: *Buckhornia carteri* n. gen., n. sp.

The bivalve fauna of the Carboniferous-aged Buckhorn Asphalt deposit of southern Oklahoma is unique, containing shells that are mineralogically and geochemically unaltered and very close to the shell condition at the time of death of the organisms. Shell material accumulating on the seafloor was sealed in asphalt shortly after death of the organisms, keeping the skeletal remains isolated from pore fluids and protecting them from diagenetic alteration. Partly mineralized skeletal material such as ligament is preserved for many species (Heaney & Yancey 1991) and even the organic matrix of shell (sheaths around crystallites) is preserved (Heaney & Yancey 1993). The exceptional preservation of Buckhorn shell material reveals minute shell characters normally obscured in Palaeozoic fossils and provides unaltered shell for documentation of major and minor element chemistry (Crick & Ottensman 1983) and stable isotope geochemistry (Brand 1987; Hewitt *et al.* 1989).

This bivalve fauna is unusual in containing a dominance of small shells, preserving shells down to submillimetre size range (the smallest is a larval shell of 180 μm maximum dimension). Some of these are juveniles of common Desmoinesian bivalves, but many are shells of small taxa that have not been previously described. Bivalves of this size are not normally preserved in ancient sediments and have not been studied, although small bivalves are common in modern biotas and taxa of this size range were probably also a common component of Palaeozoic biotas.

The most unusual of the Buckhorn bivalves is the focus of this study: a small, inflated, moderately ribbed bivalve with strong angulations on the shell and a large, distinct juvenile shell stage. The most distinctive feature is the change that occurs during transition from the juvenile to the adult growth stage, when the juvenile stage is rotated relative to the orientation of the adult shell. These ontogenetic changes and large juvenile stage indicate a relationship with bivalves of the Late Palaeozoic genus *Lunulacardium* Münster 1840 and mid-Palaeozoic genus *Patrocardia* Fischer 1887.

The exceptional preservation of Buckhorn material allows accurate documentation of the shell ontogeny, microstructure, ligament and dentition for this type of lunulacardiid, and provides much new data for the Order Praecardioida. Although praecardioids are usually described as being edentulous, the Buckhorn praecardioids possess stout hinge teeth. Despite this apparent discrepancy in hinge character, other shell characters indicate that the Buckhorn taxon is properly assigned to the Family Lunulacardiidae.

Location

The material studied is from asphalt-impregnated sediments of the Boggy Formation within the Buckhorn Asphalt Quarry south of Sulphur, Murray

From: HARPER, E. M., TAYLOR, J. D. & CRAME, J. A. (eds) *The Evolutionary Biology of the Bivalvia*. Geological Society, London, Special Publications, **177**, 291–301. 1-86239-076-2/00/$15.00

County, Oklahoma (SE 1/4, sec. 23, T. 1S., R. 3E., Indian Baseline and Meridian) (Fig. 1). The Boggy Formation, of Middle Pennsylvanian (Desmoinesian Stage, Late Carboniferous) age, is part of the Deese Group and contains a diverse and abundant molluscan fauna. The shells are small and fragile and were obtained by dissolving the asphalt matrix with carbon tetrachloride in a Soxhlet™ hydrocarbon extraction apparatus. Well-preserved specimens were mounted on stubs and examined with JEOL T330 and JEOL T6400 scanning electron microscopes (SEM) for documentation of surficial and microstructural details.

Biotic assemblage

The Buckhorn praecardioids occur in a biotic assemblage that is infrequently preserved in Palaeozoic strata, i.e. an assemblage characterized by the occurrence of common orthocone cephalopods and 'edentulous' bivalves. This unusual assemblage is best known in the Silurian and Devonian cephalopod limestones of Bohemia described by J. Barrande, and at many other sites along the margins of the Proto-Tethys Sea (Kříž 1979), in the Late Devonian Naples fauna of New York described by Clarke (1904) and in the early Carboniferous Culm facies of Europe (Amler 1998). These assemblages occur in fine-grained sediments deposited in shallow environments (Kříž 1984), within basins having connections to deep-marine environments. The cephalopod shell concentrations are interpreted to originate from empty shells floated into the area and sinking to the seafloor (Kříž 1984), although an alternative interpretation is that the cephalopods occur in death assemblages formed after mating activity in shallow waters.

Sediments of the Buckhorn Asphalt deposit accumulated in moderately shallow-water conditions, adjacent to a small land mass near deep-marine waters of an open ocean basin. These sediments are sandier than sediments containing the Silurian and Devonian assemblages cited above, and are part of a cyclothem sequence deposited under conditions of systematic variation in sea level causing regular changes in water depth and migration of shorelines. Consequently, there is a greater diversity of molluscs and non-molluscan species in the Buckhorn biota than in older praecardioid-bearing assemblages.

The Buckhorn Asphalt biota is preserved in mixed terrigenous and calcareous sediment within the upper part of a cyclothem. Based on the presence of plant remains and scour channels in some of the beds exposed at the Buckhorn Quarry site, Squires (1973) suggested that the asphalt-impregnated Buckhorn Quarry sediments were deposited in a nearshore marine environment under

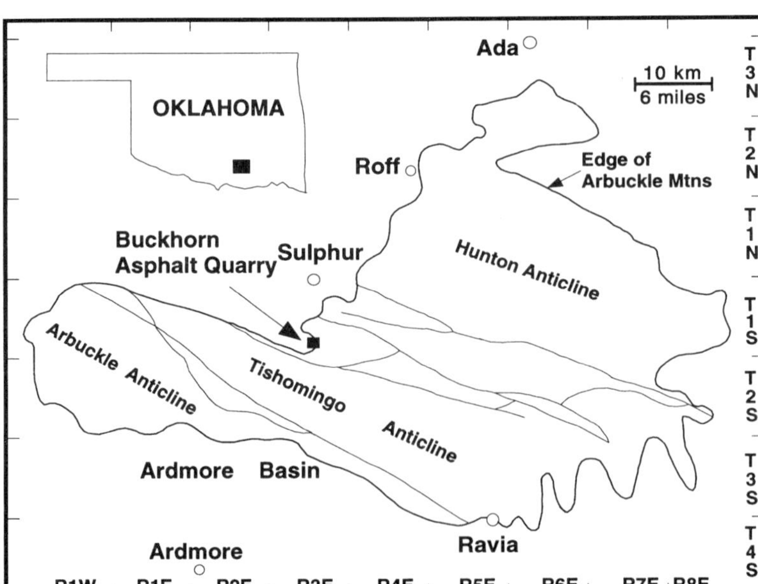

Fig. 1. Location of the Buckhorn Asphalt Quarry relative to structural units of the Arbuckle Mountains, Oklahoma. Tickmarks indicate boundaries of townships in Indian Baseline and Meridian grid system of Oklahoma. Modified from Squires (1973).

moderately turbulent conditions. Crick & Ottensman (1983) postulated a depositional setting below a fairweather wave base in a depositional environment characterized by episodic disturbances. The occurrence of abundant shell fragments, along with some wood fragments and bitumen clasts in sands containing the fossils, suggests deposition under storm conditions, which is compatible with the conclusions of Crick & Ottensman (1983). Regional analysis by Hewitt *et al.* (1989) showed that the Buckhorn area lay adjacent to structural highs on the Ardmore Block, but near deep basinal waters of the Anadarko Basin and the oceanic basin between North America and Gondwana.

Deposition of the Buckhorn sediments within a depth range of 10–20 m can be inferred from sediment characteristics and the microbored condition of the shells. Buckhorn Asphalt Quarry sediments were deposited below the fairweather wave base and above the storm wave base. The depth of the fairweather wave base at the Buckhorn site provides a means of estimating the minimum water depth, once the wave condition of ocean waters in this region is inferred. A modern analogue is available in the northwestern Gulf of Mexico, where the fairweather wave base lies at 10 m water depth or less, as determined by changes in seafloor sediments (Snedden & Nummedal 1991). Many gastropods and bivalves in the Buckhorn Asphalt deposit have their outer surfaces scarred with microborings characteristic of the upper photic zone, providing a means to estimate a lower water-depth limit. The microborings consist of tiny tubules penetrating the shell, similar to borings produced by cyanobacteria in the upper photic zone microboring assemblage of Budd & Perkins (1980) on the shelves around Puerto Rico. The depth boundary between the upper photic zone and lower photic zone around modern Puerto Rico is 20 m, although waters in Desmoinesian marine environments of southern Oklahoma may have been less clear than waters around Puerto Rico, so a reasonable lower depth limit for the Buckhorn deposits is 15–20 m. Thus, sedimentological indicators suggest a depth range of 10–20 m for the Buckhorn deposit.

Valve orientation

In most descriptions of lunulacardiid bivalves (e.g. Clarke 1904; Kříž & Serpagli 1993) the flattened or truncated portion of the shell ['lunule' of Münster (1840)] is presumed to be anterior. This convention is based on the assumption that the gape on this surface enclosed a byssus and the flattening was functional in positioning the shell against the sediment surface or substrate for byssal attachment. In this convention, the beaks of the shell are directed posteriorly (opisthogyrous). The position of the ligament on *Buckhornia* gen. nov. indicates that the beaks are prosogyrous. Several specimens preserve fibrous ligament within a groove on the juvenile shell (Fig. 2) and on the interarea of the early adult. Orientation is apparent on the juvenile shell because it is veneriform, with a ligament contained within an elongate groove located adjacent to the beaks. This is a postero-dorsal (opisthodetic) position (Newell & Boyd 1987). On some specimens it is possible to trace the extension of the juvenile ligament onto the interarea of the adult shell (Fig. 5c). The flattened or truncated portion (sical surface) of the adult shell (bounded by a carina) with its gape is thus posterior. This provides confirmation of Rathmann & Amler's (1992) opinions, and also a minority of other workers, that lunulacardiid bivalves are prosogyrous. The prosogyrous orientation may also apply to other praecardioids with a veneriform juvenile stage. In the following discussions of lunulacardiids, anterior is the direction in which beaks are pointed, and posterior is the direction where the sical surface and large gape are located.

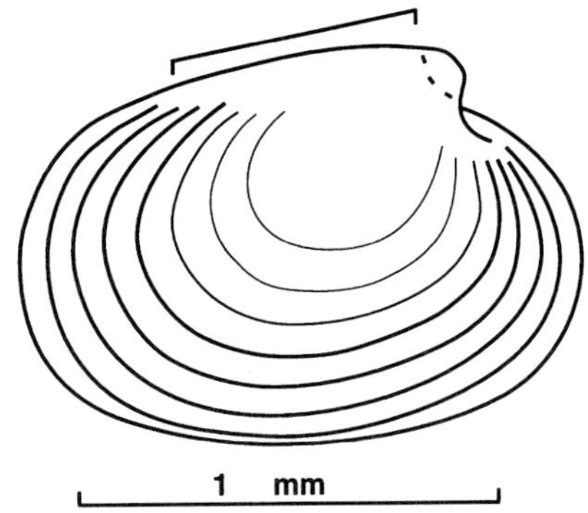

Fig. 2. Juvenile shell, showing the change in shape during growth and form of growth lines. Dashed line indicates limit of the prodissoconch growth stage. Position and extent of ligament groove indicated by bracket, although the ligament groove is not visible from this viewpoint. The crowding of growth lines on the ventral margin of the juvenile shell corresponds with development of a sharp angulation on that margin (cf. with partial juvenile shell in Fig. 4a).

A prosogyrous condition can be inferred for the lunulacardiids *Chaenocardiola* Holzapfel 1889 and *Lunulacardium* Münster 1840. The presence of a truncated margin ['lunule' of Münster (1840)] with a sical surface behind the beaks of these genera unites them with *Buckhornia* gen. nov.

(a)
(b)
(c)
(d)
(e)
(f)
(g)

Chaenocardiola haliotoidea (Roemer) has two other characters indicating a close relationship to *Buckhornia* gen. nov.: (1) a veneriform juvenile shell strongly rotated relative to the adult shell (Holzapfel 1889, pl. 7, fig. 5); (2) a wedge-shaped umbone (Holzapfel 1889, pl. 7, fig. 6b). The nature of the juvenile shell on *Lunulacardium semistriatum* Münster is not known, but the wedge-shaped umbone is apparent in Münster's original illustration of the species. Clarke (1904) noted the close relationship of *Lunulacardium* and *Chaenocardiola*, so despite limited knowledge of *Lunulacardium semistriatum* Münster, a prosogyrous condition and veneriform juvenile shell can be inferred for the genus *Lunulacardium*.

Systematic Palaeontology

Order Praecardioida Newell, 1965
Family Lunulacardiidae Fischer, 1887

Diagnosis. Praecardioids with a truncated posterior margin containing a sical surface, veneriform juvenile shell and prosogyrous beaks.

Discussion. The sical surface and truncate posterior margin of lunulacardiids is their most unusual feature and serves best to distinguish them from other praecardioids. Despite the limited knowledge of the type species of *Lunulacardium* Münster 1840, the truncate margin was the basis on which the genus was erected, making it a reliable character for family determination. The present diagnosis of the family accommodates genera of late Devonian and Carboniferous age, but several late Silurian and early Devonian praecardioid genera assigned to the Lunulacardiidae by Kříž & Serpagli (1993) may not belong in the family.

Assignment of the Lunulacardiidae to the Order Praecardioida was first suggested by Johnston & Collom (1998) in their provisional classification of cryptodont bivalves. The assignment is clearly justified by the presence of a veneriform juvenile shell, abrupt changes in growth form during ontogeny and a truncated posterior margin on genera of late Palaeozoic lunulacardiids (see discussion in Remarks section below). This combination of characters is distinctive to the praecardioids.

Genus *Buckhornia* gen. nov.

Derivation of name. In reference to the Buckhorn Asphalt Quarry, source of the specimens.

Type species. *Buckhornia carteri* sp. nov.

Diagnosis. Elongate lunulacardiids with flat, truncated posterior margin containing sical surface enclosing large gape; differences in ribbing on anterior, medial, and posterior portions of valve, separated by carinae; large veneriform juvenile shell which rotates 90° or more during transition to adult growth stage; and several stout teeth on hinge line.

Included species. *Buckhornia carteri* sp. nov. from the Buckhorn Asphalt Quarry in the Arbuckle Mountains of south-central Oklahoma.

Remarks. *Buckhornia* is similar to the genera *Lunulacardium* Münster 1840, *Chaenocardiola* Holzapfel 1889 and *Patrocardia* Fischer 1887. It is most similar to *Chaenocardiola*, but differs from it in being more elongate and in having different types of ribbing on the anterior, medial and posterior portions of the valve, separated by carinae. The nature of the *Lunulacardium* juvenile shell is unknown at this time, but assuming it also possesses a large veneriform juvenile shell, *Buckhornia* differs from *Lunulacardium* in being more elongate, having a larger posterior gape and sical surface, and differences in ribbing. *Buckhornia* differs from *Patrocardia* [which Kříž & Serpagli (1993) showed to possess a large veneriform juvenile shell] in having a large posterior gape enclosed by a well-developed sical surface and a higher degree of rotation of the juvenile shell relative to the adult shell. *Patrocardia* has a truncated posterior margin, but a sical surface has not been documented for this genus and it may not belong in the Lunulacardiidae as diagnosed in this report.

Buckhornia is the only lunulacardiid genus documented to possess stout hinge teeth, but this is not a reliable distinguishing character at this time because of uncertainty about the hinge of other lunulacardiid genera. The teeth on *Buckhornia* are fragile and post-mortem loss of teeth by breakage is common, suggesting that the supposed edentulous condition of other genera may be based on observations of incomplete specimens.

Fig. 3. *Buckhornia carteri* gen. nov., sp. nov.. (**a**)–(**d**) USNM 509755, holotype, right valve; posterior, lateral, anterior and dorsal views, respectively; scale bar, 1 mm; specimen covered with asphaltic residue, uncoated. (**e**) and (**f**) USNM 509762, paratype, right valve; (e) hinge structure; scale bar, 250 μm; (f) oblique lateral view; scale bar, 500 μm. (**g**) Paratype USNM 509763, oblique posterior view, showing sical surface and dentition; scale bar, 500 μm. Specimens (e)–(g) gold–palladium coated. Illumination on specimens (b) and (c) exaggerates the commarginal ribbing and obscures radial ribbing; commarginal ribbing becomes inconspicuous on the medial portions of larger adult shells.

Sical surface. The posterior portion (shell located posterior of the sharp carina) of the adult shell of *Buckhornia* consists of a flat surface aligned perpendicular to the plane of commissure, conspicuously different from the strongly inflated medial and anterior portions of the shell. This flat surface contains a large gape bounded by a sickle-shaped shell area on both valves. Clarke (1904) proposed the term sical surface (or sica) for this feature. The sical surface contains narrow, crowded growth lines and is bounded by a sharp ridge, in contrast to the dominant radial ribbing present on the remainder of the adult shell. Clarke (1904) interpreted the sical surface gape as a byssal gape, similar to the interpretation of Kříž (1985) for slavids and other praecardioids, but the sical surface is located on the posterior slope of *Buckhornia*, opposite the anterior position of a byssal gape.

Age and distribution. Desmoinesian Stage, Late Carboniferous, Oklahoma.

Species *Buckhornia carteri* sp. nov.

Derivation of name. In honour of Dr Joseph Carter, University of North Carolina.

Type material. Holotype. US National Museum of Natural History (USNM) 509755; paratypes USNM 509756–509776. From assorted blocks of asphaltic sandstone in Buckhorn Asphalt Quarry.

Type locality. The Buckhorn Asphalt Quarry, in the Arbuckle Mountains, south of town of Sulphur, Murray County, south-central Oklahoma. Township & Range reference: SE 1/4, sec. 23, T. 1S., R. 3E., Indian Baseline and Meridian.

Diagnosis. Lunulacardiid with large veneriform juvenile shell, which rotates anteriorly 90° during growth to adult; shell form changes from veneriform juvenile shell to subtrigonal adult shell; three to four stout, wedge-like teeth on adult hinge; ligament changes from elongate opisthodetic to wide monovincular; wide posterior sical surface enclosing a large gape extending along the length of shell; different fields of ribbing on anterior, medial, and posterior portions of valve, separated by carinae.

Description. Juvenile (nepioconch) shell veneriform in outline; reaches 1.2 mm maximum dimension; moderately inflated; initially smooth, gradually acquiring ornamentation of widely spaced fine commarginal ribs, best developed on outer zone of juvenile shell (Fig. 2); small, smooth prodissoconch.

Adult shell small, subtrigonal, well inflated; hinge line short; anterior margin rounded, separated from medial shell by a small carina; posterior margin with flat sical surface and large gape, separated from the medial portion of the shell by sharp, well-developed carina (Fig. 3d); the posterior carina is acutely angled and raised to form a ridge on adult shells; anterior and medial portions of shell ornamented with closely spaced, flat-topped radial ribs (three to four per millimetre) and broader commarginal rugae; commarginal rugae present only between anterior and posterior carinae; radial ribs grow to *c.* 0.2 mm in width and become flattened on top; new radial ribs appear by intercalation; growth lines on ventral margin of young adult shells nearly linear and oblique to direction of shell elongation, becoming rounded on mature shells (Fig. 6); sical surface with narrow, sharp commarginal growth lines and ridges (Fig. 3f); anterior and medial shell margins strongly denticulate (Fig. 5a).

Hinge. Juvenile hinge teeth unknown; juvenile shell with elongate opisthodetic ligament confined to a narrow groove along postero-dorsal margin of shell.

Adult shell with stout teeth, wedge-shaped in cross-section; three teeth on right valve and three or four teeth on left valve, with the posterior tooth on each valve smaller than other teeth; teeth attached to inner edge of valve, which lacks a distinct hinge plate (Fig. 3e), hence are weakly attached to shell, resulting in post-depositional breakage and loss of teeth in many specimens; early secreted portions of teeth occur as thick curved ridges adhering to interior of umbonal cavity; adult shell with large broad trapezoid ligament interarea, truncated on posterior margin and triangular on anterior margin (Fig. 3f); adult ligament monovincular(?), consists of fibrous layer extending between the ligamental areas; abrupt transition from juvenile to adult ligament type; ligament fibres of juvenile shell attach perpendicular to surface of ligostracum in ligament groove (Fig. 5e); ligament fibres of adult attach at oblique angle to curved surface of ligament area (Fig. 4c).

Shell microstructure. Prismato-nacreous, with very thin outer prismatic layer and thick nacreous inner layers; a thin layer of fibrous prismatic ligostracum present in the ligament groove of juvenile shell (Fig. 5E).

Ontogeny. During ontogeny, the shell of *Buckhornia* passes through three distinct growth stages separated by two abrupt developmental transformations. The first ontogenetic change occurs at the transition from prodissoconch to juvenile growth stage [named nepioconch by Kříž (1997, 1984, 1985)] and the second, and greatest, occurs at the transition from juvenile to adult growth stage (Fig. 3b and c).

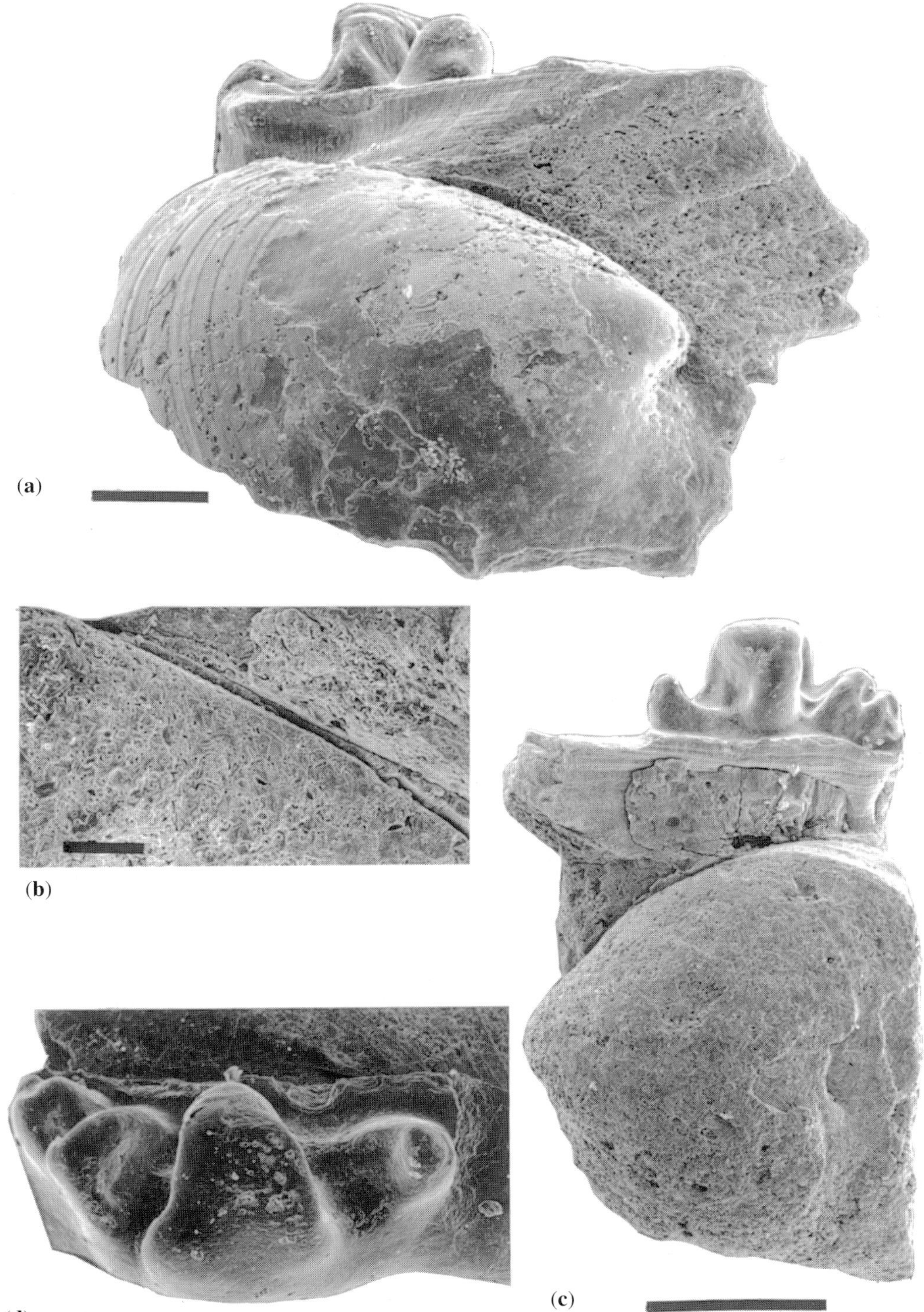

Fig. 4. *Buckhornia carteri* gen. nov., sp. nov.; paratypes. (**a**) USNM 509758, right valve, beak view of veneriform juvenile shell with commarginal ribbing; scale bar, 200 µm. (b)–(d) USNM 509759, left valve; (**b**) view of ligament groove on juvenile shell; scale bar, 100 µm; (**c**) beak area showing rotated juvenile shell and portion of fibrous ligament on interarea of adult shell; scale bar, 500 µm; (**d**) hinge teeth of adult; scale bar, 200 µm. All specimens gold–palladium coated.

The prodissoconch–dissoconch transition changes the shell from a small (100 μm) typical D-shaped shell [of a type illustrated by Lutz (1985)] to a distinctive juvenile shell of inequilateral shape. The juvenile growth stage is veneriform (Fig. 4a) and not morphologically intermediate between the morphology of the prodissoconch and adult shell, either in shell form or ornamentation. Instead, it contains fine commarginal ribbing over the middle and outer portions of this growth stage, prosogyrous beaks, an opisthodetic ligament confined to a narrow ligament groove along the postero-dorsal margin (Fig. 4b) and moderately inflated valves.

The greatest change in form occurs at the juvenile–adult transition, when the shell changes from a veneriform shape to a conocardioid shape. At this transition, the ventral margin of the juvenile shell becomes sharply bent, forming an acute angulation (Fig. 3f), with new shell growth producing a shell surface perpendicular to the plane of commissure. Shell growth on this margin causes the valves to become highly inflated. Concurrently, the hinge axis shifts 90° posteriorly, changing the posterior margin of the juvenile shell to the dorsal margin of the adult shell. Ornamentation changes from entirely commarginal ribbing on the juvenile shell to dominantly radial ribbing on the adult shell (Figs 3b and 4a). The acute posterior angulation develops into a sharp carina, separating the inflated, ribbed anterior and medial portions of the shell from the flat posterior sical surface. With subsequent growth, a minor carina appears on the anterior portion of the shell, separating an anterior area of shell with dominant commarginal ribbing from the medial area with dominant radial ribbing.

A profound change in the hinge occurs at the juvenile–adult transition, with development of a monovincular(?) ligament [a ligament type defined by Johnston & Collom (1998)]. The opisthodetic ligament of the juvenile (Fig. 4b) abruptly expands out of the ligament groove of the juvenile shell to form a broad planar ligament of trapezoidal outline (Fig. 4a). Additionally, three to four stout peg-like teeth grow beneath the ligament area (Fig. 3e–g), although this portion of the shell lacks a well-developed hinge plate to hold the teeth.

Among modern bivalves, when major changes in shell shape occur during ontogeny, it correlates with metamorphosis and change from larval growth stage at the time of benthic settlement, as shown by Ito (1999) for a modern pholadid. However, the juvenile growth stage of *Buckhornia* reaches a size much larger than modern bivalve larval shells and secretes an elongate postero-dorsal ligament that extends posteriorly with growth instead of a resilifer (expected on a larval shell). These features cannot readily be associated with larval life. Interpreting the veneriform growth stage as a juvenile instead of a prodissoconch is a change from determinations presented in earlier studies of lunulacardiids (Clarke 1904; Kříž & Serpagli 1993) and other praecardioid families with veneriform early growth stages. However, the presence of a postero-dorsal ligament instead of a resilifer justifies this change in interpretation.

Remarks. The different fields of ribbing on anterior, medial and posterior portions of the valve, and the large veneriform juvenile shell, are the most distinctive features of *Buckhornia*. The veneriform juvenile shell attains a size of 1.2 mm. At the time the shell loses its veneriform shape, the ornamentation changes from entirely commarginal ribbing to dominantly radial ribbing. The largest specimen measures 1.5 cm across the medial and posterior width of the shell, and would be *c.* 2 cm in height and nearly 2 cm wide if complete.

Buckhornia is closely related to Late Devonian specimens referred to as *Lunulacardium clymeniae* Clarke. Clarke (1904, figs 2–6) illustrates individuals with a large veneriform juvenile shell that rotates 90° during transition to the adult growth stage, possessing the same truncated adult shell with a large gape and sical surface, and having a large interarea on the hinge, differing only in the apparent lack of teeth on the hinge. However, the loss of hinge teeth by post-mortem breakage is common in *Buckhornia*, which suggests the same fate for the material illustrated by Clarke (1904). Further study is needed to determine if *L. clymeniae* should be placed in *Buckhornia* or *Chaenocardiola*.

Buckhornia is unusual in having an adult shell form similar to that of several conocardioid rostroconchs, although the presence of strong teeth and a large ligament reveal it to be a bivalve. This homeomorphy extends to the outline of the shell, shell ornament, the presence of gape and the development of a strong angulation separating a flat

Fig. 5. *Buckhornia carteri* gen. nov., sp. nov., paratypes. (**a**) USNM 509760, nacreous inner shell layers on denticulate ventral margin; scale bar, 50 μm. (**b**) USNM 509758, thin prismatic outer shell layer on juvenile shell; scale bar, 10 μm. (**c**) USNM 509762, transition of juvenile ligament groove (lower) to adult ligament interarea (upper); scale bar, 100 μm. (**d**) USNM 509759, detail of fibrous ligament on interarea; scale bar, 100 μm. (**e**) USNM 509762, ligament fibres of juvenile shell, attached to fibrous prismatic ligostracum shell layer; scale bar, 10 μm. (**f**) USNM 509763, ligament fibres in juvenile ligament; scale bar, 2 μm. All specimens gold–palladium coated.

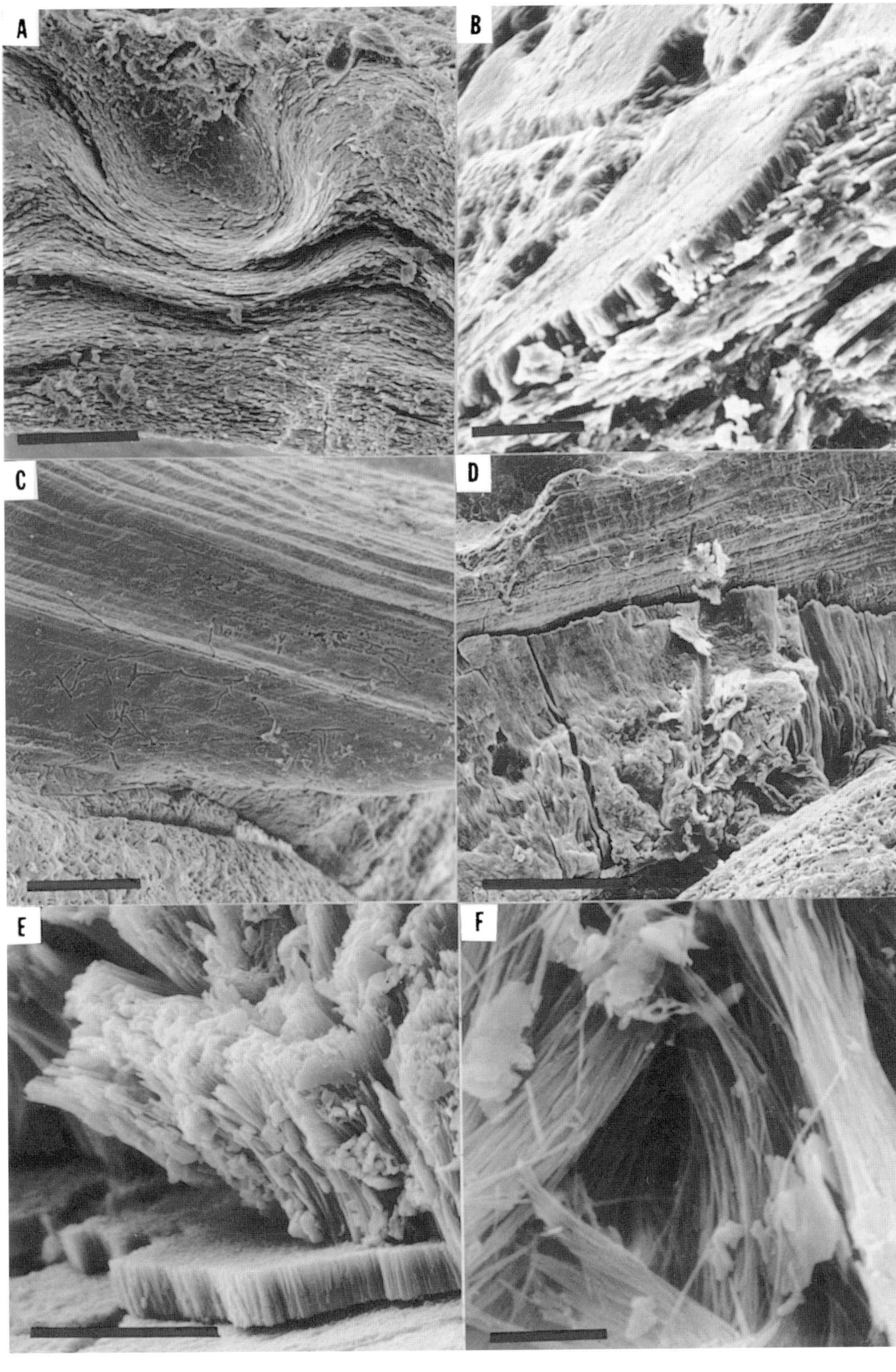
A
B
C
D
E
F

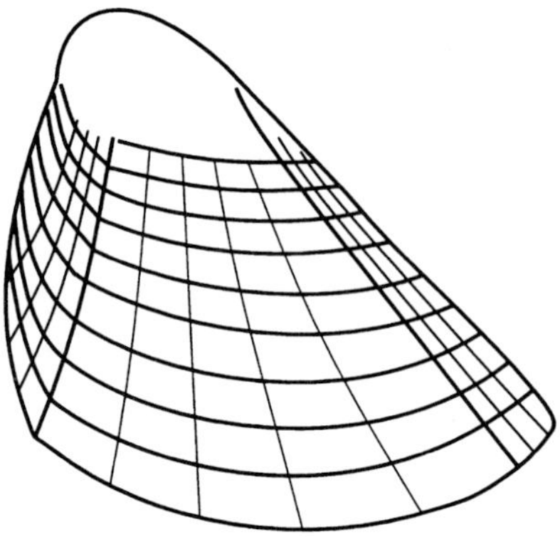

Fig. 6. Reconstruction of adult shell, showing the change in shape of growth lines on the medial portion of shell, from oblique and nearly linear to well-rounded on later growth stages. The actual density of growth lines and ribbing is greater than shown on the diagram, which is stylized for clarity. The posterior carina on the largest portion of shell becomes strongly produced. The sical surface is not visible from this shell orientation. Shells attain dimensions of *c*. 2 cm.

surface from the more rounded portions of the shell. The degree of homeomorphy is great enough that, without hinge characters, *Buckhornia* would be misidentified as a conocardioid; during the present authors' early studies it was confused with the conocardioids. Yates (1962) placed (with question) the closely related *Chaenocardiola* in the family Conocardiidae.

Age and distribution. Desmoinesian Stage, Late Carboniferous, Oklahoma.

Homeomorphy with conocardioids

The adult morphology of *Buckhornia* resembles some species of conocardioid rostroconchs and the two groups are difficult to distinguish in the absence of knowledge of hinge characters. The truncated posterior margin is very similar to the rostrum of conocardioid rostroconchs. Although the conocardioid rostrum may have other features (e.g. carinae extended into a hood, small snout), the rostrum has the distinctive characters of a flat surface on the conjoined valves aligned perpendicular to the plane of commissure, bounded by sharp carinae and enclosing a distinct gape. While there are other differences that can be used to distinguish *Buckhornia* from conocardioids (especially the large juvenile shell on *Buckhornia*), the homeomorphy of adult shells is remarkable.

This strong homeomorphy suggests that the two groups lived very similar lifestyles at comparable growth sizes. The inflated shells indicate an inability to dig rapidly through sediment, implying a shell position at or near the sediment–water boundary. Bonem (1982) discussed probable conocardioid lifestyles, concluding that for rostroconchs like *Pseudoconocardium* the antero-ventral gape opposite the rostrum lay embedded in sediment or lay along the sediment surface, and the rostral gape opened into overlying waters. A praecardioid bivalve oriented in this manner would the have the flat posterior sical surface aligned along, or close to, the sediment surface.

SEM images were produced at the Electron Microscopy Center of Texas A&M University and scanning light photography was produced by Larry Wadsworth of the Health Sciences Department. Samples from the Buckhorn Asphalt Quarry were collected by E. L. Grossman, Texas A&M University, and Rex Crick, University of Texas, Arlington. Discussions with Michael Amler, Paul Johnston and Thomas Waller have helped develop some ideas and interpretations presented in this study. This is a publication of Paleoecology Research Program of Texas A&M University.

References

AMLER, M. R. W. 1998. Early Carboniferous bivalves of the central European Culm Facies. *In*: JOHNSTON, P. A. & HAGGART, J. W. (eds) *Bivalves: An Eon of Evolution*. University of Calgary Press, Calgary, 51–67.

BONEM, R. 1982. Morphology and paleoecology of the Devonian rostroconch genus *Bigalea*. *Journal of Paleontology*, **56**, 1362–1374.

BRAND, E. 1987. Biogeochemistry of nautiloids and paleoenvironmental aspects of Buckhorn seawater (Pennsylvanian), Southern Oklahoma. *Palaeoecology, Palaeoclimatology, Palaeogeography*, **62**, 255–264.

BUDD, D. A. & PERKINS, R. D. 1980. Bathymetric zonation and paleoecological significance of microborings in Puerto Rican shelf and slope sediments. *Journal of Sedimentary Petrology*, **50**, 881–904.

CLARKE, J. M. 1904. Naples fauna in western New York, Part 2. *New York State Museum Memoirs*, **6**, 199-454.

CRICK, R. E. & OTTENSMAN, V. M. 1983. Sr, Mg, Ca, and Mn, chemistry of skeletal components of a Pennsylvanian and Recent nautiloid. *Chemical Geology*, **39**, 147–163.

FISCHER, P. H. 1887. Manuel de conchyliologie et de paléontologie conchyliologique. *Histoire naturelle des mollusques vivants et fossiles*, F. Savy (Paris), **10**, 897–1008.

HEANEY, M. J. & YANCEY, T. E. 1991. Exceptional preservation of bivalved molluscs in the Buckhorn Asphalt deposit (Pennsylvanian) of Oklahoma. *Geological Society of America Abstracts with Programs*, **23**(**5**), 166–167.

—— & —— 1993. Preserved organic matrix from Pennsylvanian aged Buckhorn asphalt Mollusca.

Geological Society of America Abstracts with Programs, **25**(6), 55.

HEWITT, R. A., DOKAINISH, M. A., EL AGHOURY, M. & WESTERMANN, G. E. G. 1989. Bathymetric limits of a Carboniferous orthoconic nautiloid deduced by finite element analysis. *Palaios*, **4**, 157–167.

HOLZAPFEL, E. 1889. Die cephalopoden führenden Kalkedes unteren Carbon von Erdbach–Breitscheid bei Herborn. *Palaontologische Abhandlungen*, N. F., 1.

ITO, Y. 1999. Ontogenetic changes in boring behavior by the rock-boring bivalve, *Barnea manilensis* (Pholadidae). *Veliger*, **42**, 157–168.

JOHNSTON, P. A. & COLLOM, C. J. 1998. The bivalve heresies – Inoceramidae are Cryptodonta, not Pteriomorphia. *In*: JOHNSTON, P. A. & HAGGART, J. W. (eds) *Bivalves: An Eon of Evolution*. University of Calgary Press, Calgary, 347–360.

KŘÍŽ, J. 1979. Silurian Cardiolidae (Bivalvia). *Sborník Geologických Véd [Journal of Geological Sciences], Paleontologie*, **22**.

—— 1984. Autecology and ecology of Silurian Bivalvia. *Special Papers in Palaeontology*, **32**, 183–195.

—— 1985. Silurian Slavidae (Bivalvia). *Sbornik Geologických Véd [Journal of Geological Sciences], Paleontologie*, **27**, 47–111.

—— & SERPAGLI, E. 1993. Upper Silurian and lowermost Devonian Bivalvia of Bohemian type from south-western Sardinia. *Bollettino della Società Paleontologica Italiana*, **32**, 289–348.

LUTZ, R. A. 1985. Identification of bivalve larvae and postlarvae: a review of recent advances. *American Malacological Bulletin Special Editions*, **1**, 59–78.

MÜNSTER, G. GRAF AU 1840. Die Versteinerungen des Ubergangskalkes mit Clymenien und Orthoceratiten von Oberfranken. *Beiträge zur Petrefaktenkunde*, **3**, 33–121.

NEWELL, N. D. 1965. Classification of the Bivalvia. *American Museum Novitates*, **2206**.

—— & BOYD, D. W. 1987. Iteration of ligament structures in pteriomorphian bivalves. *American Museum Novitates*, **2875**.

RATHMANN, S. D. & AMLER, M. R. W. 1992. Bivalven aus dem Unter-Karbon von Aprath (Wuppertal, Bergisches Land). *Geologica et Palaontologica*, **26**, 35–71.

SNEDDEN, J. W. & NUMMEDAL, D. 1991. Origin and geometry of storm-generated sand beds in modern sediments of the Texas continental shelf. *International Association of Sedimentologists Special Publications*, **14**, 283–308.

SQUIRES, R. L. 1973. *Burial environment, diagenesis, mineralogy, and Mg & Sr contents of skeletal carbonates in the Buckhorn Asphalt of Middle Pennsylvanian age, Arbuckle Mountains, Oklahoma*. PhD Thesis, California Institute of Technology, Pasadena, California.

YATES, P. J. 1962. The palaeontology of the Namurian rocks of Slieve Anierin Co. Leitrim, Erie. *Palaeontology*, **5**, 355–443.

Evolutionary significance of fossil larval shell characters: a case study from the Ostreoidea (Bivalvia: Pteriomorphia)

NIKOLAUS MALCHUS

Institut für Paläontologie, Freie Universität Berlin, Malteserstr. 74-100, Haus D, 12249 Berlin, Germany (e-mail: nikom@cc.uab.es)

Abstract: Prodissoconchs of Middle Jurassic oysters possess key characters for a new understanding of the hinge evolution within the Ostreoidea. The ancestral larval hinge most likely evolved from a mytilid-like provinculum through accelerated growth in an antero-ventral direction along a flat, helico-spiral trajectory. The posterior denticles moved into the centre of the hinge axis while the central and anterior denticles were reduced in size and eventually disappeared. The remaining, now central, posterior denticles evolved further into a secondary symmetrical hinge. In most oysters, except for example *Tiostrea*, the larval ligament maintained its relative position between posterior and anterior denticles, even though this is not obvious in the Ostreidae because of the loss of the anterior denticles. Larval and adult ligaments are continuous. The examined specimens provide further evidence for the plesiomorphic state of planktotrophy within the superfamily. Although additional direct evidence is lacking, consideration of all new available data favour the hypothesis that the postero-dorsal notch is uniquely derived by oysters. The ideas on character evolution put forward in this study are consistent with phylogeny hypotheses based on palaeontological–biological data and recent genetical studies.

Larval shells and nepioconchs of fossil bivalves provide an excellent source of information for phylogenetic inferences and decisions on character polarity (Malchus 1998, 1999, 2000*a*, *b*). Nevertheless, early shell stages are only known from a very limited number of fossil taxa, among which oysters have attracted most attention. Malchus (1995) found evidence for the presence of the postero-dorsal notch (pd-notch) (Waller 1981) of the prodissoconch II (P II) in fossil oysters and suggested that this character may be uniquely derived in the Ostreoidea. Furthermore, small prodissoconch I (P I) sizes and low P I:P II ratios were taken as indirect evidence that planktotrophy is the ancestral reproductive strategy within the superfamily. However, this conclusion contradicts biological interpretations of Andrews (1979) and Chanley & Dinamani (1980). Based on a detailed comparison of the larval hinges (provinculum), Malchus (1995) suggested that the pycnodonteinid (Gryphaeidae) type represents the ancestral state from which the other ostreoid types evolved. But the palaeontological evidence was insufficient, because provincula of ancient Gryphaeidae were unknown at that time. The discovery of larval and early post-metamorphic oyster shells from the Middle Jurassic of Poland now permits a reassessment of the previously drawn conclusions. In particular, the material allows a model for the evolutionary changes of the larval oyster hinge to be proposed and some of its so far enigmatic characteristics to be understood.

Materials and methods

Most of the new fossil material referred to in this study comes from an old core through Bajocian–Callovian sediments recovered in 1937 near the village of Klemmen (now Kleby) *c.* 50 km northeast of Szczecin, northwestern Poland. The material is owned by the Bundesanstalt für Geowissenschaften und Rohstoffe (BGR) in Berlin, Germany. Other specimens were collected by the author in Middle Jurassic (probably Bathonian) sediments of a fresh clay pit near Faustianka, north central Poland. Further specimens from the Kimmeridgian of southern England, and the Eocene of France and the USA (Alabama) were available from a previous study (Malchus 1995). Live specimens for comparison were collected from coastal waters in the northwestern Mediterranean and could be partly reared to post-metamorphic stages at the Marine Laboratory in Banyuls s/M, France. The Polish and Mediterranean material is housed in the BGR in Berlin, Germany (labelled BGR X*xxxxx–x*).

From: HARPER, E. M., TAYLOR, J. D. & CRAME, J. A. (eds) *The Evolutionary Biology of the Bivalvia*. Geological Society, London, Special Publications, **177**, 303–312. 1-86239-076-2/00/$15.00

Specimens labelled IRSNB-TI are housed in the Institut royal des Sciences naturelles de Belgique, Brussels.

The Kleby material was found already processed and the depth carefully noted with each sample. Some of the remaining fragments of the core were recently processed for examination of gastropod protoconchs and age determination with ostracodes (Gründel 1997, 1999). The preserved oysters comprise numerous larval shells [left valves (LV) and right valves (RV)] and predominantly nepionic RV up to a size of *c.* 4000 µm. Most of the shells are Middle Callovian in age.

All specimens were mounted on stubs and coated with gold for examination under a scanning electron microscope (SEM). Sample processing and handling under SEM followed Malchus (1995).

Description and determination of early shell stages

The description concentrates on distinctive shell features of the superfamily and features with a bearing on character evolution. More detailed descriptions will be given elsewhere.

Overall shell characters and determination

Left prodissoconch valves show the prominent growth track of the pd-notch and the inflection of the notch itself in the shell margin of the P II (e.g. Figs 1a, c, d and f and 2a). Right prodissoconchs are less easily identified because the notch is only weakly expressed. However, the prodissoconchs of many oysters (especially Ostreidae) are more asymmetrical than those of most bivalves, and they generally possess a distinctly wavy (non-euclidean) rather than flat commissural plane (e.g. Figs. 1d and 2g). Post-metamorphic shells are characterized by an irregular morphology with rugose growth increments and an almost entirely calcitic foliaceous microstructure. All other co-occurring shells with similar prodissoconchs lack the pd-notch and are nacro-prismatic as adults. Many of the studied nepioconchs (all RV) still carry a well-preserved larval shell as shown in Fig. 1b.

The pronouncedly opisthogyrate umbo of the oyster prodissoconchs and comparisons with Upper Jurassic *Liostrea plastica* from England (Palmer 1989) suggest that the specimens from Kleby belong to the Exogyrinae and/or Liostreinae (Figs 1b, d, e and g). The smaller, not fully grown, larval shells from the Faustianka clay pit have thicker shells and relatively weaker opisthogyrate umbones (Figs 1c and 2a). These specimens are tentatively assigned to the Gryphaeinae. The P II stages of all specimens possess a rather strongly developed sculpture of concentric growth welts (Fig. 1a–c). The variable shape of the valves, as well as different P I sizes, strongly suggest the presence of different species. Nevertheless, more specific determinations are currently problematic because of the lack of post-nepionic shells with unequivocal characters.

Shell dimensions

All specimens possess P I sizes of 65–86 µm (length) and 49–70 µm (height), and the P II sizes of fully grown shells range from 300 to 400 µm for both length and height. The P I:P II ratios are therefore always < 0.35. The significance of these values will be discussed below.

The provinculum

The presence of larval shells without post-metamorphic overgrowth permits an unobstructed view onto the provincular dentition and ligament. All examined specimens share the same general arrangement. This consists of a posterior and anterior wing of denticles separated by the larval, fibrous ligament, which grows anteriorwards,

Fig. 1. Early shell stages of Middle Jurassic Ostreoidea from Poland. All scale bars in µm. (**a**) Dorsal view of a LV prodissoconch of an exogyrinid(?) oyster. The P II shows the postero-dorsal notch and growth track (arrow) characteristic of LV. Kleby, BGR X10849-5. (**b**) Dorsal view of a RV nepioconch (with predatory drillhole) and well-preserved prodissoconch of an exogyrinid(?) oyster. Kleby, BGR X10849-3. (**c**) Dorsal view of a LV prodissoconch, probably not fully grown because of the small size (138 × 134 µm), of a gryphaeinid(?) oyster. Arrow indicates position of the pd-notch. Faustianka, BGR X10845-8. (**d**) Internal view of a LV-P II of an exogyrinid(?) oyster. Arrow indicates position of the pd-notch. Kleby, BGR X10849-5. (**e**) Close-up of hinge of (d) showing the posterior denticles in a pseudo-central position below the P I, somewhat reduced central and anterior denticles, and the ligament groove in between. (**f**) Internal view of a LV–P II of an exogyrinid(?) oyster. Arrow indicates position of the pd-notch. Kleby, BGR X10846-12. (**g**) Close up of hinge of (f). (**h**) Close up of P I and part of the P II hinge of (f). The posterior denticles (the middle ones of which are largest) are visible. The arrow points to the position of three, rather weakly developed, originally central denticles above the ligament groove.

subparallel to the interior shell margin, and some small, central denticles above it (Figs 1e, g, h, 2b and 3b). The posterior wing rests below the larval umbo and the P I, which itself is delimited ventrally by its straight hinge line (Fig. 1h). There are about five to eight larger denticles below the umbo. These may be subequal in size or first increase and then decrease, again towards the anterior end. The central denticles above the ligament and those of the anterior wing show a gradual increase in size

anteriorwards but without attaining the dimension of the smallest posterior denticle (e.g. Fig. 1e). There are about three to eight central and up to 12 anterior denticles in almost fully grown P II. The anterior row reaches far beyond the anterior front of the ligament.

Discussion

Taxonomic problems

Early post-metamorphic (= nepionic) shells of bivalves often differ in their shape, sculpture and internal features from the adult shell on which

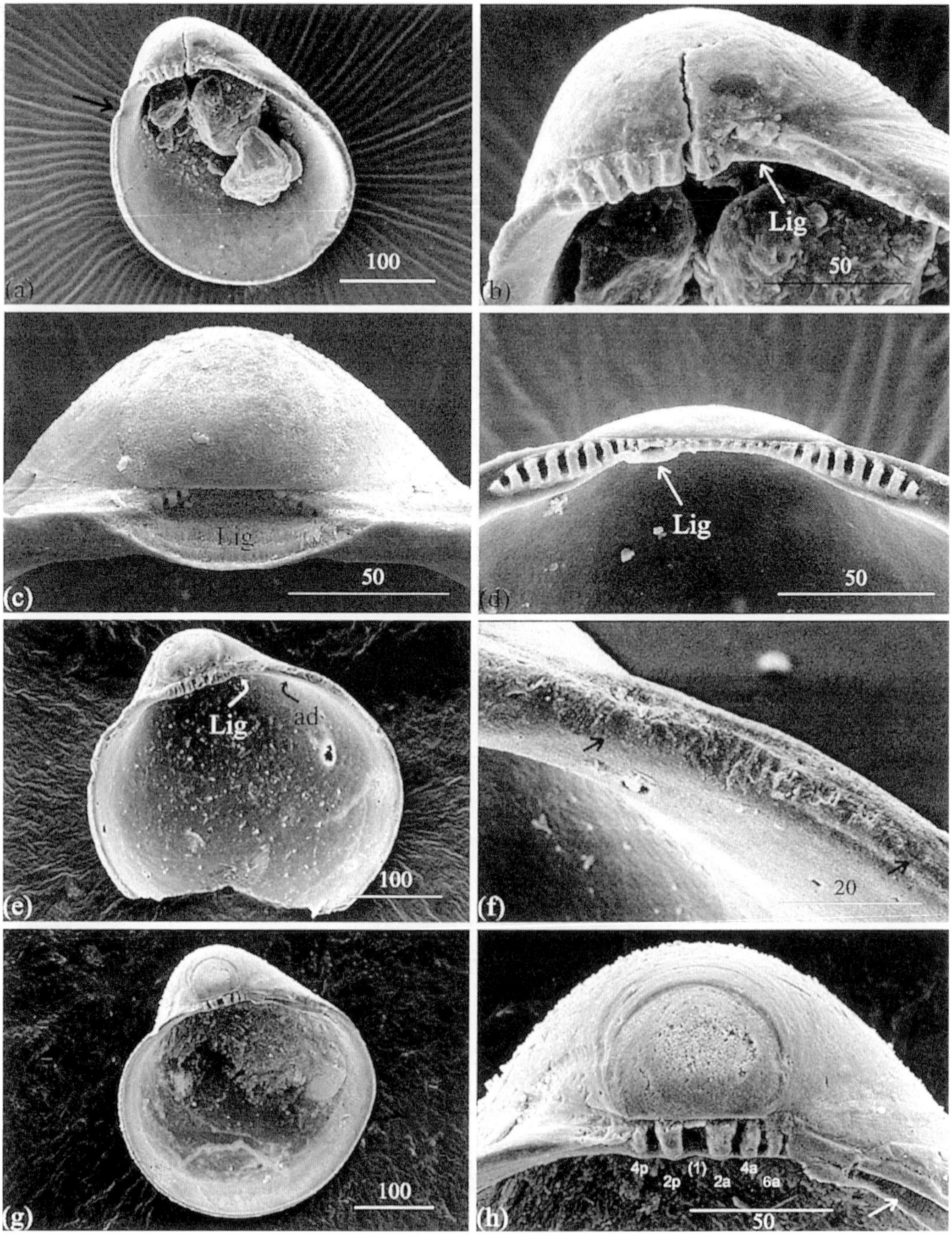

almost all diagnoses are based. In addition, knowledge of larval and nepionic shell characters of fossils is restricted to very few taxa (e.g. Palmer 1989; Malchus 1995, 1999, 2000*b*; Yancey & Heaney, pers. comm.). Thus, even the subfamily assignment of the Middle Jurassic specimens remains rather vague. However, this fact is of minor importance in the present context.

Origin and taxonomic distribution of the postero-dorsal notch

In a comprehensive study on the development on veliger larvae of *Ostrea edulis* L., Waller (1981) demonstrated that the pd-notch marks the extrusion site of a ciliated organ (of unknown function) of the P II veliger stage. So far, the notch and its growth track have only been found in the P II shells of all living oysters, except in *Tiostrea* spp. (e.g. Chanley & Dinamani 1980; Malchus 1995 and refs cited therein) and a number of fossil species belonging to the Liostreinae, Crassostreinae–Ostreinae (Ostreidae) and Pycnodonteinae (Gryphaeidae) (Malchus 1995). Veligers of other living bivalves apparently lack this shell character (Waller 1981). The same is true of recently studied prodissoconchs of Jurassic *Atreta*, *Juranomia*, *Oxytoma*, *Camptonectes*, *Grammatodon*, *Nicaniella*, *Pressastarte*, a trigoniid, numerous nacro-prismatic pterioids (Palmer 1989, Malchus 1999, 2000*b*) and some Carboniferous species (Yancey & Heaney, pers. comm.).

The taxonomic distribution strongly suggests that the notch is symplesiomorphic for the Gryphaeidae and Ostreidae. Its absence in *Tiostrea* is related to the unusually extended brooding time (Malchus 1995; Josefowicz & O'Foighil 1998) which effectively suppresses the development of the P II shell stage; it is thus a derived character state. It remains to be shown whether the notch is also present in the larval shells of the extinct Palaeolophidae which, according to Malchus (1990), are not ancestral to the modern Lophinae (Ostreidae). This latter view is supported by recent genetic studies of O'Foighil & Taylor (2000) and Hammer & Steiner (pers. comm.), which independently suggest that the Lophinae and Ostreinae are derived from the Crassostreinae. In this study, the presence of a pd-notch and its growth track was used, together with indirect evidence from other characters and a comparison with the accompanying fauna, to identify isolated larval shells as oysters.

Polarity of reproductive strategies

All living oysters with a completely planktic–planktotrophic larval life cycle, i.e. the Crassostreinae, possess P I lengths of 45–75 μm and small P I:P II ratios of between 0.25 and 0.35. A ratio <0.4 is considered indicative of non-brooding bivalves (Berkman *et al.* 1991). Most ostreoids with parental care, i.e. Ostreinae and Lophinae, have P I lengths ≥ 120 μm and P I:P II ratios >0.4. One exception is *Cryptostrea permollis*, with an average P I length of 117 μm and a P I:P II ratio of just 0.4 (Buroker 1985, table 3). According to Berkman *et al.* (1991), a ratio of 0.6 is a lower 'benchmark' in brooding bivalves. However, the correlation between length ratio and brooding in living Ostreidae suggests a threshold value of *c.* 0.4 within this family. The reproductive mode of living Gryphaeidae (Pycnodonteinae) appears to be unknown (Harry 1985, table 3); their

Fig. 2. Early shell stages of fossil oysters from Poland and Alabama and recent non-oysters from the Mediterranean. All scale bars in μm. (**a**) Internal view of LV prodissoconch of a gryphaeinid(?) oyster. Arrow indicates position of pd-notch. Faustianka (Poland), BGR X10845-4. (**b**) Close-up of hinge of (a), LV. Posterior denticles of subequal size in a pseudo-central position below P I. The anterior denticles are reduced in size. The triangular groove between the two denticle rows is the site of the ligament (arrow). (**c**) Hinge of an early post-metamorphic, recent limid (*Limatula*? sp.). The ligament emerges from the centre of the provinculum which still shows three denticles on each side just below the straight hinge. Mediterranean, BGR X10848-3. (**d**) Hinge of a LV P II of a recent mytilid (*Mytilus* sp.). The ligament (arrow) emerges posteriorly between a row of posterior denticles (left), and a row of central (below P I) and anterior denticles. Mediterranean, BGR X10847-5. (**e**) Interior view of a LV P II of *Pycnodonte* sp. 2 (Malchus 1995). Pseudo-central denticles are more or less equal in size and equidistant. The white arrow indicates the position of the ligament, the black arrow points at the vestiges of few anterior denticles (ad). Middle Eocene, Alabama. [Originally figured in Malchus (1995, pl. 7, fig. F)]. Specimen IRSNB-TI 6170. (**f**) Close-up of dorso-anterior shell margin of a LV P II of *Pycnodonte* sp. 2 [not identical to (e)]. Shell segment between arrows shows about ten very weakly developed vestiges of anterior denticles. Middle Eocene, Alabama, specimen IRSNB-TI 6171 [same specimen as in Malchus (1995, pl. 7, fig. G, H) but different detail]. (**g**) Internal view of a LV P II of *Cubitostrea sellaeformis*, Middle Eocene, Alabama. [Originally figured in Malchus (1995, pl. 4, fig. F)]. Specimen IRSNB-TI 6153. (**h**) Close-up of hinge of (g). Denticles are numbered after Malchus (1995), except for (1), which here symbolizes the socket for denticle 1 of the corresponding RV. The ligament (arrow) emerges in the lower right (anterior) corner of the pseudo-central, posterior row of denticles. [Originally figured in Malchus (1995, pl. 4, fig. D)].

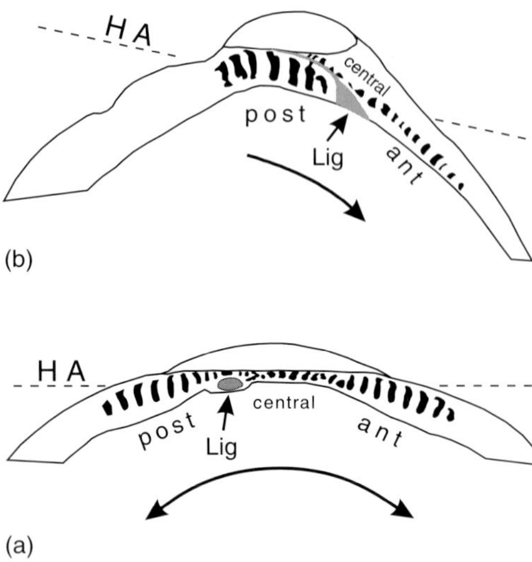

Fig. 3. Hypothesis of hinge evolution of ancient larval oysters from a mytilid-like hinge. Schematic drawings of two LV (not to scale); denticle sockets in black. ant, Anterior; HA, hinge axis; Lig, ligament; post, posterior. (**a**) Shell growth is more or less equal in anterior and posterior directions (bidirectional, double-headed arrow) and has a minimal (otherwise the shell would be flat) ortho-spiral growth component. In consequence, all denticles are near the (pivotal) hinge axis and their arrangement is almost symmetrical. Only the ligament is in a posterior position (arrow). (**b**) A change towards an anteriorly accelerated growth (unidirectional arrow) with a (rather compressed) helico-spiral component leads to opisthogyry. The posterior denticles come to lie in a central position with respect to the hinge axis, while the ligament grows anteriorly and becomes squeezed between the posterior and central denticles. Both central and anterior denticles are at an angle with the hinge axis and become reduced.

P I lengths are slightly larger than those of the Crassostreinae, but still rather small (70–100 μm) (e.g. Ranson 1960, 1967*a, b*; Waller 1981), and the P I:P II ratio is < 0.4. Presumably, they are also non-brooders. Anatomical differences to brooding ostreids are not decisive in this respect unless the presence of a supramyal passage is functionally related to the reproductive mode (see Harry 1985, table 3). However, there is no evidence for such a relation.

The larval shell lengths and ratios of fossil ostreid and gryphaeid species examined by Malchus (1995) compare well with those of Recent crassostreinid and pycnodonteinid species (Fig. 2e). Malchus (1995) therefore concluded that parental care should be the phylogenetically derived mode of reproduction within the superfamily. 'Crassostreinid' dimensions are also characteristic of all but one Middle Jurassic species described here (Fig. 1a, c and h); the exceptional specimen shown in Fig. 2a has a 'pycnodonteinid' P I length of 86 μm, which is consistent with the P I lengths of the Eocene and living Pycnodonteinae. Thus, the new data support the previous interpretation of planktotrophy as being the ancestral reproductive strategy in oysters. The genetic studies of O'Foighil & Taylor (2000) and Hammer & Steiner (pers. comm.) provide additional evidence, as both suggest that the brooding Ostreinae and Lophinae are derived from the non-brooding Crassostreinae.

Evolution of oyster provincula

Many pteriomorphian provincula pertain to one of two morphotypes. One is rather short in length and more or less symmetrical, with two or three equally sized taxodont denticles on each side of a centrally emerging, larval fibrous ligament (Fig. 2c). The other type is longer, with three to eight taxodont denticles on each side and often a central row of much smaller denticles which connects the two wings. The larval ligament emerges either in a central or a posterior position from beneath or on the provincular ledge. In a posterior position, the ligament separates the denticles asymmetrically in a posterior unit and a central row plus anterior unit. A mytilid larval hinge may serve as a model for this type (Figs 2d and 3a). Mytilids later develop some hinge characters unique to this taxon (e.g. Bernard 1896; Le Pennec 1978), but this is irrelevant in the present context.

The provinculum type of the Middle Jurassic specimens can be interpreted as a mytilid-like arrangement modified due to accelerated growth of the antero-ventral parts of the soft body along a shallow helico-spiral trajectory (Fig. 3b). Hence, the larval shell is opisthogyrate. Naturally, growth of the provinculum follows the same trajectory so that the posterior denticle wing glides into a central position with respect to both the shell and the pivotal axis, and the ligament expands exclusively anteriorly and subparallel to the dorso-anterior shell margin (Figs 1e and g, 2b and 3). Thus, with view into the LV, the entire arrangement, including the hinge axis, appears to have rotated clockwise with respect to the mytilid-like condition. As a further consequence, the small, central denticles of the model provinculum of Fig. 3a become wedged between the posterior wing plus ligament gutter and the straight hinge of the P I (Fig.1h). Beyond these, the anterior denticles emerge without abrupt changes in size (Fig. 3b). In both wings, new denticles are added only anteriorly. The anterior denticles apparently become progressively larger in this direction. In contrast, those of the posterior

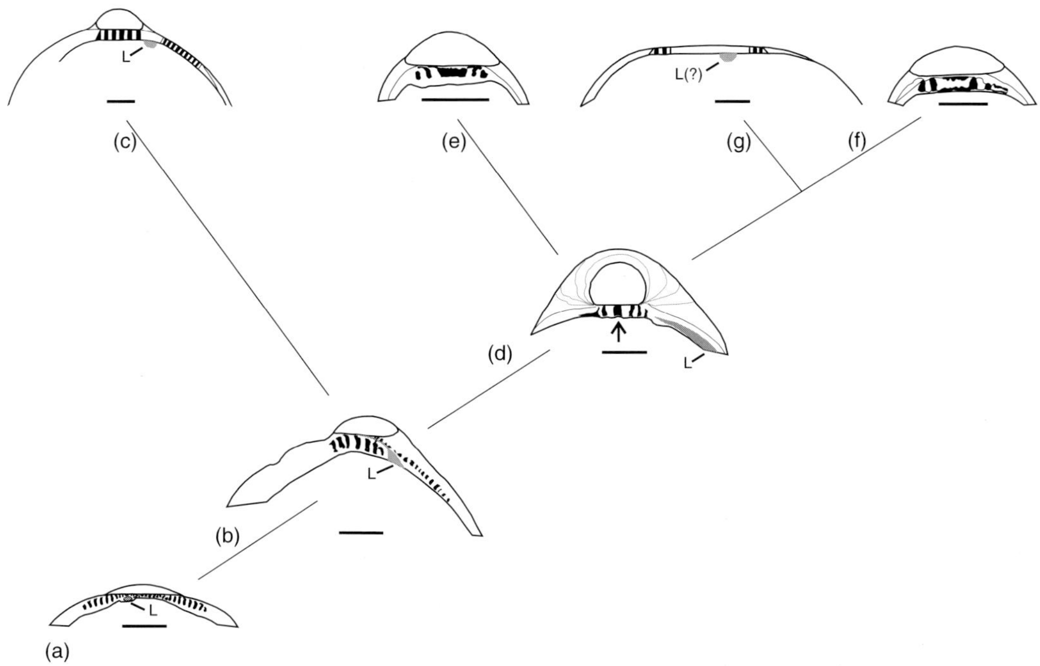

Fig. 4. Hypothesis of hinge evolution within the Ostreoidea. All schematic drawings represent LV; scale bars, 50 μm. L, Ligament. (**a**) Mytilid-like hinge. (**b**) Opisthogyrate shell with ancestral ostreoid hinge (e.g. exogyrinids). The step from (a) to (b) is explained in Fig. 3. (**c**) Pycnodonteinid hinge [redrawn from Ranson (1960, fig. 9)]. The central denticles are completely reduced but anterior denticles are still present as vestiges. The ligament sits between the two denticle rows. (**d**) *Cubitostrea* hinge: central and anterior denticles are completely reduced, and the posterior denticles become secondarily symmetrical with a small central socket in the LV (arrow). The ligament maintains its original position, although this is not obvious because of the loss of the anterior denticles as a reference (cf. Fig. 2h). (**e**) Crassostreinid hinge: similar to (d) but the central socket of the LV, and thus ridge in the RV, become much longer, probably due to fusion of the most central denticles. (**f**) Ostreinid hinge: similar to (d) and (e) but instead of a central socket/ridge, a central platform evolves in both valves that is longer than the central ridge of (e). Again, this enlargement is probably due to fusion of more central denticles. (**g**) *Tiostrea* hinge: this type has a longer platform than (f) and only weakly developed denticles at its distant ends [redrawn from Ranson (1960, fig. 133)]. The ligament apparently shifted away from its ancestral position.

wing first increase and then decrease in size (e.g. Fig. 1g and h) because the dorsal and ventral boundary of the posterior provincular ledge converge in front of the ligament gutter. The relatively larger size of the posterior denticles compared with the anterior ones is related to their central position with respect to the pivotal axis which emphasizes their functional importance.

The provinculum type just described is interpreted here as ancestral for the Ostreoidea, or it is at least very similar to the ancestral condition (Fig. 4a and b). Using this as a starting point, further evolution within the Ostreoidea followed two independent paths. One leads to the pycnodonteinid, the other to the ostreid hinge type. Characteristically, Pycnodonteinae (Gryphaeidae) possess a posterior wing of more or less equally sized and equidistant denticles 4–7 μm in width. Alternatively, the most central denticles of the posterior wing may be slightly larger, such as in Fig. 4b. In both cases, the originally central denticles and the most posterior denticles of the anterior wing are lost entirely, while vestiges of the anterior wing are still present [Fig. 4c; see Ranson (1960, 1967*a, b*,) for numerous living species]. Re-examination of two Eocene specimens figured in Malchus (1995, pl. 7, figs F and G) confirms the presence of anterior vestiges in these fossil representatives too (Fig. 2e and f).

The second ostreid path is characterized by the entire loss of all central and anterior denticles. Further evolutionary changes of the posterior hinge led to a secondary symmetrical reorganization of its denticles. This pseudo-symmetry was the basis for a descriptive hinge formula developed by Malchus (1995) for the Ostreidae. The oldest ostreid type currently known is that of the (predominantly) Eocene genus *Cubitostrea* (Figs 2h and 4d). Here, LV possess a central socket of similar dimension to its flanking denticles [the latter were coded 2a,p in Malchus (1995)]. This is filled by the central ridge [denticle 1 of Malchus (1995)] of the RV provinculum. The three central denticles are *c.* 6–13 μm in length. The hinge of living Crassostreinae shows principally the same arrangement, but the size of the central socket and ridge in the LV and RV range between 13 and 23 μm, respectively (Fig. 4e; Hu *et al.* 1993). Denticles 2a,p are sometimes ill-defined.

The central ridge/socket arrangement of *Cubitostrea* and the Crassostreinae is substituted by a long ridge or platform in both valves of many living Ostreinae and Lophinae (Ranson 1960, 1967*a, b*; Carriker & Palmer 1979; Waller 1981; Hu *et al.* 1993). These platforms meet (approximately) within the commissural plane and normally measure 30–70 μm in length (Fig. 4f), but they extend > 120 μm in *Tiostrea* (Fig. 4g). In *Ostrea*, the platforms have rugged surfaces (e.g. Pascual 1972; Waller 1981) which give the impression of ill-formed denticles. In addition, adjacent denticles [coded 2a and 2p in Malchus (1995)] are also sometimes incompletely separated from them. Thus, Malchus (1995) suggested that the platforms form by fusion of several central denticles. This probably applies to denticle 1 (RV) of the Crassostreinae as well, with respect to denticle 1 of *Cubitostrea*. In consequence, denticles 1, 2a and 2p, and those following, in these genera may not be strictly homologous. In *Tiostrea*, the central platform is apparently smooth and not rugged, and there are at most one or two very small denticles at each distal end (Fig. 4g), which is another consequence of extended brooding as is the lack of a pd-notch (Josefowicz & O'Foighil 1998). Most of these changes did not affect the ancestral position of the ligament anterior to the (posterior) hinge denticles. Only in *Tiostrea* and in a few other ostreid species described by Ranson (1960, 1967*a, b*) is this arrangement apparently lost, with the ligament emerging from beneath the anterior part of the platform (Fig. 4g).

The gross scheme of character evolution as outlined above is consistent with all currently available palaeontological–biological data, as well as with the most recent genetic studies of O'Foighil & Taylor (2000) and Hammer & Steiner (pers. comm.). They all indicate monophyly of the Gryphaeidae (Fig. 4c) and the Ostreidae (Fig. 4e–g), and a common ancestor for both families (Fig. 4b). However, Fig. 4 presents only one of two possible evolutionary paths within the Ostreidae. The figured version suggests the independent evolution of the crassostreinid and ostreinid hinges from the *Cubitostrea*-type hinge (Fig. 4d). Alternatively, hinge types 4g and 4f of *Tiostrea* and the Ostreinae–Lophinae, respectively, may have evolved from the *Crassostrea* type (Fig. 4e). This ambivalence occurs in all data sets. Palaeontological evidence only indicates that the *Cubitostrea*-type hinge (Fig. 4d) evolved earlier and that the (apparently extinct) genus belongs to the stem line of the Ostreinae–Lophinae and Crassostreinae; but it does not necessarily contain the stem species. Larval shell characters link *Cubitostrea* with modern Crassostreinae while the adult shell is neither typically crassostreinid nor ostreinid. For obvious reasons, the soft anatomy or genetics of extinct taxa cannot be examined. However, according to Malchus (1995), there may still exist some relict species of the *Cubitostrea* lineage. But these were not considered in the respective analyses.

The new data on oyster hinge evolution allow a reassessment of some earlier interpretations. Thus, the vestigial anterior denticles of the Pycnodonteinae and the pseudo-central denticles of all oysters were not recognized as homologues of the anterior and posterior wing, respectively, of other bivalve larval hinges. Instead, the pycnodonteinid anterior denticles were interpreted as chomata by Stenzel (1971, figs J39.1 and 2). It was therefore not understood that the larval ligament in oysters originally occupied a central or posterior position within the hinge. A posterior position, as depicted in Fig. 4a, would better explain the presence of small denticles between the P I and the ligament (Fig. 1h). Both conditions must be considered plesiomorphic as both are found in larval shells of other pteriomorphian bivalves. Irrespective of these details, the larval ligament of oysters is not a central, pointlike structure as indicated by Tanaka (1960, figs 3 and 5) and Stenzel (1971, fig. J39), a fact that had been already pointed out by Le Pennec (1978) and Waller (1981). But the respective descriptions of these two latter authors seem to insinuate that oysters lack a larval ligament or that it is not continuous with the adult ligament. Both assumptions now appear incorrect. The position of the primordial adult ligament in late prodissoconchs is identical to the position of the larval ligament between posteriorly present and anteriorly missing hinge denticles [cf. Figs 1e and g with Fig. 2h; see also Le Pennec (1978) and Carriker & Palmer (1979)]. Therefore, there is no reason to assume that the two are discontinuous. Technically, the larval oyster hinge was already well described by Bernard (1898) as a 'demivinculum', although he was probably unaware of all of the connotations of this term.

Conclusions

Firstly, earlier studies strongly suggest that the pd-notch and its growth track are symplesiomorphic for the Gryphaeidae and the Ostreidae (Malchus 1995). Here, this combined character is used to identify some larval shells as LV of oysters; therefore, these shells present no further direct evidence for the hypothesis. However, some additional observations, i.e. the wavy commissure, the lack of a notch in co-occurring, morphologically very similar prodissoconchs with post-metamorphic nacro-prismatic shells, and its lack in all other bivalves from the same sample (Malchus 1998, 2000*a*, *b*) are at least compatible with the hypothesis. Secondly, the observed P I sizes and P I:P II ratios of the Middle Jurassic oysters confirm the view of planktotrophy as being the ancestral reproductive mode within the Ostreoidea. Thirdly, the ancestral oyster hinge is a natural derivative from a mytilid-like provinculum. The transformation is thought to be mainly due to mechanical constraints that are related to the helico-spiral growth of the larval shell. The same constraints should act on all bivalves with a strong spiral growth component, be it the larval, nepionic or adult shell stage, and independent of an anterior or posterior growth direction. Thus, the resulting modifications outlined above may also help to understand evolutionary changes of other bivalve hinges. Fourthly, the hypothesized polarities in the evolution of larval characters are consistent with phylogeny hypotheses based on palaeontological–biological data and results from genetic studies of O'Foighil & Taylor (2000) and Hammer & Steiner (pers. comm.). This fact reciprocally supports the independently derived conclusions and emphasizes the utility of larval shell characters for phylogenetic inferences and polarity decisions in studies of character evolution.

I am indepted to M. Machalski (Inst. Paleobiologii PAN, Warszawa) for his logistic support and company on a field trip to Jurassic outcrops in north-central Poland. The final manuscript benefited from the critical comments of P. Skelton (Open University) and an unknown reviewer. This study was supported by the Deutsche Forschungsgemeinschaft (DFG; research grant MA 1259/3) which is gratefully acknowledged.

References

ANDREWS, J. D. 1979. Pelecypoda: Ostreidae. *In*: GIESE, A. C. & PEARSE, J. S. (eds) *Reproduction of marine invertebrates, V*. Academic Press, New York, 293–341.

BERKMAN, P. A., WALLER, T. R. & ALEXANDER, S. P. 1991. Unprotected larval development in the Antarctic scallop *Adamussium colbecki* (Mollusca: Bivalvia: Pectinidae). *Antarctic Science*, **3**(2), 151–157.

BERNARD, F. 1896. Troisième note sur le développement et la morphologie de la coquille chez les lamellibranches (Anisomyaires). *Bulletin de la Societé géologique de France*, **24**(3), 412–449.

—— 1898. Recherches ontogenetiques et morphologiques sur la coquille des lamellibranches, 1 partie: taxodontes et anisomyaires. *Annales des Sciences naturelles, Zoologie et Paléontologie*, **8**, 1–208.

BUROKER, N. E. 1985. Evolutionary patterns in the family Ostreidae: Larviparity vs. oviparity. *Journal of Experimental Marine Biology and Ecology*, **90**, 233–247.

CARRIKER, M. R. & PALMER, R. E. 1979. Ultrastructural morphogenesis of prodissoconch and early dissoconch valves of the oyster *Crassostrea virginica*. *Proceedings of the National Shellfisheries Association*, **69**, 103–128.

CHANLEY, P. E. & DINAMANI, D. P. 1980. Comparative descriptions of some oyster larvae from New Zealand and Chile, and a description of a new genus of oyster, *Tiostrea*. *New Zealand Journal of Marine and Freshwater Research*, **14**(2), 103–120.

GRÜNDEL, J. 1997. Zur Kenntnis einiger Gastropoden-Gattungen aus dem französischen Jura und allgemeine Bemerkungen zur Gastropodenfauna aus dem Dogger Mittel- und Westeuropas. *Berliner geowissenschaftliche Abhandlungen, Reihe E*, **25**, 69–129.

—— 1999. Truncatelloidea (Littorinimorpha, Gastropoda) aus dem Lias und Dogger Deutschlands und Nordpolens. *Berliner geowissenschaftliche Abhandlungen, Reihe E*, **30**, 89–119.

HARRY, H. W. 1985. Synopsis of the supraspecific classification of living oysters (Bivalvia: Gryphaeidae and Ostreidae). *Veliger*, **28**(2), 121–158.

HU, Y., FULLER, S. C., CASTAGNA, M., VRIJENHOEK, R. C. & LUTZ, R. A. 1993. Shell morphology and identification of early life history stages of congeneric species of *Crassostrea* and *Ostrea*. *Journal of the Marine Biological Association of the UK*, **73**, 471–496.

JOZEFOWICZ, C. J. & O'FOIGHIL, D. 1998. Phylogenetic analysis of southern hemisphere flat oysters based on partial mitochondrial 16S rDNA gene sequences. *Molecular Phylogenetics and Evolution*, **10**, 426–435.

LE PENNEC, M. 1978. *Génèse de la coquille larvaire et postlarvaire chez divers bivalves marins*. PhD Thesis, Brest (France).

MALCHUS, N. 1990. Revision der Kreide-Austern (Bivalvia: Pteriomorphia) Ägyptens (Biostratigraphie, Systematik). *Berliner geowissenschaftliche Abhandlungen, A*, **125**, 1–231.

—— 1995. Larval shells of Tertiary *Cubitostrea* Sacco, 1897, with a review of larval shell characters in the subfamilies Ostreinae and Crassostreinae (Ostreoidea, Bivalvia). *Bulletin de l'Institut royal des Sciences naturelles de Belgique*, **65**, 187–239.

—— 1998. Comparing early ontogeny shells of phylogenetically related Jurassic and Recent bivalves: a step toward taxonomic and paleoenvironmental interpretations. *In*: BIELER, R. & MIKKELSEN, P. M. (eds) *World Congress of Malacology, Washington DC, 1998*, Unitas Malacologica, Abstracts, 203.

—— 1999. Identification of larval bivalve shells by means of simple statistics. *Berliner geowissenschaftliche Abhandlungen, Reihe E*, **30**, 147–160.

—— 2000*a*. Post-larval and larval shells of *Juranomia* Fürsich and Werner, 1989, and *Anomia* Linnaeus, 1758 (Anomiidae, Bivalvia). *Paläontologische Zeitschrift*, in press.

—— 2000*b*. Larval shells of Middle Jurassic Oxytomidae (Bivalvia: Monotoidea) from Poland. *Journal of Molluscan Studies*, **66**(2), 289–292.

O'FOIGHIL, D. & TAYLOR, D. J. 2000. Evolution of parental care and ovulation behavior in oysters. *Molecular Phylogenetics and Evolution*, in press.

PALMER, C. P. 1989. Larval shells of four Jurassic bivalve molluscs. *Bulletin of the British Museum of Natural History (Geology)*, **45**(1), 57–69.

PASCUAL, E. 1972. Estudio de las conchas larvarias de *Ostrea stentina*, Payr. y *Ostrea edulis* L. *Investigación Pesqueras*, **36**(2), 297–310.

RANSON, G. 1960. Les prodissoconques (coqilles larvaires) des ostréidés vivants. *Bulletin de l'Institut océanographique de Monaco*, **1183**, 1–41.

—— 1967*a*. Les espèces d'huîtres vivant actuellement dans le monde, définies par leurs coquilles larvaires ou prodissoconques. Etude des collections de quelques – uns des grands musées naturelles. *Revue des traveaux de l'Institut des pêches maritimes*, **31**(2), 127–199.

—— 1967*b*. Les espèces d'huîtres vivant actuellement dans le monde, définies par leurs coquilles larvaires ou prodissoconques. Etude des collections de quelques – uns des grands musées naturelles, part 2. *Revue des traveaux de l'Institut des pêches maritimes*, **31**(3), 205–274.

STENZEL, H. B. 1971. Oysters. *In*: MOORE, R. C. (ed.) *Treatise on Invertebrate Paleontology. Part N. Bivalvia, Volume 3*. Geological Society of America and University of Kansas, N953–N1224.

TANAKA, Y. 1960. Identification of larva of *Saxostrea echinata* (Quoy & Gaimard). *Venus*, **21**, 32–38.

WALLER, T. R. 1981. Functional morphology and development of veliger larvae of the European oyster, *Ostrea edulis* Linné. *Smithsonian Contributions to Zoology*, **328**, 1–70.

Morphodynamics of *Bryopa* and the evolution of clavagellids

ENRICO SAVAZZI

Hagelgränd 8, 75646 Uppsala, Sweden, (e-mail: enrico.savazzi@usa.net)

Abstract: Typical clavagellids are either tube dwellers in soft sediments or facultative semiendolithic borers. *Bryopa* deviates from this pattern by being fully and obligatorily endolithic. The left valve in *Bryopa* is permanently attached to the wall of the borehole. In spite of this, the bivalve migrates forwards within the substrate throughout growth. These seemingly incompatible feats are achieved by continuously elongating the shell in the anterior direction and sliding forwards the soft parts, hinge and right valve within the shell. While the left valve becomes elongated, the posterior region of the right valve is continuously destroyed by resorption. The resulting strongly inequivalve condition is unique among clavagellids, as well as endolithic bivalves.

In the lack of direct evidence, the evolution of clavagellids from less specialized stocks remains open to alternative hypotheses. Probable parallel evolution within the clavagellid stock further complicates the problem. Evolution of tube-dwelling clavagellids directly from burrowing ancestors is as likely as their evolution from a hypothetical, endolithic protoclavagellid. In the latter case, however, *B. lata* is probably too specialized to reflect the adaptations of such an ancestor.

The superfamily Clavagelloidea d'Orbigny, 1844 belongs in the Subclass Anomalodesmata (Dall, 1889), Order Pholadomyoida Newell, 1965 (see Vokes 1967; Keen & Smith 1969). Clavagellid bivalves are remarkable in several respects. In adult individuals, one or both valves are fused to a calcareous envelope, called a crypt (Savazzi 1982), which surrounds the organism. A posterior opening in the crypt allows exposure of the siphonal tips. The anterior region of the crypt typically bears numerous openended tubules, often arranged in distinctive patterns. Softbottom clavagellids are believed to burrow by employing a hydraulic mechanism without external moving parts (Savazzi 1999, and refs cited therein). The mode of life and growth mechanisms of Recent clavagellids have attracted the attention of several modern authors, including Purchon (1956, 1960), Smith (1978), Savazzi (1982) and Morton (1984*a*, *b*). The genus *Bryopa* deviates from this pattern in being endolithic in calcareous substrates and in cementing its left valve to the substrate.

The Cretaceous clavagellid *Ascaulocardium armatum* (Morton, 1833) possesses such bizarre morphological specializations that Pojeta & Sohl (1987, p. 1) called it 'the ultimate variation on the bivalve paradigm'. The present paper deals with *Bryopa*, which displays extreme morphological adaptations to a life habit diametrically opposite, in several respects, to that of *Ascaulocardium*.

Materials and methods

Collection of the specimens

Bryopa lata (Broderip 1834) is a Recent species. However, the present author was unable to collect living specimens. This is due to a combination of factors: endolithic habits, scarcity of specimens, and a siphonal opening of the borehole that is usually located on the underside of coral boulders and is easy to confuse with that of other bivalves. In addition, collecting or breaking living corals in the study area is prohibited. On the other hand, dead specimens of *B. lata* can be collected with relative ease (see below). For these reasons, an actuo-palaeontological approach was followed.

Along the eastern coast of central Cebu Island, the Philippines, boulders are routinely dredged from intertidal and shallow subtidal flats and used to construct dry piers. This provides easy access to coral material and to their endolithic molluscs. Corals and their endolithic hosts range from well-preserved and devoid of epibionts to heavily encrusted and, in a few instances, subfossil. In most cases, the coral had been dead and exposed to seawater for an extended time, and several generations of borers and nestlers were present.

Coral boulders from dry piers were fragmented with a sledgehammer. Most of the collected clavagellid boreholes were partly exposed on old fracture surfaces. A quantitative measure of the

From: Harper, E. M., Taylor, J. D. & Crame, J. A. (eds) *The Evolutionary Biology of the Bivalvia*. Geological Society, London, Special Publications, **177**, 313–327. 1-86239-076-2/00/$15.00

frequency of endolithic clavagellids in the screened material is not available. However, their rarity is indicated by the fact that only six individuals were collected in this way, together with hundreds of specimens of other endolithic bivalves.

Taxonomy

In spite of revisions of the Clavagellidae by Smith (1962*a*) and Smith (1976), the taxonomy of *Bryopa* at the species level is still very confused. Several of the early authors [especially Broderip (1834), Gray (1858) and Lamy (1923)] described numerous species so summarily that they are impossible to characterize, although they can be recognized as probably belonging to *Bryopa*. Most of the literature lacks illustrations of *Bryopa*, or provides inadequate figures.

Majima (1994) identified three right valves from the Recent of Okinawa as *Clavagella (Bryopa) lata* Broderip. This paper provides probably the best illustration of this species available in the literature. Although the material described in the present paper does not agree in all respects with these illustrations, and Majima had only right valves available, there can be little doubt that the specimens from the Philippines available for the present paper (see below) belong to the same species, or to a closely related one. Since the shells of clavagellids are notoriously variable in size and outline, the Recent material available for this study is referred to *B. lata*.

While *Bryopa* is generally regarded as a subgenus of *Clavagella* (above references), in this paper it is treated as a genus, since its morphological characters and life habits (see below) are conspicuously different.

Materials

A single, dead and fragmentary specimen of *Bryopa lata* (specimen n. 1, Figs 5c and d) was extracted from a dead coral boulder at a depth of *c.* 0.5 m below the low watermark along the edge of a bank of coral rubble in front of Ibo, on the northwestern coast of Mactan Island, the Philippines (10° 20′30″ N, 123° 59′0″ E).

Six individuals of *Bryopa lata* (Figs 2–5a and b) were collected from dry piers (see above) near Tayud, on the eastern coast of Cebu Island, the Philippines (10° 21′30″N, 123° 59′0″ E). Specimen n. 2 is a cluster of three individuals, with two of the right valves present (Figs 2b, d and f, 3a and d, and 4c and d). Specimen n. 3 is an almost complete crypt and shell (Figs 2a, c and e, 3b, c and g). Specimen n. 4 is a fragmentary crypt and left valve (Fig. 3e and f). Specimen n. 5 is a small, isolated right valve (Fig. 5a and b).

A specimen of *Bryopa*, hereafter referred to as *B.* sp. 1, from Tat Tomna, Malta (Depiru Bed, Upper Coral Limestone, Miocene) was provided by Richard Bromley (Fig. 6).

Unless otherwise indicated in the figure captions, all illustrated material is in the possession of the author.

Preparation of the specimens

After photography, boreholes of the above specimens were filled with slowsetting silicone rubber under vacuum, in order to obtain artificial moulds. Since the substrate is highly porous, it proved impossible to mechanically extract the hardened rubber from the boreholes and the substrate had to be etched away with diluted acetic acid.

Substrate preferences

Bryopa is a borer in calcareous substrates (references below). A Recent specimen from the Red Sea, illustrated by Soliman (1971, fig. 1a), has the siphonal canal surrounded by living coral, but this is most likely due to the coral overgrowing a dead substrate. Other specimens of the same species were found in rock (Soliman 1971). *Bryopa* sp. 1 from the Miocene of Malta (Fig. 6) was a rock borer in biogenic calcirudite. Illustrations and descriptions of other species of *Bryopa* in the literature (below) confirm that this habit is typical of the taxon.

None of the available specimens of *B. lata* shows overgrowth by the coral around the siphonal opening of the crypt. This suggests boring of a dead substrate. In addition, *B. lata* is associated with a rich assemblage of boring and nestling organisms. Among these, the mytilid *Lithophaga* is especially useful for inferring the life habits of *Bryopa*. Several studies (e.g. Gohar & Soliman 1963; Kleemann 1980 and refs cited therein) agree that certain species of *Lithophaga* are always associated with living coral, while other species are constantly found in dead coral or other calcareous substrates. No species is known to occur in both living and dead coral.

The specimen of *B. lata* in Fig. 2e has the anteriormost region of the crypt very close to a large borehole containing a shell of *Lithophaga teres* (Philippi). This species, as observed by the present author in Ibo (Mactan Island, the Philippines), is a borer of rock and dead coral. The thin diaphragm of substrate separating the two boreholes (Fig. 2e) is riddled with small borings, probably produced by an endolithic sponge. The proximity of an empty *Lithophaga* borehole allowed this sponge to live because the surface of the boulder was too far away to allow direct communication with the open water and no other cavities are present in the surrounding substrate. The wall of the *Bryopa* borehole in proximity to the sponge borings is sealed by a calcareous lining deposited by the bivalve, most likely as a reaction to the presence of the sponge. This shows that the sponge could not depend on an empty *Bryopa* borehole for access to water. Therefore, the sponge, together with the *Lithophaga* borehole on which the sponge depended for respiration, antedates the arrival of *Bryopa* in the same region of substrate, proving that it did not bore in a living coral.

In conclusion, evidence shows that *Bryopa* bores exclusively in calcareous rock or dead coral.

Morphology, growth and adaptations

Shell morphology

The left valve of *Bryopa* is cemented to the substrate by its entire outer surface. The left valve is elongate, with the body cavity constituting half to two-thirds of the valve length. The posterior region of the inner surface of the left valve is a raised platform carrying coarse growth lines and irregular lines running in the antero-posterior direction (Figs 2b, d and f, 3 a and c, and 5d). This platform is referred to as the siphonal platform. Two indistinct, shallow depressions sometimes run in the antero-posterior direction along the surface of the platform, in correspondence of the inhalant and exhalant siphons. The distal portion of the platform often appears to have been secondarily eroded in order to allow the siphonal portion of the borehole to be enlarged for accommodating the growth of the siphons (Figs 2b and d).

The outer surface of the left valve, as visible in external shell moulds and fragments detached from the substrate (Figs 5c and 6c), carries coarse growth lines.

The right valve (Figs 4, 5a and b, and 6a and b) is much shorter than the left one, and is roughly of the same size and shape as the body cavity. The margin of the posterior region of this valve (Figs 4, and 5a and b) has a variable geometry and lacks a siphonal platform. The outer surface (Figs 4a and c, 5a, and 6d) carries coarse growth lines, comparable with those of the left valve.

The right valve does not form a closely fitting commissure against the left one. In fact, there is a broad, well-defined and permanent antero-ventral gape and an ill-defined siphonal or posteroventral gape.

The hinge is devoid of teeth and laterally asymmetrical, with the ligament inserted onto a platform projecting from the left valve across the medial plane (Fig. 2a and e). This platform causes the ligament plane to form an angle of 45–60° with the commissural plane. The hinge platform is too poorly differentiated from the rest of the shell to be characterized as a ligamental nymph. A similar arrangement occurs in at least some species of *Clavagella (Stirpulina)* (Fig. 1d; note that this is an internal mould of the crypt and left valve). The ligament in *B. lata* is located slightly anteriorly to the centre of the body cavity.

The musculature is dimyarian and strongly anisomyarian, with the posterior adductor scar from four to eight times larger than the anterior one. The anterior muscle scar is located near the anterior shell margin, while the posterior is just behind the hinge and roughly at the centre of the anteroposterior axis. The muscle scars in *B.* sp. 1 (Fig. 6a and b) are deeply incised, while in *B. lata* they are lightly incised or flush with the internal valve surface.

Both muscle scars are close to the dorsal shell margin and cannot therefore exert a strong leverage on the shell. Since both adductors are located below the hinge line, a diductor function comparable to that of pholadids (e.g. see Röder 1977) is not feasible. The function of the laterally asymmetrical ligament is not clear. This arrangement displaces the hinge line in the dorsal direction and therefore allows the adductor muscles to insert onto the shell in a rather dorsal position, and still retain an adducting function. In turn, this frees most of the shell interior for the placement of other organs.

Bryopa differs from the large majority of bivalves (including all other clavagellids) in that its ligament is not located near the umbo. In fact, the adult *Bryopa* has no umbo. The apical portion of the left (cemented) valve is usually eroded (see above).

In the right valve, a quick comparison of the external and internal surfaces (and, in particular, of the growth lines on the outer surface with the placement of internal features) shows that up to half of the original shell surface, including the apical and juvenile portions, is missing (Figs 4a and c, and 5a).

This can be demonstrated more rigorously by a simple geometric construct (Fig. 7a), in which homologous points (in Fig. 7a, the anteriormost and the ventralmost points along the commissure) on different growth lines are connected by straight lines. This reconstruction is based on the assumptions that the right valve grows approximately isometrically (as shown by the left valve in Fig. 6c) and that shell curvature does not introduce an appreciable error (which is reasonable, because curvature in the antero-posterior direction is only slight). Under these conditions, homologous points are aligned along a straight line passing through the umbo. Thus, the intersection of two lines passing through two separate sets of homologous points indicates the original position of the umbo. The rightmost portion of the illustrated specimen is not used because of its obvious deviation from isometric growth (perhaps due to gerontic effects).

When the above construct is made, it becomes evident that the posteroventral region of the commissure of the right valve is the site of secondary shell resorption (Fig. 7b), while it is the site of shell secretion in the opposite valve (i.e. the posterior platform). As a result of this process, the hinge and ligament in *Bryopa* continually migrate in the anterior direction along the dorsal shell margin. Thus, the ligament in *Bryopa* is strongly prosodetic with reference to the umbones. The ligament in all other clavagellids (as well as in the large majority of bivalves) is opisthodetic. A small amount of shell resorption also takes place in the left valve, posterior to the ligamental projection.

Illustrations of the left valve of *Bryopa* in the literature (e.g. Soliman 1971; Majima 1994) show clear evidence of secondary resorption of early shell portions (since the apex and juvenile region are missing, as described above). Apparently, however, the occurrence of secondary shell resorption escaped the attention of these authors.

Crypt morphology

The anterior region of the borehole extends well beyond the reach of the right valve (Figs 2a, and 3e and f). In one specimen (Fig. 6), the right valve became disarticulated after the death of the organism and moved to the anterior portion of the borehole (as shown by the offset between right and left muscle scars in Fig. 6b). This gives the false impression of a right valve reaching the bottom of the borehole. In turn, this feature implies that, at least in the anterior region of the borehole, the boring process was carried out by a mechanism other than shell abrasion.

In the material available to the present author, there are no tubules in the anterior region of the crypt (Fig. 3e and f). On the contrary, one specimen (Fig. 2a and e; see also above) exhibits a smooth and thick calcareous lining deposited on the anteriormost region of the borehole. There is no reason to suppose that all the specimens available for this study [as well as all specimens described by Soliman (1971)] are juveniles [this is made further unlikely by the fact that juvenile clavagellids are unknown in the literature, with a single exception: Smith (1910) and Morton (1984*a*)]. It is not possible to decide whether the specimen in Fig. 5a and b is a juvenile or a small adult. Thus, the lack of tubules and the facultative lining appear to be normal adult features in *B. lata*. The single available specimen of *B.* sp. 1 is also devoid of tubules. A few other species, placed in *Bryopa* and described in the literature as possessing tubules, but not available for direct inspection, are discussed below.

Unlike other species of *Bryopa* and other clavagellid genera, *B. lata* (at least in the material available in this study) builds no siphonal chimney projecting above the surface of the substrate. However, it deposits facultatively a thin calcareous lining in the siphonal region of the borehole (Fig. 3g). It can also seal the boreholes of other endolithic organisms encountered while boring by building irregular calcareous diaphragms (arrow in Fig. 2f). Judging from the haphazard aspect of these diaphragms, which contrast with the smooth and regular linings deposited in direct contact with the substrate (Fig. 2a and e), it appears that *B. lata* can exert only poor control over the shape of a lining not supported by the substrate. This suggests that the capability of building a free or semi-endolithic crypt comparable to that of typical clavagellids is lost in this species.

In *B. lata*, the secretion of a lining onto the walls of the borehole seems to take place only in the presence of larger cavities in the substrate. The natural porosity of the coral substrate does not trigger the formation of a lining. The wall of the borehole facing the right valve bears a relief matching the valve sculpture (Fig. 2c – the matching valve is shown in Fig. 4a). Repaired damage to the shell margin (Fig. 4a and c) suggests a mechanical boring process. Growth of the right valve in the anterior direction, coupled with resorption of its posterior region (see above), gradually renews the outer valve surface and results in less mechanical wear on its outer surface than could otherwise be expected. On the other hand, the anterior region of the crypt cannot be reached by the right valve (in particular, see Fig. 3e and f), indicating an exclusively chemical boring mechanism, at least in this region. A combined chemical–mechanical action in *Bryopa* is therefore likely.

The available material of *B. lata* contains a cluster of at least three boreholes (Fig. 3a and d). Coupled with the overall scarcity of individuals, this suggests gregarious habits. The material described by Soliman (1971) strengthens this assertion. These habits may be compared with gregariousness in the coral-boring coralliophilid gastropod *Leptoconchus* (Savazzi 1999, and refs cited therein). The boreholes in these specimens are conspicuously bent, implying that *B. lata* is capable of steering the direction of the borehole during growth. This character is widespread in boring bivalves (Savazzi 1999, and refs cited therein) and adaptive in avoiding the boreholes of other endolithic organisms. In a gregarious species, it relieves crowding of the substrate.

Growth

Boring a calcareous substrate, either by chemical or mechanical means, is not a fast process and, as a consequence, boring molluscs tend to grow slowly. As a result of the shift to an endolithic habitat, *B. lata* was forced to abandon the growth process of typical clavagellids – rapid juvenile growth, followed by construction of the crypt and an extended adult stage without further changes in crypt size (except for secondary growth of its siphonal extremity) (Savazzi 1999 and refs cited therein).

The presence of a non-growing adult stage in *B. lata* cannot be excluded. Indeed, other endolithic bivalves (especially pholadids) are characterized by a non-growing adult stage (Savazzi 1999, and refs cited therein). Unlike tube-dwelling clavagellids, this stage cannot be achieved by 'hurrying' through a non-optimal adaptive stage devoid of a crypt. On the other hand, the endolithic environment provides essentially the same protection to juvenile and adult *B. lata* alike, so fast juvenile growth is not essential. Therefore, *B. lata* was able to revert to the

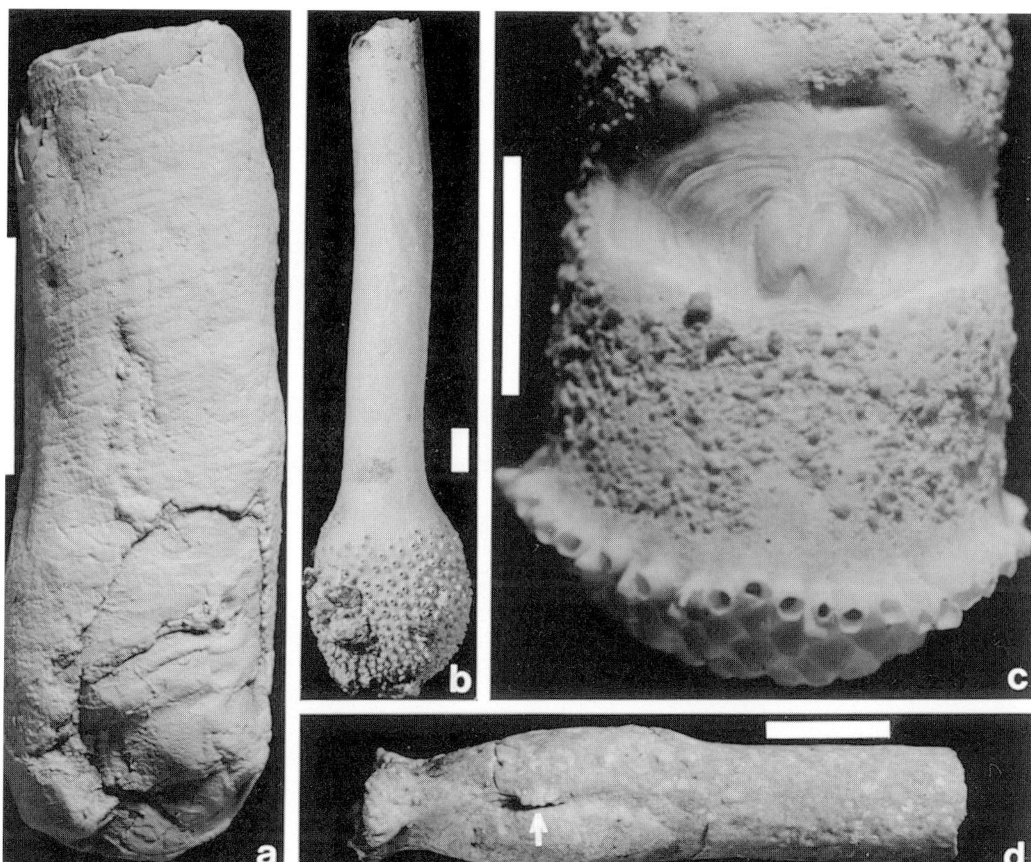

Fig. 1. (**a**) *Clavagella (Stirpulina)* cf. *oblita* Michelotti, Upper Eocene, Costalunga near Possagno, Treviso province, Italy, (Institute of Geology, University of Padova, Italy, n. 26126). The crypt is broken in the anterior region, exposing the right valve. (**b**) *Brechites (Foegia) veitchi* Smith, Recent, Tropical Pacific (Institute of Geology and Palaeontology, University of Tübingen, Germany, n. 1563/21), ventral view of crypt. (**c**) *Humphreyia* (*Nipponoclava*) *gigantea* (Sowerby), Recent, Red Sea (Swedish Museum of Natural History, Stockholm, Sweden), anterior region of crypt with valves. (**d**) *Clavagella (Stirpulina) veronensis* Savazzi, holotype, Middle Eocene, Negrar, Verona province, Italy (Institute of Geology, University of Padova, Italy, n. 26124), internal mould of crypt. Arrow points to the projecting ligamental process of the left valve (lowermost), visible as a cavity in the umbonal region of the mould (compare with Fig. 3c). Scale bars, 10 mm.

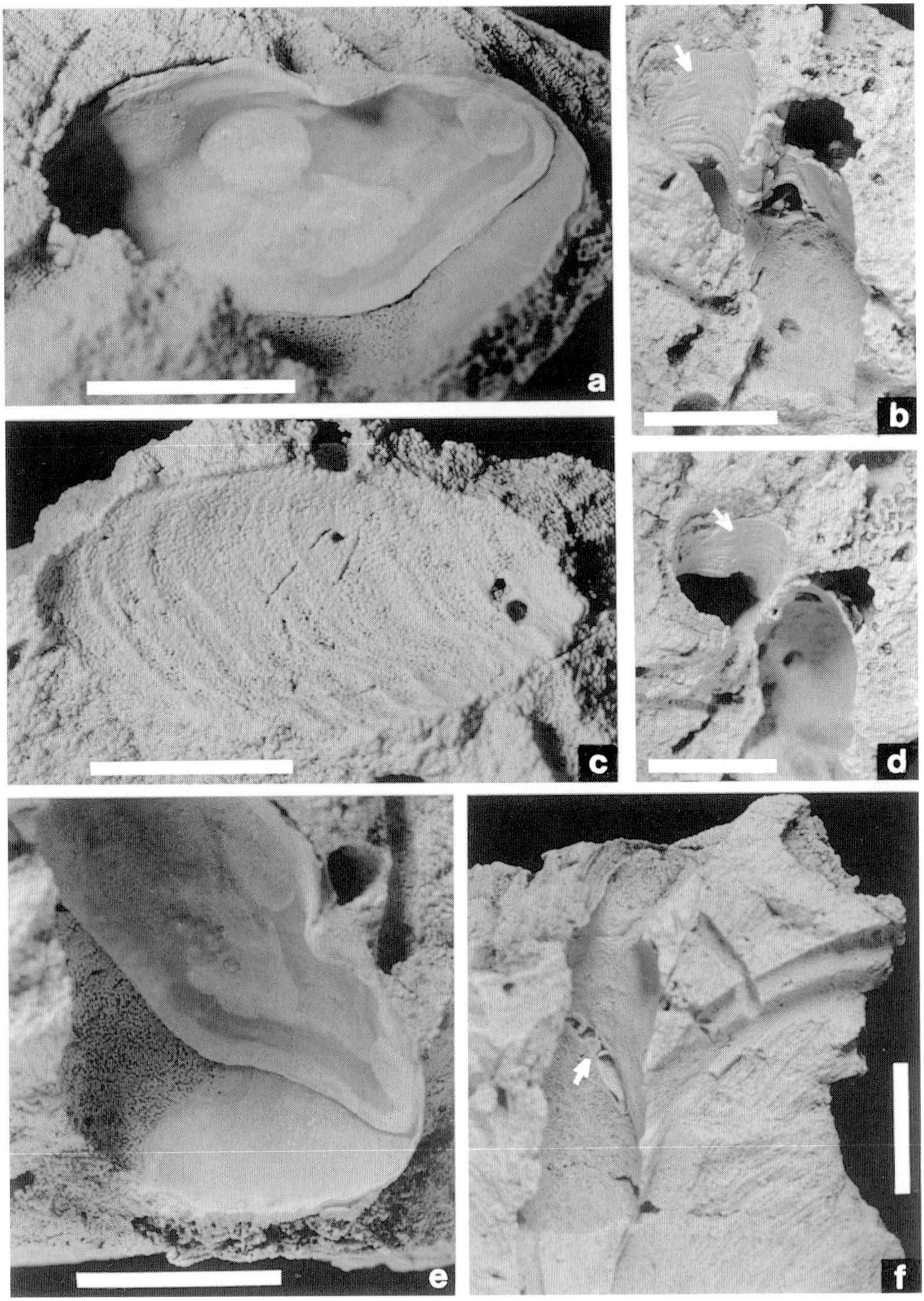

Fig. 2. *Bryopa lata* (Broderip), Recent, Tayud, Cebu Island, the Philippines. (**a**) Interior of left valve of specimen 3. (**b**) and (**d**) Oblique views of siphonal regions of specimen 2. The siphonal platform (arrows) is evident. (**c**) Borehole wall facing right valve of specimen 3 (the right valve is shown in Fig. 4a). (**e**) Anterior region of borehole of specimen 3, showing a thick calcareous lining. (**f**) Interior of borehole of specimen 2. Arrow points to irregular lining deposited by the bivalve in the borehole of another endolithic organism. The siphonal region of the borehole of another individual is visible on the right. Scale bars, 5 mm.

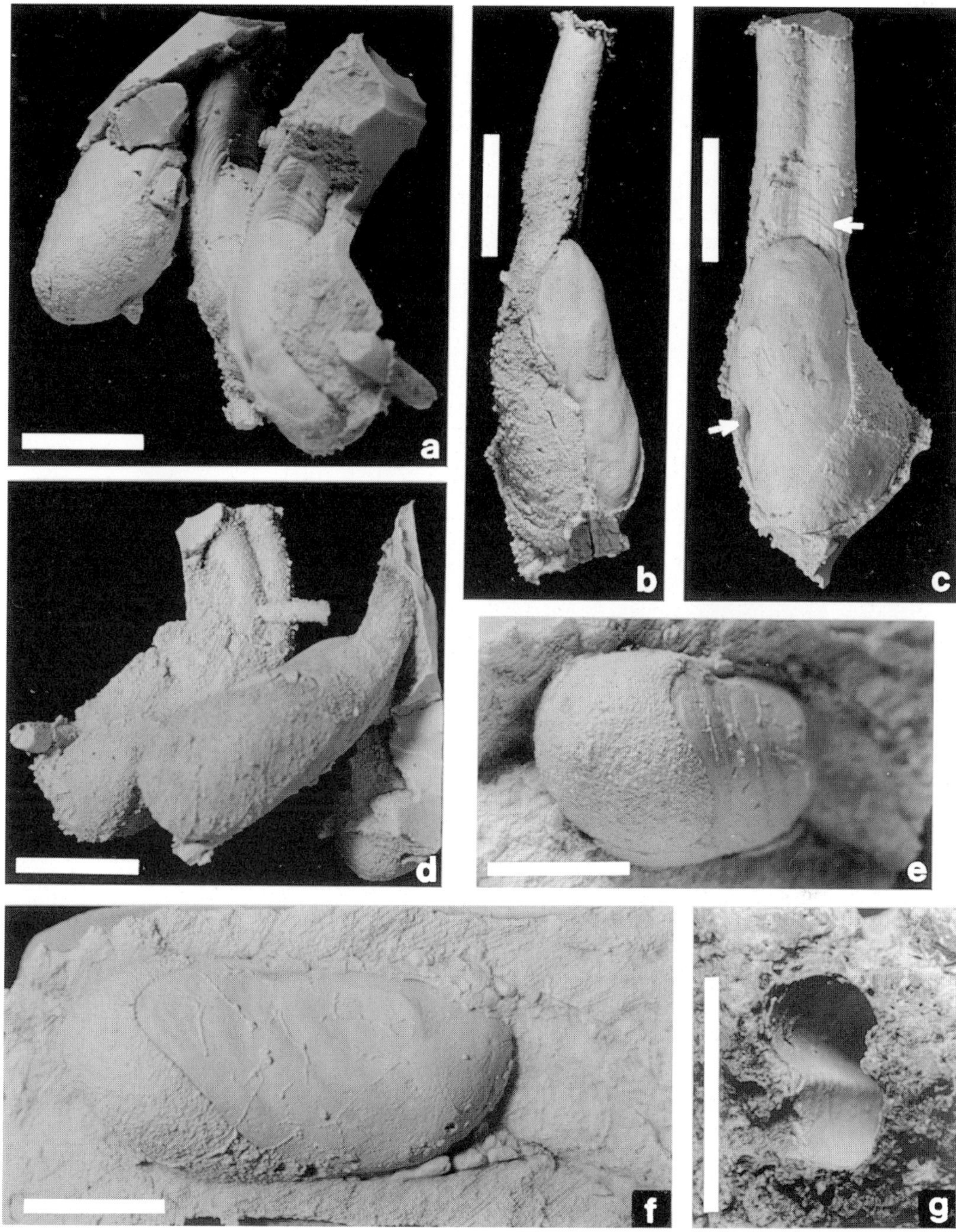

Fig. 3. *Bryopa lata* (Broderip), Recent, Tayud, Cebu Island, the Philippines. (**a**) and (**d**) Artificial mould of boreholes of specimen 2. (**b**) and (**c**) Artificial mould of borehole of specimen 3, in dorsal (**b**) and left lateral (**c**) views (the upper arrow points to the siphonal platform and the lower arrow to the ligamental process). (**e**) and (**f**) Oblique anterior (**e**) and left lateral (**f**) views of artificial mould of borehole of specimen 4. The shell and borehole walls are secondarily bored by other organisms. (**g**) Siphonal opening of borehole of specimen 3, showing a thin lining. Scale bars, 5 mm.

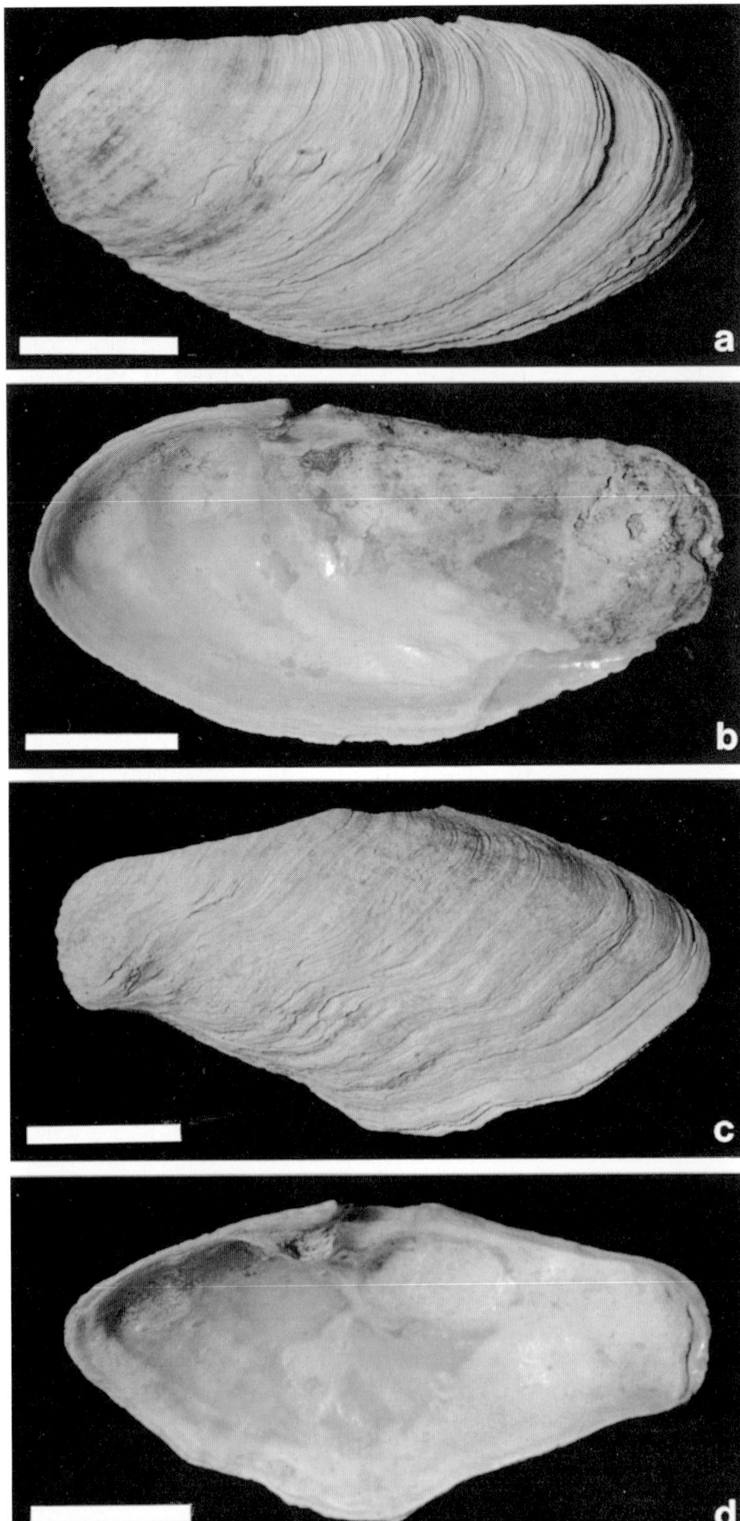

Fig. 4. Right valves of *Bryopa lata* (Broderip), Recent, Tayud, Cebu Island, the Philippines. (**a**) and (**b**) Specimen 3. (**c**) and (**d**) Specimen 2. Scale bars, 2 mm.

Fig. 5. *Bryopa lata* (Broderip), Recent, Tayud, the Philippines. (**a**) and (**b**) Cebu Island, right valve, specimen 5; (**c**) and (**d**) Ibo Island, posterior region of left valve, specimen 1. Scale bars, 2 mm.

continuous growth process (at least for the initial part of the life span) typical of bivalves. This growth process is recorded in the growth lines of *Bryopa* shells and in the ontogenetic migration of their shell cavity.

Several cemented and crevice-nestling bivalves are convergent with *Bryopa* in having a body cavity that is consistently smaller than the cemented valve and migrating during growth. Many ostreids, spondylids and chamids display these characteristics (either facultatively or as a constant feature). Several soft-bottom bivalves (e.g. *Lithiotis*, *Cochlearites*) possess similar characteristics (Chinzei 1982, and refs cited therein). In all these forms, the body cavity migrates in the ventral direction during growth and the hinge is located in a dorsal position. *Bryopa*, however, is unique in that the body cavity and hinge migrate in the anterior direction. In this genus, a migration in the ventral direction is obviously impossible, as this would prevent the siphons exiting the shell and borehole in a direction diametrically opposite to the direction of migration.

In this context, it may be pointed out that it is possible to conceive a morphogenetic process providing an alternative to resorption of the posterior region of the right valve in *Bryopa*. For instance, as in the bivalves discussed above, the right valve of *Bryopa* could follow the migration of the hinge and remain smaller than the left valve. However, this would result in two problems: (1) widely different rates of shell growth of the two valves in the area of ligamental attachment would subject the ligament to a differential stress. In other bivalves, this is largely avoided by the fact that the hinge migrates in the ventral, rather than anterior, direction and that the growth rates of the two valves along an axis perpendicular to the direction of migration are essentially the same. (2) This growth mechanism would not allow the gradual renewal of the surface of the right valve, thus making it more vulnerable to abrasion against the substrate (see above).

Morphology and evolution of the Clavagellidae

Taxonomy and stratigraphic ranges

The Clavagellidae are characterized by the construction of a calcareous envelope, or crypt, with the left, or both, valves forming part of the

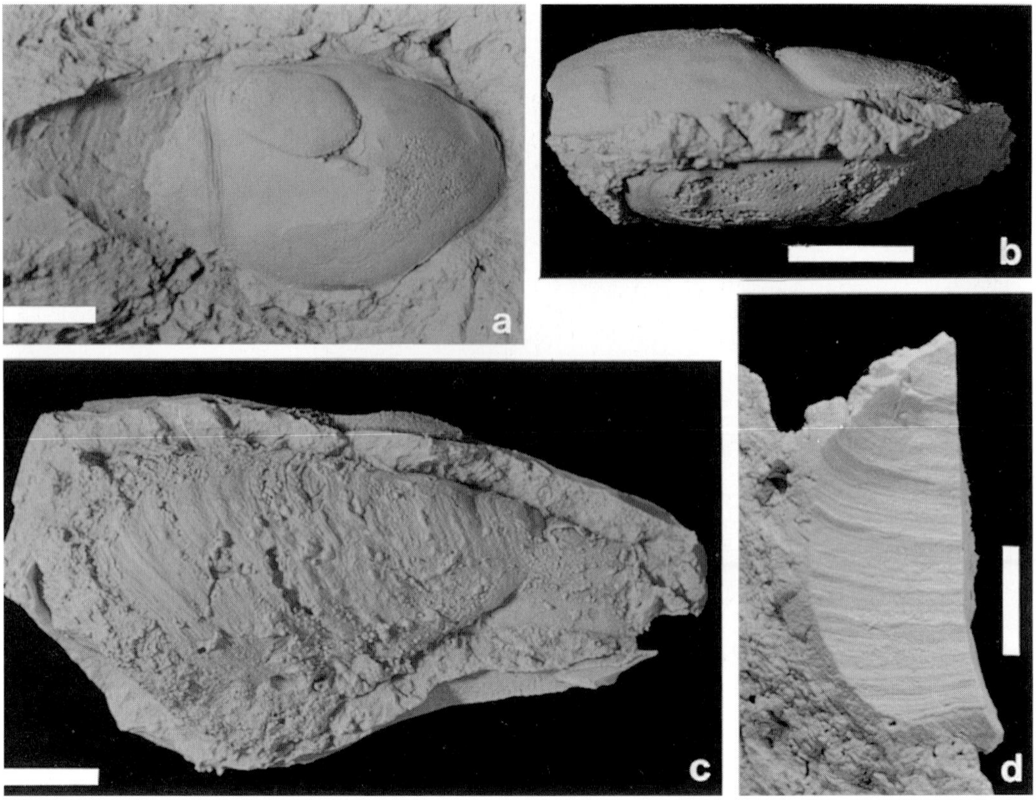

Fig. 6. *Bryopa sp. 1*, Miocene, Tat Tomna, Malta. (**a**) Composite external and internal mould, with interior of right valve visible. (**b**) Internal mould of shell in dorsal view. The offset of the muscle scars is a postmortem artefact. (**c**) Artificial replica of borehole mould showing the outer surface of the left valve, originally cemented to the substrate, and part of the siphonal canal. (**d**) Fragment of outer mould of right valve showing the shell sculpture. The borehole is filled with a fine-grained material, which stands out in contrast with the coarser material of the rock substrate. Scale bars, 5 mm (a)–(c) and 2 mm (d).

crypt which is visible on its outer surface. The posterior region of the crypt ends in a siphonal opening and the anterior region bears a number of open tubules. Three distinct morphologic groups can be recognized within this family, described below.

The *Clavagella* group (including the genera and subgenera *Clavagella*, *Stirpulina*, *Stirpuliniola*, *Parastirpulina*, *Dacosta* and *Ascaulocardium*). This group is characterized by tube dwelling as a normal habit, although *Clavagella* and *Dacosta* can be facultative semi-endolithic borers. The latter can be an extension of the juvenile habit of attaching to, or boring into, small calcareous substrates in order to stabilize the shell in soft-bottom environments. Specimens of *Clavagella* completely embedded in the substrate are rare and can be explained as a result of juveniles settling in pre-existing boreholes (e.g. Savazzi 1982). *C. mullerae* (Kilburn 1974) is described as a crevice dweller.

Common to the *Clavagella* group is the left valve being part of the crypt and visible from the outside, and the right valve articulated within the crypt (Fig. 1a and d). Shell morphology and crypt length are variable. The crypt always bears anterior tubules, either arranged into a circular fringe, or distributed uniformly on the whole anterior region. The siphonal portion of the crypt may be smooth, or bear a series of annular fringes. The shell is roughly circular to elongated in the antero-posterior direction, with the valves of roughly similar sizes.

Ascaulocardium has only four, exceptionally long, anterior tubules and a ring-shaped 'pipe' which surrounds the posterior region of the shell

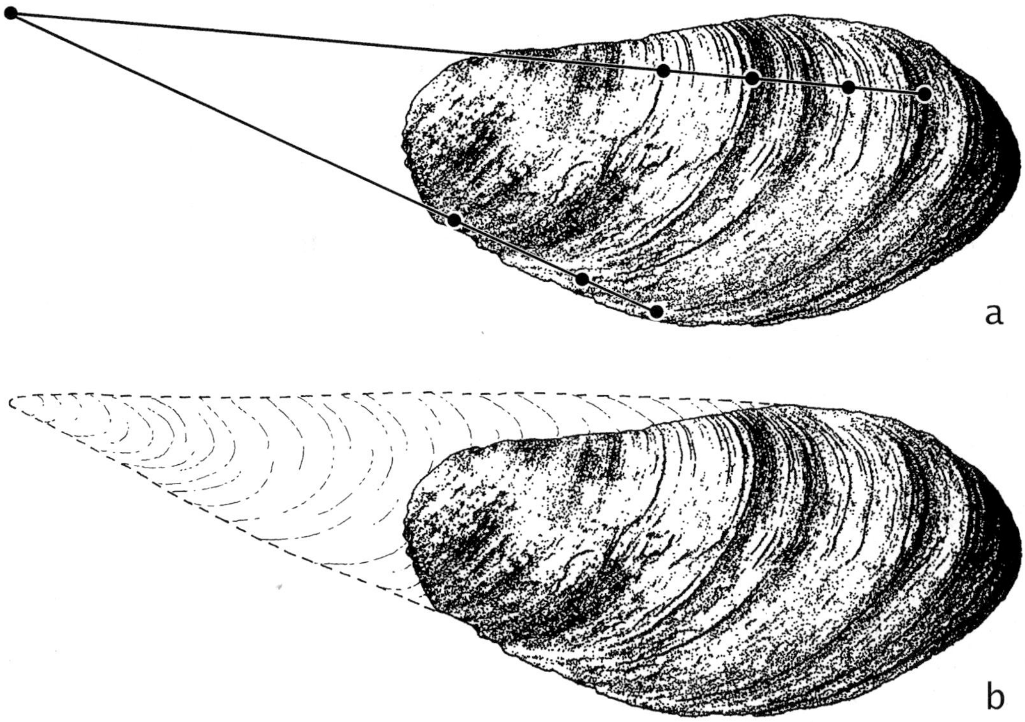

Fig. 7. (**a**) Geometric reconstruction of the original position of the umbo, using homologous points (the anteriormost and ventralmost commissure points) on different growth lines. (**b**) Schematic drawing of right valve of *Bryopa lata*, showing the extent of secondary shell resorption. The dashed outline marks the original shell margin. The solid outline marks the adult shell margin.

chamber and bears several shorter tubules (Pojeta & Sohl 1987). This pipe was apparently built by a bifurcated tentacle inserted onto the posterior region of the mantle and wrapping around the exterior surface of the crypt.

The *Clavagella* lineage probably appeared in the Upper Cretaceous of the Tethys (Smith 1962*b*), with early representatives described from the United States (including the Turonian of California; Stallwood 1995), Europe and Africa. In the Palaeocene, the lineage became restricted to Europe. In the Eocene, it migrated westwards to Florida (Pojeta & Sohl 1988; Jones & Nicol 1989), and in the Oligocene eastwards to the Indo-Pacific region, reaching Japan and the Philippines in the Miocene and Australasia in the Quaternary.

The Penicillus *group (including* Penicillus *or* Brechites, Humphreyia, Nipponoclava *and* Foegia)

This group is characterized by having both valves cemented to the crypt and visible from the outside (Fig. 1c), and by obligatory tube-dwelling habits. Members of this lineage can cement clasts (including non-calcareous materials) to the outer surface of the crypt (Fig. 1b and c) but are apparently incapable of boring.

Representatives of this lineage are typically larger than other Clavagelloidea and may reach 325 mm in crypt length. A maximum crypt length of 1 m has been repeatedly reported in the literature (usually as a quotation from earlier literature). These reports may be due to confusion with crypts of the Caenozoic teredinid *Kuphus*, which can exceed 1 m in length. For differences between *Kuphus* and clavagellids, see Savazzi (1982, 1999).

This lineage probably appeared in the Early Oligocene of the northern Tethys (Smith 1962*b*) and subsequently migrated into the Indo-Pacific region, of which it is exclusive since the Neogene.

The Bryopa *group [including the genus (see above)* Bryopa]

This group is distinguished by obligatory boring habits. There is disagreement in the literature on whether *Bryopa* is devoid of anterior tubules (Broderip 1834; Lamy 1923; Soliman, 1971; Appukuttan 1974) or whether it possesses small and short ones (e.g. Gray 1858; Smith 1962*a*; Smith 1976). However, at least some of the authors who have reported the presence of tubules may have been citing earlier literature, rather than reporting an observed feature. There may also have been some confusion between species of *Bryopa*, *Dacosta* and *Clavagella*.

Usually, *Bryopa* is described as having a secondarily elongated siphonal sheath which bears one or more reflected fringes. *B. lata*, as discussed in the present paper, and an unnamed species described by Soliman (1971) (which may well be a morph of *B. lata*) are devoid of a siphonal sheath and of tubules in the anterior portion of the crypt. *B. lata* and *B.* sp. 1 also displays a variable amount of ontogenetic migration of the soft parts in the anterior direction within the shell, accompanied by secondary resorption of the posterior slope of the right valve (see above). This also applies to specimens described in the literature, albeit to a smaller degree than the specimens available for this study.

At present, it is not possible to tell whether this group is homogeneous and the representatives that possess tubules are closely related to *B. lata* and *B.* sp. 1, or they should be moved to the *Clavagella* group. However, it is clear that the morphological and functional differences between *B. lata* and typical *Clavagella* are at least as important as those existing between the *Clavagella* and the *Penicillus* lineages.

Dacosta was listed above as a member of the *Clavagella* group. It differs from *Bryopa* in several respects, especially the semi-endolithic habit, the well-developed crypt with anterior tubules (Morton 1984*b*), and an apparent lack of migration of the soft parts within the left valve and secondary resorption of the right valve. It is similar to *Bryopa* in its anisomyarian condition. Morton (1984*a*) mentions other anatomical differences between *B. lata* and *Dacosta australis*.

On the basis of morphology alone, it is tempting to regard *Dacosta* as the member of the *Clavagella* group with the closest affinities to *Bryopa*. However, evolution of *Bryopa* from *Dacosta* is made unlikely by the stratigraphic and geographic distribution of the latter taxon, which is from the Australasian (Smith 1971) and Japanese Holocene (Majima 1994). *Dacosta* may therefore be a late offshoot from *Clavagella*, partly convergent with *Bryopa*.

Bryopa appeared in the Late Oligocene of southern Europe (Smith 1962*b*) and subsequently migrated to the tropical Indo-Pacific. It either survived the basal Neogene and Pleistocene crises in the Mediterranean, or subsequently re-entered the Mediterranean on both occasions, and its occurrence in this area is documented from the Miocene and Recent. In the Indo-Pacific, it is recorded from the Recent of several localities, principally the Red Sea, southern India, the Philippines and Japan.

Evolution within the Clavagellidae

The three groups described above appear to represent monophyletic lineages: no forms intermediate between these are known to the present author. Tubule-carrying *Bryopa* could constitute such an intermediate form, but the present author has been unable to inspect material, and literature descriptions and illustrations are insufficient to reach a decision. Each group has remained essentially constant in morphology throughout its known stratigraphic range. The disappearance of *Ascaulocardium* and *Parastirpulina* (see Pojeta & Johnson 1995) in the Upper Cretaceous, which reduced morphological diversity within the *Clavagella* group, is the only remarkable event in evolution within these lineages. The emergence of *Dacosta* is only a slight modification of the habits of *Clavagella* (s.s.).

The compactness and evolutionary conservatism within the three groups should be acknowledged in the taxonomic hierarchy, and may justify the introduction of three subfamilies. Such an idea is not new, since Lamy (1923) already did so in a short study of Recent clavagellids from the Red Sea. Modern authors (Smith 1962*a*; Smith 1971, 1976, 1978), however, did not recognize such a need.

This compactness and conservatism are an obstacle to understanding the evolution of clavagellids from an earlier stock, as well as the evolution within the three clavagellid lineages. With a lack of objective data, one is forced to formulate a hypothetical evolutionary process that must remain unverified until new material becomes available. Not surprisingly, two mutually incompatible pathways have been proposed for the evolution of the Clavagellidae (below).

Evolution of the Clavagellidae from earlier stocks

Savazzi (1982) envisaged the evolution of tube-dwelling *Clavagella* from an obligatory borer not unlike *Bryopa*. He based his hypothesis on the description of an unnamed species by Soliman (1971). Note that this evolutionary mechanism does not necessarily imply clavagellids evolved from a non-clavagellid borer (which is rather unlikely). Rather, the boring habit may have appeared early in the evolution of the Clavagellidae but after their differentiation from a pre-clavagellid stock.

According to Savazzi (1982), *Penicillus* and related genera represent a further evolutionary step, originating as an offshoot from the *Clavagella* group. Within this framework, cementation of both valves to the crypt caused the sudden divergence of this group from *Clavagella*. The process was, of necessity, a saltation, because there are no feasible intermediates. On the other hand, cementation of the left valve was already part of the clavagellid *Bauplan* and the muscular mantle could take over the function of the mobile right valve (see Savazzi 1982). Thus, all the necessary morphogenetic building blocks were available and this sudden evolutionary event therefore seems plausible. A difficulty with this evolutionary pathway is that it finds no confirmation in the fossil record, since *Bryopa* appears to be a fairly recent stock (see above).

Pojeta & Sohl (1987), on the other hand, suggested that tube-dwelling clavagellids (the *Clavagella* group in particular, being older than the others) evolved from a burrowing form morphologically similar to *Entovalva*, which envelopes the shell within the mantle, and the mantle in this genus does resemble the anterior region of a clavagellid crypt when extended. However, a difficulty with this view is that the mantle in *Entovalva* is bilaterally symmetric and encloses both valves, whereas the *Clavagella* mantle is conspicuously asymmetric [see above and Savazzi (1982, 1999)]. The recognition of such an ancestor in the fossil record is most likely precluded by the lack of preservable diagnostic characters.

An intermediate form between a 'naked' pre-clavagellid and a tube-dwelling clavagellid might have possessed a crypt only partially enveloping the organism. Such a form could conceivably be preserved, but it would be difficult to distinguish it from an incompletely grown clavagellid (incidentally, there is no fossil or Recent record of clavagellid specimens in the early phases of crypt construction). Alternatively, an intermediate form might have possessed an organic or lightly calcified crypt, possibly strengthened by agglutinated sediment particles. Such a form could stand a better chance of being recognized in the fossil record but its preservation is unlikely.

Without evidence, neither of the two competing hypotheses is completely satisfactory. In conclusion, both the appearance of the Clavagellidae and the divergence of the *Penicillus* lineage from the clavagellid stock should be regarded as unsolved problems. Also, it cannot be excluded that the *Clavagella* and *Penicillus* lineages evolved independently from pre-clavagellids, and that their similarities are only the result of convergence and similar preadaptations. On the other hand, if one accepts the fossil record as indicative of the real stratigraphic ranges of the clavagellid groups, the *Bryopa* lineage is relatively easy to explain. Facultative boring in the *Clavagella* lineage (see Savazzi 1982) became obligatory in *Bryopa*, with the consequent reduction or loss of the anterior tubules.

The two-stage growth mechanism of tube-dwelling clavagellids (i.e. rapid growth of a 'naked' stage, followed by crypt construction and no further increase in size, except for elongation of the siphonal sheath) is unsuitable in the endolithic environment, which requires the capability of compensating for substrate erosion and/or over-growth, as well as a slow growth to accommodate the necessarily slow boring process. Migration of the soft parts within the shell, coupled with secondary resorption of the right valve, provided the capability of compensating for substrate erosion by moving within the substrate. This also allowed juveniles to settle on the surface of the substrate and to become endolithic during growth, without having to rely on pre-existing cavities. The capability of building a calcareous crypt was reduced to the facultative deposition of a lining on the walls of the borehole. Construction of a siphonal sheath was retained in some species of *Bryopa* as a mechanism to fight encrustation and fouling of the substrate.

No other boring bivalve has the shell permanently cemented to the substrate. This is the factor that triggered the evolution of the unique shell morphology of *Bryopa*. The shift in habit from cemented-epifaunal to cemented-endolithic is extremely unusual (it is unknown in other bivalves and occurred just once in gastropods; unpublished observations).

The scarce and patchy fossil record of endolithic clavagellids is not unique among other such bivalves. Other examples are the arcid *Litharca* (Nicol & Jones 1986) and the pectinid *Pedum* (Savazzi 1998, and refs cited therein). Their shell morphology is highly specialized and is therefore potentially recognizable in fossils. In spite of this, both genera lack a fossil record. In both cases, this has been attributed to quick evolution from ancestors with different morphologies and life habits. While this may be true (and also apply to the sudden appearance of *Bryopa*), the fossil record for *Bryopa* shows that this genus has not substantially changed since its appearance. The extreme rarity of fossil *Bryopa* (even when compared with other clavagellids) may be due to poor preservability (their shallow boreholes are likely to be secondarily occupied and modified by other organisms, and/or to be destroyed by erosion of the substrate) coupled with individual scarcity.

Constructional morphology v. morphodynamics

In the terminology of constructional morphology (Seilacher 1970) and biological morphodynamics (Seilacher 1991), the preadaptations of *Clavagella* (juvenile cementation of the left valve to clasts; construction of a crypt; a facultative semi-endolithic habit) can be placed among phylogenetic factors. Cementation of the left valve to the substrate, originally part of the phylogenetic tradition, results in a constructional constraint that triggered the evolution of a unique morphology. The switch in the immediate surrounding environment, implicit in a habitat change from the soft bottom to the endolithic, may have constituted the main event that triggered morphological specialization in *Bryopa*.

Conclusions

A large amount of ontogenetic migration of the hinge and soft parts in the anterior direction, accompanied by extensive secondary shell resorption in the right valve, characterize *Bryopa lata* and *B.* sp. 1. Judging from literature illustrations, these characters seem to be incipient in other species of *Bryopa*. Both characters are adaptive in the context of boring in substrates subjected to surface erosion. Thus, *B. lata* seems to be adapted to an endolithic life to a higher degree than any other clavagellid. This is achieved without a substantial departure from the building plan of other species of *Bryopa*: the only 'innovative' characters of *Bryopa* are migration of the soft parts within the left valve, selective resorption of the right valve, reverting to a slower and more gradual growth process, and reduction of the crypt to a facultative lining of the borehole. This combination of adaptations results in a functioning paradox: a bivalve that cements one valve to the substrate, and yet continues to bore and move through it during growth. The stepping-stone to this complex of co-adaptations is the habit of juvenile *Clavagella* of cementing the left valve to clasts, for stabilization, on soft bottoms (Savazzi 1999, and refs cited therein).

Evolution of the Clavagellidae, and within the Clavagellidae, is at present insufficiently documented. Several key forms linking the clavagellids with their ancestors, as well as intermediate forms between the three known clavagellid lineages, still await discovery and description. In the meantime, two alternative hypotheses can be formulated on the origin and evolutionary development of this family, but there are no compelling grounds for preference of one over the other.

References

Appukuttan, K. K. 1974. Rediscovery of *Clavagella (Bryopa) lata* (Clavagellidae, Bivalvia) from the Gulf of Mannar, Southeast coast of India. *Journal of the Malacological Society of Australia*, **3**, 19–24.

Broderip, W. J. 1834. On *Clavagella*. *Proceedings of the Zoological Society of London*, **2**, 115–117.

Chinzei, K. 1982. Morphological and structural adaptations to soft substrates in the Early Jurassic monomyarians *Lithiotis* and *Cochlearites*. *Lethaia*, **15**, 179–197.

Gohar, H. A. F. & Soliman, G. N. 1963. On three mytilid species boring in living corals. *Publications of the Marine Biological Station Ghardaqa*, **12**, 65–98.

Gray, J. E. 1858. On the families of Aspergillidae, Gastrochaenidae, and Humphreyadae. *Proceedings of the Zoological Society of London*, **26**, 307–318.

Jones, D. S. & Nicol, D. 1989. Eocene clavagellids (Mollusca: Pelecypoda) from Florida: the first documented occurrence in the Cenozoic of the western hemisphere. *Journal of Paleontology*, **63**, 320–323.

Keen, M. & Smith, L. A. 1969. Superfamily Clavagellacea. *In*: Moore, R. C. (ed.) *Treatise on Invertebrate Paleontology. Part N. Mollusca 2.* Geological Society of America and University of Kansas, 857–859.

Kilburn, R. N. 1974. A new species of *Clavagella s.s.* (Bivalvia: Clavagellidae) from Natal, South Africa. *Journal de Conchyliologie*, **111**, 89–92.

Kleemann, K. H. 1980. Boring bivalves and their host corals from the Great Barrier Reef. *Journal of Molluscan Studies*, **46**, 13–54.

Lamy, E. 1923. Les Clavagelles et Arrosoirs de la Mer Rouge (d'apres les materiaux recueillis par le Dr. Jousseaume). *Bulletin du Muséum national d'Histoire Naturelle*, **19**, 104–107.

Majima, R. 1994. Clavagellidae (Mollusca; Bivalvia) in Japan. *Bulletin of the Japanese National Science Museum Series C*, **20**, 13–43.

Morton, B. 1984*a*. Adventitious tube construction in *Brechites vaginiferus* (Bivalvia: Anomalodesmata: Clavagellacea) with an investigation of the juvenile of '*Humphreyia strangei*'. *Journal of Zoology*, **204**, 461–484.

—— 1984*b*. The biology and functional morphology of *Clavagella australis* (Bivalvia: Anomalodesmata). *Journal of the Zoological Society of London*, **202**, 489–511.

Nicol, D. & Jones, D. S. 1986. *Litharca lithodomus* and adaptive radiation in arcacean pelecypods. *The Nautilus*, **100**, 105–109.

Pojeta, J. Jr & Johnson, R. 1995. *Parastirpulina sohli*: a new Cretaceous clavagellid pelecypod. *The Geological Association of New Jersey*, **12**, 14–23.

—— & Sohl, N. F. 1987. *Ascaulocardium armatum* (Morton, 1833), new genus (Late Cretaceous): the ultimate variation on the bivalve paradigm. *The Paleontological Society Memoir*, **24**, 1–77.

—— & —— 1988. Eocene clavagellids from Florida. *Journal of Paleontology*, **62**, 8–26.

PURCHON, R. D. 1956. A note on the biology of *Brechites penis* (L.) (Lamellibranchia). *Journal of the Linnean Society of London, Zoology*, **43**, 43–53.

—— 1960. A further note on the biology of *Brechites penis* (L.) (Lamellibranchia). *Proceedings of the Malacological Society of London*, **34**, 19–23.

RÖDER, H. 1977. Zur Beziehung zwischen Konstruktion und Substrat bei mekanisch bohrenden Bohrmuscheln (Pholadidae, Teredinidae). *Senckenbergiana maritima*, **9**, 105–213.

SAVAZZI, E. 1982. Adaptations to tube dwelling in the Bivalvia. *Lethaia*, **15**, 275–297.

—— 1998. Constructional morphology of the bivalve *Pedum*. *In*: JOHNSTON, P. A. & HAGGART, J. W. (eds) *Bivalves: an Eon of Evolution*. University of Calgary Press, Calgary, 413–421.

—— 1999. Boring, nestling and tube-dwelling bivalves. *In*: SAVAZZI, E. (ed.) *Functional Morphology of the Invertebrate Skeleton*. Wiley & Sons, 205–237.

SEILACHER, A. 1970. Arbeitskonzept zur Konstruktions-Morphologie. *Lethaia*, **3**, 393–396.

—— 1991. Self-organizing mechanisms in morphogenesis and evolution. *In*: SCHMIDT-KITTLER, N. & VOGEL, K. (eds) *Constructional Morphology and Evolution*. Springer-Verlag, Berlin, 251–271.

SMITH, B. J. 1971. A revision of the family Clavagellidae (Pelecypoda, Mollusca) from Australia, with descriptions of two new species. *Journal of the Malacological Society of Australia*, **2**, 135–161.

—— 1976. Revision of the Recent species of the family Clavagellidae (Mollusca, Bivalvia). *Journal of the Malacological Society of Australia*, **3**, 187–209.

—— 1978. Further notes on the Clavagellidae, with speculation on the process of tube growth. *Journal of the Malacological Society of Australia*, **4**, 77–79.

SMITH, E. A. 1910. Note on the very young stage of the genus *Humphreyia*. *Proceedings of the Malacological Society of London*, **9**, 23–25.

SMITH, L. A. 1962*a*. Revision of the Clavagellacea. *Veliger*, **4**, 167–174.

—— 1962*b*. Historical zoogeographic study of the Clavagellacea. *Veliger*, **5**, 15–19.

SOLIMAN, G. N. 1971. On a new clavagellid bivalve from the Red Sea. *Proceedings of the Malacological Society of London*, **39**, 389–397.

STALLWOOD, R. B. 1995. A Turonian clavagellid (Bivalvia) from the Ladd Formation of southern California. *Journal of Paleontology*, **69**, 84–88.

VOKES, H. E. 1967. Genera of the Bivalvia. *Bulletins of American Paleontology*, **51**, 105–393.

Calcium concretions in the interstitial tissues of the Australian freshwater mussel *Hyridella depressa* (Hyriidae)

MARIA BYRNE

Department of Anatomy and Histology F13, University of Sydney, NSW 2006, Australia (e-mail: mbyrne@anatomy.usyd.edu.au)

Abstract: The Australian freshwater mussel *Hyridella depressa* (Lamarck, 1819) has extensive deposits of calcium phosphate granules in its interstitial connective tissues. The structure, distribution and elemental profile of these granules were documented by light and electron microscopy. For the elemental study, granules in cryo-prepared mantle tissue were examined by X-ray microanalysis. The granules were bright orange and formed a conspicuous cover over the mantle and palps, and were also abundant in the visceral mass. By contrast, they were not common in the gills and foot. The granules were typically distributed in discrete clusters and, in places, dominated the tissue space. Iron was a particularly important component of the granules and may account for their colour. The granules also contained other common elements, including Mg, Mn and Al, and trace elements, including Cu, Zn and Pb. Considering the important position of the Hyriidae in understanding the evolution and phylogeny of the Unionoida, emphasis was placed on comparison of the calcium granules of *H. depressa* to those in the Margaritiferidae and Unionidae. Granule distribution in *H. depressa* was most similar to that described for margaratiferids and contrasted with that described for unionids. The impressive capacity to accumulate extensive calcium deposits in their tissues is a unique feature of the Unionoida, but the rationale underlying production of these excess calcium stores is not understood. It is suggested that the granules may be a by-product of biomineralization processes in the Unionoida associated with the highly efficient calcium uptake system these bivalves evolved in conjunction with colonization of freshwater environments.

The colonization of freshwater by the marine ancestors of the Unionoida was followed by extensive radiation of these bivalves into a wide variety of freshwater environments. Due to their local abundance and large capacity for filtration, unionoids play major ecological roles in element and nutrient cycling, and in influencing the water quality of aquatic systems (Green 1980; Nalepa *et al.* 1991). As part of their adaptation to living in freshwater, these bivalves have evolved an impressive ability to sequester the calcium required for shell formation in low-calcium, freshwater environments. They also sequester calcium in small granular concretions deposited in their tissues (Pynnönen *et al.* 1987; Silverman 1988; Byrne & Vesk 1996, 2000; Pekkarinen & Valovirta 1997; Vesk & Byrne 1999). Calcium is deposited in these granules as relatively insoluble phosphates such as orthophosphate, pyrophosphate or a mixture of these two anions (Silverman *et al.* 1983; Jeffree *et al.* 1993; Mason & Jenkins 1995; Langston *et al.* 1998). The granules often dominate freshwater mussel tissues and comprise up to half the dry tissue weight (Silverman *et al.* 1985; Pynnönen *et al.* 1987). In *Anodonta* species granules are particularly abundant in the connective tissue of the gills (Silverman *et al.* 1985; Pynnönen *et al.* 1987).

Like their northern hemisphere relatives, Australian hyriid mussels deposit calcium phosphate granules in their tissues (Jeffree & Simpson 1984; Byrne & Vesk 1996, 2000; Adams *et al.* 1997; Vesk & Byrne 1999). The granules of *Hyridella depressa* (Lamarck, 1819) sequester a range of elements from the environment and several studies have highlighted their application for monitoring metal pollution (Jeffree & Simpson 1984; Byrne & Vesk 1996, 2000; Adams *et al.* 1997; Vesk & Byrne 1999). In *H. depressa*, the calcium in the granules is bound to pyrophosphate and/or hydrogen phosphate (Jeffree *et al.* 1993). This mussel is an important component of the macrobiota of aquatic systems in southeastern Australia, and aspects of its life history and conservation biology have been documented in several studies (Jupiter & Byrne 1997; Byrne 1998; Walker *et al.* 2000). In the present study, the distribution, structure and elemental content of the calcium granules in *H. depressa* is described by

From: HARPER, E. M., TAYLOR, J. D. & CRAME, J. A. (eds) *The Evolutionary Biology of the Bivalvia*. Geological Society, London, Special Publications, **177**, 329–337. 1-86239-076-2/00/$15.00

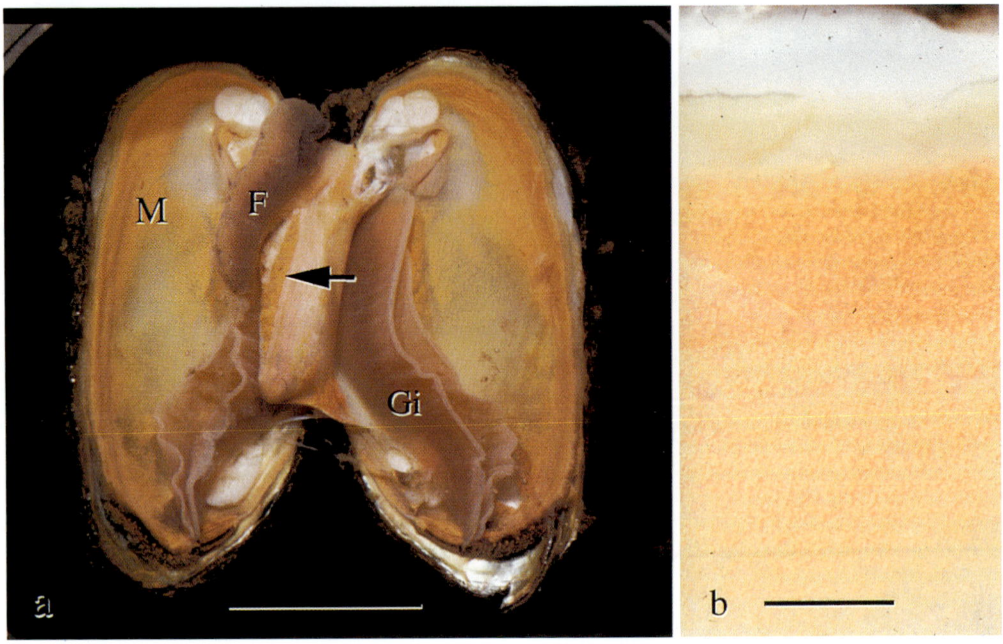

Fig. 1. (**a**) The granules form a conspicuous orange cover over the mantle (M) of *Hyridella depressa*. The visceral mass has been dissected (arrow) to show the high density of granules below the epithelium. (**b**) Detail of mantle edge showing granule clusters which are seen as orange flecks. F, foot; GI, gill. Scale bars: (a) 2.0 mm; (b) 1.0 mm.

light and electron microscopy. The elements contained in the granules were characterized in cryo-prepared mantle tissue by X-ray microanalysis. This method allowed direct analysis of granules preserved in a near-native state (Vesk & Byrne 1999) and avoided the problems of elemental loss and redistribution associated with the aqueous techniques used in previous studies (Jeffree & Simpson 1984; Silverman *et al.* 1985, 1987). Considering the important role of the Hyriidae in understanding the evolution and phylogeny of the Unionoida (Graf 2000; Walker *et al.* 2000), emphasis is placed on comparison of the granules and their distribution in the tissues of *H. depressa* with those documented for the Margaritiferidae and Unionidae. While granule accumulation is a synapomorphic feature of the Unionoida, this phenomenon in *H. depressa* exhibited several features different from those seen in their northern hemisphere counterparts.

Materials and methods

Specimens of *Hyridella depressa* were collected from three sites at Lake Burragorang, a major water storage located *c.* 60 km west of Sydney, New South Wales [see map in Byrne (1998)]. The sites were Kedumba (33°51′S; 150°20′E), Pocket Creek (33°56′S; 150°25′E) and Ripple Creek (33°54′S; 150°33′E). At each site mussels were collected from within the major size class present and shell

Fig. 2. LM of *Hyridella depressa*. (a), (c)–(e), (h) and (i) Paraffin sections (6.0 μm thick); (b), (f) and (g) semi-thin sections (0.5 μm thick). (**a**) Grazing section of the mantle showing clusters of granules (G) throughout the interstitium between the inner (left) and outer (right) epithelia (E). (**b**) Mantle section showing granules (G) aligned below the epithelium (E) and surrounding haemocoel (H) spaces. (**c**) Granule clusters in mantle interstitium. (**d**) Massive deposition of granules in the visceral.mass. (**e**) Granules in visceral mass near foot. Note the absence of granules in the foot muscle (M) region. (**f**) Cluster of granules among spermatogenic ascini. S, sperm. (**g**) Single large granule (top) and small granules (bottom) in the gill. (h) and (i) Histologically, a positive reaction of the granules to Alizarin red (**h**) and Perl's stain (**i**), indicated the presence of Ca and Fe, respectively. Scale bars: (a) 0.5 mm; (b) 5.0 μm; (c), (d) and (i) 50 μm; (e) and (h) 125 μm; (f) 30 μm; (g) 10 μm.

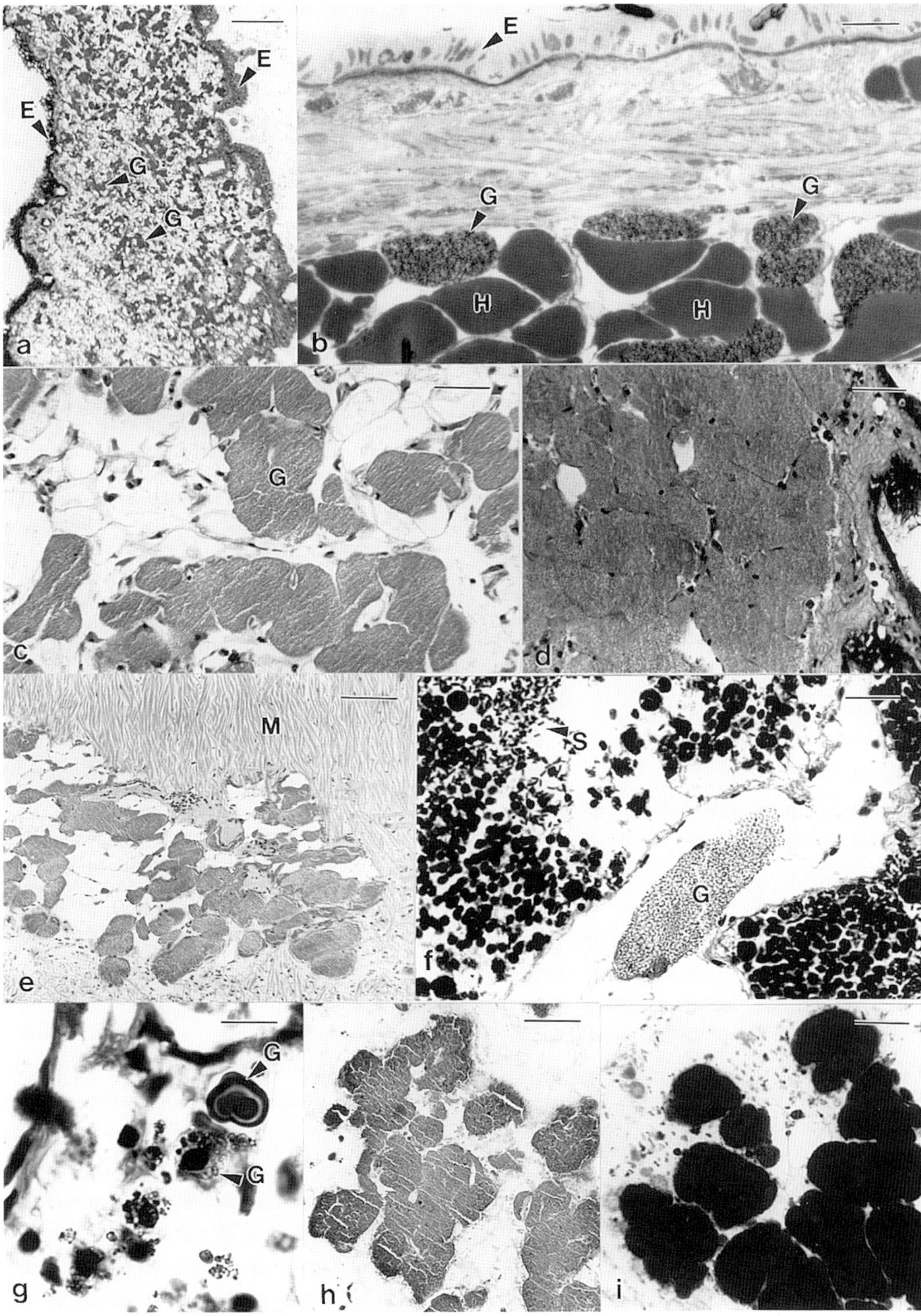
E
E
E
G
G
a
E
G
G
H
H
b
G
c
d
M
e
S
G
f
G
G
g
h
i

lengths were measured with vernier calipers. The mussels from Kedumba were 50–65 mm (shell length), while those from Pocket and Ripple Creeks were 40–58 mm (shell length). *H. depressa* were collected during the non-breeding periods of 1995 and 1996 (between April and July).

For light (LM) and transmission electron (TEM) microscopy, small pieces of tissue were placed in 2.5% glutaraldehyde in 0.02 M HEPES and post-fixed in 2% OsO_4 in 0.01 M HEPES, dehydrated in ethanols and placed in 100% acetone. The tissues were embedded in Spurr's resin and sectioned. Ultrathin sections were stained with uranyl acetate and lead citrate, then viewed in a Phillips 400 TEM or a JOEL 1010 TEM. Semi-thin sections (0.5 μm) were stained with toluidine blue. For scanning electron microscopy (SEM), mantle and palp tissues, fixed as described above, were dehydrated, critical-point dried, mounted on stubs and coated, then examined with a JOEL 35C SEM.

Tissue samples for histology were fixed in Bouin's fluid, dehydrated in graded ethanols, embedded in paraffin, sectioned (6 μm thick) and stained with haematoxylin and eosin, or with the Alizarin Red S and Perl's stain for Ca and Fe, respectively (Kiernan 1981).

Methods for X-ray microanalysis have been described in detail (Byrne & Vesk 1996; Vesk & Byrne 1999) and will only be outlined here. Mantle tissue, cryo-fixed in liquid nitrogen, was freeze-substituted with acetone and embedded in Spurr's resin. Dry-cut sections (0.5 μm thick), placed on nickel slot grids, were analysed by X-ray microanalysis in a Philips CM12 scanning transmission electron microscope (STEM) operated at 120 kV and equipped with an Edax energy dispersive X-ray detector.

Results

Granule distribution

The granules of *Hyridella depressa* were bright orange and formed a conspicuous cover over the surface of the mantle and palps (Fig. 1a and b). They were not evident on surface views of the gills or foot. An incision of the visceral mass revealed abundant granules just below the epithelium (Fig. 1a). In paraffin sections, the granules were distributed in clusters throughout the interstitial connective tissue of the mantle, palps and visceral mass (Fig. 2a–f). These clusters varied in size and, in places, the granules completely dominated the tissue space (Fig. 2d). The granules were refractile and, in unstained sections, were a gold-yellow colour. They exhibited a positive response to Alarizin Red A and Perl's stain, indicating the presence of Ca and Fe, respectively (Fig. 2h and i). Their prominence in the mantle and palps was evident with the scanning view, where each cluster was comprised of tightly packed spherical granules (Fig. 3a–d). Sections of the mantle shows groups of granules aligned below the epithelium and scattered throughout the interstitium between the inner and outer mantle epithelia (Fig. 2a and b). Although granules were encountered in the gill, they were not abundant and were usually distributed singly or in small groups (Fig. 2g). Their general absence from the foot appears to be due to the paucity of extracellular space in this highly muscular tissue (Fig. 2e).

Ultrastructure

The connective tissue of the mantle and visceral mass contains amoebocyte-like (connective tissue) cells, large vesicular cells, muscle, collagen and a sinusoidal network of vascular spaces (Figs 3a and b, 4c,d and g). The connective tissue cells had a complex profile and gave rise to numerous filipodial-like cell processes that pervaded the tissue (Fig. 4d and g). These processes contained a well-developed cytoskeleton of microfilaments and microtubules. They provided a supportive network for the interstitial tissues and surrounded vascular spaces and groups of granules (Fig. 4c and g). The vesicular cells contained large vesicles filled with amorphous material. They also gave rise to elongate processes that appeared to play a supportive role. The vesicular cells appear similar to the storage cells characteristic of bivalve tissues.

In ultrathin sections, the calcium granules were round, electron opaque and ranged in diameter from 0.5 to 2.5 μm (Fig. 4a–f). Compound granules were also evident. Although the granules were often found near a haemal space, or associated with cellular processes, many were not associated with other tissue structures. In chemically fixed mantle and visceral tissue prepared for TEM, some granules were distinctly annulated while others were not (cf. Fig. 4e and f). The annulated form usually had an electron-dense core surrounded by dense and clear annulations (Fig. 4f and h). Granules in frozen mantle tissue prepared for STEM, however, did not have an annulated structure (Fig. 4b), indicating that annulations in mantle granules may be an artefact of chemical fixation. A typical energy dispersive X-ray spectrum from microanalysis of one calcium granule from mantle tissue shows dominant P, Ca and Fe peaks (Fig. 5). They also contained variable amounts of Mn, Mg, Al, Cu, Pb, Zn and Ba. Although it is not known how the calcium granules form, the initial stages of their development appeared to occur in the connective tissue cells. Annulated granules in various stages of synthesis or

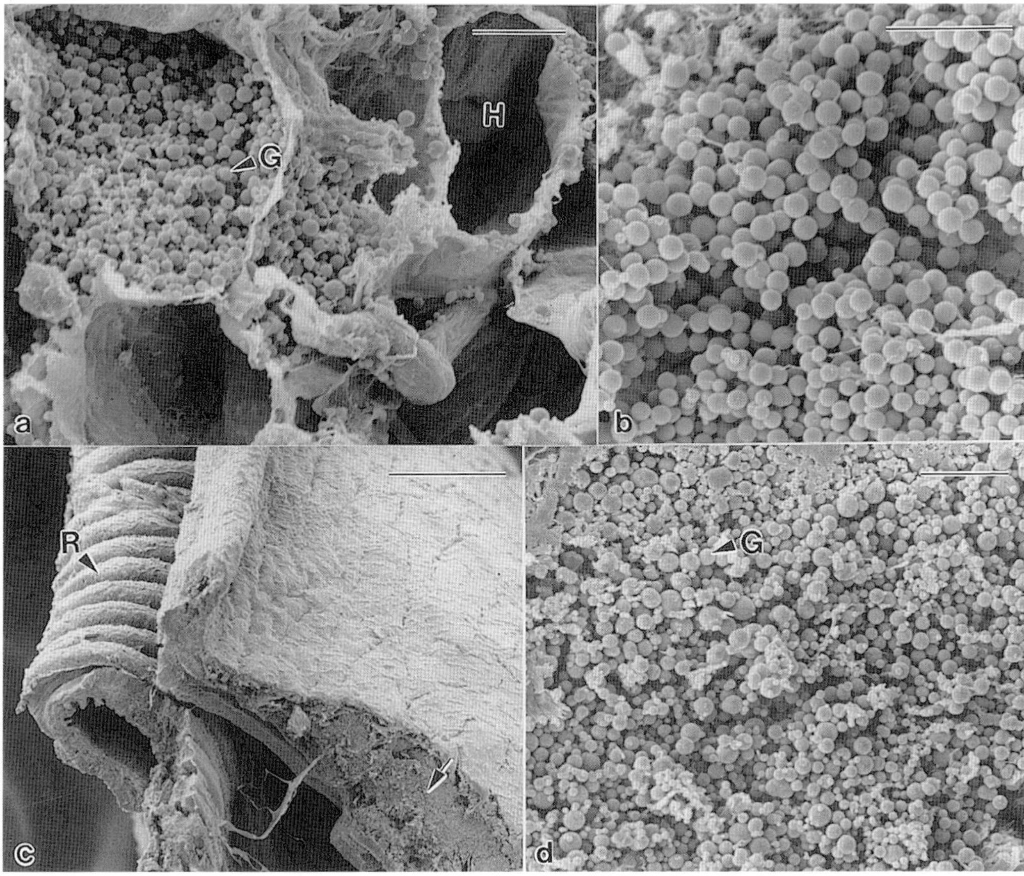

Fig. 3. SEM of *Hyridella depressa*. (**a**) and (**b**) Mantle; clusters of closely packed granules (G) in the interstitial tissue and adjacent haemal (H) spaces. (**c**) and (**d**) Palp, detail of the area indicated by the arrow in (**c**) shows granules (G) dominating the interstitial space. R, ciliated ridges. Scale bars: (a) and (b) 10 μm; (c) 250 μm; (d) 12 μm.

degradation were present in the cytoplasm of these cells (Fig. 4g and h). Connective tissue cells also served a phagocytic function in granule turnover.

In addition to calcium granules, the connective tissue cells also contained rough endoplasmic reticulum, Golgi bodies, lysosomes and residual bodies which varied in appearance. Microanalysis of some of these lysosomal structures revealed a dominant S peak and elevated levels of Cu, Pb and Zn, suggesting that they may be derived from metallothioneins (Vesk & Byrne 1999). Amorphous S-rich structures were also encountered in the extracellular tissue (Fig. 4a).

Discussion

The abundance of calcium granules in the mantle, palps and visceral mass of *Hyridella depressa* was readily visualized by macroscopic examination due to their striking orange colour. This has not been previously reported for the Unionoida, although may be common in Australian hyriids (Allison & Simpson 1989; Colville 1994; pers. obs.). The orange colour and high Fe content of the granules suggests that they contain iron oxides. High Fe concentrations are characteristic of the aquatic environment inhabited by *H. depressa* and this has been attributed to local lithology (Standard 1969; Markich & Brown 1998; Byrne & Vesk 2000). The orange colour of the granules in *H. depressa* from Lake Burragorang may be due to water chemistry rather than being a species-specific character (Byrne & Vesk 2000). Examination of specimens from Fe-poor environments is required to confirm this. Interestingly, the Fe-poor gill and Fe-rich midgut granules in the unionid *Anodonta cygnea zellensis* (Gmelin, 1791) are white and dark brown, respectively (Pynnönen *et al.* 1987). High Fe as

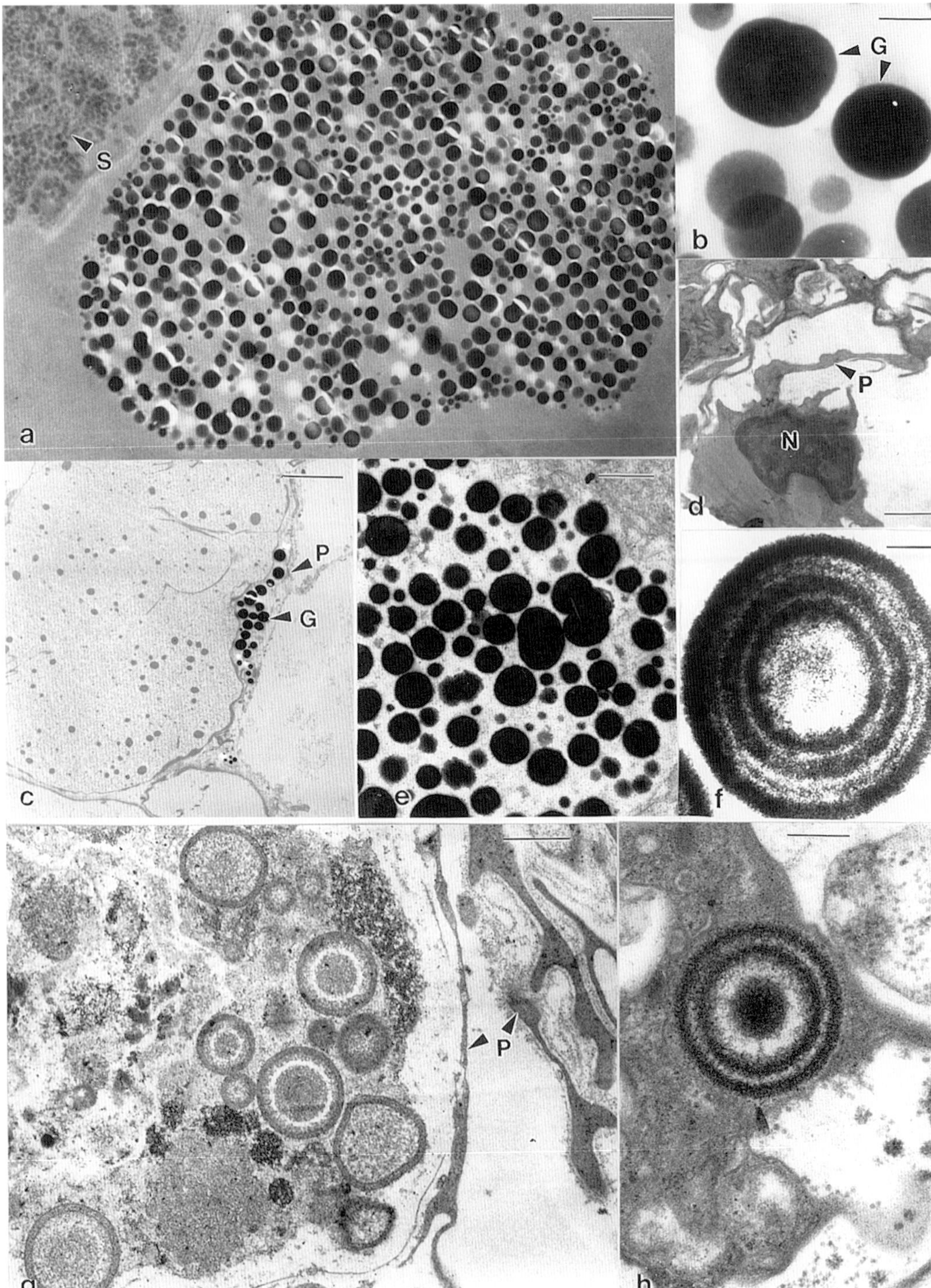

Fig. 4. STEM of cryo-prepared mantle tissue [(a) and (b)] and TEM of chemically fixed mantle [(c), (e) and (h)] and gonad [(f) and (g)] tissue. (**a**) and (**b**) Cluster of electron-dense calcium phosphate granules and adjacent S-rich (S) granules. (**c**) Small group of granules (G) surrounded by cell processes (P). (**d**) Connective tissue cells give rise to processes (P) that extend through the interstitial space. N, nucleus. (**e**) Dense and (**f**) annulated granules in fixed tissue. Intracellular annulated granules being: (**g**) broken down; or (**h**) synthesized for release. Scale bars: (a) 5.0 µm; (b) 0.4 µm; (c) 4.0 µm; (d) 1.0 µm; (e) 0.8 µm; (f) and (h) 0.2 µm; (g) 0.5 µm.

seen in the granules of *H. depressa*, however, does not appear to be typical of tissue granules in the Unionoida.

The presence of calcium granules in the interstitial tissues of *Hyridella depressa* and other hyriids, including two *Velesunio* species (Jeffree & Simpson 1984; Colville 1994; Vesk & Byrne 1999), is similar to that described for unionid and margaritiferid mussels (Petit *et al.* 1980; Machado *et al.* 1988). Like *H. depressa*, *Margaritifera margaritifera* (Linnaeus, 1758) has extensive aggregations of calcium granules in the mantle and few granules in the gill (Istin & Masoni 1973; Pekkarinen & Valoverta 1997). This differs from the pattern described for several unionids, including *Anotonta* and *Ligumia* species, where granules occur in their highest concentrations in the gills and are not abundant in the mantle (Istin & Masoni 1973; Silverman *et al.* 1985; Pynnönen *et al.* 1987; Silverman 1988). In unionids, calcium granules account for 20–50% of the dry tissue weight of the gills and granule calcium stores are mobilized for larval shell formation (Silverman *et al.* 1985; Pynnönen *et al.* 1987; Silverman 1988). By contrast, the paucity of granules in the gills of *H. depressa* suggests that they are not a significant source of calcium for hyriid glochidia. A similar conclusion was reached for *M. margaritifera* (Pekkarinin & Valoverta 1997).

It is not known if the similarities and differences in granule distribution in hyriids, margaritiferids and unionids are taxon specific but, if so, the granules may be a useful character in phylogenetic analyses. In recent times the phylogenetic relationships in the Unionoida have been examined using a number of molecular and morphological characters, and inclusion of the Hyriidae has been important in assessing clade relationships (Graf 2000).

The calcium granules of *Hyridella depressa* varied in the extent to which annulations were evident in chemically fixed tissues prepared for TEM, while annulations were not evident in cryo-prepared tissues prepared for STEM. Paired X-ray microanalyses of granules in fixed and cryo-prepared mantle tissue from individual mussels revealed that the granules loose a substantial proportion of their elements, particularly Ca, on exposure to an aqueous medium and gain other elements, by transfer, in the waterbath (Vesk & Byrne 1999). This suggests that some of the differences in calcium granule structure in chemically fixed and cryo-prepared tissues may be an artifact due to element loss or redistribution. The lighter annulations seen in granules in fixed tissues may be due to the loss of electron-opaque elements from regions where they are less tightly bound. In a previous study, granules in chemically fixed gonad tissue of *H. depressa* were noted to be more distinctly annulated compared with those in the mantle (Adams *et al.* 1997). A comparative study of the structure and elemental profile of calcium granules in cryo-prepared gill, gonad and mantle tissues is required to determine if they exhibit tissue-specific differences. Considering the different physiological roles of these tissues, some differences between their granules might be expected.

The form of phosphate varies in different types of calcium phosphate granules and is crucial to

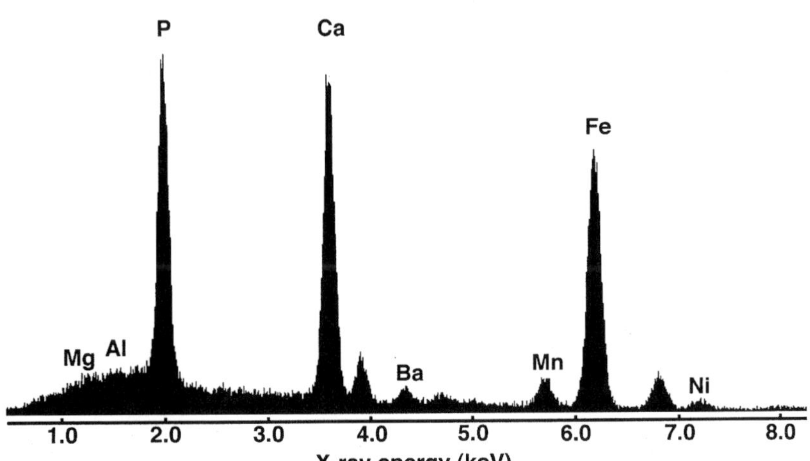

Fig. 5. Typical energy dispersive X-ray spectrum from a calcium phosphate granule in *Hyridella depressa* mantle tissue. Note characteristic P, Ca and Fe peaks (the Ni peak is due to the support grid).

granule solubility, and their physiological roles as calcium stores and detoxificatory sinks for trace elements (Jeffree *et al.* 1993; Mason & Jenkins 1995; Langston *et al.* 1998). In *Hyridella depressa*, all the mantle granules examined had a similar elemental profile and appeared to comprise a single type of granule. Although the abundance of granules in the mantle with its extensive blood supply suggests that they may function as soluble calcium stores, the presence of a range of other elements in the granules, including trace metals, suggests that they also play a role in detoxification. The granules of *H. depressa* are known to contain orthophosphate and/or hydrogen phosphate (Jeffree *et al.* 1993), which suggests that they are unlikely to be remobilized (Langston *et al.* 1998). It would be useful to determine the phosphate species in the mantle, gill and gonad granules of *H. depressa* to gain insights into their potential functions.

The presence of granules in various stages of synthesis and degradation in the cytoplasm of connective tissue cells of *Hyridella depressa* is similar to that reported for several unionids (Petit *et al.* 1980; Silverman *et al.* 1989). These cells have been variously called concretion-forming cells, calcium cells, phagocytes and amoebocytes (Petit *et al.* 1980; Machado *et al.* 1988; Silverman *et al.* 1989). Like these cells in *H. depressa*, granule-containing cells in other freshwater mussels are variable in form and have numerous processes. They are suggested to be derived from an amoebocyte-like stem cell (Silverman *et al.* 1989). In *H. depressa*, the granule-containing connective tissue cells appear to play a major role as a site for nucleation and initial development of calcium granules. They also function in the phagocytic breakdown of the granules. After newly formed granules are released into the extracellular compartment, their composition and size is likely to change in response to the elemental environment in the tissues and to the ionic exchange mechanisms. The different sizes of the calcium granules in the tissues of *H. depressa* suggests that their post-release phase involves changes in size and composition in a dynamic interaction between them and the surrounding tissues. Details of the processes underlying granule formation, release and post-release change, however, were not evident in the sectioned tissues of *H. depressa*. As seen here for *H. depressa*, the connective tissue cells of freshwater mussels often contain a number of other electron-dense inclusions which differ in structure from the calcium granules (Machado *et al.* 1988).

The extensive contribution of calcium granules to the body mass of freshwater mussels and their ability to sequester a range of elements has generated interest in the granules for studies of metal pollution (Jeffree & Simpson 1984; Pynnönen *et al.* 1987; Silverman *et al.* 1987; Byrne & Vesk 1996, 2000; Vesk & Byrne 1999). Although less abundant, S-rich lysosomal granules also occurred in the tissues of *Hyridella depressa*. The high S content of these structures and their incorporation of trace metals suggests they contain metallothioneins (metal-binding proteins) which function in detoxification (Mason & Jenkins 1995; Vesk & Byrne 1999). Both the calcium- and S-rich granules function as element-uptake sites in *H. depressa*, and provide focal structures for analyses of element dynamics and accumulation in the tissues (Byrne & Vesk 1996; Vesk & Byrne 1999).

Although the presence of extensive calcium granule deposits is a unique feature of biomineralization in the Unionoida, the rationale underlying production of what appear to be excess calcium stores by these bivalves is not understood. Suggestions for the function of the granules include calcium homeostasis, biomineralization, detoxification of harmful elements and as calcium stores for glochidial shell formation (Petit *et al.* 1980; Silverman *et al.* 1985, 1987; Mason & Jenkins 1995). They may also be a by-product of ionic regulation or biomineralization processes in the Unionoida associated with the highly efficient calcium-uptake system these bivalves evolved in conjunction with colonization of freshwater environments.

P. Vesk, A. Cerra, S. Adams and P. Selvakumaraswamy provided assistance in the laboratory and in the field. R. Smith assisted with photography. Staff of the Electron Microscope Unit, University of Sydney, are thanked for expert advice and provision of facilities, particularly Dr C. Nockolds. Special thanks to I. Wright and G. Capararo, Catchment Services, Sydney Water Corporation, for providing boat access and field support. Numerous dive buddies assisted and special thanks goes to G. Hunter. This study was supported by the New South Wales Government through its Environmental Trusts and the Australian Research Council.

References

ADAMS, S. M., SHOREY, C. D. & BYRNE, M. 1997. An ultrastructural and microanalytical study of metal-ion content in granular concretions of the freshwater mussel *Hyridella depressa*. *Micron*, **28**, 1–11.

ALLISON, H. E & SIMPSOM, R. D. 1989. *Element concentrations in the freshwater mussel,* Velesunio angasi, *in the Alligator Rivers Region.* Technical Memorandum 25. Supervising Scientist for the Alligator Rivers Region, Australian Government Publishing Services, Canberra.

BYRNE, M. 1998. Reproduction of river and lake populations of *Hyridella depressa* (Unionacea: Hyriidae) in New South Wales: Implications for their conservation. *Hydrobiologia*, **389**(38), 29–43.

—— & VESK, P. A. 1996. Microanalysis of elements in

granules in *Hyridella depressa* (Bivalvia): multivariate analysis and biomonitoring potential. *Australasian Journal of Ecotoxicology*, **12**, 91–97.

—— & —— 2000. Elemental composition of mantle tissue granules in *Hyridella depressa* (Unionida) from the Hawkesbury–Nepean River system Australia: Inferences from catchment chemistry. *Marine and Freshwater Research*, **51**, 183–192.

COLVILLE, A. E. 1994. *Comparison of granular structures and chemistry, and growth rates, in the freshwater bivalves Hyridella depressa, Hyridella australis and Velesunio ambiguus from the Nepean River*. MSc Thesis, University of Technology, Sydney.

GRAF, D. L. 2000. The Etheriodea revisited: a phylogenetic analysis of hyriid relationships (Mollusca: Bivalvia: Unionoida) of North America. *Journal of Molluscan Studies*, in press.

GREEN, R. H. 1980. Role of a unionid clam population in the calcium budget of a small artic lake. *Canadian Journal of Fisheries and Aquatic Sciences*, **37**, 219–224.

ISTIN, M. & MASONI, A. 1973. Absorption et redistribution du calcium dans le manteau des lamellibranches en relation avec la structure. *Calcified Tissue Research*, **11**, 151–162.

JEFFREE, R. A. & SIMPSON, R. D. 1984. Radium-226 is accumulated in calcium granules in the tissue of the freshwater mussel, *Velesunio angasi*: support for a metabolic analogue hypothesis. *Comparative Biochemistry and Physiology*, **79A**, 61–72.

——, MARKICH, S. J. & BROWN, P. L. 1993. Comparative accumulation of alkaline earth metals by two freshwater mussel species from the Nepean River, Australia – consistency and a resolved paradox. *Australian Journal of Marine and Freshwater Research*, **44**, 609–634.

JUPITER, S. D. & BYRNE, M. 1997. Light and scanning electron microscopy of the embryos and glochidia larvae of the Australian freshwater bivalve *Hyridella depressa* (Hyriidae). *Invertebrate Reproduction and Development*, **32**, 177–186.

KIERNAN, J.A. 1981. *Histological and Histochemical Methods: Theory and Practice*. Pergamon Press, New York.

LANGSTON, W. J., BEBIANNO, M. J. & BURT, G. R. 1998. Metal handling strategies in molluscs. *In*: LANGSTON, W. J. & BEBIANNO, M. J. (eds) *Metal Metabolism in Aquatic Environments*. Chapman & Hall, London, 219–283.

MACHADO, J., CASTILHO, F., COIMBRA, J., MONTEIRO, E., SA, C. & REIS, M. 1988. Ultrastructural and cytochemical studies in the mantle of *Anodonta cygnea*. *Tissue & Cell*, **20**, 797–807.

MARKICH, S. J. & BROWN, P. L. 1998. Relative importance of natural and anthropogenic influences on the fresh surface water chemistry of the Hawkesbury–Nepean River, south-eastern Australia. *The Science of the Total Environment*, **217**, 201–230.

MASON, A. Z. & JENKINS, K. D. 1995. Metal detoxification in aquatic organisms. *In*: TESSIER, A. & TURNER, D. R. (eds) *Metal Speciation and Bioavailability in Aquatic Systems*. John Wiley & Sons, New York, 479–589.

NALEPA, T. F., GARDNER, W. S. & MALCZYK, J. M. 1991. Phosphorus cycling by mussels (Unionidae: Bivalvia) in Lake St. Clare. *Hydrobiologia*, **219**, 239–250.

PEKKARINEN, M. & VALOVERTA, I. 1997. Histochemical and x-ray studies on tissue concretions and shells of *Margaritifera margaritifera* (Linnaeus). *Journal of Shellfish Research*, **16**, 169–177.

PETIT, H., DAVIS, W. L, JONES, R. G. & HAGLER, H. K. 1980. Morphological studies on the calcification process in the fresh-water mussel *Amblema*. *Tissue & Cell*, **12**, 13–28.

PYNNÖNEN, D. A., HOLWERDA, D. I. & ZANDEE, D. I. 1987. Occurrence of calcium concretions in various tissues of freshwater mussels, and their capacity for cadmium sequestration. *Aquatic Toxicology*, **10**, 101–114.

SILVERMAN, H. 1988. Form and function of calcium concretions in unionids. *In*: CRICK, R. E. (ed.) *Origin, Evolution and Modern Aspects of Biomineralization in Plants and Animals*. Plenum Press, New York, 367-384.

——, MCNEIL, J. W. & DIETZ, T. H. 1987. Interaction of trace metals Zn, Cd and Mn, with Ca concretions in the gills of freshwater unionid mussels. *Canadian Journal of Zoology*, **65**, 828–832.

——, STEFFENS, W. L. & DIETZ, T. H. 1985. Calcium from extracellular concretions in the gills of a freshwater unionid mussel is mobilized during reproduction. *Journal of Experimental Zoology*, **236**, 137–147.

——, RICHARD, P. E., GODDARD, R. H. & DIETZ, T. H. 1989. Intracellular formation of calcium concretions by phagocytic cells in freshwater mussels. *Canadian Journal of Zoology*, **67**, 198–207.

STANDARD, J. C. 1969. Hawkesbury sandstone. *Journal of the Geological Society of Australia*, **16**, 407–415.

VESK, P. A. & BYRNE, M. 1999. Metal levels in tissue granules of the freshwater bivalve *Hyridella depressa* (Unionida) for biomonitoring: the importance of cryopreservation. *The Science of the Total Environment*, **225**, 219–229.

WALKER, K. F., BYRNE, M., HICKEY, C. W. & ROPER, D. S. 2000. Freshwater mussels (Hyriidae, Unionidae) of Australasia. *In*: BAUER, G. & WÄCHTEL, K. (eds) *Ecology and Evolution of the Freshwater Mussels Unionoida*. Springer, Berlin, in press.

Teranota and its implications on anomalodesmatan phylogeny

NICOLE S. ROGALLA & MICHAEL R. W. AMLER

Institut für Geologie und Paläontologie der Philipps-Universität Marburg, Abteilung Invertebraten-Paläontologie, Hans-Meerwein-Strasse, 35032 Marburg, Germany (e-mail: rogalla@mailer.uni-marburg.de & amler@mailer.uni-marburg.de)

Abstract: Bedded carbonaceous siltstones from the Buechel Subformation of the Bergisch-Gladbach–Paffrath Syncline, Germany, of Middle Devonian (Middle Givetian) age have yielded a remarkable sample of extremely elongated, articulated bivalves preserved in life orientation. The specimens are associated with a single left valve embedded horizontally in the bedding plane and further isolated but articulated shells. Combining the information given by the specimens and the palaeobiological interpretations allows the reconstruction of the complete morphology and probable life habits. The very distinct morphological features led to the erection of a new taxon, *Teranota ebbighauseni* Rogalla & Amler, 2000, provoking discussion on habitats, life habits and evolutionary trends in anomalodesmatan bivalves. A combination of characters typical of the orthonotids and the modiomorphids, as well as the preserved life position at an angle of some 60–70° relative to the bedding plane, suggests that these specimens were part of a minor branch off the main evolutionary lineages within the Anomalodesmata. It is proposed that these animals represent a convergent line of endobenthic bivalves distantly related to true siphonate forms.

A large slab of carbonaceous siltstone has yielded specimens of elongated bivalves which have been described as the new genus and new species *Teranota ebbighauseni* Rogalla & Amler, 2000 (Fig. 1). The extraordinarily well-preserved specimens are embedded in life orientation and allow interpretation of life habits and evolution (Rogalla & Amler 2000). Accordingly, discussion on the systematic position of orthonotid bivalves within higher systematic categories, previously debated in Pojeta (1971, 1978, 1986, 1987), Runnegar (1974), Pojeta & Gilbert-Tomlinson (1977), Pojeta & Runnegar (1985), Bradshaw & McCartan (1991), Morris *et al.* (1991), Johnston (1993), Amler (1999) and Bradshaw (1999), is reinvestigated in this study.

Fossil material

Altogether, six specimens of differing preservation (internal and external moulds, shelly material, all incomplete, two conjoined) were collected by Dr Volker Ebbighausen, Odenthal, Germany, from fine-grained sandstones and siltstones of the Buechel Subformation of the Bolsdorf Formation of late Middle Givetian age. The locality is a temporary outcrop near Bergisch-Gladbach (R77140, H52580, geological map 1:25 000, Sheet 4908 Burscheid), situated in the Bergisch-Gladbach–Paffrath Syncline of the Rheinisches Schiefergebirge northeast of Cologne (Köln). The material is kept in the collection of the Naturkunde Museum Berlin (collection no. MB.M.1169–1171).

Morphology

The combination of all the information given by the specimens in life position accompanied by isolated but articulated shells allowed the reconstruction of the complete morphology [Figs 2–6; see also Rogalla & Amler (2000)]. The most distinctive morphological feature of *Teranota ebbighauseni* is the unusual length of the shells of > 15 cm. A height:length ratio of 1:10 anteriorly and 1:5 posteriorly indicates an increase in height towards the posterior, where the valves gape permanently. This gape is accentuated as the valves become concave towards the posterior margin. The ventral margin, otherwise straight, bears a very shallow sinus antero-ventrally. The minute umbones are placed in the anterior sixth of the dorsal margin, which is very long and straight. The adductor muscle scars are interpreted as being anisomyarian. The anterior scar forms a deep impression in the shell and the posterior scar is apparently larger, but weakly developed. The hinge margin is edentulous and not thickened as a hinge plate; externally, an

From: HARPER, E. M., TAYLOR, J. D. & CRAME, J. A. (eds) *The Evolutionary Biology of the Bivalvia*. Geological Society, London, Special Publications, **177**, 339–346. 1-86239-076-2/00/$15.00

Fig. 1. Slab of carbonaceous siltstone from the Buechel Subformation (Middle Givetian, Germany) with three specimens of *Teranota ebbighauseni* Rogalla & Amler, 2000, preserved in life position. Scale bar, 5 cm.

elongate opisthodetic, parivincular ligament is developed. The external shell ornamentation consists of sharp commarginal fila, thickened in the posteriormost region only, forming a few coarse commarginal rugae. Additionally, one slight anterior ridge, a few obscure radial furrows and oblique radiating lines are present. The radial anterior ridge forms the posterior border of a shallow depression which leads into the antero-ventral sinus. Thus, in external view the posterior shell region is slightly raised above the anterior. In cross-section the shell is separated into an umbonally inflated but ventrally narrowing anterior portion and a wider but flattened posterior one (see Figs 3 and 6).

Palaeobiology

Some of the studied specimens are preserved in life orientation but there is only limited evidence about

Fig. 2. *Teranota ebbighauseni* Rogalla & Amler, 2000, holotype, right valve, internal mould and remains of shell, damaged anteriorly and posteriorly. Scale bar, 5 cm.

Fig. 3. *Teranota ebbighauseni* Rogalla & Amler, 2000, paratype. (**a**) Right valve lateral view; (**b**) dorsal view with preserved parivincular ligament. Scale bar, 5 cm.

the former burrowing depth of the animal (Fig. 1). Nevertheless, this type of preservation, combined with morphological reconstruction, can yield information about relationships between the animals and the substrate–water interface (Fig. 7).

The specimens found embedded in life position are orientated at angles of 60–70° to the substrate surface, inferring an endobenthic mode of life. Most Palaeozoic endobenthic taxa are recognized as possessing elongated valves, a byssal sinus, a reduced but rounded anterior area, a tendency towards anisomyarian musculature and a lack of ventral flattening (Stanley 1972). In *Teranota*, the very shallow sinus in the antero-ventral margin also indicates byssal attachment. As there is no indication of a byssal slit within the ventral commissure, it can be assumed that the development of the byssus, as in many endobyssate forms, may have been quite weak. The shallow ventral sinus further narrows the tapering anterior shell portion, hence, the foot was rather small, and probably incapable of advanced and rapid burrowing. The shape of such elongate endobenthic taxa does not allow rotational movements during burrowing as in ovate taxa like modern veneroids. The possession of a permanent anterior shell gape, e.g. as in Recent members of *Ensis*, allows the animal to burrow parallel to its long axis (Stanley 1970). In contrast, *Teranota* does not possess an anterior gape and, additionally, the reduced anterior region, as well as the attachment site of the anterior adductor muscle, suggests that the foot extruded obliquely to the long axis of the shell. This feature accounts for the observed life orientation and agrees with the observations of Stanley (1972) on postulated life habits in modiomorphids, pholadomyoids and semi-infaunal mytiliids. Possession of a byssus would have stabilized the shell in the substrate and supported the weak foot in the burrowing process (Runnegar 1974). The narrow anterior shell cross-section may have been advantageous for the animal in cutting through the substrate. Where the margins

Fig. 4. *Teranota ebbighauseni* Rogalla & Amler, 2000, paratype right valve internal mould. Scale bar in mm.

Fig. 5. *Teranota ebbighauseni* Rogalla & Amler, 2000, paratype left valve shell specimen, partly crushed. Scale bar in mm.

of the valves are subparallel and become concave to form the pronounced posterior gape, that part of the shell must have protruded from the sediment surface as it cannot have been sealed by the periostracum (Morris *et al.* 1991). Due to the risks of engulfment and, conversely, the risk of being winnowed out, and its weak burrowing action, it is suggested here that the environment where *Teranota* lived was calm and that sediment input and movement were reduced, so consequently, no strong currents occurred. This opinion is supported by the grain size of the surrounding sediment and the overall sedimentary sequence.

Pojeta (1978) and Bradshaw & McCartan (1991) argued that orthonotids could have possessed short siphons. The present authors do not subscribe to this point of view, especially for the endobyssate taxa. On the one hand, endobyssate life habits and the possession of siphons are mutually exclusive (Stanley 1972), on the other hand, siphons are produced through mantle fusion and this did not appear until later in the Palaeozoic (Stanley 1968, 1972; P. A. Johnston pers. comm.). Although the presence of very short siphons is not necessarily indicated by a sinuate pallial line, there is no single evidence of definite sinupalliate orthonotids in the fossil record.

Systematic position and discussion

Any attempt to link the specimens with previously described species of elongated Palaeozoic bivalves must include a general discussion of the morphology and evolution of 'sword-shaped' bivalves in the Early and Middle Palaeozoic. Palaeozoic posteriorly elongated bivalves have

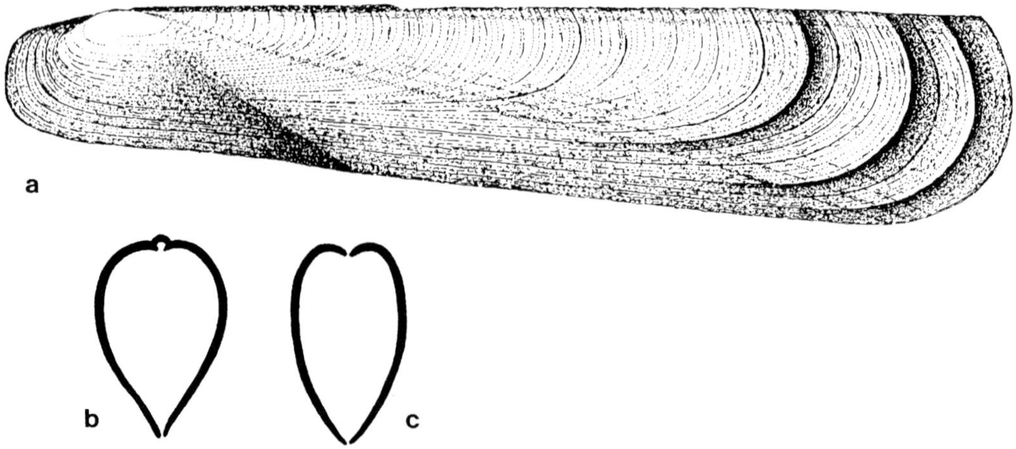

Fig. 6. *Teranota ebbighauseni* Rogalla & Amler, 2000. (**a**) Reconstruction of left valve morphology and external ornamentation; (**b**) cross-section through complete shell in anterior area; (**c**) cross-section in posterior area.

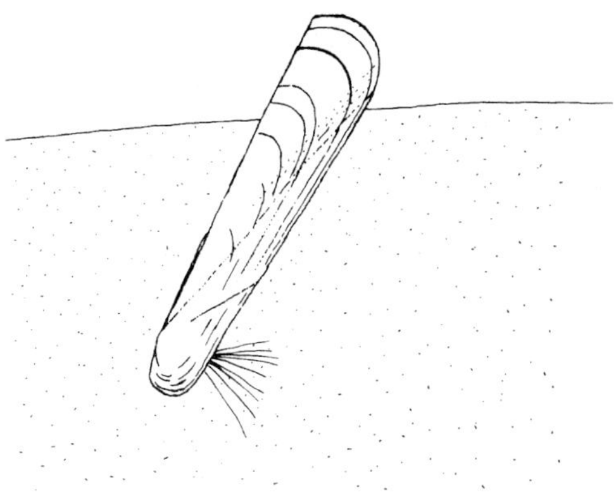

Fig. 7. Life orientation of *Teranota ebbighauseni* Rogalla & Amler, 2000, reconstructed after preserved specimens shown in Fig. 1.

been classified as being members of the Isofilibranchia, the Heterodonta or the Anomalodesmata (e.g. Pojeta 1971, 1978, 1986, 1987; Runnegar 1974; Morris *et al.* 1991). The systematic position and rank of these taxa is difficult to unravel because most higher taxa are not characterized by generally accepted autapomorphies and, therefore, lack adequate differential diagnoses. Additionally, morphological characters have been weighted differently by various authors, leaving difficulties in phylogenetic relationships unresolved.

Teranota seems to combine characters of orthonotids and elongate modiomorphids. The development of a ventral sinus and the distinct posterior increase in shell height are modiomorphid characters (see also Pojeta 1987), whereas the elongate shell form, the anterior shell shape, the ligament and the external ornament, consisting of radial and diagonal lines and furrows, are features exhibited by the orthonotids.

Based on the morphological features described above, the preserved life orientation and phylogenetic lineages, *Teranota* is classified as a member of the Family Orthonotidae Miller, 1877 in the Superfamily Pholadomyoidea (King, 1844) Gray, 1847 [Order Pholadomyoida Newell, 1965, Superorder Anomalodesmata Dall, 1889 (1899)] [for general bivalve classification see Amler (1999) and Cope (1996, 1997 and refs cited therein)].

The position of the orthonotids has been a matter of debate for decades. Newell (1969*a*) placed orthonotids, e.g. the genera *Orthonota* and *Palaeosolen*, in the Anomalodesmata; *Cymatonota* was regarded as member of the Modiomorphidae (Newell 1969*b*). Pojeta (1971), in his review of Ordovician bivalves, followed this assignment. In contrast, Runnegar (1974) rejected this opinion and excluded orthonotids from the Anomalodesmata, considering the soleniform genera to be ancestral Solenoidea, thus belonging to the Heterodonta; this was supported by Pojeta & Gilbert-Tomlinson (1977). Pojeta (1978) accepted the separate position of the soleniform orthonotids and erected the Subclass Orthonotia Pojeta, 1978, including the genera *Cymatonota*, *Orthodesma*, *Palaeosolen*, *Psiloconcha*, *Orthonota*, *Solenomorpha*, *Prothyris* and a new, unnamed genus. Later, he included the group in the Subclass Isofilibranchia (Pojeta 1986, 1987). This procedure was not followed by later authors and most recent classifications place the Orthonotidae in the Anomalodesmata [Amler 1999; see Johnston (1993) and Bradshaw (1999) for extensive discussion]. Morris *et al.* (1991) argued that placing orthonotids within the Anomalodesmata would only be a matter of convention, as they share no synapomorphies with this group. Consequently, those authors regarded the group as a separate order; the Orthonotoida.

At the generic level, *Teranota* is morphologically and phylogenetically closely linked with *Orthonota* Conrad, 1841 (type species *O. undulata* Conrad, 1841). Both share the elongate shell shape and similar shell proportions. The relatively higher shell of *Teranota* has led to differences in the position of ventral and dorsal margins compared with *Orthonota*: whereas ventral and dorsal margins run parallel in *Orthonota*, in *Teranota* they diverge in a posterior direction. Further differences are the stronger development of diagonal ridges and a narrow sulcus in several species of *Orthonota*, but those may serve as diagnostic features at species or subgenus levels only.

The shell ornamentation of *Cymatonota* Ulrich, 1893 (type species *C. typicalis* Ulrich, 1893) consists of growth lines only (Pojeta 1971). This total lack of radial elements differs from the above mentioned taxa. The main diagnostic feature is the presence of a broad lateral sulcus leading to a central ventral sinus; both are generally absent in *Orthonota* Conrad, 1841 and have a completely different form in *Teranota*. In addition, both ventral and dorsal margins run nearly parallel, and the umbones are broad and prominent. Pre-Devonian members of *Cymatonota* possess posteriorly diverging dorsal and ventral margins which may be ancestral. Some authors have assigned *Cymatonota* as ancestral to *Orthonota* (Pojeta 1971; J. G. Carter pers. comm.).

A generally accepted diagnosis is not, as yet, present for species of *Palaeosolen* Hall, 1885 (type species *Orthonota siliquoidea* Hall, 1870). The various species are characterized by an elongate shell shape and nearly terminally situated, flattened umbones (see also Beushausen 1895). Quite distinct from *Teranota* is the truncated anterior shell portion and the anteriorly and posteriorly gaping valves (Driscoll 1965; Bradshaw & McCartan 1991). The dorsal and ventral margins run parallel to each other. The external shell ornamentation consists of commarginal growth lines which coarsen posteriorly. In addition, *P. chapmani* and *P. costatus* bear external radial elements, whereas *P. frontisocurvus* shows radial elements only interiorly. In *P. quadrangularis* a single, plate-like cardinal tooth is developed in each valve, ventrally supported by a buttress.

Shells of the genus *Solenomorpha* Cockerell, 1903 (type species *S. minor* M'Coy, 1844) are distinctly elongate and differ from all other taxa as well as *Teranota* by their posteriorly tapering shells. The nasute anterior shell portion with a concave antero-dorsal margin, e.g. in *S. minor*, and a convex ventral margin, e.g. in *S. dorsocurva* and *S. scalpriformis*, distinguish them from *T. ebbighauseni*. Johnston (1993) postulated the presence of posteriorly gaping valves.

Superficially, members of the following genera may resemble the studied specimens in gross morphology, but differ largely in their shell shape by being more compact rather than strongly elongate. Members of *Sphenosolen* Pojeta & Gilbert-Tomlinson, 1977 (type species *S. draperi* Pojeta & Gilbert-Tomlinson, 1977) show a shell posteriorly increasing in height like *Teranota*, but shell proportions are not similar and no external radial ornamentation is developed.

The generic name *Spathella* Hall, 1885 (type species *S. typica* Hall, 1885) has been used for bivalves with elongated cylindrical shells and distinctly developed commarginal ornamentation, but without an increase in shell height in the posterior direction. Detailed shell characters are largely unknown in this taxon, which presently is placed in the modiomorphids (Amler 1996).

Sanguinolites M'Coy, 1844 (type species *S. discors* M'Coy, 1844), recently revised by Morris *et al.* (1991), exhibits at least one very prominent diagonal ridge where external shell sculpture bends and strengthens postero-dorsally. Further radial and commarginal ribs and fila may be developed. Some elongated taxa previously classified as *Sanguinolites* are now grouped in *Pleurophorella* Girty, 1904 (type species *P. papillosa* Girty, 1904) and *Siliquimya* Morris, Dickins & Astafieva-Urbaitis, 1991 (type species *Sanguinolaria plicata* Portlock, 1843), but these differ markedly from *Teranota* due to their 'sword-shaped' outline.

Although considered by some authors to be closely related to the above (Runnegar 1974; Pojeta 1978; Bradshaw 1999), the genera *Prothyris* Meek (1869), 1871 and *Citothyris* Ruzicka & Rehor, 1964 (see Amler 1996) are here considered to belong to a separate branch of anomalodesmatans. These taxa have a quite distinct synapomorphy – a narrow anterior lobe of unknown function.

Contrasting morphology is shown by modiomorphid taxa, which are here considered to be members of the Palaeoheterodonta (see Amler 1999). With *Teranota* they share only two characters, the distinct antero-ventral sinus in the shell margin and a posterior increase in shell height. However, all known modiomorphids are generally more compact without elongation. Only members of the genus *Liromytilus* LaRoque, 1950 are somewhat elongated but, as in other modiomorphoids (and mytiloids), slightly curved (mytiliform, myaliniform). These features are here considered to be either convergent to *Teranota* or symplesiomorphies.

Evolutionary relationships

Evolutionary relationships within the early Anomalodesmata are poorly understood. Based on morphology, *Teranota* is interpreted here to be related to *Orthonota* and *Cymatonota*. Pojeta (1987) figured an unconvincing *Cymatonota* from the Ordovician which could form an early ancestor. No direct descendents of Late Devonian or Early Carboniferous age are known for *Orthonota* and *Teranota*. Consequently, *Teranota* is here interpreted as a minor, low diverse branch off main evolutionary lines within the Orthonotidae. In contrast to most other anomalodesmatans of the Palaeozoic, *Teranota* reaches an enormous size, which is not, *per se*, a diagnostic feature but is worth mentioning, as it is a convergent character shared with *Solenomorpha* (Anomalodesmata) and, to a lesser extent, *Liromytilus* (?Isofilibranchia). The presence of the byssus is here interpreted as plesiomorphic retention of this structure. *Teranota* probably became extinct near the end of the Middle Devonian or during the Frasne–Famenne Crisis (Kellwasser Event), as there are no similar forms known from the highly diverse Famennian or Lower Carboniferous faunas (de Koninck 1885; Beushausen 1895; Hind 1896–1900; Amler 1996). This contrasts with the successful survival of *Solenomorpha* into the Carboniferous. Apart from *Solenomorpha*, all other elongated Carboniferous anomalodesmatans are either edmondiid, sanguinolitid or siliquimyoid in overall shell shape, differing from *Teranota* in several characteristics (as described above). Apparently, some of them developed siphons (e.g. *Wilkingia* Wilson, 1959) as a more advanced aid to deep burrowing. Thus, the evolutionary pathway of *Teranota* is interpreted here as leading into a blind alley, as it enabled neither faster or deeper burrowing, nor stronger byssal attachment. It may be regarded as one of many evolutionary experiments of the anomalodesmatan bloom of the late Palaeozoic.

Conclusion

The specimens studied were described as a new taxon, *Teranota ebbighauseni* Rogalla & Amler, 2000, because they bear a combination of morphological characters not known from other taxa. Although some modiomorphoid features are present, the genus *Teranota* is placed within the Anomalodesmata. The distinct combination of characters support the hypothesis of a common ancestor for the Anomalodesmata and modiomorphoids. Based on external and internal morphology, *Teranota* is placed in the Family Orthonotidae and is closely related to the Genera *Orthonota* and *Cymatonota*. With reference to hinge morphology and ligament, *Teranota* also supports the placement of the Orthonotidae in the Superorder Anomalodesmata, confirming the assumption of previous authors, although most other morphological characters of orthonotids are rather plesiomorphic. The main evolutionary lines within the Anomalodesmata followed one of two pathways – epibenthic or endobenthic. *Teranota* is interpreted as an endobenthic strategist, suggesting that these evolutionary pathways were also followed by minor groups. These extremely elongated shells, which did not occur (again) in the Late Palaeozioc, are interpreted as a (finally) failed early attempt to develop anatomical structures comparable with true siphons, which occurred convergently some millions of years later.

We would like to thank the followings: V. Ebbighausen provided us with the material and transferred the specimens to the Naturkundemuseum Berlin, K. Bieber and K. Schaumann undertook careful preparation; and A. Weisbrod carried out the photographic work. Finally, Chris Peel improved the final language of the manuscript. Both anonymous referees are thanked for their queries and suggestions.

References

AMLER, M. R. W. 1996. Die Bivalvenfauna des Oberen Famenniums West-Europas, 2. Evolution, Paläogeographie, Paläoökologie, Systematik 2. Palaeotaxodonta und Anomalodesmata. *Geologica et Palaeontologica,* **30**, 49–117.

—— 1999. Synoptical classification of fossil and Recent Bivalvia. *Geologica et Palaeontologica,* **33**, 237–248.

BEUSHAUSEN, L. 1895. Die Lamellibranchiaten des rheinischen Devon mit Ausschluss der Aviculiden. *Abhandlungen der Königlich Preussischen geologischen Landesanstalt, Neue Folge,* **17**, 1–514.

BRADSHAW, M. A. 1999. Lower Devonian bivalves from the Reefton Group, New Zealand. *Memoirs of the Association of Australasian Palaeontologists,* **20**, 1–171.

—— & MCCARTAN, L. 1991. Palaeoecology and systematics of Early Devonian bivalves from the Horlick Formation, Ohio Range, Antartica. *Alcheringa,* **15**, 1–42.

COPE, J. C. W. 1996. The early evolution of the Bivalvia. *In*: TAYLOR, J. (ed.) *Origin and Evolutionary Radiation of the Mollusca.* Oxford University Press, Oxford, 361–370.

—— 1997. The early phylogeny of the class Bivalvia. *Palaeontology,* **40**, 713–746.

DRISCOLL, E. G. 1965. Dimyarian Pelecypods of the Mississippian Marshall Sandstone of Michigan. *Paleontographica Americana,* **5**(35), 67–128.

HIND, W. 1896–1900. A monograph of the British Carboniferous Lamellibranchiata, 1. *Palaeontographical Society Monographs,* **50–54**, 1–476.

JOHNSTON, P. A. 1993. Lower Devonian Pelecypoda from southeastern Australia. *Memoirs of the Association of Australasian Palaeontologists,* **14**, 1–134.

KONINCK, L. G. DE 1885. Faune du Calcaire Carbonifère de la Belgique, 5. Lamellibranches. *Annales du Musée Royal d'Histoire Naturelle de Belgique,* **11**, 1–283.

MORRIS, N. J., DICKINS, J. M. & ASTAFIEVA-URBAITIS, K. 1991. Upper Palaeozoic Anomalodesmatan Bivalvia. *Bulletin of the British Museum (Natural History) Geology,* **47**(1), 51–100.

NEWELL, N. D. 1969*a*. Family Orthonotidae. *In:* MOORE, R. C. (ed.) *Treatise on Invertebrate Paleontology. Part N. Mollusca 6, Volume 2, Bivalvia.* Geological Society of America and University of Kansas Press, N818–819.

—— 1969*b*. Order Modiomorphoida. *In:* MOORE, R. C. (ed.) *Treatise on Invertebrate Paleontology. Part N. Mollusca 6, Volume 1, Bivalvia.* Geological Society of America and University of Kansas Press, N393–N401.

POJETA, J. JR 1971. Review of Ordovician pelecypods. *US Geological Survey Professional Paper,* **695**, 1–46.

—— 1978. The origin and early taxonomic diversification of pelecypods. *Philosophical Transactions of the Royal Society of London, Series B,* **284**, 225–246.

—— (ed.) 1986. Devonian Rocks and Lower and Middle Devonian Pelecypods of Guangxi, China, and the Traverse Group of Michigan. *US Geological Survey Professional Paper,* **1394**, 1–108.

—— 1987. Class Pelecypoda. *In:* BOARDMANN, R. S., CHEETHAM, A. & ROWELL, A. J. (eds) *Fossil Invertebrates.* Blackwell, Palo Alto, 365–435.

—— & GILBERT-TOMLINSON, J. 1977. Australian Ordovician Pelecypod Molluscs. *Bureau of Mineral Resources, Geology and Geophysics Bulletin,* **174**, 1–64.

—— & RUNNEGAR, B. 1985. The early evolution of diasome Molluscs. *In:* TRUEMAN, E. R. & CLARKE, M. R. (eds) *The Mollusca, Volume 10, Evolution.* Academic Press, Orlando, 295–336.

ROGALLA, N. S. & AMLER, M. R. W. 2000. *Teranota* n.

gen. (Bivalvia; Anomalodesmata) aus dem Mittel-Devon (Givetium) des Rheinischen Schiefergebirges. *Paläontologische Zeitschrift*, **74**, 69–73.

RUNNEGAR, B. 1974. Evolutionary history of the bivalve subclass Anomalodesmata. *Journal of Paleontology*, **48**, 904–939.

STANLEY, S. M. 1968. Post-Paleozoic adaptive radiation of infaunal bivalve molluscs – a consequence of mantle fusion and siphon formation. *Journal of Paleontology*, **42**, 214–229.

—— 1970. Relation of shell form to life habits of the Bivalvia (Mollusca). *Memoirs of Geological Society of America*, **125**, 1–496.

—— 1972. Functional morphology and evolution of byssally attached bivalved mollusks. *Journal of Paleontology*, **46**, 165–212.

The nature and origin of taxonomic diversity gradients in marine bivalves

J. A. CRAME

British Antarctic Survey, High Cross, Madingley Road, Cambridge CB3 0ET, UK

Abstract: Bivalves have been fundamental in developing the understanding of large-scale biodiversity patterns in the marine realm. A new study based on 29 regional bivalve faunas indicates that both latitudinal and longitudinal gradients are not so regular in form as was once imagined. The northern hemisphere latitudinal gradient has a marked step, or inflection, at 20–30°N, and in the southern hemisphere Australia forms a distinct diversity hotspot. A longitudinal gradient runs between a large tropical high diversity focus in the southern China–Indonesia–Australia region and a somewhat smaller one in the Panamic–Caribbean region.

It is unlikely that the marked asymmetry in both latitudinal and longitudinal gradients can be explained entirely by either the geometric patterns of species ranges in relation to geographical boundaries (the mid-domain effect) or the operation of contemporary, or equilibrium, factors. The role of history must be considered too, and in some cases this could involve a timescale of at least 60 Ma. For example, Australia has moved progressively northwards throughout the Cenozoic era and this seems to have led to the development of a unique assemblage of both tropical and temperate taxa around the continent. When Australia eventually collided with southeast Asia, some 15 Ma ago, three previously distinct tropical marine faunas were brought into close juxtaposition.

However, something else seems to have been involved in order to generate a marked pan-tropical Neogene diversification event in bivalves and other marine invertebrate taxa. It is striking how this event coincided with a prolonged interval of global cooling and the suggestion has been made that glacioeustatic cycles may have promoted the formation of a tropical species diversity pump. In concert with regular and profound sea-level changes, new taxa were created in marginal tropical regions and then concentrated in certain core regions. These same glacioeustatic cycles may also have served to accentuate the latitudinal limits of many tropical taxa.

As the knowledge of large-scale taxonomic diversity gradients in the marine realm has increased, so too has the realization that they are at least the partial products of major historical events.

Bivalves have proved to be one of the most successful groups for investigating the nature and origin of taxonomic diversity gradients in the marine realm. Widely distributed in all the world's oceans, they have been the subject of intense taxonomic investigations and are known to have an extensive fossil record. Early studies, such as those by Stehli *et al.* (1967, 1969) and Stehli (1968), illustrated their considerable potential in this respect, and this has been confirmed by the more recent work of both Flessa & Jablonski (1995, 1996) and Crame (2000*a*). Marine bivalves show clear and strong latitudinal diversity gradients in each hemisphere, as well as a pronounced pattern of longitudinal variation in the equatorial region (e.g. Stehli *et al.* 1967, fig. 3).

In a recent companion paper, a major revision of global patterns in bivalve species diversity was undertaken (Crame 2000*a*). The principal aim here was to investigate systematic changes in the taxonomic composition of bivalve faunas with respect to latitude. Do all the main groups show a regular decrease in numbers with increasing latitude, or is the decline more dramatic in some groups than in others? In addition, can the taxonomic composition of bivalve faunas tell us something about the relative ages of bivalve latitudinal gradients? Twenty-nine regional bivalve faunas were used in this study (see below) and, as a prelude to the taxonomic investigations, some general comments were made about the form of the latitudinal gradient in each hemisphere. A degree of hemispheric asymmetry was noted but its full significance was not discussed. Similarly, this new study displayed, but did not examine in detail, a pronounced longitudinal gradient.

From: Harper, E. M., Taylor, J. D. & Crame, J. A. (eds) *The Evolutionary Biology of the Bivalvia*. Geological Society, London, Special Publications, **177**, 347–360. 1-86239-076-2/00/$15.00

It is the intention of this paper to pay closer attention to the form of taxonomic diversity gradients in marine bivalves on a global scale, and what this may, in turn, be able to tell us about how they originated. Using the same basic data set as in the previous study (Crame 2000*a*), an attempt will be made to characterize the latitudinal gradient in each hemisphere and the overall longitudinal gradient. If the formation of large-scale taxonomic diversity gradients is being controlled by essentially contemporaneous, or equilibrium, factors, then it might be expected that they are regular in form. On the other hand, if historical (i.e. evolutionary or non-equilibrium) processes have prevailed then it might be expected that both types of gradient are less regular. The relative roles of equilibrium v. non-equilibrium factors in the generation and maintenance of global biodiversity patterns is currently under intense scrutiny (e.g. Ricklefs 1995).

It should also be emphasized here that an important recent series of null models have demonstrated how significant taxonomic diversity gradients can be generated in the absence of environmental gradients on either ecological or evolutionary timescales (Colwell & Lees 2000). These models are based on a phenomenon known as the mid-domain effect, an emergent statistical property of taxonomically defined biotas whereby there is an increasing overlap of species ranges towards the centre of a shared geographical domain. The latitudinal gradient predicted by such models is typically unimodal and bilaterally symmetrical (Colwell & Lees 2000).

Methods

Twenty-nine major marine bivalve faunas were selected to cover most of the principal shallow-water regions on the Earth's surface (Table 1). Region 1 is the Arctic, region 29 is the Antarctic, and in between are 27 lower latitude regions within the Pacific, Atlantic and Indian Oceans. Somewhat inevitably, these regions vary substantially in size, but there are no systematic increases or decreases in region area with either latitude or longitude. For each region a full systematic list of living bivalve species was constructed from a variety of taxonomic sources. Full details are given in Crame (2000*a*).

For this study, the number of species was plotted v. the mid-point latitude and longitude of each region (Figs 1 and 2). To judge the form of latitudinal diversity gradients, simple curves were fitted by eye to an outer envelope of points in each hemisphere (with some points in the northern hemisphere being excluded for reasons given below) (Fig. 1). In addition, classical linear regressions were performed on the relationship between the number of species and the latitude in each hemisphere. Longitudinal gradients were assessed by dividing the data points into three basic groups: northern temperate (> 32°N), tropical (32°N–25°S) and southern temperate (> 25°S) (Table 1 and Fig. 2). A three-dimensional mesh plot was also constructed to assist with the interpretation of both types of gradient (Fig. 3).

The form of the latitudinal gradient

Northern hemisphere

Initial investigations indicated what appeared to be a regular decline in bivalve diversity with increasing latitude in the northern hemisphere (Stehli *et al.* 1967; Schopf 1970; Valentine 1973). They were heavily influenced by the seminal study of Stehli *et al.* (1967), which employed a series of contour-fitting techniques to raw data from 26 regional localities. These very neatly picked out two peaks of tropical high diversity that were almost exactly coincident with the Equator, and then regular northward reductions in the numbers of species, genera and families (Stehli *et al.* 1967, figs 2–8). These latitudinal gradients seemed to accord well with those known in a number of other marine and terrestrial groups (Stehli 1968).

In a major revision of bivalve diversity patterns, Flessa & Jablonski (1995) plotted the global distribution of *c.* 600 genera from 14 superfamilies. Their northern hemisphere gradient, which was based on data from over 60 localities, strongly suggested that the decline in numbers of taxa with increasing latitude is not in fact as regular as was previously imagined. Clearly, there is a very steep fall in values that is approximately coincident with the northern margin of the tropics (i.e. 30°N), and thereafter a much gentler decline (Flessa & Jablonski 1995, fig.1). Stehli (1968, figs 13B, 33 and 50) had earlier commented on two comparatively small marine gastropod taxa, the cypraeids and strombids (Genus *Strombus*), that seemed to show steep diversity declines at the tropical margins, and Roy *et al.* (1994, 1996) found a decidedly stepped profile for a combined bivalve and gastropod gradient along the western coast of North America.

In this study, a statistically significant decline in the number of species with increasing latitude was found in the northern hemisphere, but it is again apparent that this is somewhat irregular (Fig. 1). The outer envelope of points used to assess its form was not placed through the very large value (1176 species) for the East China Sea region (32°N; Table 1) as this diversity hotspot has been displaced substantially northwards by the warm-water Kuro

Table 1. *Estimate of the number of living marine bivalve species in 29 major regions of the world*

Region	No. of species	Mid-point Latitude	Mid-point Longitude
Arctic	199	70°N	–
Norwegian Sea	268	63°N	2°W
Aleutian	173	58°N	150°W
Far-eastern Russia	261	55°N	160°E
British Isles	319	54°N	3°W
NW Atlantic	293	48°N	65°W
Oregonian	215	45°N	127°W
Mediterranean	318	38°N	22°E
East China Sea	1176	32°N	124°E
California	277	30°N	117°W
Eastern Arabia	384	23°N	54°E
Red Sea	416	20°N	43°E
Caribbean	658	20°N	64°W
Indonesia–Philippines	1211	14°N	120°E
Panamic	846	10°N	97°W
Andaman Sea	369	9°N	98°E
Sri Lanka	349	8°N	78°E
West Africa	360	4°S	6°W
East Africa	255	5°S	40°E
Brazil	391	12°S	43°W
Queensland–Northern Territory	911	20°S	142°E
Peru–Chile	177	24°S	78°W
Western Australia	658	25°S	119°E
South Africa	356	31°S	24°E
Eastern South America	148	33°S	53°W
South Australia	757	35°S	139°E
New Zealand	381	41°S	173°E
Magellan	119	49°S	67°W
Antarctica	162	70°S	–

Mid-point latitude and longitude of each region also shown. Data taken from Crame (2000*a*).

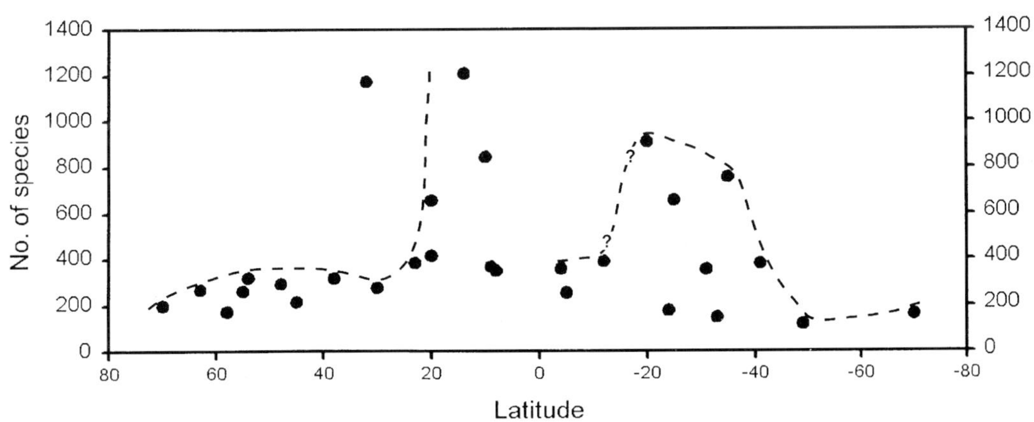

Fig. 1. A comparison of marine bivalve latitudinal gradients in species diversity in the northern and southern hemispheres. Data points taken from the 29 localities listed in Table 1 and curves fitted by eye to, essentially, an outer envelope of points (see text for further details). Latitudinal range from 80°N to 80°S. Statistical analysis of the gradients undertaken by performing classical linear regressions on data from the northern and southern hemispheres, respectively. In each case, latitude was held as the independent variable and the number of species (log transformed) as the dependent one. Results – expressed as percentage variance accounted for (i.e. the adjusted R^2 value × 100) and a t test on the null hypothesis that the regression coefficient (β) = 0 – are as follows: northern hemisphere, 38.4%, t = −3.31*; southern hemisphere, 6.5%, t = −1.33 (*, significant at p = 0.05). It should be noted that these results could be affected by the data being, at least to some extent, spatially dependent (or autocorrelated).

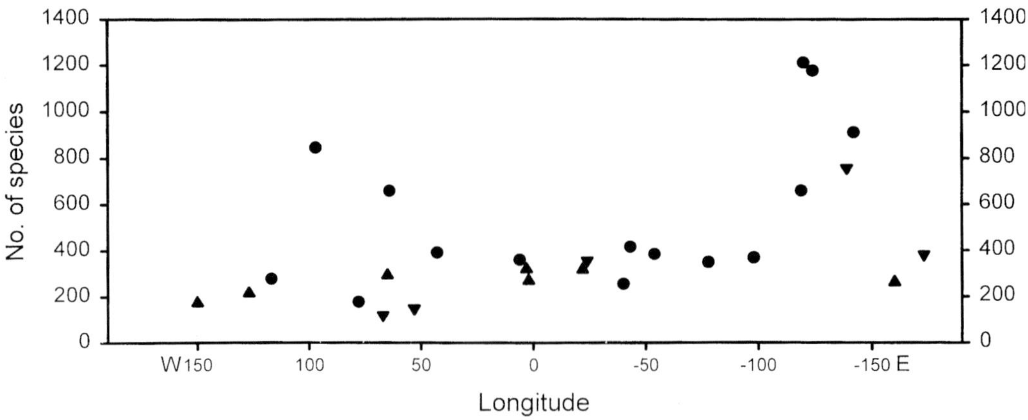

Fig. 2. A composite longitudinal gradient in species diversity for marine bivalves. Data points taken from Table 1, with the exceptions of of the Arctic and Antarctic, for which no suitable mid-point longitudes can be established. ▲, Northern temperate localities (> 32°N); ●, tropical (32°N–25°S); ▼, southern temperate (> 25°S).

Shio Current. Instead, the line of the curve has been set back closer to that for Indonesia–Philippines (14°N, 1211 species) (Fig. 1); whether it should then pass around this point and actually decrease into the lowest northern latitudes is uncertain.

What is more apparent is that, from a point of origin somewhere between the East China Sea and Indonesia–Philippines localities, the curve falls very steeply at *c.* 20°N. At *c.* 20–25°N it then levels off abruptly, giving it a distinct stepped, or bench-

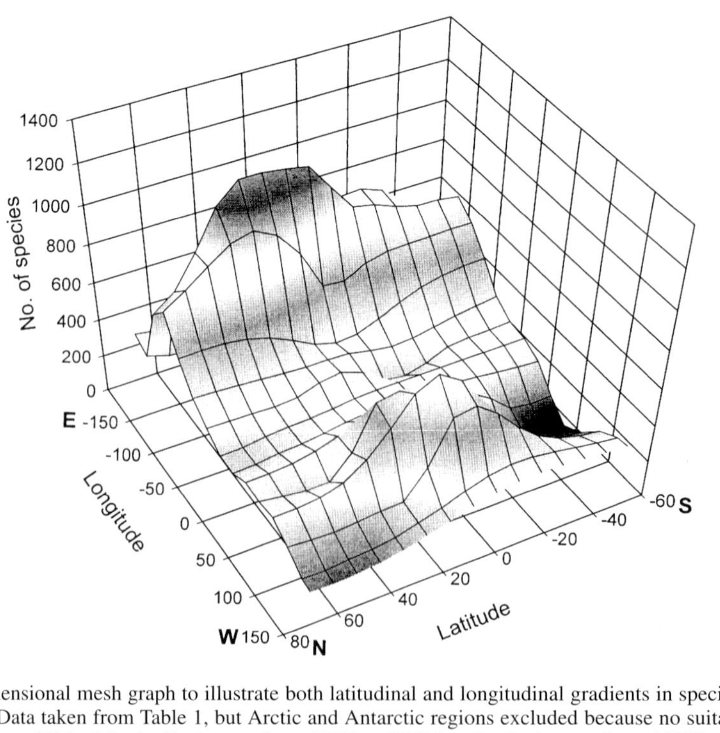

Fig. 3. Three-dimensional mesh graph to illustrate both latitudinal and longitudinal gradients in species diversity in marine bivalves. Data taken from Table 1, but Arctic and Antarctic regions excluded because no suitable mid-point longitudes can be established. Latitudinal range from 80°N to 60°S; longitudinal range from 150°E to 150°W.

like, profile overall (Fig. 1). Such a form agrees well with that described for bivalves by Flessa & Jablonski (1995, fig. 1), and for marine gastropods by Taylor & Taylor (1977, figs 1 and 2) and Roy *et al.* (1998, fig, 1). It should also be noted that a sharp, step-like rise in scleractinian coral diversity occurs at 21–23°N in the Atlantic Ocean and at 16–19°N in a more global compilation of data (Rosen 1981, figs 9.3 and 9.4).

There are some indications that the marked inflection seen at 20–30°N in bivalves and other marine taxa also occurs in the terrestrial realm. Stehli (1968, fig. 15) demonstrated its presence in lizards, and it may also be characteristic of amphibians and snakes too. Bats also show a very marked reduction in number of species at *c.* 24°N (Rozenzweig 1992, fig. 1a) and there is a distinct step-like fall in the numbers of New World birds at 20–25°N (Gaston & Blackburn 1996, fig. 1b). Taken together, these marine and terrestrial data sets are beginning to suggest that revision of the concept of the latitudinal diversity gradient in the northern hemisphere is needed. There are signs of a consistent and very steep drop in diversity that is more or less coincident with the edge of the tropics (20–30°N) and thereafter a much gentler decline into higher latitudes. Data for marine bivalves (Fig. 1) indicate that the reduction in taxa, at the species level, across this tropical step could be by as much as 66%. It is also apparent that this tropical–temperate divide marks a point of major turnover in the composition of biotas. Valentine (1973), for example, estimated an 80% turnover in the species composition of marine molluscan faunas along the western coast of North America at this level.

Southern hemisphere

Although comprehensive data sets have always been at a premium, there has been a general impression that southern hemisphere latitudinal diversity gradients were broadly similar in form to those known in the north. This was true of both the terrestrial and marine realms, and led to a widely held view that gradients were approximately symmetrical about the Equator (Stehli *et al.* 1967; Stehli 1968). In many respects this is still held to be the canonical view of global biodiversity (e.g. Ricklefs 1995). It is only comparatively recently, as many southern hemisphere biotas have come to be better known, that this concept has, at least to some extent, been brought into question. For a number of contrasting taxonomic groups there is a general impression that diversity tends to be slightly to considerably higher in the south (Gaston 1996, and refs cited therein), and this is confirmed by the latitudinal gradients established for New World birds (Gaston & Blackburn 1996), deep-sea isopods and gastropods (Rex *et al.* 1993), and deep-sea benthic foraminifera (Culver & Buzas 2000), respectively. However, it should be pointed out that in these studies there is generally less information available from the highest southern latitudes.

In contrast to these previous investigations, the curve constructed through the outer envelope of data points in this study suggests that diversity values for marine bivalves are somewhat lower in the southern hemisphere (Fig. 1). It also indicates a considerable degree of hemispheric asymmetry. From the maximum values obtained in the Queensland–Northern Territory region (20°S, 911 species) there is a steep, but by no means vertical, drop in values to the mid- and high-latitude regions. The pronounced step, or platform, is only weakly developed here, but it has to be remembered that, in comparison with the north, there are far fewer shallow-shelf localities.

It would also appear that there may be a genuine drop in diversity values from the Queensland–Northern Territory region into the three lowest latitude southern hemisphere regions (West Africa, 4°S, 360 species; East Africa, 5°S, 255 species; Brazil, 12°S, 391 species). This part of the curve is only tentatively completed in Fig. 1 because collection failure cannot be ruled out as a reason for these abnormally low values (Crame in 2000*a*). However, this may be, at best, only a partial explanation for this phenomenon and it is becoming increasingly apparent that the three Australasian regions form a distinct diversity 'bulge' (or hotspot) in the southern hemisphere (Figs 1 and 3). In contrast to the northern hemisphere, the decline in number of species with increasing latitude in the southern hemisphere is not statistically significant (Fig. 1).

The huge regional variation in southern hemisphere bivalve faunas is brought home when the overall latitudinal gradient is broken down into its component parts (Fig. 4). There is a shallow gradient along the western coast of South America, a somewhat steeper one along the eastern coast and then, in both cases, a slight rise from the Patagonian region into Antarctica. Nevertheless, it has to be pointed out that all the Antarctic data have been pooled into one very large region (Crame 2000*a*) and if smaller subunits, such as the Weddell or Ross Seas, had been used, then diversity would continue to drop into the highest southern latitudes. There is a moderately steep fall in bivalve numbers from West Africa, via South Africa, into Antarctica, but by far the most profound latitudinal gradient in the southern hemisphere is that from the Queensland–Northern Territory region, through South Australia and New Zealand, to Antarctica (Fig. 4). The very marked difference in diversity decline with

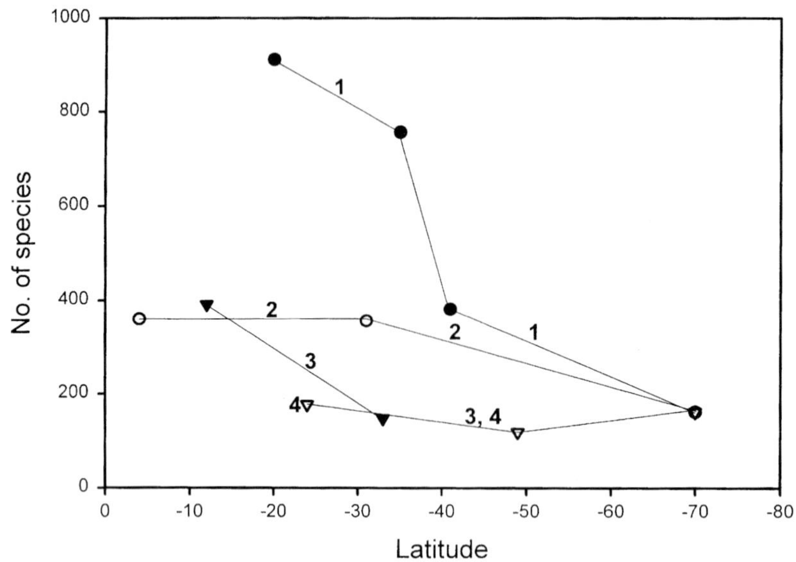

Fig. 4. Four bivalve latitudinal diversity gradients in species diversity in the southern hemisphere. 1, eastern Australia gradient between Queensland–Northern Territory, South Australia, New Zealand and Antarctica; 2, western Africa gradient between West Africa, South Africa and Antarctica; 3, eastern South America gradient between Brazil, Eastern South America, Magellan region and Antarctica; 4, western South America gradient between Peru–Chile, Magellan region and Antarctica. All regions and data points taken from Table 1.

increasing latitude between the eastern and western sectors of the southern hemisphere is also shown in the three-dimensional plot given in Fig. 3.

The form of the longitudinal gradient

Although much less often discussed than their latitudinal counterparts, longitudinal gradients are real phenomena that occur on a variety of spatial scales in both the terrestrial and marine realms (e.g. Simpson 1964; Stehli 1968; Ormond & Roberts 1997; Williams *et al.* 1997; Ellison *et al.* 1999). That produced in this study again picks out two tropical peaks or foci: a southern China–Indonesia–Australia one between *c.* 110 and 150°E, and a Caribbean–Panamic one between *c.* 65 and 110°W (Figs 2 and 3). The longitudinal gradient is particularly steep either side of these two peaks but then remarkably flat between *c.* 50°W and 100°E (i.e. much of the Atlantic and Indian Oceans).

Of course, this two-humped pattern could also be attributed to a taxonomic artefact produced by understudy of the South American and African low-latitude regions, and extensive oversplitting of faunas in both the Indonesian and central American regions. Nevertheless, it is again thought unlikely that this would have been on such a scale as to produce the observed features. It is interesting to note that similar tropical high diversity foci have also been observed in zooxanthellate corals (Rosen 1988; Veron 1995), mangroves (Ricklefs & Latham 1993; Ellison *et al.* 1999), and certain terrestrial groups such as palms, mosquitoes and parrots (Stehli 1968). Briggs (1996) has used a compendium of coral, mollusc, echinoderm and crustacean data to show that the Indo-West Pacific focus is consistently richer in taxa than the Caribbean–Panamic one. Such a conclusion is confirmed by the results obtained in this study (Figs 2 and 3).

The causes of latitudinal gradients

Some general considerations

At present, large-scale data sets, such as these from the marine realm, have not been tested against an appropriate geometric null model. It is entirely possible that the mid-domain effect (Colwell & Lees 2000) could be influencing bivalve latitudinal gradients, but on what scale is simply unknown. However, by analogy with studies that have been made in the terrestrial realm, it might be expected that the mid-domain richness peak would be essentially symmetrical about the Equator. The degree of asymmetry noted here would suggest that something else must also be involved.

Northern Hemisphere. There would seem to be a simple, but fundamental, link between latitudinal diversity gradients and temperature. Within the shallow-marine realm there is a high thermal gradient from *c.* 30°C at the Equator to < 0°C in the polar regions; a contour map of sea surface temperatures, assuming winter in both hemispheres, shows a strong latitudinal pattern (Stehli 1968, fig. 2). In turn, higher tropical temperatures generate the higher levels of primary productivity that, in some way, underpin more diverse ecosystems. There is currently a range of powerful species-energy hypotheses to account for large-scale diversity patterns in both the marine and terrestrial realms (e.g. Wright *et al.* 1993; Brown & Lomolino 1998).

It is important to emphasize that, within the tropics as a whole, patterns of primary productivity are very uneven. Many oceanic islands and atolls are surrounded by nutrient-poor, oligotrophic waters that support comparatively small numbers of suspension-feeding organisms such as bivalves. In marked contrast, many high islands and continental margins are bathed by nutrient-rich waters in which bivalve species abundance and diversity can be extremely high (Taylor 1997). A prime example here would be the East China Sea and the Indonesia–Philippines regions, which benefit from heavy pulses of nutrients generated by monsoonal rainfall on the margins of the southeast Asian continent (Crame 2000*a*). It must also be stressed that the precise mechanism whereby this high productivity is translated into high taxonomic diversity is uncertain. High tropical productivity should in fact only lead to a latitudinal gradient in biomass, so something else is required to generate a large number of tropical species. Here, two quite distinct processes are being dealt with and it may be illogical to assume that both are subject to the same set of controls (Blackburn & Gaston 1996).

It is also far from clear why diversity should drop so steeply in bivalves and other marine taxa at the edge of the tropics (i.e. at *c.* 23°N; Figs 1 and 3). It is known that there is a broad belt, *c.* 50° latitude in width, of homogenous temperatures within the tropics (Rosenzweig 1992, fig. 2) but thereafter temperature declines uniformly with increasing latitude. The sharp inflection in northern hemisphere latitudinal diversity gradients is not matched in temperature profiles (Rosen 1981, figs 9.3–9.6).

As an alternative to temperature *per se*, water density may be important, for it is known that the tropical pycnocline is a unique feature in all oceans that runs from 20°N to 20°S (Longhurst 1998). In the North Atlantic the transition between the strong tropical pycnocline and the weak subtropical one occurs at 22–23°N, coincident with the margin of the Northern Equatorial Current. In marked contrast to this, the Southern Equatorial Current is considerably broader and more diffuse than its northern equivalent, and there is no matching boundary at 22–23°S (Longhurst 1998). Here, then, may be at least a partial explanation for asymmetry in tropical–subtropical diversity patterns, but it again has to be stressed that the precise mechanism whereby water density exerts a control on taxonomic diversity patterns is unknown.

A further change in the stability of oceanic water masses occurs at 40° latitude. At latitudes lower than this level, near-surface waters (i.e. < 100 m) remain thermally stratified throughout the year, but above it wintertime cooling results in convective overturn and mixing, resupplying nutrients back into the surface waters. This means that, although mean annual primary productivity is generally higher in the high-latitude and polar regions, it is markedly seasonal. This change from annual to seasonal patterns of primary productivity at 40° latitude may have been a control on the taxonomic diversification of a number of pelagic and benthic marine groups (Taylor & Taylor 1977; Angel 1996, 1997; Culver & Buzas 2000).

Southern hemisphere. It has been argued elsewhere (Crame 2000*a*) that the Australian 'hotspot' in bivalve diversity may be directly related to a distinctive set of oceanographic conditions in the southern hemisphere. The imposition of cold boundary currents along the western coasts, and major river system outflows on the eastern ones, has severely restricted the development of coral reef and associated sedimentary environments in South America and Africa. Only along the north-eastern and northwestern coasts of Australia are reefs extensively developed, with the richest scleractinian coral faunas anywhere in the southern hemisphere occurring in the northern Great Barrier Reef, and moderately diverse ones being associated with the Northwestern Shelf Reefs (Veron 1995). Carbonate muds are still widely distributed along much of the South Australian coast and overall there is thought to be a much greater habitat heterogeneity for benthic marine organisms, such as bivalves, around the entire Australian continent (Crame 2000*a*).

The effects of history

Within the last decade there has been a dramatic increase in the understanding of the ways in which regional processes can structure natural communities (e.g. Huston 1999; Ricklefs 1999, and refs cited therein). These regional processes can be divided into two categories: (1) those that are directly attributable to the physical properties of a large area, such as its size, climate, geology and age; (2) those that are due to some form of contingent event (such as the sudden imposition of a barrier to dispersal or a mass extinction event). In both cases, such processes are reflected in changes in immigration or emigration and speciation or extinction rates. Regional processes can occur on timescales of up to millions, or even tens of millions, of years (e.g. Caley & Schluter 1997).

Northern hemisphere. If an obvious contemporary explanation cannot be found for the very marked step in diversity profiles more or less coincident with the edge of the tropics, then perhaps there is some form of historical explanation. Rosen (1981), for example, has pointed out how oceanic islands and smaller land areas within the tropics may have been particularly susceptible to a combination of climatic and tectonic events throughout the Cenozoic era. He was particularly concerned with zooxanthellate corals, noting how their normal depth range was often exceeded by the scale of glacioeustatic sea-level fluctuations, and how they were sharply bounded by a minimum temperature (normally the 20°C isotherm) and thus a maximum latitude. Rosen (1981, p. 125) argued that there must be a sudden latitudinal rise in the susceptibility of corals to climate–tectonic events. During the late Neogene (i.e. the last *c.* 15 Ma), a marked intensification of glacioeustatic cycles would have left reefs repeatedly and abruptly stranded. This has served to accentuate their northern limits, especially in the western Atlantic and western Pacific Oceans.

It is quite likely that other tropical marine invertebrate taxa were also affected by climate–tectonic events. In a study of the effects of Late Cenozoic sea-level fluctuations on tropical oceanic islands, Paulay (1990) has shown that bivalves restricted to inner-reef habitats were particularly liable to extinction. This is where the soft substrates predominate on most tropical islands and atolls, and during a sea-level regression significantly greater numbers of infaunal rather than epifaunal taxa were prone to at least local extinction. Serial restriction of coral reef and associated sedimentary habitats would have served to repeatedly limit the latitudinal range of a variety of marine invertebrate taxa.

Southern Hemisphere. Not only may Australia be unique among the southern continents in its oceanographic setting at the present day, but there is some evidence to suggest that it may have been so through the past too. Since the mid- to Late Cretaceous dispersion of the component Gondwana continents, it is evident that Australia has moved considerably further north, relative to Antarctica, than either South America or Africa (e.g. Lawver *et al.* 1992). Approximately 100 Ma ago Australia was still attached to Antarctica in a high-latitude position. It was characterized by a distinctive, temperate marine fauna that comprised part of an austral, or Weddellian, province (Zinsmeister 1982). By the end of the Cretaceous period (65 Ma ago) there were probably shallow seas between Australia and Antarctica, and by 50 Ma ago (i.e. late Early Eocene) full deep-water separation had been achieved (Crame 1999, and refs cited therein). Throughout the Cenozoic era Australia moved progressively northwards into lower, and warmer, latitudes and by the early Middle Miocene (*c.* 17 Ma ago) Indo-Pacific (i.e. tropical) molluscan genera reached a maximum development in southeastern Australia (Darragh 1985). Thereafter the percentage of tropical taxa in South Australia has declined, and today the Australian molluscan fauna can be split into essentially northern tropical and southern warm-temperate components; these are separated by extensive zones of overlap on the southeastern and southwestern coasts, respectively (e.g. Ponder & Wells 1998, fig. 1.87).

This strong admixture of tropical and temperate faunal elements may well be a prime reason for high regional diversity values around the Australian continent. As it has moved steadily northwards through time, it has literally carried a number of originally cool-temperate taxa into progressively warmer climatic belts. Australia at the present day seems to represent a zone of genuine overlap between several major biogeographic realms. It should be possible to test this hybrid biogeographic province hypothesis using a combination of biological and palaeobiological data. Already it is striking how the relict bivalve genus *Neotrigonia*, as well as representatives of taxa such as *Eucrassatella* and *Talabrica* (both members of the 'ancient' family Crassatellidae), can be found right around the continent (Lamprell & Whitehead 1992; Darragh 1998; Slack-Smith 1998).

The causes of longitudinal gradients

Some general considerations

In theory, the mid-domain effect should apply just as much to longitudinal gradients as to latitudinal ones (Colwell & Lees 2000). However, even less is known about how it may have influenced longitudinal gradients in the marine realm, and crucial to the construction of null models in this context will be the definition of geographically distinct domains. If the vast Indo-Pacific province constitutes a single, discrete domain, then it is possible that the Indo–West Pacific high diversity focus represents some form of mid-domain diversity peak. Nevertheless, if this is the case then it is not nearly so easy to explain the presence of the second, Panamic–Caribbean peak. Clearly, this is an issue that needs considerable further investigation.

The existence of bivalve longitudinal gradients may also be explained entirely by recourse to maintenance, or ecological, processes. The Indo-Pacific faunal realm is very much larger than the Atlantic one, and there is a much greater concentration of oceanic islands in the Indo-West Pacific high diversity focus than in the Caribbean. On species area grounds alone, the southern China–Indonesia–Australia diversity peak would be expected to be larger than the Caribbean–Panamic one (Figs 2 and 3). A consideration of tropical current flow and dispersal patterns would suggest that taxa should accumulate on the western margins of both the Pacific and Atlantic Oceans (e.g. Ladd 1960), and it is noticeable how some of the monsoonal coasts of the Indian Ocean have rather low diversity values.

Nevertheless, Rosen (1984) has questioned whether ecological factors alone are sufficient to account for the very marked disparity in coral faunas between the two regions (with there being 24 coral genera in the Atlantic as opposed to 87 in the Indo Pacific). This is especially so when the geological histories of the two faunas are considered and it is found that in the early Cenozoic (i.e. Palaeocene and Eocene) the global centre of coral diversity was very much in the Atlantic, Caribbean and Mediterranean regions (Rosen 1984). All the available palaeontological evidence suggests that, at least for scleractinian corals, the later Cenozoic (i.e. Neogene) was a period of relative decline for Atlantic–Caribbean faunas and one of spectacular development for Indo-West Pacific ones (Rosen 1988).

The effects of history

Further evidence of the potential importance of historical events in determining the form of present-day longitudinal diversity gradients comes from cladistic analyses of tropical coral faunas. Recent studies by both Rosen & Smith (1988) and Pandolfi (1992) have revealed a series of distinctive overlapping distribution patterns that seem to imply the presence of essentially north–south trending physical barriers. It would appear that a once homogenous coral fauna, that has been repeatedly subdivided by both climatic and tectonic events, is being observed.

It is important to emphasize that some form of physical differentiation within the vast tropical shallow-water tropical realm can be traced back to either the earliest part of the Cenozoic era, or even the latest Cretaceous (i.e. at least 50–60 Ma ago). Barriers to dispersal caused by tectonic events in the present-day Middle East and the deep-water eastern Pacific Ocean could be of at least this age (Newell 1971; Kay 1984; Rosen 1988). Nevertheless, it was not until the beginning of the Miocene (i.e. *c.* 20 Ma ago) that very strong differentiation between the Atlantic and Indo-Pacific provinces began. This was the time of collision between Africa–Arabia and Europe and the final closure of the Tethyan Ocean. Thereafter, the continued northward movement of Africa and Europe took the Mediterranean region out of the reef belt and left the Atlantic province progressively more isolated (Frost 1977; Rosen 1988).

However, there was also another crucially important Neogene tectonic event in the tropical ocean, involving the northward movement of the Australian plate until it collided with Indonesia in the late Middle Miocene (i.e. *c.* 15 Ma ago). Of course, this did not result in the formation of a complete terrestrial barrier and, indeed, may well have led to a period of convergence between previously semi-isolated, shallow-water faunas (Rosen & Smith 1988). This may, in turn, have accounted for the obvious mid- to late Miocene pulse of Indo-West Pacific coral diversification (see below; Rosen 1988).

Perhaps the most intriguing feature about the late Middle Miocene collision of Australia with Indonesia is that it was coincident with a marked deterioration in global climate (caused by a major expansion of the Antarctic ice sheet; Kennett *et al.* 1985). This would certainly have led to an intensification in the frequency and scale of glacioeustatic sea-level cycles, which were then imposed on a new geographical pattern of islands (Rosen 1984). At the very least, this would have triggered a significant increase in vicariant events when the Indonesian archipelago was very largely left exposed during the low stands. However, Rosen (1984) went on to suggest that the pattern of islands throughout the central Indo-Pacific province may now have been such as to create a significant taxonomic diversity pump. An outer zone of more widely spaced islands, situated in both the western Pacific and Indian Oceans, was the putative site of vicariance during repeated low stands. During the subsequent high stands, the new species created in this outer zone were pumped into the central Indonesian focus, which becomes a refuge comprising numerous sympatric species. It is the geographical pattern of islands within the central Indo-Pacific province, coupled with a prolonged sequence of glacioeustatic cycles, that is the key to its current high diversity. The same process has been less effective in the Atlantic because of a much sparser pattern of outlying islands relative to the Caribbean focus (Rosen 1984).

The composition of bivalve latitudinal gradients

In the companion study, Crame (2000*a*) demonstrated that the present-day latitudinal gradient in marine bivalves (in both hemispheres) is overwhelmingly underpinned by a tropical and low-latitude concentration of infaunal taxa. These are very largely contained within the extensive heteroconch clade (which comprises common families such as the Veneridae, Tellinidae, Cardiidae and Mactridae), but there are also concentrations of burrowing anomalodesmatans and certain arcoids (principally from the Subfamily Anadarinae). There is a latitudinal gradient in epifaunal pteriomorph bivalves at the present day (principally pectinids, ostreids and limids) but it is on nothing like the same scale (Crame 2000*a*).

It is also becoming apparent that latitudinal gradients in infaunal taxa can be traced back over very considerable periods of time. They have been detected (in both hemispheres) for the Late Jurassic Tithonian stage (i.e. *c*. 150 Ma ago), where they are underpinned by tropical and low-latitude concentrations of palaeoheteroconchs (i.e. trigoniids), older heteroconch families such as the Astartidae, and newer ones such as the Tancrediidae, Quenstedtiidae and Arcticidae (Crame 1996). There is even some evidence to suggest that infaunal gradients can be detected in the Late Palaeozoic era (i.e. *c*. 290 Ma ago; Bambach 1990).

The progressive infaunalization of the Bivalvia through geological time is, of course, by now well established (e.g. Crame 2000*b*, and refs cited therein), but why should it have been concentrated in tropical and low-latitude regions? Perhaps the simplest solution is that it can be directly attributed to a concomitant rise in various durophagous predatory groups, many of which clearly show their maximum development at the present day in the tropics (Vermeij 1978). The spectacular rise in tropical predators over the last 100 Ma has been matched by a similar rise in their infaunal bivalve prey, providing a striking example of evolutionary escalation (Vermeij 1978).

A further important point to consider here is that there may be more infaunal taxa in the tropics simply because there is more fine sediment available for them to burrow in. Whereas at the present day *c*. 50% of the area of inner continental shelves in the tropics is composed of fine mud (of mostly algal origin), the equivalent figure for 60° latitude is only 10% (there being a much higher proportion of sand, gravel and rock fragments of glacial origin) (Hayes 1967). This would indeed seem to be a tenable explanation at the present day but is perhaps less applicable further back in time. For example, in the Late Jurassic there were clear latitudinal gradients in infaunal bivalves but no polar ice caps (Crame 1996).

This leads to consideration of whether some form of physiological explanation for the low-latitude concentration of infaunal bivalves might not be more appropriate. Of the various alternatives that could be discussed here, two can be chosen which are known to vary systematically with latitude (Clarke 1993) – seasonality of food supply and rate of shell calcification. Nevertheless, it is not at all obvious why either of these processes should produce such a steep gradient at the present day in the heteroconchs and yet have comparatively little influence on other clades such as the protobranchs, lucinoids or mytiloids. What should be considered is a physiological process that is much more specific to infaunal taxa – i.e. the rate of burrowing. In a classic study based on a wide variety of taxa, Stanley (1970) showed conclusively that burrowing rates are strongly influenced by temperature. Maximum rates were observed in the 20–30°C range, and at lower values there was a very sharp decline in burrowing activity. It may be very much easier for bivalves to burrow, and thus escape the attentions of predators, in low- rather than high-latitude regions.

Discussion

There remains a great deal to be learnt about the nature and origin of the larger scale patterns of life on Earth. Following on from the classic early studies of Stehli (1968) and Stehli *et al.* (1967, 1969), there was, perhaps, a tendency to oversimplify these features; diversity was perceived to vary regularly with latitude, with patterns in the south largely mirroring those in the north. Such a concept was understandable because Stehli (1968) had employed a variety of techniques to smooth his data. He was, after all, using generalized taxonomic diversity gradients to test hypotheses of polar wandering and continental drift.

The patterns of regional bivalve diversity presented here can be added to a growing volume of evidence which suggests that latitudinal gradients in particular are much more complex in form than was originally envisaged (e.g. Gaston 1996; Gaston & Spicer 1998). In the northern hemisphere, a consistent pattern is beginning to emerge of a gradient that drops very steeply at the edge of the tropics (20–30°N) but then declines at a much lower rate in higher latitudes (Fig. 1). Quite why there should be this steep edge, or 'wall', to tropical high diversity values is uncertain. The promotion of tropical diversity by the mid-domain effect, and equilibrium processes such as high annual primary productivity and the habitat heterogeneity associated with coral reefs – both the latter two processes in particular may have been important in governing the accumulation of large numbers of infaunal, suspension-feeding heteroconch bivalves – should not be dismissed.

As the gradient is clearly not uniform, and there is no simple relationship between the tropical 'wall' and temperature (or indeed any other environmental parameter), consideration is required of whether non-equilibrium, or historical, factors may also have been important in its formation. Many tropical marine organisms are indeed adapted to narrow temperature and depth ranges, and repeated exposure during Cenozoic tectonic–climatic cycles may have served to accentuate their latitudinal limits (*sensu* Rosen 1981). This hypothesis needs to be tested further and to do this closer examination of the precise nature of the 'wall' is needed. Is it of the same scale and in the same position in all three major oceans; does it vary in intensity between major groups; and how complete is the turnover of taxa across it?

There is an impression of much stronger interregional variability in bivalve diversity in the southern hemisphere, with Australia forming a distinct hotspot (Figs 1–3). This could again be attributed largely to equilibrium processes, as northeastern and northwestern Australia exhibit the richest development of coral reefs in the southern hemisphere. Nevertheless, it is striking how South Australia bears a diverse temperate bivalve assemblage that seems to comprise an unusually high percentage of endemic taxa (e.g. Ponder & Wells 1998). This phenomenon may also be repeated in other South Australian benthic marine invertebrate taxa such as the peracarid Crustacea (Poore & Wilson 1993). It is known that, over the last 50 Ma, the continent has moved northwards from the edge of Antarctica to tropical latitudes, and this may have led to a unique juxtaposition of temperate and tropical faunas (Darragh 1985).

The same northward movement of Australia may also have led to the overlap of three once-separated tropical marine faunas (i.e. the Indian Ocean and the western Pacific and tropical Australian faunas), and thus the generation of a significant Indo-West Pacific high diversity focus. But then the question remains as to whether this process of simple overlap alone would have been sufficient to produce such a high diversity peak? Presumably, it involved some form of higher level diversity differentiation at the between-habitat and between-community levels (i.e. β and γ diversification). In addition, it is likely that the area of coral reefs in the Indonesian region must have increased substantially at this time too.

However, there is now a considerable volume of evidence to suggest that a marked Neogene pulse in species diversification was pan-tropical in nature. Besides scleractinian corals (Rosen 1988; Veron 1995), it has also been detected in bivalves (Flessa & Jablonski 1996; Crame 2000*a*), gastropods (Taylor *et al.* 1980; Kohn 1990) and planktonic foraminifera (Wei & Kennett 1986). This pulse also affected the Panamic and Caribbean provinces, although its effects here have to some extent been masked by high levels of local extinction in the Late Pliocene (Jackson *et al.* 1993; Allmon *et al.* 1996). This leads to an inference that something in addition to the major collision of the Australian plate with southeast Asia must have been involved in the amplification of these tropical high diversity foci, and it would seem only logical to turn to some sort of mechanism that involves global climate change.

In a series of stimulating papers, Rosen (1981, 1984, 1988) has suggested how a glacioeustatically driven species diversity pump may have operated over the last *c.* 15 Ma in the tropical regions. The key here is not so much the scale and regularity of temperature–sea-level changes as the geographical pattern of tropical islands upon which they were superimposed. That in the central Indo-Pacific region is thought to have been ideal for generating peripheral isolates during low sea-level stands and then 'pumping' newly formed species into a central reservoir during high stands. In this model, the Indo-West Pacific high diversity focus represents a refugium for a large number of sympatric species and this seems to fit well with the observed mosaics of overlapping distribution patterns for many reef corals (Rosen 1988; Pandolfi 1992). It is these same glacioeustatic cycles that are thought to have accentuated the latitudinal ranges of many tropical taxa, especially in the northern hemisphere.

Fewer islands, especially in the critical outlying regions, means that the postulated diversity pump may have been less effective in the Caribbean region. It is also apparent that this region was progressively isolated through the Neogene by the continued northward movement of the African–Mediterranean block and cooling of the world ocean (Rosen 1984; Rosen & Smith 1988). The final separation of the Panamic and Caribbean provinces was achieved by the Late Pliocene emergence of the Isthmus of Panama (Jackson *et al.* 1993).

Summary and conclusions

- Because of their widespread geographical distribution, both at the present day and in the geological past, bivalves are one of the most useful taxonomic groups for investigating the nature and origin of large-scale biodiversity patterns in the marine realm.

- A composite latitudinal gradient for the northern hemisphere shows a pronounced step, or inflection, more or less coincident with the edge of the tropics (20–30°N), and this is a feature repeated in a range of other marine and terrestrial groups. The database for the southern hemisphere is less complete. There is a very sharp fall in the number of bivalve species along the eastern coast of Australia into New Zealand and Antarctica, but elsewhere the gradients are much flatter. Australia seems to form a distinct diversity hotspot in the low- to mid-latitudes of the southern hemisphere.
- A striking longitudinal gradient in marine bivalves occurs between a large southern China–Indonesia–Australia high-diversity focus and a somewhat smaller Caribbean–Panamic one.
- There is no simple match between latitudinal gradients in temperature and species diversity. Some form of species-energy hypothesis may be more appropriate, but if so it must explain why taxonomic diversity is so obviously concentrated in some parts of the tropics but not in others. The pycnocline is a prominent asymmetrical feature of the tropical ocean whose influence on taxonomic diversity is unknown.
- It is possible that the sheer size of the tropical biome, coupled with the greater amount of incident energy that it receives, underpins the taxonomic diversity gradients seen at the present day. The bivalve gradients could, in turn, be enhanced by the fact that the rate of burrowing is positively correlated with temperature. However, it is unlikely that these processes on their own can explain either the very steep 'wall' to tropical diversity, the asymmetry between the two hemispheres or the asymmetry between the two tropical high diversity foci.
- Both latitudinal and longitudinal gradients in bivalve diversity could be influenced by the geometric theory of species richness known as the mid-domain effect, but to what extent is unknown. The marked asymmetry in these gradients may be an indication that it cannot be the only underlying cause.
- To fully explain the form of present-day taxonomic diversity gradients the role of historical, or contingent, processes also needs to be assessed. There would appear to be a long history of vicariant events within the tropical oceans but of particular importance are those that have occurred within the Neogene period (i.e. within the last *c.* 20 Ma). Within this timespan the Tethyan Ocean was finally closed and the Atlantic and Indo-Pacific marine faunas became sharply differentiated.
- The northward movement of Australia throughout the Cenozoic era is seen as a crucial historical event. It seems to have resulted in a genuine admixture of tropical and temperate faunas around the continent, and the generation of exceptionally high levels of endemism. When the northern edge of the Australian plate collided with southeast Asia 15 Ma ago the resulting overlap of tropical faunas in the Indonesian region may have led to a marked phase of diversification there.
- It is becoming apparent that a sharp pulse of Neogene diversification was pan-tropical in nature and not just confined to the Indonesian region. Quite why there should be a pulse of tropical diversification at this time is uncertain, especially as global climates are known to have deteriorated significantly over the last 15 Ma. One intriguing possibility is that glacioeustatic sea-level cycles acted as a type of species-diversity pump, with new taxa being formed in peripheral regions during low stands and then being 'pumped' into central refugia during high stands. The same climatic cycles that drove this process may also have served to accentuate the latitudinal limits of a wide range of tropical organisms.
- A combination of tropical vicariant events and glacioeustatic processes through the Neogene may have left an indelible imprint on the present-day, large-scale patterns of life on Earth.

I am grateful to two anonymous referees and J. D. Taylor for comments and suggestions that helped to significantly improve this paper.

References

ALLMON, W. D., ROSENBERG, G., PORTELL, R. W. & SCHINDLER, K. 1996. Diversity of Pliocene–Recent mollusks in the Western Atlantic: extinction, origination, and environmental change. *In*: JACKSON, J. B. C., BUDD, A. F. & COATES, A. G. (eds) *Evolution and Environment in Tropical America*. University of Chicago Press, Chicago, 271–302.

ANGEL, M. V. 1996. Ocean diversity. *In*: SUMMERHAYES, C. P. & THORPE, S. A. (eds) *Oceanography. An Illustrated Guide*. Manson Publishing, London, 228–243.

—— 1997. Pelagic biodiversity. *In*: ORMOND, R. F. G., GAGE, J. D. & ANGEL, M. V. (eds) *Marine Biodiversity. Patterns and Processes*. Cambridge University Press, Cambridge, 35–68.

BAMBACH, R. K. 1990. Late Palaeozoic provinciality in the marine realm. *In*: MCKERROW, W. S. & SCOTESE, C. R. (eds) *Palaeozoic Palaeogeography and Biogeography*. Geological Society, London, Memoir, **12**, 307–323.

BLACKBURN, T. M. & GASTON, K. J. 1996. A sideways look at patterns in species richness, or why there are so few species outside the tropics. *Biodiversity Letters*, **3**, 44–53.

BRIGGS, J. C. 1996. Tropical diversity and conservation. *Conservation Biology*, **10**, 713–718.

BROWN, J. H. & LOMOLINO, M. V. 1998. *Biogeography*, 2nd Edition. Sinauer Associates, Sunderland, MA.

CALEY, M. J. & SCHLUTER, D. 1997. The relationship between local and regional diversity. *Ecology*, **78**, 70–80.

CLARKE, A. 1993. Temperature and extinction in the sea: a physiologist's view. *Paleobiology*, **19**, 499–518.

COLWELL, R. K. & LEES, D. C. 2000. The mid-domain effect: geometric constraints on the geography of species richness. *Trends in Ecology & Evolution*, **15**, 70–76.

CRAME, J. A. 1996. Antarctica and the evolution of taxonomic diversity gradients in the marine realm. *Terra Antarctica*, **3**, 121–134.

—— 1999. An evolutionary perspective on marine faunal connections between southernmost South America and Antarctica. *Scientia Marina*, **63** (Suppl. 1), 1–14.

—— 2000*a*. Evolution of taxonomic diversity gradients in the marine realm: evidence from the composition of Recent bivalve faunas. *Paleobiology*, **26**, 188–214.

—— 2000*b*. Intrinsic and extrinsic controls on the diversification of the Bivalvia. *In*: CULVER, S. J. & RAWSON, P. F. (eds) *Biotic Response to Global Change: the Last 145 Million Years*. Cambridge University Press, Cambridge, 136–148.

CULVER, S. J. & BUZAS, M. A. 2000. Global latitudinal species diversity gradient in deep-sea benthic foraminifera. *Deep-Sea Research I*, **47**, 259–275.

DARRAGH, T. A. 1985. Molluscan biogeography and biostratigraphy of the Tertiary of southeastern Australia. *Alcheringa*, **9**, 83–116.

—— 1998. Order Trigonioida. *In*: BEESLEY, P. L., ROSS, G. J. B. & WELLS, A. (eds) *Mollusca: The Southern Synthesis. Part A. Fauna of Australia. Volume 5*. CSIRO Publishing, Melbourne, 294–296.

ELLISON, A. M., FARNSWORTH, E. J. & MERKT, R. E. 1999. Origins of mangrove ecosystems and the mangrove biodiversity anomaly. *Global Ecology and Biogeography*, **8**, 95–115.

FLESSA, K. W. & JABLONSKI, D. 1995. Biogeography of Recent marine bivalve molluscs and its implications for paleobiogeography and the geography of extinction: a progress report. *Historical Biology*, **10**, 25–47.

—— & —— 1996. Geography of evolutionary turnover. *In*: JABLONSKI, D., ERWIN, D. H. & LIPPS, J. (eds) *Evolutionary Paleobiology*. Chicago University Press, Chicago, 376–397.

FROST, S. H. 1977. Miocene to Holocene evolution of Caribbean reef-building corals. *Proceedings of the Third International Coral Reef Symposium*, **2**, 353–359.

GASTON, K. J. 1996. Biodiversity – latitudinal gradients. *Progress in Physical Geography*, **20**, 466–476.

—— & BLACKBURN, T. M. 1996. The tropics as a museum of biological diversity: an analysis of the New World avifauna. *Proceedings of the Royal Society of London*, **B263**, 63–68.

—— & SPICER, J. L. 1998. *Biodiversity. An Introduction*. Blackwell, London.

HAYES, M. O. 1967. Relationship between coastal climate and bottom sediment type on the continental shelf. *Marine Geology*, **5**, 111–132.

HUSTON, M. A. 1999. Local processes and regional patterns: appropriate scales for understanding variation in the diversity of plants and animals. *Oikos*, **86**, 393–401.

JACKSON, J. B. C., JUNG, P., COATES, A. G. & COLLINS, L. S. 1993. Diversity and extinction of tropical American mollusks and emergence of the Isthmus of Panama. *Science*, **260**, 1624–1626.

KAY, E. A. 1984. Patterns of speciation in the Indo-West Pacific. *Special Publications of the Bernice Pauahi Bishop Museum*, **72**, 15–31.

KENNETT, J. P., KELLER, G. & SRINIVASAN, M. S. 1985. Miocene planktonic foraminiferal biogeography and paleooceanographic development of the Indo-Pacific region. *In*: KENNETT, J. P. (ed.) *The Miocene Ocean: Paleoceanography and Biogeography*. Memoir of the Geological Society of America, **163**, 197–236.

KOHN, A. J. 1990. Tempo and mode of evolution in Conidae. *Malacologia*, **32**, 55–67.

LADD, H. S. 1960. Origin of the Pacific island molluscan fauna. *American Journal of Science*, **258**, 137–150.

LAMPRELL, K. & WHITEHEAD, T. 1992. *Bivalves of Australia. Volume 1*. Crawford House Press, Bathurst.

LAWVER, L. A., GAHAGAN, L. M. & COFFIN, M. 1992. The development of paleoseaways around Antarctica. *In*: KENNETT, J. P. & WARNKE, D. A. (eds) *The Antarctic Paleoenvironment: A Perspective on Global Change. Part One*. Antarctic Research Series, **56**, 7–30.

LONGHURST, A. 1998. *Ecological Geography of the Sea*. Academic Press, London.

NEWELL, N. D. 1971. An outline history of tropical organic reefs. *American Museum Novitates*, **2465**, 1–37.

ORMOND, R. F. G. & ROBERTS, C. M. 1997. The biodiversity of coral reef fishes. *In*: ORMOND, R. F. G., GAGE, J. D. & ANGEL, M. V. (eds) *Marine Biodiversity. Patterns and Processes*. Cambridge University Press, Cambridge, 216–257.

PANDOLFI, J. M. 1992. Successive isolation rather than evolutionary centres for the origination Indo-Pacific reef corals. *Journal of Biogeography*, **19**, 593–609.

PAULAY, G. 1990. Effects of late Cenozoic sea-level fluctuations on the bivalve faunas of tropical oceanic islands. *Paleobiology*, **16**, 415–434.

PONDER, W. F. & WELLS, F. E. 1998. Distribution and relationships of marine and estuarine fauna. *In*: BEESLEY, P. L., ROSS, G. J. B. & WELLS, A. (eds) *Mollusca: The Southern Synthesis. Fauna of Australia. Volume 5. Part A*. CSIRO Publishing, Melbourne, 77–80.

POORE, G. C. B. & WILSON, G. D. F. 1993. Marine species richness. *Nature*, **361**, 597–598.

REX, M. A., STUART, C. T., HESSLER, R. R., ALLEN, J. A., SANDERS, H. L. & WILSON, G. D. F. 1993. Global-scale latitudinal patterns of species diversity in the deep-sea benthos. *Nature*, **365**, 636–639.

RICKLEFS, R. E. 1995. 3.2 The distribution of biodiversity.

In: HEYWOOD, V. H. (ed.) *Global Biodiversity Assessment*. Cambridge University Press, Cambridge, 139–173.

—— 1999. Global patterns of tree species richness in moist forests: distinguishing ecological influences and historical contingency. *Oikos*, **86**, 369–373.

—— & LATHAM, R. E. 1993. Global patterns of diversity in mangrove floras. *In*: RICKLEFS, R. E. & SCHLUTER, D. (eds) *Species Diversity in Ecological Communities. Historical and Geographical Perspectives*. University of Chicago Press, Chicago, 215–229.

ROSEN, B. R. 1981. The tropical high diversity enigma – the corals'-eye view. *In*: FOREY, P. L. (ed.) *The Evolving Biosphere*. British Museum (Natural History), London and Cambridge University Press, Cambridge, 103–129.

—— 1984. Reef coral biogeography and climate through the late Cainozoic: just islands in the sun or a critical pattern of islands? *In*: BRENCHLEY, P. J. (ed.) *Fossils and Climate*. Wiley, Chichester, 201–262.

—— 1988. Progress, problems and patterns in the biogeography of reef corals and other tropical marine organisms. *Helgoländer Meeresuntersuchungen,* **42**, 269–301.

—— & SMITH, A. B. 1988. Tectonics from fossils? Analysis of reef-coral and sea-urchin distributions from late Cretaceous to Recent, using a new method. *In*: HALLAM, A. & AUDLEY-CHARLES, M. G. (eds) *Gondwana and Tethys*. Geological Society, London, Special Publications, **37**, 275–306.

ROSENZWEIG, M. L. 1992. Species diversity gradients: we know more and less than we thought. *Journal of Mammalogy*, **73**, 715–730.

ROY, K., JABLONSKI, D. & VALENTINE, J. W. 1994. Eastern Pacific molluscan provinces and latitudinal diversity gradient: No evidence for 'Rapoport's rule'. *Proceedings of the National Academy of Sciences of the USA*, **91**, 8871–8874.

——, —— & —— 1996. Higher taxa in biodiversity studies: patterns from eastern Pacific marine molluscs. *Philosophical Transactions of the Royal Society of London*, **B351**, 1605–1613.

——, —— & —— 1998. Marine latitudinal diversity gradients: Tests of causal hypotheses. *Proceedings of the National Academy of Sciences of the USA*, **95**, 3699–3702.

SCHOPF, T. J. M. 1970. Taxonomic diversity gradients in ectoprocts and bivalves and their geologic implications. *Geological Society of America Bulletin*, **81**, 3765–3768.

SIMPSON, G. G. 1964. Species density of North American Recent mammals. *Systematic Zoology*, **13**, 57–73.

SLACK-SMITH, S. M. 1998. Superfamily Crassatelloidea. *In*: BEESLEY, P. L., ROSS, G. J. B. & WELLS, A. (eds) *Mollusca: The Southern Synthesis. Fauna of Australia. Volume 5. Part A*. CSIRO Publishing, Melbourne, 325–328.

STANLEY, S. M. 1970. Relation of shell form to life habits of the Bivalvia (Mollusca). *Geological Society of America Memoir*, **125**, 1–296.

STEHLI, F. G. 1968. Taxonomic diversity gradients in pole location: The Recent model. *In*: DRAKE, E. T. (ed.) *Evolution and Environment*. Yale University Press, New Haven, 163–227.

——, DOUGLAS, R. G. & NEWELL, N. D. 1969. Generation and maintenance of gradients in taxonomic diversity. *Science*, **164**, 947–949.

——, MCALESTER, A. L. & HELSLEY, C. E. 1967. Taxonomic diversity of Recent bivalves and some implications for geology. *Geological Society of America Bulletin*, **78**, 455–466.

TAYLOR, J. D. 1997. Diversity and structure of tropical Indo-Pacific benthic communities: relation to regimes of nutrient input. *In*: ORMOND, R. F. G., GAGE, J. D. & ANGEL, M. V. (eds) *Marine Biodiversity. Patterns and Processes*. Cambridge University Press, Cambridge, 178–200.

—— & TAYLOR, C. N. 1977. Latitudinal distribution of predatory gastropods on the eastern Atlantic shelf. *Journal of Biogeography,* **4**, 73–81.

——, MORRIS, N. J. & TAYLOR, C. N. 1980. Food specialisation and the evolution of predatory prosobranch gastropods. *Palaeontology*, **23**, 375–409.

VALENTINE, J. W. 1973. *Evolutionary Paleoecology of the Marine Biosphere*. Prentice-Hall, Englewood Cliffs, NJ.

VERMEIJ, G. J. 1978. *Biogeography and Adaptation*. Harvard University Press, London.

—— 1987. *Evolution and Escalation*. Princeton University Press, Princeton, NJ.

VERON, J. E. N. 1995. *Corals in Space and Time. The Biogeography and Evolution of the Scleractinia*. Comstock/Cornell, London.

WEI, K.-Y. & KENNETT, J. P. 1986. Taxonomic evolution of Neogene planktonic foraminifera and paleoceanographic relations. *Paleooceanography*, **1**, 67–84.

WILLIAMS, P. H., GASTON, K. J. & HUMPHRIES, C. J. 1997. Mapping biodviersity values worldwide: combining higher-taxon richness from different groups. *Proceedings of the Royal Society of London*, **B264**, 141–148.

WRIGHT, D. H., CURRIE, D. J. & MAURER, B. A. 1993. Energy supply and patterns of species richness on local and regional scales. *In*: RICKLEFS, R. E. & SCHLUTER, D. (eds) *Species Diversity in Ecological Communities: Historical and Geographical Perspectives*. Chicago University Press, Chicago, 66–74.

ZINSMEISTER, W. J. 1982. Late Cretaceous–Early Tertiary molluscan biogeography of the southern circum-Pacific. *Journal of Paleontology*, **56**, 84–102.

Analysing the latitudinal diversity gradient in marine bivalves

DAVID JABLONSKI[1], KAUSTUV ROY[2] & JAMES W. VALENTINE[3]

[1]*Department of Geophysical Sciences, University of Chicago, 5734 South Ellis Avenue, Chicago, IL 60637, USA*

(e-mail: djablons@midway.uchicago.edu)

[2]*Department of Biology, University of California at San Diego, 9500 Gilman Drive, La Jolla, CA 92093-0116, USA*

[3]*Department of Integrative Biology, University of California, Berkeley, CA 94720, USA*

Abstract: Marine bivalves of the eastern Pacific continental shelf show a strong diversity gradient from the Arctic Ocean to the tropics. This gradient is underlain by strong diversity trends in both infaunal and epifaunal bivalves, contrary to Thorson's influential hypothesis (1952, *Verhandlungen der Deutschen Zoologischen Gesellschaft*, **1951**, 267–327), and is significantly correlated with mean sea-surface temperature; either raw data or a residuals analysis yields $p < 0.0001$. Patterns differ according to trophic group and phylogeny; however, suspension feeders conform to the general bivalve diversity gradient, as do facultative deposit feeders such as tellinids, while deposit feeding protobranchs do not. Infaunal and epifaunal diversity gradients have different slopes so that their ratio changes with latitude; data on Jurassic and Cretaceous bivalves suggests that this ratio has varied in slope and intercept over geological time.

The latitudinal diversity gradient, with maximum taxonomic richness in the tropics, is one of the most pervasive biological patterns, but its basic configuration and its temporal dynamics remain poorly known for marine organisms (e.g. Clarke 1992; Gaston 1996; Clarke & Crame 1997). Here the results of an analysis of 930 of the *c.* 960 bivalve species known from the eastern Pacific continental shelf (from the southern border of the tropics in northwest Peru to the north coast of Alaska in the Arctic Ocean) are presented. Distributional data were derived from the primary literature [including the excellent new volume by Coan *et al.* (2000)] and museum collections.

The marine bivalves as a whole show a strong latitudinal gradient, with a sharp, 50% decrease in diversity at the northern edge of the tropics (23°N) (Fig. 1a). Both infaunal and epifaunal bivalves exhibit strong diversity trends (Fig. 1b and c), contrary to Thorson's (1952, 1957, 1965) long-standing and influential hypothesis that only epifaunal groups increase in diversity towards the tropics. Thorson argued that infaunal species should be so buffered from spatial and temporal environmental variation that their diversity patterns should not be affected by global climate trends. An alternative approach, which has gained increasing support, views diversity as a positive function of available energy rather than an inverse function of environmental variability or harshness (e.g. Wright *et al.* 1993; Fraser & Currie 1996; Turner *et al.* 1996).

For bivalves as a group, and for infaunal and epifaunal bivalves taken separately, mean sea-surface temperature (SST) is a highly significant predictor of bivalve diversity – for all bivalves $r^2 = 0.70$, $p < 0.0001$; for infauna only, $r^2 = 0.65$, $p < 0.0001$; for epifauna only, $r^2 = 0.81$, $p < 0.0001$. However, because both diversity and temperature are correlated with latitude, along with many other biotically important factors, the relationship between the two variables was examined further by using a residuals analysis. Using this approach, the relationships between the residuals of a regression of temperature v. latitude and the residuals of a regression of diversity v. latitude, are highly significant ($r^2 = 0.92$, 0.91 and 0.94 for all bivalves, infauna and epifauna, respectively) (Fig. 2). These results are: consistent with species-energy hypotheses; strongly contradict Thorson's hypothesis; and corroborate a similar analysis for gastropods of the eastern Pacific and western Atlantic (Roy *et al.* 1998). Because SST is sensitive to factors ranging from solar input to seasonal upwelling, which in turn affect nutrient variability and productivity, the mechanisms that

From: HARPER, E. M., TAYLOR, J. D. & CRAME, J. A. (eds) *The Evolutionary Biology of the Bivalvia*. Geological Society, London, Special Publications, **177**, 361–365. 1-86239-076-2/00/$15.00

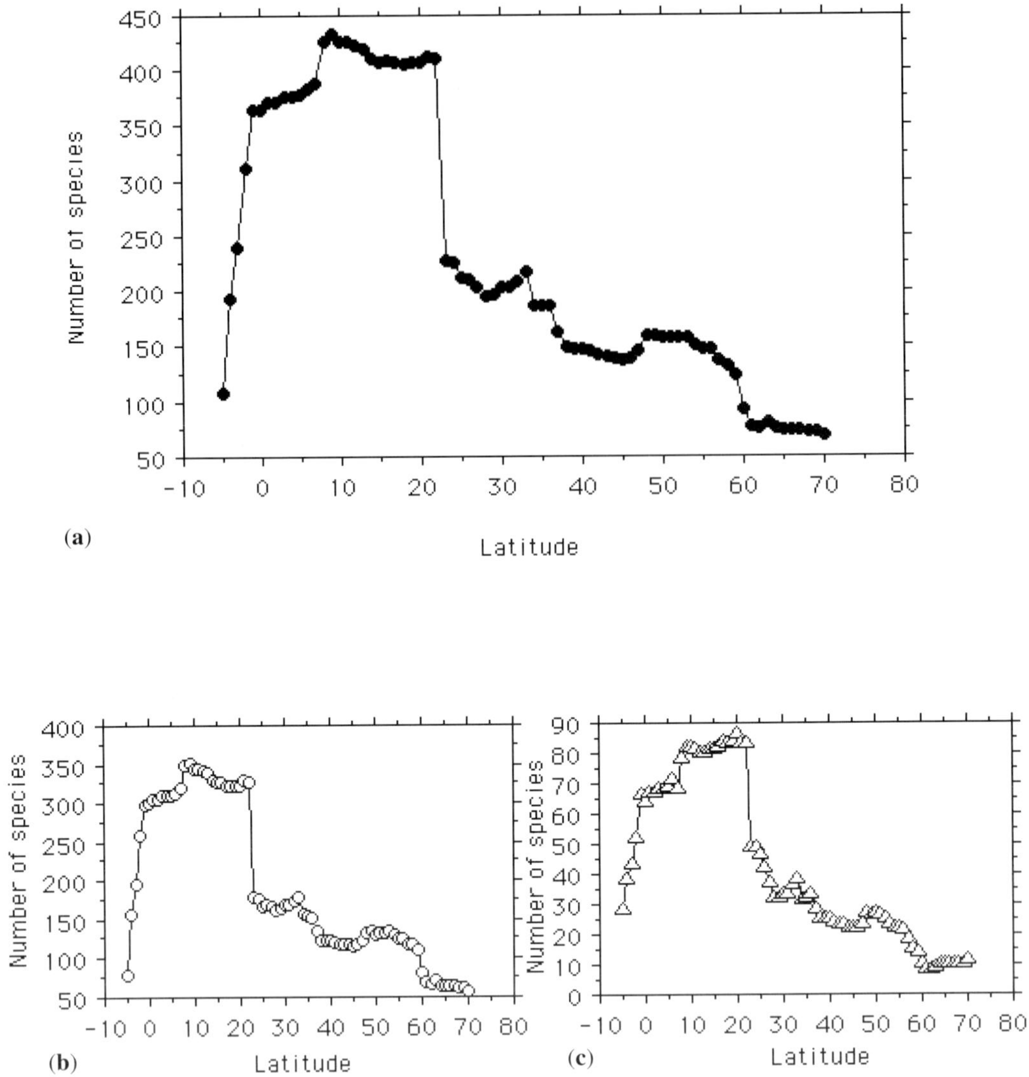

Fig. 1. Latitudinal diversity gradients for eastern Pacific continental shelf marine bivalves. (**a**) Gradient for all species combined ($N = 930$ species). (**b**) Gradient for infaunal species only ($N = 776$ species). (**c**) Gradient for epifaunal species only ($N = 154$ species).

relate SST to the diversity gradient remain uncertain (cf. Valentine 1983).

When bivalves are partitioned according to trophic groups, a more complex pattern emerges. Suspension feeders show a diversity gradient similar to that seen for bivalves as a whole, while deposit feeders exhibit two diversity plateaus (Fig. 3a and b). When the deposit feeders are partitioned phylogenetically, the tellinids, which are facultative, surface deposit feeders (Pohlo 1969; Levinton 1991; Kamermans 1994), follow a latitudinal diversity trend similar to that of suspension feeders. In contrast, the protobranchs, which are subsurface deposit feeders (e.g. Jumars *et al.* 1990), lack a latitudinal diversity gradient despite the taxonomic turnover seen at 23°N (Fig. 3c and d). The anomalous protobranch diversity trend may be related to their commitment to a relatively low fecundity, low dispersal mode of reproduction (see Valentine & Jablonski 1983).

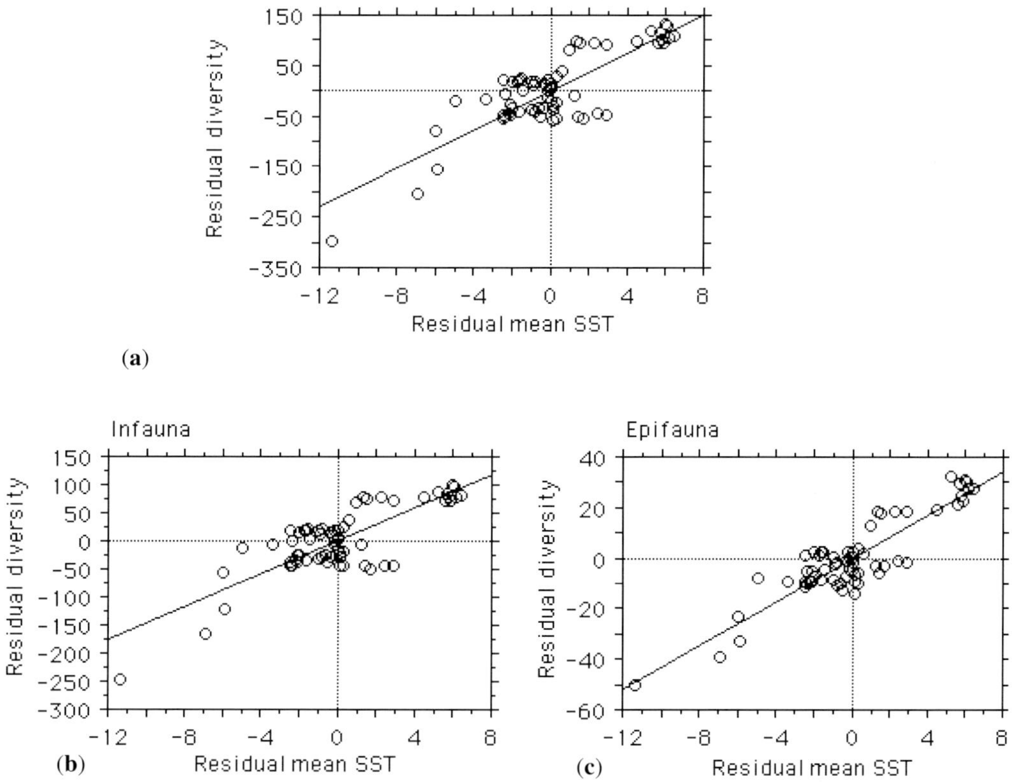

Fig. 2. Relationship between species diversity and mean sea-surface temperature (SST), using a residuals analyses (residuals from the regression of mean SST v. latitude, regressed v. the residuals from the regression of diversity v. mean SST). (**a**) All bivalves only, $r^2 = 0.92$, $p < 0.0001$. (**b**) Infaunal bivalves, $r^2 = 0.91$, $p < 0.0001$. (**c**) Epifaunal bivalves, $r^2 = 0.94$, $p < 0.0001$.

Infaunal and epifaunal bivalves both increase in diversity from poles to Equator, but they do so at different rates so that their ratio changes significantly with latitude, showing a relative increase in infauna (Fig. 4). However, Crame (1996) shows the opposite relationship with latitude, approximating a 1:1 ratio in the latest Jurassic (see also Crame 2000). The data on latest Cretaceous bivalves compiled by Raup & Jablonski (1993) and Jablonski & Raup (1995) also show an inverse trend in the infaunal:epifaunal ratio relative to modern seas, but approximating 3:1. Clearly, this is a dynamic gradient, with changes in slope and intercept driven by climate changes and differential diversification of the two functional groups. The transition to a modern trend in the infaunal: epifaunal ratio may have been driven by: (1) clade dynamics, with the ongoing diversification of siphonate infaunal suspension feeders and the relative decline of immobile suspension feeders (e.g. Stanley 1968, 1977; Skelton *et al.* 1990); (2) extinction events such as the end-Cretaceous mass extinction (although this particular extinction has not yet yielded a clear pattern of differential infaunal v. epifaunal survivorship; Jablonski & Raup 1995); and/or (3) climate change, e.g. the transition from greenhouse to icehouse worlds in the mid-Cenozoic, with accompanying changes in temperature, seasonality and other environmental factors. Comparative analysis of these trends along other latitudinal transects in modern and ancient faunas would be very interesting, and may reveal further complexity in the regional patterns that underlie global diversity gradients.

We thank the organizers for the opportunity to participate in such a stimulating meeting; the National Science Foundation for support of this research; and J. A. Crame and S. M. Kidwell for valuable comments.

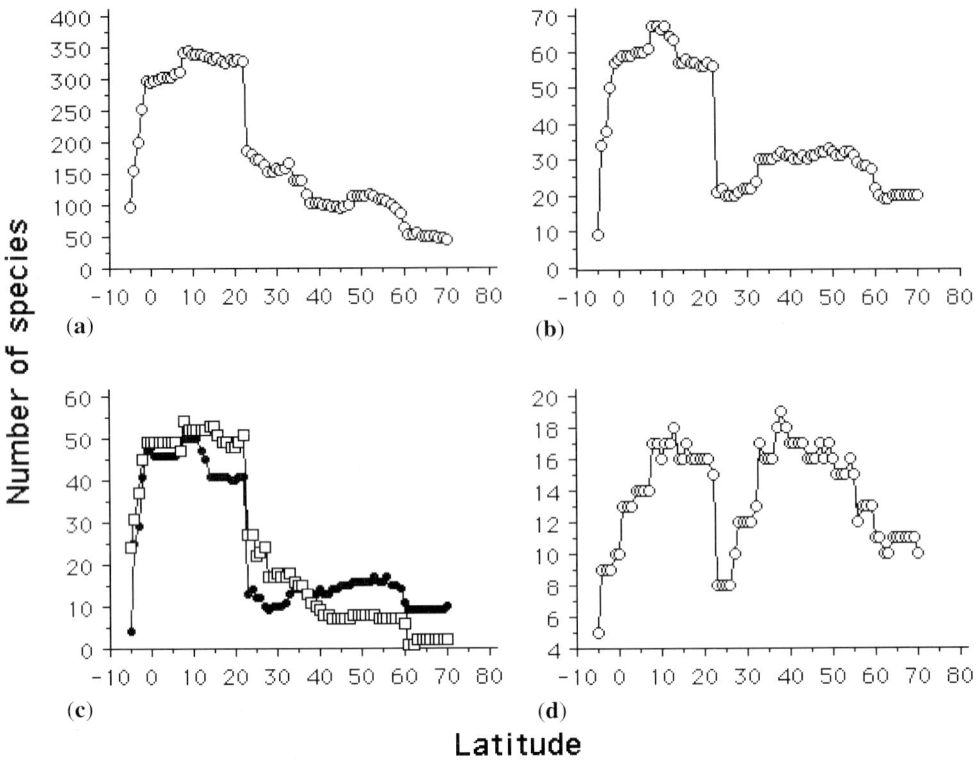

Fig. 3. Latitudinal diversity gradients in functional groups and clades of eastern Pacific bivalves. (**a**) Suspension feeders; (**b**) deposit feeders; (**c**) the facultative deposit-feeding Tellinidae (•) exhibit a diversity gradient similar to that of the suspension-feeding Veneridae (□); (**d**) the deposit-feeding protobranchs fail to conform to the overall bivalve diversity gradient. After Roy *et al.* (2000).

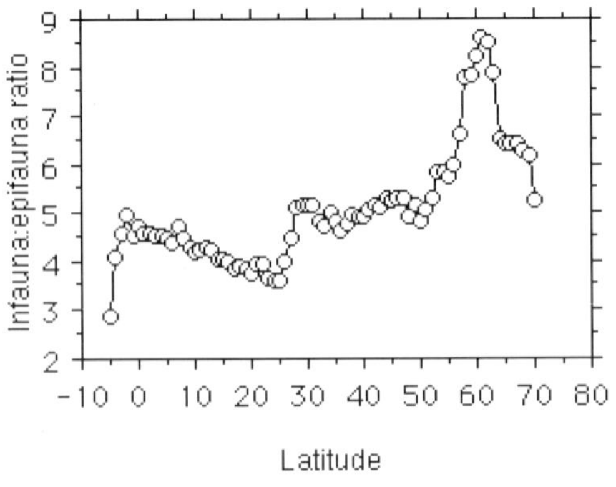

Fig. 4. The infaunal:epifaunal ratio varies with latitude in eastern Pacific bivalves. As shown in Fig. 1b and c, infauna are not *increasing* at the expense of the epifauna but are decreasing more slowly with latitude. Note, however, the downward trend within the tropics (5°S–23°N), despite the overall increase from the tropics to the poles.

References

CLARKE, A. 1992. Is there a latitudinal diversity cline in the sea? *Trends in Ecology and Evolution*, **7**, 286–287.

—— & CRAME, J. A. 1997. Diversity, latitude and time: patterns in the shallow sea. *In*: ORMOND, R. F. G., GAGE, J. D. & ANGEL, M. V. (eds) *Marine Biodiversity: Patterns and Processes*. Cambridge University Press, Cambridge, 122–147.

COAN, E. V., SCOTT, P. V. & BERNARD, F. R. 2000. *Bivalve Seashells of Western North America*. Santa Barbara Museum of Natural History Monographs, Studies in Biodiversity, **2**.

CRAME, J. A. 1996. Antarctica and the evolution of taxonomic diversity gradients in the marine realm. *Terra Antarctica*, **3**, 121–134.

—— 2000. The nature and origin of taxonomic diversity gradients in marine bivalves. *This volume*.

FRASER, R. H. & CURRIE, D. J. 1996. The species richness–energy hypothesis in a system where historical factors are thought to prevail: Coral reefs. *American Naturalist*, **148**, 138–159.

GASTON, K. J. 1996. Biodiversity – latitudinal gradients. *Progress in Physical Geography*, **20**, 466–476.

JABLONSKI, D. & RAUP, D. M. 1995. Selectivity of end-Cretaceous marine bivalve extinctions. *Science*, **268**, 389–391.

JUMARS, P. A., MAYER, L. M., DEMING, J. W., BAROSS, J. A. & WHEATCROFT, R A. 1990. Deep-sea deposit-feeding strategies suggested by environmental and feeding constraints. *Philosophical Transactions of the Royal Society of London*, **A331**, 85–101.

KAMERMANS, P. 1994. Similarity in food source and timing and feeding in deposit- and suspension-feeding bivalves. *Marine Ecology Progress Series*, **104**, 63–75.

LEVINTON, J. S. 1991. Variable feeding behavior in three species of *Macoma* (Bivalvia, Tellinacea) as a response to water flow and sediment transport. *Marine Biology*, **110**, 375–383.

POHLO, R. H. 1969. Confusion concerning deposit-feeding in the Tellinacea. *Proceedings of the Malacological Society of London*, **38**, 361–364.

RAUP, D. M. & JABLONSKI, D. 1993. Geography of end-Cretaceous marine bivalve extinctions. *Science*, **260**, 971–973.

ROY, K., JABLONSKI, D. & VALENTINE, J. W. 2000. Dissecting latitudinal diversity gradients: functional groups and clades of marine bivalves. *Proceedings of the Royal Society of London*, **B267**, 293–299.

——, ——, ——, & ROSENBERG, G. 1998. Marine latitudinal diversity gradients: tests of causal hypotheses. *Proceedings of the National Academy of Sciences USA*, **95**, 3699–3702.

SKELTON, P. W., CRAME, J. A., MORRIS, N. J. & HARPER, E. M. 1990. Adaptive divergence and taxonomic radiation in post-Palaeozoic bivalves. *In*: TAYLOR, P. D. & LARWOOD, G. P. (eds) *Major Evolutionary Radiations*. Systematics Association Special Volume, **42**, Clarendon Press, Oxford, 91–117.

STANLEY, S. M. 1968. Post-Paleozoic adaptive radiation of infaunal bivalve molluscs – a consequence of mantle fusion and siphon formation. *Journal of Paleontology*, **46**, 165–212.

—— 1977. Trends, rates, and patterns of evolution in Bivalvia. *In*: HALLAM, A. (ed.) *Patterns of Evolution*. Elsevier, Amsterdam, 209–250.

THORSON, G. 1952. Zur jetzigen Lage der mariner Bodentier-Ökologie. *Verhandlungen der Deutschen Zoologischen Gesellschaft*, **1951**, 276–327.

—— 1957. Bottom communities (sublittoral or shallow shelf). *Geological Society of America Memoirs*, **67**(1), 461–534.

—— 1965. The distribution of benthic marine Mollusca along the N.E. Atlantic shelf from Gibraltar to Murmansk. *Proceedings of the First European Malacological Congress*, 5–23.

TURNER, J. R. G., LENNON, J. J. & GREENWOOD, J. J. D. 1996. Does climate cause the global diversity gradient? *In*: HOCHBERG, M. E., CLOBERT, J. & BARBAULT. R. (eds) *Aspects of the Genesis and Maintenance of Biological Diversity*. Oxford University Press, Oxford, 199–220.

VALENTINE, J. W. 1983. Seasonality: effects in marine benthic communities. *In*: TEVESZ, M. J. S. & MCCALL, P. L. (eds) *Biotic Interactions in Recent and Fossil Benthic Communities*. Plenum Press, New York, 121–156.

—— & JABLONSKI, D. 1983. Larval adaptations and patterns of brachiopod diversity in space and time. *Evolution*, **37**, 1052–1061.

WRIGHT, D. H., CURRIE, D. J. & MAURER, B. A. 1993. Energy supply and patterns of species richness on local and regional scales. *In*: RICKLEFS, R. & SCHLUTER, D. (eds) *Species Diversity in Ecological Communities*. University of Chicago Press, Chicago, 66–74.

Marine bivalves of the Florida Keys: discovered biodiversity

PAULA M. MIKKELSEN[1] & RÜDIGER BIELER[2]

[1]*Division of Invertebrate Zoology, American Museum of Natural History, Central Park West at 79th Street, New York, New York 10024-5192, USA (e-mail: mikkel@amnh.org)*

[2]*Department of Zoology (Invertebrates), Field Museum of Natural History, 1400 S. Lake Shore Drive, Chicago, Illinois 60605-2496, USA*

Abstract: A survey of marine bivalve biodiversity in the Florida Keys, an island archipelago off southern Florida, was compiled from original collecting, museum specimens and the literature. Assembly of over 6000 records resulted in 325 species, 47% of which can be considered common to abundant in the Keys. This represents a 100% increase over the previously known fauna, largely attributable to critical review of museum specimens. Capture of species occurrences from the literature, especially when non-traditional sources (newsletters, agency reports) are excluded, is shown to be least effective, producing only 44% of the total. Bivalve distributions within the Keys show that the fauna is tropical. One-third of the species are wide ranging along the island chain; however, a latitudinal cline in faunal similarity exists from the Upper Keys southwestwards to Dry Tortugas. The fauna of Florida Bay is the most divergent within the study region and also compared to other, ecologically complex, western Atlantic tropical–subtropical regions. Limited historical records indicate little species turnover in the Keys, although population reductions along the main highway and habitat shifts (from natural to artificial substrata) are evident. These results have implications for biodiversity survey methods and, more locally, for management of the Florida Keys National Marine Sanctuary.

Estimates of biodiversity and zoogeographical comparison are only meaningful when based on empirical data. In the western Atlantic, no subtropical molluscan fauna has seen more attention by malacologists and shell collectors than southern Florida, most especially the Florida Keys.

Comprehensive knowledge of the molluscs of the Florida Keys is important for several reasons. Firstly, the Keys are of great interest zoogeographically, due to their location in a region overlapping the subtropical northwestern Atlantic, the temperate to tropical Gulf of Mexico and the tropical Caribbean Sea. Secondly, they are important malacologically as one of the best sampled regions of the eastern USA, but one whose history and relevance have never been examined. The only formal compilation of shallow- to deep-water molluscs of the Keys was produced by Lyons & Quinn (1995) as part of the preliminary Draft Management Plan of the Florida Keys National Marine Sanctuary (FKNMS). This list, compiled from the literature, the collections of the Florida Department of Environmental Protection and the authors' personal experiences, included 582 marine species, 163 of which were marine bivalves.

Thirdly, in the absence of any molluscan survey predating human development in the Keys, this analysis is sorely needed by managers of the FKNMS in helping their future attempts to assess the effectiveness of enforced preservation measures. This project involves data obtained from a multi-year effort of original field collecting by the authors, combined with critical analyses of selected museum collections and an exhaustive literature review.

The Florida Keys, at the southernmost tip of continental USA, is a chain of over 800 small, geologically young (Pleistocene Period), limestone islands mud islands, and reefs stretching 362 km from Key Largo to Key West and westwards to the Dry Tortugas Archipelago (extending *c.* 24°20′–25°21′N and 80–83°W). The populated limestone islands, *c.* 30 of which are interconnected by road bridges, reach an elevation of 6 m. To the marine biologist, the Keys are yet more extensive, comprising *c.* 10 000 km^2 of marine habitat, including hypersaline ponds, mangrove thickets, seagrass meadows, mud banks and tidal channels, sandbars, coral reefs, patch reefs, and deep sand plains, reaching into Florida Bay, the Gulf of Mexico and the Atlantic Straits of Florida. The physical environment of the Straits of Florida is dominated by the Florida Current which is known to transport Caribbean biota (e.g. coral larvae) into the Florida

From: HARPER, E. M., TAYLOR, J. D. & CRAME, J. A. (eds) *The Evolutionary Biology of the Bivalvia*. Geological Society, London, Special Publications, **177**, 367–387. 1-86239-076-2/00/$15.00

Keys (Jaap 1984). The need for conservation and management of this unique natural resource, including North America's only living barrier reef, has been well recognized. The FKNMS (est. 1990; NOAA 1996, also http://www.nos.noaa.gov/nmsp/fknms) is the newest of over a dozen active historical and ecological preservation areas in the Keys, and is the second largest marine sanctuary in the USA, encompassing the entire island chain (9100 km^2) out to the 91 m (300 ft) isobath.

Data sources/collecting history

The Florida Keys have a rich history of popular and professional mollusc collecting, spanning over a century and generating a vast accumulation of specimens in a wide variety of museum and private collections. Authors such as Pilsbry, McGinty, Rehder, Morrison, Schwengel and Houbrick, collected in the Keys and published on their molluscan discoveries. Many others (e.g. Sowerby, Dall, Bartsch, Clench, Turner, Boss, Abbott, Vokes and Petuch) have described molluscan species having type localities in the Keys. Notable among the amateur collectors frequenting the Keys (and whose collections now reside in major museum collections) were Koto, Bijur, Schilling, Feinberg and Germer. The dredgings of J. R. Black, in the 1970s, off the Dry Tortugas also supplied deep-water Keys specimens to many collectors.

Few research vessel expeditions have included the Keys to any extent. The US Coast Survey Steamer *Blake* included four deep-water dredge stations in 1877–1878 off Key West and the Dry Tortugas [molluscs published by Dall (1886, 1889)] although the expedition concentrated more extensively on the Gulf of Mexico and the Caribbean Sea. Later, throughout the 1950s, the privately owned cabin cruiser *Triton* made additional deep-water dredges for molluscs off Sombrero Key and Key West (Thompson *et al.* 1951; Pilsbry & Olsson 1953; McGinty 1955, 1962; Pilsbry 1955; McGinty & McGinty 1957). Most recently, National Oceanographic and Atmospheric Administration's (NOAA) Center for Coastal Monitoring and Assessment has sponsored a series of Macrobenthic Community Assessment cruises along the US eastern seaboard, so far including six cruises with stations off the Keys (as here defined); the reports of these cruises so far exist only as interagency reports (e.g. Vittor & Associates 1998, also http://ccma.nos.noaa.gov/datapg2.html) and many molluscan identifications are incomplete.

The most extensive sampling for molluscs in the Florida Keys was made by John B. Henderson Jr, Regent of the Smithsonian Institution. Henderson outfitted his 50 ft yacht *Eolis* specifically for dredging and sampled over 380 stations in the western Atlantic, mostly in Florida, Cuba and the Bahamas. During seven cruises in 1910–1915, over 80 stations were sampled in the Keys (as here defined), from shore to 366 m. This comprises probably the best sampling effort along the extent of the islands and could serve as an invaluable historical record of the malacofauna of the Keys in the early part of the century. Unfortunately, except for anecdotal reports of the dredging (Henderson 1911, 1913, 1914), the results and station information have never been fully assessed or published (see Abbott 1950). Fortunately, > 31 400 sorted specimen lots, although mainly incompletely unidentified, are catalogued in the National Museum of Natural History (USNM). Many *Eolis* lots have been cited in scattered systematic papers (e.g. Clench & Turner 1950; Turner & Boss 1962; Boss & Merrill 1965; Houbrick 1987), but a comprehensive treatment of this collection – which would require tray-by-tray manual extraction of *Eolis* lots from the largest mollusc collection in the world – remains to be accomplished.

Goals

The current study investigates the extent and quality of available baseline data on marine bivalves of the Florida Keys, and their application to the knowledge and assessment of regional biodiversity. Following compilation of a list of species recorded from the Florida Keys, and characterization of that fauna, these additional questions were addressed: (1) is the Florida Keys bivalve fauna a single entity (from Upper to Middle to Lower Keys to Dry Tortugas, from bayside to oceanside, from intertidal to shallow to deep water)?; (2) how much of this recorded diversity can be recaptured/verified from critical literature studies, survey of museum collections and new field collecting, and how effective are each of these data sources?; (3) can a recent historic record be reconstructed from museum and literature data and, if so, have there been changes to the bivalve fauna in the past century (conceivably associated with anthropogenic impacts to Keys ecology caused by, for example, bridge construction, the tourist industry and/or recreational diving)?; (4) how does the Florida Keys bivalve fauna compare to other subtropical western Atlantic faunas of similar ecological complexity?

Materials and methods

The Florida Keys is here defined as the waters surrounding the entire island chain from Broad Creek (*c.* 25°21′N, 80°15′W) at the northern end of

Key Largo (including Card Sound but not Biscayne Bay, southwest of, but not including, Old Rhodes Key) to the Dry Tortugas. A tangential east–west approximate midline was drawn through Florida Bay in the Upper and Middle Keys, eliminating what is more properly considered the southern end of the Florida Everglades. No oceanward limit was set. To facilitate analysis of the Keys fauna by region or zone, traditional political boundaries were employed but explicitly defined as Upper Keys (Key Largo to Craig Key), Middle Keys (Fiesta Key to the western end of Seven-Mile Bridge) and Lower Keys (Little Duck Key to Rebecca Shoal, west of the Marquesas); the Dry Tortugas Archipelago was treated separately. Delimitation of these four zones, although primarily employed by government agencies for land-planning use (Chiappone 1996*a*), is also based upon geological and hydrological boundaries (Marszalek *et al.* 1977; Shinn *et al.* 1989). While the entire Keys are founded upon an ancient coral reef known as Key Largo Limestone, that facies is exposed only in the Upper and Middle Keys, while the Lower Keys (and part of Florida Bay) are overlain by a marine oolitic bank of the Miami Limestone (Hoffmeister & Multer 1968; Mitchell-Tapping 1980). Hydrological flow patterns favour the selected boundary between the Upper and Middle Keys (N. P. Smith, pers. comm.). The Dry Tortugas is physically separated by a deep channel from Rebecca Shoal and the Lower Keys, at the western edge of the Southwest Continental Shelf (NOAA 1996). Depth categories used were based upon sampling techniques: intertidal, wading depth (*c.* 0–1 m); shallow, snorkelling–diving depth (*c.* 1–35 m); deep, beyond normal scuba depth [i.e. > 35 m or (100 ft)]; no maximum depth limit was set for this survey.

Qualitative information on bivalve occurrences in the Florida Keys was assembled from critical use of literature sources and museum lots, and from original collections. Four hundred and fifty separate publications containing Florida Keys molluscan records were compiled by scanning appropriate book publications (e.g. Johnson 1934; Abbott 1974), entire series runs (e.g. *The Nautilus, The Veliger, Bulletin of Marine Science, American Malacological Bulletin, Journal of Molluscan Studies*), shell club newsletters (e.g. *American Conchologist, Texas Conchologist*), the published papers of malacologists known to have worked in the Keys (e.g. Pilsbry, McGinty, Houbrick) and, importantly, less readily available agency reports (e.g. Lyons & Quinn 1995; Vittor & Associates 1998). Literature records are acknowledged as that portion of the data set for which species identifications could not be verified unless sufficiently illustrated. Here, 28 literature records are considered unverified (i.e. not illustrated or part of revisionary systematic work; '*' in Table 1), e.g. *Thracia corbuloides*, recorded only from the literature, claimed as 'only in Europe' (Turgeon *et al.* 1998, p. 198) and without voucher specimens for verification.

Selected museum collections were surveyed manually for Keys bivalve specimens. The American Museum of Natural History (New York; AMNH), the Field Museum of Natural History (Chicago; FMNH) and the Carnegie Museum of Natural History (Pittsburgh; CMNH) were surveyed in full; partial holdings were accessed at the Academy of Natural Sciences of Philadelphia (ANSP), the Delaware Museum of Natural History (Wilmington; DMNH), the Santa Barbara Museum of Natural History (California; SBNMH), the Bailey-Matthews Shell Museum (Sanibel Island, Florida; BMSM), the Harbor Branch Oceanographic Museum (Ft Pierce, Florida; HBOM) and the National Museum of Natural History (Washington, DC; USNM). Identifications were confirmed by the authors. Revisionary, original systematic work resulting from this survey, as well as a fully annotated compilation of the literature sources, will be published separately.

Most importantly, a five-year sampling programme of original field collections was initiated, involving hand-collecting, rock washing, shovel-and-sieving, snorkelling, scuba diving, coral-rock smashing and dredging, at over 260 'FK' stations throughout the Keys. Additional pre-survey samples by the authors in the Keys extended the original collections' portion of this work back to 1988, covering 289 stations over a total of 12 years. Original collections are vouchered at AMNH and FMNH, with partial duplicate series at DMNH and BMSM.

From these sources, a database of 6145 records was compiled (as a Microsoft Excel spreadsheet), comprising 598 records from the literature, 2223 records from museum lots and 3324 records from original collections. Within the database, each record was coded, whenever possible, for Keys zone, specific island or reef, depth zone, bayside/oceanside of the automobile highway (not coded for Dry Tortugas) and decade of collection; these allowed the data set to be subsampled and manipulated in a variety of ways to explore various questions.

Each lot was also coded as live- or dead-collected. Molluscs are one group of very few marine organisms that leave behind extensive death assemblages and it is not uncommon to find what collectors call 'fresh-dead' shells, still including organic elements (e.g. periostracum, ligament). To be conservative here, all lots were recorded as dead-collected unless soft tissue was still attached to one or more specimens, or unless the labelling indicated 'live-collected'. It is controversial, however, whether dead-collected specimens should play a role in faunal analyses. Ecological studies on benthic invertebrates generally 'count' only those individuals (including molluscs) that have been collected alive. Yet, as Kidwell (in press) has so elegantly demonstrated in her analysis of 17 separate biodiversity surveys, live-only investigations usually result in two to three times lower species richness than that represented in the death assemblage. She has corroborated the present authors' subjective impressions that empty shells are indeed reflective of the local malacofauna, and considerably more practical to assess when resources are limited, rapid assessment is mandated or when marine harvesting is restricted in protected zones. On this basis, all analyses presented herein include all specimens, whether recorded as live- or dead-collected.

Four other western Atlantic areas were selected for which comprehensive species lists could be compiled for comparison with the Florida Keys fauna: (1) the Gulf of Mexico [compiled from Steger (1962), West Florida; Haas (1940) and Gundersen (1998), Sanibel Island; Lipe (1984), Tampa Bay; Lee (1999), Cedar Key; Shelton (1997), Alabama; Lipka (1974), Flower Garden Bank; and articles in *Texas Conchologist* (1964–1999), Texas)]; (2) Cuba [from Aguayo & Jaume (1947–1948) and Espinosa *et al.* (1994)]; (3) Yucatan [from Ekdale (1974) and Vokes & Vokes (1984)] and (4) eastern Florida [from Voss *et al.* (1969), Biscayne National Monument; McGinty & Nelson (1972), Pompano Beach; Reed & Mikkelsen (1987), *Oculina* coral reefs; Lyons (1989), Hutchinson Island; Mikkelsen *et al.* (1995), Indian River Lagoon]. These areas were selected as those having similar ecological complexity, including estuarine/mangrove habitats, coral reefs and shallow- to deep-water components. Incompletely identified taxa (e.g. to genus or family level only) were excluded from these analyses.

Because of the nature of the data, only presence/absence comparisons were conducted using the Bray-Curtis (Czekanowski) similarity routine 'Cluster' in the software package PRIMER ver. 4.0 [Plymouth Routines in Multivariate Ecological Research; Carr (1997)]. Entries for each desired analysis were extracted from the original Microsoft Excel spreadsheet (a format compatible with PRIMER); duplicate records were removed manually and the data condensed, thus manually transforming the data to presence/absence format. One taxonomic group – Galeommatoidea – was omitted from all analyses in view of its extreme taxonomic uncertainty at family, generic and species levels. Results from these analyses are presented by PRIMER as dendrograms and as a table of similarity indices.

Results and discussion

Effectiveness of data sources

Of the total (6145 records, 325 species), new original collections produced more than half of all records (Fig. 1a), yet only recovered about half of all species (Fig. 1b). One hundred and thirty-four species (41%) were found by all three sources (museum, literature and original collecting). Museum collections represented only about one-third of all records, yet produced the greatest percentage of all species of any source (77%). Although locality data accompanying museum specimens are not always complete, and extraction of them is often laborious (in the absence of full computerization), the identifications of such specimens are verifiable and, if vouchers, can extend the value of published species accounts. Literature data produced the least database records (599), but recovered 237 species or *c.* 70% of the total. However, 'grey' literature played a significant role in this total; traditional literature (books and peer-reviewed journals) missed 38 species listed only by Lyons & Quinn (1995), plus another 57 species found only in agency reports or shell club newsletters, leaving a total of only 142 species (44%). This lends severe implications to studies based entirely on published species lists from this region and elsewhere. Figure 1b summarizes that, in effect, original collecting missed 45%, museum collections missed 23% and the literature missed 27% (or traditional literature 56%) of all species. Seven species were unique to the present authors' original collections and thus new records for the Keys as a result of this survey are: *Mytilopsis leucophaeata, Crassostrea virginica, Ennucula tenuis, Thyasira trisinuata, Crassinella dupliniana, Cardiomya ornatissima* and *Tivela floridana*. Sixty-two (19%) additional species were unique to museum collections; 38 species (12%) were unique to literature data.

Species composition

The Florida Keys marine molluscan species list currently comprises over 1400 species (groups other than bivalves still under review by the authors), 325 of which are bivalves (Table 1). The Florida Keys bivalve fauna represents a wide taxonomic diversity, including 60 families and 167 genera. The most diverse families were Tellinidae (41 species), Veneridae (32), Pectinidae (22) and Mytilidae (20). The species list includes 53% of the

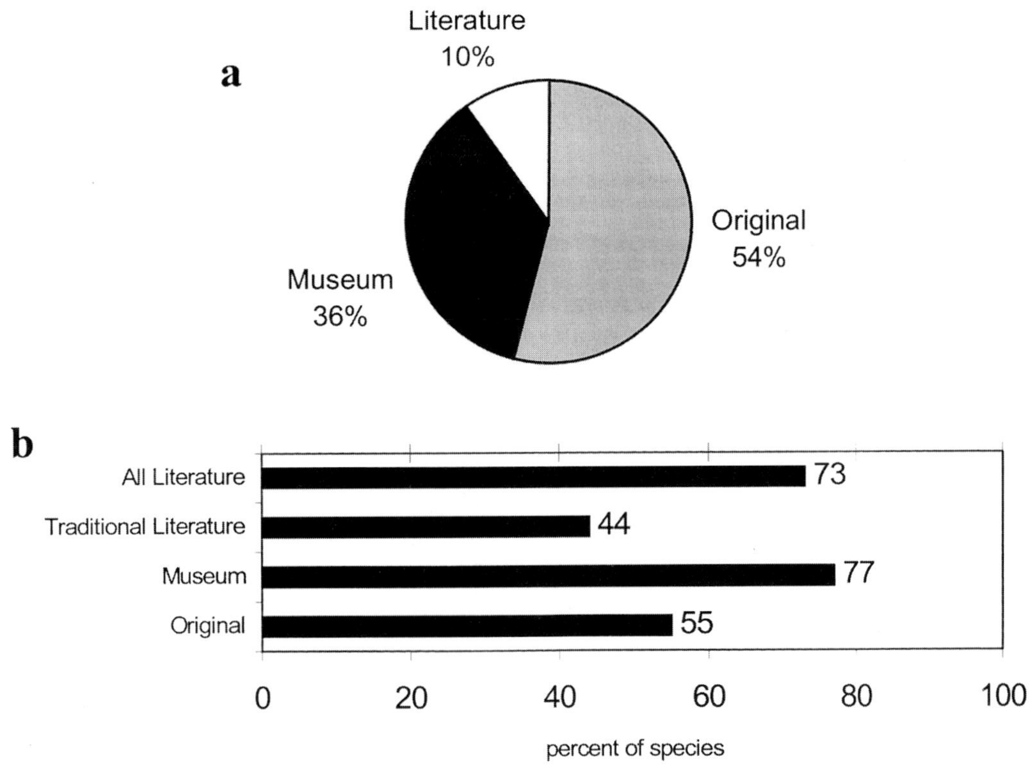

Fig. 1. Sources of Florida Keys bivalve data. (**a**) Percentage of total records (6145) from the literature (n = 598), museum lots (n = 2223) and original collections (n = 3324). (**b**) Percentage of all species (325) obtained from each source: the literature (n = 201), museum lots (n = 259) and original collections (n = 179).

marine bivalves in the western Atlantic (Turgeon *et al.* 1998) and 77% of the shallow-water bivalves of Florida (Lyons 1997). Compared with the Lyons & Quinn (1995) Florida Keys bivalve species list, this survey represents a 100% increase in the number of species (325 v. 163), an 88% increase in the number of genera and a 70% increase in the number of families. These numbers are not inflated by taxonomic 'splitting', although systematic scrutiny has played a part in developing the species list. For example, several cryptic species pairs commonly known under a single taxonomic name have been recognized as morphologically different and associated with different habitats (e.g. '*Limaria pellucida*' and '*Pinctada imbricata*' in estuarine Florida Bay v. oceanic coral reefs); these analyses are in preparation and will be published elsewhere. Some taxa are still imperfectly resolved – Cuspidariidae, Galeommatoidea, Nuculidae and Nuculanidae are groups for which few identified specimens exist, or that are based largely on unverified literature records, and, as a consequence, substantial uncertainty surrounds their species-level identifications. In view of these incomplete identifications, the species list is at worst an underestimate of the actual Florida Keys bivalve fauna.

Not surprisingly, the Keys bivalve fauna can be considered substantially tropical in character. Species ranges [mainly taken from Abbott (1974)] could be documented for 312 (96%) of the 325 species. One hundred and seventy-seven species (54%) can be considered 'widely ranging', with their distributions extending both north and south of the Florida Keys region. Only 14 species (4%) have their southern distributional limit in the Keys, while 120 species (37%), or 89% of the 'narrowly ranging' species, have their northern limit in the Keys. The Florida Keys marine bivalve fauna includes no endemic species and no threatened or endangered species. Two species of the bivalve *Mytilopsis* appear to have been recently introduced to the Keys by extension their otherwise widespread western Atlantic ranges (see below).

Although this is a qualitative study, the number of collection events, or the frequency of encounter

Table 1. *Florida Keys bivalves: summary*

Family	Species	Florida keys distribution[†]	Depth[‡]	Side[§]	Live	Recs[‖]	L&Q[¶]	No date[1]	Pre-1900	1900	1910	1920	1930	1940	1950	1960	1970	1980	1990
Anomiidae	*Anomia simplex* d'Orbigny, 1842	UMLT	ISD	OB	×	27	–	m	–	–	–	–	–	–	–	m	m	–	om
Arcidae	*Anadara baughmani* Hertlein, 1951*	T	D	–	–	1	–	l	–	–	–	–	–	–	–	–	–	–	–
	Anadara floridana (Conrad, 1869)	L	I	–	–	1	–	m	–	–	–	–	–	–	–	–	–	–	–
	Anadara notabilis (Röding, 1798)	UMLT	ISD	OB	×	60	×	ml	–	–	–	m	m	–	m	ml	m	m	om
	Anadara ovalis (Bruguière, 1789)	M	–	–	–	3	–	m	–	–	–	–	m	–	–	–	–	–	–
	Anadara transversa (Say, 1822)	UMLT	SD	OB	×	13	–	m	–	–	–	–	–	–	–	m	–	–	om
	Arca imbricata Bruguière, 1789	UMLT	ISD	OB	×	109	×	ml	l	–	–	–	m	m	m	m	m	m	om
	Arca zebra (Swainson, 1833)	UMLT	ISD	OB	×	80	×	ml	l	–	–	–	m	m	m	m	m	l	oml
	Barbatia cancellaria (Lamarck, 1819)	UMLT	ISD	OB	×	175	×	ml	–	–	–	–	m	m	m	ml	m	ml	om
	Barbatia candida (Helbling, 1779)	UMLT	ISD	OB	×	33	×	ml	–	–	–	–	m	m	m	m	m	–	o
	Barbatia domingensis (Lamarck, 1819)	UMLT	ISD	OB	×	61	×	ml	–	–	–	–	m	m	m	m	m	–	o
	Bathyarca glomerula (Dall, 1881)	T	D	–	–	1	–	–	–	–	–	–	–	–	–	–	m	–	–
Astartidae	*Astarte nana* Dall, 1886	UT	D	O	×	3	–	m	–	–	–	–	–	–	–	–	m	–	–
Cardiidae	*Americardia guppyi* Thiele, 1910	UMLT	ISD	OB	×	24	×	l	–	–	–	–	–	–	–	–	–	o	o
	Americardia media (Linné, 1758)	UMLT	ISD	OB	×	58	×	ml	l	–	–	–	–	–	m	m	m	–	ol
	Laevicardium laevigatum (Linné, 1758)	UMLT	ISD	OB	×	94	×	ml	ml	m	–	–	m	m	m	m	m	o	om
	Laevicardium mortoni (Conrad, 1830)	UMLT	ISD	OB	×	74	×	ml	–	–	–	–	m	m	m	m	–	m	oml
	Laevicardium pictum (Ravenel, 1861)	ULT	SD	OB	×	7	–	–	–	–	–	–	–	–	m	m	m	–	o
	Laevicardium sybariticum (Dall, 1886)*		–	–	–	0	×	l	–	–	–	–	–	–	–	–	–	–	–
	Nemocardium peramabile (Dall, 1881)	UL	D	O	–	6	–	ml	l	–	–	–	–	–	–	m	m	–	–
	Nemocardium tinctum (Dall, 1881)	LT	D	–	–	3	–	l	–	–	–	–	–	–	–	m	m	–	–
	Papyridea semisulcata (Gray, 1825)	UMLT	S	OB	×	16	×	ml	–	–	–	–	–	m	–	m	–	–	o
	Papyridea soleniformis (Bruguière, 1789)	UMLT	ISD	OB	–	29	×	ml	m	–	–	–	–	–	m	ml	m	l	o
	Trachycardium egmontianum (Shuttleworth, 1856)	UMLT	ISD	OB	×	31	×	ml	–	–	–	–	–	–	m	–	m	–	ol
	Trachycardium magnum (Linné, 1758)	UMLT	IS	OB	–	6	×	l	m	–	–	–	–	–	–	–	–	–	o
	Trachycardium muricatum (Linné, 1758)	UML	IS	OB	×	47	×	ml	l	–	–	–	–	–	m	m	m	–	om
Carditidae	*Carditamera floridana* Conrad, 1838	UMLT	IS	OB	×	85	×	ml	l	–	–	–	–	–	m	m	m	om	om
	Glans dominguensis (d'Orbigny, 1842)	UMLT	ISD	OB	×	11	×	ml	–	–	–	–	–	–	–	m	m	–	ol
	Pleuromeris tridentata (Say, 1826)	UMLT	ISD	OB	×	20	×	l	–	–	l	–	–	–	–	–	–	m	ol
	Pteromeris perplana (Conrad, 1841)	UMLT	SD	OB	–	8	×	ml	–	–	–	–	–	–	–	–	m	–	o
Chamidae	*Arcinella cornuta* Conrad, 1866	ULT	ISD	OB	×	27	×	ml	–	–	–	–	m	m	m	m	–	–	ol
	Chama congregata Conrad, 1833	UMLT	ISD	OB	×	64	×	ml	–	–	–	–	–	m	m	ml	m	m	ol
	Chama florida Lamarck, 1819	UMLT	ISD	OB	×	24	–	ml	–	–	–	–	–	–	m	m	–	–	o
	Chama lactuca Dall, 1886	UML	ISD	OB	–	7	×	l	–	–	–	–	–	–	ml	–	m	–	–
	Chama macerophylla Gmelin, 1791	UMLT	ISD	OB	×	110	×	ml	ml	–	–	–	–	ml	m	ml	m	ml	om
	Chama sarda Reeve, 1847	UMLT	SD	O	×	10	×	ml	–	–	–	–	–	–	–	–	–	–	o
	Chama sinuosa Broderip, 1835	UMLT	IS	OB	×	11	×	ml	–	–	–	–	–	–	m	–	–	–	o
	Pseudochama inezae Bayer, 1943	U	S	O	–	1	–	l	–	–	–	–	–	–	–	–	–	–	–
	Pseudochama radians (Lamarck, 1819)	UMLT	ISD	OB	×	44	×	ml	–	–	–	–	–	m	–	ml	m	m	o
Condylocardiidae	*Carditopsis smithii* (Dall, 1896)	UMLT	ISD	OB	×	21	×	ml	–	–	–	–	–	–	–	–	–	m	o
Corbiculidae	*Polymesoda maritima* (d'Orbigny, 1842)	UML	IS	OB	×	27	×	ml	–	–	–	–	m	–	ml	m	m	m	om

Corbulidae	*Corbula barrattiana* C. B. Adams, 1852	U	S	B	–	1	–	–	–	–	–	–	–	–	–	–	–	–	m
	Corbula caribaea d'Orbigny, 1842	MLT	ID	B	×	6	–	–	–	–	–	–	–	–	–	m	–	–	o
	Corbula contracta Say, 1822*	T	–	–	×	1	–	–	–	–	–	–	–	–	–	–	–	–	l
	Corbula dietziana C. B. Adams, 1852	ULT	D	OB	–	8	–	m	–	–	–	–	–	–	–	–	–	–	o
	Corbula swiftiana C. B. Adams, 1852	U	S	B	–	1	–	–	–	–	–	–	–	–	–	–	–	–	m
	Varicorbula limatula (Conrad, 1846)[2]	ULT	SD	OB	×	18	–	ml	–	–	–	m	–	–	–	m	–	–	ol
	Varicorbula philippii (E. A. Smith, 1885)[2]	LT	D	O	–	2	–	–	–	–	–	–	–	m	–	m	–	–	–
Crassatellidae	*Crassinella dupliniana* (Dall, 1903)	M	S	O	–	1	–	–	–	–	–	–	–	–	–	–	–	–	o
	Crassinella lunulata (Conrad, 1834)	UMLT	ISD	OB	×	18	×	l	–	–	–	–	m	m	–	m	–	–	ol
	Crassinella martinicensis (d'Orbigny, 1842)	LT	SD	B	×	5	×	l	–	–	–	–	–	–	–	–	m	–	ol
	Eucrassatella speciosa (A. Adams, 1852)	ULT	D	–	–	5	–	m	–	–	–	–	–	–	m	–	m	–	–
Cuspidariidae	*Cardiomya costellata* (Deshayes, 1830)	L	S	OB	–	2	–	–	l	–	–	–	–	–	–	–	–	–	o
	Cardiomya glypta (Bush, 1885)	T	–	–	–	1	–	m	–	–	–	–	–	–	–	–	–	–	–
	Cardiomya ornatissima (d'Orbigny, 1842)	T	D	–	–	1	–	–	–	–	–	–	–	–	–	–	–	–	o
	Cardiomya perrostrata (Dall, 1881)*	LT	D	–	×	3	–	–	l	–	–	–	–	–	–	–	–	–	l
	Cuspidaria gigantea Verrill, 1884	L	–	–	–	1	–	m	–	–	–	–	–	–	–	–	–	–	–
	Cuspidaria rostrata (Spengler, 1793)	T	–	–	–	1	–	m	–	–	–	–	–	–	–	–	–	–	–
	Leiomya claviculata (Dall, 1881)*	T	–	–	–	1	–	l	l	–	–	–	–	–	–	–	–	–	–
	Myonera limatula (Dall, 1881)*	T	D	–	–	2	–	–	l	–	–	–	–	–	–	–	–	–	–
	Plectodon granulatus (Dall, 1881)	U	–	O	×	1	–	ml	–	–	–	–	–	–	–	–	–	–	–
Cyrenoididae	*Cyrenoida floridana* (Dall, 1896)	UL	IS	B	×	2	–	–	–	–	–	–	–	–	m	–	–	–	m
Donacidae	*Donax variabilis* Say, 1822	UM	–	–	–	3	–	ml	–	–	–	–	–	–	–	–	–	–	m
	Iphigenia brasiliana (Lamarck, 1818)	U	–	–	–	1	–	–	–	–	–	–	m	–	–	–	–	–	–
Dreissenidae	*Mytilopsis leucophaeata* (Conrad, 1831)	U	I	B	×	2	–	–	–	–	–	–	–	–	–	–	–	–	o
	Mytilopsis sallei (Récluz, 1849)	U	I	B	–	2	–	l	–	–	–	–	–	–	–	–	–	–	o
Galeommatoidea	*Cymatioa* sp.	U	S	O	–	1	–	–	–	–	–	–	–	–	–	–	m	–	–
	Kellia suborbicularis (Montagu, 1803)	U	D	O	×	1	–	–	–	–	–	–	–	–	–	–	m	–	–
	Lasaea adansoni (Gmelin, 1791)	MT	IS	O	×	4	–	l	–	–	–	–	–	–	–	–	–	–	o
	Mysella planulata (Krause, 1885)	UMLT	ISD	B	×	5	–	–	–	–	–	–	–	–	–	–	–	–	ol
	Orobitella floridana (Dall, 1899)	MLT	IS	OB	×	3	–	–	–	–	–	–	–	–	–	–	–	m	ol
	Semierycina sp.	UM	IS	OB	–	2	–	–	–	–	–	–	–	–	–	–	m	–	–
Gastrochaenidae	*Gastrochaena hians* (Gmelin, 1791)	UMLT	ISD	OB	×	24	×	ml	–	–	–	–	m	m	m	m	–	–	ol
	Gastrochaena ovata Sowerby, 1834	UMLT	SD	OB	×	9	×	l	–	–	–	–	–	m	–	–	–	–	o
	Spengleria rostrata (Spengler, 1783)	UMLT	SD	O	×	11	×	ml	l	–	–	–	m	m	m	–	–	–	o
Glycymerididae	*Glycymeris americana* (DeFrance, 1829)	UT	SD	O	–	3	–	–	–	–	–	–	–	–	–	–	m	–	o
	Glycymeris decussata (Linné, 1758)	UMLT	ISD	O	–	15	×	ml	–	–	–	–	–	–	–	–	m	m	o
	Glycymeris pectinata (Gmelin, 1791)	UMLT	ISD	OB	×	111	×	ml	l	–	–	–	m	m	m	ml	m	m	oml
	Glycymeris undata (Linné, 1758)*	–	–	–	–	0	×	l	–	–	–	–	–	–	–	–	–	–	–
Gryphaeidae	*Neopycnodonte cochlear* (Poli, 1795)	UML	ISD	O	×	5	–	m	–	–	–	–	–	–	–	m	m	–	o
Hiatellidae	*Hiatella arctica* (Linné, 1767)	ULT	SD	O	×	5	–	–	–	–	–	–	–	–	–	m	m	o	ol
Isognomonidae	*Isognomon alatus* (Gmelin, 1791)	UML	IS	OB	×	60	×	ml	–	–	–	–	m	–	m	m	m	m	om
	Isognomon bicolor (C. B. Adams, 1845)	UMLT	ISD	OB	×	59	×	ml	–	–	–	–	m	m	–	m	m	m	om
	Isognomon radiatus (Anton, 1839)	UMLT	ISD	OB	×	31	×	ml	–	–	–	–	–	–	–	–	m	–	o
Limidae	*Ctenoides floridanus* Olsson & Harbison, 1953[2]	UMLT	ISD	OB	×	66	×	ml	–	–	m	–	m	m	m	m	m	m	o
	Ctenoides planulatatus (Dall, 1886)[2]	L	D	–	–	1	–	m	–	–	–	–	–	–	–	–	–	–	–
	Ctenoides sanctipauli Stuardo, 1982[2]	MLT	ID	B	×	4	–	m	–	–	m	–	–	–	–	–	m	–	–
	Ctenoides scaber (Born, 1778)[2]	UMLT	ISD	OB	×	58	×	ml	ml	–	m	–	–	m	m	m	m	ml	o

Table 1. *Continued*

Family	Species	Florida keys distribution[†]	Depth[‡]	Side[§]	Live	Recs[‖]	L&Q[¶]	No date[1]	Pre-1900	1900	1910	1920	1930	1940	1950	1960	1970	1980	1990
Limidae	*Divarilima albicoma* (Dall, 1886)	–	D	–	–	1	–	l	–	–	–	–	–	–	–	–	–	–	–
	Lima caribaea d'Orbigny, 1842[2]	UMLT	ISD	OB	×	89	×	ml	m	m	–	m	m	m	m	ml	m	m	o
	Limaria pellucida (C. B. Adams, 1846)	UMLT	ISD	OB	×	61	×	ml	l	–	–	–	m	m	m	ml	m	–	oml
Limopsidae	*Limopsis aurita* (Brocchi, 1814)	T	D	–	–	1	–	–	–	–	–	–	–	–	–	–	m	–	–
	Limopsis cristata Jeffreys, 1876	T	D	–	–	1	–	–	–	–	–	–	–	–	–	–	m	–	–
	Limopsis minuta Philippi, 1836	T	D	–	–	1	–	–	–	–	–	–	–	–	–	–	m	–	–
	Limopsis sulcata Verrill & Bush, 1898	UT	D	O	×	3	–	–	–	–	–	–	–	–	–	m	m	–	–
Lucinidae	*Anodontia alba* Link, 1807	UMLT	ISD	OB	×	33	–	m	–	–	–	–	–	–	–	m	m	–	om
	Anodontia philippiana (Reeve, 1850)	L	–	–	–	1	–	m	–	–	–	–	–	–	–	–	–	–	–
	Codakia costata (d'Orbigny, 1842)*	–	–	–	–	0	×	l	–	–	–	–	–	–	–	–	–	–	–
	Codakia orbicularis (Linné, 1758)	UMLT	ISD	OB	×	172	×	ml	ml	–	–	–	m	m	m	ml	m	oml	oml
	Codakia orbiculata (Montagu, 1808)	UMLT	ISD	OB	×	154	×	ml	–	l	l	–	m	m	m	m	m	m	om
	Codakia pectinella (C. B. Adams, 1852)	T	D	–	–	1	×	l	–	–	–	–	–	–	–	–	–	–	o
	Divalinga quadrisulcata (d'Orbigny, 1842)	UMLT	ISD	OB	–	42	×	ml	–	–	–	–	m	m	–	m	m	m	o
	Divaricella dentata (Wood, 1815)	UML	IS	OB	×	18	×	ml	m	–	–	–	m	m	m	–	m	–	o
	Lucina amianta (Dall, 1901)	UMLT	ISD	OB	×	19	×	l	–	–	–	–	–	–	–	m	–	–	oml
	Lucina floridana Conrad, 1833	L	D	O	–	1	–	–	–	–	–	–	–	–	–	m	–	–	–
	Lucina leucocyma (Dall, 1886)	UT	D	O	–	3	×	ml	–	–	–	–	–	–	–	–	m	–	o
	Lucina pectinata (Gmelin, 1791)	UML	SD	OB	×	5	×	ml	–	–	–	–	–	–	–	m	–	–	m
	Lucina pensylvanica (Linné, 1758)	UMLT	ISD	OB	×	119	×	ml	ml	–	–	–	m	m	m	ml	m	ml	oml
	Lucina radians (Conrad, 1841)	UT	D	B	×	2	–	–	–	–	–	–	–	–	–	–	m	–	l
	Lucina sombrerensis (Dall, 1886)	UMLT	ISD	OB	–	10	–	–	–	–	–	–	–	–	–	m	–	–	o
	Lucina trisulcata Conrad, 1841	UMLT	ISD	OB	×	10	×	l	–	–	–	–	–	m	–	–	m	–	ol
	Lucinisca muricata (Spengler, 1798)*	L	–	–	–	0	×	l	–	–	–	–	–	–	–	–	–	–	–
	Lucinisca nassula (Conrad, 1846)	UMLT	ISD	OB	×	61	×	ml	–	–	–	–	m	m	–	m	m	–	oml
	Lucinoma filosum (Stimpson, 1851)	T	D	–	–	1	–	–	–	–	–	–	–	–	–	–	m	–	–
	Myrtea sagrinata (Dall, 1886)*	–	D	–	–	1	–	l	–	–	–	–	–	–	–	–	–	–	–
	Parvilucina multilineata (Tuomey & Holmes, 1857)	UMLT	ISD	OB	×	22	×	l	–	–	–	–	m	m	–	–	–	–	oml
Lyonsiidae	*Entodesma beana* (d'Orbigny, 1842)	UMLT	ISD	OB	×	13	×	ml	–	–	–	–	–	m	–	m	–	–	ol
	Lyonsia floridana Conrad, 1849	ULD	IS	B	×	3	–	–	–	–	–	–	–	–	–	–	–	–	oml
Mactridae	*Anatina anatina* (Spengler, 1802)	M	–	–	–	1	–	–	–	–	–	–	–	–	–	m	–	–	–
	Mactrotoma fragilis (Gmelin, 1791)	UML	IS	OB	×	29	×	ml	–	–	–	–	–	–	–	–	m	–	om
	Raeta plicatella (Lamarck, 1818)	UL	–	–	–	2	–	m	–	–	–	–	–	–	–	–	–	–	–
	Rangia flexuosa (Conrad, 1840)*	–	–	–	–	0	–	l	–	–	–	–	–	–	–	–	–	–	–
	Spisula raveneli (Conrad, 1831)	UM	IS	OB	×	2	–	–	–	–	–	–	–	–	–	–	–	–	om
Malleidae	*Malleus candeanus* (d'Orbigny, 1842)	UMLT	SD	O	×	11	×	l	–	–	–	–	–	–	–	–	–	–	o
Myidae	*Sphenia antillensis* Dall & Simpson, 1901	U	S	B	×	1	–	–	–	–	–	–	–	–	–	–	–	–	m
Mytilidae	*Amygdalum papyrium* (Conrad, 1846)	UM	S	B	×	2	–	m	–	–	–	–	–	–	–	–	–	–	m
	Amygdalum politum (Verrill & Smith, 1880)	T	D	–	×	2	–	–	l	–	–	–	–	–	–	–	–	–	o
	Amygdalum sagittatum Rehder, 1934	LT	–	B	×	2	–	–	–	–	–	–	–	–	–	–	–	–	ml
	Botula fusca (Gmelin, 1791)	UMLT	SD	OB	×	20	×	ml	–	–	–	–	m	m	–	m	–	–	o

	Brachidontes domingensis (Lamarck, 1819)	UML	IS	OB	×	55	×	ml	–	–	–	–	–	–	m	m	m	o	o
	Brachidontes exustus (Linné, 1758)	UML	IS	OB	×	69	–	ml	–	–	m	–	–	m	–	m	m	o	oml
	Brachidontes modiolus (Linné, 1767)	UML	IS	OB	×	49	×	ml	m	–	–	–	m	m	m	m	m	m	ol
	Crenella decussata (Montagu, 1808)	UMLT	ISD	OB	×	20	×	ml	–	–	–	–	–	m	–	–	–	–	ol
	Dacrydium elegantulum hendersoni Salas & Gofas, 1997	L	D	O	–	1	–	l	–	–	–	–	–	–	–	–	–	–	–
	Geukensia granosissima (Sowerby, 1914)	M	–	–	–	2	–	m	–	–	–	–	–	–	–	–	–	–	–
	Gregariella coralliophaga (Gmelin, 1791)	UML	S	O	×	3	×	l	–	–	–	–	–	–	–	–	–	–	o
	Ischadium recurvum (Rafinesque, 1820)	UL	S	OB	×	3	–	m	–	–	–	–	m	–	–	–	–	–	m
	Lioberus castaneus (Say, 1822)	UMLT	IS	OB	×	11	×	l	–	–	–	–	–	–	–	m	–	–	oml
	Lithophaga antillarum (d'Orbigny, 1842)	UMLT	SD	OB	×	26	×	ml	–	–	–	–	–	m	m	–	–	–	o
	Lithophaga aristata (Dillwyn, 1817)	LT	SD	O	×	6	×	ml	–	–	l	–	–	m	–	–	–	–	o
	Lithophaga bisulcata (d'Orbigny, 1842)	UML	S	OB	×	19	×	ml	–	–	–	–	m	m	m	m	–	–	o
	Lithophaga nigra (d'Orbigny, 1842)	UMLT	S	OB	–	23	×	ml	–	–	–	–	–	m	–	–	–	–	o
	Modiolus americanus (Leach, 1815)	UMLT	ISD	OB	×	60	×	ml	m	–	–	–	–	m	–	m	m	m	om
	Modiolus modiolus squamosus Beauperthuy, 1967	UMLT	IS	OB	×	26	×	l	–	–	–	–	–	–	–	–	–	–	ol
	Musculus lateralis (Say, 1822)	UMLT	ISD	OB	×	27	×	l	–	–	–	–	m	m	–	m	m	–	oml
Noetiidae	*Arcopsis adamsi* (Dall, 1886)	UMLT	ISD	OB	×	118	×	ml	–	–	–	–	–	m	m	m	m	m	oml
	Noetia ponderosa (Say, 1822)	UML	–	–	–	5	–	ml	–	–	–	–	–	–	–	m	–	–	–
Nuculanidae	*Nuculana acuta* (Conrad, 1832)	UMLT	SD	OB	×	14	–	m	l	–	–	–	–	–	–	–	m	–	oml
	Nuculana concentrica (Say, 1824)	MLT	D	OB	×	8	–	–	–	–	–	–	–	–	–	–	–	–	ol
	Nuculana pusio (Philippi, 1844)*	L	D	–	–	1	–	l	–	–	–	–	–	–	–	–	–	–	–
	Nuculana solidula (E. A. Smith, 1885)*	–	D	–	–	1	–	l	–	–	–	–	–	–	–	–	–	–	–
	Nuculana verrilliana (Dall, 1886)*	–	–	–	–	0	–	l	–	–	–	–	–	–	–	–	–	–	–
Nuculidae	*Ennucula tenuis* (Montagu, 1808)	UML	IS	OB	–	5	–	–	–	–	–	–	–	–	–	–	–	–	o
	Nucula aegeensis Jeffreys, 1879*	UMLT	–	B	×	7	–	–	–	–	–	–	–	–	–	–	–	–	l
	Nucula calcicola Moore, 1977	UML	IS	OB	×	7	–	l	–	–	–	–	–	–	–	–	–	–	o
	Nucula crenulata A. Adams, 1856*	T	–	–	–	1	–	–	l	–	–	–	–	–	–	–	–	–	–
	Nucula proxima Say, 1822	UMLT	ISD	OB	×	23	–	l	–	–	–	–	–	m	–	–	–	–	o
Ostreidae	*Crassostrea rhizophorae* (Guilding, 1828)*	L	I	–	–	1	–	–	l	–	–	–	–	–	–	–	–	–	–
	Crassostrea virginica (Gmelin, 1791)	UL	IS	OB	×	7	–	–	–	–	–	–	–	–	–	–	–	–	o
	Cryptostrea permollis (Sowerby, 1871)	UML	I	OB	×	9	–	ml	–	–	–	–	–	m	m	–	m	–	–
	Dendostrea frons (Linné, 1758)	UMLT	ISD	OB	×	54	×	ml	l	–	–	–	–	m	m	m	–	m	om
	Ostreola equestris (Say, 1834)	UML	IS	OB	×	16	–	–	–	–	–	–	–	–	–	m	m	–	om
Pandoridae	*Pandora bushiana* Dall, 1886	UT	SD	O	×	2	–	l	–	–	–	–	–	–	–	–	–	–	–
	Pandora inflata Boss & Merrill, 1965	UMLT	ISD	OB	×	14	–	l	–	–	–	–	–	–	–	m	–	–	o
Pectinidae	*Aequipecten glyptus* (Verrill, 1882)	LT	D	O	–	8	–	ml	–	–	–	–	–	–	–	–	m	–	–
	Amusium laurentii (Gmelin, 1791)	LT	–	–	–	2	–	m	–	–	–	–	–	–	–	m	–	–	–
	Amusium papyraceum (Gabb, 1873)	LT	SD	–	–	7	–	m	–	–	–	–	–	–	–	m	–	–	–
	Argopecten gibbus (Linné, 1758)	UMLT	ISD	OB	×	36	×	ml	l	–	–	–	–	m	m	–	m	–	ol
	Argopecten irradians (Lamarck, 1819)	UML	ISD	OB	×	21	×	ml	–	–	–	–	m	m	–	m	–	l	om
	Argopecten lineolaris (Lamarck, 1819)	MLT	SD	O	–	11	–	m	–	–	–	–	–	–	ml	–	m	–	–
	Argopecten nucleus (Born, 1778)	UML	IS	OB	×	37	×	ml	–	–	–	–	–	m	m	m	m	–	om
	Brachtechlamys antillarum (Récluz, 1853)	UMLT	IS	OB	×	35	×	ml	–	–	–	–	m	m	m	m	m	m	o
	Caribachlamys imbricata (Gmelin, 1791)	UMLT	IS	O	–	19	×	ml	–	–	–	–	–	m	–	m	m	l	o
	Caribachlamys mildredae (Bayer, 1941)	UMLT	S	O	–	7	×	ml	–	–	–	–	l	l	–	–	–	m	–
	Caribachlamys ornata (Lamarck, 1819)	UML	IS	OB	–	9	×	ml	m	–	–	–	–	–	–	m	m	–	o

Table 1. *Continued*

Family	Species	Florida keys distribution†	Depth‡	Side§	Live	Recs‖	L&Q¶	No date[1]	Pre-1900	1900	1910	1920	1930	1940	1950	1960	1970	1980	1990
Pectinidae	*Caribachlamys sentis* (Reeve, 1853)	UMLT	ISD	OB	×	80	×	ml	–	–	–	–	m	–	m	m	ml	ml	o
	Cryptopecten phrygium (Dall, 1886)	MLT	D	OB	–	8	–	ml	–	–	–	–	–	–	l	–	m	–	–
	Euvola chazaliei (Dautzenberg, 1900)	UMLT	SD	O	–	11	×	ml	–	–	–	–	–	–	ml	m	m	–	–
	Euvola raveneli (Dall, 1898)	MLT	SD	O	–	16	–	m	–	–	–	–	–	–	m	m	m	–	–
	Euvola ziczac (Linné, 1758)	LT	SD	OB	–	27	×	ml	m	–	–	–	–	–	m	m	m	–	o
	Laevichlamys multisquamata (Dunker, 1864)*	–	–	–	–	0	×	l	–	–	–	–	–	–	–	–	–	–	–
	Lindapecten exasperatus (Sowerby, 1842)	M	S	O	–	7	×	ml	–	–	–	–	–	–	–	–	m	–	–
	Lindapecten muscosus (Wood, 1828)	UML	ISD	OB	×	35	–	ml	–	–	–	–	–	m	m	m	m	–	o
	Lyropecten kallinubilosus (Bayer, 1943)	L	S	–	–	1	–	m	–	–	–	–	–	–	–	–	–	–	–
	Nodipecten nodosus (Linné, 1758)	ULT	SD	O	–	25	×	ml	–	–	–	–	–	–	m	m	m	m	m
	Spathochlamys benedicti (Verrill & Bush, 1897)	LT	D	O	–	6	×	ml	–	–	–	–	–	–	–	–	m	–	o
Periplomatidae	*Periploma anguliferum* (Philippi, 1847)	L	–	–	–	1	–	ml	–	–	–	–	–	–	–	–	–	–	–
	Periploma tenerum Fischer, 1882	M	–	–	–	1	–	l	–	–	–	–	–	–	–	–	–	–	o
Petricolidae	*Choristodon robustum* (Sowerby, 1834)	UMLT	IS	OB	×	22	×	ml	–	–	–	–	m	–	–	m	m	–	o
	Petricola lapicida (Gmelin, 1791)	UMLT	IS	OB	×	26	×	ml	–	–	–	–	m	m	m	m	–	–	o
	Petricolaria pholadiformis (Lamarck, 1818)	L	–	–	–	1	–	–	–	–	–	–	m	–	–	–	–	–	–
Pharidae	*Ensis minor* Dall, 1900	L	–	–	–	1	–	–	m	–	–	–	–	–	–	–	–	–	–
Pholadidae	*Barnea truncata* (Say, 1822)*	–	–	–	–	0	–	l	–	–	–	–	–	–	–	–	–	–	–
	Cyrtopleura costata (Linné, 1758)	L	–	–	–	2	–	–	m	–	–	–	–	–	–	–	m	–	–
	Martesia cuneiformis (Say, 1822)	L	–	–	–	1	–	l	–	–	–	–	–	–	–	–	–	–	–
	Martesia striata (Linné, 1758)	UL	I	–	–	4	–	ml	–	–	–	–	–	–	–	–	m	–	–
	Xylopholas altanai Turner, 1972	–	D	–	–	1	–	l	–	–	–	–	–	–	–	–	–	–	–
Pinnidae	*Atrina rigida* (Lightfoot, 1786)	UML	IS	OB	×	10	×	ml	–	–	–	–	–	–	m	m	–	–	om
	Atrina seminuda (Lamarck, 1819)	UM	S	OB	×	2	×	l	–	–	–	–	–	–	–	–	–	–	om
	Atrina serrata (Sowerby, 1825)	UML	SD	OB	×	4	–	ml	–	–	–	–	–	–	–	–	–	–	m
	Pinna carnea Gmelin, 1791	UMLT	ISD	OB	×	51	×	ml	m	–	–	–	–	m	–	m	m	ml	om
	Pinna rudis Linné, 1758	–	–	–	–	0	–	m	–	–	–	–	–	–	–	–	–	–	–
Plicatulidae	*Plicatula gibbosa* Lamarck, 1801	UMLT	SD	OB	×	35	×	ml	–	–	–	–	–	–	m	m	m	–	o
Poromyidae	*Cetoconcha margarita* (Dall, 1886)	T	D	–	–	2	–	l	l	–	–	–	–	–	–	–	–	–	–
	Poromya granulata (Nyst & Westendorp, 1839)	T	D	–	–	2	–	ml	–	–	–	–	–	–	–	–	–	–	–
	Poromya rostrata Rehder, 1943	M	D	O	–	1	–	m	–	–	–	–	–	–	–	–	–	–	–
Propeamusiidae	*Cyclopecten* sp.	T	–	–	–	1	–	m	–	–	–	–	–	–	–	–	–	–	–
	Propeamussium dalli (E. A. Smith, 1885)	L	D	–	–	1	–	m	–	–	–	–	–	–	–	–	–	–	–
	Propeamussium pourtalesianum (Dall, 1886)	T	D	–	–	1	–	–	–	–	–	–	–	–	–	–	m	–	–
	Propeamussium sayanum (Dall, 1886)	T	D	–	–	1	–	–	–	–	–	–	–	–	–	–	m	–	–
Psammobiidae	*Asaphis deflorata* (Linné, 1758)	UML	IS	OB	×	12	×	ml	m	–	–	–	–	–	m	m	–	–	–
	Heterodonax bimaculatus (Linné, 1758)	–	–	–	–	0	–	ml	–	–	–	–	–	–	–	–	–	–	–
Pteriidae	*Pinctada imbricata* Röding, 1798	UML	ISD	OB	×	110	×	ml	m	–	m	–	m	m	m	m	m	ml	om
	Pinctada longisquamosa (Dunker, 1852)[2]	UMLT	ISD	OB	×	59	–	ml	–	–	–	–	m	m	m	m	–	–	om
	Pteria colymbus (Röding, 1798)	UMLT	ISD	OB	×	59	×	ml	–	m	–	m	m	m	m	ml	m	m	om
Semelidae	*Abra aequalis* (Say, 1822)	ULT	SD	OB	×	12	×	l	–	–	–	–	–	–	–	–	–	–	oml

	Abra lioica (Dall, 1881)	ML	D	O	–	2	–	m	l	–	–	–	–	–	–	–	–	–	–
	Cumingia coarctata Sowerby, 1833	UMLT	IS	OB	×	33	×	ml	–	m	–	–	–	m	–	–	–	–	ol
	Cumingia tellinoides vanhyningi Rehder, 1939	UML	IS	OB	×	37	–	ml	–	–	–	–	–	m	–	–	–	–	oml
	Ervilia concentrica (Holmes, 1860)	UMLT	ISD	OB	–	7	×	ml	–	–	–	–	–	m	–	–	–	–	ol
	Ervilia nitens (Montagu, 1806)	ML	IS	OB	–	4	×	l	–	–	–	–	–	–	–	–	–	–	o
	Ervilia subcancellata E. A. Smith, 1885	UMLT	–	–	–	7	–	l	–	–	–	–	–	–	–	–	–	–	–
	Semele bellastriata (Conrad, 1837)	UMLT	SD	OB	×	16	×	ml	–	–	–	–	–	m	–	m	m	–	ol
	Semele proficua (Pulteney, 1799)	UMLT	ISD	OB	×	30	×	ml	–	–	–	–	–	m	–	m	m	m	oml
	Semele purpurascens (Gmelin, 1791)	UMLT	ISD	OB	–	22	×	ml	l	–	–	–	–	m	m	m	m	–	o
	Semelina nuculoides (Conrad, 1841)*	LT	–	B	–	3	×	l	–	–	–	–	–	–	–	–	–	–	l
Solecurtidae	*Solecurtus cumingianus* Dunker, 1861	MLT	SD	O	–	5	–	m	–	–	–	–	–	m	–	–	m	–	o
	Tagelus divisus (Spengler, 1794)	UML	S	OB	×	6	–	–	–	–	–	–	–	m	–	–	–	–	om
	Tagelus plebeius (Lightfoot, 1786)	U	S	B	×	1	–	–	–	–	–	–	–	–	–	–	–	–	m
Solemyacidae	*Solemya occidentalis* Deshayes, 1857	UMLT	IS	B	×	12	×	ml	–	–	–	–	–	–	–	–	–	m	oml
Spondylidae	*Spondylus americanus* Hermann, 1781	UMLT	SD	O	×	19	×	ml	–	–	–	–	–	m	–	m	m	m	m
	Spondylus gussoni O. G. Costa, 1829	L	I	–	–	1	–	m	–	–	–	–	–	–	–	–	–	–	–
	Spondylus ictericus Reeve, 1856	UMLT	ISD	OB	×	43	×	ml	–	–	–	–	–	–	m	m	m	m	o
Sportellidae	*Basterotia elliptica* (Récluz, 1850)*	–	–	–	–	0	×	l	–	–	–	–	–	–	–	–	–	–	–
	Basterotia quadrata (Hanley, 1843)*	–	–	–	–	0	×	l	–	–	–	–	–	–	–	–	–	–	–
Tellinidae	*Cymatoica orientalis hendersoni* Rehder, 1939	MT	SD	OB	–	5	–	l	–	–	–	–	–	–	–	–	–	–	o
	Leporimetis intastriata (Say, 1827)	UML	IS	OB	–	10	×	ml	–	–	–	–	m	–	m	–	–	–	o
	Macoma brevifrons (Say, 1834)	UMLT	ISD	OB	×	5	×	l	–	–	–	–	–	m	–	–	m	–	oml
	Macoma cerina C. B. Adams, 1845	UMLT	ISD	OB	×	31	–	m	–	–	–	–	–	–	–	m	–	m	om
	Macoma constricta (Bruguière, 1792)	U	–	–	–	2	–	m	–	–	–	–	–	–	–	m	–	–	–
	Macoma mitchelli Dall, 1895	U	S	B	×	1	–	–	–	–	–	–	–	–	–	–	–	–	m
	Macoma tageliformis Dall, 1900	L	–	–	–	1	–	–	–	–	–	–	–	–	m	–	–	–	–
	Macoma tenta (Say, 1834)	ULT	SD	OB	×	10	–	–	–	–	–	–	–	–	–	–	m	–	om
	Strigilla carnaria (Linné, 1758)	L	–	–	–	4	×	l	ml	–	–	–	–	m	–	–	–	–	–
	Strigilla gabbi Olsson & McGinty, 1958	L	–	–	–	3	×	l	–	–	–	–	–	–	–	–	–	–	–
	Strigilla mirabilis (Philippi, 1841)	UMLT	ISD	OB	–	20	×	ml	–	–	–	–	–	m	–	–	–	–	o
	Strigilla pisiformis (Linné, 1758)	L	–	–	–	2	–	ml	l	–	–	–	–	–	–	–	–	–	–
	Tellidora cristata (Récluz, 1842)	UL	S	B	×	4	×	ml	–	–	–	–	–	–	–	–	–	–	ml
	Tellina aequistriata Say, 1824	ULT	D	O	×	8	×	ml	–	–	–	–	–	m	–	–	m	–	l
	Tellina agilis Stimpson, 1857	L	I	B	–	1	–	–	–	–	–	–	–	–	–	–	–	m	–
	Tellina alternata Say, 1822	UMLT	SD	OB	×	8	×	ml	–	–	–	–	–	–	–	–	–	–	om
	Tellina americana Dall, 1900	L	D	O	–	2	×	l	–	–	–	–	–	–	–	–	–	–	–
	Tellina angulosa Gmelin, 1791	L	–	–	–	2	×	l	m	–	–	–	–	–	–	–	–	–	–
	Tellina candeana d'Orbigny, 1842	ML	–	O	–	8	×	ml	–	–	–	–	–	m	–	–	–	–	–
	Tellina consobrina d'Orbigny, 1842	UMLT	–	O	–	7	×	l	–	–	–	–	–	–	–	–	–	–	–
	Tellina fausta Pulteney, 1799	UMLT	ISD	OB	–	85	×	ml	ml	–	–	–	m	m	m	ml	m	ml	o
	Tellina gouldii Hanley, 1846	UMLT	IS	OB	×	30	×	ml	–	–	–	–	–	m	m	–	–	–	o
	Tellina iris Say, 1822	UMLT	IS	OB	×	27	–	ml	–	–	–	–	m	m	–	–	–	–	ol
	Tellina laevigata Linné, 1758	UML	S	O	–	4	×	l	–	–	–	–	–	–	–	–	–	–	o
	Tellina lineata Turton, 1819	UL	S	B	×	9	×	ml	l	–	–	–	–	m	–	m	m	–	m
	Tellina listeri Röding, 1798	UMLT	ISD	OB	×	39	×	ml	ml	–	–	–	–	m	m	m	m	–	oml
	Tellina magna Spengler, 1798	UMLT	S	O	–	6	×	ml	–	–	–	–	–	–	m	–	–	–	–
	Tellina martinicensis d'Orbigny, 1842	LT	SD	OB	–	7	×	l	–	–	–	–	–	–	–	–	–	–	o

Table 1. *Continued*

Family	Species	Florida keys distribution†	Depth‡	Side§	Live	Recs‖	L&Q¶	No date[1]	Pre-1900	1900	1910	1920	1930	1940	1950	1960	1970	1980	1990
Tellinidae	*Tellina mera* Say, 1834	UMLT	ISD	OB	×	104	×	ml	–	–	–	–	–	–	m	m	m	m	oml
	Tellina nitens C. B. Adams, 1845	L	D	O	–	3	×	l	–	–	–	–	–	–	–	m	–	–	–
	Tellina paramera Boss, 1964	ULT	S	O	–	4	×	ml	–	–	–	–	–	–	–	–	–	–	–
	Tellina persica Dall & Simpson, 1901	L	D	O	–	1	–	–	–	–	–	–	–	–	–	m	–	–	–
	Tellina probrina Boss, 1964	MLT	SD	OB	–	15	×	l	–	–	–	–	–	–	–	–	–	–	o
	Tellina punicea Born, 1778	–	–	–	–	1	–	l	–	–	–	–	–	–	–	m	–	–	–
	Tellina radiata Linné, 1758	UMLT	ISD	O	×	39	×	ml	ml	–	–	–	m	m	m	m	m	ml	o
	Tellina similis Sowerby, 1806	UMLT	ISD	OB	×	128	×	ml	l	–	–	–	m	m	m	m	m	ml	oml
	Tellina squamifera Deshayes, 1855	UMLT	D	OB	–	11	×	ml	–	–	–	–	–	–	–	–	m	–	o
	Tellina sybaritica Dall, 1881	ULT	I	OB	×	10	×	ml	–	–	–	–	–	–	–	–	–	–	ol
	Tellina tampaensis Conrad, 1866	UL	IS	OB	×	18	×	ml	–	–	m	–	m	–	m	m	–	–	om
	Tellina texana Dall, 1900	UMLT	IS	OB	×	18	×	ml	–	–	–	–	m	–	–	–	–	–	oml
	Tellina versicolor DeKay, 1843	UMLT	ISD	OB	×	28	×	ml	–	–	–	–	–	–	–	–	–	–	oml
Teredinidae	*Bankia carinata* (Gray, 1827)	M–	–	–	×	1	–	m	–	–	–	–	–	–	–	–	–	–	–
	Nototeredo knoxi (Bartsch, 1917)	U	I	–	–	1	–	–	–	–	–	–	–	–	–	–	m	–	–
	Teredo clappi Bartsch, 1923	UL	I	–	×	2	–	l	–	–	–	–	–	–	–	–	m	–	–
Thraciidae	*Asthenothaerus balesi* Rehder, 1943	L	–	–	–	3	×	l	–	–	–	–	–	m	–	–	–	–	–
	Asthenothaerus hemphilli Dall, 1886	ULT	S	B	×	3	–	–	–	–	–	–	–	–	–	–	–	–	ml
	Thracia corbuloides Blainville, 1824*	L	SD	–	–	2	–	l	–	–	–	–	–	–	–	–	–	–	–
	Thracia phaseolina Lamarck, 1822*	–	D	–	–	1	–	l	–	–	–	–	–	–	–	–	–	–	–
	Thracia stimpsoni Dall, 1886*	LT	D	B	×	4	–	l	l	–	–	–	–	–	–	–	–	–	l
Thyasiridae	*Thyasira trisinuata* (d'Orbigny, 1842)	T	D	–	–	1	–	–	–	–	–	–	–	–	–	–	–	–	o
Trapezidae	*Coralliophaga coralliophaga* (Gmelin, 1791)	MLT	S	O	×	12	×	ml	–	–	–	–	m	m	m	–	–	–	o
Ungulinidae	*Diplodonta punctata* (Say, 1822)	UML	S	O	×	9	×	ml	–	–	–	–	–	m	–	m	–	–	o
	Diplodonta semiaspera Philippi, 1836	UMLT	IS	OB	×	9	×	ml	–	–	–	–	–	m	–	–	–	–	oml
Veneridae	*Anomalocardia auberiana* (d'Orbigny, 1842)	UMLT	ISD	OB	×	38	×	ml	–	–	–	–	m	m	–	m	m	–	oml
	Callista eucymata (Dall, 1890)	UMLT	D	O	–	13	–	ml	–	–	–	–	–	–	–	–	m	–	m
	Chione cancellata (Linné, 1767)	UMLT	ISD	OB	×	158	×	ml	ml	–	–	–	m	m	m	m	m	m	oml
	Chione mazyckii Dall, 1902	UMLT	ISD	OB	×	69	–	m	–	–	–	–	–	m	m	m	–	–	o
	Chione paphia (Linné, 1767)	LT	D	O	×	4	–	l	m	–	–	–	–	–	–	m	m	–	–
	Circomphalus strigillinus (Dall, 1902)	MLT	D	O	–	6	–	ml	–	–	–	–	–	–	m	–	m	–	–
	Cyclinella tenuis (Récluz, 1852)	UMLT	SD	OB	×	9	×	l	–	–	–	–	–	–	–	–	–	–	om
	Dosinia discus (Reeve, 1850)	MLT	–	–	×	3	–	m	–	–	–	–	–	–	–	m	–	–	l
	Dosinia elegans (Conrad, 1843)	UMLT	SD	OB	×	7	–	ml	–	–	–	–	–	–	–	m	–	–	om
	Globivenus rigida (Dillwyn, 1817)	T	–	–	–	1	×	ml	–	–	–	–	–	–	–	–	–	–	–
	Globivenus rugatina (Heilprin, 1886)	LT	D	O	–	2	–	m	–	–	–	–	–	–	–	–	m	–	–
	Gouldia cerina (C. B. Adams, 1845)	UMLT	ISD	OB	×	58	×	ml	–	–	–	–	m	m	–	m	m	–	oml
	Lirophora latilirata (Conrad, 1841)	LT	D	OB	–	15	×	ml	–	–	–	–	–	–	m	m	m	–	o
	Macrocallista maculata (Linné, 1758)	ULT	ISD	OB	×	19	×	ml	–	–	–	–	–	–	m	m	m	–	ol
	Macrocallista nimbosa (Lightfoot, 1786)	ULT	–	–	–	2	–	m	m	–	–	–	m	–	–	–	–	–	–
	Mercenaria campechiensis (Gmelin, 1791)	L	–	–	×	1	–	–	–	–	–	–	–	–	–	m	–	–	–

	Mercenaria mercenaria forma notata (Linné, 1758)	M	I	O	–	2	–	–	–	–	–	–	–	–	m	m	–	–	–
	Parastarte triquetra (Conrad, 1846)	UMLT	ISD	OB	×	13	–	–	–	–	–	–	m	–	–	–	–	–	o
	Periglypta listeri (Gray, 1838)	UMLT	IS	OB	×	58	×	ml	–	–	–	–	–	m	m	m	m	ml	om
	Pitar cordatus (Schwengel, 1951)	LT	D	OB	×	10	–	ml	–	–	–	–	–	–	m	m	m	–	o
	Pitar fulminatus (Menke, 1828)	LT	D	O	×	4	×	ml	–	–	–	–	–	–	–	–	m	–	l
	Pitar simpsoni (Dall, 1895)	UMLT	ISD	OB	×	68	×	ml	–	–	–	–	m	m	–	m	m	–	oml
	Protothaca granulata (Gmelin, 1791)*	–	–	–	–	0	–	l	–	–	–	–	–	–	–	–	–	–	–
	Puberella intapurpurea (Conrad, 1849)	UMLT	SD	OB	–	14	×	ml	–	–	–	–	–	–	m	m	m	–	o
	Puberella pubera (Bory SaintVincent, 1827)*	–	–	–	–	0	×	l	–	–	–	–	–	–	–	–	–	–	–
	Timoclea grus (Holmes, 1858)	UMLT	ISD	OB	×	17	–	–	–	–	l	–	–	–	–	m	m	–	ol
	Timoclea pygmaea (Lamarck, 1818)	UMLT	IS	OB	–	8	×	ml	–	–	–	–	–	–	–	–	–	–	o
	Tivela floridana Rehder, 1939	U	I	O	–	1	–	–	–	–	–	–	–	–	–	–	–	–	o
	Transenella conradina Dall, 1884	UMLT	ISD	OB	×	34	×	ml	–	–	–	–	–	m	–	–	–	–	ol
	Transenella cubaniana (d'Orbigny, 1842)	M	S	OB	–	4	×	l	–	–	–	–	–	–	–	–	–	–	o
	Transenella culebrana Dall & Simpson, 1901*	–	D	–	–	1	–	l	–	–	–	–	–	–	–	–	–	–	–
	Transenella stimpsoni Dall, 1902	UML	I	–	×	3	×	l	–	–	l	–	–	m	–	–	–	–	l
Verticordiidae	*Haliris fischeriana* (Dall, 1881)	T	–	–	–	1	–	m	–	–	–	–	–	–	–	–	–	–	–
	Trigonulina ornata d'Orbigny, 1842	UT	D	O	–	3	–	m	–	–	–	–	–	m	–	–	m	–	–
	Verticordia acuticostata Philippi, 1844	L	D	O	–	4	–	m	–	–	–	–	–	–	–	m	m	–	–

*unverified identification (unillustrated literature citation only, not part of a revisionary systematic work). †U, Upper Keys; M, Middle Keys; L, Lower Keys; T, Dry Tortugas. ‡i, intertidal; S, shallow; D, deep. §O, oceanside; B, bayside. ||Number of records in database. ¶Lyons & Quinn (1995). [1]m, Museum specimen record; o, original collection record; l, literature record. [2]All Nomenclature follows Turgeon *et al.* (1998) *except* for those taxa indicated by [2], which are modified according to recent systematic research by the present authors.

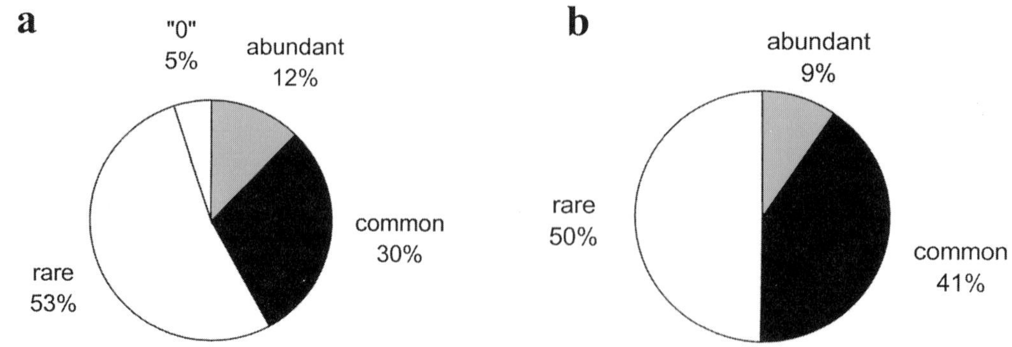

Fig. 2. Frequency of encounter of Florida Keys bivalve species from: (**a**) all sources ($n = 309$); (**b**) original collection records only ($n = 179$).

(i.e. records/species in the database), is a fair approximation of species rarity. Three hundred and nine (95%) of the 325 species were represented in the database (the remaining 16 species, with '0' records in Table 1, are known only from data indicating 'Florida Keys' and thus were not entered into the database). Accordingly, it can be said that 40 species (12%) were abundant (> 50 records), 96 species (29%) were common (10–49 records) and 172 species (53%) were rare (< 10 records) (Fig. 2a). To eliminate any possible artefact in the database caused by sampling bias of other collectors, these same statistics can be considered for original collection records only (179 species; Fig. 2b): abundant, 17 species (9%); common, 73 species (41%); rare, 89 species (50%). The most frequently collected species were *Barbatia cancellaria* (175 records, 102 original records), *Codakia orbicularis* (172 records, 100 original records), *Chione cancellata* (158 records, 116 original records) and *Codakia orbiculata* (154 records, 113 original records).

Regional similarity

Within the database of 6145 records, 6076 records (99%) could be assigned to one of the Florida Keys zones. While the Upper, Middle and Lower zones were nearly equally sampled, Dry Tortugas records represented less than half of each of the other zones (Fig. 3), probably in part the result of its inaccessibility and its much smaller area compared to the other three regions. Three hundred and six of the 325 species (94%) were coded for at least one zone; 118 species (36%) were recorded from all Florida Keys zones (UMLT in Table 1); 79 species (24%) were restricted to a single zone (Upper Keys 15; Middle Keys 10; Lower Keys 31; Dry Tortugas 23). The remaining 109 species (33%) were recorded from two or more zones.

Bray-Curtis similarity analysis of all data by Florida Keys region (Fig. 4) revealed a regional northeast–southwest gradient, with greatest similarity (82%) between the Upper and Middle Keys regions, and least similarity (72%) between the Dry Tortugas and other Keys regions. Using original data only (176 species; with Upper Keys, 132 species; Middle Keys, 139 species; Lower Keys, 138 species; and Dry Tortugas, 102 species), similarities between the Upper, Middle and Lower Keys regions formed a near trichotomy at 87%, while Dry Tortugas remained most dissimilar (67%) to the other three.

Most Florida Keys bivalves live on both sides of the island chain: 237 species (72%) could be coded for ocean- or bayside of the Keys axis; 57 species (17%) were recorded as oceanside only; 26 species (8%) were bayside only; and the remaining 154 species (47%) were recorded from both ocean- and bayside.

The close similarity between Upper and Middle

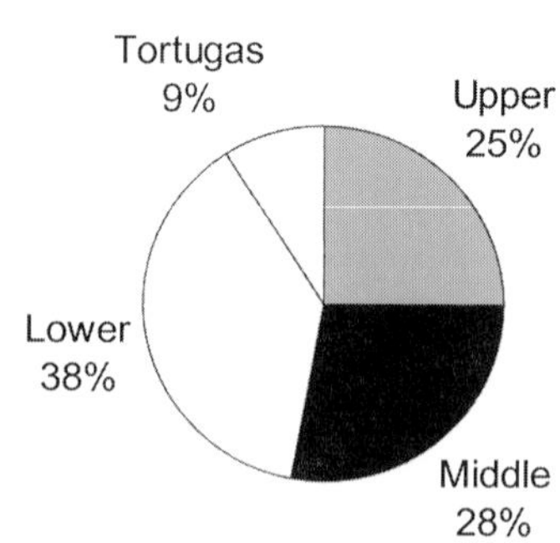

Fig. 3. Sampling effort – percentage of total bivalve records from the four Florida Keys zones: Upper Keys ($n = 1513$), Middle Keys ($n = 1688$), Lower Keys ($n = 2311$) and Dry Tortugas ($n = 565$).

	Upper	Middle	Lower
Upper			
Middle	81.818		
Lower	79.727	77.590	
DT	71.247	69.377	75.115

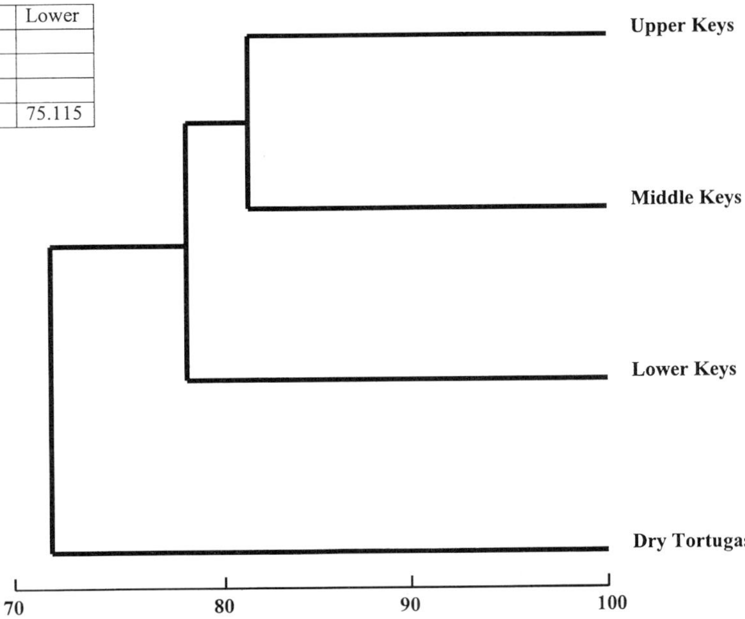

Fig. 4. Bray-Curtis similarity dendrogram of bivalve species presence/absence in the four geographical regions of the Florida Keys. Total analysis, 306 species; Upper Keys, 201 species; Middle Keys, 178 species; Lower Keys, 241 species; Dry Tortugas, 194 species.

Keys regions could be a result of the ecological differences between ocean- and bayside. Florida Bay, bayside of the Upper and Middle Keys (only), is overall shallow in depth (1.5–5 m) and experiences greater water temperature and salinity fluctuations than does the oceanside of the Upper and Middle Keys (Chiappone 1996*b*). There are 12 bivalve species unique to Florida Bay (although all are rare in frequency) and another seven are unique to bayside of the Lower Keys (outside of Florida Bay). To test whether the strong similarity of the Upper and Middle Keys is due to the influence of Florida Bay, a Bray-Curtis similarity analysis was conducted in which bay- and oceanside lists for Upper, Middle and Lower Keys were separated (Dry Tortugas remained as a single unit because its oceanic location does not allow distinction of exposures). The result of this analysis (Fig. 5) shows that the Upper and Middle bayside together form a group (71% similarity) which has lowest similarity (64%) with the remaining areas combined, including the Dry Tortugas. Bayside of the Lower Keys showed stronger similarity with the oceansides of all regions, plus the Dry Tortugas. This suggests that the Florida Bay exerts a strong influence on the bivalve fauna of the Florida Keys.

About one-quarter of Florida Keys bivalves were recorded from all depths, from intertidal to shallow to deep waters; 269 species (83%) were coded for depth zone; 85 species (26%) were recorded as ranging from shallow to deep; 14 species (4%) were recorded from only intertidal depths; 29 species (9%) were recorded from only shallow depths; and 58 species (18%) were recorded from only deep depths. Forty-three species (13%) ranged from intertidal to shallow and 38 species (12%) ranged from shallow to deep.

The dissimilarity of the Dry Tortugas to the remaining Keys regions could be an artifact of deep-water sampling – 15 of the 23 species unique to the Dry Tortugas are from deep water (the other eight are of unrecorded depth). Alternatively, the Dry Tortugas could be dissimilar due to: (1) a genuinely different fauna (associated with its lesser habitat diversity and/or recruitment by larvae from other regions); or (2) an artefact of overall poor sampling (see Fig. 3). To test whether deep-water records are the primary cause of the dissimilarity of the Dry Tortugas fauna, a Bray-Curtis similarity analysis was conducted in which all deep-water records were first eliminated. This resulted in a data set of 270 species from intertidal or shallow depths, rather evenly distributed from Upper ($n = 191$ spp.), Middle ($n = 169$), Lower ($n = 220$) and Dry Tortugas ($n = 133$) regions. The resulting dendrogram was nearly identical to that produced by all

	Upper O	Middle O	Lower O	Upper B	Middle B	Lower B
Upper oceanside						
Middle oceanside	81.081					
Lower oceanside	78.827	83.495				
Upper bayside	64.662	67.910	62.366			
Middle bayside	67.227	70.000	62.948	70.476		
Lower bayside	75.362	76.259	71.280	68.548	70.909	
Dry Tortugas	69.617	67.449	75.000	54.019	53.004	66.044

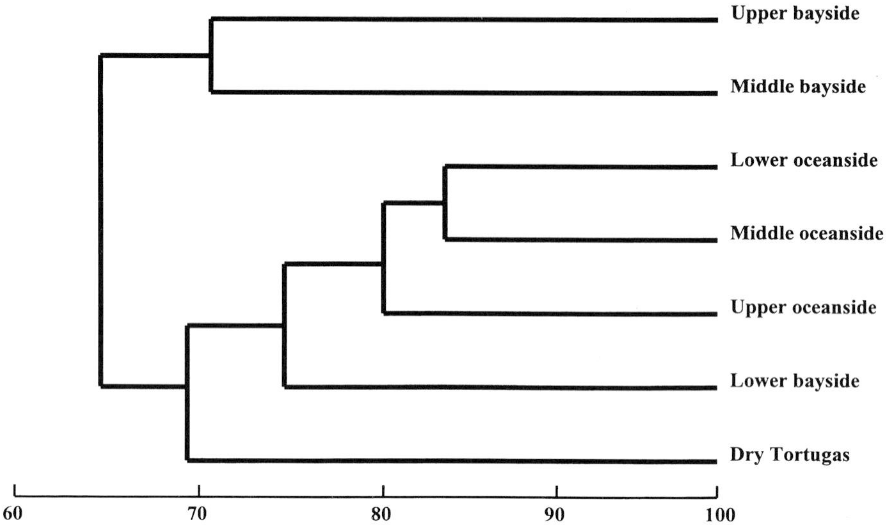

Fig. 5. Bray-Curtis similarity dendrogram of bivalve species presence/absence in the four geographical regions of the Florida Keys, with Upper, Middle and Lower Keys separated into bay- and oceanside records. Total analysis, 256 species; upper oceanside, 147 species; middle oceanside, 149 species; lower oceanside, 160 species; upper bayside, 119 species; middle bayside, 91 species; lower bayside, 129 species; Dry Tortugas, 192 species.

records (Fig. 4), with a slightly stronger affinity (82%) between Upper and Middle Keys, and a slightly weaker similarity (63%) between Dry Tortugas and all other regions. The difference in the Dry Tortugas is likely not an artefact of fewer samples overall, because this method of analysis condenses one or many records equally into a single entry. It is thus likely that the Dry Tortugas is faunistically distinct, an assumption that is also subjectively corroborated by the conspicuous apparent absence of several ubiquitous Keys bivalves (e.g. *Isognomon alatus, Brachidontes exustus, Argopecten nucleus, Pinctada imbricata, Trachycardium muricatum*). One of the factors that could play a role is the different exposure of the Dry Tortugas. There is evidence that the Gulf of Mexico Loop Current has a significant influence in this region, while other areas of the Florida Keys (such as the Lower Keys) are generally protected from the Loop Current by the blocking effect of larger islands (Shinn *et al.* 1989).

Losses and gains

Within the database of 6145 records, 5022 records (82%) representing 261 species (80%) were associated with dated collection records. Excluding the 3323 records amassed by original collections for this survey, the vast majority of the dated records were collected after 1930 (1572, or 92% of 1699 museum and literature records). Figure 6a shows this pattern clearly, with a distinct lull in data being amassed during the period from 1900 to 1929. The greatest number of records (467, or 8% of the total) was generated in the 1960s. The percentage of the total species recorded from each decade shows the same pattern, although the greatest percentage is in the 1990s, reflecting the original collections made during this study (Fig. 6b). Surprisingly, a fair number of records (63, or 1%) and species (48, or 18%) were recorded from the decades prior to the year 1900, with the earliest from the 1870s.

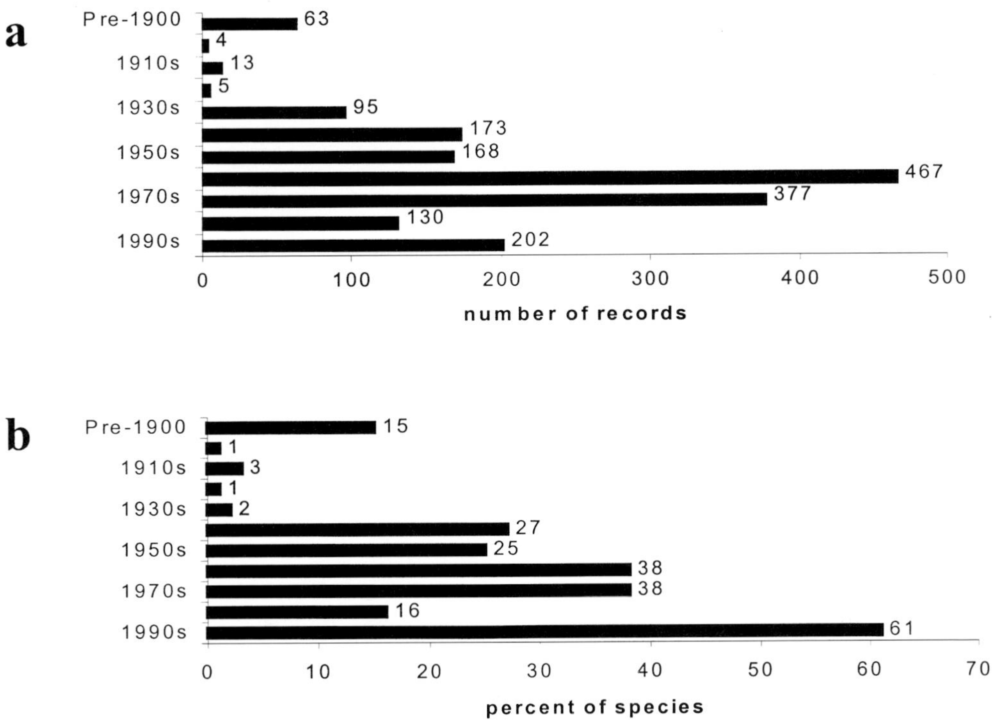

Fig. 6. Coverage by decade of collection. (**a**) Number of literature and museum records (i.e. excluding 3323 original collection records) per decade. (**b**) Percentage of total species (327) recorded from each decade: pre-1900, $n = 48$ species; 1900s, $n = 1$; 1910s, $n = 11$; 1920s, $n = 4$; 1930s, $n = 59$; 1940s, $n = 90$; 1950s, $n = 82$; 1960s, $n = 123$; 1970s, $n = 124$; 1980s, $n = 53$; 1990s, $n = 201$.

The temporal pattern of collection records correlates subjectively with accessibility of the Keys, via construction of the Flagler's extended East Coast Railroad (completed in 1912), converted to the Overseas Highway (US Route 1) in 1938 (following destruction of the railroad by the 1935 hurricane). Several interesting species anecdotes also emerge from these dated data related to the human influence in the Keys. Firstly, *Nodipecten nodosum*, the popular 'Lion's Paw' scallop, that lives on deep-water ledges and (nowadays) shipwrecks, was not recorded before the 1950s but has been collected every decade since then; this is probably strongly related to the development of scuba diving in the area. Secondly, two species of brackish-water false mussels, *Mytilopsis leucophaeata* and *M. sallei*, have only been recorded from the 1990s in the Upper Keys. Both of these species are otherwise widely distributed in the western Atlantic (Marelli & Gray 1983) but have not been previously recorded from the Keys, where marine habitats are largely oceanic in salinity. Although a direct correlation has not been proven and the present populations are small, their recent introduction could be the anthropogenic result of increased traffic of recreational boats (to which these mussels attach) and/or increased freshwater input into Florida Bay through extensive channelling in southern Florida and the Everglades (NOAA 1996).

There are no currently active commercial bivalve fisheries in Florida Keys waters. However, the impact of marine-life collectors (for the commercial and recreational aquarium trade) has been substantial through removal of large quantities of 'live rock' and rubble, and the targeting of charismatic macrofaunal species. A dramatic example was reported by Bohnsack *et al.* (1994) of annual collections of more than 50 000 'Rough Fileclams' (in this context encompassing the two limid species *Ctenoides scaber* and *C. floridanus*) during the period 1990–1992.

The lack of a collecting record does not prove that a species was in fact absent, therefore species losses or gains over time, in the absence of previous species inventories from the Keys, are difficult to

	Fla Bay	FK ocean	Tortugas	Cuba	GOM	Yucatan
Florida Bay						
FK oceanside group	69.714					
Dry Tortugas	60.550	79.012				
Cuba	46.800	61.246	56.937			
Gulf of Mexico	46.097	61.039	56.324	67.102		
Yucatan	63.102	70.796	62.005	68.439	63.750	
eastern Florida	62.234	75.330	65.245	63.245	63.551	75.314

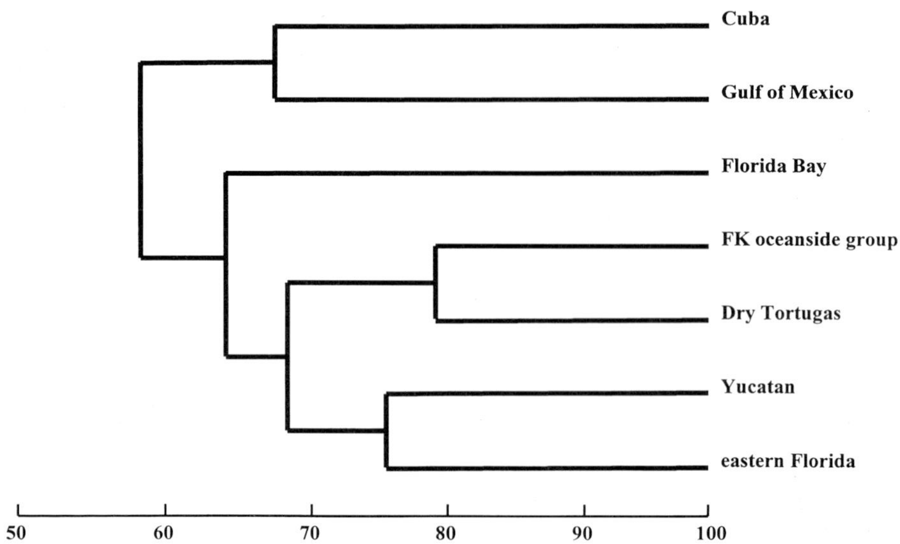

Fig. 7. Bray-Curtis similarity dendrogram of bivalve species of the Florida Keys. Florida Bay, $n = 136$ species; oceanside group, $n = 214$; and Dry Tortugas, $n = 191$ (see text). Eastern Florida, $n = 240$; Gulf of Mexico, $n = 402$; Cuba, $n = 364$; Yucatan, $n = 238$. Total analysis, 549 species.

prove. If a sufficient quantity of records from a single location (i.e. single island or reef) could be assembled, changes in the fauna might become apparent if they have occurred. In this database, restricted as it is to bivalves, the localities with the greatest number of records are Marathon (742), Key Largo (702), Key West (613), Grassy Key (274), Missouri Key (241), Lower Matecumbe Key (231) and Summerland Key (214). Unfortunately, the vast majority of records in each of these sets is from the 1990s and the spread of data across other decades is too thin to be informative. This is due in part to the deficient attention paid to Florida Keys bivalves (as opposed to gastropods) by both malacological authors and collectors. Any possible analyses of this kind must await the addition of data on other molluscan classes (e.g. gastropods – currently in preparation by the present authors).

Taking the Florida Keys as a whole, species occurrences across the decades recorded by this survey indicate little species turnover in most bivalves. All abundant species and nearly all common species range into the present decade. Of the 61 species not recorded from the 1990s, all but six are rare (fewer than ten records) in the database. Likewise, of the species not recorded since earlier decades (three not since the 1930s, four not since the 1940s, two not since the 1950s and 13 not since the 1960s), all rank as rare, and so probably do not reflect losses. Anecdotal observations relate fewer encounters along the main highway (US Route 1) in certain ‘collectible’ molluscan species (e.g. the gastropod *Nerita peloronta*, although large populations of this species occur in areas less exposed to the casual collector). Habitat shifts from natural to artificial substrata are also evident in certain species, e.g. *Isognomon alatus*, a species formerly best associated with mangrove prop roots that now successfully populates concrete seawalls and bridge abutments.

Comparative analyses

To compare the bivalve species composition of the Florida Keys with those of other western Atlantic locations, a sequence of Bray-Curtis similarity analyses was performed. The non-Keys localities in all analyses formed two consistent groups: (1) Cuba and the Gulf of Mexico (67% similarity); and (2) Yucatan and eastern Florida (75%). In the first analysis, the Florida Keys were treated as a single entity. The result showed the Keys fauna to be closer in similarity to Yucatan and eastern Florida (74%) than to Cuba and the Gulf of Mexico (71%). In the second analysis, the Keys fauna was split because earlier tests (Figs 4 and 5) showed the Keys to be composed of several different faunal components: (1) Upper and Middle Keys bayside (Florida Bay); (2) Upper, Middle and Lower Keys oceanside (plus Lower Keys bayside; oceanside group); (3) the Dry Tortugas. In this second analysis (Fig. 7), the separated Keys components retained their earlier relationships (see Fig. 5): Dry Tortugas was most similar to the oceanside group (79% similarity), with the Florida Bay less similar to the other two (60% and 70%, respectively). The final two analyses were conducted by removing either Florida Bay or both Dry Tortugas and the oceanside group; these analyses helped to elucidate the relationships of each of these two combined areas to the four non-Keys regions. The results showed the 'oceanside plus Tortugas' group with greatest similarity to Yucatan and eastern Florida (68%), while Florida Bay showed rather low similarity (54%) with the non-Keys regions. The similarity matrix for the dendrogram in Fig. 7 (the second analysis) shows that Florida Bay had only 46–47% similarity with Cuba and the Gulf of Mexico. The relationship of the Keys fauna to that of the Gulf is perplexing in view of hydrological influence of the Gulf of Mexico Loop Current (Shinn *et al.* 1989), and long-term net flow from the Gulf though tidal channels of the Middle and Lower Keys (Smith 1994). The close similarity with Yucatan, however, may relate to the northward flow of the Florida Current, past the eastern coast of Yucatan, along the Keys and north to join the Gulf Stream (NOAA 1996).

These results should not be overinterpreted because they are based on bivalve data only (rather than total molluscan faunas) and, additionally, because the comparative lists have not been as rigorously prepared as the present list for the Keys. The completeness of the latter determines the accuracy of similarity indices in comparative analyses and points to the need for robust molluscan surveys in critical tropical/subtropical regions (e.g. Bahamas, Bermuda, Great Barrier Reef). In spite of these shortcomings, several interesting indications emerge from these results: (1) the Florida Bay bivalve fauna is distinctly different, showing low similarity not only against the rest of the Keys but also against other tested western Atlantic localities; (2) the Dry Tortugas bivalve fauna, despite the differences shown earlier within the Keys, is more similar to the rest of the Keys (or more accurately, the oceanside group) than to other western Atlantic regions.

Conclusions

- Bivalves of the Florida Keys comprise 325 species, representing a 100% increase over the single previously assembled list (Lyons & Quinn 1995). According to frequency of encounter, 153 species (47%) are common to abundant. The most common species are *Barbatia cancellaria*, *Codakia orbicularis*, *C. orbiculata* and *Chione cancellata*.
- Florida Keys bivalves are mainly tropical, with 96% of the species ranging southward from the Keys into the western Atlantic tropics. The faunal list includes no endemics and no threatened or endangered species, although past overharvesting for the aquarium trade warrants continued vigilance. Two recently introduced species appear to be due to increased boat traffic and brackish habitat. No species losses over time were evident from dated records but changes in abundance along the main highway through the Keys, and in habitat (from natural to artificial substrata), are noticeable. Within the Keys, about one-third of bivalve species inhabit all Keys zones (Upper, Middle and Lower Keys, and Dry Tortugas); 72% of all species occur on both bayside and oceanside of the island chain.
- Museum collections were most successful in generating the species list, producing 77% of the total, and 62 species not found by other sources. Extensive original collecting efforts over five years captured 55% and produced the best detailed information, including live–dead distinctions. Literature sources (including newsletters and agency reports) produced 73% of the species, but an exhaustive search of the massive traditional literature (i.e. books, peer-reviewed journals) recovered only 44%. Biodiversity studies that do not take advantage of these multiple sources (each capturing taxa missed by the other methods), especially valuable existing museum collections, risk missing significant portions of the known fauna of a region. Review of traditional literature alone is viewed as the least effective approach.
- According to bivalve data, a latitudinal gradient in faunal similarity exists from the Upper Keys southwestwards to the Dry Tortugas. When bay- and oceanside are considered separately, the bayside of the Upper and Middle Keys (i.e. components of the Florida Bay) is shown to be the most divergent; the bivalve faunas of these two areas are very similar, but together differ

substantially from all other areas, bay- or oceanside. The faunistic difference in the Dry Tortugas is not due to biased or undersampling and noticeably lacks a number of otherwise ubiquitous Florida Keys bivalves.

- In comparison with four other western Atlantic regions (eastern Florida, the Gulf of Mexico, Cuba and Yucatan), the bivalves of the Florida Keys are closest to those of Yucatan and eastern Florida. Florida Bay again showed the lowest similarity, not only with other Keys areas but also with all comparative regions. The Dry Tortugas is more similar to the rest of the Keys than to other western Atlantic regions. These results are considered preliminary to: (1) improvement of comparative faunal lists; (2) completion of the Florida Keys molluscan survey.

We wish to gratefully acknowledge B. Haskell (FKNMS), D. Savage (NOAA) and M. Rice (Smithsonian Marine Station, Ft Pierce, Florida) for their interest and for facilitating our field work through permits (FKNMS 080-98) and practical assistance; M. Rice (Smithsonian Marine Station, Ft Pierce, Florida) and J. Leal (BMSM) for additional permits; T. Collins, Florida International University, and the crew of the *R/V Bartlett* for facilitating collections in the Marquesas and Dry Tortugas; W. G. Lyons (Florida Department of Environmental Protection) and B. W. Gottholm (NOAA) for unpublished data and interagency reports; G. Rosenberg (ANSP), J. Leal (BMSM), H. Chaney (SBMNH), C. Sturm (CMNH), T. Pearce and A. F. Chadwick (DMNH), D. Krumm (HBOM), and M. G. Harasewych (USNM) for access to the museum collections under their care; and A. Littman, L. Crowley, J. Slapcinsky, M. Baker, J. Karb, K. Mobley, C. Breedlove, D. Gonsalves-Jackson, C. Miles, B. Kubricht, R. Price and C. Bear for field/collections assistance and/or data entry. Field and laboratory work in Florida was facilitated in part through Visiting Scientists Awards by the Smithsonian Marine Station; M. Rice and the station staff are gratefully acknowledged for their support. Additional field support was provided over the course of this project by Harbor Branch Oceanographic Institution (Ft Pierce, Florida), DMNH and AMNH. Field collecting in the Florida Keys was supported through supplementary funding from the Bertha LeBus Charitable Trust and FMNH's Marshall Field Fund. This is Smithsonian Marine Station contribution no. 504.

References

ABBOTT, R. T. 1950. Voyages of the '*Eolis*'. *Johnsonia*, **2**, 219–220.

—— 1974. *American Seashells: the Marine Mollusca of the Atlantic and Pacific Coasts of North America*, 2nd Edition. Van Nostrand Reinhold, New York.

AGUAYO, C. G. & JAUME, M. L. 1947–1948. *Catalogo de Los Moluscos de Cuba, nos. 1–363*. Privately published, Havana, Cuba.

BOHNSACK, J. A., HARPER, D. E. & MCCLELLAN, D. B. 1994. Fisheries trends from Monroe County, Florida. *Bulletin of Marine Science*, **54**, 982–1018.

BOSS, K. J. & MERRILL, A. S. 1965. The family Pandoridae in the western Atlantic. *Johnsonia*, **4**, 181–216.

CARR, M. R. 1997. *PRIMER User Manual: Plymouth Routines in Multivariate Ecological Research*. Plymouth Marine Laboratory, Plymouth.

CHIAPPONE, M. 1996*a*. *Site Characterization for the Florida Keys National Marine Sanctuary and Environs, Volume 1. Geology and Paleontology of the Florida Keys and Florida Bay*. The Nature Conservancy & The Preserver, Zenda, Wisconsin.

—— 1996*b*. *Site Characterization for the Florida Keys National Marine Sanctuary and Environs, Volume 2. Oceanography and Shallow-Water Processes of the Florida Keys and Florida Bay*. The Nature Conservancy & The Preserver, Zenda, Wisconsin.

CLENCH, W. J. & TURNER, R. D. 1950. The genera *Sthenorytis, Cirsotrema, Acirsa, Opalia* and *Amaea* in the western Atlantic. *Johnsonia*, **2**, 221–248.

DALL, W. H. 1886. Report on the results of dredging by the United States Coast Survey Steamer 'Blake'. XXIX. Report on the Mollusca. Part I. Brachiopoda and Pelecypoda. *Bulletin of the Museum of Comparative Zoology, Harvard University*, **12**, 171–318, pls 1–9.

—— 1889. Reports on the results of dredging, under the supervision of Alexander Agassiz, in the Gulf of Mexico (1877–78) and in the Caribbean Sea (1879–80), by the U. S. Coast Survey steamer 'Blake', Lieut.-Commander C. D. Sigsbee, U. S. N., and Commander J. R. Bartlett, U. S. N., commanding. XXIX. Report on the Mollusca. Part 2. Gastropoda and Scaphopoda. *Bulletin of the Museum of Comparative Zoology*, **18**, 1–492, pls 10–40.

EKDALE, A. A. 1974. Marine molluscs from shallow-water environments (0 to 60 meters) off the northeast Yucatan coast, Mexico. *Bulletin of Marine Science*, **24**, 638–668.

ESPINOSA, J., FERNANDEZ-GARCÉS, R. & ROLÁN, E. 1994. Catalogo de los moluscos marinos actuales de Cuba. *Cuadernos Investigaciones Biologíca (Bilbao)*, **18**, 85–154.

GUNDERSEN, R. W. 1998. *The Seashells of Sanibel and Captiva Islands*. Privately published, Kenosha, Wisconsin.

HAAS, F. 1940. Ecological observations on the common mollusks of Sanibel Island, Florida. *American Midland Naturalist*, **24**, 369–378.

HENDERSON, J. B. 1911. Extracts from the log of the *Eolis*. *The Nautilus*, **25**, 71–72, 81–82.

—— 1913. Marine shells from drift on Upper Matecumbe Key, Florida. *The Nautilus*, **27**, 59–60.

—— 1914. Notes. *The Nautilus*, **27**, 120.

HOFFMEISTER, J. E. & MULTER, H. G. 1968. Pleistocene limestones of the Florida Keys. *In*: GINSBURG, R. N. (ed.) *Guidebook for Field Trip*. Geological Society of America, Miami, Florida, **1**, 57–61.

HOUBRICK, R. S. 1987. Transfer of *Cerithiopsis crystallina* Dall to the genus *Varicopeza* Gründel, family Cerithiidae (Prosobranchia: Gastropoda). *The Nautilus,* **101**, 80–84.

JAAP, W. C. 1984. *The Ecology of the South Florida Reefs: a Community Profile.* United States Fish & Wildlife Service, Washington, DC, FWS/OBS-82/08.

JOHNSON, C. W. 1934. List of marine Mollusca of the Atlantic coast from Labrador to Texas. *Proceedings of the Boston Society of Natural History*, **40**, 1–204.

KIDWELL, S. M. in press. Ecological fidelity of molluscan death assemblages. *In*: ALLER, J. Y., WOODIN, S. A. & ALLER, R. C. (eds) *Organism–Sediment Interactions*. Belle W. Baruch Symposium, University of South Carolina Press, Columbia, in press.

LEE, H. G. (WITH THE JACKSONVILLE SHELL CLUB). 1999. *Cedar Key Marine Mollusks*. World Wide Web address: http://home.sprynet.com/~wfrank/cedarkey.htm.

LIPE, R. 1984. Tampa Bay Shells. *Conchologists of America Bulletin*, **12**, 26–27.

LIPKA, D. A. 1974. *Biota of the West Flower Garden Bank*. Gulf Publishing Company, Houston.

LYONS, W. G. 1989. Nearshore marine ecology at Hutchinson Island, Florida: 1971–1974. XI. Mollusks. *Florida Marine Research Publication,* **47**, 1–131.

—— 1997. *Preliminary Check-list of Littoral, Estuarine, and Shallow-water Marine Mollusks of Florida.* Unpublished report (copy on file in reprint library of Division of Invertebrate Zoology, American Museum of Natural History, New York).

—— & QUINN, J. F., JR 1995. Appendix J. Marine and terrestrial species and algae: Phylum Mollusca. *Florida Keys National Marine Sanctuary Draft Management Plan/Environmental Impact Statement, Volume III.* United States Government Printing Office, Washington, DC, J-10–J-26.

MCGINTY, P. L. 1955. New marine mollusks from Florida. *Proceedings of the Academy of Natural Sciences of Philadelphia*, **107**, 75–85.

—— 1962. Caribbean marine shells. *The Nautilus*, **76**, 39–44, pl. 3.

—— & MCGINTY, T. L. 1957. Dredging for deep water shells in southern Florida. *The Nautilus,* **71**, 37–47.

—— & NELSON, M. M. 1972. *Mollusca Dredged off Pompano Beach, Florida, in 10 Fathoms.* Unpublished report (copy on file in reprint library of Division of Invertebrate Zoology, American Museum of Natural History, New York).

MARELLI, D. C. & GRAY, S. 1983. Conchological redescriptions of *Mytilopsis sallei* and *Mytilopsis leucophaeta* of the brackish western Atlantic. *The Veliger,* **25**, 185–193.

MARSZALEK, D., BABASHOFF, G., JR, NOEL, M. R. & WORLEY, D. R. 1977. Reef distribution in south Florida. *Proceedings of the Third International Coral Reef Symposium,* **2**, 223–229.

MIKKELSEN, P. M., MIKKELSEN, P. S. & KARLEN, D. J. 1995. Molluscan biodiversity in the Indian River Lagoon, Florida. *Bulletin of Marine Science*, **57**, 94–127.

MITCHELL-TAPPING, H. J. 1980. Depositional history of the oolite of the Miami Limestone formation. *Florida Scientist*, **43**, 116–125.

NOAA (NATIONAL OCEANIC & ATMOSPHERIC ADMINISTRATION). 1996. *Florida Keys National Marine Sanctuary, Final Management Plan/Environmental Impact Statement. Volume I–III.* United States Government Printing Office, Washington, DC.

PILSBRY, H. A. 1955. Another Floridan *Conus. The Nautilus*, **69**, 47–48, pl. 3.

—— & OLSSON, A. A. 1953. Materials for a revision of east coast and Floridan volutes. *The Nautilus,* **67**, 1–13, pls 1–3.

REED, J. K. & MIKKELSEN, P. M. 1987. The molluscan community associated with the scleractinian coral, *Oculina varicosa. Bulletin of Marine Science*, **40**, 99–131.

SHELTON, D. N. 1997. *A Systematic List of Mollusks in the Northern Gulf of Mexico off the Coast of Alabama.* World Wide Web Address: http://fly.hiwaay.net/~dwills/marine/alamarsp.html.

SHINN, E. A., LIDZ, B. H., HALLEY, R. B., HUDSON, J. H. & KINDINGER, J. L. 1989. *Reefs of Florida and the Dry Tortugas. Field Trip Guidebook T176.* American Geophysical Union, Washington, DC.

SMITH, N. P. 1994. Long-term Gulf-to-Atlantic transport through tidal channels in the Florida Keys. *Bulletin of Marine Science,* **54**, 602–609.

STEGER, D. 1962. *Check List, Bay and Beach Mollusca, West Coast of Florida, Cedar Keys to Marco Id.* Unpublished report (copy on file in reprint library of Delaware Museum of Natural History, Wilmington).

THOMPSON, A. R., MCGINTY, P. L. & MCGINTY, T. L. 1951. Dredging from the cruiser *Triton. The Nautilus,* **65**, 37–43.

TURGEON, D. D., QUINN, J. F., JR, BOGAN, A. E. *ET AL.* 1998. *Common and Scientific Names of Aquatic Invertebrates from the United States and Canada: Mollusks,* 2nd Edition. American Fisheries Society, Special Publications, **26**.

TURNER, R. D. & BOSS, K. J. 1962. The genus *Lithophaga* in the western Atlantic. *Johnsonia*, **4**, 81–116.

VITTOR (BARRY A.) & ASSOCIATES, INC. (FOR NOAA, SILVER SPRING, MARYLAND). 1998. *Florida Keys to Dry Tortugas Benthic Community Assessment*. World Wide Web Address: http://ccma.nos.noaa.gov/datapg2.html.

VOKES, H. E. & VOKES, E. H. 1984 (1983). Distribution of shallow-water marine Mollusca, Yucatan Peninsula, Mexico. *Mesoamerican Ecology Institute, Monograph 1; Middle American Research Institute, Publications,* **54**.

VOSS, G. L., BAYER, F. M., ROBINS, C. R., GOMON, M. & LAROE, E. T. 1969. *A Report to the National Park Service, Department of the Interior, On the Marine Ecology of the Biscayne National Monument.* Institute of Marine and Atmospheric Sciences, University of Miami.

Genetic relationships of *Mytilus galloprovincialis* Lamarck populations worldwide: evidence from nuclear-DNA markers

CLAIRE DAGUIN & PHILIPPE BORSA

Laboratoire Génome Populations Interactions and IRD, Station Méditerranéenne de l'Environnement Littoral, 1 Quai de la Daurade, 34200 Sète, France (e-mail: daguin@univ-montp2.fr)

Abstract: Allozyme surveys of genetic variation in *Mytilus galloprovincialis* Lamarck throughout the world have identified three groups within this species: a northeastern (NE) Atlantic group that also includes the *M. galloprovincialis* population of South Africa, a Mediterranean group that also includes the *M. galloprovincialis* populations from the eastern and the western coasts of the North Pacific, and an Australasian group. Hypotheses that have been proposed to account for the genetic differentiation patterns and disjunct, worldwide distribution of *M. galloprovincialis* include the recent introduction of this species into the southern hemisphere and the North Pacific through human agency, and an alternative hypothesis that each of the three groups is endemic. In this study, two nuclear-DNA markers (the polyphenolic adhesive protein gene *Glu-5′* and the first intron of the actin gene *mac-1*) were used to investigate in more depth the genetic relationships among *M. galloprovincialis* populations. Samples were taken between 1996 and 1999 from California, the NE Atlantic, the Mediterranean Sea, South Africa, Korea, Western Australia, Tasmania and New Zealand. NE Atlantic *M. edulis* L. were used as an outgroup. While all *M. galloprovincialis* samples were fixed, or nearly so, for the diagnostic *G* allele at locus *Glu-5′*, correspondence analysis of *mac-1* allele frequency data highlighted the genetic distinctness of Australasian mussels relative to other *M. galloprovincialis* populations. The latter consisted of two differentiated groups (NE Atlantic and Mediterranean) as formerly reported at allozyme loci. Another sample, from Chile, was nearly identical to Mediterranean *M. galloprovincialis*. Nuclear-DNA data thus enforce the idea that *M. galloprovincialis* have probably been introduced from the Mediterranean to the North Pacific (and now Chile), and from the NE Atlantic to South Africa. It is argued in this study that Australasian mussels derive from a proto- *M. galloprovincialis* population introgressed by *M. edulis*-like genes, and should be considered as a regional subspecies of *M. galloprovincialis*.

Smooth-shelled *Mytilus* spp. mussels are distributed over the temperate and subpolar regions of all oceans. Global surveys of allozyme variation in *Mytilus* spp. populations have focused on the clarification of the systematics of the genus (McDonald & Koehn 1988; Varvio *et al.* 1988; Koehn 1991; McDonald *et al.* 1991). These unambiguously showed that all smooth-shelled *Mytilus* populations in the southern and the northern hemispheres could be ascribed to one of three species – *M. edulis* Linnaeus, 1758, *M. galloprovincialis* Lamarck, 1819 or *M. trossulus* Gould, 1850. Full species status is warranted by the fact that each of these entities maintains its genetic integrity over broad geographical areas, in spite of hybridization in areas of overlap and in the face of potential for long-distance dispersal by planktotrophic larvae (Koehn 1991). Although the majority of the genetic differences were accounted for by this systematic distinction, geographical differentiation was also detected within each species. For example, McDonald *et al.* (1991) recognized Australasian (Australia, Tasmania, New Zealand) *Mytilus* as *M. galloprovincialis*, but noted some differences in allele frequency between the latter and their northern hemisphere conspecifics.

The present study is concerned with worldwide genetic variation in *Mytilus galloprovincialis*. The main objective was to test, with novel nuclear-DNA markers, the biogeographical hypotheses arising from the analysis of allozyme loci.

Allozyme surveys of genetic variation in *Mytilus galloprovincialis* along the Iberian Peninsula have demonstrated clear-cut geographical isolation between the northeastern (NE) Atlantic and the Mediterranean populations of this species (Sanjuan *et al.* 1994; Quesada *et al.* 1995). Compilation and analysis of the data set generated by the above

From: HARPER, E. M., TAYLOR, J. D. & CRAME, J. A. (eds) *The Evolutionary Biology of the Bivalvia*. Geological Society, London, Special Publications, **177**, 389–397. 1-86239-076-2/00/$15.00

allozyme studies has since clarified the world-wide picture of the geographic structure in *M. galloprovincialis* (Sanjuan *et al.* 1997). These authors examined the allele frequency data at five allozyme loci using multidimensional scaling analysis on the matrix of Nei's genetic distance estimates between populations. Unfortunately, they did not consider any outgroup, so their genetic networks were unrooted. They found evidence for three genetically differentiated groups of *M. galloprovincialis*: a NE Atlantic group that also included *M. galloprovincialis* from South Africa, a Mediterranean group that also included *M. galloprovincialis* from California and eastern Siberia, and a third group from Western Australia and Tasmania. New Zealand *Mytilus* appeared to be more closely related to North Pacific *M. galloprovincialis* than to the other Australasian populations.

Phylogenetic inference is desirable for a more in-depth understanding of the evolutionary processes, including colonization and vicariance, that shape the genetic composition of species. Figure 1 is a neighbour-joining tree, rooted by *Mytilus edulis*, inferring the phylogeny of *M. galloprovincialis* populations from allele frequency data at seven allozyme loci (McDonald *et al.* 1991; Väinölä & Hvilsom 1991; Quesada *et al.* 1995). The seven loci considered here included the five loci examined by Sanjuan *et al.* (1997). Jackknife resampling of loci, where each locus was omitted in turn from the data set, supported a principal separation between NE Atlantic–Mediterranean *M. galloprovincialis* (samples Sesimbra and Palavas) and Australasian *M. galloprovincialis* (samples Albany, Huon River and Wellington). Figure 1 also confirms that Californian *M. galloprovincialis* (Los Angeles) is genetically closer to the Mediterranean than the Atlantic population. The Wellington (New Zealand) sample occurred within an Australasian clade (i.e. together with Albany and Huon River) four to seven times, whereas in the remaining three to seven pseudotrees it diverged slightly before the node separating the Australian (Albany + Huon River) clade from the NE Atlantic–Mediterranean clade.

Kenchington *et al.* (1995) obtained the sequence of 18S rDNA for several *Mytilus* spp. samples. These included *M. galloprovincialis* from an introduced and farmed population in Puget Sound

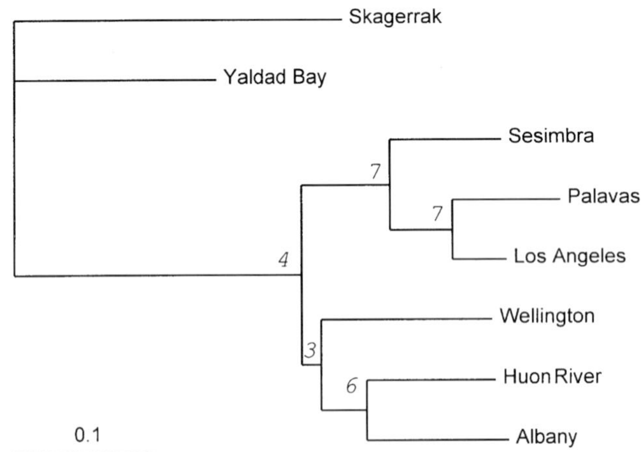

Fig. 1. *Mytilus galloprovincialis*. Neighbour-joining tree (Saitou & Nei 1987; NEIGHBOR procedure of PHYLIP: Felsenstein 1993) constructed from the matrix of absolute genetic distance estimates (Gregorius 1984), based on electromorph frequency data at seven allozyme loci (*Ap*, *Est-D*, *Gpi*, *Lap*, *Mpi*, *Odh* and *Pgm*) in samples from: Los Angeles, California (McDonald *et al.* 1991), Palavas (Mediterranean) and Sesimbra (NE Atlantic) (Quesada *et al.* 1995); Yaldad Bay, Chile, Albany, Western Australia, Huon River Estuary, Tasmania, and Wellington, New Zealand (McDonald *et al.* 1991). Numbers at a node are scores of across-locus jacknife resampling. Electromorph identities were deduced from cross-comparisons of electromorph frequencies in McDonald & Koehn (1988), McDonald *et al.* (1991), Väinölä & Hvilsom (1991) and Quesada *et al.* (1995). *M. edulis* sample Skagerrak of Väinölä & Hvilsom (1991) [which is also SWE of Varvio *et al.* (1988)] was used as outgroup to root the tree. These samples were chosen because of their geographical proximity with samples of the present study (see Fig. 2) as follows: BOD, Los Angeles; CHL, Yaldad Bay; STB, Sesimbra; SET, Palavas; AUS, Albany; TAS, Huon River; NZL, Wellington; FLO, Skagerrak. Rooting the entire tree with *M. trossulus* sample Tillamook, Oregon, of McDonald *et al.* (1991), did not change its topology (data not shown), confirming that the choice of *M. edulis* as outgroup for all *M. galloprovincialis* was appropriate. Scale bar, 0.1 unit absolute genetic distance

in the eastern North Pacific, *M. edulis planulatus* Lamarck, 1819 from southeastern Tasmania [McDonald *et al.* (1991) had concluded that *M. edulis planulatus* were *M. galloprovincialis* according to both allozyme frequency and morphometric data], and several *M. edulis* and *M. trossulus* samples. Kenchington *et al.* (1995) also analysed an '*M. galloprovincialis*' sample from Morgat (Britanny, France), but the present authors have disregarded this identification because the location lies within the European hybrid zone between *M. edulis* and *M. galloprovincialis* (Coustau *et al.* 1991; Gosling 1992), where there is no evidence for the presence of non-introgressed *M. galloprovincialis* (Coustau *et al.* 1991; C. Daguin, F. Bonhomme and P. Borsa, pers, comms). The phylogeny inferred from the 18S rDNA sequences (Kenchington *et al.* 1995) strongly suggests an early separation of northern hemisphere *M. galloprovincialis* from the other smooth-shelled *Mytilus* spp., including *M. edulis*, *M. trossulus* and Australasian *M. galloprovincialis*.

A number of hypotheses have been suggested to account for the differentiation patterns and disjunct worldwide distribution of *Mytilus galloprovincialis*. (1) *Mytilus galloprovincialis* has been introduced to South Africa, the western North Pacific, the eastern North Pacific, and perhaps Australasia, through human agency (Wilkins *et al.* 1983; Grant & Cherry 1985; McDonald & Koehn 1988; McDonald *et al.* 1991; Vermeij 1992). This hypothesis is based on the lack of evidence for *M. galloprovincialis*-like mussels in the fossil record, and their absence from the literature and museum collections in South Africa and in the North Pacific until recently (Wilkins *et al.* 1983; Grant & Cherry 1985; McDonald & Koehn 1988, and refs cited therein; Geller 1999). However, sub-fossil *Mytilus* are present in Aboriginal middens in Tasmania, southern Australia and New Zealand (McDonald *et al.* 1991, and refs cited therein). This leads to the following, modified hypothesis for the genetic affinities of Australasian *M. galloprovincialis*. (2) Introduced *Mytilus galloprovincialis* have displaced native Australasian *Mytilus* spp. [suggested, although believed to be unlikely, by McDonald *et al.* (1991)] or introgressed with the latter (Seed 1992). Sanjuan *et al.* (1997) also write that 'a human introduction of North Pacific mussels into Australia is [...] possible (Carlton 1987) [and] may explain the genetic heterogeneity of Australasian samples'. Both displacement and introgression by presumably introduced *M. galloprovincialis* have affected native *M. trossulus* in California (McDonald & Koehn 1988; Geller 1999). (3) North Pacific *Mytilus galloprovincialis* have recently been introduced from the South Pacific. Quoting Koehn (1991): 'As *M. galloprovincialis* is probably native to large areas of the South Pacific, introductions into northern Pacific sites [...] may not have originated in Europe.' However, this is contradicted by the quite large genetic dissimilarity between North Pacific and South Pacific *M. galloprovincialis*, which is at variance with the close genetic similarities between North Pacific and Mediterranean *M. galloprovincialis* (Gosling 1992). (4) Australasian, North Pacific and Mediterranean *Mytilus galloprovincialis* all derive from an ancestral Pacific stock (Sanjuan *et al.* 1997). While hypotheses (1) and (2) here above rely on the presumption that *M. galloprovincialis* is native to the NE Atlantic–Mediterranean, Sanjuan *et al.* (1997) consider the alternative hypothesis that either Australasia or the North Pacific be the geographic origin of *M. galloprovincialis*. According to this hypothesis, proto-*M. galloprovincialis* crossed the Equator in the Pacific Ocean; North Pacific *M. galloprovincialis* subsequently crossed the Arctic to colonize the NE Atlantic and the Mediterranean. This amounts to proposing that present-day Australasian, North Pacific and Mediterranean populations are endemic forms of *M. galloprovincialis*. However, the data provided by Sanjuan *et al.* (1997, table 4) failed to substantiate part of this scenario, since the genetic distance estimate between the North Pacific and the Mediterranean [$D = 0.029$ (0.011 – 0.057)] was not significantly larger than that among sub-populations within the Mediterranean [$D = 0.028$ (0.007 – 0.054)].

The allozyme-based phylogeny proposed in Fig. 1 suggests that Australasian *Mytilus galloprovincialis* differentiated early on from all the other *M. galloprovincialis*, supporting the hypothesis that Australasian *M. galloprovincialis* are endemic (Koehn 1991; McDonald *et al.* 1991). Figure 1 further shows that the divergence between NE Atlantic and Mediterranean *M. galloprovincialis* is more recent than the separation between the Australasian and the NE Atlantic–Mediterranean groups. The close genetic relationship of Californian with Mediterranean *M. galloprovincialis* (see above; Fig. 1) reflects an even more recent common origin, which is indeed compatible with the hypothesis that North Pacific *M. galloprovincialis* were introduced from the Mediterranean through human agency. Sanjuan *et al.* (1997) instead propose that 'a genetic mixing between [*M. galloprovincialis* and *M. edulis*], perhaps combined with selective pressure, may explain the larger allozyme divergence between *M. galloprovincialis* of Mediterranean vs Atlantic populations, than between Mediterranean vs North Pacific populations'. If this hypothesis were true, allele frequencies in NE Atlantic *M.*

galloprovincialis would be intermediate between those of *M. edulis* and Mediterranean *M. galloprovincialis*. This appears to be the case at only one (*Ap*), perhaps two (adding *Gpi*), out of the six loci (*Ap*, *Est-D*, *Gpi*, *Lap*, *Mpi*, *Odh*) at which the most significant differences were observed between the NE Atlantic and the Mediterranean *M. galloprovincialis* (Quesada *et al.* 1995), and which were also scored in *M. edulis* (Skibinski *et al.* 1980; Coustau *et al.* 1991; Sanjuan *et al.* 1994). All of these loci were taken into account for computing the neighbour-joining tree of Fig. 1. The distinctness of the NE Atlantic v. the Mediterranean *M. galloprovincialis* thus far appears to reflect vicariance more than differential introgression by *M. edulis* allozyme genes.

To what extent do nuclear-DNA markers enforce and refine the conclusions drawn from allozyme data?

Methods

Mytilus galloprovincialis populations were sampled in (all abbreviations following locations refer to Fig. 2): California (BOD), the NE Atlantic (STB), the Mediterranean Sea (SET), South Africa (SAF), eastern Asia (KOR), Western Australia (AUS), Tasmania (TAS) and New Zealand (NZL) between 1996 and 1999. Reference *M. edulis* samples were collected in the North Sea (GFP), the Skagerrak (FLO) and the Kattegat (GIL) in 1996 and 1997. A sample of uncertain taxonomic status was collected in southern central Chile (CHL) in 1998. All mussel shells in this sample were *M. galloprovincialis* according to morphology. *Mytilus* samples from Chile examined so far at allozyme loci and using morphometrics were *M. edulis*-like (McDonald *et al.* 1991; Fig. 1), but a sample from southern Chile recently analysed at nuclear-DNA and mitochondrial-DNA loci exhibited alleles identical to those of New Zealand *M. galloprovincialis* (Toro 1998), raising the possibility that this species might also be present in southern Chile.

The total genomic DNA was extracted from each individual and used as a template for polymerase chain reactions (PCR) using primer pairs specific to a fragment of the polyphenolic adhesive protein gene *Glu-5′*, and a fragment of intron 1 of the actin gene *mac-1*. The *Glu-5′* marker was developed by Inoue & Odo (1994) and Rawson *et al.* (1996). The *mac-1* marker was developed by Ohresser *et al.* (1997) and Daguin & Borsa (1999). Protocols for DNA extraction, PCR, gel electrophoresis and allele nomenclature at locus *Glu-5′* have been reported in Borsa *et al.* (1999). Those for locus *mac-1* have been detailed in Daguin & Borsa (1999): *mac-1* is the only non-coding locus out of the ten nuclear loci (also including seven allozyme loci, 18S rDNA and *Glu-5′*) considered in this paper.

Correspondence analysis (FCA: Benzécri 1982) was performed using the AFC procedure implemented in *BIOMECO* (Lebreton *et al.* 1990)

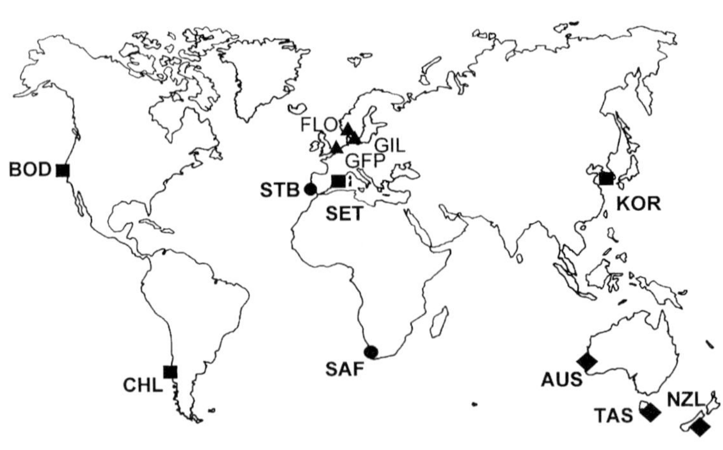

Fig. 2. Sampling sites for *Mytilus galloprovincialis*. Abbreviations for samples: BOD, Bodega Bay, California; CHL, Dichato, southern Central Chile; STB, Setubal, Portugal; SET, Sète, France [sample SETE of Daguin & Borsa (1999)]; SAF, Bloubergstrand, South Africa; KOR, western coast of South Korea; AUS, Nedlands, Western Australia; TAS, d'Entrecasteaux Channel, Tasmania; NZL, Dunedin, New Zealand. Symbols refer to the present nuclear-DNA characterization of samples: ■, Mediterranean *M. galloprovincialis* ; ●, Atlantic *M. galloprovincialis* ; ◆, Australasian *M. galloprovincialis*. ▲, *M. edulis* reference samples: FLO, Flodevigen, Skagerrak; GFP, Grand Fort Philippe, North France; GIL, Gilleleje, Kattegat [sample GILL of Daguin & Borsa (1999)].

on the matrix of allelic frequencies per sample. This method, previously used by Coustau *et al.* (1991) on allozyme data, is well-adapted to express the genetic differences present in a data set because the eigenvalues of each FCA axis are analogous to S. Wright's *Fst* (Guinand 1996).

Results and discussion

Table 1 gives the allele frequencies at locus *Glu-5′*. Allele *G* was fixed, or nearly fixed, in all *Mytilus galloprovincialis* samples [thereby extending the preliminary findings of Rawson *et al.* (1996) and Borsa *et al.* (1999)] and in the *Mytilus* sp. sample from Chile. The *Glu-5′* marker, being quasi-monomorphic, is therefore of little help in analysing the genetic relationships among *M. galloprovincialis* populations.

In contrast, preliminary surveys of genetic variation at the *mac-1* locus have demonstrated considerable polymorphism, with up to 18 size alleles in *Mytilus galloprovincialis* samples from the NE Atlantic and the Mediterranean (Daguin & Borsa 1999). Here, allele frequency data at locus *mac-1* suggested closer genetic affinities of Australasian *Mytilus* with *M. edulis* rather than *M. galloprovincialis* (Table 2): all three Australasian samples possessed *mac-1* allele *a*2 at a high frequency (this allele is found at a moderate frequency in NE Atlantic *M. edulis*, but at significantly lower frequency in *M. galloprovincialis*: Daguin & Borsa 1999; Table 2); the Tasmanian and New Zealand samples did not possess any of the alleles (*b*2, *c*2) inferred to be characteristic of Mediterranean and NE Atlantic *M. galloprovincialis* (Daguin & Borsa 1999; Table 2), although these were present at a moderate frequency in Western Australia. The three-dimensional projection of the FCA of *M. galloprovincialis* samples collected worldwide (Fig. 3) illustrates the clear distinctness of Australasian *Mytilus*, with the first axis of FCA separating *M. edulis* and Australasian mussels from all the other samples. The second axis then clearly distinguished Australasian mussels from *M. edulis*. The third axis yielded evidence of (comparatively slighter) further distinction between NE Atlantic (STB) and Mediterranean (SET) *M. galloprovincialis*. The homogeneity of *mac-1* allele frequencies within each of the two latter regions has been reported elsewhere (Daguin & Borsa 1999). South African mussels clustered together with NE Atlantic *M. galloprovincialis*, while mussel samples from California, Chile and Korea clustered together with Mediterranean *M. galloprovincialis*. The estimate of θ, the equivalent of Wright's *Fst* (Weir & Cockerham 1984), was $\hat{\theta} = 0.027$ between these two groups. This value compares with the value reported at locus *mac-1* between NE Atlantic (northwestern African) and Western Mediterranean *M. galloprovincialis* ($\hat{\theta} = 0.016$; $p = 0.02$; Daguin & Borsa 1999) and with the mean *Gst* value reported for 13 allozyme loci between samples from the western and the eastern coasts of the Iberian Peninsula ($Gst = 0.029$; $p < 0.001$; Quesada *et al.* 1995).

Thus, *mac-1* data (Daguin & Borsa 1999, this study) are in accordance with the results of former allozyme-based surveys in: (1) allowing a clear distinction between NE Atlantic and Mediterranean *Mytilus galloprovincialis* populations (Sanjuan *et al.* 1994, 1997; Quesada *et al.* 1995); (2) demonstrating the genetic affinities of South African *M. galloprovincialis* with NE Atlantic *M. galloprovincialis* (Sanjuan *et al.* 1997); (3) demonstrating the genetic affinities of *M. galloprovincialis* from both the eastern and the western coasts of the North Pacific with Mediterranean *M. galloprovincialis* (Sanjuan *et al.* 1997; Fig. 1); (4) arguing against the possibility that Australasian *M. galloprovincialis* were transported to the North Pacific (Gosling 1992; Fig. 1).

Table 1. *Allele frequencies at locus* Glu-5′ *in samples of* Mytilus galloprovincialis *from the northern and southern hemispheres, and in two* M. edulis *reference samples*

Allele*	Sample†										
	BOD	CHL	STB	SET	SAF	KOR	AUS	TAS	NZL	FLO	GIL
E	–	–	–	0.01	0.05	–	–	–	–	0.89	0.50
E'	–	–	–	–	–	–	–	–	–	0.11	0.47
E"	–	–	–	–	–	–	–	–	–	–	0.03
G	0.98	1.00	1.00	0.99	0.95	1.00	1.00	1.00	1.00	–	–
T	0.02	–	–	–	–	–	–	–	–	–	–
(*N*)‡	(23)	(48)	(19)	(56)	(65)	(19)	(46)	(25)	(77)	(35)	(16)

* Nomenclature of Borsa *et al.* (1999). † Abbreviations for samples as in legend to Fig. 2; data for GIL, BOD and SET from Borsa *et al.* (1999); additional data for SET from Rawson *et al.* (1996). ‡ Sample size.

Table 2. *Allele frequencies at locus* mac-1 *in samples of* Mytilus galloprovincialis *from the northern and southern hemispheres, and in three* M. edulis *reference samples*

Allele*	Sample†											
	BOD	CHL	STB	SET	SAF	KOR	AUS	TAS	NZL	FLO	GIL	GFP
*f*1	–	–	0.02	–	0.02	–	–	–	–	–	–	–
*f*2	–	–	–	–	0.01	–	–	–	–	–	–	–
*f*3	–	0.01	–	–	0.01	–	–	–	–	–	–	–
*b*0	–	–	–	–	0.01	–	–	–	–	–	–	–
*b*05	–	–	–	–	–	–	–	–	0.02	–	–	–
*b*2	0.04	0.05	0.04	0.05	0.04	0.02	–	–	–	–	–	–
*b*1	0.32	0.32	0.15	0.21	0.09	0.42	0.04	–	–	–	–	0.01
*b*3	–	0.01	–	–	–	0.02	–	–	–	–	–	–
*b*4	–	–	0.02	–	–	–	–	–	–	–	–	–
*b*5	–	–	0.02	–	–	–	–	–	–	–	–	–
*c*1	0.04	0.08	0.10	0.07	0.10	0.05	–	–	–	–	–	0.01
*c*12	–	0.01	–	–	–	–	–	–	–	–	–	–
*c*15	–	–	–	–	0.01	–	–	–	–	–	–	–
*c*2	0.41	0.39	0.50	0.54	0.53	0.43	0.16	–	–	–	–	–
*c*3	–	0.01	0.02	–	0.04	–	–	–	–	–	–	–
*c*4	–	–	–	–	–	–	–	–	–	0.05	–	0.02
*c*6	0.01	–	–	0.01	0.01	–	–	–	–	–	–	–
*a*0	–	–	–	–	–	–	–	–	–	0.02	–	–
*a*05	–	–	–	–	–	–	–	–	0.04	–	–	–
*a*1	–	–	0.02	–	–	–	–	–	–	0.06	0.02	0.01
*a*15	–	–	–	–	–	–	0.01	–	0.01	–	–	–
*a*2	–	0.03	0.02	–	0.02	0.02	0.62	0.99	0.92	0.10	0.15	0.20
*a*3	0.03	–	0.04	0.01	0.03	–	0.16	0.01	0.02	0.29	0.31	0.24
*a*4	–	0.01	–	–	–	–	–	–	–	0.07	0.17	0.18
*a*5	0.10	0.01	–	–	0.02	–	–	–	–	0.38	0.27	0.29
*a*6	–	0.01	–	0.01	0.02	–	0.01	–	–	–	0.08	0.04
*a*7	0.01	0.03	0.02	0.04	0.04	0.03	–	–	–	–	–	–
*a*8	0.01	0.05	0.04	0.06	0.01	0.02	–	–	–	–	–	–
*a*9	–	–	–	–	–	–	–	–	–	0.01	–	–
d	–	–	–	–	–	–	–	–	–	0.01	–	–
(*N*)‡	(34)	(76)	(26)	(68)	(62)	(30)	(38)	(40)	(79)	(47)	(26)	(42)

* Nomenclature of Daguin & Borsa (1999); alleles are presented from the slower migrating (*f*1) to the faster migrating (*d*).
† Abbreviations for samples as in legend to Fig. 2; data for GIL and SET from Daguin & Borsa (1999). ‡ Sample size.

Australasian *Mytilus* are *M. galloprovincialis* when considering allozyme data (McDonald *et al.* 1991), even though substantial differences with northern hemisphere *M. galloprovincialis* were detected using multidimensional scaling (Sanjuan *et al.* 1997) and neighbour-joining phylogenetic inference (Fig. 1). Australasian *Mytilus* remained grouped with *M. galloprovincialis* when considering *Glu-5′* data (Table 1). However, *mac-1* data (Table 2; Fig. 3) revealed strong differences between Australasian *Mytilus* and northern hemisphere *M. galloprovincialis*. The summing up of all data available regarding the genetic characterization of Australasian *Mytilus* (McDonald *et al.* 1991; Kenchington *et al.* 1995; this study) shows noticeable discrepancies among loci. For instance, if allele frequency patterns at the four loci (allozyme loci *Est-D* and *Mpi*; nuclear-DNA loci *Glu-5′* and *mac-1*) which can be considered as diagnostic between *M. galloprovincialis* and *M. edulis* are examined (Skibinski *et al.* 1983; Rawson *et al.* 1996; Daguin & Borsa 1999): Australasian *Mytilus* possess fixed, or nearly fixed, *M. galloprovincialis* alleles at *Mpi* and *Glu-5′*, while at *Est-D* the frequency of *M. edulis* alleles was zero in sample Wellington, *c.* 0.25 in sample Albany and *c.* 0.5 in sample Huon River (McDonald *et al.* 1991); at locus *mac-1*, Australasian *Mytilus* possess, at high frequency, a size allele (*a*2) which is more characteristic of *M. edulis* than of *M. galloprovincialis* (Table 2). At other allozyme loci, Australasian mussels can be considered as either closer to *M. galloprovincialis* than *M. edulis* (*Aap*, *Ap*, *Gpi* and *Lap*), or the contrary (*Pgm*), or quite different from both (*Odh*) (McDonald *et al.* 1991). Finally, 18S rDNAs of Tasmanian *Mytilus* showed

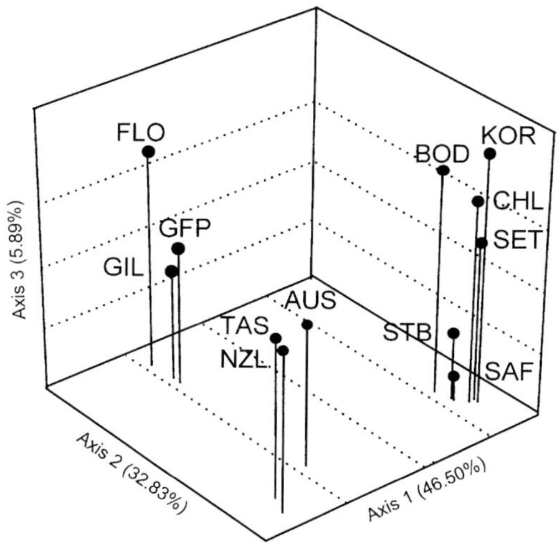

Fig. 3. *Mytilus galloprovincalis* and *M. edulis*. Three-dimensional representation of the outcome of correspondence analysis (Benzécri 1982) on the matrix of *mac-1* allelic frequencies per sample. See Fig. 2 for abbreviations.

higher sequence similarity with *M. edulis* (and *M. trossulus*) than with *M. galloprovincialis* (Kenchington *et al.* 1995).

The present results provide new insight into the biogeography of *Mytilus galloprovincialis*. Firstly, Australasian *Mytilus* now appear to be clearly distinct from all *M. galloprovincialis* populations elsewhere in the world. Therefore, the hypothesis that these mussels were recently introduced by humans (e.g. from the North Pacific), or that they may have undergone recent, substantial introgression by alien *M. galloprovincialis* (e.g. Seed 1992) is here rejected. Instead, Australasian *M. galloprovincialis* have a patchy genetic architecture, with a high frequency of *M. edulis*-like alleles at some loci and of *M. galloprovincialis*-like alleles at others. Assuming that the choice of *M. edulis* as an outgroup of *M. galloprovincialis* holds valid, this suggests that Australasian *Mytilus* originates from a proto-*M. galloprovincialis* population that underwent introgression by proto-*M. edulis*. In other words, introgression by *M. edulis*-like alleles was detected in Australasian mussels at loci *mac-1* and *Est-D* (and, perhaps, *Pgm*), but this may be ancient since at locus *mac-1* these alleles are now fixed or quasi-fixed. It is also possible that the *M. edulis*-like genes of Australasian mussels do not derive from a proto-*M. edulis* form but originate from more modern southern hemisphere *M. edulis* (South America, Falklands, Kerguelen; McDonald *et al.* 1991). It is expected that sequencing analysis will confirm whether the *mac-1* alleles present in Australasian mussels are phylogenetically closer to those from northern hemisphere, or southern hemisphere *M. edulis*. On the basis of the close morphological and allozymic resemblance borne by Australasian mussels to *M. galloprovincialis* (McDonald *et al.* 1991), it is proposed here that these be attributed a subspecific rank within *M. galloprovincialis*.

Secondly, the hypothesis, implicit in Sanjuan *et al.* (1997), that North Pacific and Mediterranean mussels are endemic forms of *M. galloprovincialis* is rejected here, because there was no evidence of genetic differentiation between populations of these two regions. The close genetic similarity at all loci, including *mac-1*, of populations separated by such geographical distances and geographical barriers (e.g. continents) conforms to the hypothesis of recent introduction by humans. Since fossil *M. galloprovincialis* have been found in the Mediterranean (Mars 1956) and have not been reported from the North Pacific (see above), it is legitimate to presume that North Pacific *M. galloprovincialis* were introduced from the Mediterranean and not the contrary. Rejecting the present existence of endemic North Pacific *M. galloprovincialis* does not necessarily imply that *M. galloprovincialis* is a Mediterranean offshoot of North Atlantic *M. edulis*, as is frequently assumed (Barsotti & Meluzzi 1968; Skibinski *et al.* 1983; Gosling 1984, 1992; Grant & Cherry 1985; Seed 1992; Rawson & Hilbish 1998). As emphasized by Vermeij (1992), 'It is possible that the original Tethyan *Mytilus* persisted in the Mediterranean regions throughout the Neogene, and that the Pleistocene and Recent *M. galloprovincialis* is derived from this purported stock rather than from the Pacific-derived *trossulus-edulis* group' – although fossil evidence for this scenario remains to be found.

Thirdly, *M. galloprovincialis* was here identified for the first time in Chile. Its high genetic similarity, at locus *mac-1*, with Mediterranean and North Pacific *M. galloprovincialis* suggests it has been recently introduced to Chile from either of these regions by maritime transport or, perhaps, by unreported, intentional transplantation. The population from which our Chilean sample originates thus appears to be different from any of the *M. edulis*-like Chilean populations sampled by McDonald *et al.* (1991), but it is perhaps the same as the one sampled by Toro (1998). In this case, the taxonomic status proposed by Toro for these mussels (a subspecies of *M. edulis*) should be disregarded.

Note added in proof: A recent mitochondrial-DNA survey aimed at elucidating the origin of the antitropical distribution pattern of *Mytilus* spp. (Hilbish *et al.* 2000, *Marine Biology*, **136**, 69–77) yielded results that confirm the main conclusions of the present study. Most Australasian *Mytilus* spp. female mitochondrial lineages clustered into a single clade (*D2*) whose closest relative was the *D* clade found in northern *M. galloprovincialis* (*D1*), demonstrating both the originality of Australasian *Mytilus* sp. and their close relatedness to *M. galloprovincialis*. However, a few Australasian mussels possessed mitochondria of the *A* type characteristic of *M. edulis*: this is consistent with our hypothesis that these Australasian mussels derive from a proto-*M. galloprovincialis* population introgressed by *M. edulis*-like genes. Mitochondrial-DNA data were also consistent with the hypothesis that North Pacific *M. galloprovincialis* originate from the Mediterranean through human agency (all mussels from San Diego, California, possess mitochondria of type *D1* or *A*, as do Mediterranean mussels). Hilbish *et al.* (2000, *Marine Biology*, **136**, 69–77) failed to detect *D1* haplotypes in their Chilean samples.

We are grateful to: N. Bierne, F. Bonhomme, B. Delay, M. Raymond and R. D. Ward for discussions; to J. Taylor, R. D. Ward and an anonymous referee for comments on an earlier manuscript; to J. Beesley, W. Borgeson, P. Boudry, P. Fréon, A. Leitao, C. Lemaire, D. Moraga, M. Ohresser, J. Panfili, C. Perrin, M. Raymond and C. Riquelme for providing samples; to V. Rolland for carefully examining the Chilean *Mytilus galloprovincialis* shells; to S. Ramos Caetano for assistance in the laboratory; to IFREMER URM 16 for research funds; to MENRT and IRD for our salaries.

References

Barsotti, G. & Meluzzi, C. 1968. Osservazioni su *Mytilus edulis* L. e *Mytilus galloprovincialis* Lamarck. *Conchiglie*, **4**, 50–58.

Benzécri, J.-P. 1982. *L'Analyse des Données. 2. L'Analyse des Correspondances*. Dunod, Paris.

Borsa, P., Daguin, C., Ramos Caetano, S. & Bonhomme, F. 1999. Nuclear-DNA evidence that northeastern Atlantic *Mytilus trossulus* mussels carry *M. edulis* genes. *Journal of Molluscan Studies*, **65**, 524–527.

Carlton, J. T. 1987. Patterns of transoceanic marine biological invasions in the Pacific Ocean. *Bulletin of Marine Science*, **41**, 452–465.

Coustau, C., Renaud, F. & Delay, B. 1991. Genetic characterization of the hybridization between *Mytilus edulis* and *M. galloprovincialis* on the Atlantic coast of France. *Marine Biology*, **111**, 87–93.

Daguin, C. & Borsa, P. 1999. Genetic characterisation of *Mytilus galloprovincialis* Lmk. in North West Africa using nuclear DNA markers. *Journal of Experimental Marine Biology and Ecology*, **235**, 55–65.

Felsenstein, J. 1993. *Phylip 3.5 (Phylogeny Inference Package)*. University of Washington, Seattle, Washington, USA.

Geller, J. B. 1999. Decline of a native mussel masked by sibling species invasion. *Conservation Biology*, **13**, 661–664.

Gosling, E. M. 1984. The systematic status of *Mytilus galloprovincialis* in western Europe: a review. *Malacologia*, **25,** 551–568.

—— 1992. Systematics and geographic distribution of *Mytilus*. *In*: Gosling, E. (ed.) *The Mussel* Mytilus*: Ecology, Physiology, Genetics and Culture*. Developments in Aquaculture and Fisheries Science, **25**, 1–20. Elsevier, Amsterdam.

Grant, W. S. & Cherry, M. I. 1985. *Mytilus galloprovincialis* in South Africa. *Journal of Experimental Marine Biology and Ecology*, **90,** 179–191.

Gregorius, H. R. 1984. An unique genetic distance. *Biometrical Journal*, **26**, 13–18.

Guinand, B. 1996. Use of a multivariate model using allele frequency distributions to analyse patterns of genetic differentiation among populations. *Biological Journal of the Linnean Society*, **58**, 173–195.

Inoue, K. & Odo, S. 1994. The adhesive protein cDNA of *Mytilus galloprovincialis* encodes decapeptide repeats but no hexapeptide motif. *Biological Bulletin*, **186**, 349–355.

Kenchington, E., Landry, B. & Bird, C. J. 1995. Comparison of taxa of the mussel *Mytilus* (Bivalvia) by analysis of the nuclear small-subunit rDNA gene sequence. *Canadian Journal of Fisheries and Aquatic Sciences*, **52**, 2613–2620.

Koehn, R. K. 1991. The genetics and taxonomy of species in the genus *Mytilus*. *Aquaculture*, **94**, 125–145.

Lebreton, J.-D., Roux, M., Banco, G. & Bacou, A.-M. 1990. *Biomeco (Biométrie-écologie), Logiciel d'Ecologie Statistique pour PC et Compatibles. Version 3.9.* Centre National de la Recherche Scientifique, Montpellier, France.

McDonald, J. H. & Koehn, R. K. 1988. The mussels *Mytilus galloprovincialis* and *M. trossulus* on the Pacific coast of North America. *Marine Biology*, **99**, 111–118.

——, Seed, R. & Koehn, R. K. 1991. Allozymes and morphometric characters of three species of *Mytilus* in the northern and southern hemispheres. *Marine Biology*, **111**, 323–333.

Mars, P. 1956. Faunes malacologiques du Pliocène et du Quaternaire de Milazzo (Sicile). *Bulletin du Muséum d'Histoire Naturelle de Marseille*, **16**, 33–52.

Ohresser, M., Borsa, P. & Delsert, C. 1997. Intron-length polymorphism at the actin gene locus *mac-1*: a genetic marker for population studies in the marine mussels *Mytilus galloprovincialis* Lmk and *M. edulis* L. *Molecular Marine Biology and Biotechnology*, **6**, 123–130.

Quesada, H., Zapata, C. & Alvarez, G. 1995. A multilocus allozyme discontinuity in the mussel *Mytilus galloprovincialis*: the interaction of ecological and life-history factors. *Marine Ecology-Progress Series*, **116**, 99–115.

Rawson, P. D. & Hilbish, T. J. 1998. Asymmetric introgression of mitochondrial DNA among European populations of blue mussels (*Mytilus* spp.). *Evolution*, **52**, 100–108.

——, JOYNER, K. L., MEETZE, K. & HILBISH, T. J. 1996. Evidence for intragenic recombination within a novel genetic marker that distinguishes mussels in the *Mytilus edulis* species complex. *Heredity*, **77**, 599–607.

SAITOU, N. & NEI, M. 1987. The Neighbor-Joining method: a new method for reconstructing phylogenetic trees. *Molecular Biology and Evolution*, **4**, 406–425.

SANJUAN, A., ZAPATA, C. & ALVAREZ, G. 1994. *Mytilus galloprovincialis* and *M. edulis* on the coasts of the Iberian Peninsula. *Marine Ecology–Progress Series*, **113**, 131–146.

——, —— & —— 1997. Genetic differentiation in *Mytilus galloprovincialis* Lmk. throughout the world. *Ophelia*, **47**, 13–31.

SEED, R. 1992. Systematics, evolution and distribution of mussels belonging to the genus *Mytilus*: an overview. *American Malacological Bulletin*, **9**, 123–137.

SKIBINSKI, D. O. F., BEARDMORE, J. A. & CROSS, T. F. 1983. Aspects of the population genetics of *Mytilus* (Mytilidae; mollusca) in the British Isles. *Biological Journal of the Linnean Society*, **19**, 137–183.

——, CROSS, T. F. & AHMAD, M. 1980. Electrophoretic investigation of systematic relationships in the marine mussels *Modiolus modiolus* L., *Mytilus edulis* L., and *Mytilus galloprovincialis* Lmk. (Mytilidae; Mollusca). *Biological Journal of the Linnean Society*, **13**, 65–73.

TORO, J. E. 1998. PCR-based nuclear and mtDNA markers and shell morphology as an approach to study the taxonomic status of the Chilean blue mussel, *Mytilus chilensis* (Bivalvia). *Aquatic Living Resources*, **11**, 347–353.

VÄINÖLÄ, R. & HVILSOM, M. M. 1991. Genetic divergence and a hybrid zone between Baltic and North Sea *Mytilus* populations. *Biological Journal of the Linnean Society*, **43**, 127–148.

VARVIO, S.-L., KOEHN, R. K. & VÄINÖLÄ, R. 1988. Evolutionary genetics of the *Mytilus edulis* complex in the North Atlantic region. *Marine Biology*, **98**, 51–60.

VERMEIJ, G. J. 1992. Trans-equatorial connections between biotas in the temperate eastern Atlantic. *Marine Biology*, **112**, 343–348.

WEIR, B. S. & COCKERHAM, C. C. 1984. Estimating *F*-statistics for the analysis of population structure. *Evolution*, **38**, 1358–1370.

WILKINS, N. P., FUJINO, K. & GOSLING, E. M. 1983. The Mediterranean mussel *Mytilus galloprovincialis* Lmk. in Japan. *Biological Journal of the Linnean Society*, **20**, 365–374.

Size: all it's shaped up to be? Evolution of shape through the lifespan of the Cenozoic bivalve *Spissatella* (Crassatellidae)

JAMES S. CRAMPTON[1] & PHILLIP A. MAXWELL[2]

[1] *Institute of Geological and Nuclear Sciences, PO Box 30-368, Lower Hutt, New Zealand (e-mail: j.crampton@gns.cri.nz.)*

[2] *Bathgates Road, RD 10, Waimate, New Zealand*

Abstract: Outline shape is a key morphological character of infaunal bivalves that reflects both phylogenetic history and function. A new method of Fourier shape analysis has been used here to study the evolution of shape and size in the Cenozoic genus *Spissatella* (Crassatellidae), using material from New Zealand. The Fourier method can be used to construct a map of morphological 'geography', to generate synthetic average or extreme morphologies, and to visualize marginal growth fields and allometries. The results demonstrate that growth in *Spissatella* was strongly allometric and that differences between individual growth stages are commonly far greater than evolutionary changes in shape spanning 20 Ma. In all taxa, allometry was dominated by the relative elongation of the posterior margin with growth, an inferred functional adaptation for life in relatively high-energy environments. Heterochrony was the dominant evolutionary mechanism and, following an initial peramorphic expansion into morphospace, both paedomorphosis and peramorphosis probably occurred in approximately equal proportions. In general, it is not possible to decouple patterns of shape and size evolution. The data reveal no directional trends in evolution of these traits. The ontogenetic pathway followed by *Spissatella* may be a key genus-level taxonomic character and apparently represents a developmental constraint that largely controlled evolution within the genus.

Relationships between ontogenetic development, ecology, selection and evolution are increasingly the focus of diverse studies that bridge the interface of developmental biology and palaeontology (e.g. Gould 1977, 1989; Atchley 1987; Foote & Cowie 1988; McKinney & McNamara 1991; Zelditch *et al.* 1992; Polly 1998). Fossil bivalve molluscs are an ideal subject for such studies because they have a readily preserved shell that retains a complete record of external traits for all post-larval growth stages. In addition, the shell possesses internal features that can be used to infer soft-part morphology, such as the pallial line and muscle scars. The present analysis is based upon the outline shape and size of the lateral profile of the bivalve genus *Spissatella* Finlay, 1926 (family Crassatellidae).

Outline shape is a key morphological character that, in most bivalves, reflects both phylogenetic history, and function and life habit. *Spissatella* was a shallow-burrowing, non-siphonate, infaunal suspension feeder. It is likely that undisturbed adults were largely sessile within the substrate. The exhalent and inhalent currents were located at the posterior margin of the shell that must have been positioned at, or just above, the sediment–water interface, as is the case for most living Crassatellidae (Slack-Smith 1998). Living crassatellids are, in some areas, heavily preyed upon by octopuses, boring gastropods and shell-crushing chondrichthyans (Slack-Smith 1998), and survival is, to a great extent, dependent upon burial depth and speed of reburial following disinterment by predators or water currents. The same must have been true for *Spissatella*, in which burial depth was limited by the overall length of the shell and, as with most infaunal bivalves, burrowing speed was determined primarily by shape, inflation and sculpture (Stanley 1970, 1975; Kondo 1987; Watters 1993). The length and shape of the shell, therefore, are likely to have been key targets of directional and/or stabilizing selection. This prediction is corroborated, in part, by the work of Stanley & Yang (1987) who studied 19 Pliocene–Recent infaunal bivalves and inferred that shape was indeed subject to stringent stabilizing selection pressure within species, whereas overall size was not. Similarly, Soares *et al.* (1998) studied the living infaunal bivalve *Donax* and attributed intraspecific shape differences to directional selection related to habitat.

From: HARPER, E. M., TAYLOR, J. D. & CRAME, J. A. (eds) *The Evolutionary Biology of the Bivalvia*. Geological Society, London, Special Publications, **177**, 399–423. 1-86239-076-2/00/$15.00

This paper examines the relationships between ontogenetic and other intraspecific variations in shape and size on the one hand, and the evolutionary history of *Spissatella* on the other. Specifically, three questions are addressed: (1) how do shape and size vary or covary within species?; (2) how do intraspecific patterns of shape and size variation vary between species?; (3) phenotypic variation, at the population level, is the raw material of evolutionary selection – to what extent is such variation controlled by internal developmental processes throughout the lifespan of this bivalve genus? As noted by Gould (1989, p. 537), 'Natural history is a science of relative frequencies'; advance in many fields of palaeontological debate requires compilation of detailed observations across diverse fossil groups and time spans. This paper represents such a contribution.

The genus *Spissatella* and material studied

Spissatella is a moderately speciose genus of small to medium-sized crassatellids that is well represented in mid-Cenozoic shallow-water sandstones and siltstones in the South Island, New Zealand (Finlay 1926; Beu & Maxwell 1990). There are very few occurrences in the North Island. The only convincing record of the genus from outside the New Zealand region is the Early Oligocene *S. maudensis* (Pritchard, 1903) from Victoria, Australia (Darragh 1965, p. 110), although it is possible that some Argentinian Miocene species belong here (Beu *et al.* 1997, table 1).

Beu & Maxwell (1990) listed ten nominal species ranging in age from Late Eocene to Middle Miocene; another species may be *Crassatellites cordiformis* Suter, 1917 (late Oligocene?), but this is based on very poorly preserved material and its relationships are obscure. The earliest typical member of the genus is *Spissatella acculta* Finlay, 1926 (Early Oligocene), but some older crassatellids (Late Eocene–Early Oligocene), including *S. media* (Marwick, 1926), probably belong here despite their much smaller sizes (Figs 1 and 2). These earliest *Spissatella*-like forms were probably derived from the genus *Eucrassatella*. Most records of *Spissatella* are from Late Oligocene to Early Miocene sites, particularly in South Canterbury and North Otago on the east coast of the South Island. Thereafter, the genus became rare and is last known from the Early Pliocene. Extinction of *Spissatella* may have been related to Late Miocene and Early Pliocene cooling. Two of the species currently assigned to *Spissatella* – *S. scopalveus* and *S. concisa* Finlay, 1926 – are more appropriately referred to *Eucrassatella*, based on details of sculpture, shape, ontogenetic development and shell thickness (unpublished data).

In addition to having the general appeal of bivalves noted previously, *Spissatella* is an attractive subject for a morphometric analysis of evolution for several reasons. Firstly, the genus has a restricted geographic range: the vast majority of species occur in, and are apparently restricted to, New Zealand. Hence it seems reasonable to assume that evolution within the genus occurred largely within the New Zealand region. Secondly, well-preserved specimens of *Spissatella* are abundant in many collections from a variety of shallow-water facies in the South Island of New Zealand. Thirdly, the genus displays obvious and intriguing changes in morphology both through ontogeny and geological time. Fourthly, these changes are difficult to describe or quantify using more traditional descriptive or morphometric techniques. Lastly, *Spissatella* has a reasonable history of description and an existing taxonomic framework. This taxonomy is based on a variety of internal and external characters in addition to the outline shape studied here, notably hinge structures and sculpture.

In total, 299 individuals were included in the morphometric study. Collection and locality information is summarized in Table 1. For ease of communication, collections are identified throughout the text using ***bold italic*** labels shown on Table 1. On all but the smallest shells, two or three growth stages of each specimen were included in the analysis. The study is, therefore, based on both longitudinal (i.e. data from several ontogenetic stages of each individual) and cross-sectional data (i.e. data from many individuals of differing size). The description of intraspecific variation in shape has been emphasized, information that allows the significance of variations between populations to be gauged. The majority of specimens (650 outlines from 254 individuals) are from 18 main collections of *Spissatella* that span the Late Eocene–Early Miocene (Table 1, Fig. 1). Where possible, approximately 20 individuals were sampled from each collection. All holotypes of described species and subspecies were included in the analysis. The remaining specimens (97 outlines from 45 individuals) represent taxa formerly included in *Spissatella* and herein referred to *Eucrassatella*, other comparative specimens of *Eucrassatella* and a collection of *Salaputium animula*, a very small species closely related to, and hitherto assumed to be derived from, *Spissatella* or *Eucrassatella*. This was originally described as *Chattonia animula*, the type species of *Chattonia* Marwick, 1929, and has been included in *Salaputium* Iredale, 1924, following Finlay's (1930) questionable synonymy of *Chattonia* with *Salaputium*.

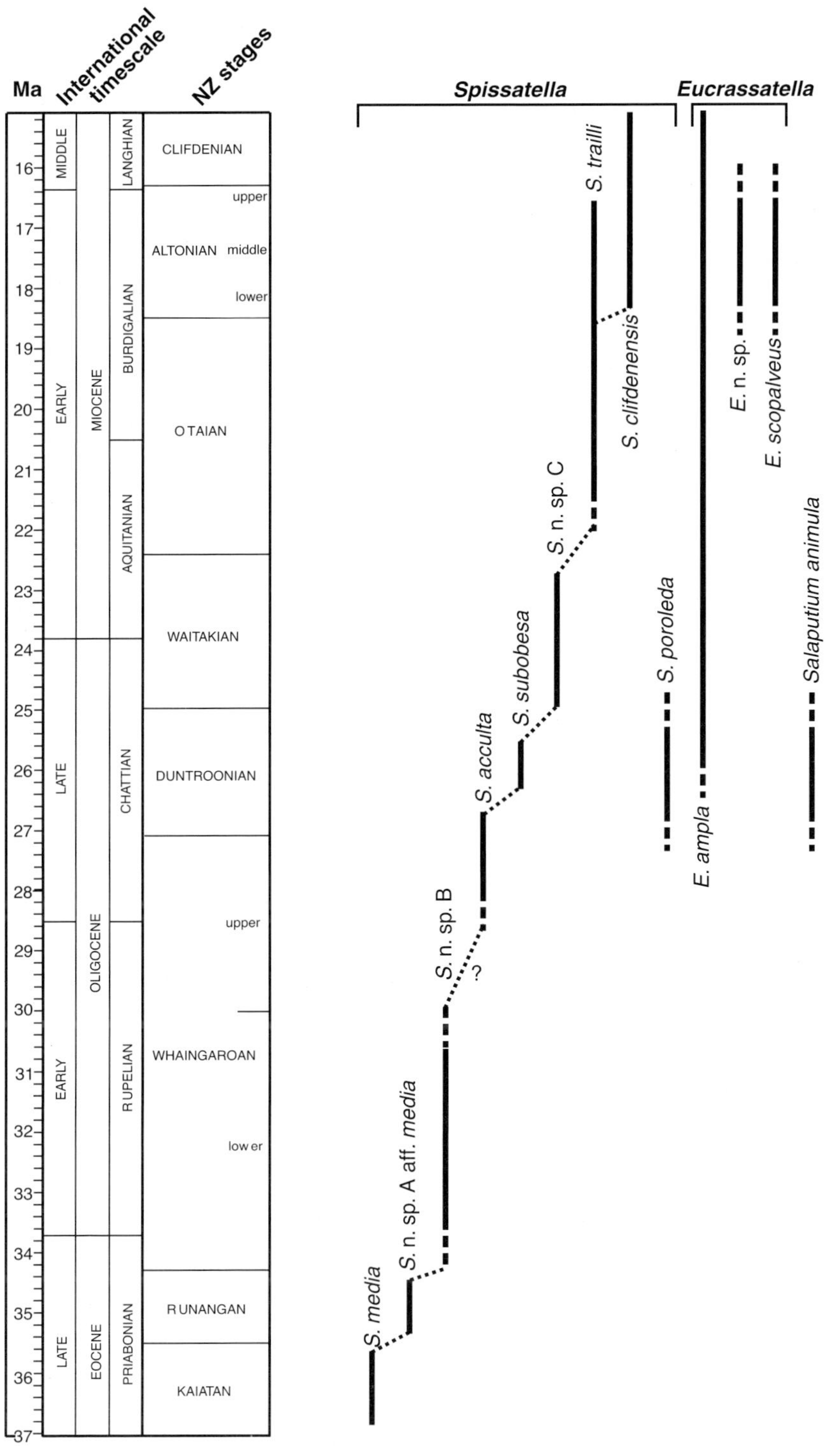

Fig. 1. Geological timescale, showing correlations between the International and New Zealand units for the study interval [after Morgans *et al.* (1996)], stratigraphic ranges of studied taxa and inferred phylogeny for *Spissatella*. All taxa for which ranges are shown are illustrated in Fig. 2.

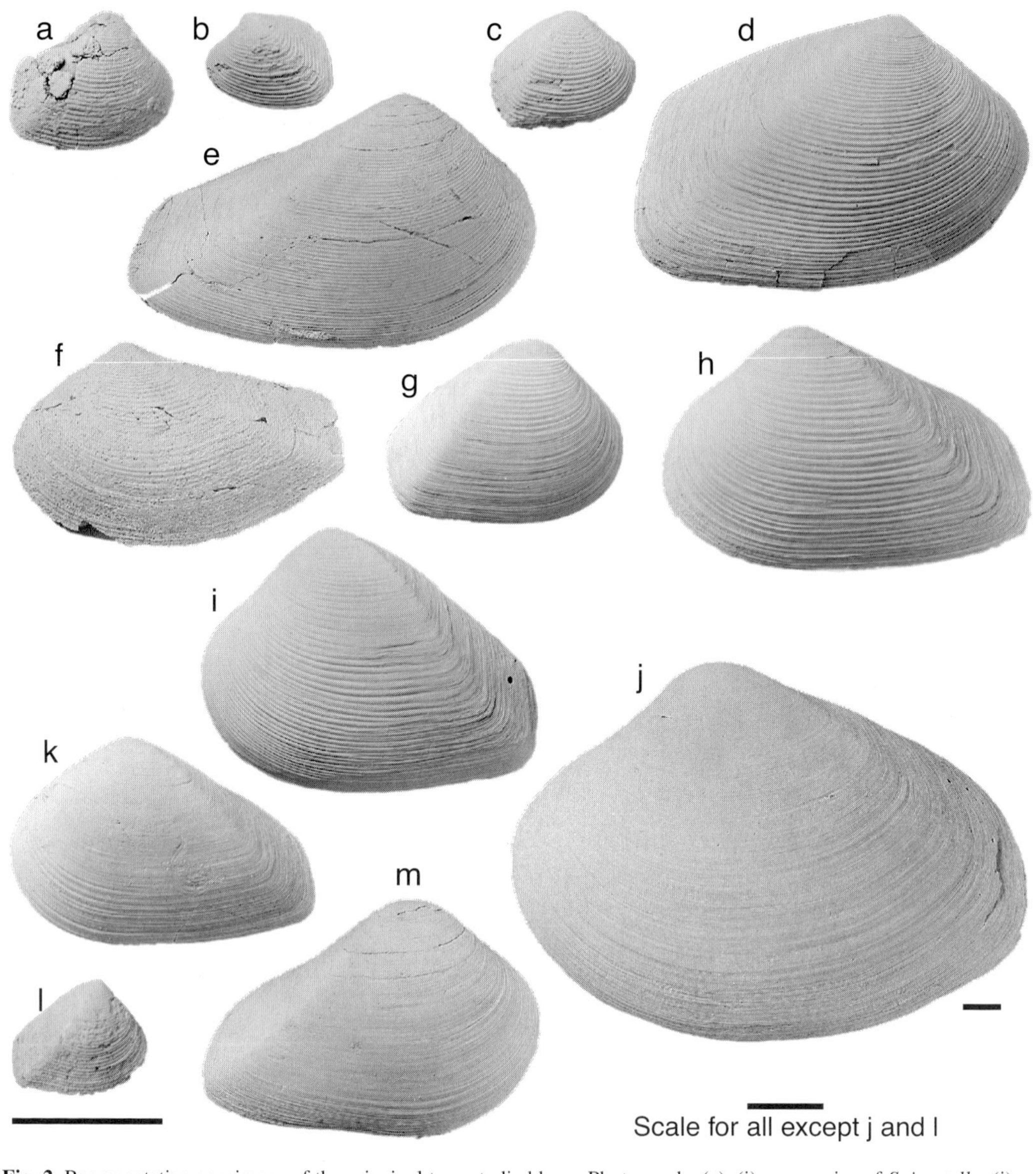

Fig. 2. Representative specimens of the principal taxa studied here. Photographs (a)–(i) are species of *Spissatella*, (j), (k) and (m) are species of *Eucrassatella*, and (l) is *Salaputium*. Locality and age details are given in Table 1. All specimens are whitened with magnesium oxide. Scale bar, 10 mm; all photographs are life size except (j) and (l). (**a**) *Spissatella media*, holotype, TM4299, right valve. (**b**) *S.* n. sp. A aff. *media*, TM7843 (GS11214), left valve. (**c**) *S.* n. sp. B, TM7844 (GS9536), right valve. (**d**) *S. acculta*, TM7845 (GS11195), right valve. (**e**) *S. subobesa*, holotype, TM4300, right valve. (**f**) *S. poroleda*, holotype, AK70725, left valve. (**g**) *S.* n. sp. C, TM7846 (SCIP041), right valve. (**h**) *S. trailli*, TM7847 (GS9685), left valve. (**i**) *S. clifdenensis*, TM7848 (GS10365), left valve. (**j**) *Eucrassatella ampla*, TM7849 (GS3600), left valve. (**k**) *E.* n. sp., TM7850 (SCIP080), left valve. (**l**) *Salaputium animula*, TM7872 (GS9806), right valve. (**m**) *Eucrassatella scopalveus*, holotype, AK70726, right valve.

Methods

Allometric heterochrony

This paper is concerned, in part, with the identification of heterochrony (e.g. McKinney & MacNamara 1991). The study of heterochrony requires, by definition, some means of calibrating the rate and timing of developmental events and growth between different individuals. Most workers have assumed that chronological time is the most appropriate or practical referent ['age-structured' studies; but see discussion in Hall (1992)].

In the present study, it has not been possible to establish the chronological age of individual specimens. The use of stable isotopes and growth line analyses to age molluscs are well established and could, potentially, allow age-structured analysis of the *Spissatella* data (e.g. Jones 1988; Brey & Mackensen 1997; Jones & Gould 1999; Tojo & Ohno 1999). Specimens of *Spissatella*, however, generally do not possess conspicuous growth lines or discontinuities in the commarginal sculpture that might represent annual increments. In some cases, growth lines are visible on the juvenile and immature parts of the shell, but these typically are absent on the adult. Hence sclerochronology would, by necessity, rely on isotopic analyses or examination of radial sections of the shell, methods that are destructive of material and were considered impractical in the present study.

In this study, therefore, size has been used as an implicit proxy for time and the data are size structured, even though the assumption of age–size equivalence is almost certainly flawed (e.g. Jones 1988). Inferences of heterochrony based on size-structured data are distinguished as allometric heterochrony (McKinney 1988). The potential pitfalls of this approach have been discussed by many workers (e.g. Blackstone & Yund 1989; Jones & Gould 1999).

Fourier shape analysis

Fourier shape analysis operates on populations of digitized outline traces and, for each outline, yields a suite of Fourier coefficients that describe a spectrum of harmonically related trigonometric curves or harmonics. These coefficients furnish the input to standard multivariate statistical analyses.

Digitized outlines were captured via a two-stage process:

- The growth line of interest was drawn using a *camera lucida* attached to a Wild binocular microscope. Specimens were arranged so that the growth line lay within the plane of view. Studied growth lines were spaced at *c.* 10 mm intervals along the axial length (*AL*) or conspicuous carina that extends from the umbo to the postero-ventral margin. Specimens were reoriented beneath the microscope before each growth line was drawn. Because of the vagaries of preservation, growth lines could not be selected at constant intervals of *AL*, although the distribution of *AL* values is strongly trimodal with modes at *c.* 8, 19 and 34 mm.
- The outline drawings were digitized automatically using the pc-based image analysis software *Optimas, v. 5.10* (Optimas Corporation 1995). Outlines were traced in a standard orientation (using the hinge line) and sense, starting from the umbo, and 2000 *xy* coordinates were written to file.

Fourier coefficients were computed and manipulated using the suite of programs Hangle, Hmatch and Hcurve, written by John Haines (Crampton & Haines 1996; Haines & Crampton 2000). Ten Fourier harmonics were generated for each outline; the first harmonic had no shape significance and was ignored during subsequent statistical analysis. Ten harmonics captured all biologically meaningful shape information in the outlines (Crampton 1995). For each harmonic, Hangle produced two Fourier coefficients.

During processing, a number of adjustments and standardizations were necessary: (1) outlines were automatically smoothed in program Hangle to eliminate the corrupting effects of high frequency pixel digitization 'noise' (Haines & Crampton 2000). (2) Outlines were automatically standardized for size in program Hangle based on perimeter length. This is appropriate given that the effects of size, if not removed, will dominate any analysis of shape. Size information was retained separately and reincorporated into the study at a later stage. In this way, size and shape were separated and could be studied independently, as recommended by Stanley & Yang (1987). (3) Right valves were automatically mirrored in program Hangle, prior to computation of the Fourier coefficients, to eliminate the effects of primitive bilateral symmetry in bivalves (Crampton 1995). Following this procedure, a perfectly equivalve left and right valve pair will appear identical during statistical analysis and separation will increase with increasing inequivalveness. (4) Outlines were standardized for starting position of the digitized trace using the program Hmatch. Previous studies have demonstrated that Fourier methods are very sensitive to the starting position (and thereby orientation) of the digitized trace (Haines & Crampton 2000). This is a problem in the *Spissatella* data set where the only landmark that could be used to start the trace, the umbo, is broadly

Table 1. *Summary of all specimens and collections included in the analysis*

Label	Species	Collections*	Age†	Locality‡	Facies and inferred§ palaeoenvironment	Number (individuals/outlines)
1	*Spissatella acculta* Finlay, 1926	**AK70722**	Whaingaroan	Wharekuri, Otago	Glauconitic sandstone; inner shelf	1/3
2	*Spissatella acculta*	GS11195	Duntroonian	I40/f349; head of Lake Waitaki, South Canterbury	Glauconitic sandstone; inner shelf	9/27
3	*S. clifdenensis* Finlay, 1926	**AK70723**	Altonian(a)	Clifden section, Waiau River, Southland	Massive siltstone; inner to mid-shelf	1/3
4	*S. clifdenensis*	GS11182	Altonian(a)	D45/f8819; Clifden section, Waiau River, Southland	Massive siltstone; inner to mid-shelf	17/44
5	*S. clifdenensis*	GS10344	Altonian(b)	D45/f8483; Clifden section, Waiau River, Southland	Sandstone; inner shelf	20/56
6	*S. clifdenensis*	GS10365, SCIP053	Altonian(c)	D45/f8598; Clifden Section, Waiau River, Southland	Muddy sandstone; mid-shelf?	10/24
7	*S. discrepans* Finlay, 1926	**AK70724**	Duntroonian–Waitakian	Lake Wakatipu, Southland	Unknown	1/3
8	*S. media* (Marwick, 1926)	**TM4299**	Kaiatan	J41/f8460; Lorne, near Weston, North Otago	Debrite; flank of seamount	1/1
9	*S. media*	GS9481	Kaiatan	J41/f8025; Lorne, near Weston, North Otago	Debrite; flank of seamount	2/3
10	*S. poroleda* Finlay, 1926	**AK70725**	Duntroonian	Shell Gully, Chatton, Southland	Shelly sandstone; inner shelf	1/3
11	*S. poroleda*	GS9806, SCIP046	Duntroonian	F45/f9668; Shell Gully, Chatton, Southland	Shelly sandstone; inner shelf	11/36
12	*S. subobesa* (Marshall & Murdoch, 1919)	**TM4300**	Duntroonian	Head of Lake Waitaki, South Canterbury	Glauconitic sandstone; inner shelf	1/6
13	*S. subobesa*	GS10837, SCIP011	Duntroonian	I40/f1; head of Lake Waitaki, South Canterbury	Glauconitic sandstone; inner shelf	24/68
14	*S. trailli* (Hutton, 1873)	GS11154	Otaian	J39/f26; Mount Horrible, Pareora River, South Canterbury	Siltstone; outer shelf	20/41
15	*S. trailli*	GS11283	Otaian	J39/f7500c; Blue Cliffs, Otaio River, South Canterbury	Massive to poorly bedded siltstone; outer shelf	9/24
16	*S. trailli*	**TM2859**	Altonian(a)	Awamoa Creek, North Otago	Poorly bedded siltstone; mid-shelf	1/3
17	*S. trailli*	GS1160, SCIP006	Altonian(a)	J41/f8499; Awamoa Creek, North Otago	Poorly bedded siltstone; mid-shelf	20/55
18	*S. trailli*	GS9685, OU10193	Altonian(b)	J41/f9499b; Pukeuri, North Otago	Siltstone; mid-shelf	6/18
19	*S.* n. sp. A aff. *media*	GS11214, SCIP084	Runangan	J42/f126; Bridge Point, North Otago	Debrite; flank of sea-mount	9/12

20	*S.* n. sp. B	GS9536	Whaingaroan	J42/f6032; Gees Point, Kakanui, North Otago	Calcareous, fine agglomerate; flank of sea-mount	6/8
21	*S.* n. sp. C	GS1473	Waitakian(a)	I40/f9519; Trig Z, Otiake, North Otago	Sandy limestone; inner to mid-shelf	20/51
22	*S.* n. sp. C	SCIP039	Waitakian(a)	I40/f345; Brothers Stream, Hakataramea valley, South Canterbury	Glauconitic sandstone; inner shelf	15/32
23	*S.* n. sp. C	GS7166, SCIP040	Waitakian(b)	I40/f7551; Brothers Stream, Hakataramea valley, South Canterbury	Glauconitic siltstone; mid-shelf?	20/55
24	*S.* n. sp. C	SCIP041	Waitakian(b)	I40/f188B; Sisters Creek, Hakataramea valley, South Canterbury	Calcareous siltstone; mid- to outer shelf	20/51
25	*Eucrassatella ampla* (Zittel, 1864)	GS3600	Waitakian	F46/f8492; Mataura River, Southland	Sandy mudstone	1/5
26	*E. australis* (Hutton, 1873)	GS9959	Bortonian	J40/f6610; Pentland Hills, Waihao River, South Canterbury	Gravely sandstone	1/4
27	*E. concisa* (Finlay, 1926)	**AK70727**	Altonian	Target Gully, Oamaru, North Otago	Shelly sandstone; mid-shelf	1/3
28	*E. kingicola* (Lamarck, 1805)	WM2982	Recent	Australia (locality unknown)		1/4
29	*E.* n. sp.	SCIP080	Altonian	J39/f244; Pareora River near end of Purves Road, South Canterbury	Shelly sandstone; inner shelf	9/22
30	*E. scopalveus* (Finlay, 1926)	**AK70726**	Altonian	Target Gully, Oamaru, Otago	Shelly sandstone; inner shelf	1/3
31	*E. scopalveus*	GS951, OU189	Altonian	J41/f8498; Target Gully, Oamaru, North Otago	Shelly sandstone	18/60
32	*Salaputium animula* (Marwick, 1929)	**M1847**	Duntroonian	Shell Gully, Chatton, Southland	Shelly sandstone; inner shelf	10/10
33	*Salaputium animula*	TM6758, GS9806	Duntroonian	F45/f9668; Shell Gully, Chatton, Southland	Shelly sandstone; inner shelf	10/10
34	*S. animula*	GS9805	Duntroonian	F44/f9501; Coal Creek, Wendon valley, Southland	Carbonaceous pebbly sandstone	10/10

*Collection numbers shown in bold indicate holotypes. All material is housed in New Zealand institutions, and specimen and collection numbers are prefixed with the following repository identifiers: AK, type Mollusca number, Auckland University, Auckland; GS, collection number, Institute of Geological and Nuclear Sciences (IGNS), Lower Hutt; M, type Mollusca number, Museum of New Zealand, Wellington; SCIP, private collections of the author (PAM); TM, type Mollusca number, IGNS; WM, world Mollusca collection, IGNS. †Ages are given in terms of New Zealand local ages; correlations with the international timescale are shown in Fig. 1. Relative positions within ages are indicated, in ascending order, by (a), (b), etc. ‡Fossil localities are registered in the Fossil Record File (FRF), a national database administered by the Geological Society of New Zealand. The file is arranged according to NZMS260 series 1:50 000 topographical map sheet areas. FRF numbers appear in the format J41/f123, where J41 refers to the map sheet area and f123 refers to a unique locality number within that sheet. §Palaeoenvironments are inferred from total fossil faunas, facies and stratigraphic considerations.

rounded and cannot be defined with precision. Other Fourier programs have standardized for starting position using properties of individual outlines. Program Hmatch, on the other hand, uses properties of the entire sample of outlines (i.e. 747 outlines) and is a consistently robust approach to standardization for starting position (Haines & Crampton 2000). 'Matching' in this way has the effect of minimizing the first eigenvalue during statistical analysis (see below).

Fourier coefficients were examined using three statistical methods, principal components analysis (PCA), discriminant function analysis (DFA) and analysis of variance (ANOVA). Statistical analyses were performed using the pc-based software NTSYS-pc, v. 1.80 (Rohlf 1993) and the Unix-based program S-plus v. 5.0 (Mathsoft 1998). Multivariate analyses (PCA and DFA) were based on the variance–covariance matrix and the Fourier coefficients were not standardized for amplitude [see explanation in Crampton (1995)]. In the DFA, the significance of differences between group means was tested using the Wilks' lambda statistic transformed to an F-statistic. Because of relatively small group sizes and the requirement that the number of individuals exceed the number of variables, in some cases it was necessary to limit the number of harmonics input to DFA to the first five or six. Tests with varying numbers of harmonics indicated that eliminating the higher order harmonics has little impact on group discrimination. The number of harmonics used is reported in the text with the results of DFA. The ANOVA used principal component scores for each outline and corresponding specimen sizes, as measured by the AL, to examine linear relationships between shape and size. In many cases correlations are non-linear and/or show scale-dependent increase of variance with size, a condition termed heteroscedasticity that contravenes an assumption of the ANOVA method. To eliminate or minimize these problems, AL was log-transformed prior to analysis (cf. Foote & Cowie 1988).

Results

Presentation of results

Interpretation was based largely on the results of PCA of the total data set. For ease of comparison, all data are shown ordinated against the first three principal component (PC) axes and all plots are homologous. The first three PC axes or eigenvectors explain 51, 16 and 11%, respectively, of the total variance in the data. Both a scree plot of eigenvalues and the broken-stick model (Joliffe 1986; Jackson 1993) suggest that only the first three eigenvectors are significant and explain more variance than expected by chance alone. Hence the remaining eigenvectors were ignored in subsequent interpretation. The PC axes are cast in units of standard deviation and centred about the overall mean. Their lengths are scaled to be proportional to the variance explained by each eigenvector and also normalized so that they are consistent with average taxonomic distances amongst a set of objects (Rohlf 1993).

Ordinated against the three PC axes of Fig. 3 are a suite of wholly synthetic outline shapes corresponding to unit standard deviation steps along each axis (± two standard deviations). The outline positioned at the origin is the overall mean shape. Figure 3 also shows a grey-scale map of size, based on AL, for the first two PC axes. Note that PCs 2 and 3 show no consistent size–shape correlations and, for simplicity, the size map is omitted from this plot. Size data are examined in more detail below using ANOVA. The outlines of Fig. 3 facilitate easy interpretation of the distributions of the real data and should be used as a reference for the following discussions. They translate an arbitrary mathematical space into the biologically meaningful 'morphospace', upon which the present interpretations are based. In combination with the size data, they transform the PC axes into a map depicting morphological 'geography'.

In the following plots, subsets of the data are summarized by 95% confidence ellipses on their group means. In other words, each group population mean has a 95% probability of lying within its confidence ellipse. Similarly, in some diagrams the distributions of individual points are summarized using 95% probability ellipses. These ellipses are expected to enclose 95% of data points for any particular group. The meaning and relationships of confidence and probability ellipses are illustrated in Fig. 4. The ellipses were generated using the method of Altman (1978), modified to the F-statistic following Sokal & Rohlf (1995). Confidence and probability ellipses are a convenient way to summarize and represent a large number of individual data points. In addition, they allow the reader to gauge (approximately) the significance of differences between group means.

It should be remembered, however, that PCA is an ordination that maximizes separation of individual points in multivariate space for the data set as a whole. It does not identify axes that optimally discriminate any particular pair of subgroups of the data (e.g. collection or size groups etc.). Hence, where relevant, more rigorous pairwise comparison of group means was achieved using DFA, results of which are reported in the text.

In order to test for patterns of allometric growth,

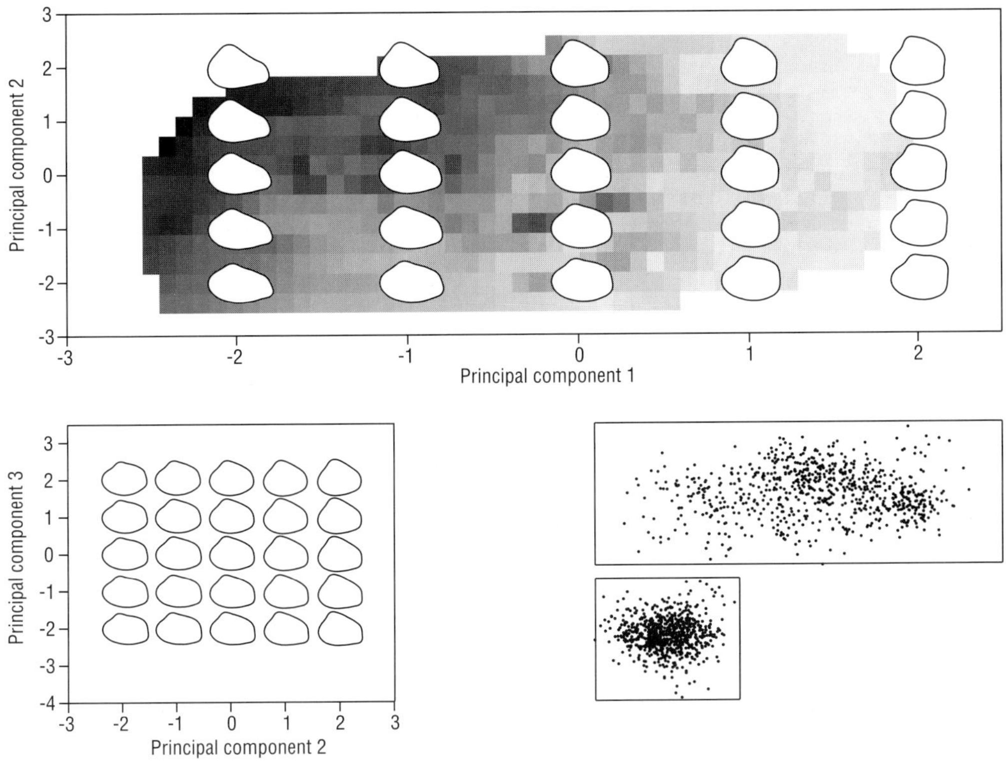

Fig. 3. First three principal component (PC) axes based on the total data set (see text for explanation). Ordinated against these axes are a suite of wholly synthetic outline shapes corresponding to unit standard deviation steps along each axis. These shapes facilitate easy interpretation of distributions of real data and should be used as a reference for subsequent plots. Synthetic outline shapes were generated using the inverse Fourier program Hcurve and the procedure outlined in Haines & Crampton (2000). Also shown v. the first two axes is a grey-scale map of average shell size (based on axial length, *AL*), calculated as the average *AL* for all data points within each rectangle of an arbitrary grid. White grid rectangles correspond to an average size of 0 mm (i.e. no data) and black rectangles correspond to an average size of 54 mm (i.e. the largest average *AL* encountered). The map reveals an approximately uniform increase in size towards the upper left corner of the plot (see text for further discussion). Note that specimens of *Eucrassatella* were omitted during computation of the size map. Size data show no consistent pattern v. PC 2 and 3 and have, for clarity, been omitted. For reference, small plots on the bottom right of the diagram show the distributions of the real data v. the first three PC axes.

data were divided into three arbitrary size classes based on the *AL* (remembering, of course, that outlines have been standardized for size *per se*). These size classes were chosen to reflect the size modes of the digitized outlines (see above) and are: (1) $AL < 15$ mm; (2) 15 mm $\leq AL < 30$ mm; and (3) $AL \geq 30$ mm. By convention, and for convenience, these are referred to as juvenile, immature and adult growth stages, respectively, with the caveat that certain size classes will not correspond to the same absolute age classes in different taxa.

Lines linking different growth stages of individual specimens or group means are termed ontogenetic trajectories (Fig. 5a). As noted above, outlines were standardized for size during processing and, therefore, two outlines that have identical shapes and differ only in size will plot at the same point in morphospace. The degree of separation is a measure of shape difference. Consequently, the degree of separation between different growth stages of a single individual is a measure of allometry. Conversely, if growth is isometric, two ontogenetic stages of an individual will occupy the same point in morphospace. Clearly, in some taxa allometric growth is pronounced and individual ontogenetic trajectories traverse much of the total morphospace occupied by the data set as a whole (Fig. 5a). Using overlain real or synthetic average outlines, it is possible to visualize changes in the location, extent or intensity

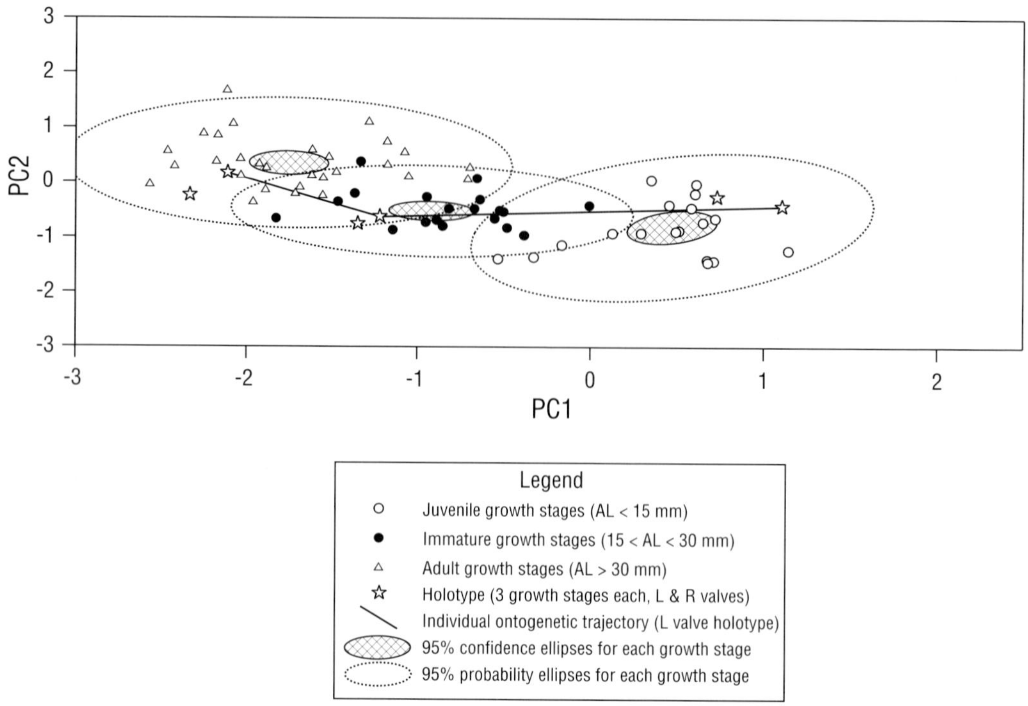

Fig. 4. First two PC axes showing all data for one collection of *S. subobesa* (***12***, ***13***) and including the holotype (left and right valves, TM4300, ***12***). Data points are distinguished according to size class and, for each size class, the 95% probability and confidence ellipses are shown. Also shown, for the left valve of the holotype, is one example of an individual ontogenetic trajectory. This plot illustrates the relationship between the actual data points and their probability and confidence ellipses, and serves as a reference for later figures.

of marginal growth fields and to thus specify shape allometries. This is illustrated in Fig. 5c, showing size-standardized outlines for three growth stages of a single individual. Regions of negative and positive relative growth are indicated using shading and arrows. This method of visualization is utilized below. The regions thus identified are empirical analogues of the aperture maps of Rice (1998) that depict the relative magnitudes of shell production at different points around the apertures of hypothetical gastropods.

Where relevant, synthetic average shapes for specific subsets of the data are shown with the statistical plots. These outlines were generated using the averaged Fourier coefficients for the group in question and the inverse Fourier program Hcurve. They allow the reader to gauge morphological differences between specific subsets of the data.

The set of 747 outlines was examined by collection, geological age, facies, a priori taxonomic groupings and size class. Due to the large number of ways the data have been explored, only significant results and relevant plots are presented here.

Interpretation of the principal component axes: the shape morphospace

Using the synthetic outlines of Fig. 3, it is possible to assign morphological meaning to the three significant PC axes. The first principal component can be explained largely by the degree of posterior elongation: individuals with low scores on PC 1 have an extended posterior margin, are strongly inequilateral and have a rostrate posterior; those with high scores have a truncated posterior margin and are relatively equilateral. PC 2 codes for several traits, most importantly the prominence and inflation of the umbo, the length of the anterior part of the shell, and the height and roundness of the posterior margin. Individuals with low PC 2 scores have relatively weakly inflated umbones, extended anterior regions and high umbonal angles approaching 180°. Shells with high PC 2 scores

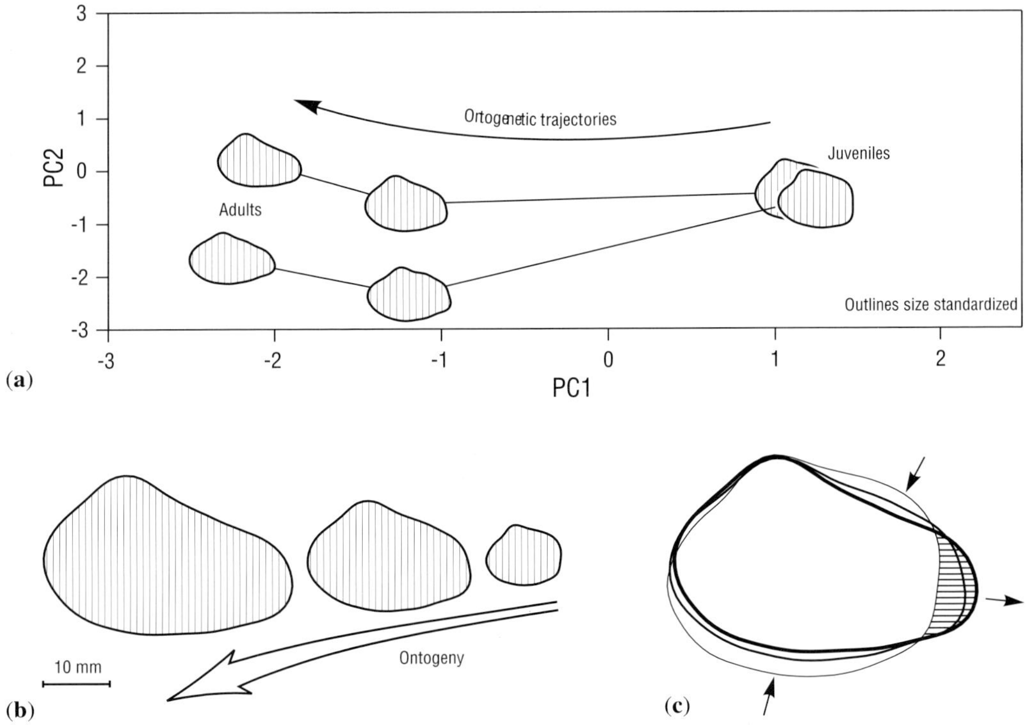

Fig. 5. (**a**) First two PC axes showing actual, size-normalized outlines for three growth stages each of two individuals. The upper outline is *S. subobesa* (holotype, TM4300, ***12***, left valve) and the lower outline is *S. poroleda* (TM7873, SCIP046, ***11***, right valve). (**b**) The same three outlines for TM4300 shown in their true size proportions. (**c**) Size-normalized outlines for TM4300 overlain to reveal regions of positive (shaded) and negative (white) marginal allometry. Arrows highlight the sense of allometry for each segment of the margin. The thinnest outline indicates juvenile growth stage and the thickest outline the adult growth stage.

have prominent umbones, short anterior regions and low umbonal angles close to 90°. With increasing PC 2, the outlines become less rectangular and increasingly triangular in appearance. PC 3 accounts largely for convexity of the ventral margin and the angularity of the postero-umbonal carina: low scores correspond to a relatively straight ventral margin and sharp carina, high scores code for a more convex margin and carina.

General taxonomic observations

Examination of the data by collection has revealed that, with almost no exceptions, individual collections occupy continuous regions of morphospace and, for any particular size class, do not define more than one discrete cluster. This suggests that, in general, each collection comprises a single taxonomic unit, based on shape alone, and that different species, if truly coeval, did not coexist at a single site but were partitioned by facies and/or palaeogeographic setting.

Morphometric and evolutionary patterns through time

At a gross scale, Fig. 6 reveals long-term patterns in the evolution of adult shape throughout the Late Eocene–Early Miocene, both in terms of morphological disparity and average form. During the latest Eocene and Early Oligocene, diversity was low, comprising a single taxon at each level, and shape is restricted to relatively high scores on PC 1. Over this period, and into the Late Oligocene, there was progressive expansion of adult form into regions of morphospace corresponding to low scores on PC 1. During the middle Late Oligocene there were at least two lineages of *Spissatella* (Fig. 1) and morphological disparity was greatest over this interval; three species are recognized and their average shapes are widely separated in morphospace. Subsequently, during the latest Oligocene–Early Miocene, there was a contraction in total morphospace occupation and group averages are restricted largely to the upper half of

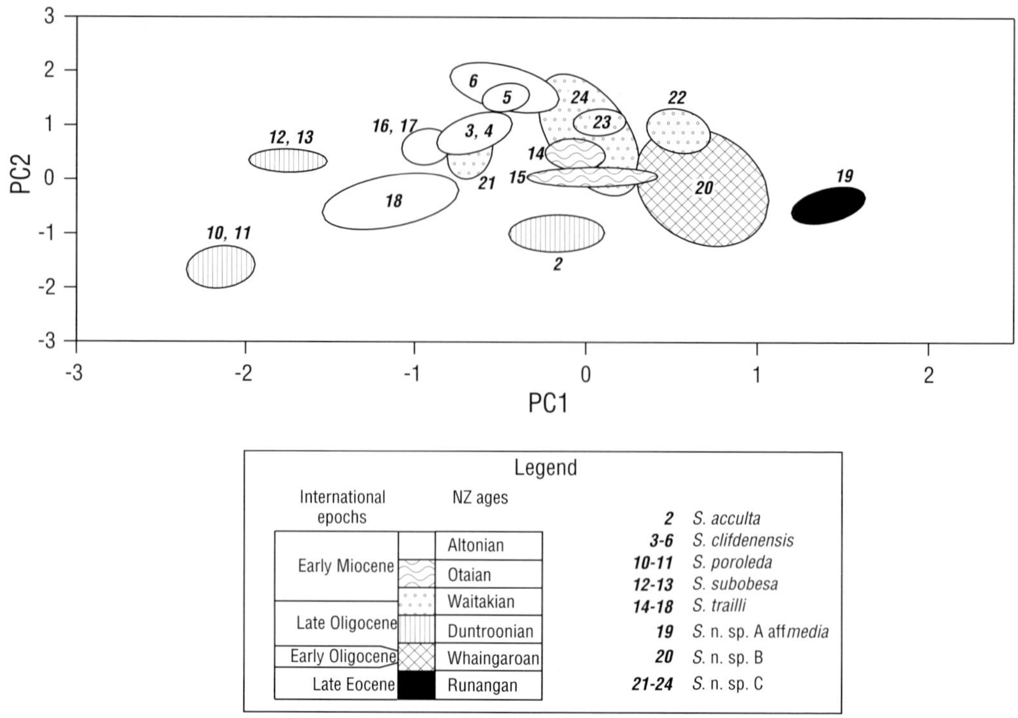

Fig. 6. First two PC axes showing 95% confidence ellipses for largest growth stage of all major collections of *Spissatella*. Confidence ellipses on PC 3 are largely overlapping and are not shown here. Ages are coded according to the New Zealand local timescale but approximate correlations with the international epochs are indicated (see Fig. 1 for more precise correlations). Ellipses are identified using the collection labels of Table 1.

the plot. For this interval of time, confidence ellipses for collections are relatively tightly clustered and, in many cases, overlapping. Overall, following the initial morphological expansion, there is no directional trend in morphospace occupation through time. Figure 6 reveals patterns of convergence and repeated occupations of particular regions of morphospace. Relationships at specific times and between particular groups are explored in more detail below, in stratigraphically ascending order.

From independent analysis of both internal and external characters, and the close stratigraphic proximity of the studied topotype collections, it is inferred that *S. acculta* (*2*) was immediately ancestral to *S. subobesa* (***12, 13***) in the middle Late Oligocene. In addition, available evidence favours evolution of latest Oligocene–earliest Miocene *S.* n. sp. C (***21***) from *S. subobesa*. Average ontogenetic trajectories for these three species are virtually parallel in three-dimensional morphospace and they differ primarily in length (Fig. 7a). This arrangement of successive trajectories can be readily explained as evolution by global heterochrony: the three forms display near-identical patterns of shape development, varying only in the rate and/or timing of progress along the ontogenetic pathway. In addition, however, they also show progressive translations of their ontogenetic trajectories towards higher PC 2 scores that correspond to increased prominence of the umbo and slightly increased inflation (cf. Fig. 3). In the case of *S.* n. sp. C, the offset is clearly significant and the increase in inflation is expressed early in ontogeny. This offset might represent a non-heterochronic, morphological innovation in early ontogeny, termed a cenogenic event by some authors (e.g. Swan 1988). Alternatively, it is very likely that increased inflation was itself a consequence of heterochronic evolution, although this has not been studied in detail here. The shape data for these three species can be recast with PC score v. size *AL* to create a series of plots equivalent to those used by McKinney (1988) to illustrate different forms of allometric heterochrony (Fig. 7b). Thus, the inferred transition from *S. acculta* to *S. subobesa* represents evolution via a combination of allometric acceleration and hypermorphosis, whereas the transition from *S. subobesa* to *S.* n. sp. C represents evolution primarily via allometric progenesis (cf. McKinney & McNamara 1991, figs 2–15). By way of contrast, the ontogenetic trajectory for middle Late Oligocene *S. poroleda* is conspicuously non-parallel to the others and has opposed polarity on PC 3. It is inferred here that *S. poroleda* either evolved from a rather different, and as yet unknown, ancestor or that its evolution from the common ancestor to *S. acculta* involved mechanisms other than heterochrony. These results can be viewed another way, using overlain, size-standardized, average outlines for each growth stage (Fig. 7c, cf. Fig. 5c). In the three species *S. acculta*, *S. subobesa* and *S.* n. sp. C, positive and negative allometric growth fields are distributed similarly around the growing margin and vary only in their magnitude in *S. subobesa*. *S. poroleda*, on the other hand, has strongly negative allometric growth at the postero-ventral extremity of the shell, a pattern not observed on the other species, and it clearly follows a rather different ontogenetic growth pattern. These differences are mirrored by its relatively low inflation and differences in hinge structures (unpublished data), suggesting that perhaps *S. poroleda* should be referred to its own subgenus.

Evolution of *Spissatella* during the latest Oligocene–earliest Miocene is rather more problematic. Four collections have been examined and these can be grouped into older and younger pairs of collections based on associated planktonic foraminifera (Table 2). Two of the collections, ***22*** and ***23***, are closely associated in the same section and span a microfossil zonal boundary. The older collections, ***21*** and ***22***, have substantially different mean shapes, although there is overlap between their immature probability ellipses (Fig. 8). The younger pair of collections, ***23*** and ***24***, are indistinguishable on the basis of shape, their means having a 63% probability of being drawn from the same population (Table 1). In all other comparisons, means are significantly different from each other at probabilities of > 99.9% (Table 2). Individual immature data points for the four groups all overlap with all others to varying degrees and together occupy a continuous region of morphospace (the 95% probability ellipses of Fig. 8). Shifts in mean positions are not consistent through time and do not define a uniform trend. Shape variability in these collections is matched by comparable variation in the strength and distribution of commarginal sculpture (unpublished data). Although the possibility that the older collections, ***21*** and ***22***, represent distinct species cannot be discounted, from available evidence it is inferred that the four collections are drawn from a single, highly variable species. Variability may have been related to ecophenotypic effects associated with facies and/or to evolution by random walk, or to evolution in response to a fluctuating selection regime. The present data do not allow these hypotheses to be tested. It is worth noting, however, that the two indistinguishable collections are derived from similar siltstone facies; the other two collections are from sandstone and limestone facies (Table 1). The influence of facies and palaeoenvironment is discussed below.

Stratophenetic patterns during the middle–late Early Miocene have been examined using seven collections of two species, *S. trailli* and *S. clifdenensis* (Table 1). Two collections of *S. trailli* of middle Early Miocene age (***14*** and ***15***) are morphologically similar to each other (Fig. 9a) and their means are not significantly different (Table 3). Although not shown, these two samples include a few larger individuals that are indistinguishable from adult *S. trailli* in the succeeding late Early Miocene collection (***17***). Topotype collections of *S. trailli* and *S. clifdenensis* from the late Early Miocene have very similar shapes (Fig. 9b, lower outlines) and there is a high degree of overlap between their populations (Fig. 9a inset, ***4*** and ***17***). Their mean shapes are, however, significantly different at the 99.5% level of confidence (Table 3). They also differ in details of the hinge line: *S. clifdenensis* has a relatively high hinge and short resilifer. Subsequently, both species diverge morphologically from the ancestral populations, moving in opposite senses on the PC plot. The geologically youngest samples (***6*** and ***18***) are widely separated, their 95% probability ellipses barely touch (Fig. 9a) and they have distinct outlines (Fig. 9b, upper outlines). The geologically youngest *S. trailli* are morphologically convergent upon immature *S. subobesa* and the two youngest collections of *S. clifdenensis* are indistinguishable on the basis of shape (Table 3). Using overlain average outlines for each growth stage (Fig. 9c), it is apparent that increasing differences between the species result from rather subtle changes in the magnitudes of growth fields on the ventral and postero-ventral margins. In particular, negative allometry on the ventral margin that is visible in all other *Spissatella* is suppressed in *S. clifdenensis* – i.e. growth on the ventral margin is isometric. Even though ontogenetic trajectories are non-parallel, these patterns may record local heterochronic changes in particular marginal growth fields that occurred without significant global change in adult size (cf. Figs 3 and 9a). Patterns of shape change during the Early Miocene, therefore, suggest the following evolutionary hypotheses (Fig. 10). First, negligible evolution of *S. trailli* through the middle Early Miocene conforms to a model of stasis. Secondly, it is inferred that *S. clifdenensis* evolved

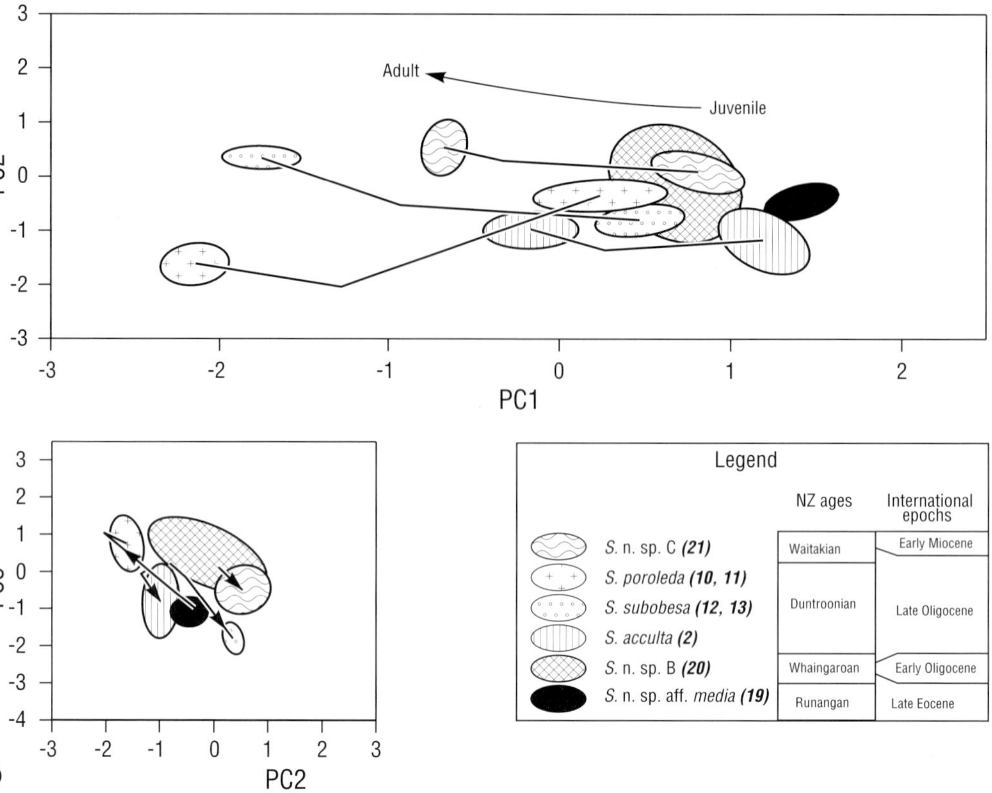

(a)

by cladogenesis from *S. trailli* at some time in the middle Early Miocene. Thirdly, morphological divergence of the two species occurred during the late Early Miocene and suggests anagenetic evolution within both lineages, although this requires thorough testing with additional, stratigraphically closely spaced samples. Fourthly, the youngest populations of *S. clifdenensis* experienced apparent stasis for an appreciable length of time. Note that the inference of anagenesis (within lineage evolution) makes no statement about evolutionary rate, for which stratigraphic resolution would need to be improved, perhaps by an order of magnitude (cf. Chiba 1996).

Overall correlations between size and shape

Looking at the data set as a whole, the most striking feature is the strong negative correlation between PC 1 score and size (Fig. 3), and the consistent separation of juvenile and adult size classes along this axis (Fig. 11). The general pattern of size change shown in Fig. 3 is interrupted by a single 'peak' in size around PC 1 = 0 and PC 2 = –1. This peak represents *S. acculta*, which is, compared to other taxa, anomalously large for its adult shape. Overall relationships between size and shape were studied in more detail using ANOVA. As explained previously, this is used here to explore linear correlations between PC scores and their corresponding log-transformed values of *AL* (size). Four nested models were fitted for each of the principal components. These models identify the relative proportions of the total variance that are explained by: (1) overall size–shape correlations; (2) differences in size–shape correlations between collections; (3) differences between individuals from each collection and the mean regression for that collection; and (4) residual or unexplained variations.

Results of ANOVA are presented in Table 4. These reveal that 63% of the shape variance on PC 1 is explained by the correlation between size and shape. In morphological terms, this correlation corresponds to increasing posterior elongation with increasing size (Fig. 3). Nearly all the remaining variance, 30%, is explained by differences between collections; the scatter of individuals about the mean regression for each collection is negligible (< 3%). The value for the residual on PC 1 (4%) is

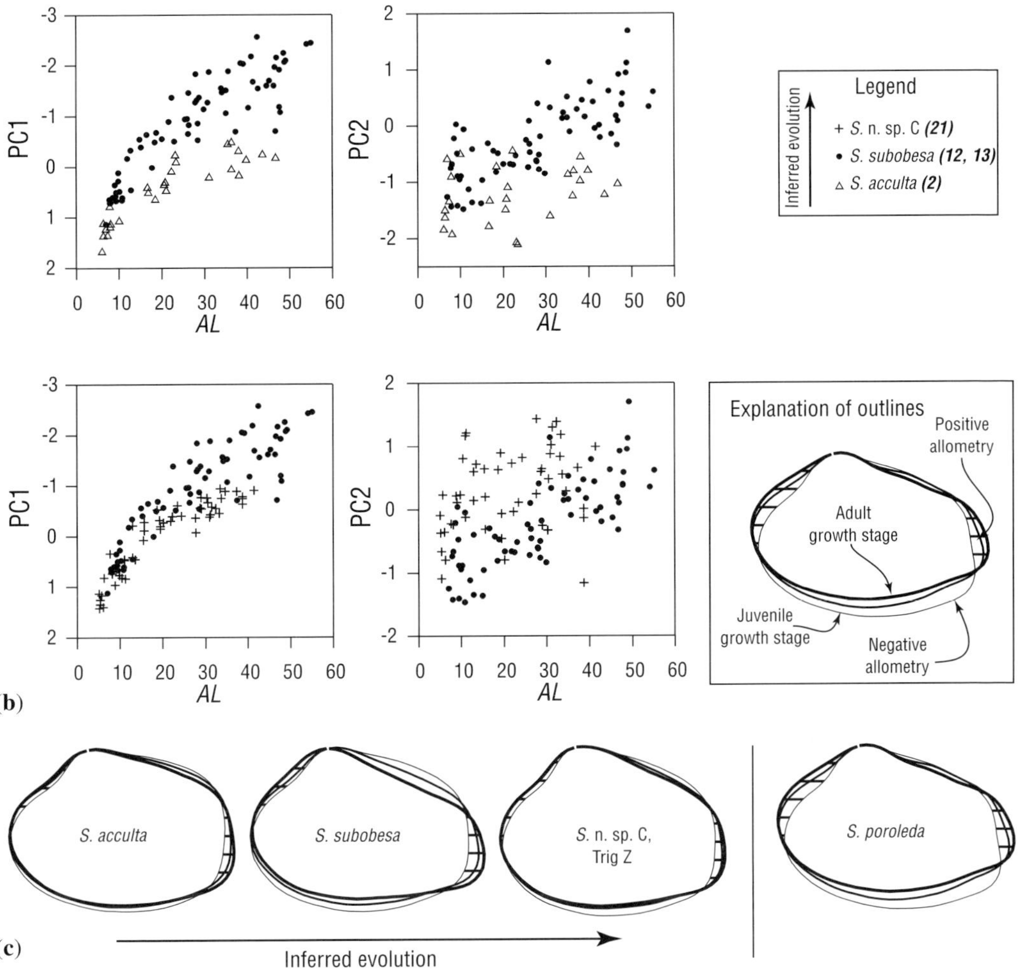

Fig. 7. (**a**, opposite) First three PC axes showing data for latest Eocene–Late Oligocene *Spissatella*. Note that, for clarity, immature ellipses are omitted from all plots and juvenile ellipses are omitted from the lower plot. (**b**) Data for *S. acculta*, *S. subobesa* and *S.* n. sp. C (collection GS1473) ordinated on PC v. *AL* (i.e. size) plots for PC 1 and 2. These three species are inferred to form an evolutionary lineage. (**c**) Overlain, size-normalized average outlines for three growth stages each of *S. acculta*, *S. subobesa*, *S.* n. sp. C and *S. poroleda*, showing regions of positive and negative allometry on the shell margin (cf. Fig. 5c).

surprisingly low and emphasizes the high explanatory importance of size. On PC 2 and 3, the greatest component of variance is explained by differences between collections (55 and 50%, respectively) and overall size–shape correlations are relatively unimportant to negligible. Taken together, the weighted sums for the three PC show that close to half (45%) of the total variance in the PCA is explained by size and over one-third (37%) relates to differences between collections. Only a minor part, < 7%, is explained by intracollection scatter between individuals. Overall, the residual or unexplained variance is low (11%). Further analysis of the residuals reveals that, for each collection, there is no significant correlation between the residual and the fitted PC score; in other words, the variance or scatter of the individual points about the size–shape regression does not change significantly with size. This both satisfies an assumption of the ANOVA and indicates that the shape of *Spissatella* does not become more or less variable with growth.

These results can be visualized by reference to Fig. 11, which shows average ontogenetic trajectories for all collections of *Spissatella*. On an ordination of PC 1 v. PC 2, all ontogenetic

Table 2. *Results of pairwise discriminant function analyses (DFA) for immature growth stages of all latest Oligocene–earliest Miocene collections, based on harmonics 2–6*

	GS1473 (***21***)	SCIP039 (***22***)	GS7166, SCIP040 (***23***)
SCIP039 (***22***)	0.000 (F = 13.872, v_1 = 10, v_2 = 23)	–	–
GS7166, SCIP040 (***23***)	0.000 (F = 5.767, v_1 = 10, v_2 = 37)	0.000 (F = 8.698, v_1 = 10, v_2 = 37)	–
SCIP041 (***24***)	0.000 (F = 6.528, v_1 = 10, v_2 = 30)	0.000 (F = 12.732, v_1 = 10, v_2 = 30)	0.626 (F = 0.804, v_1 = 10, v_2 = 44)

The null hypothesis states that any two collections are drawn from the same population. The first value in each cell is the probability that the null hypothesis is true.

Table 3. *Results of pairwise discriminant function analyses (DFA) for adult growth stages of selected collections of* S. trailli *and* S. clifdenensis

		S. trailli		*S. clifdenensis*	
		GS11154 (***14***) l. Aquitanian–e. Burdigalian	GS1160, SCIP006 (***17***) late Burdigalian(a)	GS10344 (***5***) late Burdigalian(b)	GS10365 (***6***) late Burdigalian(c)
S. trailli	GS11283 (***15***) l. Aquitanian–e. Burdigalian	0.275 (F = 1.326, v_1 = 10, v_2 = 23)	–	–	–
S. clifdenensis	GS11182 (***4***) late Burdigalian(a)	–	0.005 (F = 3.986, v_1 = 10, v_2 = 19)	0.000 (F = 5.994, v_1 = 10, v_2 = 20)	–
	GS10344 (***5***) late Burdigalian(b)	–	–	–	0.679 (F = 0.711, v_1 = 8, v_2 = 19)

Analyses based on harmonics 2–6 except for ***5*** v, ***6*** which, because of small sample sizes, is based on harmonics 2–5. The null hypothesis states that any two collections are drawn from the same population. The first value in each cell is the probability that the null hypothesis is true.

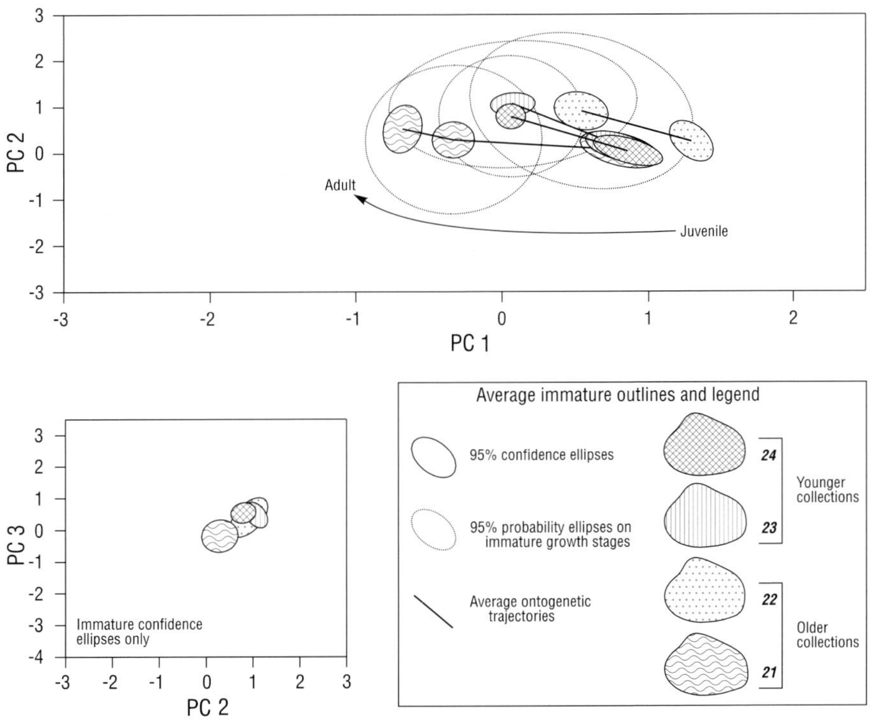

Fig. 8. First three PC axes showing data for latest Oligocene–earliest Miocene *S.* n. sp. C. Ninety-five per cent probability ellipses are shown for the immature size class on the upper plot. For clarity, probability and juvenile confidence ellipses are omitted from the lower plot. Larger specimens are lacking or uncommon in some collections and hence comparisons are based largely on the immature size class. Also shown in the legend are average immature outlines for each collection. Based on associated planktonic foraminifera, the four collections are grouped into two older and two younger samples. Collections are compared statistically in Table 2.

trajectories, except for one, are approximately parallel and show monotonic juvenile to adult trends from right to left. Juvenile confidence ellipses cluster within a relatively small region of morphospace. The oldest definite *Spissatella* examined, *S.* n. sp. A aff. *S. media* (***19***), is highlighted and lies close to the intersection point of nearly all trajectories. In contrast, on a plot of PC 2 v. PC 3, ontogenetic trajectories are widely divergent and show no consistent trend, even within trajectories.

Correlations between facies and shape

At the level of individual species, the present data cannot resolve whether or not ecophenotypic variation is an important influence on the shape of *Spissatella*. This is because it is difficult to eliminate other potential sources of variation and study facies effects in isolation. For example, in the latest Oligocene–earliest Miocene collections illustrated in Fig. 8, shape variation is correlated with both facies and geological age, and it is unclear whether either, both or neither of these factors influenced shape variation. To resolve this question would require a number of very well dated collections from a variety of facies, material that is currently not available.

At a larger scale, however, there are very clear correlations between immature–adult shape, substrate type and inferred palaeoenvironment (Fig. 12). Collections from sandy facies and/or inferred shallow-water environments display a much greater variance than those from siltstone facies and/or inferred mid- to outer-shelf settings. Furthermore, low scores on PC 1, corresponding to the largest and most inequilateral shells with elongate posterior margins, are restricted to shallow-water *Spissatella*.

Comparisons with other closely related genera

Figure 13 shows data for a number of other, closely related, fossil and Recent crassatellids that were included in the analysis for comparative purposes.

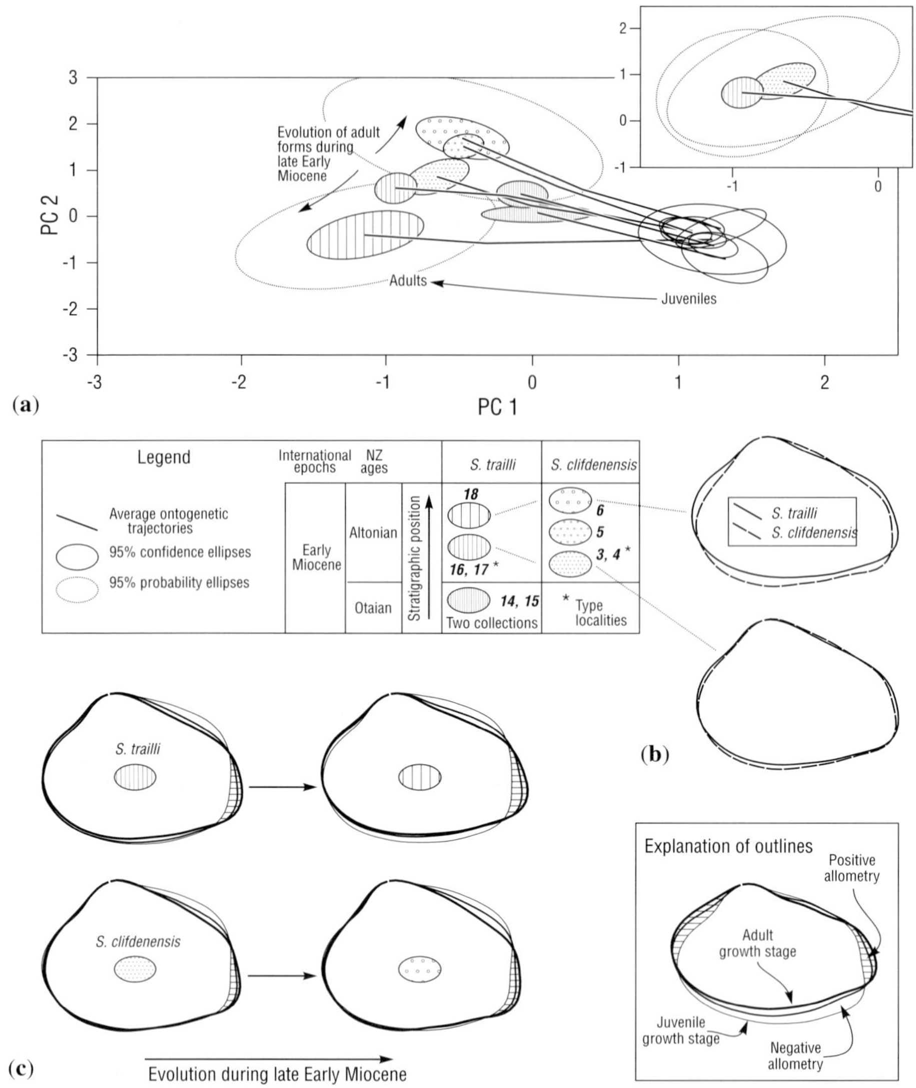

Fig. 9. (**a**) First two PC axes showing data for middle–late Early Miocene *S. trailli* and *S. clifdenensis*. Adult 95% probability ellipses are shown for the youngest collections of each species and, inset, for the topotype collections. These two species are not discriminated on PC 3 and, for simplicity, a plot of PC 2 v. PC 3 is omitted. (**b**) Overlain size-normalized average adult outlines for oldest (lower outlines) and youngest late Early Miocene collections of *S. trailli* and *S. clifdenensis*. The older outlines are very similar (although statistically different), whereas the younger outlines are quite distinct. (**c**) Overlain, size-normalized average outlines for three growth stages each of older and younger late Early Miocene *S. trailli* and *S. clifdenensis*. These reveal evolutionary changes in allometry through this interval of time.

Five of these are species of *Eucrassatella*, including the type species *E. kingicola*. Two of these, *E. scopalveus* and *E.* n. sp., were formerly included within *Spissatella* but are herein referred to *Eucrassatella* on the basis of shape, their relatively thick shells and their commarginal sculpture that becomes obsolete on the flanks at an early stage of growth. All five species of *Eucrassatella* have ontogenetic trajectories that distinguish them from *Spissatella*, although this is based on single individuals for three of the species. In *Eucrassatella*, ontogenetic trajectories originate in an area of morphospace not occupied by juvenile *Spissatella*, and the two sets of trajectories have

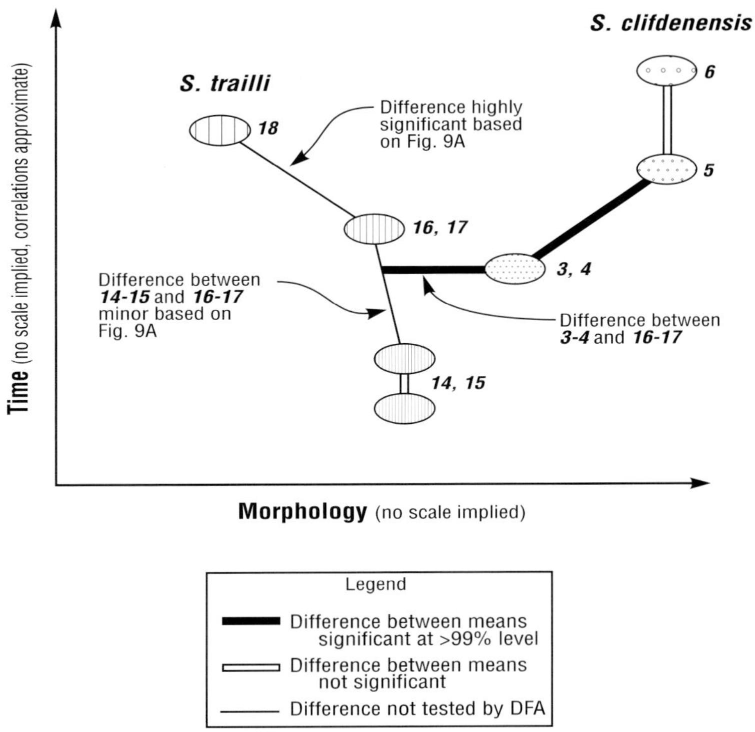

Fig. 10. Hypothesized evolutionary relationships between studied collections of *S. trailli* and *S. clifdenensis*. Collections are identified using same fill patterns as in Fig. 9a and using numeric labels of Table 1. Values of statistical significance are given in Table 3.

different orientations in morphospace (cf. Figs 11 and 13). These results suggest that genus-level differences in development apparently affect early stages of ontogeny that are not studied here (i.e. at the larval or immediately post-larval stage) and result in different 'starting conditions' for subsequent ontogenetic trajectories.

Also shown is the 95% confidence ellipse for the juvenile (and only) size class of *Salaputium animula*. This small species has been regarded as a separate clade derived by paedomorphosis from the common ancestor of *Spissatella* or *Crassatella* (Finlay 1930). *S. animula*, however, occupies morphospace that is well removed from the general ontogenetic trajectory of *Spissatella* and hence is unlikely to have been derived from a *Spissatella*-like ancestor by simple paedomorphosis acting on the shape of the growing margin. This conclusion is corroborated by differences in microsculpture between these genera (unpublished data). The evolutionary origins of *Salaputium* remain unknown, but it is possible that it evolved either from a different crassatellid group outside the New Zealand region or by innovation early in ontogeny.

Discussion

Three results demonstrate that, throughout much of the lifespan of *Spissatella*, shape variation is very highly correlated with size and is explained largely by a consistent pattern of allometric growth across all species within the genus. In other words, there was little evolution of size without concomitant evolution in adult shape, and vice versa. This pattern is expressed as approximately parallel development of individual and species-averaged ontogenetic trajectories in multidimensional morphospace. Differences between species primarily involve the magnitude of mean ontogenetic shape and size change along their average trajectories. In addition, the shapes of the geologically oldest (and smallest) species and the juvenile stages of all younger species are highly conservative and trend towards a single point in morphospace. Together, these results demonstrate a correspondence between individual and species ontogenies and the evolution of adult form, and suggest that heterochrony has been the major evolutionary mechanism throughout the lifespan of the genus. Early

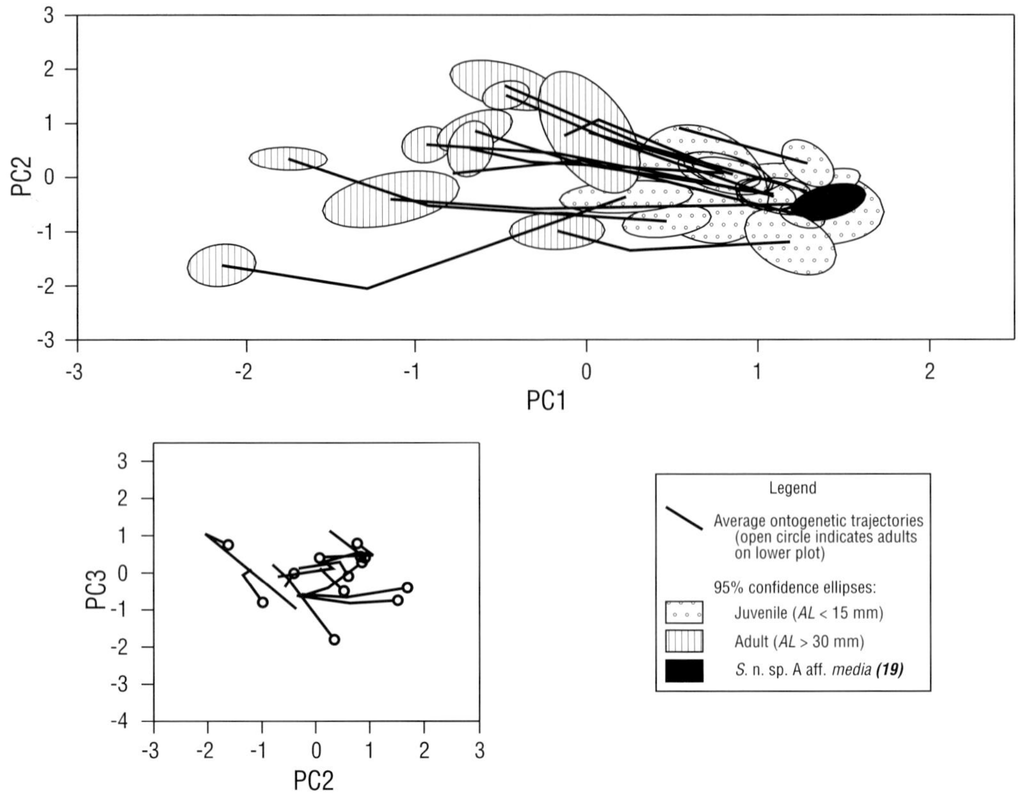

Fig. 11. First three PC axes showing average ontogenetic trajectories for all collections of *Spissatella*. Ninety-five per cent confidence ellipses for juveniles and adults (but not for the immature size class) are shown on the upper plot but have, for clarity, been omitted from the lower plot.

evolution was apparently dominated by peramorphosis. Apart from this early expansion into morphospace during the middle Late Oligocene, no other conspicuous directional trends in shape or size evolution have been identified and the genus records repeatedly convergent evolution of shape. These patterns suggest that paedomorphosis and peramorphosis occurred in approximately equal frequencies from the latest Oligocene onwards.

As noted in the introduction, the burial depth of

Table 4. *Results of analysis of variance (ANOVA) of linear relationships between shape, as measured along each of three principal component (PC) axes, and the axial length of the outline (*AL*, i.e. size)*

	Total variance %			
	Size	Collections	Individuals	Residual
PC 1	63.2	29.8	2.6	4.4
PC 2	17.7	54.7	11.0	16.6
PC 3	0.3	47.9	18.4	33.4
Weighted sum	45.2	37.4	6.5	10.9

The ANOVA was computed for *Spissatella* data only. For each PC, the analysis identifies the proportion of total shape variance that is explained by: (1) overall size–shape correlations; (2) differences in size–shape correlations between collections; (3) differences between individuals from each collection and the mean regression for that collection; and (4) residual or unexplained variations. The weighted sum is the sum across the three PCs, each component being weighted by its relative contribution to the PCA (i.e. weighted by its eigenvalue as a proportion of the sum of the first three eigenvalues). The weighted sums reveal, for the PCA as a whole, the proportions of total shape variance that are explained by size–shape correlations, differences between collections and intracollection variations.

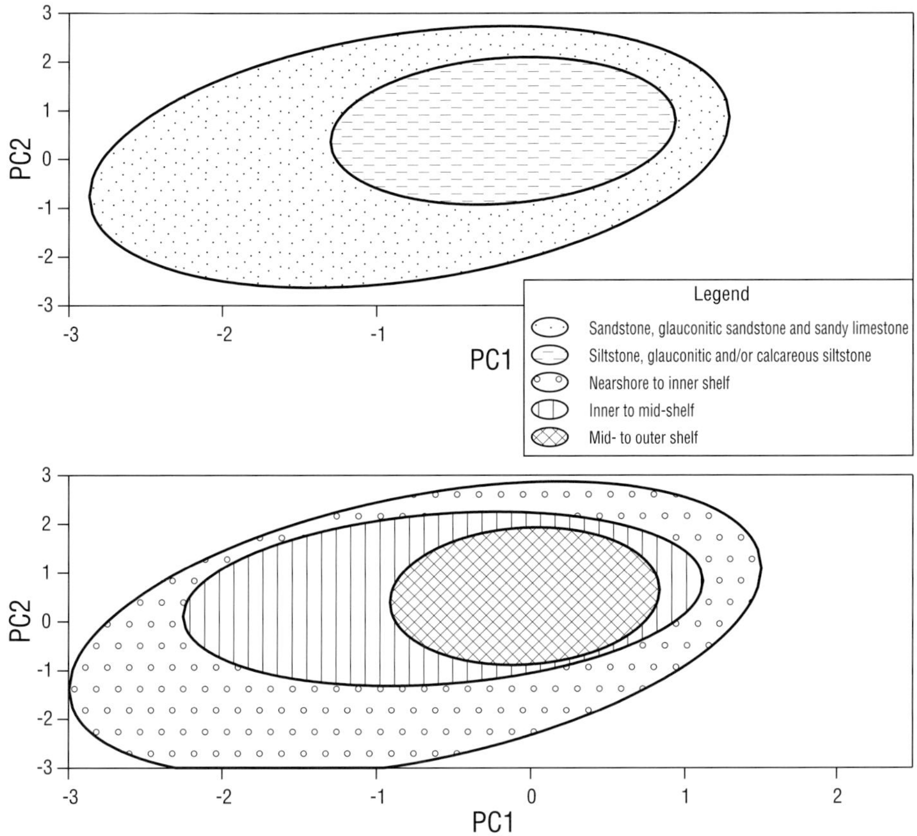

Fig. 12. First two PC axes showing 95% probability ellipses based on data for all immature and adult *Spissatella* plotted by facies (upper plot) and inferred palaeoenvironment (lower plot). Facies and inferred palaeoenvironments for each collection are summarized in Table 1.

Spissatella was limited by the overall length of the shell. Hence, the general pattern of posterior elongation with increasing size has a clear functional interpretation. Increasing the length of the posterior margin allowed the animal to bury the visceral bulk of the shell comparatively deeply whilst retaining contact with the water. This, together with an overall increase in size and weight, would have enhanced stability in shifting substrates, and under energetic conditions (Stanley 1970). Furthermore, the compressed, rostrate shape of the elongate posterior probably reduced current scour of sediment from around the exposed part of the shell (Stanley 1977, p. 881). Therefore, the dominant component of ontogenetic size–shape variation in *Spissatella* is interpreted here as an adaptation for life in coarser substrates and more energetic environments. It is inferred that selection targeted increasing size, increasing elongation, or (most probably) both traits simultaneously. This result is consistent with the observation that the largest, most elongate species were restricted to the coarsest substrates in the shallowest waters.

Clearly, however, posterior elongation was not the only successful adaptation to increasing energy, since species from sandstone facies and inferred shallow waters have the greatest overall interspecific variance, occupy the full range of *Spissatella* morphospace, and include small, equilateral, juveniles and adults. It is likely that such forms gained adaptive advantage from an ability to burrow and rebury rapidly following disinterment, as has been inferred for living inter- and subtidal *Donax* (Soares *et al.* 1998; see also Stanley 1970, p. 58).

In contrast to the coarser facies, species from siltstones that were deposited in relatively stable, mid- to outer-shelf palaeoenvironments are all comparatively small in size and occupy a restricted region of morphospace.

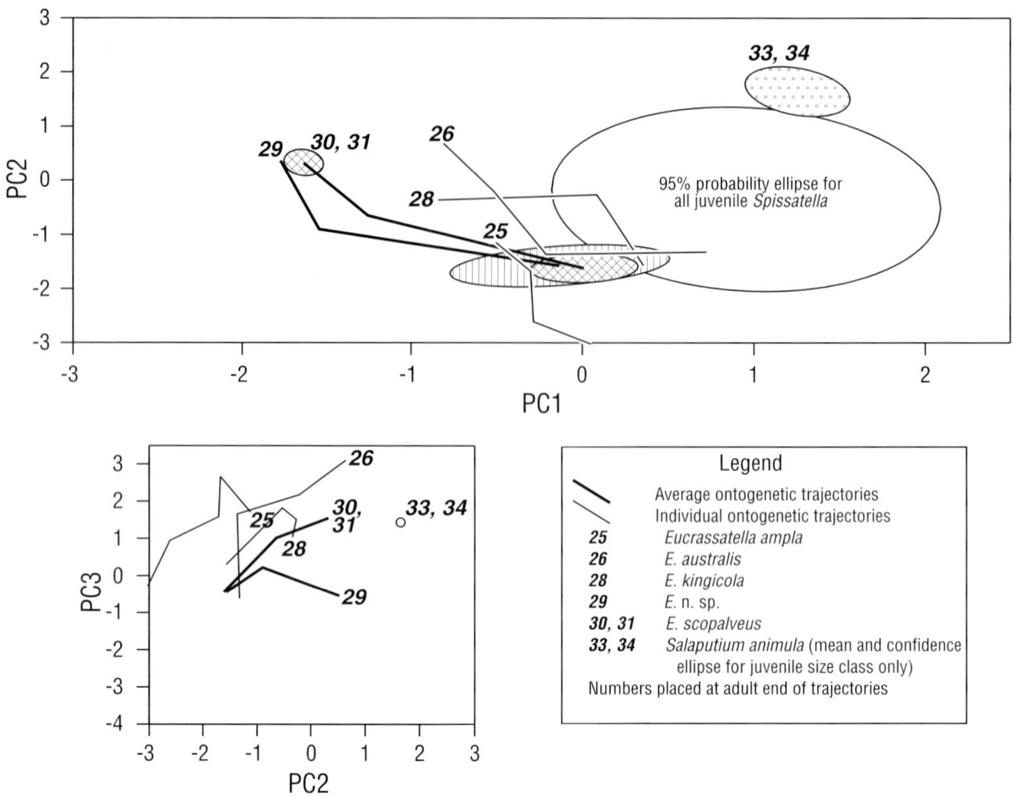

Fig. 13. First three PC axes showing data for species of *Eucrassatella* and *Salaputium animula*. Also shown is the 95% probability ellipse for all juvenile *Spissatella*. For clarity, confidence ellipses are omitted from the lower plot. *E. ampla*, *E. australis* and *E. kingicola* (the type species of the genus) are represented by a single specimen each. *E. scopalveus* and *E*. n. sp. are represented by collections. Both were formerly included within *Spissatella* and are here referred to *Eucrassatella*. Because of the small number of adult specimens, the adult confidence ellipse is omitted from the data for *E*. n. sp. (See text for discussion.)

From a study of 19 lineages of infaunal Neogene bivalves, Stanley & Yang (1987, p. 134) concluded that 'size is much more labile than shape in phyletic evolution' and that this 'constitutes a basic rule of evolution'. They recommended that these two traits be considered separately in the calculation of evolutionary rates. Likewise, Budd & Johnson (1991) inferred that size and shape were decoupled (i.e. behaved independently) during evolution in the gastropod *Strombina*. Both these studies are in accord with the elegant theoretical model of gastropod morphogenesis developed by Rice (1998). He hypothesized that alterations in the 'shape of the aperture map', or allometric changes in the relative magnitude of shell production at different points around the aperture, are more difficult to achieve in evolution than those that require a change in global rate or size parameters. In contrast to these studies, the present results demonstrate that, within Eocene–Miocene *Spissatella*, evolution of size and shape are inextricably linked, they co-vary in a predictable manner along a restricted allometric pathway and they cannot be considered independently. Only one significant evolutionary transition has been identified in this study, where shape changes occur in the absence of overall adult size changes, namely in the divergence of late Early Miocene *S. trailli* and *S. clifdenensis*. If, as implied by Stanley & Yang (1987), size and shape typically behave independently during evolution of infaunal bivalves, then *Spissatella* represents a significant departure from the norm.

Similarly, numerous authors have postulated a general correspondence between decreasing body size with increasing environmental disturbance (*r*-selection) and vice versa (*K*-selection) (e.g. McKinney 1986; Chiba 1998). Although the concept of *r–K* selection is overly simplistic, some predictions such as the general relationship of body size to environmental stability remain consistent with the recent 'intermediate disturbance model' of evolution proposed by McKinney & Allmon (1995). For molluscs, such a relationship has been inferred, for example, in Ordovician nuculoid bivalves (Snyder & Bretsky 1971) and in Miocene–Recent columbellid gastropods (Budd & Johnson 1991). Again, the present results are not in accord with this general hypothesis. The largest species of *Spissatella* were apparently restricted to the coarsest substrates in the shallowest waters: environments that are likely to have been relatively high energy and unstable.

It is considered here that the contradictions of the previous two paragraphs are inevitable consequences arising from the interplay of rigid developmental constraints combined with functional selection for life in shallow water. Maynard Smith *et al.* (1985, p. 265) defined developmental constraints as 'biases on the production of variant phenotypes or limitations on phenotypic variability caused by the structure, character, composition, or dynamics of the developmental system'. In the present context, three key developmental constraints are recognized in *Spissatella*; these are historical or phylogenetic constraints *sensu* Gould (1989) [and *local* constraints *sensu* Maynard Smith *et al.* (1985)]. The first two are the lack of a siphon in Crassatellidae and the 'primitive', approximately equilateral, shape of early *Spissatella*. These constraints define the 'starting condition' for evolution within the genus. The third and most important constraint, inferred by the present authors, is the ontogenetic pathway followed by all *Spissatella*; it is this that caused the genus to 'break the rules' described above. The ontogenetic constraint acted to channel variance. Within each species, variance orthogonal to the mean ontogenetic trajectory was negligible, as recorded by the low residuals in the ANOVA, and directional selection apparently targeted available variation parallel to the ontogenetic pathway. Given these conditions and selective pressure to adapt to shallow-water, high-energy environments, *Spissatella* 'responded' along the only pathway available: peramorphic elongation of the posterior margin with increasing size. Selection may have targeted increasing size, increasing posterior elongation or, quite probably, both traits simultaneously. In this way the present results are in accord with Jones & Gould (1999), who stress that 'constraints' may be positive in that selection may target two (or more) correlated and 'constrained' traits, each of which promotes evolutionary advantages of the other. The developmental constraint described here in *Spissatella* is certainly not universal to all non-siphonate infaunal bivalves, e.g. as evidenced by the spectacular morphological diversity of the trigoniids (e.g. Stanley 1977).

Conclusions

- The lateral outline shape of infaunal, burrowing bivalves is a key determinant of burrowing speed and depth and is, therefore, subject to strong selection pressure. This study seeks to understand major evolutionary processes and controls on outline shape throughout the lifespan of the infaunal, non-siphonate, crassatellid bivalve *Spissatella*.
- Fourier shape analysis is a useful morphometric technique for the description of outline shape in landmark-poor bivalve outlines. The method can be used to construct a morphological 'map', to project ontogenetic trajectories onto this map, to generate synthetic average and extreme morphologies, and to visualize marginal growth allometries.
- Growth in *Spissatella* was strongly allometric. Ontogenetic changes in the shapes of individual shells are, in many cases, far greater than total evolutionary changes within the genus throughout *c.* 20 Ma. Furthermore, ontogenetic trajectories for different species are essentially parallel in morphospace and vary primarily in length, i.e. they involve consistent patterns of allometric growth around the margin of the shell, but these growth fields differ in magnitude. In general, ontogenetic patterns involve: (1) moderate to weak negative allometry on the postero-dorsal margin; (2) moderate to negligible negative allometry on the ventral margin; and (3) moderate to strong positive allometry on the posterior margin. To varying degrees, all *Spissatella* display increasing elongation of the posterior and become increasingly inequilateral with growth.
- It is inferred here that posterior elongation in *Spissatella* was a functional adaptation for increased depth of burial and stability in relatively coarse substrates and in relatively high-energy, shallow marine environments.
- Preliminary comparisons with the genera *Eucrassatella* and *Salaputium* suggest that the orientation of mean ontogenetic trajectories in morphospace may be a key genus-level taxonomic character in Crassatellidae and, by inference, other bivalve groups. Genus-level differences in development apparently affect early stages of ontogeny (i.e. at the larval or immediately post-larval stage) and result in different 'starting conditions' for the ontogenetic transitions studied here.

- Based on the correspondence between species ontogenies and evolutionary changes in adult form, it is inferred here that heterochrony was the dominant evolutionary mechanism throughout the lifespan of *Spissatella*. Following an early peramorphic morphological expansion, both paedomorphosis and peramorphosis probably occurred in approximately similar proportions. The data reveal no other directional trends in size or shape evolution. Contrary to some other studies, it is generally not possible to decouple patterns of shape and size evolution in the genus.
- Stratigraphic resolution within the present data set is too imprecise to reliably quantify species evolutionary rates, and to distinguish gradual evolution from punctuated evolution and random walk [e.g. see recent discussion in Roopnarine *et al.* (1999)]. The results do, however, support models of stasis in *S. trailli* during the middle Early Miocene and in *S. clifdenensis* during the late Early Miocene. In contrast, variability within *S.* n. sp. C during the latest Oligocene–earliest Miocene may have resulted from random walk, evolution in response to a fluctuating selection regime, ecophenotypic effects, or some combination of these factors.
- It is concluded that the ontogenetic pathway followed by *Spissatella* represents a rigid developmental constraint that largely controlled morphological evolution within the genus.

This paper is Institute of Geological and Nuclear Sciences (IGNS) contribution 1781. This research was supported in part by the Marsden Fund project entitled 'Heterochrony, evolution and extinction in Mesozoic and Cenozoic Mollusca', administered by the Royal Society of New Zealand and funded by the Foundation for Research, Science and Technology. The Commonwealth Science Council generously contributed funds that enabled one of us (JSC) to attend the meeting 'Biology and Evolution of the Bivalvia' held in Cambridge, UK, September 1999. This work is based, to a great extent, on the Fourier software developed by J. Haines (University of Cambridge). For assistance with some of the statistical tests, we would like to thank D. Rhoades (IGNS), E. Abrahams (National Institute of Water and Atmospheric Research) and G. Scott (IGNS). Thanks also to G. Scott for foraminiferal age determinations. Early drafts of the manuscript were substantially improved by the thorough reviews of A. Beu (IGNS), M. Foote (University of Chicago), G. Scott and A. Swan (Kingston University). Roger Tremain digitized approximately half the outlines. The photographs of Fig. 2 were prepared by W. St George.

References

ALTMAN, D. G. 1978. Plotting probability ellipses. *Applied Statistics*, **27**, 347–349.

ATCHLEY, W. R. 1987. Developmental quantitative genetics and the evolution of ontogenies. *Evolution*, **41**, 316–330.

BEU, A. G. & MAXWELL, P. A. 1990. Cenozoic Mollusca of New Zealand. *New Zealand Geological Survey Paleontological Bulletin*, **58**, 1–518.

——, GRIFFIN, M. & MAXWELL, P. A. 1997. Opening of Drake Passage gateway and Late Miocene to Pleistocene cooling reflected in Southern Ocean molluscan dispersal: evidence from New Zealand and Argentina. *Tectonophysics*, **281**, 83–97.

BLACKSTONE, N. W. & YUND, P. O. 1989. Morphological variation in a colonial marine hydroid: a comparison of size-based and age-based heterochrony. *Paleobiology*, **15**, 1–10.

BREY, T. & MACKENSEN, A. 1997. Stable isotopes prove shell growth bands in the Antarctic bivalve *Laternula elliptica* to be formed annually. *Polar Biology*, **17**, 465–468.

BUDD, A. F. & JOHNSON, K. G. 1991. Size-related evolutionary patterns among species and subgenera in the *Strombina* group (Gastropoda: Columbellidae). *Journal of Paleontology*, **65**, 417–434.

CHIBA, S. 1996. A 40,000-year record of discontinuous evolution of island snails. *Paleobiology*, **22**, 177–188.

—— 1998. A mathematical model for long-term patterns of evolution: effects of environmental stability and instability on macroevolutionary patterns and mass extinctions. *Paleobiology*, **24**, 336–348.

CRAMPTON, J. S. 1995. Elliptic Fourier shape analysis of fossil bivalves: some practical considerations. *Lethaia*,**28**, 179–186.

—— & HAINES, A. J. 1996. Users' manual for programs HANGLE, HMATCH, and HCURVE for the Fourier shape analysis of two-dimensional outlines. *Institute of Geological and Nuclear Sciences Science Report*, **96/37**, 1–28.

DARRAGH, T. A. 1965. Revision of the species of *Eucrassatella* and *Spissatella* in the Tertiary of Victoria and Tasmania. *Proceedings of the Royal Society of Victoria*, **78**, 95–114.

FINLAY, H. J. 1926. New shells from New Zealand Tertiary beds: part 2. *Transactions of the New Zealand Institute*, **56**, 227–258.

—— 1930. Notes on recent papers dealing with the Mollusca of new Zealand. *Transactions of the New Zealand Institute*, **61**, 248–258.

FOOTE, M. & COWIE, R. H. 1988. Developmental buffering as a mechanism for stasis: evidence from the pulmonate *Theba pisana*. *Evolution*, **42**, 396–399.

GOULD, S. J. 1977. *Ontogeny and Phylogeny*. Harvard University Press, Cambridge.

—— 1989. A developmental constraint in *Cerion*, with comments on the definition and interpretation of constraint in evolution. *Evolution*,**43**, 516–539.

HAINES, A. J. & CRAMPTON, J. S. 2000. Improvements to the method of Fourier shape analysis as applied in morphometric studies. *Palaeontology*, in press.

HALL, B. K. 1992. *Evolutionary Developmental Biology*. Chapman & Hall, London.

IREDALE, T. 1924. Results from Roy Bell's molluscan collections. *Proceedings of the Linnean Society of New South Wales*, **49**, 179–278.

JACKSON, D. A. 1993. Stopping rules in principal components analysis: a comparison of heuristical and statistical approaches. *Ecology*, **74**, 2204–2214.

JOLIFFE, I. T. 1986. *Principal Component Analysis*. Springer-Verlag, New York.

JONES, D. S. 1988. Sclerochronology and the size versus age problem. *In:* MCKINNEY, M. L. (ed.) *Heterochrony in Evolution: A Multidisciplinary Approach*. Plenum Press, New York, 93–108.

—— & GOULD, S. J. 1999. Direct measurement of age in fossil *Gryphaea*: the solution to a classic problem in heterochrony. *Paleobiology*, **25**, 158–187.

KONDO, Y. 1987. Burrowing depth of infaunal bivalves – observation of living species and its relation to shell morphology. *Transactions and Proceedings of the Palaeontological Society of Japan, New Series*, **148**, 306–323.

MCKINNEY, M. L. 1986. Ecological causation of heterochrony: a test and implications for evolutionary theory. *Paleobiology*, **12**, 282–289.

—— 1988. Classifying heterochrony: allometry, size and time. *In:* MCKINNEY, M. L. (ed.) *Heterochrony in Evolution: A Multidisciplinary Approach*. Plenum Press, New York, 17–34.

—— & ALLMON, W. D. 1995. Metapopulations and disturbance: from patch dynamics to biodiversity dynamics. *In:* ERWIN, D. H. & ANSTEY, R. L. (eds) *New Approaches to Speciation in the Fossil Record*. Columbia University Press, New York, 123–183.

—— & MCNAMARA, K. J. 1991. *Heterochrony; The Evolution of Ontogeny*. Plenum Press, New York.

MARWICK, J. 1926. Molluscan fauna of the Waiarekan Stage of the Oamaru Series. *Transactions of the New Zealand Institute*, **56**, 307–316.

—— 1929. Tertiary molluscan fauna of Chatton, Southland. *Transactions of the New Zealand Institute*, **59**, 903–934.

MATHSOFT 1998. S-Plus statistical software, version 5.0. MathSoft, Inc. Seattle, Washington.

MAYNARD SMITH, J., BURIAN, R., KAUFFMAN, S. *ET AL.* 1985. Developmental constraints and evolution. *Quarterly Review of Biology*, **60**, 265–287.

MORGANS, H. E. G., SCOTT, G. H., BEU, A. G., GRAHAM, I. G., MUMME, T. C., ST GEORGE, W. & STRONG, C. P. 1996. New Zealand Cenozoic timescale (version 11/96). *Institute of Geological and Nuclear Sciences Science Report*, **96/38**, 1–12.

OPTIMAS CORPORATION 1995. *Optimas 5 Software*. Optimas, Washington.

POLLY, P. D. 1998. Variability, selection, and constraints: development and evolution in viverravid (Carnivora, Mammalia) molar morphology. *Paleobiology*, **24**, 409–429.

PRITCHARD, G. B. 1903. Contributions to the palaeontology of the Older Tertiary of Victoria. Lamellibranchs. Part 3. *Proceedings of the Royal Society of Victoria*, **15**, 87–103.

RICE, S. H. 1998. The bio-geometry of mollusc shells. *Paleobiology*, **24**, 133–149.

ROHLF, F. J. 1993. *NTSYS-pc; Numerical Taxonomy and Multivariate Analysis System; Version 1.80*. Exeter Software, Setauket.

ROOPNARINE, P. D., BYARS, G. & FITZGERALD, P. 1999. Anagenetic evolution, stratophenetic patterns, and random walk models. *Paleobiology*, **25**, 41–57.

SLACK-SMITH, S. M. 1998. Superfamily Crassatelloidea. *In*: BEESLEY, P. L., ROSS, G. J. G. & WELLS, A. (eds) *Mollusca: The Southern Synthesis. Fauna of Australia. Volume 5*. CSIRO Publishing, Melbourne, 325-328.

SNYDER, J. & BRETSKY, P. W. 1971. Life habits of diminutive bivalve molluscs in the Maquoketa Formation (Upper Ordovician). *American Journal of Science*, **271**, 227–251.

SOARES, A. G., CALLAHAN, R. K. & DE RUYCK, A. M. C. 1998. Microevolution and phenotypic plasticity in *Donax serra* Röding (Bivalvia: Donacidae) on high energy sandy beaches. *Journal of Molluscan Studies*, **64**, 407–421.

SOKAL, R. R. & ROHLF, F. J. 1995. *Biometry,* 3rd Edition. W. H. Freeman, New York.

STANLEY, S. M. 1970. Relation of shell form to life habits of the Bivalvia (Mollusca). *Geological Society of America Memoir*, **125**, 1–296.

—— 1975. Why clams have the shape they have: an experimental analysis of burrowing. *Paleobiology*, **1**, 48–58.

—— 1977. Coadaptation in the Trigoniidae, a remarkable family of burrowing bivalves. *Palaeontology*, **20**, 869–899.

—— & YANG, X. 1987. Approximate evolutionary stasis for bivalve morphology over millions of years: a multivariate, multilineage study. *Paleobiology*, **13**, 113–139.

SUTER, H. 1917. Descriptions of new Tertiary Mollusca occurring in New Zealand, accompanied by a few necessary changes in nomenclature. *New Zealand Geological Survey Palaeontological Bulletin*, **5**, 1–93.

SWAN, A. R. H. 1988. Heterochronic trends in Namurian ammonoid evolution. *Palaeontology*, **31**, 1033–1051.

TOJO, B. & OHNO, T. 1999. Continuous growth-line sequences in gastropod shells. *Palaeogeography, Palaeoclimatology, Palaeoecology*, **145**, 183–191.

WATTERS, G. T. 1993. Some aspects of the functional morphology of the shell of infaunal bivalves (Mollusca). *Malacologia*, **35**, 315–342.

ZELDITCH, M. L., BOOKSTEIN, F. L. & LUNDRIGAN, B. L. 1992. Ontogeny of integrated skull growth in the cotton rat *Sigmodon fulviventer*. *Evolution*, **46**, 1164–1180.

The Queen Scallop *Aequipecten opercularis*: a new source of information on late Cenozoic marine environments in Europe

A. L. A. JOHNSON[1], J. A. HICKSON[1], J. SWAN[1], M. R. BROWN[2], T. H. E. HEATON[3], S. CHENERY[4] & P. S. BALSON[4]

[1]*Division of Earth Sciences, University of Derby, Derby DE22 1GB, UK (e-mail: A.L.A.Johnson@derby.ac.uk)*

[2]*Division of Computing, University of Derby, Derby DE22 1GB, UK*

[3]*NERC Isotope Geosciences Laboratory, Keyworth, Nottingham NG12 5GG, UK*

[4]*British Geological Survey, Keyworth, Nottingham NG12 5GG, UK*

Abstract: Few data exist on seasonal variation in the temperature (or other aspects of the environment) of late Cenozoic shelf seas in Europe. Ontogenetic records in the shell of *Aequipecten opercularis*, a widespread, fast-growing and typically well-preserved bivalve, are a potential source. Study of modern forms has shown that oxygen stable isotopes are incorporated in equilibrium with surrounding seawater (hence providing a faithful record of temperature) and data from late Holocene *A. opercularis* of the southern North Sea Basin (SNSB) indicate that extreme winter (as well as summer) temperatures are registered. Oxygen isotope data from apparently well-preserved, mid-Pliocene shells of the SNSB indicate seasonal temperatures similar to present, whereas microgrowth increment data suggest substantially warmer conditions, in accordance with other evidence. The balance of evidence thus implies cryptic diagenetic corruption of the isotopic temperature signature in mid-Pliocene shells. However, it would be premature to discount the possibility of cooler mid-Pliocene conditions than currently recognized. Ontogenetic variation in carbon isotopic composition within shells is minor and apparently unrelated to environment, but differences between mid-Pliocene, late Holocene and modern shells probably relate to changes in atmospheric CO_2 concentration. Unlike certain taxa, seasonal variation is not evident in the strontium or magnesium contents of the *A. opercularis* shell, but may be displayed by other trace elements, hence affording (together with seasonal increment width variation) a means of independent temporal calibration of isotope profiles. Ontogenetic records of environment from *A. opercularis* can be complete (and, in the case of increment data, easily recovered) but are of short duration. They must be complemented by (less complete) records from longer lived taxa to obtain the fullest possible environmental history. Seasonal cycles in ontogenetic records afford a means of establishing age and growth rate, and can therefore provide information of value for evolutionary studies and management of living populations.

Concerns over the possible effect on climate of anthropogenic CO_2 emissions to the atmosphere have promoted interest in the extent and cause of recent natural climatic fluctuation, and have hence directed attention to the late Cenozoic interval. Variations in atmospheric CO_2 content have, with little doubt, contributed to changes in global mean temperature during the late Cenozoic, e.g. in the mid-Pliocene when atmospheric CO_2 was apparently 30–35% above the pre-industrial modern level (Kürschner *et al.*1996; Raymo *et al.* 1996) and the global mean temperature an estimated 3.6°C higher (Sloan *et al.* 1996). However, locally and, in particular, in the northeastern North Atlantic area, climatic variation may well have been determined more by changes in the supply of heat through ocean currents than by alterations in atmospheric retention of reflected solar radiation as a result of shifts in CO_2 level. Thus, reduced Gulf Stream heat supply has been plausibly invoked to explain intervals of relative cold during the Pleistocene in the European area (Broecker 1997), while during the mid-Pliocene, Gulf Stream heat supply was apparently substantially enhanced (Dowsett *et al.* 1992) and presumably, at least in part, the cause of warmer conditions on land in Europe (Thompson & Fleming 1996).

Although the geographic pattern of mid-Pliocene warming in the North Atlantic favours enhanced Gulf Stream heat supply, the reduced overall

From: HARPER, E. M., TAYLOR, J. D. & CRAME, J. A. (eds) *The Evolutionary Biology of the Bivalvia*. Geological Society, London, Special Publications, **177**, 425–439. 1-86239-076-2/00/$15.00

latitudinal temperature gradient in the oceans (Dowsett *et al.* 1996) makes this a theoretical improbability (Crowley 1996). The evidence of temperature in the mid-Pliocene North Atlantic (and of ocean temperatures generally) derives from the distribution of various elements of the microbiota and is not immune from question, being based on analogy with the temperature associations of modern representatives of taxa. The same applies to almost all estimates of sea temperature on the European continental shelf in the Pliocene (e.g. Raffi *et al.* 1985; Wood *et al.* 1993; Head 1997, 1998) and Pleistocene (e.g. Funnell *et al* 1979; Gibbard *et al.* 1991; Meijer & Preece 1995); only scant evidence is available from the alternative approach of oxygen isotope thermometry (Buchardt 1978). Lack of secure palaeotemperature estimates for the European continental shelf sea is regrettable because this area, transitional between land and the Atlantic Ocean proper, is surely key to evaluating the influence of variations in Gulf Stream heat supply on climate of the European landmass. From the fact that at present, summer and winter sea temperatures on the Atlantic seaboard of Europe are approximately equally elevated above those (uninfluenced by the Gulf Stream) at the same latitude off North America (National Oceanic and Atmospheric Administration of the United States 1999), it can be expected that enhanced Gulf Stream heat supply would be manifested in shelf temperature increases of about the same magnitude in summer and winter (and likewise diminished supply by approximately equal decreases). It has been predicted, by contrast, that, at the latitude of Europe, markedly discrepant increases in summer and winter atmospheric (and by inference oceanic) temperatures would result from elevated atmospheric CO_2 (Crowley 1991).

It is evident from the above that accurate data on seasonal sea temperature variation in the European area during the late Cenozoic would be of value both to test existing temperature estimates and to investigate the causes of secular change. Such data, together with information on other aspects of environment, are potentially recoverable from chemical and morphological ontogenetic records contained in accretionary skeletons. The Queen Scallop, *Aequipecten opercularis* (L.), has a large, rapidly accreted shell and, as a common component of Miocene–Recent marine faunas in Europe, appears to be a promising subject for research. In this paper the general suitability of *A. opercularis* for provision of environmental information from ontogenetic records is reviewed, and then the quality and implications of data from three specific sources – stable isotopes, microgrowth increments and trace elements – are discussed.

General suitability of *Aequipecten opercularis*

As a scallop, *A. opercularis* (Fig. 1) has a predominantly calcitic shell which, while not guaranteeing faithful preservation of ontogenetic records of environment, is unlikely to suffer the wholesale alteration or dissolution common in aragonitic taxa. Being epibenthic, scallops are more likely to provide a record of general water conditions than infaunal forms, whose carbon isotopic composition, at least, may reflect porewater geochemistry (Krantz *et al.* 1987). Shelf-dwelling scallops are generally fast-growing and *A. opercularis* is no exception, reaching a height of *c.* 40 mm after one year and 55 mm after two years, with growth occurring during the winter months (Ursin 1956; Hickson 1997). This provides, in principle, for a well-resolved and complete record of seasonal environmental variation.

Against these favourable growth characteristics must be set the relatively short lifespan of *A. opercularis* (rarely more than six years; Broom 1976) and the sharp decline in growth rate with age, such that if an animal reaches six years old its growth is almost zero (Taylor & Venn 1978). This precludes recovery of a long continuous history of environment. However, such limited data as exist provide no indication that any other species with an accretionary skeleton, found widely in the late Cenozoic deposits of Europe, combines sub-

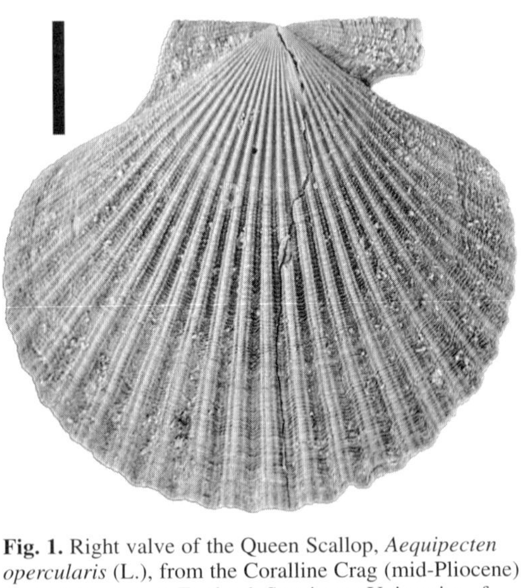

Fig. 1. Right valve of the Queen Scallop, *Aequipecten opercularis* (L.), from the Coralline Crag (mid-Pliocene) of Suffolk, eastern England. Specimen: University of Derby, Division of Earth Sciences (UD) 52795. Scale bar, 10 mm.

stantially greater longevity with significant growth at all times of year. *Mytilis edulis* is quite widespread from early Pleistocene times onward and, according to Seed (1976), lives for up to 16 years. However, winter growth is only *c.* 5 mm in the first year and even less in the second; in view of a decline in annual growth rate, winter growth must be negligible in older animals [see also Jones (1983)]. *Pinna nobilis*, locally common at present in the Mediterranean Sea, reaches 13 years and is still growing rapidly (probably in winter as well as summer) at eight years (Richardson *et al.* 1999). Unfortunately, however, this species appears to be restricted to warm waters, not extending into northern Europe even in the Pliocene (Raffi *et al.* 1985). The infaunal bivalve *Arctica islandica* reaches very much greater ages (200+ years; Jones 1983) but ceases growing in winter (Weidman *et al.* 1994). Evidence from North American species suggests that seasonal growth interruptions may be characteristic of relatively long-lived infaunal bivalves (Jones 1980; Jones *et al.* 1989; Jones & Quitmyer 1996).

A. opercularis is a common and widespread species at present, breeding populations extending from northern Norway around the Atlantic coast of Europe and as far east into the Mediterranean as the Adriatic (Waller 1991; Margus 1991). It thus occupies waters ranging in temperature from an average winter minimum of *c.* 5°C to an average summer maximum of *c.* 24°C (National Oceanic and Atmospheric Administration of the United States 1999). This tolerance predisposes *A. opercularis* to be useful as a tool for documenting temperature in that its own occurrence is unlikely to be affected by this variable, something borne out by the presence of the species in, for instance, cold-water deposits of the Pleistocene in eastern England (Funnell *et al.* 1979) and warm-water deposits of the Pliocene in southern Spain (Aguirre *et al.* 1996). *A. opercularis* also occurs over a substantial depth range (from low water to 183 m) and is tolerant of a wide variety of substrates, being abundant at present on sediments ranging from sandy mud to sandy gravel and shell hash (Tebble 1976).

Although eurytopic in the above respects, *A. opercularis* requires some current flow (Ursin 1956) and is tolerant only of quite small, short-term departures from normal marine salinity (Paul 1980). As a result of the latter, while it may be common in fully marine late Cenozoic sequences [e.g. in the southern North Sea Basin (SNSB), the Pliocene Coralline and Red Crag formations of eastern England, and the Oosterhout Formation of the Netherlands], the species is absent from sequences deposited under strong fluvial influence (e.g. much of the Pleistocene of the Netherlands; Gibbard *et al.* 1991). This feature at least ensures that salinity-related variation in the isotopic composition of ambient water can be largely ruled out as a factor influencing shell isotopic composition, hence simplifying interpretation of isotopic data (see below).

The frequency with which *A. opercularis* occurs in British Geological Survey cores of late Holocene deposits beneath the southern North Sea is in marked contrast to its rarity in surface grab samples from the same area (A. Weller, pers. comm.). Indigenous modern individuals also exhibit more interrupted growth (see below). Given that conditions of temperature, salinity, tides and substrate in modern and late Holocene times were almost identical (Cameron *et al.* 1992), it seems likely that present-day occurrence and growth characteristics in the southern North Sea reflect some form of pollution rather than a natural control.

Although affected by some natural and, apparently, anthropogenic factors, *A. opercularis* exhibits remarkable short-term hardiness which makes it highly suitable for *in vivo* experimental studies. Thus, animals used in the North Sea culture referred to below suffered no ill effects in a 12 h car journey from their source (western Scotland) merely enveloped in wet seaweed and ice. Moreover, growth occurs at a healthy rate even in specimens tightly packed within onion nets, a method of culture used experimentally off western Scotland (J. P. Mikolajunas, pers. comm.). Such tolerance suggests that experiments involving artificial manipulation of conditions to assess the fidelity of ontogenetic records (e.g. temperature control of oxygen isotopic composition and microgrowth increment width; control of shell trace element content by environmental concentration – see discussions below) could be conducted without risk of mortality.

Data from modern, late Holocene and mid-Pliocene shells

Stable isotope data

Although investigations of ontogenetic variation in the stable isotopic composition of the *A. opercularis* shell were initiated only recently (Hickson 1997; Hickson *et al.* 1999), studies of scallop species occurring outside Europe started some time ago (e.g. Krantz *et al.* 1984), and stable isotope data have been used as a source of information on seasonal variation in temperature (including water-column stratification) and the timing of upwelling events, phytoplankton blooms and freshwater influxes (e.g. Krantz *et al.* 1987, 1988; Krantz 1990; Jones & Allmon 1995). All such inferences are based on the assumption of something close to equilibrium incorporation of elemental isotopes into shell carbonate; i.e. incorporation in accordance with their relative abundance in the immediate environment and a fractionation factor which, in the case of oxygen, varies strongly with temperature. At the smallest spatial and temporal scale equilibrium is probably unavoidable, but at larger scales, e.g. where seawater surrounding a shell is taken to be the immediate environment rather than the extrapallial fluid from which it is actually precipitated, the condition does not necessarily hold. Thus, in the scallop *Pecten maximus* (Mitchell *et al.* 1994; C. A. Richardson, pers. comm.) the interposition of some effect relating to the biology of the organism, perhaps slow replenishment from seawater of ions in the extrapallial fluid (resulting in significant 'pollution' by, or reequilibration with, mantle-derived ions; cf. Hickson *et al.* 1999), results in disequilibrium between shell and seawater.

Equilibrium between shell and external environment has been demonstrated convincingly for the scallops *Placopecten magellanicus* (Krantz *et al.* 1984) and *Adamussium colbecki* (Barrera *et al.* 1990) but, given the evidence from *Pecten maximus*, cannot be assumed for other species. Indeed, the whole issue of the detailed mechanisms by which carbonate precipitates, and their effects on its isotopic composition, is a subject of active debate (Spero *et al.* 1997; Zeebe 1999). It was for these reasons that, preparatory to analysis of ancient *A. opercularis*, an investigation of the occurrence of isotopic equilibrium was conducted on modern examples. This involved analysis of shells which had grown under known environmental conditions, the animals concerned either having been maintained in monitored cultures over an autumn–winter period in the southern North Sea or collected live from sites for which environmental data were available from general monitoring programmes. The results have been published elsewhere (Hickson *et al.* 1999) and are only summarized below, attention being concentrated on the implications of data from late Holocene and mid-Pliocene shells.

Oxygen isotopes. Isotopic ratios are expressed in the δ notation, in per mil (‰), where $\delta^{18}O$ (or $\delta^{13}C$) = $[(R_{sample}/R_{standard}) - 1] \times 10^3$. R is the $^{18}O/^{16}O$ (or $^{13}C/^{12}C$) ratio, and the standards are Vienna Pee Dee Belemnite (VPDB) for carbonates and Vienna Standard Mean Ocean Water (VSMOW) for water (Coplen 1994). Under conditions of isotopic equilibrium the difference between the $\delta^{18}O$ value of calcite and the $\delta^{18}O$ value of the water from which it precipitates is dependent on temperature, and for water with a seasonally constant isotopic composition (as is usually the case for fully marine waters) the change in $\delta^{18}O_{calcite}$ as a function of temperature is *c.* –0.24‰ °C^{-1} (O'Neil *et al.* 1969).

For the investigation of modern shells (Hickson *et al.* 1999), seawater temperatures and $\delta^{18}O$ were either measured (for cultured animals) or derived from other records (for indigenous animals). Solutions of the equation of O'Neil *et al.* (1969), relating the difference between $\delta^{18}O_{calcite}$ and $\delta^{18}O_{water}$ as a function of temperature, then yielded predictions of $\delta^{18}O$ for shells growing in isotopic equilibrium. The range of $\delta^{18}O_{shell}$ values from both cultured and indigenous animals (according quite closely with the predicted values), and the pattern of change in the former (increasing $\delta^{18}O_{shell}$ with decreasing temperature), show that something (at least) very near to isotopic equilibrium existed between shells and seawater at the time of precipitation. Indigenous shells (e.g. Fig. 2a) uniformly exhibit $\delta^{18}O$ maxima (i.e. extreme winter values) slightly less than values corresponding to the minimum temperatures experienced. Rather than disequilibrium, this almost certainly reflects a brief interruption of growth during the coldest months because in each case the $\delta^{18}O$ maximum corresponds to a marked 'growth ring' (formed of a group of very small microgrowth increments – see below), which is evidently the culmination of a slowdown in growth (shown by the relatively sharp increase in $\delta^{18}O$ against shell height in Fig. 2a).

Data from a late Holocene shell from the SNSB (Fig. 2b) describe a strikingly different winter record, with an asymptotic approach to the $\delta^{18}O$ maximum which is greater (reflecting a lower temperature) than in any modern indigenous shell; no growth ring is developed. The same features are observed in isotopic profiles from the first two years of growth of other late Holocene shells, ranging in age from *c.* 1 to 3 Ka, and only rarely is there a significant growth ring associated with the maximum $\delta^{18}O$ value (Hickson 1997). This evidence, which is matched by data from cultured shells, shows that, under appropriate circumstances (evidently not abnormally warm winter conditions), *A. opercularis* does not merely exhibit growth in the overall winter period but actually grows significantly in the coldest months. Given that at present in the southern North Sea, where winter growth interruptions are observed, *A. opercularis* is uncommon (presumably as a result of some unfavourable feature of the environment; see above), it may be that, under conditions which support larger populations, growth during the coldest months is the norm, at least in the first year or two of life.

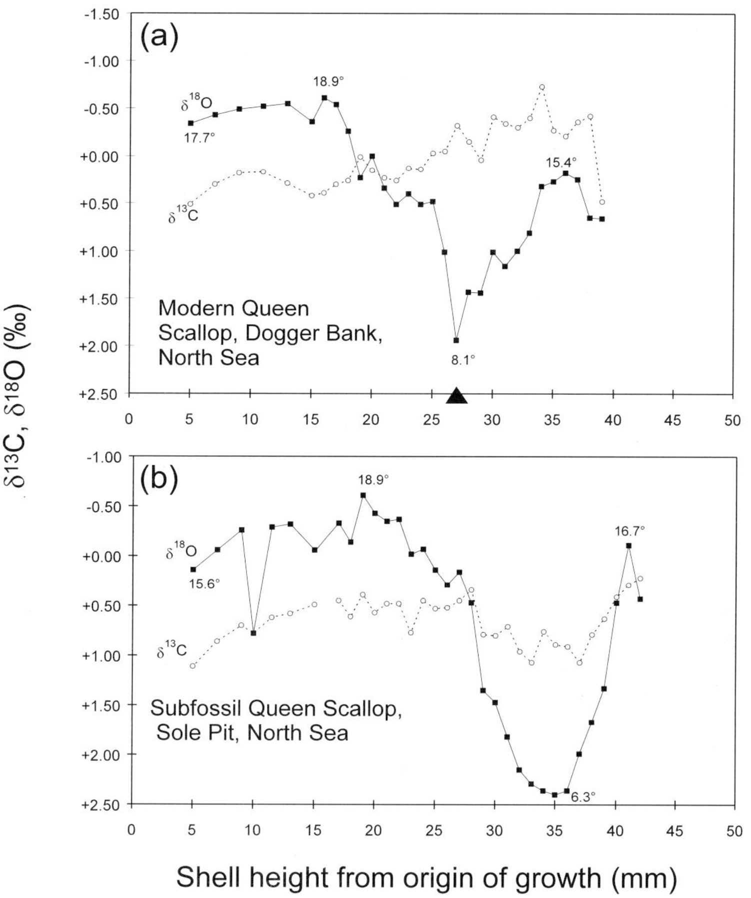

Fig. 2. Examples of oxygen ($\delta^{18}O$) and carbon ($\delta^{13}C$) stable isotopic compositions of: (**a**) indigenous modern; (**b**) subfossil (late Holocene) Queen Scallops from the southern North Sea. In accordance with convention, δ values are shown decreasing upwards on the *y*-axis. High $\delta^{18}O$ at a shell height of 10 mm in (b) almost certainly reflects an instrumental aberration. Temperatures shown were calculated from $\delta^{18}O$ values using the equation of O'Neil *et al.* (1969) and $\delta^{18}O_{water}$ = +0.1‰ (v. VSMOW). ▲, Position of growth ring (growth cessation). Uncorrected ^{14}C AMS date of late Holocene specimen: 2760 ± 50 years BP (Lab. Code AA-27134). Specimens: (a) Naturmuseum Senckenberg DOGD 25; (b) British Geological Survey Zt 9954 [PB4 of Hickson (1997)].

Data (Fig. 3) from two apparently well-preserved mid-Pliocene specimens of *A. opercularis* from the SNSB exhibit a cyclicity similar to that seen in data from younger shells, indicating a primary (temperature) control. However, whilst assemblage analysis and the few existing isotopic data (see above) suggest that mid-Pliocene temperatures were substantially higher than now, the temperatures represented by both $\delta^{18}O$ maxima and minima in Fig. 3 are not significantly greater than typical winter (5.5–7.5°C) and summer (16–18°C) extremes in the current southern North Sea (Lane & Prandle 1996). Shell UD 52797 (Fig. 3b) provides evidence of slightly warmer extreme winter temperatures (8.8°C) but the summer extreme is indistinguishable from present, and that indicated by shell UD 52796 (Fig. 3a) is, at 13.0°C, substantially below typical modern extreme summer temperatures.

These last two findings could be explained by the existence of a summer growth halt (resulting in failure to register the highest temperatures experienced), as in warm water populations of the bivalve *Mercenaria* (Jones & Quitmyer 1996). However, while the 'peaks' (in fact, relatively low $\delta^{18}O$ values) in the isotope profiles are rather narrower than those from late Holocene and modern shells (Fig. 2), such growth rings as exist are conspicuously not coincident with them (Fig. 3a). Other than the rather remote possibility that previous

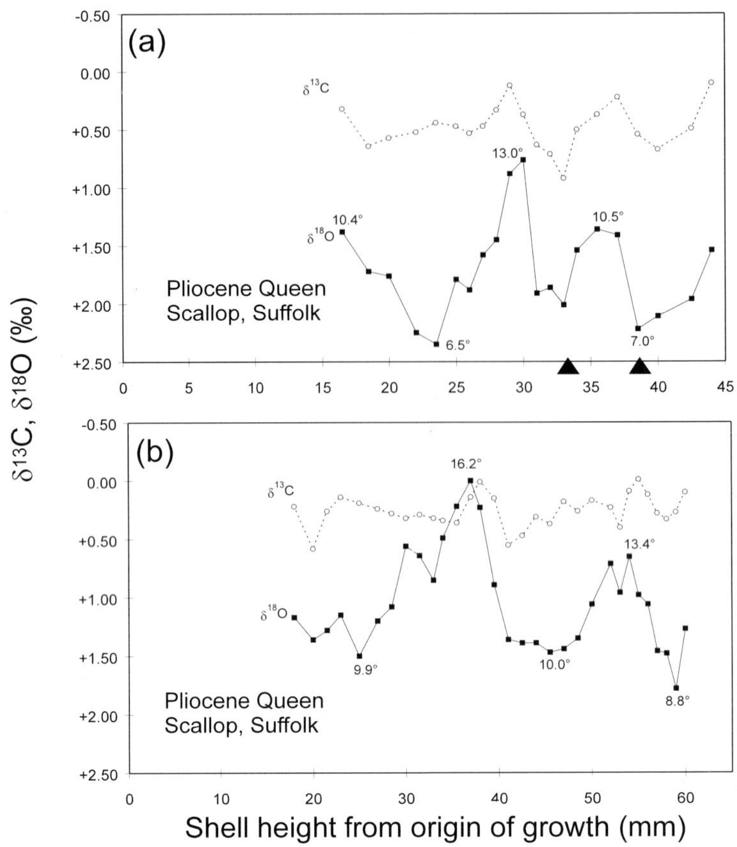

Fig. 3. Isotopic profiles from mid-Pliocene (Coralline Crag) Queen Scallops from eastern England. Format as in Fig. 2. In calculating temperatures, no correction was made for possible ice-volume effects due to a lack of consensus over the most appropriate value (cf. Jones & Allmon 1995). All suggested corrections would result in a lowering of calculated temperatures. Specimens: (a) UD 52796; (b) UD 52797.

temperature estimates are all markedly in error, two potential explanations remain: that the analysed shells lived during anomalously cool years or that they underwent cryptic diagenetic alteration, resulting in corruption of the isotopic signature. The latter possibility is supported by evidence from microgrowth increment widths (see below) which indicate warmer conditions than present.

Carbon isotopes. Isotope mass balance equations relating to ^{13}C and ^{12}C, and incorporating appropriate fractionation factors, indicate that, unlike other marine bivalve species, a maximum of 20% of carbon in the shell carbonate of *A. opercularis* is derived from metabolic sources (Hickson *et al.* 1999). This, in turn, suggests that something close to isotopic equilibrium exists at the time of precipitation between shell carbon and dissolved inorganic carbon (DIC) in seawater. There is no measurable temperature effect on $^{13}C/^{12}C$ fractionation between DIC and carbonate (Romanek *et al.* 1992), hence, given isotopic equilibrium, variations in $\delta^{13}C_{shell}$ will reflect changes in $\delta^{13}C_{DIC}$. Short-term variations of $\delta^{13}C_{DIC}$ in shelf seas relate to factors such as local phytoplankton blooms (causing increased $\delta^{13}C_{DIC}$ by removal into tissue of a relatively high proportion of ^{12}C) and influxes of freshwater or deep-marine waters (typically causing reduced $\delta^{13}C_{DIC}$ by introduction of water relatively rich in ^{12}C from the decay of organic matter). Over the longer term, changes in $\delta^{13}C$ of atmospheric CO_2 (caused partly by variations in the rate of oxidation of organic matter and hence related to the atmospheric concentration of CO_2) might be expected to influence $\delta^{13}C_{DIC}$ by air–sea exchange (Charles *et al.* 1993), although the relatively large size of the oceanic carbon reservoir would tend to damp fluctuations. That changes in atmospheric CO_2 content are indeed reflected in $\delta^{13}C_{DIC}$, and recorded in skeletal calcium carbonate, is shown by the reduction in $\delta^{13}C$ of shallow–depth foraminiferal and sponge material over the period of significant anthropogenic addition to atmospheric CO_2; apparently, preservation of the atmospheric signal is due to incomplete mixing of shallow and deeper waters (Beveridge & Shackleton 1994; Böhm *et al.* 1996).

Unlike records from modern scallops of the US Atlantic Coast (Krantz *et al.* 1988), $\delta^{13}C$ data from modern to mid-Pliocene *A. opercularis* of the SNSB exhibit little within shell variation, rarely > 1‰ over the lifespan (or period of North Sea growth) of the organism (Figs 2 and 3; Hickson 1997; Hickson *et al.* 1999). This is consistent with evidence (from direct measurement or the composition of biotic assemblages) of relatively invariant salinities at the locations of the shells concerned and the considerable distance of the southern North Sea (depth generally < 40 m) from any substantial area of significantly deeper water (hence eliminating the possibility of major upwelling effects). A spring phytoplankton proliferation occurs, but unlike many other areas (including the northern North Sea) is not immediately followed by a depletion of resources and population crash (Tett & Walne 1995). It is possible that this protracted 'bloom' causes little perturbation of $\delta^{13}C_{DIC}$ and consequently of $\delta^{13}C_{shell}$.

With respect to longer term changes, the modern shell represented in Fig. 2a and the mid-Pliocene shells represented in Fig. 3 all exhibit $\delta^{13}C$ values which are generally less than those of the late Holocene shell represented in Fig. 2b. This pattern is confirmed by comparison of mean values for all data available from modern and late Holocene (Hickson 1997) and mid-Pliocene (Fig. 3) shells, [−0.04 (n = 197), +0.79 (n = 209) and +0.34 (n = 43), respectively]. It is in accordance with, for the present time, evidence from direct measurement and, for the Pliocene, consistent proxy evidence (Kürschner *et al.* 1996; Raymo *et al.* 1996) of relatively high atmospheric CO_2 concentrations. In view of rather limited ontogenetic variation it appears, therefore, that $\delta^{13}C$ values of *A. opercularis* shell could be used to chart late Cenozoic changes in atmospheric CO_2 content, although independent evidence of phytoplankton levels (e.g. from trace element data; see below) and the extent of upwelling would be necessary to obtain an accurate picture.

Microgrowth increment data

Except in cases of extreme abrasion, the shell exterior of *A. opercularis* displays a conspicuous record of the pattern of fine-scale accretion in the form of microgrowth increments of varying width (Fig. 4). Regular examination of animals grown under semi-natural conditions has revealed that, at certain times of year (May–July), the number of increments deposited approaches, but does not exceed, the number of days elapsed (Broom & Mason 1978). Therefore, it may be that each increment corresponds to one day's growth. Even if this is so (a contrasting interpretation has been made in the case of similar increments in *Pecten maximus*; Gruffydd 1981), it cannot be the case that growth occurs every day because at other times of year there is a marked discrepancy between the number of days and increments. Broom & Mason (1978) also recorded a marked seasonal variation in the width of microgrowth increments. They suggested that this is determined by food supply as

Fig. 4. Enlarged view of the ventral part of the shell illustrated in Fig. 1, showing microgrowth increments of varying width but no growth ring (i.e. group of very small microgrowth increments marking a growth cessation). Area outlined is enlarged further and analysed in Fig. 6. Scale bar, 4 mm.

much as temperature, but their data (Broom & Mason 1978, fig. 5), which show increasing increment widths during May and June whilst the phytoplankton standing crop underwent a major reduction, suggest that temperature is the primary control, at least when food supply is above the minimum for body maintenance.

Measurement of increment widths in North Sea

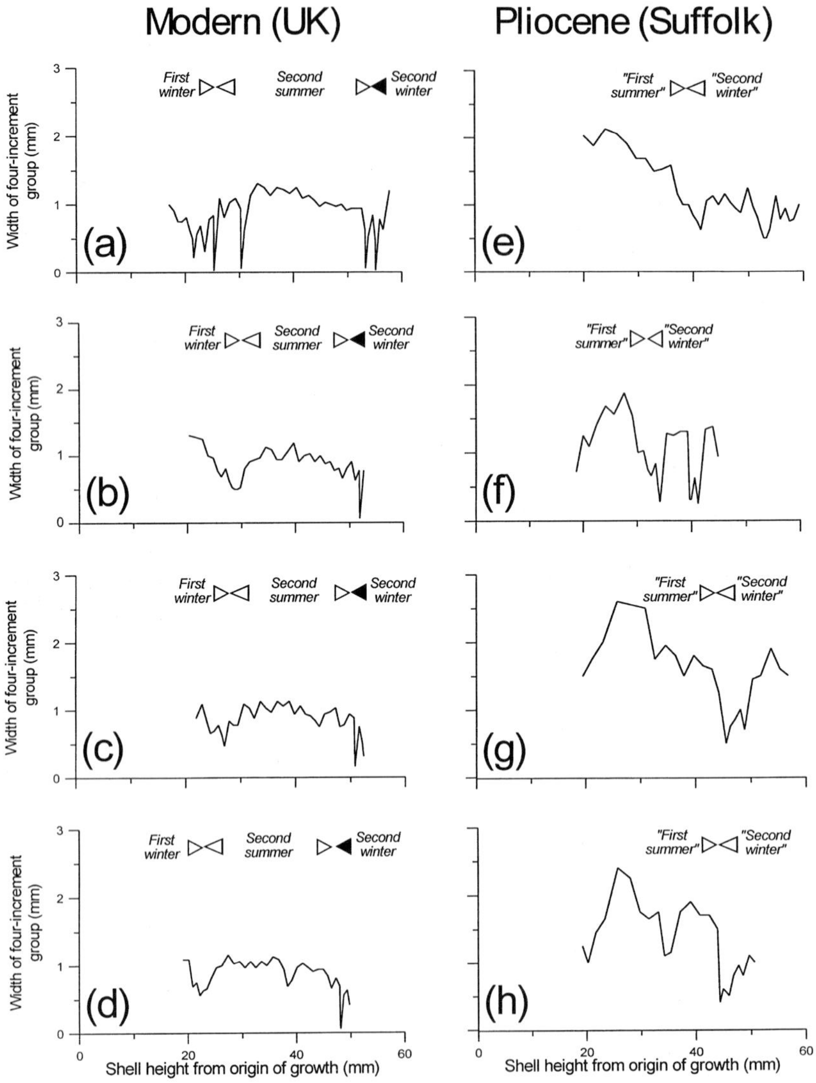

Fig. 5. Profiles of microgrowth increment width in modern [(**a**)–(**d**)] and mid-Pliocene [(**e**)–(**h**)] Queen Scallop shells, measured using a microscope fitted with an eyepiece graticule. Modern animals had been spawned in spring–summer 1993 and cultured on the west coast of Scotland until October 1994, when they were transferred to nets in the southern North Sea until February 1995, alongside animals spawned in spring–summer 1994. The pattern of growth in the latter, together with published information on growth rates (e.g. Broom & Mason 1978; Paul 1981), confirms identification of seasons represented in former prior (open symbol) to second winter (solid symbol) growth. Interpretation of seasons in mid-Pliocene shells assumes autumn spawning. Specimens (all UD): (**a**) 52791; (**b**) 52792; (**c**) 52793; (**d**) 52794; (**e**) 52796; (**f**) 52797; (**g**) 52798; (**h**) 52799. Profiles in (e) and (f) cannot be precisely compared with isotopic profiles from the same specimens (Fig. 3) because 'shell height' in increment plots is derived by addition of increment widths; hence, even a slight systematic error in measurement of the latter will result in a significant departure from true shell height at large sizes.

cultured shells (see above) confirms the existence of a seasonal cycle of variation; a similar pattern is evident amongst mid-Pliocene shells from eastern England (Fig. 5), including the two (Fig. 5e and f) for which isotope data are available. The mid-Pliocene shells, however, show markedly higher increment width maxima and minima (presumably corresponding to extreme summer and winter temperatures) and, in association, higher average increment widths (range of means for four-increment groups: 1.00–1.55 for Pliocene shells and 0.82–0.86 for modern shells). On the basis of the increment width–temperature relationship, this indicates substantially warmer conditions in the mid-Pliocene, a finding in agreement with all previous views (see above) and at variance with isotopic evidence (Fig. 3) from the shells represented in Fig. 5e and f.

While these results suggest that growth increment data from *A. opercularis* might provide an independent means of determining secular temperature change through the late Cenozoic, the approach requires testing by analysis of modern forms from different temperature regimes. The specific matter of the accuracy of increment-based temperature estimates for the mid-Pliocene could be investigated with data from Mediterranean populations, which experience temperatures similar to those generally inferred for the SNSB in the mid-Pliocene. Even if it should prove unjustifiable to attach absolute temperatures to increment width values, it seems very likely that increment widths will supply a reliable indication of season and hence provide the basis for an independent temporal scale to calibrate isotope profiles. Such calibration is of considerable value for isotope profiles from fossil shells, where it is never entirely possible to exclude the possibility of cryptic diagenesis and the existence of spurious peaks and troughs (providing misleading indications of seasonal temperature variation). Increment width data might also be used to distinguish between growth rings resulting from traumatic causes (e.g. storms or attempted predation, both of which are likely to be represented by an abrupt reduction in the width of microgrowth increments) and those growth rings due to cessations of growth associated with summer or winter extremes of temperature (likely to be preceded by gradual reduction of increment widths).

Not only are microgrowth increments in *A. opercularis* generally well preserved, width data can be recovered easily and quite quickly by direct measurement under the microscope. Even more rapid data acquisition can be envisaged by application of image-analysis software. Figure 6 shows a 'brightness' trace for the area of shell outlined in Fig. 4. The edges of increments are faithfully identified as major peaks and it is a simple matter to translate the positions of these into a plot depicting increment width variation (Fig. 7). Use of this technology [see Chavaud *et al.* (1998) and Diou *et al.* (1999) for similar approaches] offers the opportunity to obtain really large data sets, a boon for investigations of palaeotemperature where substantial variation about the mean is to be expected and a small body of data might give a

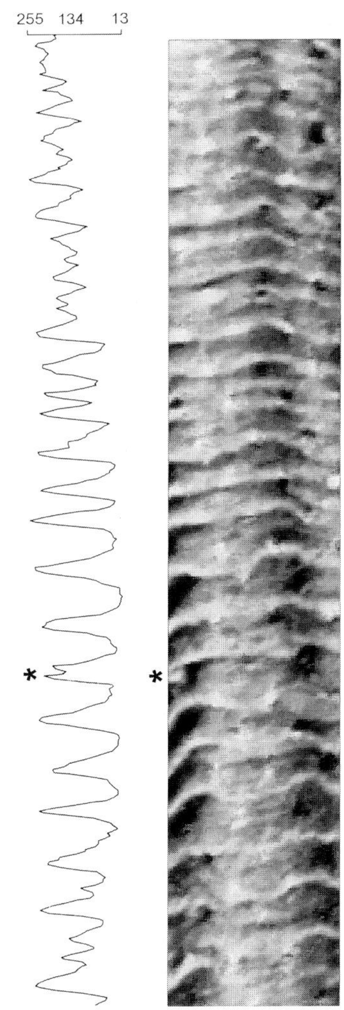

Fig. 6. Enlargement of area outlined in Fig. 4 with plot of 'brightness' (arbitrary scale) measured along left edge of display. Major peaks, in general, correspond precisely with edges of microgrowth increments. Double peak at * is caused by presence of adherent sediment particle; such 'noise' might be removed by use of an appropriate wavelength filter. Image analysis using Wit Version 5.2 (Logical Vision).

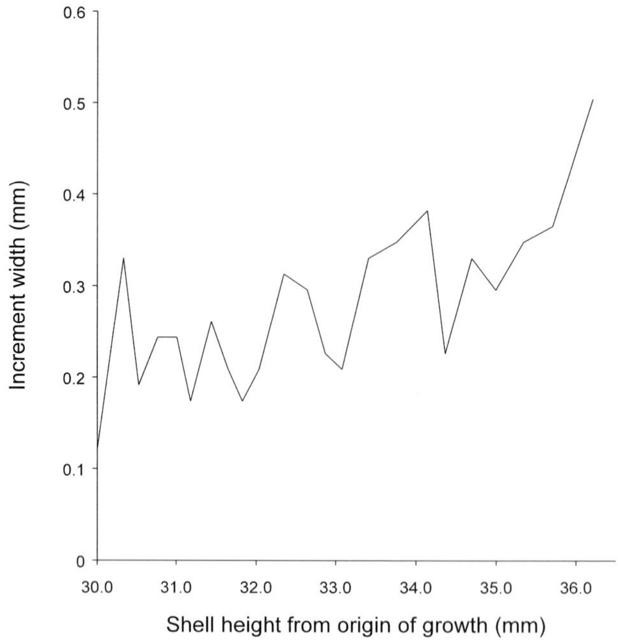

Fig. 7. Plot of increment widths for area of shell analysed in Fig. 6; produced by automatic measurement of distances between major peaks (defined here as those with a brightness value > 200). The plot clearly reflects the overall change in increment widths evident in the display in Fig. 6.

misleading impression (cf. interpretation of oxygen isotope data from mid-Pliocene *A. opercularis*).

At least on the right valve (against the substrate), the record of microgrowth increments is usually obscured by abrasion on the earliest formed *c.* 15 mm of shell, representing the first 4–8 months of life (Paul 1981). While this shortens the increment record that can be recovered from a shell, the effect is the same on an isotope record obtained (as conventional for scallops) from the outermost part of the outer shell layer. This is because material deposited beneath the shell surface, as sampled on an abraded shell, is not contemporaneous with surface material (marking the advance of the shell edge) and, in any case, may not have been precipitated in isotopic equilibrium with seawater (Hickson *et al.* 1999).

Trace element data

In certain invertebrate taxa, the concentrations in skeletal materials of the trace elements magnesium and strontium show some correlation with temperature (Rosenberg 1980, 1990), although the control is probably metabolic rate (strongly influenced by temperature in invertebrates) rather than temperature itself (Rosenberg & Hughes 1991). For taxa with accretionary skeletons in which the relationship applies there is thus the prospect of identifying seasons, if not absolute temperatures of growth. In the same way as microgrowth increment evidence of season, this would be a valuable complement to stable isotope data.

The magnesium and strontium contents of four North Sea cultured *A. opercularis* shells were investigated using the recently introduced technique of laser ablation inductively coupled plasma mass spectrometry; by reason of the tiny sample sizes involved, this facilitates construction of very detailed ontogenetic profiles (e.g. Fuge *et al.* 1993). Contrary to what had been hoped, profiles of strontium concentration show almost no fluctuation and while magnesium displays some cyclicity this is of much shorter than seasonal period (Fig. 8; cf. Leng & Pearce 1999). Evidently, therefore, strontium and magnesium concentrations in *A. opercularis* shell show no correlation with temperature and cannot provide independent temporal calibration of isotope profiles. (Note: magnesium profiles would be of some use if the cause and/or temporal basis of cyclicity, currently unknown, could be established.)

Other trace elements may yet prove to be useful indicators of the season of shell growth. The concentrations of cobalt, copper, iron, lead, manganese, nickel and zinc all vary seasonally in

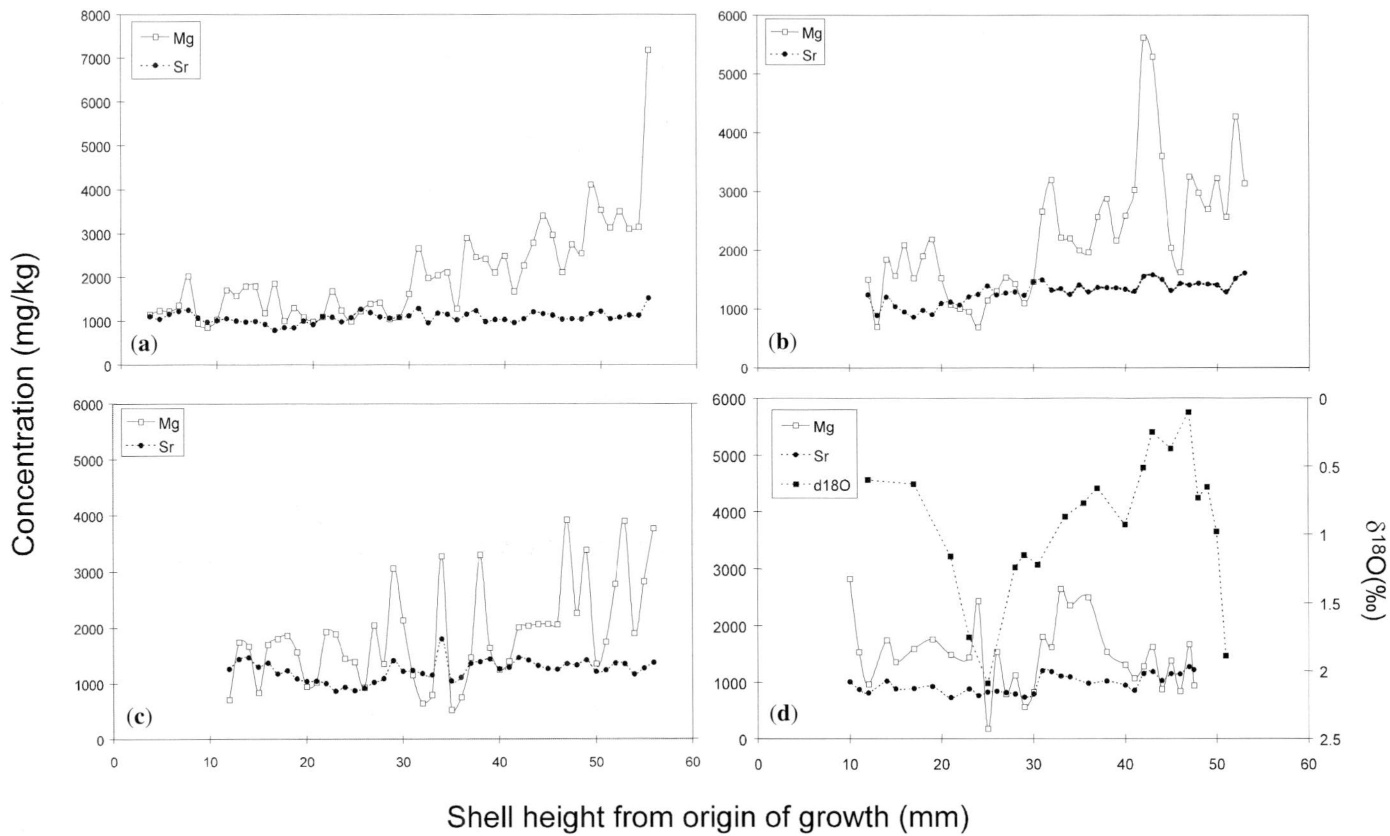

Fig. 8. Profiles of magnesium and strontium concentration for four Queen Scallops initially cultured on the west coast of Scotland and then transferred to the southern North Sea. Oxygen isotope data included in (d), showing summer–winter seasonality, demonstrate that variation in Mg content of this specimen [and those in (a)–(c)] does not have a seasonal basis. Specimens (all UD): (**a**) 52800 (S2/5); (**b**) 52801 (N4/A2); (**c**) 52802 (N4/A28); (**d**) 52803 (N4/A9). Alphanumeric codes in brackets were used by Hickson (1997) and Hickson *et al.* (1999).

soft tissues of *A. opercularis*, probably in relation to sequestering by phytoplankton (Bryan 1973), and the fact that ontogenetic profiles from shells of other scallop species exhibit seasonal variation in concentration of the trace element cadmium, again probably in relation to phytoplankton uptake (Krantz *et al.* 1988), makes it reasonable to expect that concentrations of elements in the former group would show seasonal variation in the *A. opercularis* shell.

Conclusions and wider applications

A. opercularis is well suited for investigation of late Cenozoic marine environments in Europe. The species has many propitious features, amongst the most important of which are equilibrium (or near) stable isotope incorporation, and the capacity to furnish several detailed and independent ontogenetic records of environment from an individual shell, hence enabling checks for accuracy. It is not, however, a perfect tool: it has the notable defect of a relatively short lifespan which prevents acquisition of long-term records. In this respect, it is distinctly inferior to the bivalve *Arctica islandica* which lives to over 200 years and has been dubbed 'the tree of the North Atlantic' (Thompson & Jones 1977; Weidman *et al.* 1994). However, unlike this species, *A. opercularis* exhibits rapid, year-round growth which enables recovery of well-resolved, complete records of seasonal variation in environment (principally temperature). Combined use of these species represents a means of constructing the fullest possible environmental history.

With respect to the particular major issues identified at the outset, carbon isotope data from mid-Pliocene *A. opercularis* of the SNSB are consistent with the notion of relatively high levels of atmospheric CO_2 and hence of potential 'greenhouse' warming of atmosphere and oceans. Microgrowth increment data appear to indicate higher summer and winter sea temperatures than present but, until the increment width–temperature relationship is calibrated through further research on modern forms, the relative magnitudes of summer and winter temperature increases cannot be determined, and hence the role of greenhouse heating (as v. Gulf Stream heat supply) cannot be fully evaluated. Indeed, further research on modern forms is necessary to confirm that growth increments can provide a reliable indication of differences in *general* temperature regime. Oxygen isotope data indicate mid-Pliocene sea temperatures little different from now in the SNSB. The balance of evidence suggests that these are erroneous estimates, resulting from cryptic diagenetic alteration of the original isotopic signature. However, there are no more direct indications that this is the case; hence, pending results from further studies of growth increments in modern forms, it cannot be ruled out that previous notions of relatively high sea temperatures in the SNSB during the mid-Pliocene are significantly in error.

Information on the age and growth rate of organisms is of fundamental importance for management of commercially exploited populations and for investigations of the evolutionary role of heterochrony (changes in the rate and times of onset–cessation of developmental processes). Seasonal cycles in oxygen isotopic composition of skeletal materials have been used as a basis for obtaining such information in both contexts (e.g. Jones 1988; Dare & Deith 1991). However, while seemingly reliable, the oxygen isotope approach demands substantial investment of time and money. Microgrowth increment analysis has been adopted in the management of some scallop fisheries (Dare 1995), but the recent analysis by Jones & Gould (1999) of annual increments in the oyster *Gryphaea* is one of very few instances where any kind of increment size data has been employed in evolutionary studies. At least for scallops, readily obtainable microgrowth increment data offer the prospect of answering many questions concerning the role of heterochrony in evolution (e.g. Johnson 1984).

Stable-isotope analysis was carried out at the NERC Isotope Geosciences Laboratory, Keyworth, under an award of services to ALAJ (IP/417/0994). We thank H. Sloane and C. Arrowsmith for assistance with analysis of carbonates and waters, respectively. The investigation was mainly conducted during tenure by JAH of a NERC studentship (GT4/94/322). Additional support for trace element analysis (performed at the British Geological Survey, Keyworth) was provided by the University of Derby. We thank J. P. Mikolajunas (formerly at Seafish Industry Authority, Ardtoe) for supply of specimens used in North Sea culture, R. Janssen (Naturmuseum Senckenberg, Frankfurt) for supply of North Sea indigenous specimens and R. Pouwer (Netherlands Institute of Applied Geoscience, Utrecht) for provision of information on the occurrence of Pliocene *A. opercularis* in the Netherlands. We thank A. Weller (British Geological Survey, Keyworth) and C. A. Richardson (University of Wales, Bangor) for permission to quote unpublished findings relating to *A. opercularis and Pecten maximus*, respectively.

References

AGUIRRE, J., BRAGA, J. C., JIMÉNEZ, A. P. & RIVAS, P. 1996. Substrate-related changes in pectinid fossil asemblages. *Palaeogeography, Palaeoclimatology, Palaeoecology*, **126**, 291–308.

BARRERA, E., TEVESJ, M. J. S. & CARTER, J. G. 1990. Variations in oxygen and carbon isotopic compositions and microstructure of the shell of

Adamussium colbecki (Bivalvia). *Palaios*, **5**, 149–159.

BEVERIDGE, N. A. S. & SHACKLETON, N. J. 1994. Carbon isotopes in recent planktonic foraminifera: A record of anthropogenic CO_2 invasion of the surface ocean. *Earth and Planetary Science Letters,* **126**, 259–273.

BÖHM, F., JOACHIMSKI, M. M., LEHNERT, H., MORGENROTH, G., KRETSCHMER, W., VACELET, J. & DULLO, W.-C. 1996. Carbon isotope records from extant Caribbean and South Pacific sponges: Evolution of $\delta^{13}C$ in surface water DIC. *Earth and Planetary Science Letters,* **139**, 291–303.

BROECKER, W. S. 1997. Thermohaline circulation, the Achilles heel of our climate system: Will man-made CO_2 upset the current balance? *Science*, **278**, 1582–1588.

BROOM, M. J. 1976. *Synopsis of Biological Data on Scallops* Chlamys (Aequipecten) opercularis *(Linnaeus),* Argopecten irradians *(Lamarck),* Argopecten gibbus *(Linnaeus).* Food and Agriculture Organization of the United Nations, Fisheries Synopsis, **114**.

—— & MASON, J. 1978. Growth and spawning in the pectinid *Chlamys opercularis* in relation to temperature and phytoplankton concentration. *Marine Biology,* **47**, 277–285.

BRYAN, G. W. 1973. The occurrence and seasonal variation of trace metals in the scallops *Pecten maximus* (L.) and *Chlamys opercularis* (L.). *Journal of the Marine Biological Association of the United Kingdom*, **53**, 145–166.

BUCHARDT, B. 1978. Oxygen isotope palaeotemperatures from the Tertiary period in the North Sea area. *Nature,* **275**, 121–123.

CAMERON, T. J., CROSBY, A., BALSON, P. S., JEFFREY, D. H., LOTT, G. K., BULAT, J. & HARRISON, D. J. 1992. *The Geology of the Southern North Sea*. HMSO, for the British Geological Survey.

CHARLES, C. D., WRIGHT, J. D. & FAIRBANKS, R. G. 1993. Thermodynamic influences on the marine carbon isotope record. *Paleooceanography*, **8**, 691–697.

CHAUVAUD, L., GÉRARD, T. & PAULET, Y.-M. 1998. Effects of environmental factors on the daily growth rate of *Pecten maximus* individuals in the Bay of Brest (France). *Journal of Experimental Marine Biology and Ecology*, **227**, 83–111.

COPLEN, T. B. 1994. Reporting of stable hydrogen, carbon and oxygen isotopic abundances. *Pure and Applied Chemistry*, **66**, 273–276.

CROWLEY, T. J. 1991. Modelling Pliocene warmth. *Quaternary Science Reviews*, **10**, 272–282.

—— 1996. Pliocene climates: the nature of the problem. *Marine Micropaleontology*, **27**, 3–12.

DARE, P. J. 1995. Use of external shell microgrowth patterns for determining growth and age in the scallop *Pecten maximus* (L.). *IFREMER, Actes de Colloques,* **17**, 211–217.

—— & DEITH, M. R. 1991. Age determination of scallops, *Pecten maximus* (L.), using stable oxygen isotope analysis, with some implications for fisheries management in British waters. *In*: SHUMWAY, S. E. & SANDIFER, P. A. (eds) *An International Compendium of Scallop Biology and Culture*. World Aquaculture Society, Baton Rouge, 118–133.

DIOU, A., DUMONT, C., LALLGANT, O., TOUBIN, M., TRUCHETET, F., VERRECCHIA, E. P. & ABIDI, M. A. 1999. Multiscale analysis of range image: its use for growth increment characterization. *Optical Engineering*, **38**, 2016–2021.

DOWSETT, H., BARRON, J. & POORE, R. 1996. Middle Pliocene sea surface temperatures: a global reconstruction. *Marine Micropaleontology,* **27**, 13–25.

——, CRONIN, T. M., POORE, R. Z., THOMPSON, R. S., WHATLEY, R. C. & WOOD, A. M. 1992. Micropaleontological evidence for increased meridional transport in the north Atlantic Ocean during the Pliocene. *Science*, **258**, 1133–1135.

FUGE, R., PALMER, T. J., PEARCE, N. J. G. & PERKINS, W. T. 1993. Minor and trace element chemistry of modern shells: a laser ablation inductively coupled plasma mass spectrometry study. *Applied Geochemistry, Supplement*, **2**, 111–116.

FUNNELL, B. M., NORTON, P. E. P. & WEST, R. G. 1979. The Crag at Bramerton, near Norwich, Norfolk. *Philosophical Transations of the Royal Society of London*, **B287**, 489–534.

GIBBARD, P. L., WEST, R. G., ZAGWIJN, W. H. *ET AL.* 1991. Early and early Middle Pleistocene correlations in the southern North Sea basin. *Quaternary Science Reviews*, **10**, 23–52.

GRUFFYDD, LL. D. 1981. Observations on the rate of production of external ridges on the shell of *Pecten maximus* in the laboratory. *Journal of the Marine Biological Association of the United Kingdom*, **61**, 401–411.

HEAD, M. J. 1997. Thermophilic dinoflagellate assemblages from the mid Pliocene of eastern England. *Journal of Paleontology,* **71**, 165–193.

—— 1998. Pollen and dinoflagellates from the Red Crag at Walton-on-the-Naze, Essex: evidence for a mild climatic phase during the early Late Pliocene of eastern England. *Geological Magazine*, **135**, 803–817.

HICKSON, J. A. 1997. *Stable isotope profiles in shells of the bivalve* Aequipecten opercularis *(L.) as environmental indicators*. PhD Thesis, University of Derby.

——, JOHNSON, A. L. A., HEATON, T. H. E. & BALSON, P. S. 1999. The shell of the Queen Scallop *Aequipecten opercularis* (L.) as a promising tool for palaeoenvironmental reconstruction: evidence and reasons for equilibrium stable-isotope incorporation. *Palaeogeography, Palaeoclimatology, Palaeoecology*, **154**, 325–337.

JOHNSON, A. L. A. 1984. The palaeobiology of the bivalve families Pectinidae and Propeamussiidae in the Jurassic of Europe. *Zitteliana*, **11**.

JONES, D. S. 1980. Annual cycle of shell growth increment formation in two continental shelf bivalves and its paleoecologic significance. *Paleobiology*, **6**, 331–340.

—— 1983. Sclerochronology: reading the record of the molluscan shell. *American Scientist*, **71**, 384–391.

—— 1988. Sclerochronology and the size versus age

problem. *In*: McKinney, M. (ed.) *Heterochrony in Evolution. A Multidisciplinary Approach*. Plenum, New York, 93–108.

—— & Allmon, W. D. 1995. Records of upwelling, seasonality and growth in stable-isotope profiles of Pliocene mollusk shells from Florida. *Lethaia,* **28**, 61–74.

—— & Gould, S. J. 1999. Direct measurement of age in fossil *Gryphaea*: the solution to a classic problem in heterochrony. *Paleobiology,* **25**, 158–187.

—— & Quitmyer, I. R. 1996. Marking time with bivalve shells: oxygen isotopes and season of annual increment formation. *Palaios*, **11**, 340–346.

——, Arthur, M. A. & Allard, D. J. 1989. Sclerochronological records of temperature and growth from shells of *Mercenaria mercenaria* from Narragansett Bay, Rhode Island. *Marine Biology*, **102**, 225–234.

Krantz, D. E. 1990. Mollusk-isotope records of Plio-Pleistocene marine paleoclimate, U.S. Middle Atlantic Coastal Plain. *Palaios*, **5**, 317–335.

——, Jones, D. S. & Williams, D. F. 1984. Growth rates of the sea scallop, *Placopecten magellanicus*, determined from the $^{18}O/^{16}O$ record in shell calcite. *Biological Bulletin*, **167**, 186–199.

——, Kronick, A. T. & Williams, D. F. 1988. A model for interpreting continental shelf hydrographic processes from the stable isotope and cadmium: calcium profiles of scallop shells. *Palaeogeography, Palaeoclimatology, Palaeoecology,* **64**, 123–140.

——, Williams, D. F. & Jones, D. S. 1987. Ecological and paleoenvironmental information using stable isotope profiles from living and fossil molluscs. *Palaeogeography, Palaeoclimatology, Palaeoecology*, **58**, 249-266.

Kürschner, W. M., Van der Burgh, J., Visscher, H. & Dilcher, D. L. 1996. Oak leaves as biosensors of late Neogene and early Pleistocene paleoatmospheric CO_2 concentrations. *Marine Micropaleontology*, **27**, 299–312.

Lane, A. & Prandle, D. 1996. Inter-annual variability in the temperature of the North Sea. *Continental Shelf Research*, **16**, 1489–1507.

Leng, M. J. & Pearce, N. J. G. 1999. Seasonal variation of trace element and isotopic composition in the shell of a coastal mollusk, *Mactra isabelleana*. *Journal of Shellfish Research*, **18**, 569–574.

Margus, D. 1991. Yugoslavia. *In*: Shumway, S. E. (ed.) *Scallops: Biology, Ecology and Aquaculture*. Elsevier, Amsterdam, 789–793.

Meijer, T. & Preece, R. C. 1995. Malacological evidence relating to the insularity of the British Isles during the Quaternary. *In*: Preece, R. C. (ed.) *Island Britain: a Quaternary Perspective*. Geological Society, London, Special Publications, **96**, 89–110.

Mitchell, L., Fallick, A. E. & Curry, G. B. 1994. Stable carbon and oxygen isotope compositions of mollusc shells from Britain and New Zealand. *Palaeogeography, Palaeoclimatology, Palaeoecology,* **111**, 207–216.

National Oceanic and Atmospheric Administration of the United States 1999. COADS monthly climatology. World Wide Web Address: http://ferret.wrc.noaa.gov/fbin/climate_server.

O'Neil, J. R., Clayton, R. N. & Mayeda, T. K. 1969. Oxygen isotope fractionation in divalent metal carbonates. *Journal of Chemical Physics*, **51**, 5547–5558.

Paul, J. D. 1980. Salinity–temperature relationships in the queen scallop *Chlamys opercularis* (L.). *Marine Biology*, **56**, 295–300.

—— 1981. Natural settlement and early growth of spat of the queen scallop *Chlamys opercularis* (L.), with reference to the formation of the first growth ring. *Journal of Molluscan Studies*, **47**, 53–58.

Raffi, S., Stanley, S. M. & Marasti, R. 1985. Biogeographic patterns and Plio-Pleistocene extinction of Bivalvia in the Mediterranean and southern North Sea. *Paleobiology*, **11**, 368–388.

Raymo, M. E., Grant, B., Horowitz, M. & Rau, G. H. 1996. Mid-Pliocene warmth: stronger greenhouse and stronger conveyor. *Marine Micropaleontology,* **27**, 313–326.

Richardson, C. A., Kennedy, H., Duarte, C. M., Kennedy, D. P. & Proud, S. V. 1999. Age and growth of the fan mussel *Pinna nobilis* from south-east Spanish Mediterranean seagrass (*Posidonia oceanica*) meadows. *Marine Biology*, **133**, 205–212.

Romanek, C. S., Grossman, E. L. & Morse, J. W. 1992. Carbon isotopic fractionation in synthetic aragonite and calcite: effects of temperature and precipitation rate. *Geochimica et Cosmochimica Acta*, **56**, 419–430.

Rosenberg, G. 1980. An ontogenetic approach to the environmental significance of bivalve shell chemistry. *In*: Rhoads, D. C. & Lutz, R. A. (eds) *Skeletal Growth of Aquatic Organisms*. Plenum, New York, 133–168.

—— 1990. The 'vital effect' on skeletal trace element content as exemplified by magnesium. *In*: Carter, J. G. (ed.) *Skeletal Biomineralisation: Patterns, Processes and Evolutionary Trends. Volume 1*. Van Nostrand Reinhold, New York, 567–577.

—— & Hughes, W. W. 1991. A metabolic model for the determination of shell composition in the bivalve mollusc, *Mytilus edulis*. *Lethaia*, **24**, 83–96.

Seed, R. 1976. Ecology. *In*: Bayne, B. L. (ed.) *Marine Mussels: Their Ecology and Physiology*. Cambridge University Press, 13–65.

Sloan, L. C., Crowley, T. J. & Pollard, D. 1996. Modelling of middle Pliocene climate with the NCAR GENESIS general circulation model. *Marine Micropaleontology*, **27**, 51–61.

Spero, H. J., Bijma, J., Lea, D. W. & Bemis, B. E. 1997. Effect of seawater carbonate concentration on foraminiferal carbon and oxygen isotopes. *Nature*, **390**, 497–500.

Taylor, A. C. & Venn, T. J. 1978. Growth of the queen scallop, *Chlamys opercularis*, from the Clyde Sea area. *Journal of the Marine Biological Association of the United Kingdom*, **58**, 687–700.

Tebble, N. 1976. *British Bivalve Seashells*. HMSO, Edinburgh.

Tett, P. & Walne, A. 1995. Observations and simulations of hydrography, nutrients and plankton in the southern North Sea. *Ophelia*, **42**, 371–416.

Thompson, I. L. & Jones, D. S. 1977. The ocean quahog,

Arctica islandica, 'tree' of the North Atlantic shelf. *Geological Society of America, Abstracts with Programs*, **9**, 1199.

THOMPSON, R. S. & FLEMING, R. F. 1996. Middle Pliocene vegetation: reconstructions, paleoclimatic inferences, and boundary conditions for climate modeling. *Marine Micropaleontology,* **27**, 27–49.

URSIN, E. 1956. Distribution and growth of the queen, *Chlamys opercularis* (Lamellibranchiata), in Danish and Faroese waters. *Meddelelser fra Danmarks Fiskeri- og Havundersøgelser*, **1**, 1–31.

WALLER, T. R. 1991. Evolutionary relationships among commercial scallops (Mollusca: Bivalvia: Pectinidae). *In*: SHUMWAY, S. E. (ed.) *Scallops: Biology, Ecology and Aquaculture*. Elsevier, Amsterdam, 1–73.

WEIDMAN, C. R., JONES, G. A. & LOHMANN, K. C. 1994. The long-lived mollusc *Arctica islandica*: A new palaeoeceanographic tool for the reconstruction of bottom temperatures for the continental shelves of the northern North Atlantic Ocean. *Journal of Geophysical Research – Oceans*, **99**, 18 305–18 314.

WOOD, A. M., WHATLEY, R. C., CRONIN, T. M. & HOLTZ, T. 1993. Pliocene palaeotemperature reconstruction for the southern North Sea based on Ostracoda. *Quaternary Science Reviews*, **12**, 747–767.

ZEEBE, R. E. 1999. An explanation of the effect of seawater carbonate concentration on foraminiferal oxygen isotopes. *Geochimica et Cosmochimica Acta*, **63**, 2001–2007.

The myth of metabolic cold adaptation: oxygen consumption in stenothermal Antarctic bivalves

LLOYD S. PECK & LUCY Z. CONWAY

British Antarctic Survey, High Cross, Madingley Road, Cambridge CB3 0ET, UK

(e-mail: l.peck@bas.ac.uk)

Abstract: Antarctic marine ectotherms are often described as only being capable of living in a restricted temperature range, i.e. they are stenothermal. However, few data exist demonstrating that for a given group this is the case. The Antarctic bivalve molluscs *Laternula elliptica* and *Limopsis marionensis* are similar to other Antarctic invertebrates and can only exist within a temperature window of 6–12°C. This is two to six times smaller than the range for temperate and tropical bivalves, thus demonstrating their stenothermal nature. The possibility of elevated metabolic rates of cold-water ectotherms has been a topic of debate over many years. Recently, the suggestion that metabolic rates must be elevated at low temperatures to overcome constraints has been supported by findings that mitochondrial contents of muscles in ectotherms are higher at low temperatures. Data, presented here for standard or routine metabolic rates of 41 species of bivalve mollusc from polar, temperate and tropical sites, indicate that oxygen consumption is not elevated at low temperatures. Indeed, analysis of Q_{10} coefficients between 0 and 25°C suggests that metabolic rates of polar species may be lower than would be expected by comparison with temperate bivalves.

In the early part of the twentieth century, Krogh (1916) noted that species living at low temperatures were still capable of significant activity and, comparing this with the very low metabolism and limited locomotory capabilities of goldfish held at low temperatures, postulated that metabolic rates of polar ectotherms should be elevated to compensate for the physiological constraints imposed by living near or below 0°C. In the 1950s and 1960s data were produced to support this hypothesis, and the concept of metabolic cold adaptation (MCA) was put forward by Scholander *et al.* (1953) and Wohlschlag (1964). Holeton (1974), however, showed that some of the previous data indicating elevated metabolism were probably the product of stress, and Clarke (1980, 1983) provided convincing arguments to show polar ectotherm metabolic rates are not elevated. More recent studies have continued to show low resting metabolisms for polar species (Chapelle & Peck 1995; Peck 1989, 2000; Pörtner *et al.* 1999), with a few exceptions (e.g. Wells 1987).

Low temperature physiology has been an increasing field over the last 5–10 years, and the findings that some processes are well adapted to low temperatures, e.g. the cycling of microtubules (Detrich *et al.* 1989), have led to the resurrection of the idea that metabolic costs must be elevated at low temperatures. This argument has been supported particularly by the finding that the muscles of polar fish species contain approximately twice as many mitochondria per unit mass as muscles from species living normally at 20°C (Johnston *et al.* 1994; Guderley 1998).

It is not possible to test the MCA concept adequately by using measures on single, or even a few, species. A rigorous evaluation requires data from many species at their normal habitat temperatures, so that a robust measure of the level of metabolic rates in polar species can be made in comparison with lower latitude types. There are very few taxa where there are sufficient data for this approach but, in a recent study of perciform fish, Clarke & Johnston (1999) found no elevation of metabolism at low temperatures. This paper provides a similar comparison for bivalve molluscs and is the first such data for an invertebrate taxon.

Phylogenetic constraints and environmental history are important when investigating evolutionary adaptations. Comparisons should, as far as possible, be made within a phylogenetically consistent group. The age and environmental consistency of an environment are also important because if current conditions have existed for only a short period on evolutionary timescales it is unlikely: (1) that many species will have evolved adaptations; or (2) that evolved adaptations will have reached a stable equilibrium. This is true for all

From: Harper, E. M., Taylor, J. D. & Crame, J. A. (eds) *The Evolutionary Biology of the Bivalvia*. Geological Society, London, Special Publications, **177**, 441–450. 1-86239-076-2/00/$15.00

environmental types, whether there is significant temperature variation or not, i.e. organisms need sufficient time to evolve adaptions to ambient conditions irrespective of the type of environment. In this context, data for Antarctic ectotherms are particularly valuable. The Antarctic marine environment is one of extreme thermal stability; sea temperatures vary by < 3°C annually at maritime Antarctic sites, and by < 0.2°C at some high Antarctic sites (Clarke 1988). Also, these conditions have been in existence for at least 5–10 Ma, possibly even since Antarctica became isolated from other continents 15–25 Ma ago (Clarke & Crame 1992). Data are available for metabolic rates from several species of Antarctic bivalve molluscs, and comparisons with temperate and tropical species can be expected to be robust.

Data are presented here for oxygen consumption in the Antarctic bivalves *Laternula elliptica* and *Cyclocardia astartoides*. These data are combined with previously published measures of metabolism and thermal tolerance in bivalves worldwide to show that Antarctic species are stenothermal and can in no way be construed to possess elevated metabolic rates.

Materials and methods

Experimental specimens were collected from Borge Bay, Signy Island, South Orkney Islands (60°43′S, 45°36′W). *Laternula elliptica* were collected by scuba divers from depths of 10–20 m and *Cyclocardia astartoides* were collected by Agassiz trawl from depths of 50–150 m. Specimens were transported by ship and held in the UK in refrigerated aquaria at 0° ± 1.0°C under conditions of low light. Experimental animals were kept in the UK for 3–9 months prior to measurements being made and remaining stocks were stable for a further 18 months in the aquarium system prior to them being used in other experiments. Under normal holding conditions animals were fed fortnightly by adding 2 l each of cultures of *Thalassiosira pseudonanna* and *Tetraselmis suecica* to the circulating seawater. Bivalves were not fed during experiments.

Measures of oxygen consumption were made on individuals held in acrylic or glass respirometers, using closed-bottle respirometry techniques similar to those of Chapelle & Peck (1995) and Pörtner *et al.* (1999). Water oxygen contents were obtained by couloximetry (Peck & Uglow 1990). Experimental animals were carefully placed in respirometers for the duration of each trial but were not exposed to air during this process. Behaviour was observed; the *L. elliptica* commenced pumping within minutes of being placed in respirometers and the *C. astartoides* opened soon after being moved. During trials bivalves were held in a 50 l jacketed waterbath linked to a thermocirculator. The whole system was held in a controlled-temperature room. Using this method experimental temperatures were held to 0 ± 0.2°C. Oxygen contents of respirometers were not allowed to fall below 75% of full saturation during trials.

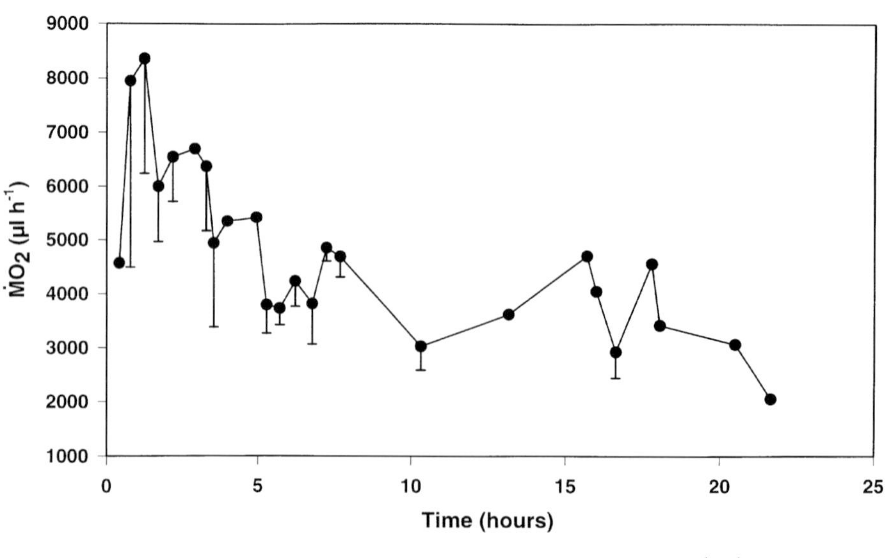

Fig. 1. *Laternula elliptica*: timecourse evaluation of oxygen consumption (µl O_2 animal^{-1} h^{-1}) over the first 24 h after animals were placed in respirometers. Data are the mean of measures for eight animals and are presented as values per standard 12.5 g tissue dry mass animal. This value was close to the mean mass for the group (12.55 g).

At the end of the experiment internal tissues were removed from the bivalves, dried to constant weight at 60°C and weighed. The mean tissue mass of the *C. astartoides* was 10.8 mg ($n = 8$, SE = 1.97) and for the *L. elliptica* 12.5 g ($n = 20$, SE = 2.80).

Results

Metabolic rates

Timecourse evaluations of oxygen consumption (MO_2) revealed that initial metabolic rates were elevated by up to three times compared with the acclimated rates for both *Laternula elliptica* (Fig. 1) and *Cyclocardia astartoides* (Fig. 2). Thus, a standard 12.5 g tissue dry mass *L. elliptica* had an oxygen consumption rate of 8400 µl O_2 h^{-1} after 1.5 h, which declined to *c.* 3000 µl O_2 h^{-1} after 20–25 h. For a standard 11 mg tissue dry mass *C. astartoides* these values were 1.51 µl O_2 h^{-1} after 2.5 h, which fell to *c.* 0.5 µl O_2 h^{-1} after 20–25 h. Converting the acclimated data (rates obtained between 15 and 24 h after the start of trials) to mass specific oxygen consumption rates produced values of 46.3 µl O_2 g dry tissue $mass^{-1}$ h^{-1} for *C. astartoides* and 239 µl O_2 g dry tissue $mass^{-1}$ h^{-1} for *L. elliptica*. Values for *L. elliptica* are high and those for *C. astartoides* are low for Antarctic species, but both are within the range of oxygen consumption rates previously reported for polar bivalve molluscs (Table 1).

There is great variability in the data presented in Table 1, with rates varying by over eight times from 35 to 292 µl O_2 g dry tissue $mass^{-1}$ h^{-1}. However, when these data are compared with those for bivalves from other latitudes, it is clear that such variability is characteristic of the group as a whole and not just of polar regions (Fig. 3), as for all temperatures there is around a ten fold range in the values reported. Some of this variability will be explained by differences in techniques and size of animal, since mass-specific data are only strictly comparable when the mass exponent is 1. Mass exponents for bivalve molluscs are *c.* 0.73 (Griffiths & Griffiths 1987) and so the variability in the data in Table 1 because of this will not be great. The size range used in each temperature regime was similar and the ecological niches represented were also broad. There is also a small temperature bias in these data. Values have been presented in volumes of oxygen consumed because most studies have used these units. However, the volume occupied by a gas varies with temperature and 1 mol or 1 g of oxygen at 30°C occupies *c.* 11% more volume than at 0°C. Comparisons between polar (0°C) and temperate (15°C) sites, and tropical (30°C) and temperate (15°C) sites would differ by 5% compared with data expressed on a molar or mass basis.

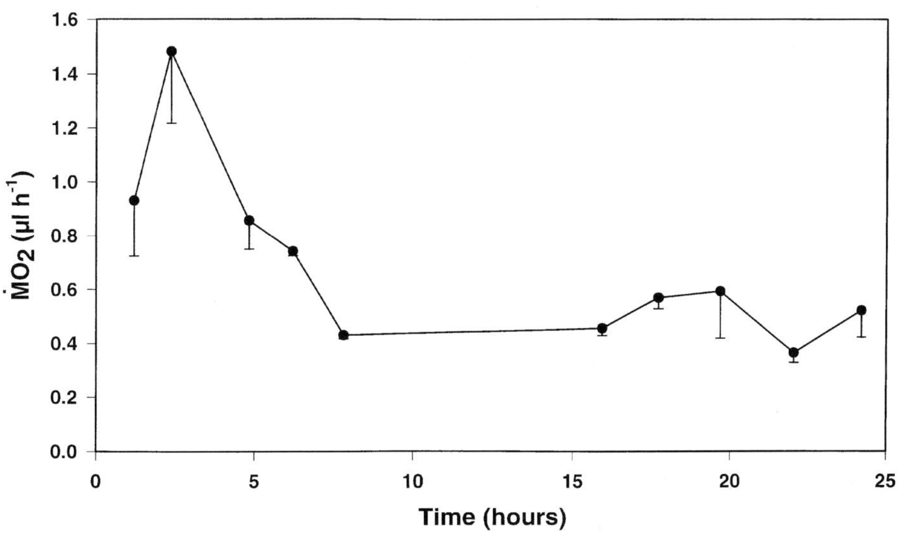

Fig. 2. *Cyclocardia astartoides*: timecourse evaluation of oxygen consumption (µl O_2 $animal^{-1}$ h^{-1}) over the first 24 h after animals were placed in respirometers. Data are the mean of measures for 20 individuals and data are presented as oxygen consumed per standard 11 mg tissue dry mass animal. This value was close to the mean mass of the group (10.81 mg).

Table 1. *Metabolic rates of polar bivalve molluscs*

Species	Temperature (°C)	MO_2* (μl g^{-1} h^{-1})	Arctic/Antarctic	Reference
Gaimardia trapesina	0	292	Antarctic	Ralph & Maxwell (1977)
Yoldia eightsi	0	164	Antarctic	Davenport (1988)
Limopsis marionensis	0	121	Antarctic	Pörtner *et al.* (1999)
Laternula elliptica	0.5	144	Antarctic	Ahn & Shim (1998)
Laternula elliptica	0	239	Antarctic	This study
Cyclocardia astartoides	0	46.3	Antarctic	This study
Chlamys islandica	2	190	Arctic	Vahl (1978)
Chlamys islandica	0	113	Arctic	Schmid (1996)
Clinocardium ciliatum	0	35–40	Arctic	Schmid (1996)

*Data are presented as μl O_2 consumed per gram of dry tissue mass per hour.

Metabolic rate varied significantly with temperature (ANOVA, $F = 2.74$, 85 df, $P < 0.001$). An Arrhenius plot of the data (ln rate v. reciprocal of absolute temperature, T) produces a relationship between metabolic rate and temperature (Fig. 4) which is significantly modelled by a linear relationship:

$$\ln MO_2 = 26.7 - 5.95(1000/T) \quad (r^2 = 0.47, F = 74.33, P < 0.001)$$

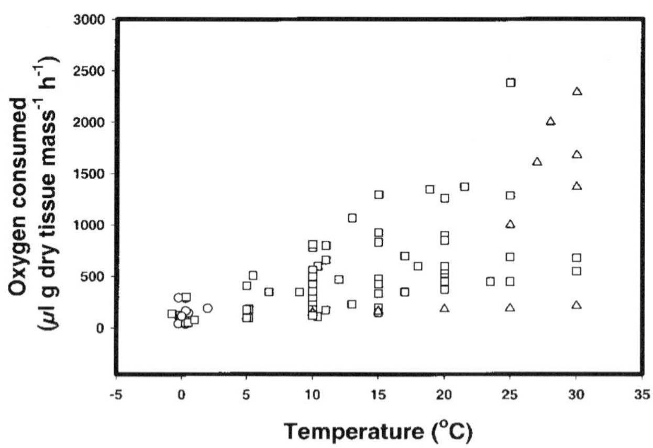

Fig. 3. Oxygen consumption values (μl O_2 g dry tissue mass^{-1} h^{-1}) for bivalve molluscs from different latitudes. Data presented are resting routine or standard rates for species at their normal temperatures. Where data were available for winter and summer rates for species inhabiting a wide temperature range, more than one value is included. ○, Polar species; □, temperate species; △, tropical species. Species and authorities are as follows: *Ruditapes philipinarum* (Kim *et al.* 1999); *Cerastoderma glaucum* (Wilson & Ekhaim 1997); *Saccostrea cucullata, Crassostrea belcheri, Crassostrea iradelei* (Davenport & Wong 1992); *Donax variabilis* (Wilson 1999); *Tellina tenuis* (Trevallion 1971); *Arctica islandica* (Taylor & Brand 1975); *Arctica islandica, Laevicardium crassum, Mytilus edulis* (Bayne 1971); *Placopecten magellanicus* (MacDonald & Thompson 1986); *Macoma balthica, Mulinia lateralis, Mya arenaria* (Kennedy & Mihursky 1972); *Scrobicularia plana* (Worral *et al.* 1983); *Geloinia ceylonica, Anadara granosa* (Bayne 1973); *Donax vittatus* (Ansell 1973); *Mya arenaria* (Lowe & Trueman 1972); *Chlamys islandica, Clinocardium ciliatum* (Schmid 1996); *Chlamys hasata, Clinocardium nuttali* (Bernard & Noakes 1990); *Chlamys opercularis* (McKlusky 1973); *Chlamys islandica* (Vahl 1978); *Polymesoda caroliniana* (Deaton 1991); *Modilous demissus, Mytilus edulis, Mytilus californianus, Perna perna* (Bayne 1976); *Chlamys varia* (Shafee 1980); *Argopecten irradians* (Bricelj *et al.* 1987); *Yoldia eightsi* (Davenport 1988); *Gaimardia trapesina* (Ralph & Maxwell 1977); *Geukensia demissa, Cardium edule, Mytilus californianus, Tridacna gigas, Cardium glaucum, Mytilus galloprovincialis* (Mingoa-Licuanan 1993); *Limopsis marionensis* (Pörtner *et al.* 1999); *Laternula elliptica* (Ahn & Shim 1998); *Crassostrea virginica* (Dame 1972); *Ostrea edulis* (Rodhouse 1978); *Mytilus chilensis* (Navarro & Winter 1982); *Laternula elliptica, Cyclocardia astartoides* (this study).

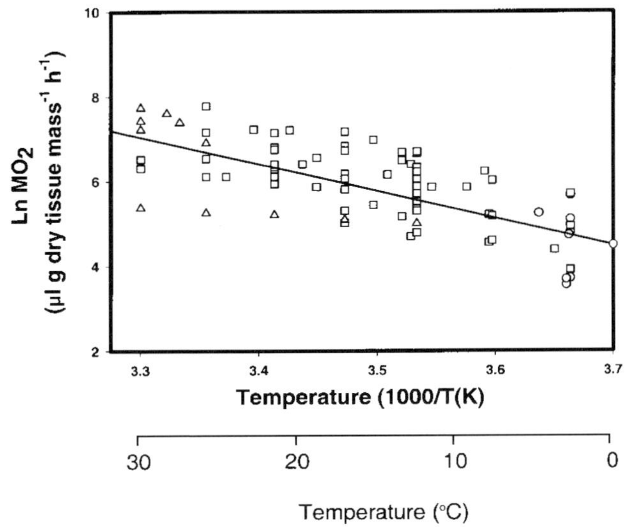

Fig. 4. Arrhenius plot of bivalve metabolic rates as ln oxygen consumed g dry tissue mass^{-1} h^{-1} v. the reciprocal of absolute temperature (1000/*T*). ○, Polar species; □, temperate species; △, tropical species. Note the reversal of the temperature axis to scale with increasing temperature to the right.

However, the form of the data does suggest some curvilinearity. Assessing whether metabolic rates at low temperatures are elevated compared to other latitudes is not easy, even from these types of data. Fitting nonlinear models, or transforming and fitting several regressions to different temperature ranges and then testing to see if the slopes of the relationships obtained are less at low temperature, is fraught with problems. Fitting nonlinear models and testing for significant changes in slopes of tangents is technically non-trivial, requiring very large amounts of data throughout the range so that estimates of the variance of tangents can be made. A similar problem exists when analysing different regressions for subsections of the overall relationship; because *x*-axis ranges are restricted, and fewer data are used in each regression, comparisons are perforce coarser.

An alternative approach is to split the data into 5°C blocks and compare means at each temperature step using Q_{10} coefficients. If rates are elevated at low temperatures, Q_{10} coefficients should progressively decrease as temperature is reduced. When the bivalve oxygen consumption data in Fig. 3 are treated this way no significant relationship between Q_{10} and temperature is obtained over the 0–30°C temperature range [Pearson correlation coefficient (PCC) = –0.18, $P = 0.77$] and the mean Q_{10} value is 2.56 (Fig. 5). However, if the analysis is restricted to 0–25°C a significant relationship is obtained (PCC = –0.987, $P = 0.013$). This would indicate that Q_{10} values are higher at polar latitudes than in temperate or tropical zones, and hence that metabolic rates are lower at low temperatures than would be predicted by extrapolation from warmer sites.

Temperature tolerances

Data on the temperature ranges that Antarctic bivalve molluscs can survive are scarce. In parallel with the present oxygen consumption study, measures of survival in relation to elevated temperature were made on *L. elliptica* (L. S. Peck, H.-O. Pörtner & I. Hardewig, unpublished data). Specimens were held in the temperature-controlled system for respirometry and temperatures progressively elevated from 0 to 3°C, and then 6–9°C. Specimens were allowed one week to acclimate to each new temperature before temperatures were raised again. This time was allowed because Peck (1989) found that long-term acclimation occurred in similar trials with the gastropod *Nacella concinna*. At 6°C and above siphons were dramatically extended and had much less turgor than previously. After 4 days at 9°C, 50% of the specimens died. For short periods of time the *L. elliptica* could survive higher temperatures than this, because it took several days for them to die at 9°C. However, such short-term exposures would not be representative of the species true upper lethal limits. The only other data on upper lethal temperature limits for an Antarctic bivalve mollusc are for *Limopsis marionensis*,

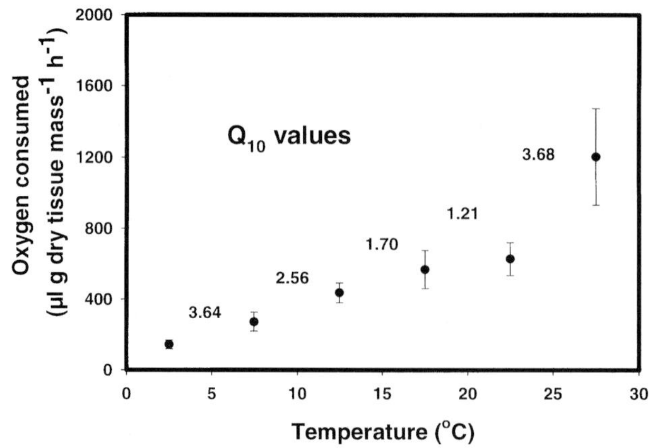

Fig. 5. Q_{10} values for the relationship between bivalve metabolic rate (µl O_2 g dry tissue mass^{-1} h^{-1}) and temperature (°C). Data have been grouped into 5°C blocks and means (± SE) plotted. Q_{10} values shown are for differences between means in successive 5°C blocks.

which had an upper lethal temperature of +4.5°C (Pörtner *et al.* 1999). These values are similar to those for other Antarctic groups, but are two to five times less than for temperate and tropical bivalves (Fig. 6).

Discussion

In *L. elliptica* and *C. astartoides* oxygen consumption rates are similar to previously reported values for polar bivalve molluscs, and are low compared to species from temperate and tropical localities. Data for 41 bivalve species across the 0°–30°C temperature range showed a significant temperature effect on metabolism. It has been argued that an Arrhenius relationship is the best descriptor of the relationship between whole-animal metabolism and temperature (Clarke & Johnston 2000). The Arrhenius relationship establised in this study indicates that a rise in temperature from 0 to 30°C causes an elevation in metabolism of 8.6 times. This would be equivalent to a Q_{10} of 2.05 over the whole range, which is well within the normally expected range.

The idea of metabolic cold adaptation (MCA)

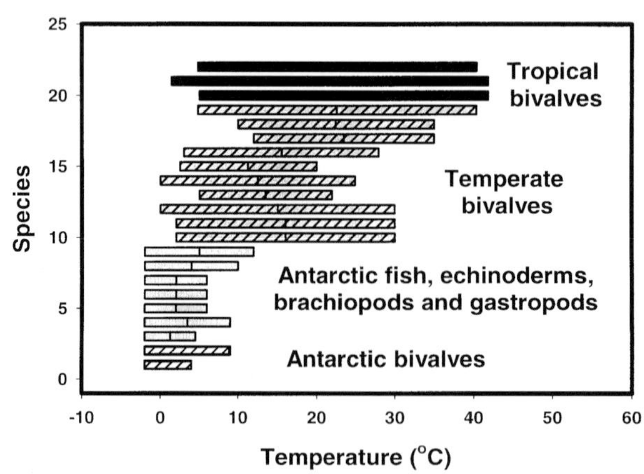

Fig. 6. Temperature tolerance windows for Antarctic bivalves, fish, echinoderms, brachiopods and gastropods, compared with temperate and tropical bivalves. Data from Pörtner *et al.* (1999), Peck (1989), Somero & De Vries (1967), Davenport & Wong (1992), Rodhouse (1978), Kennedy & Mihursky (1972), MacDonald & Thompson (1986) and Bricelj *et al.* (1987).

was originally based on within-species relationships between metabolism and temperature for eurythermal temperate taxa and extrapolating those relationships to polar temperatures. Even using this philosophy, a body of data have been produced suggesting that metabolic rates of polar species are not elevated compared to extrapolation of data for temperate species. However, several authors still purport to have data supporting the MCA concept (Forster *et al.* 1987; Wells 1987; Torres & Somero 1988). Hardewig *et al.* (1998) have also recently produced data indicating polar eelpout have metabolic rates which are relatively elevated compared to expectations from measures on temperate species. These data are compelling when viewed in isolation. However, as pointed out by both Zimmerman & Hubold (1998) and Clarke & Johnston (1999), the best method for testing the hypothesis of elevated metabolism at polar latitudes is to analyse standard or resting metabolic rates of species at their normal habitat temperatures across a wide temperature range. This was the procedure used by Clarke & Johnston (1999) for fish and was repeated here for bivalve molluscs. It should be emphasized here that neither the data for fish nor the present data provide any support for significant elevation of whole-organism metabolism at low temperatures. Indeed, the higher Q_{10} values at lower temperatures for bivalve molluscs suggests that resting and basal metabolic rates may even be lower than would be expected from extrapolation from temperate groups. This is because higher Q_{10} coefficients indicate a greater difference between the low and high temperature metabolic rates, i.e. a greater slope on the rate–temperature graph. The present data, the first on an invertebrate taxon, strongly support the conclusion of Clarke & Johnston (1999) for fish that 'metabolic cold adaptation (*sensu* Krogh 1916) does not exist'.

Clearly, polar ectotherms are adapted to low temperatures in that they survive and carry out all the necessary biological functions. Also, changes in enzyme kinetic properties and quantities have been demonstrated (Somero *et al.* 1998) and mitochondrial densities are elevated (Johnston *et al.* 1994, 1998; Guderley 1998). Pörtner *et al.* (1998) have argued persuasively that even though individual mitochondrial costs will be reduced under long-term permanently cold conditions, such as those in Antarctica, mitochondrial proliferation must inevitably lead to some elevation of resting metabolism because of enhanced requirements for mitochondrial synthesis and maintenance. They correctly point out that a major component of baseline metabolic costs is the maintenance of ionic gradients and, in this respect, the compensation for H^+ leakage across mitochondrial membranes will mean that they are more costly to maintain per unit volume than other cellular elements. The cost of mitochondrial maintenance has been estimated to be 45% of standard metabolic costs in animals (Brand 1990), which emphasizes this point.

The reconciliation of the fact that mitochondrial proliferation must carry an enhanced cost, in contrast with the finding here that there is no detectable increment in whole-animal metabolism at low temperatures (which may even be lower than expected in bivalves) can be explained by one of three mechanisms. (1) *Mitochondrial maintenance costs are reduced at low temperatures.* This is certainly not the case for within-species comparisons over adaptational timescales, as pointed out for *Arenicola marina* by Pörtner *et al.* (1998), who cited data showing mitochondrial proliferation and elevated low-temperature metabolism from North Sea to White Sea populations. However, within-species data are contradictory, with many species showing lowered metabolisms in lower temperature populations (Clarke & Johnston 1999). Data do exist which show that mitochondria function more slowly at low temperatures (Guderley 1998; Johnston *et al.* 1998). However, the same problem exists here as for evaluations of whole-animal metabolism; namely that data on mitochondrial maintenance costs are needed from many species over a wide temperature range before evolutionary implications can be assessed. The present data on bivalves indicate that, over evolutionary time, reduced mitochondrial maintenance costs should not be ruled out. (2) *Other physiological functions may be so reduced in low temperature species that the increment in mitochondrial cost is balanced.* This runs contrary to the Krogh hypothesis, in that if this is the case, low temperature reduces the costs of functions, such as nervous conduction, ion exchange and protein turnover, to such an extent that overall costs are reduced. There are insufficient data to draw conclusions in this respect; however, as pointed out by Clarke & Johnston (1999), resting metabolism does represent a cost to the animal which evolution should be expected to minimize, and lowered temperatures allow organisms to reduce such costs. (3) *Whole animal maintenance costs are elevated at low temperatures but the increment is too small to measure.* The data on bivalves produce a mean Q_{10} value over the 10–30°C temperature range of 2.29 (SE = 0.54); the value for 0°C–5°C is 3.64 and for this value to be lowered to 2.29 would require the resting metabolic rates of the polar bivalves in this study to be raised by 25%. For a significant elevation in metabolic costs above that predicted by a Q_{10} of 2.29, published oxygen consumption rates for polar species would have to be underestimated by 50–100%. It is possible that the metabolic rates of the eight polar bivalves presented here are all underestimated by 25–100%, but it is unlikely.

Temperature tolerances

Although polar ectotherms are often described as stenothermal, data evaluating the extent of their stenothermality are scarce, possibly restricted to a handful of published reports (Somero & DeVries 1967; Arnaud 1985; Peck 1989; Pörtner *et al.* 1999). Few, if any, data exist comparing temperature tolerance envelopes for polar, temperate and tropical species within a given taxonomic group. Polar bivalve molluscs live within a temperature range of –2–+4 to +10°C, an envelope of 6–12°C (Fig. 5). Conservative estimates for temperate species indicate an envelope of 15–30°C and the limited data for tropical bivalves suggest they may be able to survive over a range of 30–40°C. Thus, the temperature window in which polar species can exist is *c.* 2.5 to three times smaller than temperate and tropical species, and may be up to six times smaller. These limits for survival have been associated with critical temperatures (Sommer *et al.* 1997) which are characterized by a transfer to anaerobic metabolism, and oxygen supply has been strongly implicated in setting upper and lower temperature limits (Pörtner *et al.* 1998, 1999, 2000).

The restriction in temperature envelope and the low metabolic rates are evolutionary adaptations. In studies of this type one should be aware of the past history of the environment. Data showing elevated metabolism at low temperatures within species are investigating adaptational changes where costs may be higher because species colonizing lower temperature regimes need to adapt to reduced temperatures prior to losing characters associated with warmer habitats. In this context, data from phylogenetically related species which have inhabited differing thermal regimes over long evolutionary timescales will be of more value than others. The long timescale over which the Antarctic marine environment has existed is possibly one of the best such cases and species from such systems should be given precedence in the examination of the evolution of physiological characters.

We thank the staff of the British Antarctic Survey Signy Island Station, especially the diving officer Rob Wood and the marine assistant Simon Brockington. Jon Ward gave invaluable assistance in the maintenance of stock in the UK.

References

AHN, I.-Y. & SHIM, J. H. 1998. Summer metabolism of the Antarctic clam, *Laternula elliptica* (King and Broderip) in Maxwell Bay, King George Island and its implications. *Journal of Experimental Marine Biology and Ecology*, **224**, 253–264.

ANSELL, A. D. 1973. Oxygen consumption by the bivalve *Donax vittatus* (Da Costa). *Journal of Experimental Marine Biology and Ecology*, **11**, 311–328.

ARNAUD, P. M. 1985. Essai de Synthèse des particularités éco-biologiques (adaptations) des invetrébrés benthiques antarctiques. *Oceanis*, **11**, 117–124.

BAYNE, B. L. 1971. Oxygen consumption by three species of lamellibranch mollusc in declining ambient oxygen tension. *Comparative Biochemistry and Physiology*, **40A**, 955–970.

—— 1973. The responses of three species of bivalve mollusc to declining oxygen tension at reduced salinity. *Comparative Biochemistry and Physiology*, **45A**, 793–806.

—— 1976. *Marine Mussels: Their Ecology and Physiology*. Cambridge University Press, Cambridge.

——, WIDDOWS, J. & NEWELL, R. I. E. 1977. Physiological measurements on estuarine bivalve molluscs in the field. *In*: KEEGAN, B. F., CEIDIGH, P. O. & BOADEN, P. J. S.(eds) *Biology of Benthic Organisms, Proceedings of the 11th European Symposium on Marine Biology*. Pergamon Press, Oxford, 57–68.

BERNARD, F. R. & NOAKES, D. J. 1990. Pumping rates, water pressure and oxygen use in eight species of marine bivalve molluscs from British Columbia. *Canadian Journal of Fisheries and Aquatic Science*, **47**, 1302–1306.

BRAND, M. D. 1990. The contribution of the leak of protons across the mitochondrial inner membrane to standard metabolic rate. *Journal of Theoretical Biology*, **145**, 267–286.

BRICELJ, V. M., EPP, J. & MALOUF, R. E. 1987. Comparative physiology of young and old cohorts of bay scallop *Argopecten irradians irradians* (Lamarck): mortality, growth and oxygen consumption. *Journal of Experimental Marine Biology and Ecology*, **112**, 73–91.

CHAPELLE, G. & PECK, L. S. 1995. The influence of acclimation and substratum on the metabolism of the Antarctic amphipods *Waldeckia obesa* (Chevreux 1905) and *Bovallia gigantea* (Pfeffer 1888). *Polar Biology*, **15**, 225–232.

CLARKE, A. 1980. A reappraisal of the concept of metabolic cold adaptation in polar marine invertebrates. *Biological Journal of the Linnean Society*, **14**, 77–92.

—— 1983. Life in cold water: the physiological ecology of polar marine ectotherms. *Oceanography Marine Biology Annual Reviews*, **21**, 341–453.

—— 1988 Seasonality in the Antarctic marine environment. *Comparative Biochemistry and Physiology*, **90B**, 461–473.

—— & CRAME, J. A. 1992. The Southern Ocean benthic fauna and climate change: a historical perspective. *Philosophical Transactions of the Royal Society of London*, **338B**, 299–309.

—— & JOHNSTON, N. 1999. Scaling of metabolic rate and temperature in teleost fish. *Journal of Animal Ecology*, **68**, 893–905.

DAME, R. F. 1972. The ecological energetics of growth,

respiration and assimilation in the intertidal American oyster, *Crassostrea virginica*. *Marine Biology*, **17**, 243–250.

DAVENPORT, J. 1988. Oxygen consumption and ventilation rate at low temperatures in the Antarctic protobranch bivalve mollusc *Yoldia (= Aequiyoldia) eightsii* (Courthouy). *Comparative Biochemistry and Physiology*, **90A**, 511–513.

—— & WONG, T. M. 1992. Effects of temperature and aerial exposure on three tropical oyster species, *Crassostrea belcheri*, *Crassostrea iradelei* and *Saccostrea cucullata*. *Journal of Thermal Biology*, **17**, 135–139.

DEATON, L. E. 1991. Oxygen uptake and heart rate of the clam *Polymesoda carliniana* Bosc in air and in seawater. *Journal of Experimental Marine Biology and Ecology*, **147**, 1–7.

DETRICH, H. W. III, JOHNSON, K. A. & MARCHESE-RAGONA, S. P. 1989. Polymerisation of Antarctic fish tubulins at low temperature: energetic aspects. *Biochemistry*, **28**, 10 085–10 093.

FORSTER, M. E., FRANKLIN, C. E., TAYLOR, H. H. & DAVISON, W. 1987. The aerobic scope of an Antarctic fish, *Pagothenia borchgrevinki*, and its significance for metabolic cold adaptation. *Polar Biology*, **8**, 155–159.

GRIFFITHS, C. L. & GRIFFITHS, R. J. 1987. Bivalvia. *In*: PANDIAN, T. J. & VERNBERG, F. J. (eds) *Animal Energetics, Volume 2, Bivavlia through Reptilia*. Academic Press, San Diego, 1–88.

GUDERLEY, H. 1998. Temperature and growth as modulators of the metabolic capacities of fish muscle. *In*: PÖRTNER, H.-O. & PLAYLE, J. (eds) *Cold Ocean Physiology. Society for Experimental Biology Seminar Series, 66*. Cambridge University Press, Cambridge, 58–87.

HARDEWIG, I., VAN DIJK, P. L. M. & PÖRTNER, H. O. 1998. High energy turnover at low temperatures: recovery from exercise in Antarctic and common celpont (Zoarcidae). *American Journal of Physiology*, **274**(R), 1789–1796.

HOLETON, G. F. 1974. Metabolic cold adaptation of polar fish: fact or artefact? Physiological Zoology, **47**, 137–152.

JOHNSTON, I. A., CALVO, J., GUDERLEY, H., FERNANDEZ, D. & PALMER, L. 1998. Latitudinal variation in the abundance and oxidative capacities of muscle mitochondria in perciform fishes. *Journal of Experimental Biology*, **201**, 1–12.

——, GUDERLEY, H. E., FRANKLIN, C. E. CROCKFORD, T. & KAMUNDE, C. 1994. Are mitochondria subject to evolutionary temperature adaptation. Journal of Experimental Biology, **195**, 293–306.

KENNEDY, V. S. & MIHURSKY, J. A. 1972. Effects of temperature on the respiratory metabolism of three Chesapeake Bay bivalves. *Chesapeake Science*, **13**, 1–22.

KIM, W. S., HUH, H. T., LEE, J.-H., RUMOHR, H. & KOH, C. H. 1999. Endogenous circatidal rhythm in the manila clam *Ruditapes philipinarum* (Bivalvia: Veneridae). *Marine Biology*, **134**, 107–112.

KROGH, A. 1916. *Respiratory Exchange of Animals and Man*. Longmans, Green, London.

LOWE, G. A. & TRUEMAN, E. R. 1972. The heart and water flow rates of *Mya arenaria* (bivalvia: mollusca) at different metabolic levels. *Comparative Biochemistry and Physiology*, **41A**, 487–494.

MACDONALD, B. A. & THOMPSON, R. J. 1986. Influence of temperature and food availability on the ecological energetics of the giant scallop *Placopecten magellanicus*. *Marine Biology*, **93**, 37–48.

MCKLUSKY, D. S. 1973. The effect of temperature on the oxygen consumption and filtration rate of *Chlamys (Aequipecten) opercularis* (L.) (Bivalvia). *Ophelia*, **10**, 141–154.

MINGOA-LICUANAN, S. S. 1993. Oxygen consumption and ammonia excretion in juvenile *Tridacna gigas* (Linne, 1785): effects of emersion. *Journal of Experimental Marine Biology and Ecology*, **171**, 119–137.

NAVARRO, J. M. & WINTER, J. E. 1982. Ingestion rate, assimilation efficiency, and energy balance in *Mytilus chilensis* in relation to body size and different algal concentrations. *Marine Biology*, **67**, 255–266.

PECK, L. S. 1989. Temperature and basal metabolism in two Antarctic marine herbivores. *Journal of Experimental Marine Biology and Ecology*, **127**, 1–12.

—— 2000. Ecophysiology of Antarctic marine ectotherms: limits to life. Keynote contribution, SCAR Ecology of the Antarctic Sea–Ice Zone, Bremerhaven, June 1999. *Polar Biology*, in press.

—— & UGLOW, R. F. 1990. Two methods for the assessment of the oxygen content of small volumes of seawater. *Journal of Experimental Marine Biology and Ecology*, **141**, 53–62.

PÖRTNER, H. O., HARDEWIG, I., SARTORIS, F. J. & VAN DIJK, P. L. M. 1998. Energetic aspects of cold adaptation: critical temperatures in metabolic, ionic and acid–base regulation. *In*: PÖRTNER, H.-O. & PLAYLE, J. (eds) *Cold Ocean Physiology. Society for Experimental Biology Seminar Series, 66*. Cambridge University Press, Cambridge, 88–120.

——, PECK, L. S., ZIELINSKI, S. & CONWAY, L. Z. 1999. Intracellular pH and energy metabolism in the highly stenothermal Antarctic bivalve *Limopsis marionensis* as a function of ambient temperature. *Polar Biology*, **22**, 17–30.

——, VAN DIJK, P., HARDEWIG, I. & SOMMER, A. 2000. Levels of metabolic cold adaptation: tradeoffs in eurythermal and stenothermal ectotherms. *In:* DAVISON, W. & HOWARD WILLIAMS, C. (eds) *Antarctic Ecosystems: Models for Wider Ecological Understanding*. Caxton Press, Christchurch, New Zealand, in press.

RALPH, R. & MAXWELL, J. H. G. 1977. The oxygen consumption of the Antarctic lamellibranch *Gaimardia trapesina trapesina* in relation to cold adaptation in polar marine invertebrates. *British Antarctic Survey Bulletin*, **45**, 41–46.

RODHOUSE, P. R. G. 1978. Energy transformation by the oyster *Ostrea edulis* L. in a temperate estuary. *Journal of Experimental Marine Biology and Ecology*, **34**, 1–22.

SCHOLANDER, P. F., FLAGG, W., WALTERS, V. & IRVING, L. 1953. Climatic adaptation in Arctic and tropical poikilotherms. *Physiological Zoology*, **26**, 67–92.

SCHMID, M. K. 1996. On the distribution and oxygen consumption of ecologically important benthic animals in the waters around Svalbard. *Berichte zur Polarforschung*, **202**, 1–92.

SHAFEE, M. S. 1980. *Ecophysiological studies on a temperate bivalve* Chlamys varia *(L) from Lanveoc (Bay of Brest)*. PhD Thesis, Université de Bretagne Occidentale.

SOMERO, G. N. & DE VRIES, A. L. 1967. Temperature tolerance of some Antarctic fishes. *Science*, **156**, 257–258.

——, FIELDS, P. A., HOFMANN, G. E., WEINSTEIN, R. B. & KAWALL, H. 1998. Cold adaptation and stenothermy in Antarctic notothenioid fishes: what has been gained and what has been lost? *In*: DI PRISCO, G., PISANO, E. & CLARKE, A. (eds) Fishes of Antarctica. A biological overview, Springer-Verlag, Milan, 97–109.

SOMMER, A., KLEIN, B. & PÖRTNER, H. O. 1997. Temperature induced anaerobiosis in two populations of the polychaete worm *Arenicola marina*. *Journal of Comparative Physiology*, **167B**, 25–35.

TAYLOR, A. C. & BRAND, A. R. 1975. Effects of hypoxia and body size on the oxygen consumption of the bivalve *Arctica islandica* (L.). *Journal of Experimental Marine Biology and Ecology*, **19**, 187–196.

TORRES, J. J. & SOMERO, G. N. 1988. Metabolism, enzymic activities and cold adaptation in Antarctic mesopelagic fishes. *Marine Biology*, **98**, 169–180.

TREVALLION, A. 1971. Studies on *Tellina tenuis* Da Costa. III. Aspects of general biology and energy flow. *Journal of Experimental Marine Biology and Ecology*, **7**, 95–122.

VAHL, O. 1978. Seasonal changes in oxygen consumption of the Iceland scallop (*Chlamys islandica*(O.F. Muller)) from 70°N. *Ophelia*, **17**, 1–24.

WELLS, R. M. G. 1987. Respiration of Antarctic fishes from McMurdo Sound. *Comparative Biochemistry and Physiology*, **88A**, 417–424.

WILSON, J. G. 1999. Population dynamics and energy budget for a population of *Donax variabilis* (Say) on an exposed South Carolina beach. *Journal of Experimental Marine Biology and Ecology*, **239**, 61–83.

—— & EKAIM, B. 1997. Seasonal and geographical differences in oxygen consumption with temperature of *Cerastoderma glaucum* and a comparison with *C. edule*. *Estuarine, Coastal and Shelf Science*, **45**, 571–577.

WOHLSCHLAG, D. E. 1964. Respiratory metabolism and ecological characteristics of some fishes in McMurdo Sound, Antarctica. *Antarctic Research Series of the American Geophysical Union*, **1**, 33–62.

WORRAL, C. M., WIDDOWS, J. & LOWE, D. M. 1983. Physiological ecology of three populations of the bivalve *Scrobicularia plana*. *Marine Ecology Progress Series*, **12**, 267–279.

ZIMMERMAN, C. & HUBOLD, G. 1998. Respiration and activity of Arctic and Antarctic fish with different modes of life: a multivariate analysis of experimental data. *In*: DI PRISCO, G., PISANO, E. & CLARKE, A. (eds) *Fishes of Antarctica. A Biological Overview*. Springer-Verlag, Milano, 163–174.

Phenotypic plasticity of burrowing depth in the bivalve *Macoma balthica*: experimental evidence and general implications

PIM EDELAAR

Netherlands Institute for Sea Research (NIOZ), PO Box 59, 1790 AB Den Burg, The Netherlands, (e-mail: edelaar@nioz.nl) and Centre for Ecological and Evolutionary Studies (CEES), University of Groningen, PO Box 14, 9750 AA Haren, The Netherlands

Abstract: Depth of burrowing in bivalves presents the individual with a trade-off: burrowing more deeply increases safety against predators but shallow burrowing increases food intake. Large variation in burrowing depth exists both between and within species. This study investigated whether part of this variation in depth in the tellinid *Macoma balthica* can be attributed to phenotypic plasticity. A predatory crab and food availability were either offered or withheld in separate aquaria and the response in burrowing depth measured. Individuals burrowed more deeply in the presence of a crab but only to shallow depths in the presence of food. This suggests that individuals can assess their environment and subsequently make decisions on their burrowing depth.

Phenotypic plasticity like this flexible anti-predation behaviour in *M. balthica* can have far-ranging effects. In the presence of predators burrowing depth increases, reducing energy uptake and subsequent growth and reproduction, whereas increased burrowing depths reduces mortality by predation, further affecting prey and predator population dynamics.

Many bivalves burrow in relatively soft substrata, like sand, mud, peat or even wood. Rapid burrowing probably evolved in the Silurian and became increasingly common, especially during the Cretaceous (Vermeij 1993). It is now widely distributed amongst several bivalve families (e.g. Stanley 1970). Burrowing reduces predation risk by decreasing visibility and accessibility to predators. Decreased visibility lowers the rate of detection and decreased accessibility increases handling times, making buried prey less profitable (e.g. Zwarts *et al.* 1996). Burrowing deeply may even provide a refuge against (some) predators (e.g. Blundon & Kennedy 1982; Zwarts & Wanink 1993).

But burrowing is not just beneficial; among the conceivable disadvantages, being buried deeply within the sediment reduces the access to food resources like suspended and deposited food particles. These resources are typically found in the water column or at the substrate–water interface, i.e. some distance away from the buried animal (Kamermans 1994). If buried deeper in the sediment, food uptake rates for a suspension feeder are likely to be reduced by the increased resistance of the currents through the outstretched siphons. For a deposit feeder, the sediment surface area available for deposit feeding will decrease owing to the limitations in extension of the siphon. When individual *M. balthica* were experimentally fixed at certain depths in the sediment, flesh growth decreased with increasing burrowing depth (Fig.1: linear regression; proportion flesh growth = 0.15 – 0.049 × burrowing depth (cm), $n = 126$, $R^2 = 0.11$, $p = 0.0005$ for constant, $p = 0.0001$ for slope). Zaklan & Ydenberg (1997) found that food intake decreased with an increase in burrowing depth when suspension-feeding *Mya arenaria* were positioned at fixed depths in the sediment, whereas De Goeij & Luttikhuizen (1998) found the same effect for growth in *M. balthica*.

Burrowing depth therefore can be regarded as a trade-off between predation risk (cost) and food intake (benefit). For successful reproduction an animal has to survive and have energy available for gamete production; lifetime reproductive success in a sequential reproducer is the summation of the reproductive output at age *i* times the probability to survive until age *i*, over all ages *i* (Stearns 1992). A rather generalized model shows this graphically (Fig. 2). Given general assumptions about the relationship between burrowing depth and food intake or survival (each monotonously decreasing and increasing, respectively, with increasing depth), reproductive output can be maximal at

From: HARPER, E. M., TAYLOR, J. D. & CRAME, J. A. (eds) *The Evolutionary Biology of the Bivalvia*. Geological Society, London, Special Publications, **177**, 451–458. 1-86239-076-2/00/$15.00

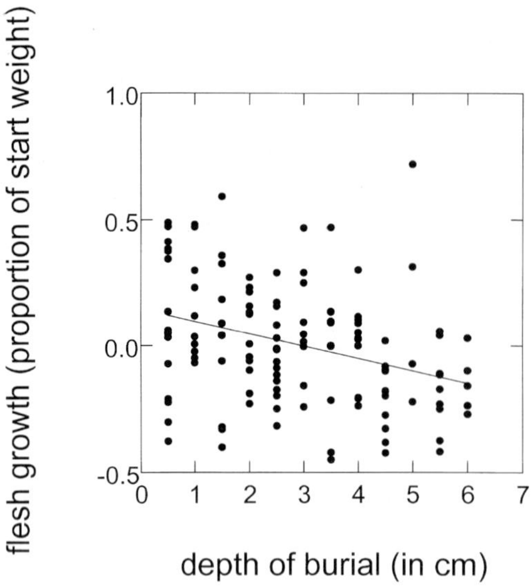

Fig. 1. The effect of experimentally fixed burrowing depth on flesh growth in *Macoma balthica*. Individuals collected in the field were positioned in a wire mesh frame and subsequently buried on an intertidal mudflat from 28 April 1999 until 18 May 1999. The frame enabled them to reach the sediment surface with their siphons for respiration and foraging, and to move horizontally, but prevented them from moving vertically, thus fixing the depth in the sediment at which they lived. Flesh growth was estimated by measuring the final ash-free dry weight (AFDW), and subtracting the estimated initial AFDW (based on a regression equation of AFDW measurements of subsamples taken at the start of the experiment v. shell length). Flesh growth was defined as the change in AFDW divided by the estimated initial AFDW. See text for statistics.

some intermediate burrowing depth. As a result of natural selection, burrowing depth may have evolved towards those values that maximize fitness.

Figure 2 reflects the situation in a constant environment. More realistically, environmental conditions vary both on ecological and evolutionary timescales. The trait value that maximizes fitness will therefore vary too. The question of interest is whether individual organisms can adjust to changes in optimum trait value by modifying their phenotype (like depth of burrowing). Following the framework of quantitative genetics, the phenotype of an individual is usually divided into the following additively contributing components: genotypic value (all additive, dominant and epistatic genetic effects); common environmental effects (general effects working on all individuals of a population equally); genotype × environment interactions (not all genotypes respond identically to a particular environment); specific environmental effects (random deviation due to individual uniqueness in time and space; i.e. unexplained variation) (Lynch & Walsh 1998). Phenotypic adjustment to environmental variation is possible for an individual if the common environmental effects and/or genotype × environment interaction statistically explain part of the variation in phenotype between individuals. This can be summed under the term phenotypic plasticity. If such plasticity does not exist, changes in the environment can only be tracked by natural selection on genotypic values. Thus, phenotypic plasticity can be viewed as any individual response to environmental change with respect to morphology, physiology or behaviour (also including phenotypic flexibility; Piersma & Lindström 1997).

Appropriate adjustments can only be made if the means to sense and process environmental information and the decision-making tools have evolved (Dusenbery 1992). It is known that burrowing depth differs widely between species (e.g. Stanley 1970), but to what extent has phenotypic plasticity in this behaviour evolved? Can bivalves respond to environmental changes that affect the costs and benefits of this behaviour?

M. balthica, a small tellinid bivalve, is found mainly in intertidal areas and the shallow subtidal

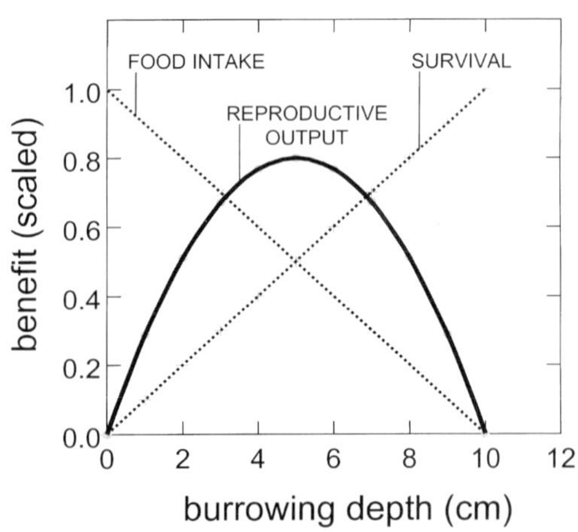

Fig. 2. Hypothetical relationship between burrowing depth and survival, food intake and reproductive output. Burrowing increases survival by reduction of predation risk but decreases food intake. Reproductive output is a multiplication of survival and energy investment, so reproductive output is non-linearly related to burrowing depth. The burrowing depth that maximizes reproductive output will be at some intermediate value.

(at least in Europe). It burrows in soft sediments, usually muddy sands, and feeds both on suspended food particles, mainly microalgae, as well as on deposited food particles and diatoms growing on the sediment surface (Kamermans 1994). Predators on adults include fishes, crabs and birds. Predation is often linked to the size of the predators, with smaller predators taking smaller *M. balthica* due to preference or even foraging constraints (e.g. Zwarts & Wanink 1993; Seed & Hughes 1995). Burrowing depth shows extreme variability in the field, both between and within sites (Zwarts & Wanink 1989; Piersma *et al.* 1994). It is not clear to what extent this phenotypic variation represents individual responses to environmental fluctuations (phenotypic plasticity).

In the section 'Experimental evidence' (below), this study first explores whether the burrowing depth of *M. balthica* represents flexible behaviour, responsive to food and predators, or behaviour that is constant for individuals. Implications of plasticity in depth of burrowing for shell formation, population dynamics and (in general) evolution of traits and species will then be discussed in 'General implications'.

Experimental evidence

Methods

Individual *M. balthica* were collected at five locations in the Dutch Western Wadden Sea and were kept for three months for acclimatization in outdoor containers filled with sand and supplied with unfiltered seawater. Suspended food was thus available, but predators were excluded.

At the beginning of the experiment, thin nylon threads, length of 10 cm and with individually numbered labels, were glued to the right valve of a total of 480 individuals with shell lengths of 10–25 mm, following the method of Zwarts (1986). Subsequently, four randomly assigned individuals of each of the five populations were released in each of 24 aquaria ($40 \times 50 \times 30$ cm), containing 15 cm of sandy sediment and 15 cm of filtered (food-free) water. No renewal of water occurred during the experiment but the water was oxygenated continuously. The individuals were free to burrow in the sand. Burrowing depth was measured by gently pulling the thread until it was perpendicular to the sediment surface and under tension, and then pulling the individual out of the sand and measuring the length of thread which was buried underneath the sediment. The same individuals were used during two manipulations, the first with predators present, and then with food present. Individuals that failed to burrow successfully in either of the two manipulations were excluded from the analyses (33 out of 480, i.e. < 7% – this was due to poor condition, damage during handling or the valves becoming inadvertently glued together).

Comparing the first and second depth of burrowing of each individual in the two controls (i.e. no predators or food present), it is clear that it is very consistently maintained in these completely extracted individuals (Fig. 3: Pearson's $r = 0.86$, $n = 111$, $p < 0.001$). This indicates that this method of measurement does not cause disturbance and that the measured depth yields relevant information concerning the depth in the sediment that an individual occupies at any point in time during the experiment.

The predator manipulation consisted of exposing the buried *M. balthica* to cues from feeding shorecrabs (*Carcinus maenas*). Separate small mesh cages with a closed floor were placed in the aquaria and were stocked with either a crab or no crab, yielding $n = 12$ replicate aquaria for each manipulation. Water was free to move between the crab cage and the rest of the aquarium, but the crab was prevented from reaching the bivalves. All crabs were fed with two *M. balthica* each day. Excrement and prey remains were prevented from falling onto the sediment surface as the closed floor of the crab cage was fitted with a small ledge. Burrowing depth was first measured after seven days.

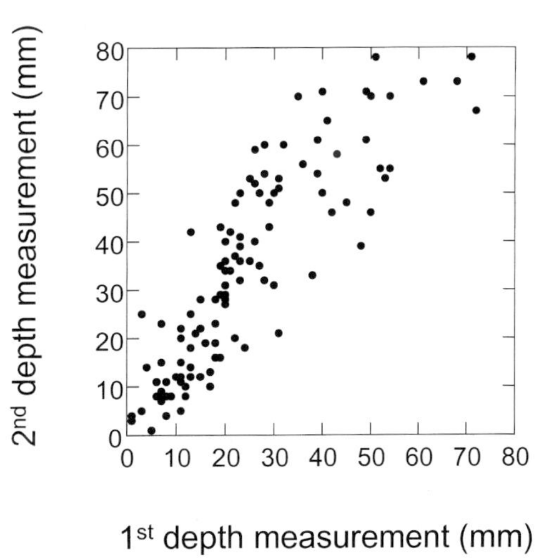

Fig. 3. Repeatability of burrowing depth in *Macoma balthica*. Only individuals that were not subjected to food or a crab in both manipulations were selected. Individuals were free to burrow twice for seven days and burrowing depth was measured by extracting the individuals completely. See text for statistics.

Subsequently, the crabs and cages were removed and the bivalves were replaced in their aquarium and allowed to re-burrow. Water was completely replaced and flowed freely in order to remove potential chemical predation cues. After six days, food was provided to the *M. balthica* in 12 out of the 24 aquaria (balanced with regards to the previous manipulation to avoid unwanted correlation between the food and predator manipulations). Food was provided by adding 0.5 g of finely pulverized dried spinach to 100 mL of filtered seawater and pouring this suspension into the aquarium. Pilot experiments showed that *M. balthica* uses these fine particles as food: the water was cleared, and inspection of the intestines showed that the spinach was accepted as food, and not just removed as pseudo-faeces. The control manipulation consisted of the addition of 100 mL of filtered seawater only. After two days, burrowing depth was measured and all individuals were stored frozen for subsequent analyses.

The statistical unit used for two-factor ANOVA was the burrowing depth averaged over all individuals in a single aquarium. This is necessary because individuals were grouped within a single aquarium and the measured burrowing depths cannot be considered as independent. Inspection of the mean depths revealed that their distributions were skewed. Transforming the data by calculating the logarithm resulted in very acceptable normal distributions and these transformed values were used for subsequent parametric statistical analyses. Testing for the effect of crab and food was performed by two-factor ANOVA on the mean burrowing depths of both manipulations. Both factors, crab and food, were included in the test of each of the two manipulations in order to use the appropriate degrees of freedom. Additionally, testing for the effect of food was performed by testing for a difference between the two categories of the food manipulation in the change in average burrowing depth. This change in depth was used since it represents a much more sensitive measurement. Because burrowing depth was highly correlated between the two depth measurements in individuals (Fig. 3), the same is true for the mean burrowing depths of the aquaria, but variation between aquaria was large. By taking the change in burrowing depth as the response variable, this large but consistent variation is accounted for (as in a paired *t*-test).

Results

The average depths of the first and second measurements for each aquarium are plotted in Fig. 4. The output of the two-factor ANOVA on the first depth measurement, the second measurement and the change in burrowing depth between the two measurements are given in Table 1. It should be noted that in the ANOVA of the first depth measurement, the effect of food is expected to be non-significant since this was not manipulated; it tests for systematic errors in the initial design that might explain later significant results of the food manipulation.

In the first measurement, individuals burrowed more deeply when in the presence of a crab (mean = 38.7 mm) compared to no crab present (mean = 24.6 mm) (Fig. 4: triangles v. circles). In

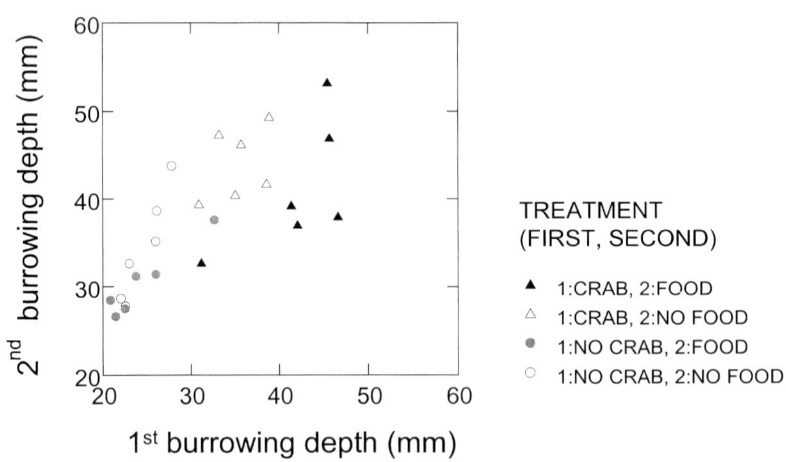

Fig. 4. Aquarium mean burrowing depths – first depth measurement v. second depth measurement for each of the four treatments. In the first manipulation, crabs were either present (triangles) or absent (circles) . In the second manipulation, food was either present (closed symbols) or absent (open symbols). See text for statistics.

Table 1. F- *and* p-*values of two-factor ANOVA on first and second depth measurement, and on the change in burrowing depth* (n = *24 for all tests*)

Treatment	First burrowing depth	Second burrowing depth	Change in depth
Crab	$F_{1,20} = 73.6$ $p < 0.0001$	$F_{1,20} = 20.8$ $p = 0.0002$	$F_{1,20} = 20.3$ $p = 0.0002$
Food	$F_{1,20} = 2.18$ n.s.	$F_{1,20} = 2.63$ n.s.	$F_{1,20} = 19.4$ $p = 0.0003$
Crab × food	$F_{1,20} = 2.85$ n.s.	$F_{1,20} = 0.11$ n.s.	$F_{1,20} = 3.06$ n.s.

n.s., non significant.

the second measurement, *M. balthica* overall burrowed more deeply (i.e. 92 out of 111 for the individuals of Fig. 3). There was no significant effect of the food manipulation on burrowing depth but, despite the fact that the crabs had been removed, the effect of crab presence was still significant. When the change in burrowing depth was tested for the effect of food, it was found that the change in depth was not identical for individuals from the two feeding categories; individuals burrowed less deeply in the presence of food compared to food not being present (Table 1 and Fig. 4: closed v. open symbols).

Discussion

Variation in burrowing depth can be as large as two orders of magnitude (Fig. 3). Most of this variation is probably explained by differences in size and condition, as is true in field samples (Zwarts & Wanink 1993; pers. obs.). Since the individuals were randomly assigned to the aquaria, no systematic differences between manipulation categories were expected and indeed they were statistically far from significant in the analyses. More interesting is the fact that burrowing depth is highly consistent (Fig. 3). Some mechanism may be constraining burrowing depths or perhaps a preferred depth for each individual exists. When comparing the second measurement of burrowing depth with the first one, it was found that most individuals burrowed more deeply the second time. Burrowing depth of the first measurement was therefore not (completely) constrained, indicating that depth is open to individual decision making. It is possible that individuals burrowed more deeply the second time as a result of the experience of being completely extracted from the sediment and handled outside the water. Such experiences may indicate risk of predation or erosion.

These experiments show that, through varying their burrowing depth, individuals respond to exposure to food and predators. Shallow burrowing became beneficial as food was added, with the animals staying closer to the surface. Adding a predator to the aquarium increased the (perceived) costs of shallow burrowing, and the animals buried, on average, 57% deeper. Is it difficult to say how much the increase in survival rate would be as depth increases from 25 to 39 mm, since this is completely dependent on the foraging behaviour of the predator – a complex function of profitability and density of many classes of potential prey types (e.g. Krebs & Davies 1987). However, it seems highly likely that it will increase survival rate, as was found in other studies concerned with burrow depth and predation risk of various predators (e.g. Blundon & Kennedy 1982; Zwarts & Wanink 1993; Zwarts *et al.* 1996). This is especially so as shorecrabs forage mainly in the upper few centimetres of the sediment (pers. obs.).

However, the presence of a crab feeding on *M. balthica* may not be directly causally related to the increased burrowing depth. For instance, the condition of the water in the tank may have changed in biotic or abiotic parameters. As feeding remains and crab excrement were prevented from falling onto the sediment, it is unlikely that *M. balthica* will have fed on them. It is possible that bacterial loads increased, but since bacteria are also utilized as food source by *M. balthica* (Kamermans 1994), a decrease in burrowing depth would have been expected (i.e. as in the food addition manipulation). In contrast, an increase in burrowing depth when crabs were present was observed, and therefore this possibility was dismissed. Similar reasoning is valid for potential confounding by abiotic variables such as oxygen saturation and detrimental chemicals. Due to the aeration, oxygen would not have been limiting. It is known that *M. balthica* is very able to cope with stressful situations without large changes in survival or functioning (Jahn *et al.* 1997; Modig & Olafsson 1998). Moreover, when the chemical environment deteriorates the usual response is to reduce burrowing depth, or even to leave the sediment completely (pers. obs.), and not to increase it.

Individuals of *M. balthica*, as has been demonstrated before in many other species (Lima & Dill 1990), seem to have the capacity to adaptively respond to changes in the environment that affect the trade-off between food and predation risk. Without dismissing the possibility that this trade-off is open to natural selection on a longer, evolutionary, timescale (Stearns 1992), the experiments show that this trade-off is open to individual decision making (i.e. a behavioural response to environmental input; Krebs & Kacelnik 1991). This may seem unexpected for a 'sluggish' creature like a bivalve. However, there is one other study documenting increased burrowing depth under stimuli of increased risk of predation. Here, *Mya arenaria* responded to various stimuli that mimicked increased risk of predation by red rock crabs (Zaklan & Ydenberg 1997). Exactly how bivalves sense (chemically?) and process the cues presented to them is unclear (cf. Dusenbery 1992). Also, it is not clear whether *M. balthica* responded solely to the presence of crabs, the presence of crushed and eaten conspecifics, or a combination of the two. This distinction is not an issue here as each of these sources of information indicates an increased risk. Rather then speculating about the mechanism, further discussion will focus on the repercussions of behavioural plasticity on ecological and evolutionary processes.

General implications

Flexibility in anti-predation behaviour resulting in, for example, increased burrowing depth may have far ranging ecological consequences. If individuals find themselves exposed to greater danger, continued increased burrowing depth results in decreased food intake (Zaklan & Ydenberg 1997). With less energy available, growth slows down (Fig. 1; De Goeij & Luttikhuizen 1998). In species with growth rings spaced in time, this effect should be visible, even after the death of the individual. Decreased growth can affect shape and thickness of the shell (Vermeij 1978). Moreover, shell size, shape and thickness have been found to be flexible anti-predation traits too (e.g. Appleton & Palmer 1988; Crowl & Covich 1990; Trussell 1996). Palaeontologists and neontologists should avoid interpreting such phenotypic differences between populations as genetic differences without due acknowledgment of the possibility of phenotypic plasticity. Nor should they attribute such differences to the wrong environmental variables, such as a low food availability that constrains growth instead of high predator density resulting in the choice to grow slowly but remain safe. Information on predator abundance is usually collected directly (e.g. catches, fossil counts) or indirectly (e.g. by determining the proportion of predated shells). Yet, the level of predator-induced damage on shells is not necessarily a reliable estimator of predator density, as a few predators may be very successful in a situation where bivalves do not benefit in taking refuge at the cost of reduced food intake.

Besides the effects of flexibility in behaviour on individuals, population dynamics can also change (Komers 1997). Adaptive anti-predation behaviour changes outcomes of simple population dynamic models as a consequence of changed predation mortality rates (Fryxell & Lundberg 1997). In more complex models, even without any substantial predation mortality, a population can even become extinct if reduced reproductive rates due to reduced food intake cannot compensate adult mortality (Matsuda & Abrams 1994). Especially in situations where fertilization, larval survival, settlement, spat and adult survival, or other demographic variables, are density dependent, adaptive changes in anti-predation behaviour can seriously change prey and/or predator population dynamics (e.g. Matsuda *et al.* 1993; Jennings 1997; McCann *et al.* 1998). In summary, modellers of predator–prey interactions that have included flexibility of behaviour have stressed its importance in model outcomes.

Unfortunately, hardly any field measurements and experiments have focused on these longer-term population consequences of behavioural flexibility. Long-term experiments are hard to fund and prone to failure, and it is almost impossible to design an experiment that only affects the behaviour of a focal organism and not all other interacting organisms too. Untangling the total response of the populations into the causal components is thus extremely difficult; measuring only the final overall change is hardly enlightening. Here lies a major task with which to further integrate the fields of behavioural and community ecology.

Finally, is there an effect of phenotypic plasticity on the evolution of biodiversity? Plasticity helps organisms to cope with changes in the environment (Via *et al.* 1995). If plasticity came without costs and limits, life would still have evolved but not as we know it (DeWitt *et al.* 1998). Costs and limits can be found and measured in organisms in genetic, ontogenetic and energetic terms. In a completely stable environment, adaptive plasticity is expected to be selected against, as maintaining the genetic coding for all aspects of plasticity, as well as maintaining the sensory systems for monitoring potential change in the environment, would be energetically wasteful (DeWitt *et al.* 1998). The mere presence of plasticity in an organism therefore already points out that its environment varies sufficiently and predictably enough to 'justify' its evolution (e.g. Pijanowska *et al.* 1993; Via *et al.*

1995; Storfer & Sih 1998). Indeed, plasticity is higher in species that live in more variable environments (Komers 1997).

Plasticity can increase adaptiveness in more then one environment (Via *et al.* 1995). When plasticity is perfect, variation in time or space of a particular potential selective force does not result in evolution by means of natural selection. In effect, the fossil record documents that species with (assumed) high levels of adaptive phenotypic plasticity are more resilient and less responsive to changes in selective regimes (Sheldon 1990). Highly plastic species are diverging at a lower rate and are evolutionary longer lived then less plastic species. Intriguingly, in a recent review, Jackson & Cheetham (1999) found that new species of marine biota arise not by gradual change but by 'punctuated equilibria'; quick shifts in morphology with no intermediate types found in the fossil record. This indicates that speciation comes about by sudden changes in the environment, like the occurrence of geographic barriers with subsequent allopatric speciation, whereas the inevitable minor environmental changes apparently did not result in speciation. This is in line with the thought that plasticity prevents speciation. The study of plasticity of traits (by means of experimental manipulations) of living taxa might therefore explain some of the evolutionary patterns of stasis v. adaptive radiation between different groups (Soares *et al.* 1998). As many species of bivalves have a larval phase with a potentially very high dispersal capacity, high levels of plasticity may indicate realized high levels of dispersal and gene flow (e.g. Parsons 1997, 1998), thus preventing local adaptation and speciation.

I would like to thank J. Drent, P. Luttikhuizen, T. Piersma, J. Van der Meer, D. Welink, W. Wolff, the editors and two anonymous referees, for comments and discussion on earlier versions. This study was supported by a PIONIER grant of the Netherlands Organization for Scientific Research (NWO) to T. Piersma. This is NIOZ publication 3482.

References

APPLETON, R. D. & PALMER, A. R. 1988. Water-borne stimuli released by predatory crabs and damaged prey induce more predator-resistant shells in a marine gastropod. *Proceedings of the National Academy of Sciences, USA*, **85**, 4387–4391.

BLUNDON, J. A. & KENNEDY, V. S. 1982. Refuges for infaunal bivalves from the Blue crab, *Callinectus sapidus* (Rathbun), predation in Chesapeake Bay. *Journal of Experimental Marine Biology and Ecology,* **65,** 67–81.

CROWL, T. A. & COVICH, A. P. 1990. Predator-induced life-history shifts in a freshwater snail. *Science,* **247,** 949–951.

DE GOEIJ, P. & LUTTIKHUIZEN, P. 1998. Deep-burying reduces growth in intertidal bivalves: field and mesocosm experiments with *Macoma balthica. Journal of Experimental Marine Biology and Ecology,* **228,** 327–337.

DEWITT, T., SIH, A. & WILSON, D. S. 1998. Costs and limits of phenotypic plasticity. *Trends in Ecology & Evolution,* **13,** 77–81.

DUSENBERY, D. B. 1992. *Sensory Ecology. How Organisms Acquire and Respond to Information.* Freeman, New York.

FRYXELL, J. M. & LUNDBERG, P. 1997. *Individual Behaviour & Community Dynamics.* Chapman & Hall, London.

JACKSON, J. B. C. & CHEETHAM, A. H. 1999. Tempo and mode of speciation in the sea. *Trends in Ecology & Evolution,* **14,** 72–77.

JAHN, A., JANAS, U., THEEDE, H. & SZANIAWSKA, A. 1997. Significance of body size in sulphide detoxification in the Baltic clam *Macoma balthica* (Bivalvia, Tellinidae) in the Gulf of Gdansk. *Marine Ecology Progress Series,* **154,** 175–183.

JENNINGS, S. 1997. Aquatic life cycle strategies: survival in a variable environment. *Trends in Ecology & Evolution,* **12,** 384–385.

KAMERMANS, P. 1994. Similarity in food source and timing of feeding in deposit- and suspension-feeding bivalves. *Marine Ecology Progress Series,* **104,** 63–75.

KREBS, J. R. & DAVIES, N. B. 1987. *An Introduction to Behavioural Ecology*. Blackwell, Oxford.

—— & KACELNIK, A. 1991. Decision making. *In*: KREBS, J. R. & DAVIES, N. B. (eds) *Behavioural Ecology, An Evolutionary Approach*. Blackwell, Oxford.

KOMERS, P. E. 1997. Behavioural plasticity in variable environments. *Canadian Journal of Zoology,* **75,** 161–169.

LIMA, S. L. & DILL, L. M. 1990. Behavioural decisions made under the risk of predation: a review and prospectus. *Canadian Journal of Zoology,* **68,** 619–640.

LYNCH, M. & WALSH, B. 1998. *Genetics and Analysis of Quantitative Traits*. Sinauer Associates, Inc., Sunderland.

MCCANN, K., HASTINGS, A. & HUXEL, G. R. 1998. Weak trophic interactions and the balance of nature. *Nature,* **395,** 794–798.

MATSUDA, H. & ABRAMS, P. A. 1994. Timid consumers: self-extinction due to adaptive change in foraging and anti-predator effort. *Theoretical Population Biology,* **45,** 76–91.

——, —— & HORI, M. 1993. The effect of adaptive anti-predator behaviour on exploitative competition and mutualism between predators. *Oikos,* **68,** 549–559.

MODIG, H. & OLAFSSON, E. 1998. Responses of Baltic benthic invertebrates to hypoxic events. *Journal of Experimental Marine Biology and Ecology,* **229,** 133–148.

PARSONS, K. E. 1997. The role of dispersal ability in the phenotypic differentiation and plasticity of two marine gastropods. I. Shape. *Oecologia,* **110,** 461–471.

—— 1998. The role of dispersal ability in the phenotypic differentiation and plasticity of two marine gastropods. II. Growth. *Journal of Experimental Marine Biology and Ecology,* **221,** 1–25.

PIERSMA, T. & LINDSTRÖM, Å. 1997. Rapid reversible changes in organ size as a component of adaptive behaviour. *Trends in Ecology & Evolution,* **12,** 134–138.

——, VERKUIL, Y. & TULP, I. 1994. Resources for long-distance migration of knots *Calidris canutus islandica* and *C. c. canutus*: how broad is the temporal exploitation window of benthic prey in the western and eastern Wadden Sea? *Oikos,* **71,** 393–407.

PIJANOWSKA, J., WEIDER, L. J. & LAMPERT, W. 1993. Predator-mediated genotypic shifts in a prey population: experimental evidence. *Oecologia,* **96,** 40–42.

SEED, R. & HUGHES, R. N. 1995. Criteria for prey size-selection in molluscivorous crabs with contrasting claw morphologies. *Journal of Experimental Marine Biology and Ecology,* **193,** 177–195.

SHELDON, P. R. 1990. Shaking up evolutionary patterns. *Nature,* **245,** 772.

SOARES, A. G., CALLAHAN, R. K. & DE RUYCK, A. M. C. 1998. Microevolution and phenotypic plasticity in *Donax serra* Röding (Bivalvia: Donacidae) on high energy sandy beaches. *Journal of Molluscan Studies,* **64,** 407–421.

STANLEY, S. M. 1970. *Relation of Shell Form to Life Habits of the Bivalvia (Mollusca).* The Geological Society of America, Boulder, CO.

STEARNS, S. C. 1992. *The Evolution of Life Histories.* Oxford University Press, Oxford.

STORFER, A. & SIH, A. 1998. Gene flow and ineffective antipredator behavior in a stream-breeding salamander. *Evolution,* **52,** 558–565.

TRUSSELL, G. C. 1996. Phenotypic plasticity in an intertidal snail: the role of a common crab predator. *Evolution,* **50,** 448-454.

VERMEIJ, G. J. 1978. *Biogeography and Adaptation.* Harvard University Press, Cambridge, MA.

—— 1993. *A Natural History of Shells.* Princeton University Press, Princeton, New Jersey.

VIA, S., GOMULKIEWICZ, R., DE JONG, G., SCHEINER, S. M., SCHLICHTING, C. D. & VAN TIENDEREN, P. H. 1995. Adaptive phenotypic plasticity: consensus and controversy. *Trends in Ecology & Evolution,* **10,** 212–217.

ZAKLAN, S. D. & YDENBERG, R. 1997. The body size–burial depth relationship in the infaunal clam *Mya arenaria. Journal of Experimental Marine Biology and Ecology,* **215,** 1–17.

ZWARTS, L. 1986. Burying depth of the benthic bivalve *Scrobicularia plana* (Da Costa) in relation to siphon cropping. *Journal of Experimental Marine Biology and Ecology,* **101,** 255–289.

—— & WANINK, J. H. 1989. Siphon size and burying depth in deposit- and suspension-feeding benthic bivalves. *Marine Biology,* **100,** 227–240.

—— & —— 1993. How the food supply harvestable by waders in the Wadden Sea depends on the variation in energy density, body weight, biomass, burying depth and behaviour of tidal-flat invertebrates. *Netherlands Journal of Sea Research,* **31,** 441–476.

——, ENS, B. J., GOSS-CUSTARD, J. D., HULSCHER, J. B. & LE V. DIT DURELL, S. E. A. 1996. Causes of variation in prey profitability and its consequences for the intake rate of the Oystercatcher *Haematopus ostralegus. Ardea,* **84A,** 229–268.

Herring gulls feeding on a recent invader in the Wadden Sea, *Ensis directus*

GERHARD C. CADÉE

Netherlands Institute for Sea Research (NIOZ), PO Box 59, 1790 AB Den Burg, The Netherlands (e-mail: Cadee@nioz.nl)

Abstract: The American razor clam *Ensis directus* has been a successful non-indigenous invader in the Wadden Sea since the late 1970s. It often shows poorly understood mass mortalities in the winter half-year, which are also known from its native area. During such mass mortalities *Ensis directus* partly leaves its burrow and, as it is unable to re-burrow, is an easy prey for herring gulls during low tide. During a mass mortality in March 1999, *Ensis* consumption by herring gulls was studied. Shell fragmentation was achieved by shaking the shell vigorously, not by hammering. This caused characteristic shell fragmentation ranging from small fragments broken from one or both valves near the middle, to both valves broken near the middle, but still connected by the ligament. About one-quarter of the shells remained undamaged. No other predators are known to produce similar *Ensis* shell fragments. In taphonomic studies the importance of predators in fragmenting shells should be taken into account.

Invasions by exotic species into coastal zones have escalated in recent years, mainly due to increased human activities such as ballast-water transport (Carlton & Geller 1993). In the new environment these invaders often show population explosions due to the absence of their natural predators, parasites and diseases. Their success may disturb the native ecosystem, e.g. as illustrated by the invasion of the ctenophore *Mnemiopsis leidyi* in the Black Sea leading to a collapse of pelagic fisheries (Gesamp 1997). Although no reports exist on extinction of native marine species related to invasions, successful invaders caused disappearance of the original communities. As a result, natural biodiversity at the community and ecosystem level decreased (Carlton 1996). Apart from these negative effects, invaders offer new opportunities to study aspects of species richness in ecosystem processes, a topic of discussion in biodiversity studies (e.g. Lawton 1997), and also relations between invasions and extinction of native species (e.g. Vermeij 1989). Finally, these invasions offer opportunities to study the evolution of predator–prey relations. For example, Vermeij (1982) studied possible increase in shell thickness of *Littorina littorea* shells after introduction of the predatory crab *Carcinus maenas* on American shores.

In this paper, the reaction of one predator, the herring gull (*Larus argentatus* Pontoppidan), to a recent invader, the American razor clam *Ensis directus* (Conrad), in the Dutch Wadden Sea, is reported. Whether this predator produces characteristic shell fragments, useful in taphonomic studies that try to differentiate between biological and physical agents in the production of shell fragments (Parsons & Brett 1991; Cadée 1968, 1994, 1995, and refs cited therein), was also studied.

The American razor clam *Ensis directus* has been a successful ballast-water invader in the North Sea coastal area since the late 1970s. Its dissemination has been studied regularly and more papers are now written on this species in its new habitat than in its native area (Beukema & Dekker 1995; Armonies & Reise 1999). *E. directus* shows characteristic mass mortalities in the winter half-year and such mortalities are poorly understood (Mühlenhardt-Siegel *et al.* 1983; Cadée *et al.* 1994; Armonies & Reise 1999); they have been related, amongst other things, to: parasites; the influence of cold periods; energy depletion after spawning. Comparable mass mortalities in spring also occur in its native area (e.g. in the Gulf of St Lawrence; E. Kenchington pers. comm.). During such mass mortalities *E. directus* protrudes at least half its length above the sediment and is unable to re-burrow (Cadée & Cadée-Coenen 1994; Armonies & Reise 1999, fig. 3). This makes them an easy prey for herring gulls, which otherwise would be unable to collect this normally rapidly burrowing bivalve (Cadée &

From: HARPER, E. M., TAYLOR, J. D. & CRAME, J. A. (eds) *The Evolutionary Biology of the Bivalvia*. Geological Society, London, Special Publications, **177**, 459–464. 1-86239-076-2/00/$15.00

Cadée-Coenen 1994). Probably only some oystercatchers specialized for feeding on *E. directus* are able to catch healthy specimens (Swennen *et al.* 1985).

Methods

Observations on herring gulls consuming *Ensis directus* were made in early March 1999 along the Wadden Sea dyke of the southern part of the island of Texel, the Netherlands. Between 1 and 5 March 1999 all shells that accumulated along the dyke studied (and still containing parts of the adductor muscles) were collected. In this way, a total of 750 *Ensis* specimens were collected for the study of fragmentation. Of these, 200 were used for length measurements (with vernier callipers) and age determination (using annual rings). The picture of the *Ensis* shells (Fig. 2) was made by a direct scan method using a Hewlett Packard Flatbed scanner (Scanjet 6200 C). Direct scanning has also proved to be very efficient method for objects that are not completely flat, such as *Ensis* shells (see also Bromley & Richter 1999).

Results

Processing Ensis *by herring gulls*

Herring gulls were observed to collect *Ensis* during low tide. They swam around in areas of some 40 cm or less water depth, looking for protruding *Ensis*. Once they had discovered one, they tried to collect the specimen by diving. Herring gulls are not true divers and cannot disappear entirely below the water surface. To get momentum they first lift themselves somewhat out of the water and then dive for the *Ensis*. Once they had collected a specimen they brought it to the nearest dry place, flying with the *Ensis* square in the beak. Along the Wadden Sea dyke this dry place was the foot of the dyke (Fig. 1). After arrival they started shaking the *Ensis* in their beak for some seconds, subsequently dropping it and trying to consume the flesh protruding from the shell. This process of shaking and dropping was repeated several times. In general, 1–2 min were sufficient to consume the *Ensis*, leaving only the empty shell. If it took longer this was usually due to disturbance by other herring gulls. The whole process of shaking and dropping the prey is very similar to their handling of shorecrabs (*Carcinus maenas*).

Fig. 1. Herring gull (just arrived) at the foot of the Wadden Sea dyke with *Ensis directus* square in its beak and starting to shake *Ensis* to free the flesh; specimen still undamaged. In the background are empty specimens of *Ensis* left over by herring gulls. [Drawing kindly made by Sytske Dijksen (Texel)].

Fragmentation

Shaking of the shells by herring gulls caused damage in the middle part of the valves. Gulls were observed also to hack at the shells when they had dropped them on the ground after shaking, but this was mainly to cut the adductor muscles or remove flesh attached to the shells, not to break the shells. The amount of resulting fragmentation was variable, with four grades of damage being discernable (Fig. 2 and Table 1). About one-quarter of the shells showed no damage at all; these were apparently specimens which opened easily and which had lost the flesh early in the handling (shaking) process, probably because they had been dead for some time. In the remaining three-quarters, however, shell damage was apparent; this varied from small damage to one or both shells, via one valve broken in the middle, to both valves broken in the middle. In all cases the anterior parts of the valves were still connected by the ligament (Fig. 2).

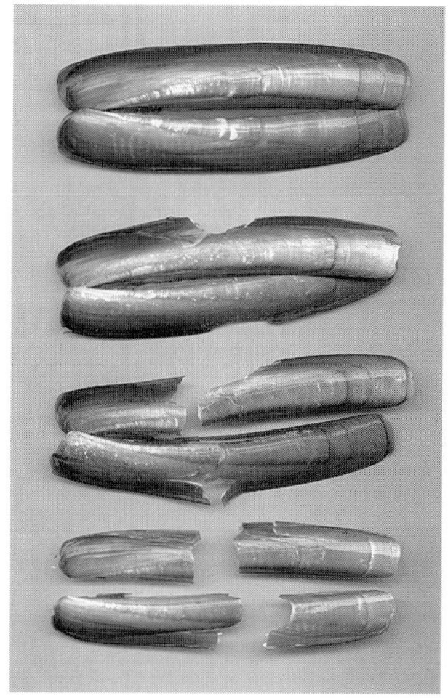

Fig. 2. The four grades of shell damage discerned (from top to bottom): undamaged; small damage in the middle part of one or both valves, one valve broken; both valves broken. Length of top specimen, 142.2 mm.

Size–frequency distribution and growth

The average length of 200 specimens of *Ensis* consumed was 139.5 mm (standard deviation 13.4 mm); the range was 82.8–168.5 mm (Fig. 3). Based on the counts of the annual rings (not clearly visible in all specimens, due to the formation of disturbance rings), the majority consisted of specimens from the 1996 (45.5%) and 1994 cohorts (34%) – their size–frequency distributions are given separately in Fig. 3. As found in most earlier mass mortalities of *Ensis directus*, this mortality was mainly restricted to older animals (Mühlenhardt-Siegel *et al.* 1983; Cadée *et al.* 1994; Armonies & Reise 1999); this also applies to mortalities in their native lands in the Gulf of St Lawrence (E. Kenchington pers. comm.). Only Armonies & Reise (1999) report mass mortality in cohorts of < 1 year old observed in the lower intertidal of Königshaven (Sylt, Germany) in March 1980 and February 1982 (but apparently not observed since).

By measuring annual rings on the shells, growth of the 1994 and 1996 cohorts could be estimated, and in Table 2 these data are compared with growth data for other cohorts of *Ensis directus* from the same tidal-flat area bordering the southern part of Texel, the Schanserwaard and elsewhere. It can be seen that there is a large variation in growth in the first year, but this then diminishes in later years. Growth data fall within the range observed in European coastal waters (Dörjes 1992; Beukema & Dekker 1995; Rasmussen 1996). It seems high, however, when compared with the few published data on growth (only given for the first year) in its native land, of *c.* 30 mm (McDermott 1976; Kenchington *et al.* 1998).

Table 1. *Type and presence of damage in 750* Ensis directus *specimens consumed by herring gulls*

Grade of damage	Number	Percentage
No damage	198	26.4
One or two valves slightly damaged	156	20.8
One valve broken in the middle	196	26.1
Both valves broken in the middle	200	26.7
Total	750	100.0

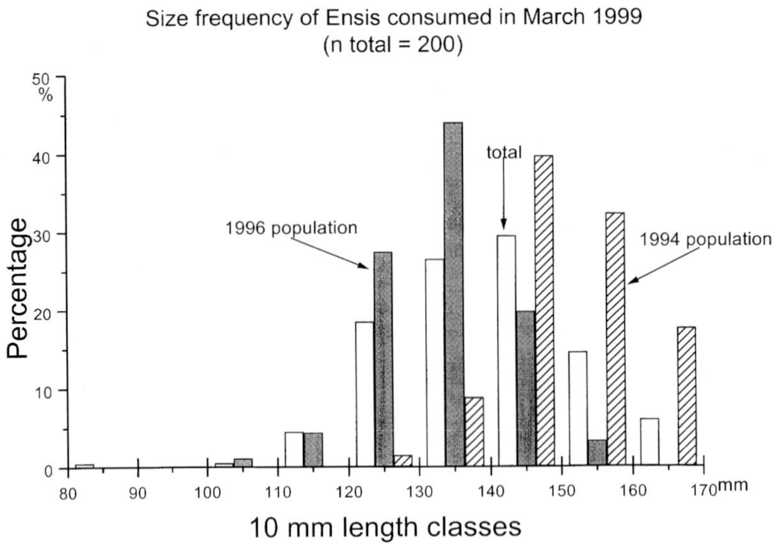

Fig. 3. Size–frequency distribution of 200 specimens of *Ensis directus* consumed by herring gulls between 1 and 5 March 1999. Size–frequency distribution of 1994 ($n = 68$) and 1996 ($n = 91$) cohorts (the bulk of the sample) are also given separately.

Discussion

In the Wadden Sea, the optimal zone for *E. directus* is the lower, rarely emerging, region of the intertidal and the upper subtidal zones, in relatively exposed areas with almost silt-free sediments (1% or less < 60 µm). This area is generally poor in macrobenthos and thus may have provided an 'empty' niche for this rapidly burrowing bivalve adapted to maintain its position in an exposed habitat (Beukema & Dekker 1985). It is apparently also relatively undisturbed here by predators. Only some oystercatchers are specialized in feeding on *Ensis* (Swennen *et al.* 1985), but they can reach only the intertidal specimens. Herring gulls, and probably also eiderducks, feed only on protruding specimens during mass mortalities (Cadée & Cadée-Coenen 1994; this study). Herring gulls either collect such specimens themselves during low tide (the main method observed in March 1999) or by stealing them from eiderducks who had collected them by diving (observed frequently in 1994; Cadée & Cadée-Coenen 1994). Predation on these dying *Ensis* will not influence density of the population. The omnivorous herring gull, and more specialized molluscan predators such as oystercatcher and eiderduck (Swennen *et al.* 1985), rapidly added this invader to their diet.

Invasion of *E. directus* started in a period of increasing phytoplankton primary production in the Wadden Sea, doubling in the late 1970s and remaining high in the 1980s and 1990s (Cadée 1986; Cadée & Hegeman 1993; Cadée pers. obs.). Besides its occupation of an empty niche, this

Table 2. *Position of annual rings in various cohorts* (first column) of* Ensis directus

Year-class	*n*	1st ring mm (SD)	2nd ring mm (SD)	3rd ring mm (SD)	4th ring mm (SD)	5th ring mm (SD)	Reference
1984	200	48.4 (11.1)	98.5 (15.3)	124.2 (11.6)	133.7 (10.9)		Cadée 1989
1988	31	26.4 (4.5)	80.4 (6.3)	124.5 (5.4)	144.6 (5.9)		Swennen 1992
1991	100	59.2 (10.5)	120.5 (7.1)	134.5 (7.1)			Cadée *et al.* 1994
1994	68	73.5 (13.1)	120.7 (8.8)	133.4 (7.9))	147.2 (8.3)	151.2 (8.9)	Present study
1996	91	49.0 (15.3)	112.4 (11.9)	133.5 (8.8)			Present study

*All specimens from the same tidal flat (Schanserwaard) bordering the southern part of the island of Texel, based on published data and this study. *n*, Number of shells used; SD, standard deviation of measurements.

might have favoured its successful development in the Dutch coastal area, where it would have to compete with other suspension feeders such as mussels and cockles. However, using their long-term monitoring programmes (1970–1990) Beukema & Cadée (1997) were able to indicate that food limitation for macrozoobenthos in the Balgzand, a well-studied area of the western Wadden Sea, only occurred in the middle part of the intertidal zone and not in the lower part inhabited by *E. directus,* where exposure to currents and wind limited macrobenthos.

E. directus shell fragments produced by herring gulls are characteristic for this predator. *Ensis* consumed by oystercatchers only showed damage to the posterior part of the shell (C. Swennen, pers. comm.). This adds another example to the list of characteristic predator marks produced on shells (e.g. Vermeij 1978, 1987; Cadée *et al.* 1997 and refs cited therein). The broken *Ensis directus* shells reported by Rasmussen (1996) from the Isefjord (Denmark) were more heavily fragmented than those consumed by oystercatchers, and it is likely that this fragmentation was also due to herring gulls. The specimens were collected in April 1994, when an enormous mass mortality of *Ensis* also occurred in the Wadden Sea (Cadée *et al.* 1994; de Wolf & Cadée 1995), causing much higher numbers of dead *Ensis* than in March 1999.

This study indicates the importance of predators in producing shell fragments. Biological factors, and in particular predation, are more important in shell fragmentation in the Wadden Sea than are physical factors (Cadée 1994, 1995). In taphonomic studies the role of predators in shell fragmentation should be taken into account.

Although in the past, 26 non-indigenous macrobenthic species, including *E. directus*, have been introduced by humans, successfully invading German and Dutch coastal waters, this has not yet led to drastic changes in the functioning of the ecosystem (Nehring & Leuchs 1999). There are, however, indications that the native *Ensis* species in the coastal zone of the open North Sea, not discussed in this study, are declining, as fresh shells are becoming rarer in beach drift, whereas *E. directus* has become an abundant shell on the beaches.

This is part of a larger low-budget project in shell damage and shell repair of the Paleobiology department of NIOZ; it is NIOZ publication no. 3485. Without the help of herring gulls, alerting me to the mass mortality of *Ensis directus* and also providing me with the *Ensis* material, this study would have been impossible. I am further grateful to Sytske Dijksen for the drawing of the herring gull handling an *Ensis directus,* and to Jan Beukema, John Davenport, Alistair Crame and an anonymous referee for critically reading the manuscript.

References

ARMONIES, W. & REISE, K. 1999. On the population development of the introduced razor clam *Ensis americanus* near the island of Sylt. *Helgoländer Meeresuntersuchungen*, **52**, 291–300.

BEUKEMA, J. J. & DEKKER, R. 1995. Dynamics and growth of a recent invader into European coastal waters: the American razor clam, *Ensis directus*. *Journal of the Marine Biological Association of the UK*, **75**, 351–362.

—— & CADÉE, G. C. 1997. Local differences in macrozoobenthic response to enhanced food supply caused by mild eutrophication in a Wadden Sea area: food is only locally a limiting factor. *Limnology and Oceanography*, **42**, 1424–1435.

BROMLEY, R. G. & RICHTER, B. 1999. Direkte skanning: Ny technik til illustrering af geologiske materiale. *Geologisk Tidsskrift,* **1998**(4), 1–10.

CADÉE, G. C. 1968. Molluscan biocoenoses and thanatocoenoses in the Ria de Arosa, Galicia, Spain. *Zoologische Verhandelingen Leiden*, **95**, 1–121.

—— 1986. Increased primary production in the Marsdiep. *Netherlands Journal of Sea Research*, **20**, 285–290.

—— 1989. Size-selective transport of shells by birds and its palaeoecological implications. *Palaeontology,* **32**, 429–437.

—— 1994. Eider, shelduck, and other predators, the main producers of shell fragments in the Wadden Sea: palaeoecological implications. *Palaeontology* **37**, 181–202.

—— 1995. Birds as producers of shell fragments in the Wadden Sea, in particular the role of the Herring gull. *Géobios, Mémoir Spécial*, **18**, 77–85.

—— & CADÉE-COENEN, J. 1994. Hoe Zilvermeeuwen Amerikaanse zwaardscheden (*Ensis directus*) vangen. *Correspondentieblad der Nederlandse Malacologische Vereniging*, **278**, 64–67.

—— & HEGEMAN, J. 1993. Persisting high levels of primary production at declining phosphate concentrations in the Dutch coastal area (Marsdiep). *Netherlands Journal of Sea Research*, **31**, 147–152.

——, CADÉE-COENEN, J. & WITTE, J. IJ. 1994. Massale sterfte van *Ensis directus* op de Schanserwaard en elders blijft raadselachtig. *Correspondentieblad der Nederlandse Malacologische Vereniging*, **279**, 86–93.

——, WALKER, S. E. & FLESSA, K. W. 1997. Gastropod shell repair in the intertidal of Bahia la Choya (N. Gulf of California). *Palaeogeography, Palaeoclimatology, Palaeoecology*, **136**, 67–78.

CARLTON, J. T. 1996. Marine bioinvasions: the alteration of marine ecosystems by nonindigenous species. *Oceanography*, **9**, 36–43.

—— & GELLER, J. B. 1993. Ecological roulette: the global transport of nonindigenous marine organisms. *Science,* **261**, 78–82.

DE WOLF, P. & CADÉE, G. C. 1995. Shells of *Ensis directus* on the beaches of Texel. *Annual Report 1994 Netherlands Institute for Sea Research,* 74–75.

DÖRJES, J. 1992. Die amerikanische Schwertmuschel *Ensis directus* (Conrad) in den Deutschen Bucht. III. Langzeitentwicklung nach 10 Jahren. *Senckenbergiana Maritima*, **22,** 29–35.

GESAMP (IMO/FAO/UNESCO–IOC/WMO/WHO/IAEA/UN/UNEP JOINT GROUP OF EXPERTS ON THE SCIENTIFIC ASPECTS OF MARINE ENVIRONMENTAL PROTECTION). 1997. Opportunistic settlers and the problem of the ctenophore *Mnemiopsis leidyi* invasion in the Black Sea. *GESAMP Reports and Studies*, **58**, 1–84.

KENCHINGTON, E., DUGGAN, R. & RIDDELL, T. 1998. Early life history characteristics of the razor clam (*Ensis directus*) and the moonsnail (*Euspira* spp.) with applications to fisheries and aquaculture. *Canadian Technical Report of Fisheries and Aquatic Sciences*, **2223**, 1–32.

LAWTON, J. H. 1997. The role of species in ecosystems: aspects of ecological complexity and biological diversity. *In*: ABE, T., LEVIN, S. A. & HIGASHI, M. (eds) *Biodiversity: An Ecological Perspective.* Springer, New York, 215–228.

MCDERMOTT, J. J. 1976. Predation of the razor clam *Ensis directus* by the nemertean worm *Cerebratulus lacteus. Chesapeake Science*, **17,** 299–301.

MÜHLENHARDT-SIEGEL, U., DÖRJES, J. & VON COSEL, R. 1983. Die amerikanische Schwertmuschel *Ensis directus* (Conrad) in der Deutschen Bucht. II. Populationsdynamik. *Senckenbergiana Maritima,* **15**, 93–110.

NEHRING, S. & LEUCHS, H. 1999. Neozoa (Makrozoobenthos) an der deutschen Nordseeküste – Eine Übersicht. *Bundesantalt für Gewässerkunde, Koblenz, Bericht BfG*, **1200**, 1–131.

PARSONS, K. M. & BRETT, C. E. 1991. Taphonomic processes and biases in modern marine environments: an actualistic perspective on fossil assemblage preservation. *In*: DONOVAN, S. K. (ed.) *The Processes of Fossilization.* Belhaven Press, London, 22–65.

RASMUSSEN, E. 1996. Nyt om den amerikanske knivmuslin (*Ensis americanus* Gould 1870 (= *E. directus* (Conrad)) i danske farvande. *Flora og Fauna,* **101**, 53–60.

SWENNEN, C. 1992. De groei van de Amerikaanse zwaardschede (*Ensis americanus*) in de Waddenzee. *Het Zeepaard*, **52**, 129-131.

——, LEOPOLD, M. F. & STOCK, M. 1985. Notes on growth and behaviour of the American razor clam *Ensis directus* in the Wadden Sea and the predation on it by birds. *Helgoländer Meeresuntersuchungen*, **39**, 255–261.

VERMEIJ, G. J. 1978. *Biogeography and Adapatation.* Harvard University Press, Cambridge, MA.

—— 1982. Environmental change and the evolutionary history of the periwinkle (*Littorina littorea*) in North America. *Evolution*, **36**, 561–580.

—— 1987. *Evolution and Escalation.* Princeton University Press, Princeton NJ.

—— 1989. Invasion and extinction: the last three million years of North Sea pelecypod history. *Conservation Biology*, **3**, 274–281.

Marine mussels, their evolutionary success, ecological significance and use as chronometers of environmental change

R. SEED, C. A. RICHARDSON & K. SMITH

School of Ocean Sciences, University of Wales–Bangor, Menai Bridge, Anglesey LL59 5EY, UK (e-mail: r.seed@bangor.ac.uk)

Abstract: The evolution of the heteromyarian form and neotenous retention of the byssus into the adult stage were central to the colonization of hard substrata and the subsequent evolutionary success of epibyssate mytilids. In addition, these mussels are supreme filter feeders and exhibit many structural, behavioural and life-history traits which make them particularly successful colonizing species. As competitively dominant species, mussels can potentially monopolize epibenthic communities with a consequent reduction in diversity of the primary space-occupying species. However, mussel patches themselves are extremely diverse, offering numerous resources which are exploited by a range of associated organisms from most animal phyla. Mussel beds can influence large-scale coastal and estuarine processes such as productivity and eutrophication, and can significantly alter sediment biogeochemistry. Annual growth lines and tidal bands in mussel shells have been used to estimate age, and to investigate the effects of environmental factors such as seawater and air temperature, spring–neap lunar tidal cycles and exposure to chlorination regimes in power station culverts. Analysis of the elemental composition of the incremental record in long-lived mytilids has been used to assess the effects of anthropogenic inputs into coastal waters.

Mytilid mussels, especially those belonging to the genus *Mytilus*, are amongst the most intensively researched marine organisms with entire books devoted solely to their study (e.g. Bayne 1976; Stefano 1990; Gosling 1992). There are many reasons for such scientific interest; e.g. these mussels are widely distributed and exceedingly abundant throughout the world's oceans; they are important ecologically as dominant space-occupying organisms, particularly in coastal and estuarine waters (e.g. Seed & Suchanek 1992); they are important economically as food and fouling organisms (e.g. Hickman 1992). Moreover, from as early as 1976, when the Mussel Watch Monitoring Program was initiated, mussels have been widely used as sentinels or biomonitors of coastal water quality (e.g. Goldberg 1986; Widdows & Donkin 1992). Some species, such as the common blue mussel (*Mytilus edulis* Linnaeus), have also proved to be model organisms in studies of physiology, biochemistry and population genetics (see relevant chapters in Gosling 1992). Moreover, mussel beds are amongst the most productive assemblages on Earth, often rivalling the productivity of tropical rain forests and kelp beds, whilst in historic and prehistoric times *Mytilus* shells were used extensively for tools by native North American Indians as well as by the Pilgrim settlers (Suchanek 1985).

Amongst the marine mussels, species of *Mytilus* are particularly widespread throughout the cooler waters of both the northern and southern hemispheres, attached by means of byssus threads to rock and other hard or semi-consolidated surfaces [see McDonald *et al.* (1991) and Suchanek *et al.* (1997) for current taxonomy and global distributions of these species]. In tropical and subtropical latitudes, however, this genus is replaced by other dominant zone-forming genera such as *Perna* and *Septifer*. Whilst most marine mussels occur in shallow inshore waters, dense populations of highly specialized species have been reported from cold-seep areas and sites of hydrothermal vent activity in the deep sea (e.g. Hessler *et al.* 1988; Jahnke *et al.* 1995). These bathymodiolids filter particulate material from the water column in the usual way but, remarkably, can also tap other nutrient sources in the form of sulphur- and/or methane-oxidizing bacteria contained within their own tissues (e.g. Reid 1990; Kochevar *et al.* 1992; Fisher *et al.* 1993). There are now over 20 known species of *Bathymodiolus*, the largest of which (*B. boomerang* n. sp.) can attain a shell length of 360 mm and is probably the largest recorded living

From: Harper, E. M., Taylor, J. D. & Crame, J. A. (eds) *The Evolutionary Biology of the Bivalvia*. Geological Society, London, Special Publications, **177**, 465–478. 1-86239-076-2/00/$15.00

mussel (von Cosel & Olu 1998). In addition to these bathymodiolids, species of *Idas* and *Adipicola* live associated with sunken decaying wood and other organic remains in deep water (Horikoshi & Tsuchida 1984; Warén 1991); *Idas simpsoni* (Marshall) curiously occurs on whale skeletons or in drifting sperm whale blubber – these mussels may also house symbiotic bacteria. Other unusual mussels include those which construct byssal nests (e.g. *Amygdalum*), live within the tests of ascidians (e.g. *Musculus*) or bore into the substratum (e.g. *Adula*).

Thus, although the marine mytilids exhibit a remarkably wide diversity of lifestyles, this paper focuses only on those epibyssate colonizers of hard intertidal and shallow subtidal substrata. Firstly, some of the adaptive features that underpin the success of epibyssate mytilids will be explored, then the ecological significance of this success will be examined, in particular their competitive ability and the importance of mussel beds as islands of enhanced biodiversity. Finally, the potential value of mytilids as chronometers of environmental change and as environmental biomonitors will be considered.

Evolution and adaptation

Of the many factors that are undoubtedly responsible for the success of mytilid mussels, the evolution of the heteromyarian form associated with the neotenous retention of the byssus is of particular significance [see Seed (1983) and Morton (1992) for reviews]. The secretion of byssus threads by a special gland at the base of the foot provided an effective means of attachment onto hard surfaces and enabled various bivalve classes, including mussels, to become independent of the soft sediments inhabited by their ancestors. Thus, the neotenous retention into adult life of a post-larval characteristic, originally used for temporary attachment during metamorphosis, created the opportunity for a marked shift in evolutionary direction. The broadly triangular heteromyarian form of the 'typical' mussel, coupled with secure byssal attachment throughout life, proved to be the ideal solution for bivalves like epibyssate mussels which live in high population densities on hard or moderately consolidated substrata. Indeed, so successful was this solution that it has been evolved independently amongst otherwise unrelated bivalve lineages, most notably the freshwater dreissenid mussels and their marine counterparts the mytilids, thus providing a particularly clear and elegant example of convergence.

Factors other than shell morphology and byssal attachment have contributed to the success of mytilid mussels. One such factor is their supreme efficiency as filter-feeding organisms. Individual mussels can filter several litres of water per hour and are capable of extracting extremely small particles, such as bacteria; this is one of the reasons why they can be a potentially serious threat to public health (e.g. Shumway 1989). However, this filtering efficiency has also been put to use for clearing polluted dockland sites: e.g. in the two years following the introduction of mussels to Liverpool docks, water quality and oxygen content of both the water column and the sediment improved significantly as bacteria and other suspended particulate material were removed by the mussels (Allen & Hawkins 1993).

Marine mytilids, as well as some brackish and freshwater bivalves such as the zebra mussel, *Dreissena polymorpha* (Pallas), possess several life-history traits which make them especially effective as colonizing species. For example; they are highly fecund; they occur at high population densities, producing large numbers of planktotrophic larvae which ensure rapid and widespread dispersal (further facilitated by bysso-pelagic drifting of juveniles); they are tolerant of extended periods of aerial exposure and thus easily transported, often accidentally, from one locality to another [for general reviews see Bayne (1976) and Gosling (1992)]. Once established they can, by virtue of their competitive superiority (see later), quickly become the dominant space-occupying members of the community, with the potential to completely eliminate most other sessile species. This colonizing ability of mussels can create potentially serious economic problems for industry, in particular to nuclear power stations, if mussel larvae present in the cooling water are allowed to settle and foul the vulnerable condenser tubes (Thompson *et al.* 1997).

Specific adaptations to conditions prevailing at opposite ends of an environmental gradient are exhibited by *Perna viridis* (Linnaeus) and *Septifer virgatus* (Wiegmann). Although these two common mytilids are widely distributed throughout the coastal waters of Hong Kong, only rarely do their local distribution patterns overlap, *Perna* occurring predominantly in sheltered, often heavily polluted, waters and *Septifer* on cleaner, wave- exposed shores of the outer coast. These mussels have evolved contrasting suites of structural and behavioural traits which correlate well with a presumed advantage for each species within its preferred habitat (Seed & Richardson 1999). Thus, whereas *Septifer* possesses a more robust, ventrally flattened shell and stronger byssal attachment than *Perna*, characteristics well suited for life on high-energy rocky shores, *Perna* produces copious amounts of mucus and has unusually large labial palps with strong ciliary rejection tracts, features more appropriate for coping with the high sediment loadings often associated with sheltered, low-energy habitats. Moreover, it has been suggested that excessive mucus production may be associated with the elimination of toxic trace metals such as copper (Sze & Lee 1995). *Perna* also possesses a larger, more mobile foot and crawls more readily than *Septifer*. In particular, it exhibits a marked propensity to climb vertically upwards, a response which effectively elevates this mussel above any accumulated sediment; *Septifer*, by contrast, does not exhibit this behavioural trait. Similar adaptations have been previously reported for *Mytilus californianus* (Conrad) and *M. edulis* which occupy wave-swept and sheltered habitats, respectively, along the Pacific coast of North America (Harger 1972).

Ecological significance and biodiversity

An important organizing principle in many space-limited communities, such as those associated with the rocky intertidal zone where most of the common species are sedentary, is the existence of an approximate hierarchy in the competitive abilities of the primary space-occupying species. Such hierarchies, if left undisturbed, can eventually lead to extreme dominance (monopolization) by the competitively superior species. However, moderate levels of disturbance, either physical (e.g. waves) or biological (e.g. predation), operating on species high in the hierarchy can promote species richness by interrupting the competitive process and creating patches of bare rock which can then be exploited, at least temporarily, by lower ranking opportunistic species. Low levels of disturbance may be insufficient to prevent competitive monopolization whilst high levels may reduce species richness by acting adversely on all species. Keystone predators are those which consume competitively dominant species. On rocky shores, mussels are often simultaneously the competitively dominant species *and* the preferred prey for the keystone predator, typically a starfish, strongly suggesting that common mechanisms serve to organize rocky shore communities in different geographical areas (Boaden & Seed 1985). The diversity promoting role of predation has been elegantly demonstrated in the *Mytilus–Pisaster* interaction on the wave-exposed shores of Washington State (Paine 1974). At sites where the starfish was regularly removed the number of primary space-occupying species within the community (*c*. 25) declined and after a period of approximately ten years only mussels remained; at the undisturbed control sites, where starfish were left to forage on the mussels, species richness remained high throughout the experimental period. The striking results of these predator exclusion experiments provide a clear and unambiguous demonstration of an ecological principle operating in nature.

The tidal restriction of most marine predators often results in distinct predation lines below which established prey populations are rare. Thus, on the east coast of England, predatory starfish [*Asterias rubens* (Linnaeus)] and dogwhelks [*Nucella lapillus* (Linnaeus)] effectively remove *Mytilus edulis* from much of the lower shore (Seed 1969). Predation in the upper shore, by contrast, is minimal and mussels occurring there can often be exceptionally long lived (> 20 years). Similarly, experiments carried out on the Pacific Coast of North America have shown that *Pisaster ochraceus* (Brandt) effectively controls the distribution of its main prey, *Mytilus californianus*, on the lower shore. When this starfish was experimentally removed from a section of shore, over a period of five years, the lower edge of the mussel zone gradually extended downshore, but when the starfish was allowed to re-enter the system the lower limit of mussels gradually returned to the previous state (Paine 1974). Similar downward shifts in the distribution of *Perna canaliculus* (Gmelin) in New Zealand and *Perumytilus purpuratus* (Lamarck) in Chile have been observed following the removal of major predatory starfish *Stichaster* and *Heliaster*, respectively (Paine *et al.* 1985).

The impact of the large predatory gastropod *Concholepas concholepas* (Bruguière) on the distribution of mussels has been demonstrated on the rocky coastline of central and southern Chile where the intertidal zone is typically dominated by *Perumytilus purpuratus*. When the gathering of *Concholepas*, a prized commercial species, was halted, by excluding fishermen from a designated Marine Reserve, the intertidal zone switched dramatically from one dominated by mussels to one dominated by barnacles. No such change occurred outside the reserve where *Concholepas* was still heavily harvested. Moreover, when *Perumytilus* was experimentally protected from *Concholepas* by cages deployed inside the reserve the mussel population soon became re-established (Moreno *et al.* 1986). Thus, the act of simply protecting an important natural predator caused the entire landscape of the intertidal zone to change, and it therefore appears that the mussel-dominated community normally observed along much of the Chilean coast is in an artificially altered state which is largely shaped and maintained by human factors.

A relatively new and exciting field of research involving mussels concerns the extent to which these densely aggregated, filter-feeding organisms act as processors of inshore water. This research involves the use of large (*c.* 10 m long) plastic tunnels which are placed across the mussel bed and secured firmly to the substratum. Water is sampled as it enters and leaves the tunnel whilst monitors, placed within the tunnel, enable the flow dynamics of water over the mussel bed to be established. Any changes in the quality of the inflowing and outflowing water can then be attributed to the activities of the mussels. Results have shown that mussel beds can remove large quantities of suspended particulate organic matter and also release dissolved metabolites back into the environment where they are readily utilized by the phytoplankton (Fig. 1). Biodeposits (faeces and pseudo-faeces) produced by the mussels are incorporated into the sediment and enter the mineralization process. These tunnel experiments have demonstrated that mussel beds (and oyster reefs) can have a profound influence on large-scale coastal and estuarine processes such as productivity and eutrophication, and that they can also alter significantly the biogeochemistry of the underlying sediment (Dame 1996).

It is clear from the above that, in the absence of adequate levels of disturbance, mussels, as superior spatial competitors, can effectively monopolize large areas of the rocky intertidal zone by denying access to the weaker more opportunistic species. Such loss of species richness, however, applies only to those species that compete for the primary resource – attachment space – on the rock surface. In contrast, the complex three-dimensional structure of the mussel patches themselves provide a suitable habitat for rich assemblages of associated organisms, including representatives from most of the major animal phyla (e.g. Suchanek 1979; Tsuchiya & Nishihira 1986; Ong Che & Morton 1992).

Aggregations of mussels can drastically modify the local environment through enhanced water retention, biodeposition of faecal and pseudofaecal material (= 'mussel mud'), and the provision of additional attachment surface and shelter by the mussels themselves – features which serve to encourage species enrichment in habitats wherever mussels are present in abundance. Mussels are thus effective 'ecosystem engineers', a term used by Lawton & Jones (1995) to describe species that either directly or indirectly modulate the availability of resources to other species by causing physical state changes in biotic or abiotic materials, thereby modifying, maintaining and creating habitats. Mussel patches support rich assemblages of associated organisms (Seed 1996). Some of these organisms live attached to the mussel shells (= epibiota), others typically live amongst the rich sediments and shell fragments which accumulate at the base of the bed (= infauna), whilst mobile organisms rove freely through the complex matrix of shells and interconnecting byssus threads.

Table 1 documents the number of macroinvertebrate taxa associated with several rocky shore mytilids from wave-exposed rocky shores together with some of the dominant species found within these communities. A striking feature is that similar taxa, often from the same genus, regularly recur within these communities worldwide. Moreover, with the exception of *Mytilus californianus* – a much larger bodied mussel which forms thick multilayered beds – the number of associated taxa is broadly comparable between species. Such observations suggest that the number of niches available within these structurally similar 'habitats'

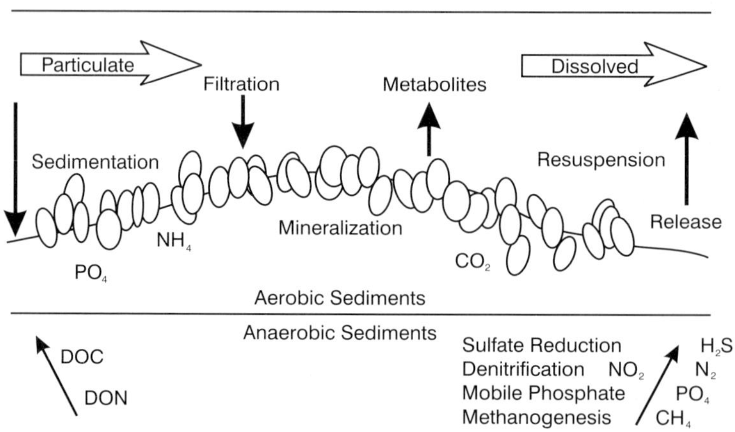

Fig. 1. A conceptual summary of the processes occurring in and around dense systems of filter-feeding bivalves such as mussels [after Dame (1996)].

Table 1. *Macroinvertebrate fauna associated with various species of rocky shore mytilids [after Seed (1996)]*

Mussel species (location)	No. taxa	Dominant species within community: Mollusca	Crustacea	Polychaeta
*Septifer virgatus** (Hong Kong)	52	*Lasaea nipponica* *Hormomya mutabilis* *Isognomon legumen*	*Hyale* sp. *Chthamalus sinensis*	*Typosyllis* sp.
Mytilus edulis† (North Wales)	56	*Lasaea rubra*	*Semibalanus balanoides*	
Brachidontes rostratus‡ (Southeast Australia)	56	*Lasaea australis* *Xenostrobus pulex*	*Chthamalus antennatus* *Allorchestes compressa*	*Typosyllis* sp. *Perinereis* sp.
M. californianus§ (North America)	*c*. 270	*Lasaea subviridis* *Barleeia sanjuanensis*	*Hyale* sp. *Semibalanus cariosus* *Chthamalus dalli*	*Typosyllis adamanteus* *Syllis* sp.
Perumytilus purpuratus (Chile)	65	*Lasaea petitiana* *Semimytilus algosus* *Siphonaria lessoni*	*Hyale* sp. *Chthamalus scabrosa* *Jehlius cirratus*	*Typosyllis* sp. *Perinereis* sp. *Pseudonereis* sp.

*Ong Che & Morton (1992); †Lintas & Seed (1994); ‡Peake & Quinn (1993); §Suchanek (1979).

is limited and that when the associated communities are at, or near, equilibrium these are occupied by taxonomically and functionally similar species. It seems therefore that the pattern of parallel communities on rocky shores may be replicated on a much finer scale within mussel patches (Seed 1996). Despite their generally high levels of biodiversity, which can vary both temporally (e.g. Peake & Quinn 1993) and spatially according to the age, size and structural complexity of the patch, the degree of aerial and tidal exposure and the amount of accumulated sediment (e.g. Tsuchiya & Nishihira 1986; Lintas & Seed 1994), intertidal mussel communities are typically dominated by a few very abundant species. Figure 2 shows the rank abundance of the fauna associated with three species of rocky shore mytilids. The overall distributions are similar for all three communities and indicate high dominance by a few common species.

The macroinvertebrate communities associated with rocky intertidal mussel patches, particularly when these patches experience similar physical conditions and are of similar size and structural complexity, often exhibit a remarkable degree of conformity when compared using simple univariate statistics. However, such techniques, which typically collapse the full set of species counts into a single coefficient, tend to grossly oversimplify complex systems and reveal little about community organization (Schulter & Ricklefs 1993). Thus, although this group's, as yet unpublished, data show that the communities associated with *Perumytilus purpuratus* in central Chile and *Mytilus edulis* in the UK are indistinguishable in terms of their species richness, diversity and evenness, significant differences in community structure do emerge (Fig. 3) when multivariate techniques are applied to these same communities. For example, whilst the *P. purpuratus* community supports surprisingly large numbers of sea anemones and grazing gastropods, these taxa are only rarely encountered within the *M. edulis* community which, in turn, supports large numbers of meiofaunal taxa including nematodes, nemerteans and mites. The preliminary data on *P. purpuratus* further suggests that dominance by relatively few species is most pronounced in those communities that experience enhanced levels of anthropogenic stress. As biodiversity typically reflects environmental health, mussel patch communities could prove to be valuable bioindicators of coastal and estuarine water quality.

Mussels as chronometers of environmental change

In mussels, as in most bivalves, changes in growth are permanently recorded in the shell in the form of rings or bands. A well-established technique which enables the internal structure of mussel shells to be examined for the presence of annual growth checks and microgrowth patterns involves the preparation of acetate peel replicas of cut, polished and etched radial shell sections (Rhoads & Lutz 1980;

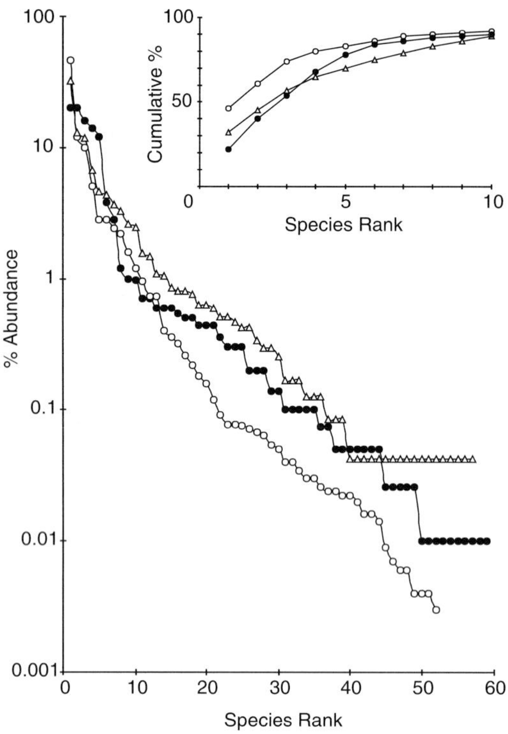

Fig. 2. Rank abundance plots showing the diversity of the macroinvertebrate fauna associated with *Mytilus edulis* from North Wales (●); *Septifer virgatus* from Hong Kong (○); and *Perumytilus purpuratus* from Central Chile (△). Inset: Cumulative abundance of the ten most common species within each community. *M. edulis* and *S. virgatus* data from Lintas & Seed (1994) and Ong Che & Morton (1992), respectively.

Richardson 1989, 1990). When these peels are viewed under a light microscope a clearly defined pattern of growth bands or growth lines can be seen in various regions of the shell. Figure 4a shows a diagrammatic section through a generalized mussel shell. The shell typically consists of a periostracum, an outer prismatic shell layer of calcite or aragonite, an inner nacreous layer and an umbo. Prominent growth lines can be seen as a pattern of light and dark lines (arrows) in the internal structure of the umbo and in the nacreous layer (Fig. 4b and c). In *Mytilus edulis* (Lutz 1976; Richardson *et al.* 1990) and *Geukensia demissa* (Dillwyn) (Lutz & Castagna 1980) these inner nacreous lines have been shown to be deposited annually. This fact has enabled determination of the age and growth rates of *M. edulis* (e.g. Bayne & Worrall 1980; Rodhouse *et al.* 1984), thereby also allowing individual linear growth rates to be determined, providing that the relationship between growth of the nacreous lines and shell length has first been established (Seed & Richardson 1990). A potentially useful technique for validating the position of the annual rings in temperate water mussels, and other bivalve species, utilizes the ratio of naturally occurring oxygen isotopes ($^{18}O/^{16}O$) within the calcite of the shell. This ratio is temperature dependent and samples taken at known intervals along the shell provide a temperature profile which may reflect seasonal variations in growth rate (Margosian *et al.* 1987). Dodd (1965) demonstrated a strong negative correlation between strontium concentrations in the nacre and temperature (13–19°C) in *Mytilus edulis*

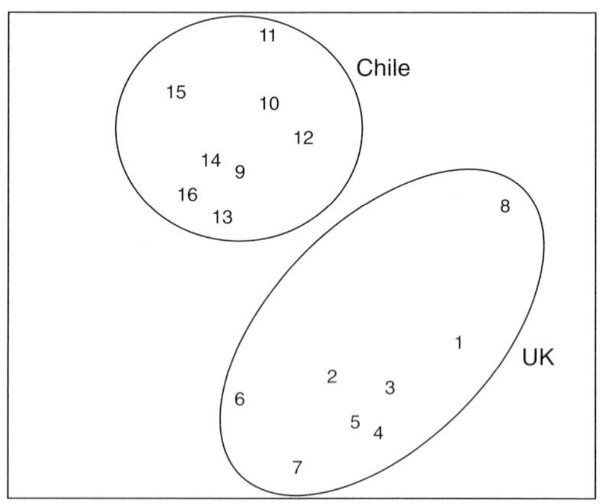

Fig. 3. Non-metric multidimensional scaling (MDS) plot for intertidal mussel bed communities at eight sites in the UK and eight in Chile. Stress value, 0.1; ANOSIM Global R, 0.886 ($p < 0.05$); these values indicate significant differences in the communities at these two major geographical sites.

Fig. 4. (**a**) Schematic diagram of a shell section of *Mytilus edulis chilensis* to illustrate the layers of the shell: p, periostracum; pl, prismatic layer; n, nacreous layer; u, umbo. (b)–(g) Photomicrographs of acetate peels of sections of *Mytilus*: (**b**) annual growth lines (arrows) in the umbone and (**c**) annual growth lines (arrows) in the nacreous layer of *M. e. chilensis* from the Falkland Islands; (**d**) position of the winter growth check (arrow) in the shell of *M. e. chilensis*; (**e**) a disturbance check (arrow) in the prismatic layer of the shell of *M. e. chilensis*; (**f**) very faint bands deposited during a period of non-chlorination in the shell of *Mytilus edulis*; (**g**) disturbance to the normal band pattern (arrow) in *M. edulis* exposed to chlorination from a power station culvert. Scale bars, 100 μm.

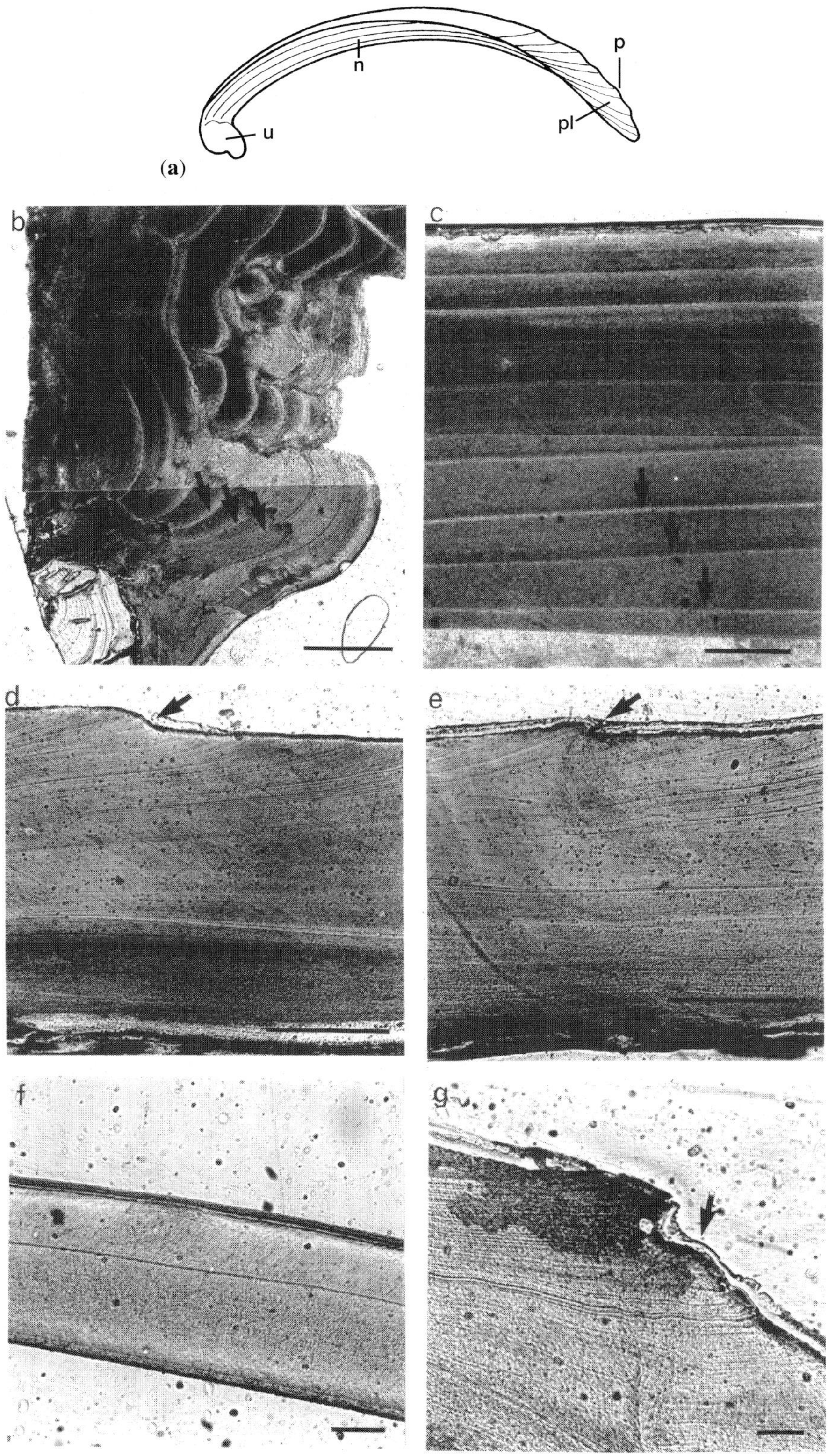
n
p
pl
u
(a)
b
c
d
e
f
g

and *Mytilus californianus* collected from Californian waters. Magnesium concentrations in the outer prismatic layer of *Mytilus edulis* increased markedly with decreasing salinity and showed a weak positive correlation with temperature (Dodd 1965, 1967).

Magnesium/calcium ratios determined in sequentially drilled samples in the shell of *Mytilus trossulus* from British Columbia, Canada, were found to provide an accurate measure of seawater temperature (Klein *et al.* 1996). Moreover, it has recently been suggested that the actual volume of calcite deposited in *Mytilus* shells is inversely related to water temperature (Carter & Seed 1998), a relationship which offers some potential for estimating sea-surface palaeotemperatures from fossil material. Variation in the $^{13}C/^{12}C$ ratios in the shell of *Mytilus californianus* has been shown to correspond to upwelling events, thus providing an opportunity for determining the frequency of palaeoupwelling events in fossil mussel shells (Killingley & Berger 1979).

The outer shell layer of *Mytilus edulis* contains a series of fine microscopic alternating light and dark bands (Fig. 4d). Dark bands have a semi-diurnal periodicity of formation, each band being deposited when the mussel is emersed at low tide; the wider lighter bands are laid down during immersion when the mussel is actively feeding (Richardson 1989). Under continuously immersed conditions the bands are weaker in definition and reflect an endogenous rhythm in shell deposition (Richardson 1989). Tidal bands in the outer layer have been used to determine the annual rates of growth of several mytilid species. Growth increments vary along the shell in response to seasonal environmental changes. During the summer (May–August in northern latitudes, December–March in southern latitudes), when seawater temperatures are maximal, shell growth is rapid and wide increments are deposited. With the onset of autumn, shell formation gradually slows and the increments become progressively narrower. This seasonal pattern in the narrowing of the increments results in annual growth checks in the outer shell layer (Fig. 4d) and the distance between successive checks can be measured directly from the peel, thus allowing the growth rate of individual mussels, and hence the population growth rate, to be determined. However, the presence of seasonal patterns in increment width may not necessarily be associated with seasonal changes in water temperatures. Richardson *et al.* (1995) showed that in *Septifer virgatus*, Hong Kong, two groups of narrowly spaced tidal bands, representing periods of reduced linear growth, were deposited each year. One of these groups appeared to be associated with low seawater temperatures during winter (February–March) whilst the other appeared to be related to a period of post-reproductive stress aggravated by high temperatures during summer (July–August). A confounding factor in estimating age is the presence of growth disturbances in the shell (Fig. 4e) resulting from, for example, removal of cage-held mussels during cage cleaning (Richardson *et al.* 1990). These checks arise due to withdrawal of the mantle edge during disturbance and a change in the direction of shell growth, which could be misinterpreted as an annual growth check. A sudden interruption to the normal banding pattern distinguishes these growth checks from the seasonally induced changes in shell growth.

Shell growth rates can be obtained either by measuring the width of the most recently deposited growth increment, corresponding to one tides growth (Fig. 5a), or the width of a number of increments equivalent to several days (tides) growth. Intertidal mussels show variations in the definition of the growth bands and differences in shell growth rate during the spring–neap lunar tidal cycle. Mussels occupying the lower intertidal zones deposit dark bands when they are emersed during spring low tides, and weakly defined bands and enhanced shell growth rates during neap tides when they are continuously immersed (Fig. 5a and b). Mussels growing in the middle and upper zones of the shore display a pronounced spring–neap lunar banding pattern with narrowly spaced increments and strongly defined bands laid down at neap tides when they are emersed for longer periods than at spring tides. These consistent and obvious differences in band strength and increment width, documented in mussel shells from different tidal levels worldwide (Richardson *et al.* 1990, 1995; Gray 1997), could have palaeoecological significance.

Growth patterns in fossil or subfossil mussel shells, or those from death assemblages, for instance, could be used to reconstruct tidal conditions and determine the shore level which the mussels had occupied: weakly defined patterns would indicate subtidal growth whilst an obvious 14 day spring–neap lunar variation in increment width, with pronounced narrowing of increments together with the presence of dark growth bands, would be indicative of a high shore existence. Differences in the nature of the amplitude of the tides might be discerned from variations in increment width and definition of adjacent tidal bands. *Mytilus edulis* exposed to a semi-diurnal tide of similar amplitude (emersion twice daily) in the Menai Strait, North Wales, deposit tidal bands of uniform definition (Richardson 1989), whereas *M. e. chilensis* (Hupé) inhabiting a mixed semi-diurnal tidal regime, e.g. in the Falklands (Fig. 5c), where there is unequal tidal amplitude at spring tides

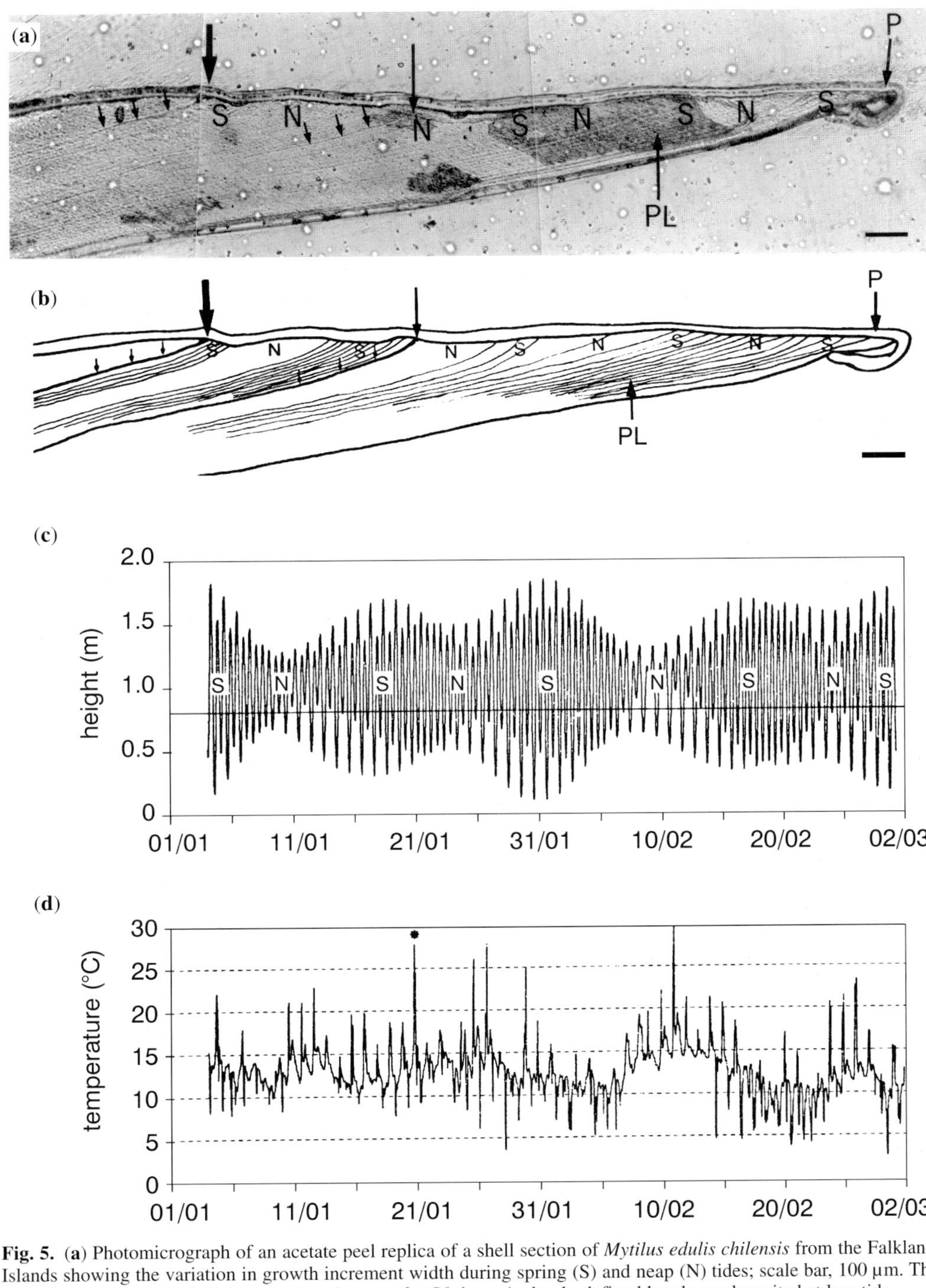

Fig. 5. (**a**) Photomicrograph of an acetate peel replica of a shell section of *Mytilus edulis chilensis* from the Falkland Islands showing the variation in growth increment width during spring (S) and neap (N) tides; scale bar, 100 μm. The mussel was date marked (large arrow) and grown for 56 days. A clearly defined band was deposited at low tide during an anomalously high air temperature (small arrow). P, periostracum; PL, prismatic layer. (**b**) Camera lucida drawing of the positions of the bands in the prismatic layer in (a). (**c**) Predicted tidal cycle during the experimental period; horizontal line marks the tidal height where the mussel was growing. (**d**) Combined records of seawater temperatures during immersion and air temperatures during emersion; *, the anomalously high air temperature which occurred when the mussel was emersed for several hours during spring low tide [data from Gray (1997)].

(daily emersion), deposit one dark widely spaced band during emersion and two narrow bands a day during emersion at neap tides (emersion twice daily) (Gray 1997). Pannella (1976) has demonstrated that suitably preserved fossil bivalve shells may be used to reconstruct the tidal regime of past oceans, and evidence from experimentally marked and transplanted mussels (Richardson 1989; Gray 1997) indicates that it is feasible to determine the animal's tidal elevation on the shore from the shell banding patterns.

Since the tidal increments can be used to measure variations in shell growth rates they can also be used to investigate the effects of environmental factors on shell growth. Peels of shells of *M. edulis* collected from power station culverts (Fig. 4f), unlike those from naturally occurring mussel populations, show only a very faint banding pattern. These culvert mussels annually receive, between June and November, continuous exposure to chlorination designed to reduce biofouling within the culvert. Mussels that survive chlorination leave evidence of this event as a series of dark prominent bands which result from reduced growth when the shell valves remain tightly closed (Fig. 4g). Chlorinated water discharged from the power station into coastal waters may also affect the natural mussel populations. Examination of the internal structure of mussel shells from within the culvert for the presence of prominent growth bands has enabled the efficiency of chlorination in retarding shell growth to be assessed (Thompson *et al.* 2000). The technique could be extended to investigate the impact of chlorinated water discharges on the natural mussel populations in the vicinity of the culvert.

Using experimentally marked *M. e. chilensis*, transplanted for 56 days onto the shore at Camilla Creek, Falklands, Gray (1997) demonstrated that an unusually distinct band was deposited in the shell which coincided with an anomalously high air temperature during emersion on one spring low tide. The growth band pattern, in which each increment and band was assigned a date, was analysed and compared with continuous *in situ* hourly records of seawater and air temperatures. On 20 January 1996, when the mussels were emersed during spring low tide for several hours they experienced a temperature of 28°C which resulted in the formation of a distinct band (Fig. 5a and d). During neap tides, however, although unusually high air temperatures were recorded, the mussels were not emersed and no distinct bands were deposited.

Mussels have many attributes which have led to their use as 'sentinel' or 'indicator' organsisms in environmental monitoring programmes (Phillips 1980; Widdows & Donkin 1992). They are, as was demonstrated earlier, dominant members of coastal and estuarine communities and have a wide geographical distribution; they are sedentary and therefore better than mobile species as integrators of chemical contamination in a given area; they process large volumes of water and concentrate many chemicals in their tissues and, as they are commercially important, contamination of their tissues is of concern to public health. Most metals occur in forms which are highly soluble in seawater and are taken up directly from solution into the soft tissues. Metals can also be absorbed and incorporated into the shell, thus providing the potential for long-term records of changes in environmental contamination. Metals such as lead are highly concentrated in the periostracum covering the shell (Sturesson 1976). However there does not appear to be a relationship between metal levels in the whole shell and the tissues (Bryan & Uysal 1978), although Bourgoin (1990) recently showed that nacre lead levels in *Mytilus edulis* were strongly correlated with the tissue lead concentrations; nacre contained one-tenth of the lead in the tissues. Using modern analytical tools, e.g. electron probe (Rosenberg & Jones 1975) and, more recently, laser ablation inductively coupled plasma-mass spectrometry (LA–ICP–MS) (Gray 1982), it has been possible to analyse the chemical composition of the shell, particularly with respect to environmentally sensitive elements such as lead, copper, cadmium and zinc, and to biogenic elements such as calcium, strontium and magnesium which comprise the natural structure of the shell. Since mytilids contain annual patterns of linear growth within their shells and some, e.g. the horse mussel *Modiolus modiolus*, attain ages of > 50 years (Anwar *et al.* 1990), the possibility exists for studying both annual and seasonal variations in a range of elements in the chemical record of shell growth.

Using electron microprobe step-scan analysis across the inner nacreous layer of a sectioned *Mytilus edulis* shell, Lutz (1981) concluded that there was no correlation between strontium and the inner nacreous growth lines. Using LA–ICP–MS, Perkins (1992) demonstrated temporal variations in magnesium, and geographic variations in copper, lead and zinc in shells of *M. edulis* from around the coastline of Wales. Richardson *et al.* (2000) analysed the chemical composition of the incremental growth record in *M. modiolus* shells collected in 1984 from two sites in the North Sea: one of these, a 'waste dump' (impacted) site to the east of the Humber Estuary, had historically received a mixture of sewage sludge and industrial sludge (1.1×10^5 wet tonnes and 5×10^3 wet tonnes, respectively, in 1986), whereas the other site, off the Norfolk coast, was distant from any

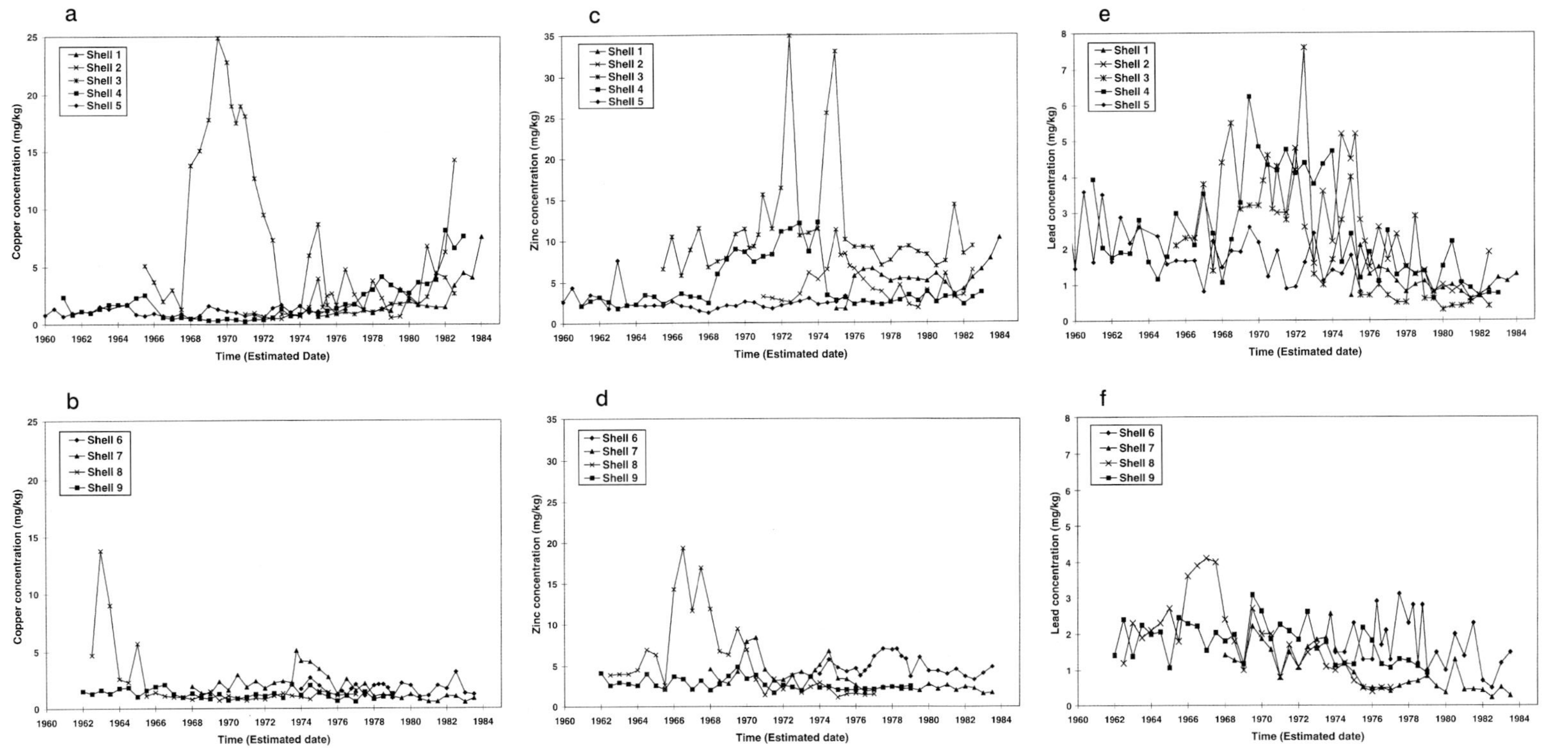

Fig. 6. Contaminant element concentrations in horse mussel, *Modiolus modiolus*, shells (1–5) from an impacted site and shells (6–9) from a control site between 1962 and 1984. (**a**) and (**b**) copper levels; (**c**) and (**d**) zinc levels; (**e**) and (**f**) lead levels (from Richardson *et al.* 2000).

known point-source inputs (control site). The age of the shells was determined from the alternating pattern of light (summer) and dark (winter) lines present in the shells (Anwar *et al.* 1990), and a Spectron frequency quadrupled Nd:YAG laser-ablation microprobe operating in the far-UV (266 nm) and linked to a light microscope was used to ablate 40 μm diameter, 50 μm depth samples. Up to 22 summer and winter lines (= 22 years of mussel life) in the oldest shells, covering the period of 1962–1984, were sampled (Richardson *et al.* 2000). ICP–MS analysis of ablated shell material demonstrated that the shells from the impacted site contained elevated levels of copper, zinc and especially lead between 1968 and 1976 (Fig. 6a, c and e), coinciding with a change in contaminant inputs between the early 1960s and 1984 when the mussels were collected. During the same period, concentrations of these elements in control shells remained around background levels (Fig. 6b, d and f). In 1974 the Dumping at Sea Act was enacted, which subsequently became the Food and Environment Protection Act, and Richardson *et al.* (2000) speculated that introduction of this legislation was likely to have had an impact on trace element concentrations in the shells between 1964 and 1984. The observed levels of heavy metals in shells from the impacted site during a period of dumping of sewage sludge and industrial sludge from the Humber Estuary, followed by a decline in metal concentrations in the shells after the mid-1970s, suggested that the shells were indeed providing a record of metal contamination at this particular site.

References

ALLEN, J. R. & HAWKINS, S. J. 1993. Can biological filtration be used to improve water quality? *In*: WHITE, K. N., BELLINGER, E. G., SAUL, A. J., SYMES, M. & HENDRY, K. L. (eds) *Urban Waterside Regeneration: Problems and Prospects*. Ellis Horwood Press, 377–385.

ANWAR, N. A., RICHARDSON, C. A. & SEED, R. 1990. Age determination, growth rate and population structure of the horse mussel *Modiolus modiolus*. *Journal of the Marine Biological Association of the UK*, **70**, 441–457.

BAYNE, B. L. (ed.) 1976. *Marine Mussels: Their Ecology and Physiology*. Cambridge University Press.

—— & WORRALL, C. M. 1980. Growth and production of mussels *Mytilus edulis* from two populations. *Marine Ecology Progress Series*, **3**, 317–328.

BOADEN, P. J. S. & SEED, R. 1985. *An Introduction to Coastal Ecology*. Blackie.

BOURGOIN, B. P. 1990. *Mytilus edulis* shell as a bioindicator of lead pollution: considerations on bioavailability and variability. *Marine Ecology Progress Series*, **61**, 253–262.

BRYAN, G. W. & UYSAL, H. 1978. Heavy metals in the burrowing bivalve *Scrobicularia plana* from the Tamar estuary in relation to environmental levels. *Journal of the Marine Biological Association of the UK*, **58**, 89–108.

CARTER, J. G. & SEED, R. 1998. Thermal potentiation and mineralogical evolution in *Mytilus* (Mollusca; Bivalvia). *In*: JOHNSTON, P. & HAGGART, J. W. (eds) *Bivalves: An Eon of Evolution – Paleobiological Studies Honoring Norman D. Newell*. University of Calgary Press, 87–117.

COSEL, R. VON & OLU, K. 1998. Gigantism in Mytilidae. A new *Bathymodiolus* from cold seep areas on the Barbados accretionary prism. *Comptes Rendus de l'Academie des Sciences Serie III, Sciences de la Vie*, **321**, 655–663.

DAME, R. F. 1996. *Ecology of Marine Bivalves: An Ecosystem Approach*. CRC Marine Science Press.

DODD, J. R. 1965. Environmental control of strontium and magnesium in *Mytilus*. *Geochimica et Cosmochimica Acta*, **29**, 385–398.

—— 1967. Magnesium and strontium in calcareous skeletons: a review. *Journal of Paleontology*, **41**, 1313–1330.

FISHER, C., BROOKS, J. M., VODENICHAR, J. S., ZANDE, J. M., CHILDRESS, J. J. & BURKE, R. A. 1993. The co-occurrence of methanotrophic sulfur-oxidising bacterial symbionts in a deep-sea mussel. *Marine Ecology*, **14**, 277–289.

GOLDBERG, E. D. 1986. The mussel watch concept. *Environmental Monitoring and Assessment*, **7**, 91–103.

GOSLING, E. M. (ed.) 1992. *The Mussel Mytilus: Ecology, Physiology, Genetics and Culture*. Elsevier.

GRAY, A. L. 1982. Solid sample introduction by laser ablation for inductively coupled plasma source mass spectrometry. *Analyst*, **10**, 551–556.

GRAY, A. P. 1997. *The ecology of the mussel* Mytilus edulis chilensis *from three sites in the Falkland Islands*. PhD Thesis, University of Wales, Bangor.

HARGER, J. R. E. 1972. Competitive coexistence: maintenance of interacting associations of sea mussels, *Mytilus edulis* and *Mytilus californianus*. *Veliger*, **14**, 387–408.

HESSLER, R. R., SMITHEY, W. M., BOUDRIAS, M. A., KELLER, C. H., LUTZ, R. A. & CHILDRESS, J. J. 1988. Temporal changes in megafauna at the Rose Garden hydrothermal vent (Galapagos Rift; eastern tropical Pacific). *Deep Sea Research*, **35**, 1681–1709.

HICKMAN, R.W. 1992. Mussel Cultivation. *In*: GOSLING, E. M. (ed.) *The Mussel Mytilus: Ecology, Physiology, Genetics and Culture*. Elsevier, 465–510.

HORIKOSHI, M. & TSUCHIDA, E. 1984. Distribution, mode of life and allometric growth of a deep-sea mytilid bivalve, *Adipicola longissima* (Thiele and Jaekel). *Venus, Japanese Journal of Malacology*, **43**, 86–100.

JAHNKE, L. L., SUMMONS, R. E., DOWLING, L. M. & ZAHIRALIS, K. D. 1995. Identification of methanotrophic lipid biomarkers in cold-seep mussel gills: chemical and isotopic analysis. *Applied and Environmental Microbiology*, **61**, 576–582.

KILLINGLEY, J. S. & BERGER, W. H. 1979. Stable isotopes

in a mollusk shell: detection of upwelling events. *Science*, **205**, 186–188.

KLEIN, R. T., LOHMANN, K. C. & THAYER, C. W. 1996. Sr/Ca and $^{13}C/^{12}C$ ratios in skeletal calcite of *Mytilus trossulus*: covariation with metabolic rate, salinity, and carbon isotopic composition of seawater. *Geochimica et Cosmochimica Acta*, **60**, 4207–4221.

KOCHEVAR, R. E., CHILDRESS, J. J., FISHER, C. R. & MINNICH, E. 1992. The methane mussel: roles of symbiont and host in the metabolic utilization of methane. *Marine Biology*, **112**, 389–401.

LAWTON, J. H. & JONES, C. G. 1995. Linking species and ecosystems: organisms as ecosystem engineers. *In*: JONES, C. G. & LAWTON, J. H. (eds) *Linking Species and Ecosystems*. Chapman & Hall, 141–150.

LINTAS, C. & SEED, R. 1994. Spatial variation in the fauna associated with *Mytilus edulis* on a wave-exposed rocky shore. *Journal of Molluscan Studies*, **60**, 165–174.

LUTZ, R. A. 1976. Annual growth layers in the shell of *Mytilus edulis*. *Journal of the Marine Biological Association of the UK*, **56**, 723–731.

—— 1981. Electron probe analysis of strontium in mussel (Bivalvia, Mytilidae) shells: feasibility of estimating water temperature. *Hydrobiologia*, **83**, 377–382.

—— & CASTAGNA, M. 1980. Age composition and growth rate of a mussel (*Geukensia demissa*) population in a Virginia salt marsh. *Journal of Molluscan Studies*, **46**, 106–115.

MCDONALD, J. H., SEED, R. & KOEHN, R. K. 1991. Allozymes and morphometric characters of three species of *Mytilus* in the northern and southern hemispheres. *Marine Biology*, **111**, 323–335.

MARGOSIAN, A., TAN, F. C., CAI, D. & MANN, K. H. 1987. Seawater temperature records from stable profiles in the shell of *Modiolus modiolus*. *Estuarine, Coastal and Shelf Science*, **25**, 81–89.

MORENO, C. A., LUNECKE, K. M. & LEPEZ, M. I. 1986. The response of an intertidal *Concholepas concholepas* (Gastropoda) population to protection from man in southern Chile and the effects on benthic assemblages. *Oikos*, **46**, 359–364.

MORTON, B. 1992. The evolution and success of the heteromyarian form in the Mytiloida. *In*: GOSLING, E. M. (ed.) *The Mussel Mytilus: Ecology, Physiology, Genetics and Culture*. Elsevier, 21–52.

ONG CHE, R. G. & MORTON, B. 1992. Structure and seasonal variations in abundance of the macroinvertebrate community associated with *Septifer virgatus* (Bivalvia: Mytilidae) at Cape d'Aguilar, Hong Kong. *Asian Marine Biology*, **9**, 217–233.

PAINE, R. T. 1974. Intertidal community structure: experimental studies on the relationship between a dominant competitor and its principal predator. *Oecologia*, **15**, 93–120.

——, CASTILLA, J. C. & CANCINO, J. 1985. Perturbation and recovery patterns of starfish dominated intertidal assemblages in Chile, New Zealand and Washington State. *American Naturalist*, **125**, 679–691.

PANNELLA, G. 1976. Tidal growth patterns in recent and fossil mollusc bivalve shells: a tool for the reconstruction of paleotides. *Naturwissenchaften*, **63**, 539–543.

PEAKE, J. & QUINN, G. P. 1993. Temporal variation in species–area curves for invertebrates in clumps of an intertidal mussel. *Ecography*, **16**, 269–277.

PERKINS, W. T. 1992. Role of inductively coupled plasma mass spectrometry in natural environment research. *Journal of Analytical Atomic Spectrometry*, **7**, 25–34.

PHILLIPS, D. J. H. (ed.) 1980. *Quantitative Aquatic Biological Indicators: Their Use to Monitor Trace Metal and Organochlorine Pollution*. Applied Science Publishers Ltd, London.

REID, R. G. B. 1990. Evolutionary implications of sulphide-oxidizing symbioses in bivalves. *In*: MORTON, B. (ed.) *The Bivalvia: Proceedings of a Memorial Symposium in Honour of Sir Charles Maurice Yonge (1899–1986), Edinburgh, 1986*. Hong Kong University Press, 127–140.

RHOADS, D. & LUTZ, R. A. (eds) 1980. *Skeletal Growth of Aquatic Organisms*. Plenum Press, New York.

RICHARDSON, C. A. 1989. An analysis of the microgrowth bands in the shell of the mussel *Mytilus edulis* (L.). *Journal of the Marine Biological Association of the UK*, **69**, 477–491.

—— 1990. Tidal rhythms in the shell secretion of living bivalves. *In*: BROSCHE, P. & SUNDERMANN, J. (eds) *Earth's Rotation from Eons to Days*. Springer-Verlag, 215–226.

——, SEED, R. & NAYLOR, E. 1990. Use of internal growth bands for measuring individual and population growth rates in *Mytilus edulis* from offshore production platforms. *Marine Ecology Progress Series*, **66**, 259–265.

——, ——, BROTOHADIKUSUMO, N. A. & OWEN, R. 1995. Age, growth and allometric relationships in *Septifer virgatus* (Bivalvia: Mytilidae). *Asian Marine Biology*, **12**, 39–52.

RICHARDSON, C. A., CHENERY, S. R. N. & COOK, J. M. 2000. Assessing the history of trace metal (Cu, Zn, Pb) contamination in the North Sea through laser ablation–ICP–MS of horse mussel, *Modiolus modiolus* shells. Marine Ecology Progress Series, in press.

RODHOUSE, P. G., RODEN, C. M., BURNELL, G. M., HENESEY, M. P., MCMAHON, T., OTTWAY, B. & RYAN, T. J. 1984. Food resource, gametogenesis and growth of *Mytilus edulis* on the shore and in suspended culture: Killary Harbour, Ireland. *Journal of the Marine Biological Association of the UK*, **64**, 513–529.

ROSENBERG, G. D. & JONES, C. B. 1975. Approaches to chemical periodicities in molluscs and stromatolites. *In*: ROSENBERG, G. D. & RUNCORN, S. K. (eds) *Growth Rhythms and the History of the Earth's Rotation*. John Wiley and Sons, 223–242.

SCHULTER, D. & RICKLEFS, R. E. 1993. *Species Diversity: An Introduction to the Problem*. University of Chicago Press.

SEED, R. 1969. The ecology of *Mytilus edulis* (Lamellibranchiata) on exposed rocky shores. 2. Growth and mortality. *Oecologia*, **3**, 317–350.

—— 1983. Structural organisation, adaptive radiation and classification of molluscs. *In*: HOCHACHKA, P. W. (ed.) *The Mollusca I. Metabolic Biochemistry and Molecular Biomechanics.* Academic Press, 1–54.

—— 1996. Patterns of biodiversity in the macro-invertebrate fauna associated with mussel patches on rocky shores. *Journal of the Marine Biological Association of the UK*, **76**, 203–210.

—— & RICHARDSON, C. A. 1990. *Mytilus* growth and its environmental responsiveness. *In*: STEFANO, G. B. (ed.) *The Neurobiology of Mytilus edulis.* Manchester University Press, 1–37.

—— & —— 1999. Evolutionary traits in *Perna viridis* (Linnaeus) and *Septifer virgatus* (Wiegmann) (Bivalvia: Mytilidae). *Journal of Experimental Marine Biology and Ecology*, **239**, 273–287.

—— & SUCHANEK, T. H. 1992. Population and community ecology of *Mytilus*. *In*: GOSLING, E. M. (ed.) *The Mussel Mytilus: Ecology, Physiology, Genetics and Culture*. Elsevier, 87–169.

SHUMWAY, S. E. 1989. A serious threat to shellfish aquaculture. *World Aquaculture*, **20**, 65–74.

STEFANO, G. B. (ed.) 1990. *The Neurobiology of* Mytilus edulis. Manchester University Press.

STURESSON, U. 1976. Lead enrichment in shells of *Mytilus edulis*. *Ambio*, **5**, 253–256.

SUCHANEK, T. H. 1979. *The* Mytilus californianus *community. Studies on the composition, structure, organization and dynamics of a mussel bed.* PhD Thesis, University of Washington, Seattle.

—— 1985. Mussels and their role in structuring rocky shore communities. *In*: MOORE, P. G. & SEED, R. (eds) *The Ecology of Rocky Coasts.* Hodder & Stoughton, 79–96.

——, GELLER, J. B., KREISER, B. R. & MITTON, J. B. 1997. Zoogeographic distribution of the sibling species *Mytilus galloprovincialis* and *Mytilus trossulus* (Bivalvia, Mytilidae) and their hybrids in the North Pacific. *Biological Bulletin*, **193**, 187–194.

SZE, P. W. C. & LEE, S. Y. 1995. The potential role of mucus in the depuration of copper from the mussels *Perna viridis* (Linnaeus) and *Septifer virgatus* (Weigmann). *Marine Pollution Bulletin*, **31**, 390–393.

THOMPSON, I. S., RICHARDSON, C. A., SEED, R. & WALKER, G. 2000. Quantification of mussel (*Mytilus edulis*) growth from power station cooling waters in response to chlorination procedures. *Biofouling*, in press.

——, SEED, R., RICHARDSON, C. A., HUI, L. & WALKER, G. 1997. Effects of low level chlorination on the recruitment, behaviour and shell growth of *Mytilus edulis* Linnaeus in power station cooling water. *Scientia Marina*, **61**, 77–85.

TSUCHIYA, M. & NISHIHIRA, M. 1986. Islands of *Mytilus* as habitat for small intertidal animals: effect of *Mytilus* age structure on the species composition of the associated fauna and community organization. *Marine Ecology Progress Series*, **31**, 171–178.

WARÉN, A. 1991. New and little known Mollusca from Iceland and Scandinavia. *Sarsia*, **76**, 53–124.

WIDDOWS, J. & DONKIN, P. 1992. Mussels and environmental contaminants: bioaccumulation and physiological aspects. *In*: GOSLING, E. M. (ed.) *The Mussel Mytilus: Ecology, Physiology, Genetics and Culture*. Elsevier, 383–424.

Index of subjects

Page numbers in *italics* refer to figures; those in **bold** type to tables.

Systematic Index

Page numbers in *italics* refer to figures.